AF713439

The Medicinal Chemist's Guide to Solving ADMET Challenges

Drug Discovery Series

Editor-in-chief
David Thurston, *King's College London, UK*

Series editors:
David Fox, *Vulpine Science and Learning, UK*
Ana Martinez, *Centro de Investigaciones Biologicas-CSIC, Spain*
Hong Shen, *Roche Innovation Center Shanghai, China*
Ian Storer, *AstraZeneca, UK*

Editorial advisor:
Corey Hopkins, *University of Nebraska Medical Center, USA*

Titles in the Series:
1: Metabolism, Pharmacokinetics and Toxicity of Functional Groups
2: Emerging Drugs and Targets for Alzheimer's Disease; Volume 1
3: Emerging Drugs and Targets for Alzheimer's Disease; Volume 2
4: Accounts in Drug Discovery
5: New Frontiers in Chemical Biology
6: Animal Models for Neurodegenerative Disease
7: Neurodegeneration
8: G Protein-Coupled Receptors
9: Pharmaceutical Process Development
10: Extracellular and Intracellular Signaling
11: New Synthetic Technologies in Medicinal Chemistry
12: New Horizons in Predictive Toxicology
13: Drug Design Strategies: Quantitative Approaches
14: Neglected Diseases and Drug Discovery
15: Biomedical Imaging
16: Pharmaceutical Salts and Cocrystals
17: Polyamine Drug Discovery
18: Proteinases as Drug Targets
19: Kinase Drug Discovery
20: Drug Design Strategies: Computational Techniques and Applications
21: Designing Multi-target Drugs
22: Nanostructured Biomaterials for Overcoming Biological Barriers
23: Physico-Chemical and Computational Approaches to Drug Discovery
24: Biomarkers for Traumatic Brain Injury
25: Drug Discovery from Natural Products
26: Anti-inflammatory Drug Discovery
27: New Therapeutic Strategies for Type 2 Diabetes: Small Molecules
28: Drug Discovery for Psychiatric Disorders
29: Organic Chemistry of Drug Degradation
30: Computational Approaches to Nuclear Receptors

31: Traditional Chinese Medicine
32: Successful Strategies for the Discovery of Antiviral Drugs
33: Comprehensive Biomarker Discovery and Validation for Clinical Application
34: Emerging Drugs and Targets for Parkinson's Disease
35: Pain Therapeutics; Current and Future Treatment Paradigms
36: Biotherapeutics: Recent Developments using Chemical and Molecular Biology
37: Inhibitors of Molecular Chaperones as Therapeutic Agents
38: Orphan Drugs and Rare Diseases
39: Ion Channel Drug Discovery
40: Macrocycles in Drug Discovery
41: Human-based Systems for Translational Research
42: Venoms to Drugs: Venom as a Source for the Development of Human Therapeutics
43: Carbohydrates in Drug Design and Discovery
44: Drug Discovery for Schizophrenia
45: Cardiovascular and Metabolic Disease: Scientific Discoveries and New Therapies
46: Green Chemistry Strategies for Drug Discovery
47: Fragment-based Drug Discovery
48: Epigenetics for Drug Discovery
49: New Horizons in Predictive Drug Metabolism and Pharmacokinetics
50: Privileged Scaffolds in Medicinal Chemistry: Design, Synthesis, Evaluation
51: Nanomedicines: Design, Delivery and Detection
52: Synthetic Methods in Drug Discovery: Volume 1
53: Synthetic Methods in Drug Discovery: Volume 2
54: Drug Transporters: Role and Importance in ADME and Drug Development
55: Drug Transporters: Recent Advances and Emerging Technologies
56: Allosterism in Drug Discovery
57: Anti-aging Drugs: From Basic Research to Clinical Practice
58: Antibiotic Drug Discovery: New Targets and Molecular Entities
59: Peptide-based Drug Discovery: Challenges and New Therapeutics
60: Drug Discovery for Leishmaniasis
61: Biophysical Techniques in Drug Discovery
62: Acute Brain Impairment Through Stroke: Drug Discovery and Translational Research
63: Theranostics and Image Guided Drug Delivery
64: Pharmaceutical Formulation: The Science and Technology of Dosage Forms
65: Small-molecule Transcription Factor Inhibitors in Oncology
66: Therapies for Retinal Degeneration: Targeting Common Processes
67: Kinase Drug Discovery: Modern Approaches
68: Advances in Nucleic Acid Therapeutics
69: MicroRNAs in Diseases and Disorders: Emerging Therapeutic Targets
70: Emerging Drugs and Targets for Multiple Sclerosis

71: Cytotoxic Payloads for Antibody–Drug Conjugates
72: Peptide Therapeutics: Strategy and Tactics for Chemistry, Manufacturing, and Controls
73: Anti-fibrotic Drug Discovery
74: Protein Degradation with New Chemical Modalities: Successful Strategies in Drug Discovery and Chemical Biology
75: Artificial Intelligence in Drug Discovery
76: New Tools to Interrogate Endocannabinoid Signalling: From Natural Compounds to Synthetic Drugs
77: Phenotypic Drug Discovery
78: Protein–Protein Interaction Regulators
79: The Medicinal Chemist's Guide to Solving ADMET Challenges

How to obtain future titles on publication:
A standing order plan is available for this series. A standing order will bring delivery of each new volume immediately on publication.

For further information please contact:
Book Sales Department, Royal Society of Chemistry, Thomas Graham House, Science Park, Milton Road, Cambridge, CB4 0WF, UK
Telephone: +44 (0)1223 420066, Fax: +44 (0)1223 420247
Email: booksales@rsc.org
Visit our website at books.rsc.org

The Medicinal Chemist's Guide to Solving ADMET Challenges

Edited by

Patrick Schnider
Rejuveron Life Sciences AG, Switzerland
Email: patrick.schnider@alumnibasel.ch

ROYAL SOCIETY OF CHEMISTRY

Drug Discovery Series No. 79

Print ISBN: 978-1-78801-227-0
PDF ISBN: 978-1-78801-641-4
EPUB ISBN: 978-1-83916-049-3
Print ISSN: 2041-3203
Electronic ISSN: 2041-3211

A catalogue record for this book is available from the British Library

© The Royal Society of Chemistry 2021

Reprinted 2022

All rights reserved

Apart from fair dealing for the purposes of research for non-commercial purposes or for private study, criticism or review, as permitted under the Copyright, Designs and Patents Act 1988 and the Copyright and Related Rights Regulations 2003, this publication may not be reproduced, stored or transmitted, in any form or by any means, without the prior permission in writing of The Royal Society of Chemistry or the copyright owner, or in the case of reproduction in accordance with the terms of licences issued by the Copyright Licensing Agency in the UK, or in accordance with the terms of the licences issued by the appropriate Reproduction Rights Organization outside the UK. Enquiries concerning reproduction outside the terms stated here should be sent to The Royal Society of Chemistry at the address printed on this page.

Whilst this material has been produced with all due care, The Royal Society of Chemistry cannot be held responsible or liable for its accuracy and completeness, nor for any consequences arising from any errors or the use of the information contained in this publication. The publication of advertisements does not constitute any endorsement by The Royal Society of Chemistry or Authors of any products advertised. The views and opinions advanced by contributors do not necessarily reflect those of The Royal Society of Chemistry which shall not be liable for any resulting loss or damage arising as a result of reliance upon this material.

The Royal Society of Chemistry is a charity, registered in England and Wales, Number 207890, and a company incorporated in England by Royal Charter (Registered No. RC000524), registered office: Burlington House, Piccadilly, London W1J 0BA, UK, Telephone: +44 (0) 20 7437 8656.

For further information see our website at www.rsc.org

For general enquiries, please contact books@rsc.org

For EU product safety enquiries, please email books@rsc.org or contact Royal Society of Chemistry Worldwide (Germany) GmbH, Römischer Hof, Unter den Linden 10, 10117 Berlin.

Preface

Small molecules constitute the most versatile modality in drug discovery since they can be designed by medicinal chemists to be amenable to all routes of administration and to reach any tissue within the body. With the introduction of early absorption, distribution, metabolism, excretion and toxicity (ADMET) profiling later-stage failures due to inadequate pharmacokinetics and a number of previously dreaded safety liabilities could be markedly reduced. During the last two to three decades an impressive body of knowledge has been accumulated on design strategies and tactics to solve common ADMET issues, which have been collected and diligently summarized in this handbook by a distinguished global group of experienced medicinal chemists and ADMET experts across academia and the pharmaceutical industry.

Medicinal chemistry experience and knowledge of how best to modify molecular structure to solve ADMET issues is difficult and tedious to retrieve from the literature, public databases and even corporate data warehouses. *The Medicinal Chemist's Guide to Solving ADMET Challenges* addresses this gap by presenting the most proven strategies to solve challenges associated with a comprehensive set of ADMET parameters which are commonly subject to medicinal chemistry optimization. These tactics are exemplified with a curated collection of concrete examples displayed in a highly visual "table-of-contents" style format, allowing readers to rapidly identify the most promising approaches applicable to their own challenges.

Each chapter of this handbook is dedicated to a single or a group of related ADMET properties. The sections summarizing the mitigation strategies and the examples are preceded by a concise yet comprehensive introduction of the parameter. This comprises a discussion of the relevance of the parameter, a description of state-of-the-art screening strategies including key assays and *in silico* prediction methods, structural and mechanistic aspects and,

Drug Discovery Series No. 79
The Medicinal Chemist's Guide to Solving ADMET Challenges
Edited by Patrick Schnider
© The Royal Society of Chemistry 2021
Published by the Royal Society of Chemistry, www.rsc.org

for enzymes and transporters, an overview of the protein family, expression, activity and substrates. A major remit is to point readers who would like to learn about some of these aspects in more detail to high-quality literature sources. Out of these around half a dozen key references are highlighted in a box, visually offsetting the discussion and exemplification of mitigation strategies from the introductory part.

The unique and uniform format of the chapters of this handbook is in part inspired by Wikipedia, which provides a quick and concise overview of a topic while providing links and references for further reading. Wikipedia is also a source of knowledge about ADMET properties, however, on a more superficial level and generally lacking information on mitigation strategies. Yet there is a great tradition of hugely popular and useful printed reference works on finding solutions which many chemists like to have at their fingertips, such as *Protective Groups in Organic Synthesis* by Theodora W. Greene and Peter G. M. Wuts or Jerry March's *Advanced Organic Chemistry*. In the field of small molecule drug discovery, *The Handbook of Medicinal Chemistry* edited by Andrew Davis and Simon E. Ward has become a landmark oeuvre, which encyclopedically covers the entire R&D process. *The Medicinal Chemist's Guide to Solving ADMET Challenges* is meant to complement the latter by compendiously providing medicinal chemistry practitioners and drug discovery scientists across small and large pharmaceutical companies as well as students interested in entering the field with answers and solutions to their daily challenges pertaining to the universal task of optimizing ADMET properties in search of candidate drugs with a well-balanced overall profile.

I would like to conclude by giving my sincere thanks to all chapter authors for their outstanding work and dedication to bringing this handbook to a successful completion. Furthermore, I would also like to extend my thanks to Alexander Mayweg and Andrew Thomas for their key roles and support in the inception of this book. Last but not least my special thanks go to Sarah Skerratt for her enthusiasm and invaluable contributions to the initiation of this project.

Patrick Schnider

Contents

Drug Discovery Series No. 79
The Medicinal Chemist's Guide to Solving ADMET Challenges
Edited by Patrick Schnider
© The Royal Society of Chemistry 2021
Published by the Royal Society of Chemistry, www.rsc.org

CHAPTER 1

Overview of Strategies for Solving ADMET Challenges

PATRICK SCHNIDER*

Rejuveron Life Sciences AG, Wagistrasse 18, 8952 Schlieren, Switzerland
*E-mail: patrick.schnider@alumnibasel.ch

1.1 Introduction

This chapter provides a high-level overview of all the strategies for solving challenges arising during the optimization of absorption, distribution, metabolism, excretion and toxicity (ADMET) properties in small molecule drug discovery as described in this book. While this is meant as a quick reference for the reader's convenience, it is highly recommended to consult the pertinent chapters for a detailed yet concise discussion, including concrete examples of how these strategies were employed previously.

Each chapter describes mitigation strategies for a single ADMET property. It is clear that hit and lead optimization with a focus on just few parameters is overly simplistic, since each structural modification targeted at changing a particular property in the desired direction will inevitably confer a change on all of a compound's properties. In drug discovery it is therefore advisable to adopt a holistic perspective from the beginning, with efficacious dose predictions ultimately being the most holistic metric. Applying dose predictions early on can be an effective way to identify critical drug properties and guide further optimization.[1–5] While Lipinski's "rule-of-five", which predicts that poor absorption or permeation is more likely when there are more than five

Drug Discovery Series No. 79
The Medicinal Chemist's Guide to Solving ADMET Challenges
Edited by Patrick Schnider
© The Royal Society of Chemistry 2021
Published by the Royal Society of Chemistry, www.rsc.org

hydrogen-bond donors, more than ten hydrogen-bond acceptors, the molecular weight is greater than 500 and the calculated partition coefficient (logP) is greater than five,[6] is an early version of an integrated approach, the parameters of this and similar mnemonics[7] should be used for guidance rather than as strict cut-offs. This is equally true for the techniques for how to mitigate challenges associated with individual ADMET properties described in the chapters of this book. As the overview Table 1.1 in Section 1.3 reveals, the attempt to design a universally perfect drug is elusive. The medicinal chemist's task is rather to come up with candidate drugs with a good balance of molecular properties.[8,9] In order to find this balance, it is essential to develop a solid understanding of the target pharmacokinetic–pharmacodynamic (PK–PD) profile and an adequate ratio of therapeutic benefits to safety risks for a given indication.[1,3,5,10–14]

While a holistic consideration of molecular properties is indispensable during the hit-to-lead and lead optimization phases, medicinal chemists often encounter situations in which particular attention needs to be given to the improvement of single ADMET properties. It is the purpose of this book to provide efficient access to the most promising mitigation strategies and tactics to improve individual properties while offering a uniquely comprehensive overview of how the modifications employed as part of such optimization efforts might affect other ADMET parameters. The properties featured in this book comprise those which are most commonly and universally addressed during hit and lead optimization in small molecule drug discovery programs.

It should be mentioned that there are no chapters dedicated to two fundamental pharmacokinetic parameters, volume of distribution and plasma protein binding, since it has been recommended not to target these parameters for optimization in drug design *per se*, or only with caution.[15–17] However, both can be modulated by changing physicochemical properties; the main trends are therefore included in Table 1.1 in Section 1.3. Lipophilic positively charged molecules have a tendency to partition into biological membranes due to a positive interaction with the phospholipid bilayer, which results in high volumes of distribution. Neutral compounds have no electrostatic interaction with the surface of membranes. Their ability to partition into membranes will thus mainly be driven by increasing lipophilicity. Negatively ionized compounds have a very low affinity for membranes and consequently low volumes of distribution but tend to bind strongly to serum albumin, the most abundant plasma protein.[17] The strongest determinant of albumin binding for neutral and basic molecules is again lipophilicity.[18]

Lipophilicity (or hydrophobicity) is the molecular property which has the single most profound impact on ADMET properties (Table 1.1).[7,19–22] The effective lipophilicity is expressed as logD, the logarithm of the distribution coefficient, D, which is the ratio of equilibrium concentrations of all ionized and unionized forms of a solute in a mixture of a hydrophobic phase such as *n*-octanol and an aqueous phase at a given pH, usually 7.4. The partition coefficient, P, refers to the equilibrium concentration ratio of a hypothetical unionized compound; logP is termed intrinsic lipophilicity since it is independent of the pH of the aqueous phase. Together with logP the acid dissociation constant (pK_a) values of all ionizable groups of a compound therefore determine the effective lipophilicity logD (Chapter 2).[19,23] It has been suggested that an optimal lipophilicity range lies between logD 1 and 3.[20] While logD values above 3 are associated with low solubility and an overall increased safety risk, low logD values below 0–1 will entail poor membrane permeability and increased renal clearance. Due to the prominent influence of lipophilicity on ADMET properties, the experimentally determined logD[21,24] is invaluable to effectively guide medicinal chemistry optimization. For drug design calculated measures of lipophilicity are of great use. The most common and popular methods to estimate logP are based on the summation of lipophilicity contributions of fragments. Estimation of logD from calculated logP and pK_a is intrinsically unreliable due to the uncertainty of predictions, which is exacerbated by the propagation of errors.[20] Advances in computational methods, such as machine learning using large corporate data sets, have enabled the generation of robust models, even for the prediction of logD. However, these still need to be trained continuously and are not broadly available to the community. A recent compilation of logD contributions of commonly used substituents based on experimental logD data and a molecular matched pairs analysis constitutes a practical tool for compound design.[25]

Beyond its crucial effect on logD, charge is also an important determinant of numerous ADMET properties in its own right (Table 1.1).[7] For this reason optimization efforts often require careful fine-tuning of the pK_a of ionizable groups, which highlights the need to continuously expand the knowledge base of how ionization constants of acids and bases can be modulated by structural modifications.[26–28]

The size of a compound, *e.g.* expressed as molecular weight (MW) as a simple and intuitive measure, is another fundamental molecular property. With some exceptions, striving to keep molecular weight low – ideally below 500 or even better 400 Da[7] – will broadly benefit ADMET properties, in particular membrane permeability, but also contribute to achieving high ligand binding efficiency.[29] While these trends have been firmly established, it should also be mentioned, though, that there are a number of orally absorbed drugs with MW >500 Da in the "beyond rule of five" (bRo5) space.[30]

In the quest to gain an ever better understanding of how molecular properties influence ADMET properties, numerous molecular descriptors have been evaluated. It should be pointed out that many descriptors of molecular shape and counts or summations of various structural features, such as number of rotatable bonds or even (topological) polar surface area [(T)PSA] are to some degree correlated with molecular weight. This is also true for hydrogen bond acceptor and donor counts. Nevertheless, these parameters are featured in Table 1.1 since hydrogen bond acceptors and, in particular, donors play a prominent role pertaining to certain ADMET properties, such as membrane permeability and phase 2 metabolism.

Contacts with aromatic rings are among the most frequent non-covalent protein–ligand interactions, which impressively underlines the importance of aryl groups in drug discovery.[31,32] It is therefore to be expected that the number, nature and positioning of aromatic rings may have an effect on ADMET properties which involve a protein–ligand interaction, such as inhibition of cardiac ion channels, plasma protein binding or cytochrome P450 (CYP) inhibition. It has been shown that detrimental effects on these ADMET parameters are mainly due to the number of carboaromatic rings.[33] It should be pointed out, though, that carboaromatic ring count also tends to correlate with lipophilicity, which is a key determinant of these properties. Nevertheless, it has been suggested that aromatic ring count is an important parameter in its own right and that limiting the sum of logD and the number of aromatic rings serves as a good predictor of developability.[21] It is also worth mentioning that aromatic ring count does not adequately describe the relevance of aromaticity for some ADMET properties; *e.g.* photosensitivity, or genotoxicity due to intercalation or the formation of reactive metabolites such as nitrenium ions may be mitigated by breaking conjugation. The formation of electrophilic epoxide or quinone metabolites may be suppressed by replacement of an electron-rich aryl ring by a more electron-deficient one. In contrast, replacement of electron-deficient (aza-)heterocycles by less electron-deficient rings may be a successful strategy to disfavor DNA intercalation or reduce inhibition of CYP1A2.

1.2 Strategies by ADMET Properties

1.2.1 Tactics to Improve Solubility (Chapter 2)

by Robert J. Young

- Reduce lipophilicity
- Introduce charge
- Introduce polar substituents
- Replace aromatic CH by N or O
- Reduce crystal packing and melting point
  - Reduce aromatic ring count or increase $sp^3 : sp^2$ ratio
  - Reduce hydrogen and halogen bonding
  - Reduce intramolecular hydrogen bonding
- Salt forms

1.2.2 Optimisation of Passive Permeability for Oral Absorption (Chapter 3)

by Andy Pike and R. Ian Storer

- Decrease molecular weight
- Increase logD: increase logP and/or modulate pK_a of ionizable centers

- Reduce PSA: lower TPSA or experimental PSA (ePSA) by reduction of hydrogen bond donor (HBD) and/or hydrogen bond acceptor (HBA) count or introduction of intramolecular hydrogen bonds, *via* small rings in acyclic molecules and across larger macrocycles
- Reduce HBD count
- Prodrugs: a derivative compound with improved physicochemical characteristics for absorption which can undergo facile chemical or metabolic degradation to the pharmacologically active species

1.2.3 Targeting Gastrointestinal Uptake Transporters (Chapter 4)

by Simone H. Stahl, Katherine S. Fenner, M. Raymond V. Finlay, Ravindra V. Alluri, Beth Williamson, Johan X. Johansson and Jason Kettle

- Oligopeptide transporter 1 (PepT1, SLC15A1) mediated uptake
  - Modification of a compound to mimic key features of the pharmacophore of the natural substrates of PepT1, dipeptides and tripeptides, *e.g.* by addition of an amino acid
- Sodium-dependent multivitamin transporter (SMVT, SLC5A6) mediated uptake
  - Drug conjugation with biotin (vitamin B7) or pantothenic acid (vitamin B5), the endogenous substrates of SMVT
- Apical sodium-dependent bile acid transporter (ASBT, SLC10A2) mediated uptake
  - Drug conjugation with a bile acid
- Monocarboxylate transporter 1 (MCT1, SLC16A1) mediated uptake
  - A monocarboxylate structure is the key prerequisite; other ionisable groups need to be masked, *e.g.* as prodrugs
- Organic cation transporter OCTN2 (SLC22A5) mediated uptake
  - Drug conjugation with OCTN2's endogenous substrate L-carnitine
- Organic cation transporter OCT1 (SLC22A1) and OCT3 (SLC22A3) mediated uptake (*cf.* Chapter 6)
  - Substrates are largely small hydrophilic compounds ranging from approximately 60 to 350 Da in size, with at least one positively charged group
- Organic anion transporting polypeptide (OATP, SLCO, subtypes 1A2 and 2B1) mediated uptake
  - Relevance is unclear; substrates including a number of marketed drugs are lipophilic acids; endogenous substrates include prostaglandins and sulfate-conjugated steroids
- Nucleoside transporter [concentrative nucleoside transporters (CNTs, SLC28) and equilibrative nucleoside transporters (ENTs, SLC29)] mediated uptake
  - Substrates are nucleosides and nucleobases as well as derivatives

1.2.4 Drug Efflux Transporters: P-gp and BCRP (Chapter 5)

by Peter Bungay and Sharan Bagal

- Maximizing oral absorption
  - MW <500
  - logP <5
  - PSA <120–140 Å^2
  - Hydrogen bond donor count <5
  - Hydrogen bond acceptor count <10
- Maximizing brain penetration by minimizing P-glycoprotein (P-gp) and breast cancer resistance protein (BCRP) mediated efflux at the blood–brain barrier
  - MW <400
  - logP 2–5
  - PSA <70–90 Å^2
  - Hydrogen bond donor count <2
  - Charge: attenuate basic pK_a (<8–8.5), avoid negative charge

1.2.5 OATs and OCTs: The SLC22 Family of Organic Anion and Cation Transporters (Chapter 6)

by Pär Matsson and Maria Karlgren

- Substrates
  - Transport usually requires a positive [organic cation transporter (OCT)] or negative [organic anion transporter (OAT)] charge
  - Molecular weight rarely exceeding 400 Da
  - Lipophilic drugs tend to be poor OCT and OAT substrates
- Inhibitors
  - Inhibitors are usually positively (OCT) or negatively (OAT) charged
  - Greater molecular size is associated with a higher likelihood of inhibition
  - Reducing lipophilicity tends to decrease the probability of inhibition

1.2.6 OATPs: The SLCO Family of Organic Anion Transporting Polypeptide Transporters (Chapter 7)

by Maria Karlgren and Pär Matsson

- Substrates
  - The majority of OATP1B1/1B3 substrates are negatively charged, but some neutral compounds have been reported as weaker substrates
  - Substrates tend to have a relatively high molecular weight in the 400–900 Da range
  - Lipophilic drugs tend to be less efficient OATP substrates

- Inhibitors
  - Inhibitors are usually negatively charged but may also be neutral
  - The likelihood of inhibition correlates with descriptors of increasing molecular size, including molecular weight, volume, number of hydrogen bond acceptors and (topological) polar surface area
  - Reducing lipophilicity leads to a decreased likelihood of inhibition

1.2.7 Bile Salt Export Pump (BSEP) Inhibition (Chapter 8)

by Alexander Treiber and Martin H. Bolli

- Reduce lipophilicity
- Decrease molecular weight
- Cationic and zwitterionic compounds tend to be less potent inhibitors
- Tendency toward decreased inhibition with increasing number of hydrogen bond donors

1.2.8 Cytochrome P450 Metabolism (Chapter 9)

by Antonia F. Stepan and R. Scott Obach

- Increasing metabolic stability
  - Reduce lipophilicity
  - Modify the site of metabolism: remove or block the metabolic soft spot, or disfavor binding to the catalytic site
  - Add fluorine

1.2.9 Cytochrome P450 Induction (Chapter 10)

by Hua Lv, Wei Zhu and Hong Shen

- Strategies to mitigate CYP3A4 induction mediated by pregnane X receptor (PXR) activation; the pharmacophore of human PXR (hPXR) agonists comprises an essential H-bond acceptor and at least two flanking (preferably aromatic) hydrophobic groups
  - Introduce a polar substituent to the hydrophobic group
  - Remove or replace the key hydrophobic group with a less hydrophobic group
  - Introduce steric hindrance or rigidify the structure

1.2.10 Strategies to Mitigate CYP450 Inhibition (Chapter 11)

by Alexander G. Dossetter, Marcel J. de Groot and Sarah E. Skerratt

- Strategies applicable to all CYP isoforms to impede binding of the nitrogen lone pair of an azaheterocycle to the heme group
  - Add a flanking group (*e.g.* a methyl group) next to an aromatic nitrogen
  - Change the heterocycle

- 1A2
  - Increase molecular weight
  - Reduce aromaticity
- 2C9
  - Avoid or reduce negative charge
  - Reduce lipophilicity
- 2C19
  - Reduce lipophilicity
- 2D6
  - Avoid or reduce positive charge
  - Reduce lipophilicity
- 3A4
  - Decrease molecular weight
  - Reduce lipophilicity
  - Add charge; negative charge is most promising

1.2.11 Aldehyde and Xanthine Oxidase Metabolism (Chapter 12)

by David C. Pryde, Dharmendra B. Yadav and Rajib Ghosh

- Preventing oxidation of azaheterocycles featuring an aromatic carbon–hydrogen bond adjacent to an aromatic nitrogen atom
  - Remote functionalization
  - Alternative heterocycles
  - Add a blocking group adjacent to the aromatic nitrogen atom

1.2.12 Glucuronidation (Chapter 13)

by Yue Pan

- Preventing glucuronidation
  - Remove or block the glucuronidation site
  - Use bioisosteres to replace the susceptible moiety
  - Sterically or electronically decrease glucuronidation rate
  - Reduce lipophilicity
  - Sterically disrupt the substrate's binding to uridine 5′-diphospho-glucuronosyltransferase (UGT)
  - Protect the soft spot as a prodrug

1.2.13 Sulfation (Chapter 14)

by Yue Pan

- Preventing sulfation
  - Remove or block the sulfation site
  - Use bioisosteres to replace the susceptible moiety
  - Sterically or electronically decrease the sulfation rate

    - Reduce lipophilicity
    - Increase the size of the molecule to disrupt binding to sulfotransferase (SULT)

1.2.14 Reactive Metabolites (Chapter 15)

by Amit S. Kalgutkar

- Mitigation of epoxidation of (hetero-)aromatic ring and double or triple bonds
    - Introduce innocuous metabolic soft spots
    - Replacement
    - Disfavor metabolic oxidation by reducing electron density
- Mitigation of electron-deficient double (and triple) bonds, including quinones, quinone-methides, quinone-imines, imine-methides, diimines, classical Michael acceptors and iminium ions
    - Iminium ions
        - Structural modifications to prevent formation
        - Electronically disfavor stabilization
    - Quinones, quinone-imines, quinone-methides, diimines, imine-methides
        - Prevent formation by blocking metabolic soft spot (*e.g. para* or *ortho* position of six-membered rings, or benzylic position)
        - Disfavor metabolic oxidation by reducing electron density
    - Michael acceptors
        - Decrease electrophilicity by increasing electron-density
        - Prevent formation by metabolic oxidation followed by elimination reaction (*e.g.* by blocking oxidation of the β-position or replacing a proton in the α-position)
- Reduce reactivity of acyl glucuronides of aliphatic carboxylic acids towards nucleophiles by alkyl substitution at the α-carbon

1.2.15 Genotoxicity (Chapter 16)

by Stephan Kirchner and Patrick Schnider

- Avoiding the formation of aryl nitrenium ions
    - Reduce electron density of the aromatic ring to reduce nitrenium ion stabilization
    - Break or reduce conjugation
    - Impede metabolism of the amino or nitro group by steric shielding or remote substitution
    - Introduce innocuous metabolic soft spots
    - Prevent the release of an aromatic amine (*e.g.* from an amide or *N*-aryl heterocycle) by *N*-substitution, electronic or steric modifications of the aryl group, or replacement or modification of the amide or heterocycle

- Avoiding alkylating agents
  - Avoid alkyl or acyl moieties substituted with good leaving groups
    - Strain in three- and four-membered rings increases reactivity
    - Ensure complete removal of alkylating agents used during synthesis
  - Prevent the metabolic formation of epoxides, Michael-type acceptors and iminium ions (*cf.* Chapter 15)
- Avoiding intercalation and minor groove binding
  - Reduce planarity or aromaticity
  - Introduce bulky substituents
  - Increase electron density of the aryl system
  - Reduce positive charge
  - Disrupt hydrogen bonding (mainly minor groove binders)
- Optimization against binding to the ATP site of kinases regulating the cell cycle
  - Modify hydrogen bond (donor–)acceptor(–donor) hinge-binding motif

1.2.16 Drug-induced Photosensitivity (Chapter 17)

by Jean-François Fournier

- Decrease intrinsic property forecast index (iPFI): Reduce lipophilicity and number of aromatic rings
- Break conjugation
- Remove an aryl halogen atom
- Introduce an intramolecular radical scavenger
- Subtle structural modifications, *e.g.* change positional isomers

1.2.17 Drug-induced Phospholipidosis (Chapter 18)

by Laura Goracci and Gabriele Cruciani

- Reduce basicity
- Reduce lipophilicity
- Reduce amphiphilicity
- Modulation of metabolism
  - Improve metabolism which decreases the potential of a drug to induce phospholipidosis by reduction of overall lipophilicity and basicity
  - Avoid the formation of metabolites that induce phospholipidosis more strongly (and may have a lower clearance and consequently a tendency to accumulate) than the parent, *e.g.* a secondary amine metabolite from a tertiary amine or an amine metabolite from deacylation

1.2.18 Cardiac Ion Channel Inhibition (Chapter 19)

by Cinzia Bordoni, Daniel J. Brough, Gemma Davison, James H. Hunter, J. Daniel Lopez-Fernandez, Kate McAdam, Duncan C. Miller, Pasquale A. Morese, Alexia Papaioannou, Stefan Schunk, Mélanie Uguen, Paul Ratcliffe, Nikolay Sitnikov and Michael J. Waring

- Voltage-gated sodium channel 1.5 ($Na_V1.5$ channel) inhibition
  - Reduce lipophilicity
  - Reduce or eliminate basicity
  - Modify (hetero)aromatic rings and/or (hetero)aromatic substitution pattern
  - Disrupt binding by introduction of steric clashes
- Voltage-gated calcium channel 1.2 ($Ca_V1.2$ channel) inhibition
  - Reduce lipophilicity
  - Reduce or eliminate basicity
  - Modify (hetero)aromatic rings and/or (hetero)aromatic substitution pattern
- Human ether-à-go-go-related gene (hERG) potassium channel inhibition
  - Reduce lipophilicity
  - Reduce or eliminate basicity
  - Introduce acidic centres
  - Reduce the number of aromatic rings
  - Modify (hetero)aromatic rings and/or (hetero)aromatic substitution pattern
  - Disrupt binding by introduction of steric clashes and changes in conformation

Key References

- M. P. Gleeson, *J. Med. Chem.*, 2008, **51**, 817–834.
  - *Guidance on the impact of lipophilicity, molecular weight and ionization state on key ADMET properties based on an analysis of large data sets.*
- M. J. Waring, *Expert Opin. Drug Discovery*, 2010, **5**, 235–248.
  - *Comprehensive review of lipophilicity concluding that issues and risks related to ADMET properties are minimized best in the logD range between ca. 1 and 3.*
- R. J. Young, D. V. S. Green, C. N. Luscombe and A. P. Hill, *Drug Discovery Today*, 2011, **16**, 822–830.
  - *Discussion of the relative influence of intrinsic and effective lipophilicity on key ADMET properties and of the sum of logP/D and aromatic ring count as a composite index to predict developability.*
- M. M. Hann and G. M. Keserü, *Nat. Rev. Drug Discovery*, 2012, **11**, 355–365.
  - *Guide on how to effectively apply existing knowledge of key trends and principles in a holistic way to find the "sweet spot".*
- M. V. Varma, S. J. Steyn, C. Allerton and A. F. El-Kattan, *Pharm. Res.*, 2015, **32**, 3785–3802.
  - *Introduction of the Extended Clearance Classification System (ECCS) for the prediction of the predominant clearance mechanism based on physicochemical properties and passive permeability.*
- T. S. Maurer, D. Smith, K. Beaumont and L. Di, *J. Med. Chem.*, 2020, **63**, 6423–6435.
  - *Overview of the opportunities and challenges associated with dose prediction, the most holistic metric reflecting a compounds potential to become a drug.*

1.3 Strategies by Molecular Properties

Table 1.1 High-level overview of the effects of key molecular properties on ADMET properties.

ADMET property		Lipophilicity	Charge	Molecular weight	Aromatic/planar rings	HBD	HBA
Solubility	↑	↓	↑		↓ planarity/ring count	↑/↓	↑/↓
Plasma protein binding	↓	↓ (positive, neutral)	↓ negative	↓	↓		
Volume of distribution	↓	↓ (positive, neutral)	↓ positive (↑negative)	(↓)			
Passive permeability	↑	↑	↓ negative	↓		↓	(↓)
P-glycoprotein (and BCRP)	↓ Efflux	(↑)	↓	↓		↓	(↓)
OCTs	↓ Transport	↑	↓ positive	↑			
	↓ Inhibition	↓	↓ positive	↓			
OATs	↓ Transport	↑	↓ negative	↑			
	↓ Inhibition	↓	↓ negative	↓			
OATPs	↓ Transport	↑	↓ negative	↓			
	↓ Inhibition	↓	↓ negative	↓			↓
BSEP	↓ Inhibition	↓	(↓ negative, ↑ positive)	↓		↑	
CYP450 metabolism	↓ Metabolism	↓					
CYP450 induction	↓ 3A4 induction	(↓)			(↓ ring count)		
CYP450 inhibition	↓ 1A2 inhibition			↑	↓		
	↓ 2C9 inhibition	↓	↓ negative				
	↓ 2C19 inhibition	↓					
	↓ 2D6 inhibition	↓	↓ positive				
	↓ 3A4 inhibition	↓	↑ (negative)	↓			

Table 1.1 High-level overview of the effects of key molecular properties on ADMET properties. (continued)

ADMET property		Lipophilicity	Charge	Molecular weight	Aromatic/planar rings	HBD	HBA
Glucuronidation	↓ Conjugation	↓				(↓ block/ remove/protect glucuronidation site)	
Sulfation	↓ Conjugation	↓		↑		(↓ block/ remove/ protect sulfation site)	
Reactive metabolites	↓ Metabolism	↓			prevent aromatic ring oxidation: replace/modify		
Genotoxicity	↓ Reactive metabolite formation	↑/↓		↑ (especially aromatic amines)	↓ break or reduce conjugation (especially aromatic amines)		
	↓ Intercalation/ minor groove binding	↑/↓	↓ positive	(↑)	↓ reduce planarity or aromaticity	↓ (especially groove binding)	(↓)
	↓ Kinase inhibition					modify hinge-binding (donor–)acceptor (–donor) motif	
Photosensitivity	↓ Photosensitivity	↓			↓ reduce number of rings, break aromaticity		
Phospholipidosis	↓ Phospholipidosis	↓	↓ positive				
Cardiac ion channels	↓ $Na_V1.5$ inhibition	↓	↓ positive		modify rings/ substitution		
	↓ $Ca_V1.2$ inhibition	↓	↓ positive		modify rings/ substitution		
	↓ hERG inhibition	↓	↓ positive, ↑ negative		↓ number of rings, modify rings/ substitution		

References

1. J. Gabrielsson, H. Dolgos, P.-G. Gillberg, U. Bredberg, B. Benthem and G. Duker, *Drug Discovery Today*, 2009, **14**, 358–372.
2. N. A. Meanwell, *Chem. Res. Toxicol.*, 2011, **24**, 1420–1456.
3. P. Ballard, P. Brassil, K. H. Bui, H. Dolgos, C. Petersson, A. Tunek and P. J. H. Webborn, *Drug Metab. Rev.*, 2012, **44**, 224–252.
4. T. T. Wager, B. L. Kormos, J. T. Brady, Y. Will, M. D. Aleo, D. B. Stedman, M. Kuhn and R. Y. Chandrasekaran, *J. Med. Chem.*, 2013, **56**, 9771–9779.
5. T. S. Maurer, D. Smith, K. Beaumont and L. Di, *J. Med. Chem.*, 2020, **63**, 6423–6435.
6. C. A. Lipinski, F. Lombardo, B. W. Dominy and P. J. Feeney, *Adv. Drug Delivery Rev.*, 1997, **23**, 3–25.
7. M. P. Gleeson, *J. Med. Chem.*, 2008, **51**, 817–834.
8. T. W. Johnson, K. R. Dress and M. Edwards, *Bioorg. Med. Chem. Lett.*, 2009, **19**, 5560–5564.
9. M. M. Hann and G. M. Keserü, *Nat. Rev. Drug Discovery*, 2012, **11**, 355–365.
10. J. J. Sutherland, J. W. Raymond, J. L. Stevens, T. K. Baker and D. E. Watson, *J. Med. Chem.*, 2012, **55**, 6455–6466.
11. M. V. Varma, S. J. Steyn, C. Allerton and A. F. El-Kattan, *Pharm. Res.*, 2015, **32**, 3785–3802.
12. A.-K. Sohlenius-Sternbeck, J. Janson, J. Bylund, P. Baranczewski, A. Breitholtz-Emanuelsson, Y. Hu, C. Tsoi, A. Lindgren, O. Gissberg, T. Bueters, S. Briem, S. Juric, J. Johansson, M. Bergh and J. Hoogstraate, *Curr. Drug Metab.*, 2016, **17**, 253–270.
13. M. K. Bayliss, J. Butler, P. L. Feldman, D. V. S. Green, P. D. Leeson, M. R. Palovich and A. J. Taylor, *Drug Discovery Today*, 2016, **21**, 1719–1727.
14. M. C. Wenlock, *MedChemComm*, 2017, **8**, 571–577.
15. D. A. Smith, L. Di and E. H. Kerns, *Nat. Rev. Drug Discovery*, 2010, **9**, 929–939.
16. X. Liu, M. Wright and C. E. C. A. Hop, *J. Med. Chem.*, 2014, **57**, 8238–8248.
17. D. A. Smith, K. Beaumont, T. S. Maurer and L. Di, *J. Med. Chem.*, 2015, **58**, 5691–5698.
18. S. Endo and K.-U. Goss, *Chem. Res. Toxicol.*, 2011, **24**, 2293–2301.
19. G. Caron, G. Ermondi and R. A. Scherrer, in *Comprehensive Medicinal Chemistry II*, ed. J. B. Taylor and D. J. Triggle, Elsevier, Oxford, 2007, pp. 425–452.
20. M. J. Waring, *Expert Opin. Drug Discovery*, 2010, **5**, 235–248.
21. R. J. Young, D. V. S. Green, C. N. Luscombe and A. P. Hill, *Drug Discovery Today*, 2011, **16**, 822–830.
22. R. R. Miller, M. Madeira, H. B. Wood, W. M. Geissler, C. E. Raab and I. J. Martin, *J. Med. Chem.*, 2020, **63**, 12156–12170.
23. R. A. Scherrer and S. M. Howard, *J. Med. Chem.*, 1977, **20**, 53–58.
24. B. Wagner, H. Fischer, M. Kansy, A. Seelig and F. Assmus, *Eur. J. Pharm. Sci.*, 2015, **68**, 68–77.
25. M. L. Landry and J. J. Crawford, *ACS Med. Chem. Lett.*, 2020, **11**, 72–76.

26. M. Morgenthaler, E. Schweizer, A. Hoffmann-Röder, F. Benini, R. E. Martin, G. Jaeschke, B. Wagner, H. Fischer, S. Bendels, D. Zimmerli, J. Schneider, F. Diederich, M. Kansy and K. Müller, *ChemMedChem*, 2007, **2**, 1100–1115.
27. D. T. Manallack, R. J. Prankerd, E. Yuriev, T. I. Oprea and D. K. Chalmers, *Chem. Soc. Rev.*, 2013, **42**, 485–496.
28. P. Schnider, C. Dolente, H. Stalder, R. E. Martin, V. Reinmüller, R. Marty, C. Wyss Gramberg, B. Wagner, H. Fischer, A. M. Alker and K. Müller, *ChemBioChem*, 2020, **21**, 212–234.
29. A. L. Hopkins, G. M. Keserü, P. D. Leeson, D. C. Rees and C. H. Reynolds, *Nat. Rev. Drug Discovery*, 2014, **13**, 105–121.
30. V. Poongavanam, B. C. Doak and J. Kihlberg, *Gen. Therm.*, 2018, **44**, 23–29.
31. L. M. Salonen, M. Ellermann and F. Diederich, *Angew. Chem., Int. Ed.*, 2011, **50**, 4808–4842.
32. R. Ferreira de Freitas and M. Schapira, *MedChemComm*, 2017, **8**, 1970–1981.
33. T. J. Ritchie, S. J. F. Macdonald, R. J. Young and S. D. Pickett, *Drug Discovery Today*, 2011, **16**, 164–171.

CHAPTER 2

Tactics to Improve Solubility

ROBERT J. YOUNG*

Blue Burgundy Consulting, Formerly, GlaxoSmithKline, Medicines Research Centre, Stevenage, SG1 2NY, UK
*E-mail: rob@blue-burgundy.co.uk

2.1 Introduction: Relevance of Solubility for Drug Disposition

A drug needs to be solubilised at appropriate concentrations in aqueous environments to enable efficacious actions. This encompasses, *inter alia*, the compound being solvated to suitable levels in bodily fluids to enable delivery to and engagement with the target, dissolution in the gastrointestinal (GI) tract to enable absorption into the general circulation for oral drugs and/or high solubility to enable intravenous administration.

There is an abundant literature on poor physicochemical properties in drug discovery,[1] characterised by excessive lipophilicity[2] and the consequent low solubility,[3] which contribute to lower chances of success.[4,5]

Other pertinent parameters, such as wettability and dissolution rate, both of which can also be influenced by the crystalline form or micro manipulation of the solid, also have roles in drug disposition.[6] Whilst they are important, they are beyond the scope of this chapter that focusses on tactics to influence inherent solubility.

It is a cliché in drug development that solubility almost invariably worsens as more stable polymorphs of higher purity compounds are produced during

Drug Discovery Series No. 79
The Medicinal Chemist's Guide to Solving ADMET Challenges
Edited by Patrick Schnider
© The Royal Society of Chemistry 2021
Published by the Royal Society of Chemistry, www.rsc.org

progression[7] and scale up.[8] To maximize chances of success it is thus better to start in a good place in early discovery.[9]

2.2 Definitions

As defined by the International Union of Pure and Applied Chemistry (IUPAC), solubility is the analytical composition of a saturated solution expressed as a proportion of a solute in a designated solvent. In drug discovery, biorelevant solvents are simulated by buffered aqueous media with various surfactants and additives (such as lecithin and taurocholic acid) at pH values that mimic environments in the gastro-intestinal tract and body compartments.

Most drugs and molecules in discovery programmes are "very slightly soluble or practically insoluble" according to European Pharmacopoeia classifications (Table 2.1). The volume of liquid in the stomach is typically around 300 mL, which gives an indication of why keeping the dose below 100 mg is a sensible aspiration.[10]

2.2.1 Which Solubility Measure is Most Pertinent?

Aqueous buffers at different pH values (*e.g.* gastric, approximately pH 2; intestinal, pH 5.5 to 6.5; plasma, pH 7.4) are commonly used for measurements, and all are valuable in understanding drug behaviour. In common medicinal chemistry practice, kinetic solubility in pH 7.4 buffer is a very good benchmark for appraising the solubility of experimental molecules and is commonly employed with increasingly high throughput and automation;[11,12] such precipitative methods use as little as few tens of microlitres of 10 mM dimethyl sulfoxide (DMSO) stock.

As indicators of likely behavior in the GI tract for oral drugs solubilities are commonly measured in fasted or fed state simulated (gastric or intestinal) fluids (FaSSGF, FeSSGF, FaSSIF or FeSSIF; Table 2.2).

Table 2.1 European Pharmacopoeia solubility definitions. Reproduced from ref. 13 with permission from Springer Nature, Copyright 2014.

Descriptive term	Solubility (mg mL^{-1})	Molar solubility MW 400	log *S*	Approximate volume (mL) to dissolve 100 mg
Very soluble	>1000	>2.5 M	>0.4	0.1
Soluble	33 to 100	83 to 240 mM	−1.1 to −0.6	1.0 to 3.0
Sparingly soluble	10 to 33	25 to 83 mM	−1.6 to −1.1	3.0 to 10
Slightly soluble	1 to 10	2.5 to 25 mM	−2.6 to −1.6	10 to 100
Very slightly soluble	0.1 to 1	0.25 to 2.5 mM	−3.6 to −2.6	100 to 1000
Practically insoluble	<0.1	<250 μM	<−3.6	>1000 (or 1 litre)

Table 2.2 Simulated gastrointestinal fluids used in drug discovery. Reproduced from ref. 13 with permission from Springer Nature, Copyright 2014.[a]

Fluid	Solvent and pH	Key components	Aqueous acid/ base added
FaSSGF	Water 1.6	Sodium taurocholate, lecithin, NaCl	HCl, qs
FeSSGF	1 : 1 milk : water 5	NaCl, acetic acid, sodium acetate	HCl, NaOH qs
FaSSIF	Water 6.5	Sodium taurocholate, lecithin, maleic acid, NaCl (dilute)	NaOH qs
FeSSIF	Water (oil?) 6.5 (or lower)	Sodium taurocholate, lecithin, maleic acid, NaCl (concentrated)	NaOH qs

[a]qs = *quantum satis* – the amount needed.

2.2.2 Solubility, Lipophilicity and pK_a

The time taken in measuring solubility is important. Typically, shorter experiments (often using a precipitative method from a carrier solvent, such as DMSO) give a kinetic solubility, representing the maximum solubility of the fastest precipitating species of the compound.[12] Longer times furnish thermodynamic solubility data, representing equilibrium in the dissolution process, whereby the dissolved compound is in equilibrium with the undissolved material (*e.g.* a stable polymorph) in excess.[12]

Solubility is one of three physicochemical properties most pertinent to drug discovery, along with lipophilicity and acid dissociation constant (pK_a); the three are inexorably linked.[13]

Lipophilicity, the inherent affinity of a molecule for a lipophilic environment, is the most important single parameter in understanding drug disposition and interactions. It is of course the antithesis of hydrophilicity, the affinity for water. In drug discovery the relative affinity for the two contrasting environments is usefully measured by partition and distribution coefficients, based on octanol/aqueous buffer measurements (usually expressed as log P or log D at a given pH).

Log P is a constant for any given compound; the only figure for compounds with no ionisable centres or the asymptotic value for the unionised form if there are potential charged centres.

$$P = [\text{solute}]\ \text{octanol}/[\text{solute}]\ \text{water}$$

Log D_{pH} is the effective hydrophobicity of a compound and measures distribution of *all* species at a stated pH (pH 7.4 commonly quoted in drug discovery). This varies with pH, given profound differences (typically 500- to 10 000-fold, represented by ΔP) in the distribution of ionized and unionized forms; the difference between the pK_a and pH will influence the degree of ionization.

$$D_{pH} = [\text{HA}] + [\text{A}^-]\ \text{octanol (acids)}/[\text{HA}] + [\text{A}^-]\ \text{water}$$

$$D_{pH} = [\text{B}] + [\text{BH}^+]\ \text{octanol (bases)}/[\text{B}] + [\text{BH}^+]\ \text{water}$$

The pK_a refers to the pH at which an ionisable centre, be it an acid or a base, is present with equal proportions of the charged and uncharged forms, *i.e.* 50% ionised. It is derived from the Henderson–Hasselbalch equation:

$$pK_a = pH - \log([A^-]/[HA]) \text{ for an acid and}$$

$$pK_a = pH - \log([BH^+]/[B]) \text{ for a base}$$

such that when the species is 50% ionised then $[A^-] = [HA]$ (or $[BH^+] = [B]$) and the pH = pK_a

The interrelationship between log P and log D_{pH} can be computed, given measured or calculated distribution values and pK_a, using the equation:[14]

$$\log D_{pH} = \log(10^{(\log P)} + 10^{[\log P - \Delta P + ch\times(pK_a - pH)]}) - \log(1 + 10^{[ch\times(pK_a - pH)]})$$

Where:

ch (charge) = +1 for all bases, −1 for all acids
ΔP = difference in partition of ionised and neutral species

ΔP is variable and represents a 500-fold to 10 000-fold difference in partition between the neutral and fully charged forms of the ionisable molecule.

Representative ΔP values for ionophores are as follows: acids, 4.2; aliphatic amines, 3.1; aromatic amines (*e.g.* aniline), 2.5; aromatic embedded nitrogen (*e.g.* pyridine) 4.0.[13]

When (pK_a − 2) < pH for a monobasic compound and (pK_a + 2) > pH for a monoacidic compound, the equation to estimate the log P or log D_{pH}, respectively, gets simplified as follows:[15,16]

$$\text{Monobasic: } \log D_{pH} = \log P - \log(1 + 10^{(pK_a - pH)})$$

$$\text{Monoacidic: } \log D_{pH} = \log P - \log(+10^{(pH - pK_a)})$$

Clearly, the calculations are more complex for multiple charges, but there are no substitutes for quality measured values![17]

2.2.3 General Solubility Equation (GSE)

The essence of solubility is captured in the general solubility equation,[18] developed by Yalkowsky and co-workers,[19] which is a useful predictive model originally derived for the solubility of "weak electrolytes". The GSE relates aqueous solubility to a combination of inherent affinity for water *versus* lipid

Table 2.3 Computed and visualised GSE solubility predictions. Reproduced from ref. 13 with permission from Springer Nature, Copyright 2014.[a]

Computed log *S* values (from GSE) in table		$\log_{10}$ (partition coefficient)				
		1	2	3	4	5
Melting point values (°C)	50	−0.75	−1.75	−2.75	−3.75	−4.75
	100	−1.25	−2.25	−3.25	−4.25	−5.25
	150	−1.75	−2.75	−3.75	−4.75	−5.75
	200	−2.25	−3.25	−4.25	−5.25	−6.25
	250	−2.75	−3.75	−4.75	−5.75	−6.75
	300	−3.25	−4.25	−5.25	−6.25	−7.25

[a]Predicted solubility: >200 µM; 30–200 µM; <30 µM.

environments (estimated by the log of the partition coefficient, log *P*) and the lattice energy of the solid form, estimated by the melting point (MP). It is worth noting that for compounds within the log *P* range of 2 to 3 (the median range for drugs) the GSE predicts that only the lowest melting ones have solubility >250 µM ($\log_{10}$ Sol > −3.6) (Table 2.3). As an aide memoire it is useful to consider avoiding (high melting) *brick dust* and (overtly lipophilic) *grease balls* in drug discovery.[20]

General solubility equation:

$$\log[\text{molar aqueous solubility}] = 0.5 - 0.01(\text{MP} - 25) - \log P$$

or

$\log S_{\text{pHx}} = 0.5 - 0.01(\text{MP} - 25) - \log D_{\text{pHx}}$ is a useful guide and approximation

2.2.4 Other Predictors: *In Silico*

Many *in silico* packages to predict solubility are available;[21] most are based on lipophilicity[22] and yield varying predictivity, largely because the form and energy of the crystal lattice (which affect the melting point) are very difficult to predict.

The Lipinski rule of five was proposed as a guide to predict the likelihood of oral absorption based on chances of achieving both acceptable permeation and solubility.[23] Reasonable oral absorption can be expected when a given molecule does not violate two or more of the following, molecular weight >500; log *P* >5, number of hydrogen bond acceptors >10, number of hydrogen bond donors >5. The GSE would concur with the excessive lipophilicity constraint, and there is evidence that higher hydrogen bond acceptor (HBA) or hydrogen bond donor (HBD) counts influence increased melting points. Molecular weight does not affect solubility *per se*,[17] but often correlates with factors that affect permeability, such as size.[17]

Key References

- N. A. Meanwell, Improving drug candidates by design: a focus on physicochemical properties as a means of improving compound disposition and safety. *Chem. Res. Toxicol.*, 2011, **24**, 1420–1456.
  - *Review of physicochemical properties and their influence on progression.*
- R. J. Young, D. V. S. Green, C. N. Luscombe and A. P. Hill, Getting physical in drug discovery II: the impact of chromatographic hydrophobicity measurements and aromaticity. *Drug Discovery Today*, 2011, **16**, 822–830.
  - *Where octanol–water lipophilicity measurements break down and developability parameters are better resolved with chromatographic methods, plus the orthogonal effects of aromatic ring count.*
- R. J. Young, Physical properties in drug design, in *Tactics in Contemporary Drug Design*, Springer Berlin Heidelberg, 2014, pp. 1–68.
  - *Broad chapter on properties, what they are, how they can be measured, predicted and employed in the design of better molecules.*
- C. A. S. Bergström, C. M. Wassvik, K. Johansson and I. Hubatsch, Poorly soluble marketed drugs display solvation limited solubility. *J. Med. Chem.*, 2007, **50**, 5858–5862.
  - *The concepts of brick dust and grease balls.*
- S. H. Yalkowsky and S. C. Valvani, Solubility and partitioning I: Solubility of nonelectrolytes in water. *J. Pharm. Sci.*, 1980, **69**, 912–922.
  - *Seminal work relating solubility to lipophilicity and melting point.*
- C. A. S. Bergström and A. Avdeef, Perspectives in solubility measurement and interpretation. *ADMET & DMPK*, 2019, **7**, 88–105.
  - *Looking deeper into methods and pH dependency.*
- L. Di, P. V. Fish and T. Mano, Bridging solubility between drug discovery and development. *Drug Discovery Today*, 2012, **17**, 486–495.
  - *An introduction to concepts beyond the scope of this review.*

2.3 Mitigation Strategies

2.3.1 Decrease Lipophilicity

It is logical, as embodied in the GSE, to thus consider various tactics that lower lipophilicity as a means of improving solubility; as discussed in the introduction, this will also reduce other attrition-related risks. The following subsections consider various tactics that can help to reduce lipophilicity.

2.3.2 Charge

Reducing lipophilicity by the incorporation of a charged (basic or acidic) centre into a molecule is a commonly employed tactic to reduce lipophilicity. This may be either a recognition feature in the molecule or a non-binding appendage (or solubilising motif). It should be noted that, *e.g.* for a basic compound with a pK_a of 7.4 the solubility at pH 7.4 is only twice that of the unionised form of the molecule, illustrated in Figures 2.1 and 2.2. The same, but opposite – due to charge reversal, effects apply to acidic compounds.

The effective solubility at a given pH (S_{pH}) can, like the effective distribution, be related to intrinsic solubility (S_0) and degree of ionisation by the Henderson–Hasselbalch equation:[24]

$$\text{Monobasic compounds: } \log S_{pH} = \log S_0 + \log(10^{(+pK_a-pH)} + 1)$$

$$\text{Monoacidic compounds: } \log S_{pH} = \log S_0 + \log(10^{(-pK_a+pH)} + 1)$$

The pH changes going through the GI tract should be considered in this context in understanding solubility behaviour, although it can be complicated by transit times and the occurrence of supersaturation, which is typically highly variable since it is liable to food interference and is not a reliable means of achieving desired solubility levels.[25] Predicting and understanding GI solubility behaviour of charged compounds, complicated further for zwitterions and multiply charged molecules, is a science of its own and published references give a good introduction.[24,25]

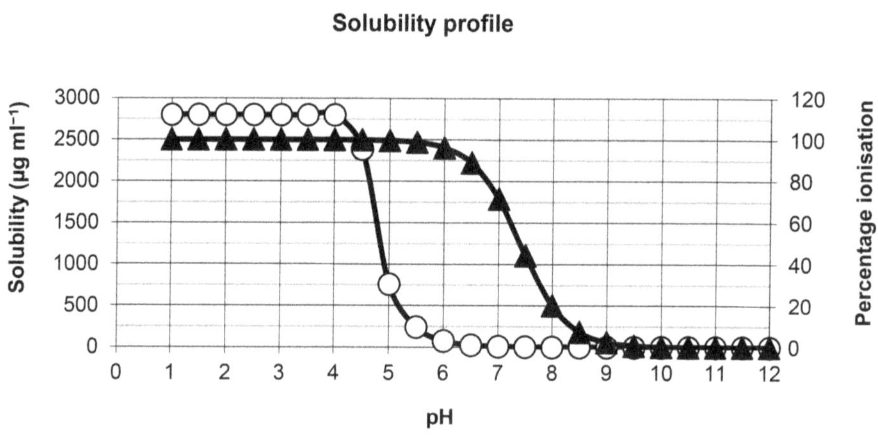

Figure 2.1 Example solubility (white circles) and ionisation (black triangles) *vs.* pH profiles for a basic compound with pK_a of 7.4. This assumes an intrinsic solubility of the free base of 3 µg mL^{-1} and solubility in SGF (pH 1.5) of 2800 µg mL^{-1}. At pH 7.4 the solubility is 6 µg mL^{-1} (2 × 3 µg mL^{-1} as when pH = pK_a, equal concentrations of ionised and unionised forms are explicit, the rest of the unionised form precipitates). Reproduced from ref. 13 with permission from Springer Nature, Copyright 2014.

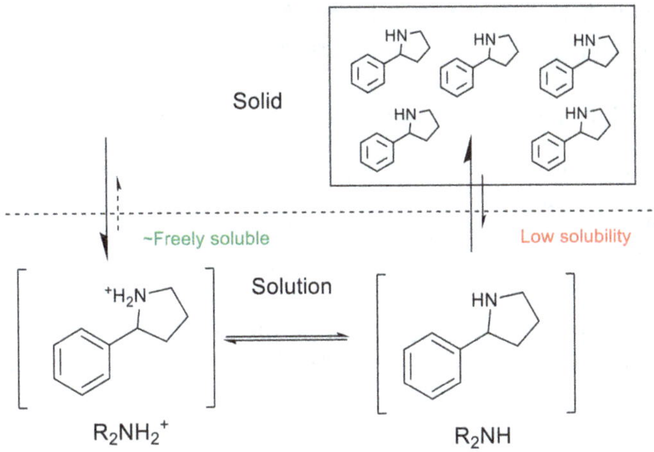

Figure 2.2 Visualising the solubility of a basic compound with low intrinsic solubility. At the pK_a, then $[R_2NH_2^+] = [R_2NH]$, but solubility is only twice that of the low solubility unionised form.

2.3.2.1 *Charge – Not Always a Quick Fix*

While charged groups are clearly hydrophilic and thus contribute to increased solubility, they can reduce binding affinity since the energetic cost of desolvation of associated water molecules is rarely offset by the accrued energy from electrostatic and/or hydrogen bonds. Solubilising groups are best left in bulk water outside of the binding pocket.

There are risks and shortcomings with charged molecules, discussed in other chapters of this book. Promiscuity (interactions with many targets) tends to be driven by intrinsic lipophilicity (or log *P*, the partitioning of the neutral form),[17] frequently exacerbated by a positive charge, *e.g.* human ether-à-go-go related gene (hERG) and general promiscuity.[26] Basic molecules also tend to bind to negatively charged membrane phospholipids, thus increasing volume of distribution and reducing plasma concentration – or they may catalyse the breakdown of these phospholipids, causing phospholipidosis. Acidic molecules often tend to be poorly permeable and bind to basic regions on plasma proteins, resulting in reduced unbound drug concentrations at sites of action.

2.3.3 Introduction of Polar Substituents

Introduction of polar functional groups based on heteroatoms or their combinations, particularly based on oxygen, nitrogen and sulphur, can be effective ways to increase solubility by the reduction of lipophilicity. However, this effect may in some cases be offset by an increase in lattice energy, due to intermolecular contacts, such a hydrogen bonds, occurring in crystal lattices. It is vitally important that polarity is introduced without compromising activity.[27] The magnitudes of these heteroatom effects are dependent on the molecular conformation, where the effects of neighbouring polar groups can be significant, and whether the substitution leads to a decrease of a basic pK_a, the net effect of which may be an increase in lipophilicity. Replacement of hydrogen by fluorine in aliphatic chains also tends to decrease lipophilicity[28] and can improve solubility, whilst having the most profound effect on conformation and pK_a modulation.[29]

2.3.4 Replacement of Aromatic CH by N or O

Benzenoid structures are inherently very lipophilic, and replacement of aromatic CH by one or more nitrogens is a common and useful solubility enhancer[30] and can introduce moderate basicity to reduce effective hydrophobicity (log *D*) or aid with salt formation. Additionally, five-membered heterocyclic rings are common replacements with reduced lipophilicity, and various connectivities involving combinations of N, O and S are feasible. However, such modifications with particular topologies might also lead to increases in lattice energy, with a negative effect on solubility.

2.3.5 Reduce Crystal Packing and Melting Point

2.3.5.1 Reducing Aromatic Ring Count or Increasing sp^3 : sp^2 Ratio

Reducing aromatic ring count[31] or increasing sp^3:sp^2 ratio[32] in molecules ("escape from flatland") are principles that will not only improve solubility but also increase the chance of success for drug molecules in general. The property forecast index[17] (PFI, originally the solubility forecast index) is a useful guide for assessing the chances of achieving favourable developability outcomes, based on the summation of log *P* (or log D_{pH}) plus aromatic ring count. PFI < 6 (on the chromatographic log *D* scale) indicates a high probability of achieving reasonable solubility. The effect of flatness on solubility is probably manifested both by π-stacking and other aromatic interactions in the solid state and lesser entropy release with the phase change of the more rigid structures (Figure 2.3).

The diagonal line separates the distribution of generally good *vs.* generally poor solubility – representing PFI = 7.

Beyond reduction of flatness and disruption of hydrogen bonding networks, lattice energy can in some cases be reduced effectively by – sometimes even subtle – modifications of shape or conformation, such as stabilisation of bent conformations, increase of dihedral angles, or reduction in symmetry.[33] A commonly employed tactic toward this end is methylation, whereby the slight increase in lipophilicity may be outweighed by the reduction of crystallinity.

2.3.5.2 Reducing Hydrogen and Halogen Bonding

In addition to effects noted above with aromatic rings there is an increasing appreciation of the influence of wider crystal packing phenomena, whereby, for example, networks of intermolecular hydrogen bonds can influence the supramolecular structure and thus lattice energy. The number of hydrogen-bond donors as well as adjacent hydrogen-bond donor–acceptor motifs are strongly associated with high crystallinity, and matched molecular pair analyses demonstrated the effects of other motifs such has halogens on melting point.[34] The use of single molecule X-ray crystal structures can aid in the identification of positions to ameliorate these effects (*e.g.* see Example 3 in Section 2.4.2).

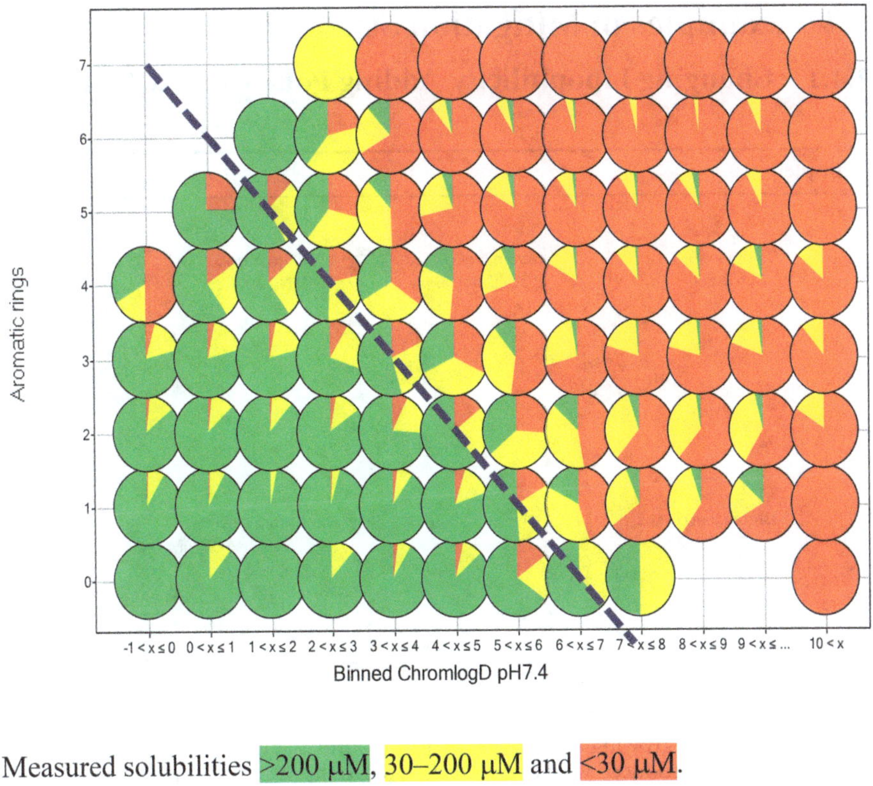

Figure 2.3 Distribution of measured kinetic solubilities *versus* binding measured chromatographic log $D_{7.4}$ and aromatic ring count. Reproduced from ref. 13 with permission from Springer Nature, Copyright 2014.

2.3.5.3 Reducing Intramolecular Hydrogen Bonding

Rigidification of structures through intramolecular hydrogen bonding (*e.g.* Example 1 in Section 2.4.1) can reduce solubility as it lowers the effective hydrophobicity by shielding hydrophilic functionality. Such effects are very important in the purported chimeric behaviours of macrocycles that can show hydrophilic or hydrophobic faces depending on intermolecular hydrogen bonding and the local environment.[35,36]

2.3.6 Salt Forms

Different salts, hydrates and inclusion complexes can be used to improve solubilities; these are usually employed in later stages of discovery or development. The focus here is on improving the intrinsic solubility of the parent molecule, although salts can increase both dissolution rate and solubility. High solubility levels of protonated compounds in the highly acidic environment of the stomach can enable increased absorption due to supersaturation, which maintains the state of dissolution irrespective of the pH rising above the pK_a of the compound in the duodenum that would reduce the intrinsic solubility.

2.4 Examples of Mitigation Strategies

2.4.1 Changing Lipophilicity, Adding Polar Groups, Twists

Example

1

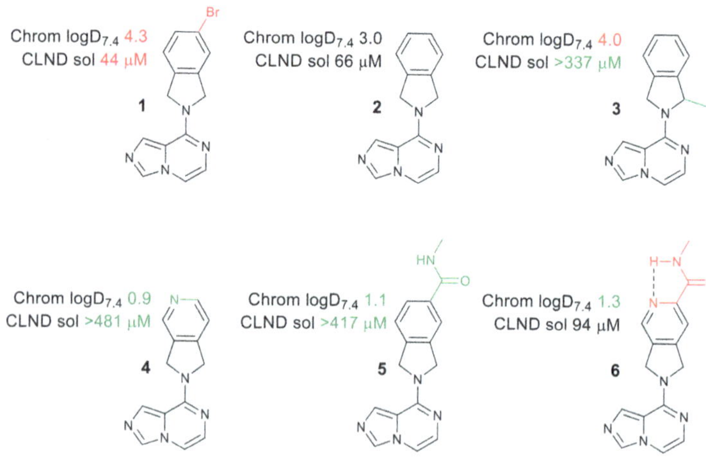

In the process of growing a fragment hit **1**, profound changes in lipophilicity and solubility were produced with minimal changes to the structure. Methylated **3** disrupted planarity and improved solubility in a kinetic assay with chemiluminescent nitrogen detection (CLND) in spite of increased lipophilicity. Introducing a heterocycle (**4** *vs.* **2**) or a polar carboxamide substituent (**5** *vs.* **2**) both reduced lipophilicity approximately 100-fold and increased solubility. However, combining these changes in **6** was not additive and decreased solubility as a significant reduction of the basicity of the pyridine nitrogen along with a reduction of accessibility of the pyridine nitrogen and the amide NH by water due the intramolecular hydrogen bonding increased the lipophilicity. Higher crystallinity, as a consequence of increased planarity effected by the intramolecular hydrogen bond, may further contribute to the drop in solubility.[37]

2.4.2 Solid State Structure Manipulation to Reduce Melting Point

---

Example

2

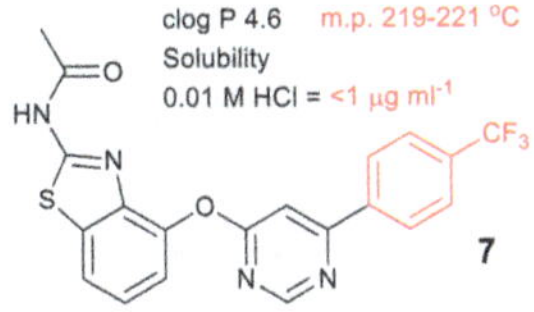

clog P 4.6 m.p. 219-221 °C
Solubility
0.01 M HCl = <1 μg ml^{-1}

7

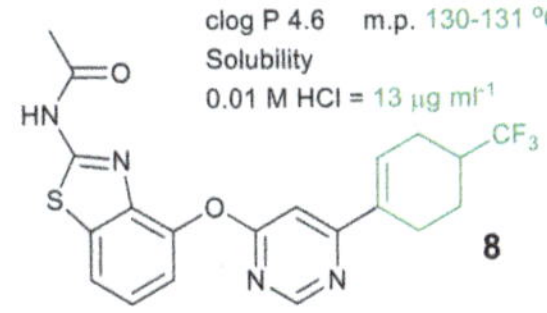

clog P 4.6 m.p. 130-131 °C
Solubility
0.01 M HCl = 13 μg ml^{-1}

8

Partial saturation of a distal ring markedly reduced the melting point of 8, producing a marked improvement of solubility of 7 in spite of similar lipophilicity.[38]

3

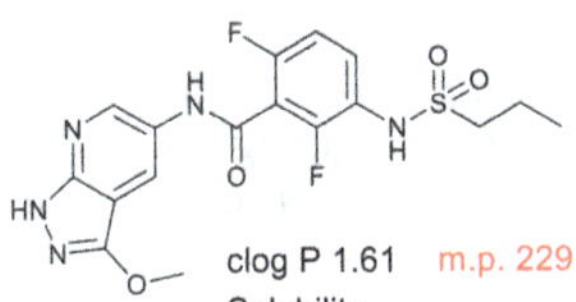

9

clog P 1.61 m.p. 229 °C
Solubility
pH 6.5 = 4 μg ml^{-1}
pH 7.4 = 6 μg ml^{-1}

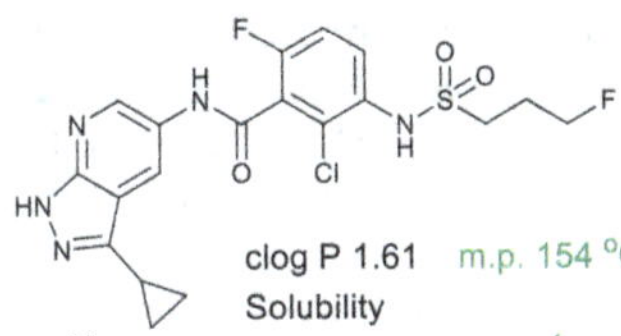

11

clog P 1.61 m.p. 154 °C
Solubility
pH 6.5 = 130 μg ml^{-1}
pH 7.4 = 265 μg ml^{-1}

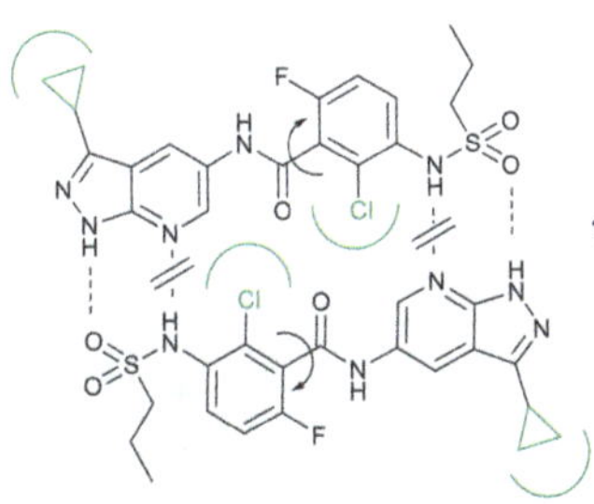

10

clog P 1.61 m.p. 164 °C
Solubility
pH 6.5 = 54 μg/ml
pH 7.4 = 127 μg ml^{-1}

The poor solubility of the lead compound **9** was attributed to its high melting point; a single-crystal X-ray structure indicated that the compound formed head to tail dimers of an extended conformation through hydrogen bonding interactions of the pyrazolopyridine with the sulphonamide groups. Similar analysis of a second compound (**10**) revealed no head-to-tail dimer formation, influenced by the extra steric demands of the chlorinated central ring, which prevented the formation of head to tail dimers in the solid state and resulted in the formation of a different crystal form with fewer π-stacking interactions. The aliphatic fluorine in **11** further lowered lipophilicity.[39]

2.4.3 Modulating Intramolecular Interactions – Reducing Melting Point

Example

4

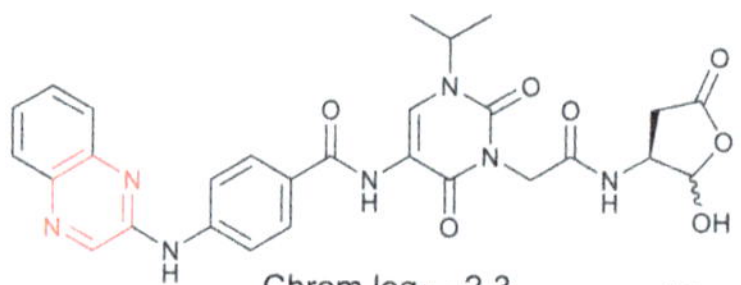

Chrom $\log_{7.4}$ 2.3
Aq solubility 8.3 μM
PG/EtOH sol 5.2 mM

12

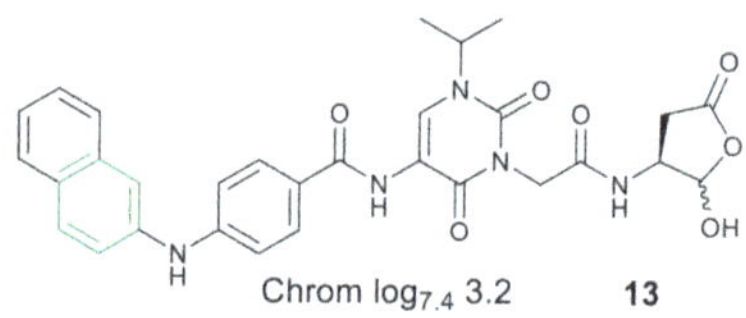

Chrom $\log_{7.4}$ 3.2
Aq solubility 52 μM
PG/EtOH sol 36 mM

13

This is an interesting example from a paper which considers the effects of crystal packing and melting point on solubility of compounds for topical administration, illustrated by the counterintuitive (and not generally recommended, due to toxicity of *beta*-naphthylamine) change of quinoxaline **12** to naphthalene **13**, which led to improved solubility in spite of an increase in log $D_{7.4}$. The observation that the quinoxaline **12** is yellow, while the naphthalene analogue **13** is colourless indicates that the former is a more rigid, higher melting point structure and thus has lower solubility in both water and propylene glycol or ethanol.[40]

2.4.4 Reducing Planarity

Example

5

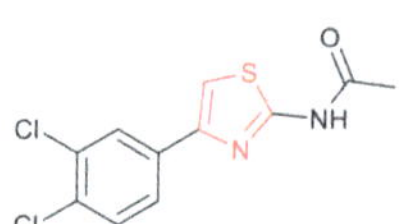

14

Chrom $\log D_{7.4}$ 5.7
PFI 7.7
CLND sol 1 μM

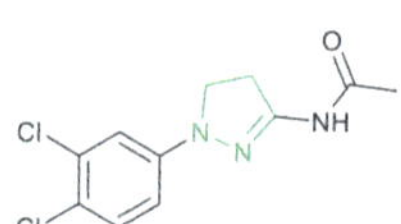

15

Chrom $\log D_{7.4}$ 5.0
PFI 6.0
CLND sol 31 μM

Despite its small size, the lead amino thiazole **14** was highly lipophilic and thus poorly soluble in a kinetic assay with CLND. Fragment hybridisation led to the less planar dihydropyrazole series with considerably improved solubility for **15** in spite of similar lipophilicity.[41]

2.4.5 Heterocyclic Switches, Reducing Aromaticity

Example

6

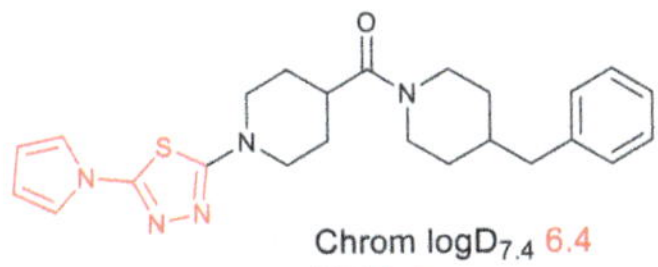

16

Chrom log$D_{7.4}$ 6.4
PFI 9.4
CLND sol <1 μM

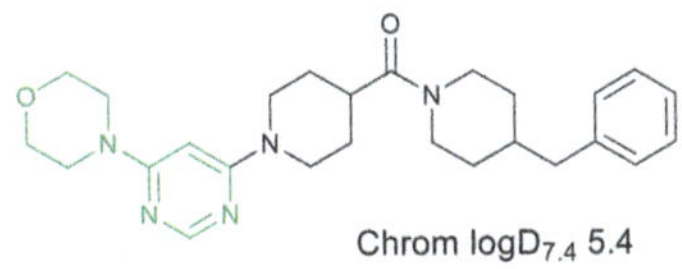

17

Chrom log$D_{7.4}$ 5.4
PFI 7.4
CLND sol 280 μM

The pyrrolopyrazole **16** was a known antitubercular compound, disclosed by GSK to foster collaborations to enable target identification and optimisation. This compound, shown to inhibit DprE1,[42] was one of those where "attention to decreasing lipophilicity and/or minimizing aromatic ring count would be a sensible approach in lead optimization".[43] Similarity searching identified morpholino pyrimidine as a suitable replacement in **17**, retaining activity and reducing both lipophilicity and aromatic ring count, resulting in considerably improved solubility and excellent *in vivo* activity in the series from oral dosing.[44]

2.4.6 Heterocyclic Switch

Example

7

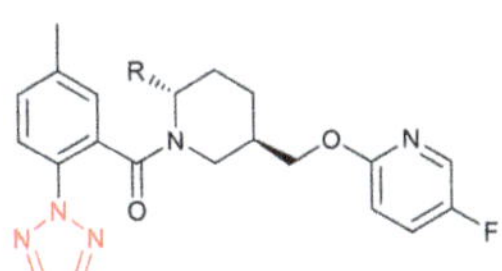

18

R = Me - more potent than H
Datylight clogP 4.3
Sol pH 7 <10 μM

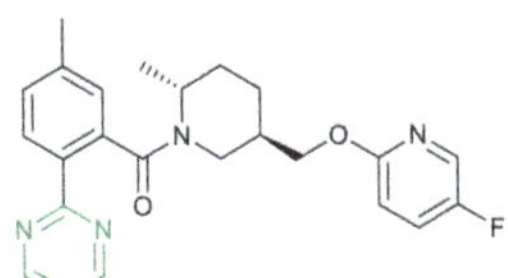

19

Filorexant
Daylight clogP 3.5
Sol pH 7 192 μM

In the Merck Orexin antagonist programme, addition of a trans methyl **18** enforced an axial arrangement of the 3-methyleneoxy enhancing potency, but this compound had poor aqueous solubility (<10 μM) (data above include Daylight calculated log partition coefficient (clog *P*). Optimization of the distal aryl moiety identified the pyrimidine **19** with excellent potency and enhanced aqueous solubility (192 μM), thus filorexant (MK-6096) was selected for progression into clinical trials.[45]

2.4.7 Disruption of Planarity

Example

8

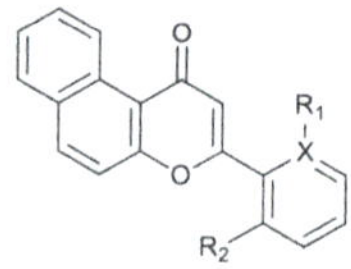

R_1	R_2	X	Solubility (μg mL^{-1})	Melting point °C	Calculated dihedral angle	clogP
OMe	H	C	46	192–193	18.5	4.1
H	H	C	85	165–167	17.8	4.7
F	H	C	153	157	9.1	4.8
F	F	C	248	150	40.5	4.9
H	Me	C	262	135–137	37.9	4.9
—	H	N	299	187–188	0	3.4
Me	Me	C	1270	92	70.0	5.1

The results of this study indicated that increased dihedral angle in the analogues decreased melting point and increased solubility; the pyridyl analogue, with increased melting point showed the effect of the planarity of this compound counteracting the reduced lipophilicity of the compound.[33]

2.4.8 Simple Changes, Big Differences – Crystal Packing and Effect of a Twist

Example

9

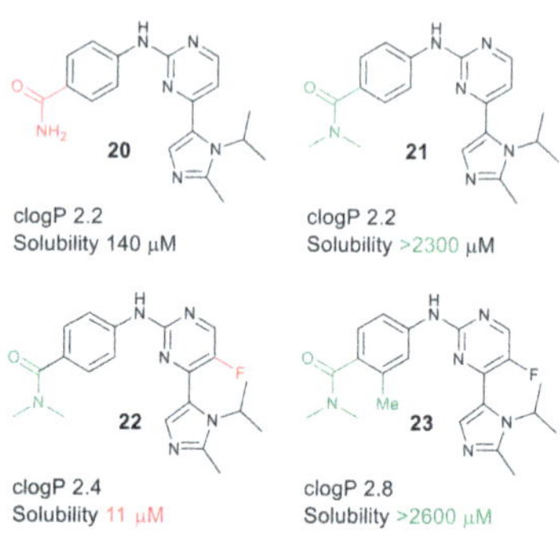

Some interesting effects to consider are described in this paper, although it would be interesting to have more measured lipophilicity (data above are Daylight clogP) and melting point data to hand. Starting from **20**, what would appear to be a relatively high melting point compound has markedly increased solubility by addition of the dimethyl amide **21** (reduced intermolecular hydrogen bonding in crystals?). Modification to the 5-fluoropyrimidine **22** improved activity (attributed to the "increased H-bond acidity of the NH"), but solubility markedly fell due to "increased lipophilicity". The marked level of solubility returned with ortho-methylation in **23**, presumably disrupting crystal packing by turning the amide out of plane in the even more lipophilic molecule.[46]

2.4.9 Introducing Polar Clashes to Induce a Twist

Example

10

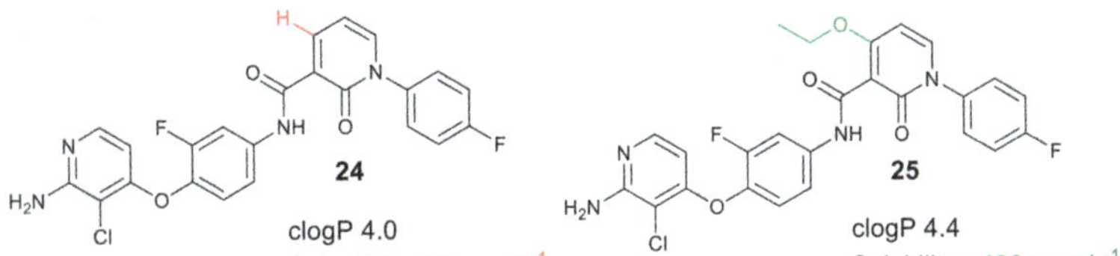

24
clogP 4.0
Sol, pH 1 <10 μg mL^{-1}

25
clogP 4.4
Solubility >400 μg mL^{-1}

The ethoxy substituent, whilst giving a small net increase in lipophilicity, nonetheless showed a marked increase in solubility, attributed to "a deviation from planarity induced by the ethoxy substituent which results in disruption of crystal packing forces". It should be noted that the amide and the pyridone are fully in plane in an X-ray crystal structure of **25** bound to the Met kinase.[47]

2.4.10 Profound Effect of Methyl Substituents

Example

11

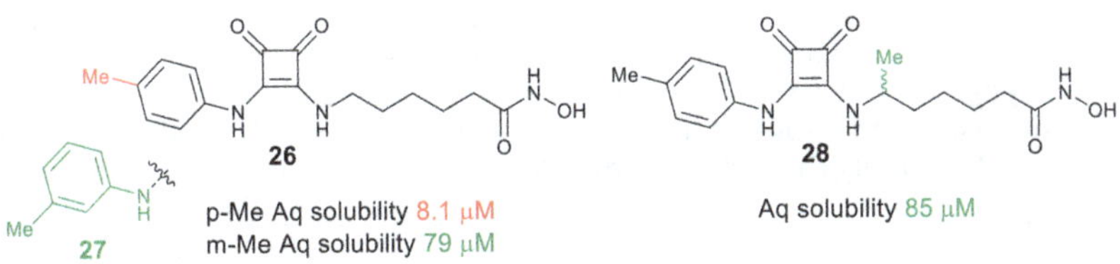

26
27
p-Me Aq solubility 8.1 μM
m-Me Aq solubility 79 μM

28
Aq solubility 85 μM

Addition or placement of methyls can have a profound effect on solubility, most probably *via* crystal packing, given lipophilicity changes, illustrated here by the differences in solubility of meta **26** and para tolyl **27** exemplars from a series of histone deacetylase (HDAC) inhibitors and the tenfold increase achieved by methylation of the more active analogue alpha to the other amino linkage **28**.[48]

2.4.11 Additions of a Solubilising Group

Examples

12

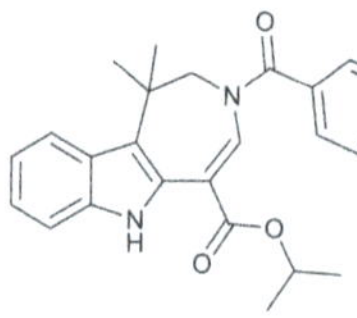

29

No useful DMPK exposure

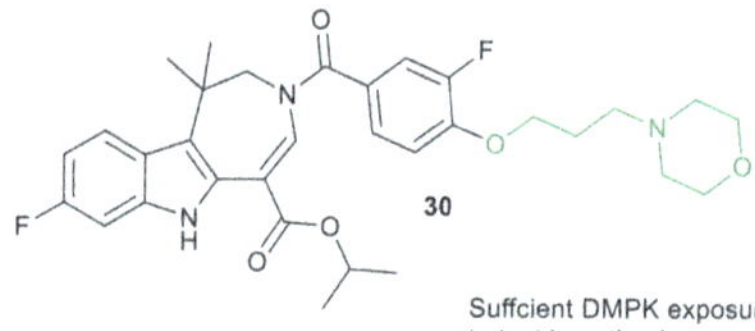

30

Suffcient DMPK exposure
to test hypothesis

A small chain substituent on **29** linked to a mildly basic morpholine was employed to enhance aqueous solubility in a farnesoid X receptor (FXR) candidate **30**. This improved oral exposure without resorting to complex formulations to achieve exposure in preclinical model species.[49]

13

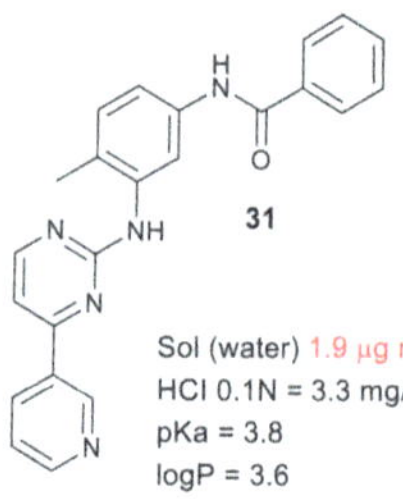

31

Sol (water) 1.9 μg ml^{-1}
HCl 0.1N = 3.3 mg/ml.
pKa = 3.8
logP = 3.6

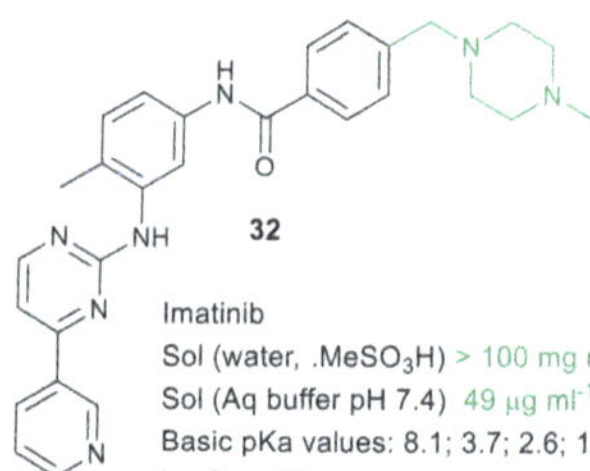

32

Imatinib
Sol (water, .$MeSO_3H$) > 100 mg ml^{-1}
Sol (Aq buffer pH 7.4) 49 μg ml^{-1}
Basic pKa values: 8.1; 3.7; 2.6; 1.5
Log$D_{7.4}$ 2.0

The use of a solubilising group was an important element that led to the first licenced kinase inhibitor imatinib, which is used against an number of cancers based on inhibition of the tyrosine kinases breakpoint cluster region–Abelson kinase (Bcr–Abl), c-kit and platelet derived growth factor receptor (PDGF-R). An early lipophilic and weakly basic lead **31** was poorly soluble except under low-pH conditions; incorporation of the solubilising piperazine in **32**, imatinib, gave much improved solubility that led to the mesylate salt as the commercialised form.[50,51]

2.4.12 Salt Formation

Example

14

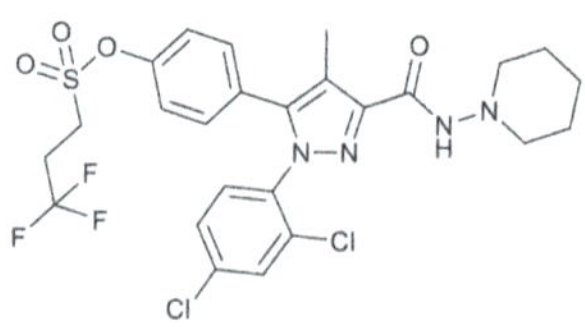

Parent pKa 3.2, solubility 0.01-0.02 μg/ml
Salts made:
.H^+Cl^-
.$H^+O^-SO_2OH$
.$H^+O^-SO_2$(1,5-Napth)SO_2OH

This paper gives an informative review of the issues around salt formation, highlighting the need to have >2 log difference in pK_a values of the partners, the use of a salt screen and how the stability, dissolution rate and solubility all need attention in identifying the final form. In this example, chloride, hemisulphate and hemi-1,5-naphthalenedisulphonic acid gave crystalline forms and the latter was most stable. All increased dissolution rate, solubility and oral absorption.[52]

References

1. P. D. Leeson and B. Springthorpe, *Nat. Rev. Drug Discovery*, 2007, **6**, 881–890.
2. M. M. Hann, *MedChemComm*, 2011, **2**, 349–355.
3. A. P. Hill and R. J. Young, *Drug Discovery Today*, 2010, **15**, 648–655.
4. P. D. Leeson and A. M. Davis, *J. Med. Chem.*, 2004, **47**, 6338–6348.
5. N. A. Meanwell, *Chem. Res. Toxicol.*, 2011, **24**, 1420–1456.
6. *Handbook of Pharmaceutical Salts Properties, Selection and Use*, ed. P. H. Stahl and C. G. Wermuth, Wiley-VCH, Weinheim, Germany, 2011.
7. L. I. Di and E. H. Kerns, in *Solvent Systems and Their Selection in Pharmaceutics and Biopharmaceutics*, ed. P. Augustijns and M. E. Brewster, Springer New York, New York, NY, 2007, pp. 111–136.
8. L. Di, P. V. Fish and T. Mano, *Drug Discovery Today*, 2012, **17**, 486–495.
9. R. J. Young and P. D. Leeson, *J. Med. Chem.*, 2018, **61**, 6421–6467.
10. M. K. Bayliss, J. Butler, P. L. Feldman, D. V. Green, P. D. Leeson, M. R. Palovich and A. J. Taylor, *Drug Discovery Today*, 2016, **21**, 1719–1727.
11. E. H. Kerns, *J. Pharm. Sci.*, 2001, **90**, 1838–1858.
12. T. Sou and C. A. S. Bergström, *Drug Discovery Today: Technol.*, 2018, **27**, 11–19.
13. R. J. Young, in *Tactics in Contemporary Drug Design*, ed. N. A. Meanwell, Springer Berlin Heidelberg, Berlin, Heidelberg, 2014, pp. 1–68.
14. G. Caron, G. Ermondi and R. A. Scherrer, in *Comprehensive Medicinal Chemistry II*, ed. J. B. Taylor and D. J. Triggle, Elsevier, Oxford, 2007, p. 425–452.
15. R. A. Scherrer and S. M. Howard, *J. Med. Chem.*, 1977, **20**, 53–58.
16. A. Hersey, A. P. Hill, R. M. Hyde and D. J. Livingstone, *Quant. Struct.-Act. Relat.*, 1989, **8**, 288–296.
17. R. J. Young, D. V. Green, C. N. Luscombe and A. P. Hill, *Drug Discovery Today*, 2011, **16**, 822–830.
18. S. H. Yalkowsky and S. C. Valvani, *J. Pharm. Sci.*, 1980, **69**, 912–922.
19. N. Jain and S. H. Yalkowsky, *J. Pharm. Sci.*, 2001, **90**, 234–252.
20. C. A. S. Bergström, C. M. Wassvik, K. Johansson and I. Hubatsch, *J. Med. Chem.*, 2007, **50**, 5858–5862.
21. J. S. Delaney, *Drug Discovery Today*, 2005, **10**, 289–295.
22. S. B. Bunally, C. N. Luscombe and R. J. Young, *SLAS Discovery*, 2019, **24**, 791–801.
23. C. A. Lipinski, F. Lombardo, B. W. Dominy and P. J. Feeney, *Adv. Drug Delivery Rev.*, 1997, **23**, 3–25.
24. C. A. S. Bergström and A. Avdeef, *ADMET & DMPK*, 2019, **7**, 88–105.
25. A. Avdeef, E. Fuguet, A. Llinàs, C. Ràfols, E. Bosch, G. Völgyi, T. Verbić, E. Boldyreva and K. Takács-Novák, *ADMET & DMPK*, 2016, **4**, 117–178.
26. J.-U. Peters, P. Schnider, P. Mattei and M. Kansy, *ChemMedChem*, 2009, **4**, 680–686.
27. E. Freire, *Chem. Biol. Drug Des.*, 2009, **74**, 468–472.
28. K. Müller, *Chimia*, 2014, **68**, 356–362.

29. E. P. Gillis, K. J. Eastman, M. D. Hill, D. J. Donnelly and N. A. Meanwell, *J. Med. Chem.*, 2015, **58**, 8315–8359.
30. T. J. Ritchie, S. J. Macdonald, R. J. Young and S. D. Pickett, *Drug Discovery Today*, 2011, **16**, 164–171.
31. T. J. Ritchie and S. J. Macdonald, *Drug Discovery Today*, 2009, **14**, 1011–1020.
32. F. Lovering, J. Bikker and C. Humblet, *J. Med. Chem.*, 2009, **52**, 6752–6756.
33. M. Ishikawa and Y. Hashimoto, *J. Med. Chem.*, 2011, **54**, 1539–1554.
34. S. Schultes, C. de Graaf, H. Berger, M. Mayer, A. Steffen, E. E. J. Haaksma, I. J. P. de Esch, R. Leurs and O. Krämer, *MedChemComm*, 2012, **3**, 584–591.
35. A. T. Bockus, K. W. Lexa, C. R. Pye, A. S. Kalgutkar, J. W. Gardner, K. C. Hund, W. M. Hewitt, J. A. Schwochert, E. Glassey, D. A. Price, A. M. Mathiowetz, S. Liras, M. P. Jacobson and R. S. Lokey, *J. Med. Chem.*, 2015, **58**, 4581–4589.
36. M. Withnall, H. Chen and I. V. Tetko, *ChemMedChem*, 2018, **13**, 599–606.
37. B. C. Whitehurst, *Bioorg. Med. Chem. Lett.*, manuscript in preparation.
38. H.-L. Wang, J. Katon, C. Balan, A. W. Bannon, C. Bernard, E. M. Doherty, C. Dominguez, N. R. Gavva, V. Gore, V. Ma, N. Nishimura, S. Surapaneni, P. Tang, R. Tamir, O. Thiel, J. J. S. Treanor and M. H. Norman, *J. Med. Chem.*, 2007, **50**, 3528–3539.
39. S. Wenglowsky, D. Moreno, J. Rudolph, Y. Ran, K. A. Ahrendt, A. Arrigo, B. Colson, S. L. Gloor and G. Hastings, *Bioorg. Med. Chem. Lett.*, 2012, **22**, 912–915.
40. J.-F. Fournier, L. Clary, S. Chambon, L. Dumais, C. S. Harris, C. Millois, R. Pierre, S. Talano, É. Thoreau, J. Aubert, M. Aurelly, C. Bouix-Peter, A. Brethon, L. Chantalat, O. Christin, C. Comino, G. El-Bazbouz, A.-L. Ghilini, T. Isabet, C. Lardy, A.-P. Luzy, C. Mathieu, K. Mebrouk, D. Orfila, J. Pascau, K. Reverse, D. Roche, V. Rodeschini and L. F. Hennequin, *J. Med. Chem.*, 2018, **61**, 4030–4051.
41. G. L. Simpson, S. M. Bertrand, J. A. Borthwick, N. Campobasso, J. Chabanet, S. Chen, J. Coggins, J. Cottom, S. B. Christensen, H. C. Dawson, H. L. Evans, A. N. Hobbs, X. Hong, B. Mangatt, J. Munoz-Muriedas, A. Oliff, D. Qin, P. Scott-Stevens, P. Ward, Y. Washio, J. Yang and R. J. Young, *J. Med. Chem.*, 2019, **62**, 2154–2171.
42. S. M. Batt, M. Cacho Izquierdo, J. Castro Pichel, C. J. Stubbs, L. Vela-Glez Del Peral, E. Perez-Herran, N. Dhar, B. Mouzon, M. Rees, J. P. Hutchinson, R. J. Young, J. D. McKinney, D. Barros Aguirre, L. Ballell, G. S. Besra and A. Argyrou, *ACS Infect. Dis.*, 2015, **1**, 615–626.
43. L. Ballell, R. H. Bates, R. J. Young, D. Alvarez-Gomez, E. Alvarez-Ruiz, V. Barroso, D. Blanco, B. Crespo, J. Escribano, R. Gonzalez, S. Lozano, S. Huss, A. Santos-Villarejo, J. J. Martin-Plaza, A. Mendoza, M. J. Rebollo-Lopez, M. Remuinan-Blanco, J. L. Lavandera, E. Perez-Herran, F. J. Gamo-Benito, J. F. Garcia-Bustos, D. Barros, J. P. Castro and N. Cammack, *ChemMedChem*, 2013, **8**, 313–321.

44. J. A. Borthwick, C. Alemparte, I. Wall, B. C. Whitehurst, A. Argyrou, G. Burley, P. de Dios-Anton, L. Guijarro, M. C. Monteiro, F. Ortega, C. J. Suckling, J. C. Pichel, M. Cacho and R. J. Young, *J. Med. Chem.*, 2020, **63**, 2557–2576.
45. P. J. Coleman, J. D. Schreier, C. D. Cox, M. J. Breslin, D. B. Whitman, M. J. Bogusky, G. B. McGaughey, R. A. Bednar, W. Lemaire, S. M. Doran, S. V. Fox, S. L. Garson, A. L. Gotter, C. M. Harrell, D. R. Reiss, T. D. Cabalu, D. Cui, T. Prueksaritanont, J. Stevens, P. L. Tannenbaum, R. G. Ball, J. Stellabott, S. D. Young, G. D. Hartman, C. J. Winrow and J. J. Renger, *ChemMedChem*, 2012, **7**, 415–424.
46. C. D. Jones, D. M. Andrews, A. J. Barker, K. Blades, K. F. Byth, M. R. V. Finlay, C. Geh, C. P. Green, M. Johannsen, M. Walker and H. M. Weir, *Bioorg. Med. Chem. Lett.*, 2008, **18**, 6486–6489.
47. G. M. Schroeder, Y. An, Z.-W. Cai, X.-T. Chen, C. Clark, L. A. M. Cornelius, J. Dai, J. Gullo-Brown, A. Gupta, B. Henley, J. T. Hunt, R. Jeyaseelan, A. Kamath, K. Kim, J. Lippy, L. J. Lombardo, V. Manne, S. Oppenheimer, J. S. Sack, R. J. Schmidt, G. Shen, K. Stefanski, J. S. Tokarski, G. L. Trainor, B. S. Wautlet, D. Wei, D. K. Williams, Y. Zhang, Y. Zhang, J. Fargnoli and R. M. Borzilleri, *J. Med. Chem.*, 2009, **52**, 1251–1254.
48. J.-F. Fournier, Y. Bhurruth-Alcor, B. Musicki, J. Aubert, M. Aurelly, C. Bouix-Peter, K. Bouquet, L. Chantalat, M. Delorme, B. Drean, G. Duvert, N. Fleury-Bregeot, B. Gauthier, K. Grisendi, C. S. Harris, L. F. Hennequin, T. Isabet, F. Joly, G. Lafitte, C. Millois, R. Morgentin, J. Pascau, D. Piwnica, Y. Rival, C. Soulet, É. Thoreau and L. Tomas, *Bioorg. Med. Chem. Lett.*, 2018, **28**, 2985–2992.
49. J. T. Lundquist, D. C. Harnish, C. Y. Kim, J. F. Mehlmann, R. J. Unwalla, K. M. Phipps, M. L. Crawley, T. Commons, D. M. Green, W. Xu, W.-T. Hum, J. E. Eta, I. Feingold, V. Patel, M. J. Evans, K. Lai, L. Borges-Marcucci, P. E. Mahaney and J. E. Wrobel, *J. Med. Chem.*, 2010, **53**, 1774–1787.
50. J. Zimmermann, E. Buchdunger, H. Mett, T. Meyer, N. B. Lydon and P. Traxler, *Bioorg. Med. Chem. Lett.*, 1996, **6**, 1221–1226.
51. J. Zimmermann, E. Buchdunger, H. Mett, T. Meyer and N. B. Lydon, *Bioorg. Med. Chem. Lett.*, 1997, **7**, 187–192.
52. K. Sigfridsson, M. L. Ulvinge, L. Svensson and A. K. Granath, *Drug Dev. Ind. Pharm.*, 2019, **45**, 202–211.

CHAPTER 3

Optimisation of Passive Permeability for Oral Absorption

ANDY PIKE*[a] AND R. IAN STORER*[b]

[a]DMPK, Oncology R&D, AstraZeneca, Cambridge, UK; [b]Medicinal Chemistry, Discovery Sciences, Biopharmaceuticals R&D, AstraZeneca, Cambridge, UK

*E-mail: andrew.pike@astrazeneca.com, ian.storer@astrazeneca.com

3.1 Introduction

3.1.1 Potential Mechanisms of Membrane Permeation

To reach their site of action following oral administration, drugs must first be absorbed from the gastrointestinal tract. Therefore, design of compounds capable of passing through the gut wall is a key aspect of drug discovery.

While this chapter focuses on passive permeability, it should be acknowledged that there are several mechanisms by which compounds can reach the systemic circulation from the gastrointestinal tract (Figure 3.1):

1. Paracellular absorption: compounds are absorbed *via* inter-cellular spaces. Generally, this route of absorption requires compounds to have low logD and molecular weight (typically <300 Daltons) and therefore is extremely difficult, if not impossible, to design against in many drug discovery projects due to the constraints imposed by the target

Drug Discovery Series No. 79
The Medicinal Chemist's Guide to Solving ADMET Challenges
Edited by Patrick Schnider
© The Royal Society of Chemistry 2021
Published by the Royal Society of Chemistry, www.rsc.org

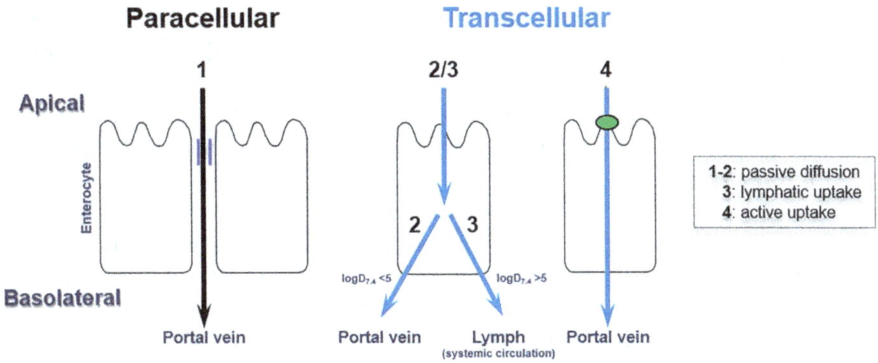

Figure 3.1 Mechanisms of membrane permeability.

pharmacophore. Additionally, the paracellular route is known to be species-dependent *e.g.* dogs are known to show higher absorption by this route.[1] The potential to modulate paracellular absorption *via* the use of formulation excipients has been reported,[2] although it remains unclear if this approach will modify the physicochemical space to which it can be applied effectively.

2. Transcellular absorption: compounds are absorbed through the gut epithelial cells requiring passage through the apical and basolateral membranes. The transcellular route is by far the most common mechanism of drug absorption and this chapter discusses in detail various strategies for optimizing passive permeability to enable this process.
3. Lymphatic uptake: the process by which lipids are absorbed into the lymphatic system, which requires the association of molecules with chylomicrons and other triglyceride-rich lipoproteins to promote lymphatic uptake. There are relatively few well-established examples of drugs which are absorbed by this route and the structural and physicochemical determinants remain poorly understood, although high lipophilicity is clearly required. However, with increasing interest in developing drugs in beyond rule-of-five space this is potentially an area of specific need that may offer specific advantages such as bypassing first pass metabolism.[3] Examples of compounds with a lymphatic contribution to their absorption include the B-cell lymphoma 2 (BCL-2) inhibitors navitoclax (ABT-263) and venetoclax (ABT-199).[4]
4. Active uptake: the body utilizes a wide range of membrane transporters for nutrient uptake and in instances where drug pharmacophores overlap those of endogenous substrates, or alternatively it is possible to add a targeting motif, then these processes can be targeted to enable drug absorption. Specific examples of active uptake are discussed in

Chapter 4 of this book. The potential negative effect of gut efflux transporters on drug absorption are likewise discussed in detail in Chapter 5. However, it should be noted in terms of design tactics that increasing passive permeability will have the effect of reducing the net effect of efflux transporters independently of any change in specific transporter interaction.

While potentially important for specific compound classes, paracellular, lymphatic and transporter mediated uptake impose significant constraints on physicochemical properties and/or requirement for the inclusion of a specific pharmacophore making them difficult to design for in many, if not most, situations. Therefore, when designing for oral absorption this usually means optimising passive permeability to enable absorption by the transcellular route.

3.1.2 Strategies for Screening for Passive Permeability

Screening for compound permeability is ultimately a trade-off between throughput/resource (including cost) and data content/direct relevance and therefore different methodologies tend to be employed at different points in the screening cascade.

Due to the ready availability of computational tools in the modern drug discovery environment, and increasingly large historical data sets, *in silico* approaches are often employed at the design stage to avoid making compounds with a low probability of success. The application of computational approaches based on physicochemical characteristics, once in place, requires minimal resources and is easily incorporated into design strategies. Such computational models range in sophistication from simple checks on rule-of-five (Ro5) compliance through to machine learning approaches. It is likely given the continuing increases in processing power, sophistication of machine learning approaches and refinements in the application of physicochemical descriptors that the role of *in silico* predictions will continue to grow.

Chromatographic techniques, such as immobilised artificial membranes (IAM) and experimental polar surface area (ePSA), provide a rapid and low cost means of screening large numbers of compounds and avoid the potential solubility limitations of concentration-based methods.

IAM is a chromatographic technique employing a phospholipid bilayer bonded to the surface of the stationary phase of a liquid chromatography column to simulate the lipid bilayer environment. The principle being that partitioning onto the stationary phase will correlate to the propensity of a molecule to enter and pass a lipid bilayer and therefore chromatographic retention will provide an index of compound permeability.[5]

In contrast to the calculated topological polar surface area (TPSA), a two-dimensional parameter, ePSA is measured experimentally using supercritical fluid chromatography.[6,7] Using this method, the potential effects of three-dimensional conformation of the molecule and any ability to form polarity masking intramolecular interactions can be accounted for, thus enabling identification of compounds with potentially good passive permeability despite apparently unfavourable two-dimensional physicochemical or structural characteristics.[6]

The parallel artificial membrane permeability assay (PAMPA), which uses a lipid impregnated filter between donor and receiver compartments to provide an artificial membrane substitute for the lipid bilayer of the cell, was developed as a low cost, robust alternative to cell monolayers which could provide a correlate of *in vivo* permeability.[8] The non-biological nature of the system also enables testing under conditions which can't be applied in cell-based assays due to viability concerns *e.g.* across a wide pH range or long time period. However, it should be noted that the PAMPA membrane does not share the structure of the cell membrane *e.g.* it is much thicker (approximately 125 μm) than a lipid bilayer (approximately 5 nm) or cell monolayer (approximately 20 μm) and is not structured in a bilayer but instead has been suggested to be composed of reverse micelle structures.[9] The effects of these differences on the measured value may not always be predictable and therefore establishing correlation with permeability data from a cellular system within a given chemical series may be beneficial.

The use of confluent cell monolayers grown in permeable transwell culture plates remains the "gold-standard" for *in vitro* permeability assessment by measuring flux of a compound from donor to receiver compartments. The most widely used cell line for this purpose is Caco-2, derived from a human adenocarcinoma, although efflux transporter inhibitors are required to estimate intrinsic permeability in isolation.[10] The use of other cell lines in the same assay format has also been described, *e.g.* Madin–Darby canine kidney (MDCK). Alternatives, such as the Ralph Russ canine kidney (RRCK), a low P-glycoprotein expressing sub-clone of MDCK, may simplify the assessment of passive permeability in isolation due to the reduced efflux transporter activity removing the need for chemical inhibitors of efflux transporters.[11–13] Potential human stem cell derived culture systems have also recently been reported in the literature, although it is not yet clear what advantage is offered over the established cell lines for assessment of passive permeability.[14] Disadvantages of cell-based transwell systems can include solubility dependence, non-specific binding to labware and the potential for compound trapping in the cell monolayer.[15] However, it seems likely that such cell-based assays will remain key tools for medium throughput assessment of permeability, although refinements in predictive *in silico* models may provide better triage to reduce costs.

For complete mechanistic evaluations, including potentially in human tissue, *ex vivo* tissue-based approaches, *e.g.* Ussing chambers or inverted gut sacs, can be used.[16] However, these techniques are inherently limited in throughput due to their greater complexity and, particularly for human, by availability of donor tissue.

Ultimately, any strategy requires validation by the demonstration of *in vivo* oral bioavailability/exposure consistent with a high fraction absorbed (F_a). However, effective screening can minimise requirements for *in vivo* studies and lead to a higher success rate.

3.1.3 Influence of Physicochemical Properties on Passive Permeability

The principal function of the cell membrane is to prevent free diffusion of ions, metabolites and cell proteins. While the structure of the cell membrane is complex (precise lipid composition, incorporation of proteins, *etc.*), for the purposes of understanding passive permeability the cell membrane can be considered to consist of a double layered sheet of amphiphilic phospholipids with hydrophilic phosphate head groups orientated to the aqueous environment and fatty acid tails forming a hydrophobic interior. As such, diffusion through the membrane requires the compound to pass through the hydrophobic environment of the lipid bilayer interior, and understanding the physicochemical constraints on this process has been instrumental in designing orally absorbed drugs.

The rate of flux through a membrane is a function of permeability, concentration gradient and available surface area. Therefore, for optimal absorption a drug will ideally be both highly soluble (in the context of the dose), to maximise the concentration gradient, and highly permeable, to enable membrane diffusion. Balancing requirements for solubility and permeability is a key factor in physicochemical rules-based approaches to finding oral drugs. Probably the most well-known and influential such approach is the Ro5 introduced by Lipinski, *et al.* to describe oral drugs.[17] The Ro5 stipulates that "poor absorption or permeation is more likely when":

1. Molecular weight (MW) >500
2. logP >5
3. There are more than five hydrogen bond donors (HBD)
4. There are more than ten hydrogen bond acceptors (HBA) and HBD

Other such physicochemical rules-based approaches have since been proposed using alternative descriptor combinations, and it is beyond the scope of this chapter to provide a comparison of their individual characteristics or value to compound design.[18–33] However, it is worth noting some general points about such models and the implications for compound properties and design.

Firstly, solubility and permeability can't simultaneously be optimised to an arbitrarily high degree as increasing capacity for aqueous solvation comes at the cost of increasing the desolvation penalty associated with passing into the lipophilic environment of the membrane. While impossible to generalize across all discovery projects, it should be noted that solubility requirements can also be mitigated by minimizing dose requirement through increased potency, reduction of unbound clearance or, in many – but not all – instances, addressed by formulation approaches. Therefore, improving permeability is often the primary focus of design.

Secondly, many physicochemical descriptors are to some degree inherently correlated, potentially risking spurious correlations being made; *e.g.* changes in MW may also affect lipophilicity, polar surface area (PSA), rotatable bond count, *etc.* Therefore, in a drug design sense, the effects of compound modifications should be considered holistically.

Lastly, rules-based approaches are generally seeking to identify molecules with high oral bioavailability, of which passive permeability is only one component. Poor oral bioavailability can also result from low solubility, efflux or high first-pass metabolism, and rules-based approaches do not distinguish contributions of individual factors.

While understanding of *in vivo* F_a is the aim, using the model systems previously described, the passive permeability component of absorption can be studied in a more controlled fashion to enable quantitative structure–activity relationships (QSAR) to be built independently of other pharmacokinetic properties. There are many such published studies of permeability, often reaching subtly different conclusions about the effects of various descriptors. These discrepancies are probably due to analysis of different data sets or subjective analysis decisions such as data binning giving rise to "correlation inflation" type effects for specific parameters.[34] However, across the literature three key themes emerge:

1. Lipophilicity: logD (logP and ionization state)
2. TPSA/ePSA
3. HBD count

Lipophilicity is an important characteristic of drug molecules and has become firmly embedded in the concept of "drug-likeness". It is commonly quantified as the log of the partition coefficient (logP), which is defined as the ratio of concentrations of the neutral form of a molecule between two immiscible phases, typically water and octanol. Increased lipophilicity has been established to correlate with improved potency, driven by well-tolerated non-directional protein–ligand interactions and hydrophobic effects.[35] However, excessive lipophilicity can also negatively affect absorption, distribution, metabolism and excretion (ADME), selectivity and toxicological properties, leading to compound attrition. Given its potentially broad effects, metrics such as lipophilic efficiency (LipE or LLE) have become a key part of drug design, with the aim to avoid pursuing potency through excessive lipophilicity which will probably also result in high unbound clearance, pharmacological promiscuity, *etc.*[22]

In terms of passive permeability, the benefit of increasing logP can clearly be rationalised by the lowering of desolvation energy enabling subsequent diffusion into the lipid bilayer. Although, for oral bioavailability, an upper limit will clearly be imposed by a corresponding negative effect on solubility and rate of metabolism, which is consistent with the Ro5 stipulation of logP <5. This behavior has been summarized in a number of reviews which highlight trends towards a balanced lipophilicity 'sweet-spot' for optimal oral bioavailability.[18] However, while increased logP will improve the permeability of the neutral form, many drugs are ionized at the pH of the small intestine; ionization state is therefore also a factor in drug absorption. As early as the 1950s it was proposed that acidic drugs could be rapidly absorbed if their pK_a was >3 whereas strong acids showed very little absorption at all.[36] Indeed, results of a recent analysis indicated an overall trend to acidic compounds having a lower F_a than bases, which is proposed to reflect a higher likelihood of being in an ionized state at the pH of the small intestine.[37] Conversely, basic drugs were proposed to be rapidly absorbed if their pK_a was <8.[36] These observations are consistent with the assumption that a high fraction unionized at a physiologically relevant pH for the gastrointestinal tract is beneficial for absorption due to the higher permeability of the unionized form. An experimental example, using alfentanil and cimetidine, determining the permeability in Caco-2 cells across a pH range supports this hypothesis.[38] Permeability was shown to have a linear relationship with the fraction of drug unionized and the permeabilities of the unionized species of alfentanil and cimetidine were estimated to be 150- and 30-fold higher than those of the ionized forms. However, notably, it was concluded that at a fraction unionized below 0.1, this proportion could potentially still represent a significant contribution to overall flux.

Therefore, to incorporate charge state into lipophilicity optimization the log of the distribution coefficient (logD) at a given pH is often used as the preferred measure of lipophilicity. The logD is defined as the ratio of concentrations of a molecule between two immiscible phases measured at a defined pH, typically applying pH 7.4 buffer and octanol. Design strategies can then target increased logD by either increasing logP or by appropriate modulation of the pK_a for ionizable species.

The calculated TPSA of a molecule is defined as the molecular surface composed of all polar atoms (primarily oxygen and nitrogen), effectively incorporating the HBD and HBA count, and correlates with the desolvation energy cost in entering the lipid membrane. It has been suggested that PSA in isolation can effectively predict F_a, with compounds showing a F_a >90% typically having a PSA ≤60 Å^2 while those with a PSA >140 Å^2 tend towards F_a <10%.[39] However, this is not an absolute cut off, and increasingly calculated TPSA values are being supplemented with ePSA values which can account for effects, not always obvious from the two-dimensional structure, such as intramolecular hydrogen bonds (IMHBs) that can effectively mask polarity.[40]

PSA, in combination with the number of rotatable bonds (#Rot B), a measure of molecular flexibility, has been suggested to effectively predict oral bioavailability.[41] However, it is not clear if this relates specifically to the F_a or also influences rate of metabolism.[37]

As stated previously, HBD count contributes to PSA and therefore may be partially mitigated by consideration of that descriptor in design. However, a further subtlety to the interpretation of TPSA is the relative effects of HBA/D on permeability. The Ro5 implies a higher weighting of HBD compared with HBA in determining oral bioavailability, setting a lower limit for HBD *vs.* combined count. Interestingly, a recent examination of oral drugs in beyond Ro5 (bRo5) space indicated that while overall PSA was correlated with molecular weight up to 1000 Da, this was principally due to an increase in HBA count while the apparent limit for HBD count was only increased from five to six, indicating that permeability is more sensitive to the presence of HBD than polar surface area *per se*.[42]

3.1.4 Future Directions: Beyond Rule-of-five

Application of the principles of the Ro5, and similar rules-based approaches, has provided an effective strategy for identifying oral drugs. However, changes in pharmacological targets *e.g.* protein–protein interactions or modalities, such as protein targeting chimeras (PROTACs), mean that in the future strict adherence to the Ro5 is likely to be too restrictive in many instances. Results from several analyses have indicated that increasing numbers of drugs are being developed in bRo5 space and, furthermore, there is a shift within this space from natural products to designed molecules.[43,44] Therefore, understanding how to design drugs in this space is likely to become increasingly important.

These new design pressures have precipitated a re-examination of the physicochemical limits of drug absorption, leading some authors to question if some aspects of Ro5 *e.g.* the molecular weight limit, are in fact overly simplistic or at least can be broken providing other conditions are met.[20,44]

It is reasonable to assume that, as molecules grow larger, additional considerations may come into play for optimal permeability as the potential for flexibility and three-dimensional shape differences increases. Therefore, physicochemical properties, such as number of rotatable bonds or radius of gyration, as correlates of flexibility and three-dimensional shape respectively, may begin to play a bigger role.[45–47] Likewise, the potential for secondary structure, such as the formation of intramolecular hydrogen bonds leading to unexpectedly high permeability of compounds such as daclatasvir, or compounds showing so-called chameleonic properties due their potential to exist in distinct conformations will need to be considered.[48] A new metric of lipophilic permeability efficiency was recently proposed to try to reconcile the opposing effects of lipophilicity on membrane permeability and solubility on examples of peptidic and non-peptidic macrocycles, indicating that chameleonic properties may allow the two to be at least partially decoupled.[49]

Data in bRo5 space is still extremely sparse and heavily biased by large numbers of closely related compounds against a limited range of targets. As a result, it may be too soon to derive consistent broad design principles in this space. However, the potential to move considerably beyond the Ro5 limits has been established for various classes of compounds and is likely to be an area of growing interest.

3.1.5 Relevance

For many reasons, most importantly patient convenience, but also including other factors, such as cost of goods, in the short term at least the preferred route of drug administration is likely to remain oral. Therefore, it is desirable where possible to identify compounds with intrinsically good passive permeability and solubility as the simplest and most effective means of minimizing development risk and cost-of-goods.

Key References

- C. A. Lipinski, *J. Pharmacol. Toxicol. Methods*, 2000, **44**, 235.
  - *Derivation of the rule-of-five.*
- M. M. Hann and G. M. Keserue, *Nat. Rev. Drug Discovery*, 2012, **11**, 355.
  - *Review of small molecule physical property metrics.*
- D. F. Veber, S. R. Johnson, H. Y. Cheng, B. R. Smith, K. W. Ward and K. D. Kopple, *J. Med. Chem.*, 2002, **45**, 2615.
  - *Alternative rules-based approach to oral bioavailability based on non-rotatable bonds and polar surface area.*
- V. Poongavanam, B. C. Doak and J. Kihlberg, *Curr. Opin. Chem. Biol.*, 2018, **44**, 23.
  - *Guidelines for discovery of orally absorbed compounds in beyond rule-of-five space.*
- K. Beaumont, R. Webster, I. Gardner and K. Dack, *Curr. Drug Metab.*, 2003, 4, 461.
  - *Prodrug review.*

3.2 Mitigation Strategies

Many physicochemical parameters (molecular weight, ionization state, lipophilicity, HBD, number of rotatable bonds, *etc.*) have been shown to correlate to passive permeability, although in some cases it remains debatable if they are truly causative or simply correlated to a more limited set of key parameters.

While some of these parameters *e.g.* molecular weight or ion class, can be important considerations in series selection for various reasons, it is relatively rare that they can be arbitrarily changed during the optimization process due to the constraints of the target pharmacophore. Therefore, while providing some examples of such approaches, we have primarily concentrated on strategies involving modulation of logD, polar surface area and HBD count, with additional consideration of prodrug-based strategies.

1. Lowering molecular weight.
2. Increasing logD: increasing logP and/or modulating pK_a of ionizable centres.
3. Lowering PSA: lowering TPSA or ePSA by reduction of HBD and/or HBA count or introduction of IMHBs, *via* small rings in acyclic molecules and across larger macrocycles.
4. Lowering of HBD count.
5. Pro-drugs: a derivative compound with improved physicochemical characteristics for absorption which can undergo facile chemical or metabolic degradation to the pharmacologically active species. Such compounds are interesting in a design sense as they may result in simultaneous modification of lipophilicity, TPSA and charge state.[50]

3.3 Examples of Mitigation Strategies

3.3.1 Lower Molecular Weight (MW)

Although molecular weight (MW) is well established to correlate with permeability this often proves challenging to implement as a design tactic during lead optimization since reduction in MW tends to lead to loss of desired pharmacological potency. As a result, molecular weight is more often used as a metric to assist in early series selection decisions, often in the form of a ligand efficiency metric.

Examples

1	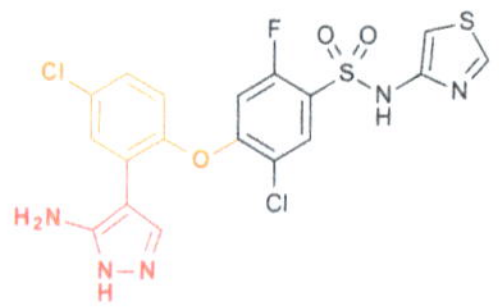	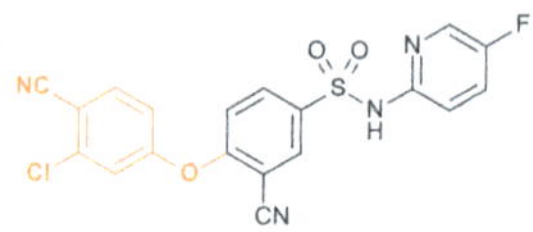	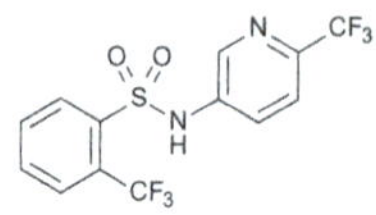
RRCK P_{app} (×10^{-6} cm s^{-1})	3.8	16	27
MW	500	429	370
HBA/HBD (count)	6/4	6/1	3/1
logD/clogP	2.6/4.2	1.8/3.7	1.4/3.3
pK_a (base)	6.3	5.8	5.6
TPSA (Å^2)	123	124	67
Rat *F* (%)	27	100	98

Design tactics:

- Reduce MW
- Reduce HBD

A clinical voltage activated sodium channel, $Na_V1.7$, compound was discovered to also be a weak urate transporter 1 (URAT1) inhibitor. Subsequent compounds were optimized for URAT1 activity and improved permeability and bioavailability by lowering molecular weight and reducing the numbers of hydrogen bond donors.[51–53]

3.3.2 Increase logD

Tactics to increase logD largely comprise addition of more lipophilic groups to increase logP, removal of a polar group, or modulation of the pK_a of an ionizable centre to increase the non-ionized fraction presented at the apical membrane interface, thereby increasing the productive concentration gradient. However, care needs to be taken as, despite increased lipophilicity leading to improved passive permeability, it is also well established to correlate with other negative outcomes, such as reduced solubility and selectivity.

Examples			
1	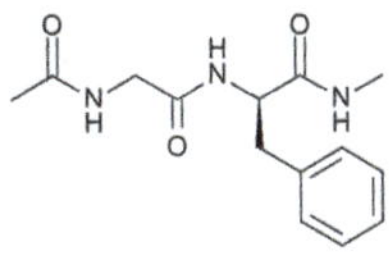	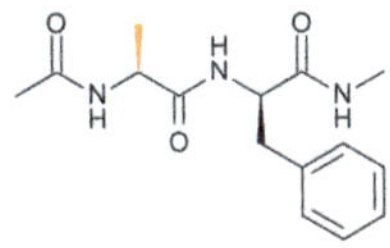	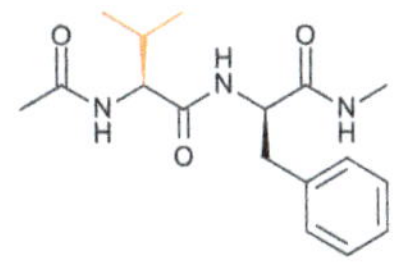
Caco-2 P_{app} ($\times 10^{-6}$ cm s^{-1})	0.17	0.24	0.91
ACD clogP	0.03	0.38	1.26
	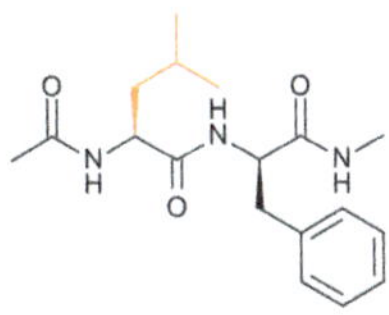	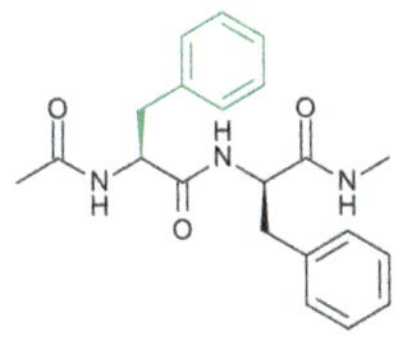	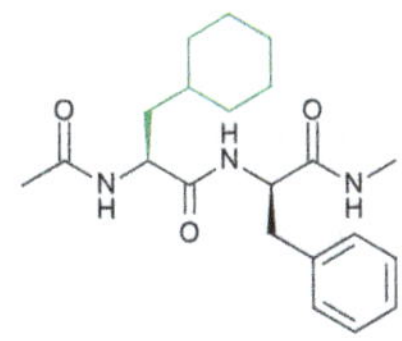
Caco-2 P_{app} ($\times 10^{-6}$ cm s^{-1})	1.41	2.75	7.20
ACD clogD/clogP	1.79	2.22	2.97

Design tactics:

- Increase logP

Replacing a glycine in a peptide chain with other amino acids that bear lipophilic side chains resulted in increased permeability correlated with lipophilicity.[54]

2	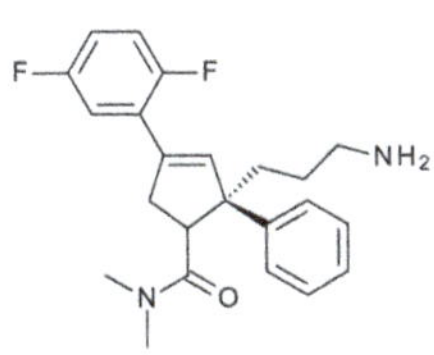	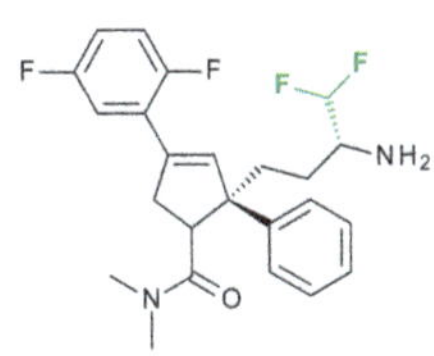
MDCK P_{app} ($\times 10^{-6}$ cm s^{-1})^{a}	4	38
logD/clogP	−1.7/1.2	3.2/3.2
pK_a (base)	10.3	7.0
TPSA (Å^2)	46	42

Design tactics:

- Increase logP
- Introduce electron-withdrawing group to reduce pK_a of ionizable basic group (reduce fraction ionized)

For a series of kinesin spindle protein (KSP) inhibitors addition of group adjacent to the basic amine led to an increase in permeability. This can be rationalized by an increased logD through both an increase in lipophilicity and lowering of the basic pK_a. Also, there is a possibility that a five-membered IMHB could be formed between the CHF_2 group and the NH_2 to mask the effect of a HBD to further enhance permeability.[55,56]

(*continued*)

Examples

3

	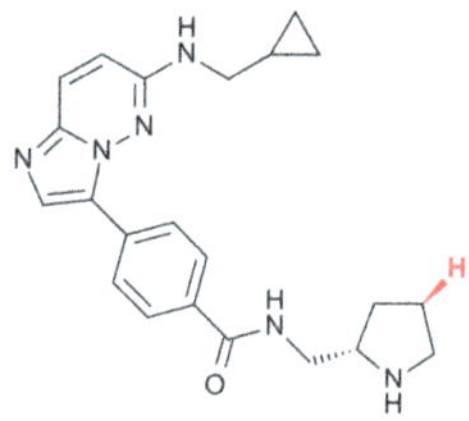	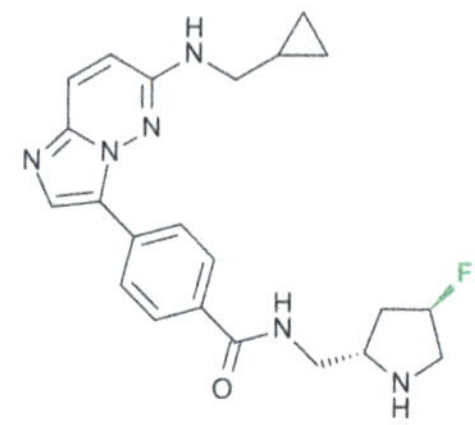
PAMPA P_{app} ($\times 10^{-6}$ cm s^{-1})	0.84	36
ACD clogD/clogP	−1.53/1.29	0.08/1.12
ACD cpK_a	10.2	8.4

Design tactics:

- Reduce pK_a of ionizable basic group (reduce fraction ionized) which increases logD

For a series of Imidazol[1,2-*b*]pyridazines as inhibitor of nuclear factor-κB kinase (IKK_β) inhibitors, addition of a fluorine on a pyrrolidine ring led to a significant improvement of permeability through reducing the pK_a of the basic group, which in turn led to an increase in logD despite a lower clogP.[57,58]

4

	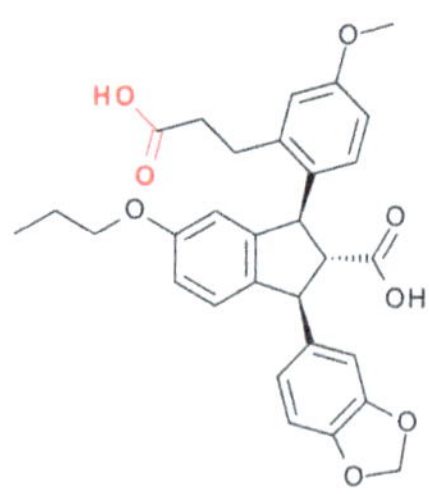	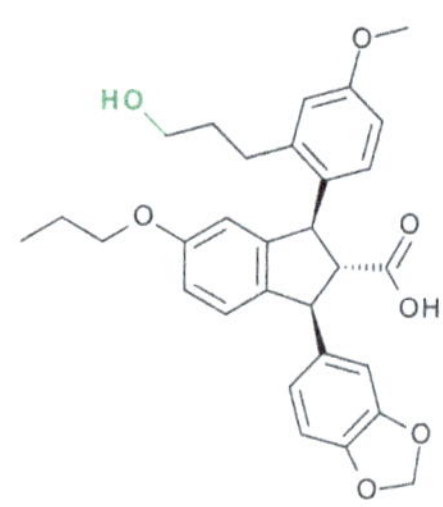
Caco-2 P_{app} ($\times 10^{-6}$ cm s^{-1})	2.1	56.8
***F* (rat; %)**	4	66
ACD clogD/clogP	−0.14/5.78	2.64/5.82
ACD cpK_a (acids)	4.2/4.7	4.2
TPSA (Å^2)	118	100

Design tactics:

- Remove an ionizable group (acid) which increases logD
- Lower TPSA

A carboxylic acid in a series of diacid endothelin receptor A antagonists was replaced with a primary alcohol. This change reduced the TPSA, removed an ionizable centre and led to an increase in logD which overall resulted in a significant increase in permeability and rat oral bioavailability (F).[59,60]

Examples		
5	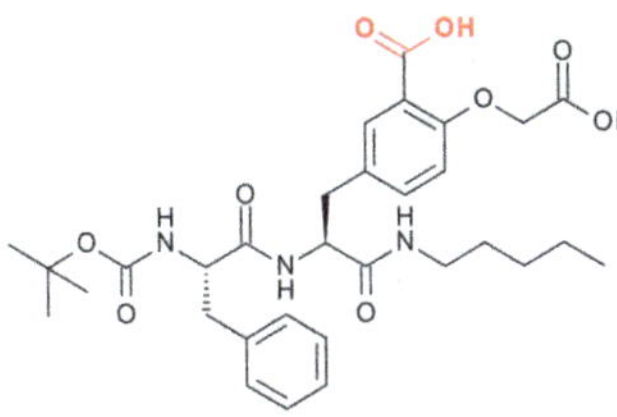	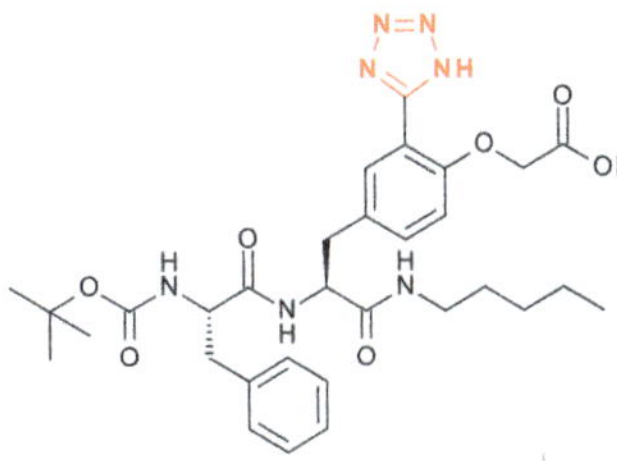
Caco-2 P_{app} (×10^{-6} cm s^{-1})	<0.10	0.19
ACD clogD/clogP	−3.4/4.8	−2.0/4.7
ACD cpK_a (acids)	3.0/3.5	3.0/5.3
TPSA (Å^2)	198	211
Cellular activity	No	Yes

Design tactics:

- Replace carboxylic acid with less acidic tetrazole isostere
- Increase logD

Replacement of a carboxylic acid in a series of protein tyrosine phosphatase 1B (PTP1B) inhibitors with a less acidic bioisosteric tetrazole led to an overall increase in logD which in turn gave a modest improvement in permeability despite a higher TPSA.[59,61,62]

[a]Passive permeability in MDCK-MDR1 cell line with P-glycoprotein (P-gp) inhibited.

3.3.3 Lower Effective Polar Surface Area

Examples		
1	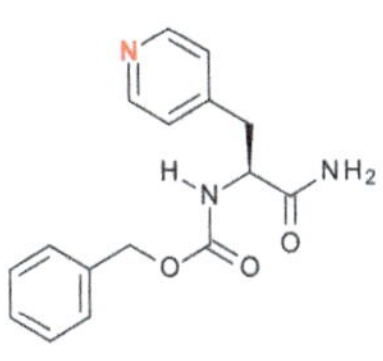	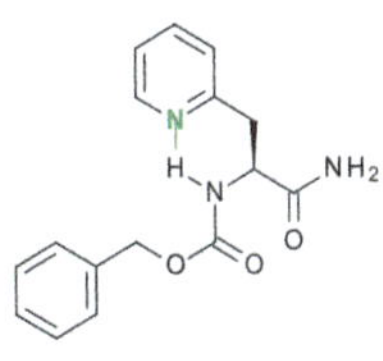
Caco-2 P_{app} (×10^{-6} cm s^{-1})	4.3	17.7
MW	299	299
HBA/HBD (count)	6/3	6/3
ACD clogD/clogP	1.2/1.2	1.2/1.1
TPSA (Å^2)	98	98

Design tactics:

- Introduce IMHB

In a systematic permeability study of peptidic small molecules the effect of introducing an IMHB was explored whilst minimizing other concomitant changes to lipophilicity *etc.* The introduction of an IMHB proved highly successful at raising passive permeability.[55,63]

(continued)

Examples

2

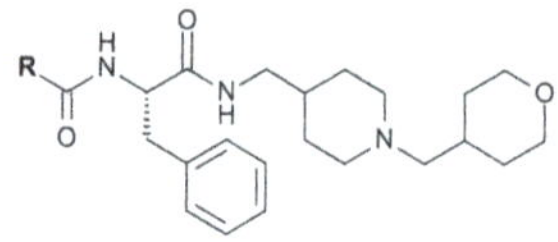

R =

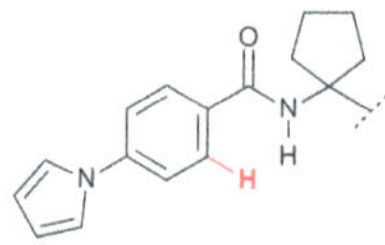

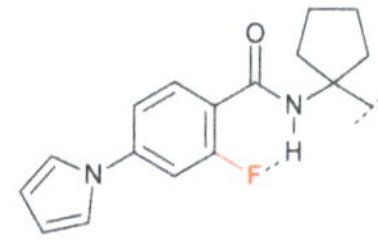

PAMPA P_{app} (×10^{-6} cm s^{-1})	n/a	0.66
Caco-2 P_{app} (×10^{-6} cm s^{-1})	<1.0	13.8
MW	640	658
ACD clogP/clogD	4.7/3.0	4.5/3.3
HBA/HBD (count)	9/3	9/3
TPSA (Å^2)	109	109

R =

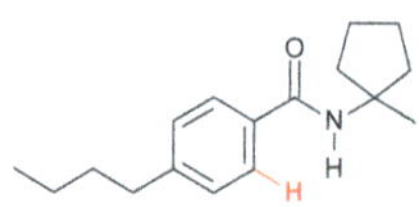

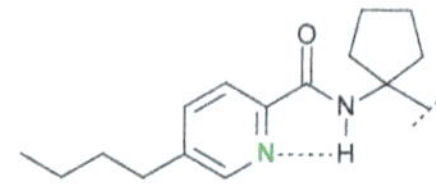

PAMPA P_{app} (×10^{-6} cm s^{-1})	1.6	7.2
Caco-2 P_{app} (×10^{-6} cm s^{-1})	9.3	17.6
MW	631	632
ACD clogP/clogD	5.4/3.8	5.0/3.5
HBA/HBD (count)	8/3	9/3
TPSA (Å^2)	109	118

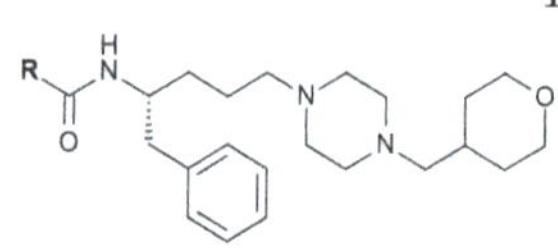

R =

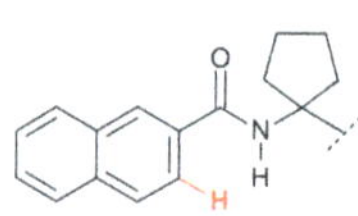

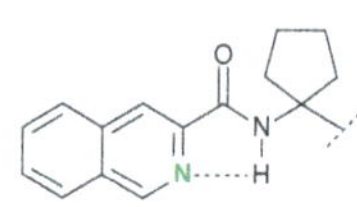

PAMPA P_{app} (×10^{-6} cm s^{-1})	6.0	15.5
Caco-2 P_{app} (×10^{-6} cm s^{-1})	14.6	25.2
MW	611	612
ACD clogP/clogD	4.3/4.1	3.9/4.1
HBA/HBD	7/2	8/2
TPSA (Å^2)	77	87

Design tactics:

- Introduce IMHB (*e.g.* halogen interactions)

In a study into the effects of hydrogen bonding on permeability various groups, including halogens and nitrogen acceptors, were added to form five- and six-membered IMHBs. These all led to a modest improvement in permeability.[55,64]

Examples

3

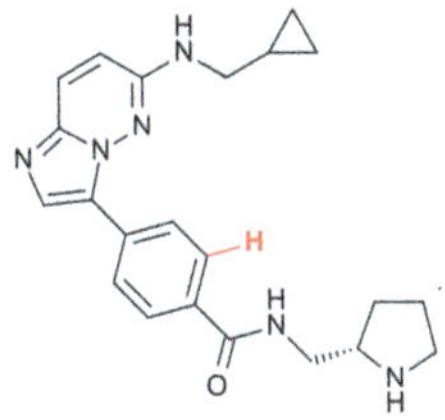

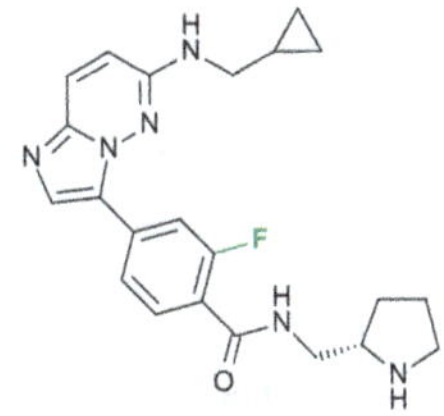

PAMPA P_{app} ($\times 10^{-6}$ cm s^{-1})	0.84	19
MW	391	409
ACD clogP/clogD	1.3/−1.5	0.9/−1.9
HBA/HBD	7/3	7/3
TPSA (Å^2)	85	85

Design tactics:
- Introduce IMHB (*via* halogen interactions)

Introduction of halogens to make six-membered IMHBs to an ortho amide group of Imidazo[1,2-*b*]pyridazines as IKK$_{\beta}$ inhibitors led to significant improvements in permeability.[57]

4

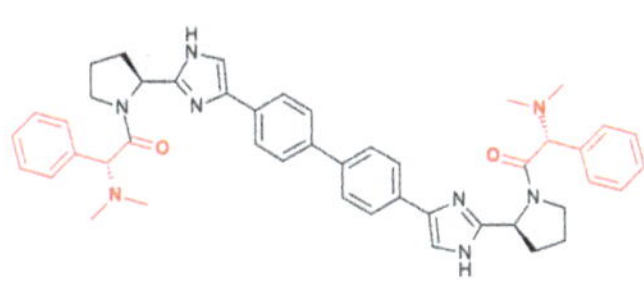

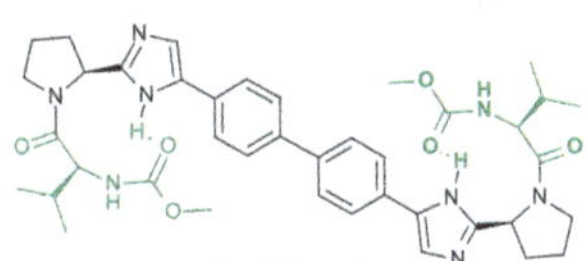

Daclatasvir

RRCK P_{app} ($\times 10^{-6}$ cm s^{-1})	0.84	19
MW	747	739
logD	3.2	2.6
HBA/HBD	6/2	6/4
TPSA (Å)	105	175
ePSA (Å)	128	103

Design tactics:
- Introduce IMHB to lower ePSA

Research on the hepatitis C virus nonstructural protein 5A (NS5A) inhibitor Daclatasvir enabled significant understanding about the design of chameleonic compounds outside of traditional Ro5 space. Incorporation of amino acid groups, despite increasing the 2D TPSA and polarity of the molecules, led to significantly increased permeability. This has been rationalized and is believed to result from the carbamate group making IMHBs with an imidazole in the linker, thereby reducing the ePSA.[65,66]

(continued)

Examples

5

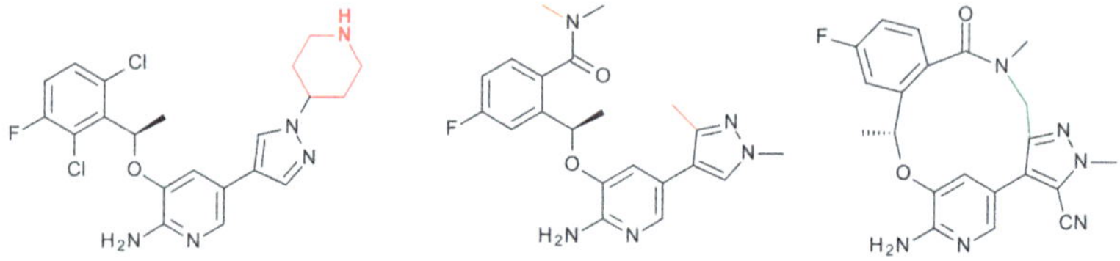

Crizotinib Lorlatinib

MDCK AB P_{app} ($\times 10^{-6}$ cm s^{-1})	0.28	2.5	19.3
MDCK BA P_{app} ($\times 10^{-6}$ cm s^{-1})	12.5	18.8	28.0
ACD clogP	4.7	2.1	2.4
logD	2.0	2.1	2.3
HBA/HBD (count)	6/3	7/2	8/2
TPSA (Å^2)	75	80	99

Design tactics:

- Macrocyclise to restrict conformation, favour IMHB formation and reduce solvent-accessible surface area

The design of the anaplastic lymphoma kinase (ALK) inhibitor lorlatinib as a central nervous system (CNS) penetrant kinase inhibitor illustrated the potential value of macrocyclization as a strategy to increase permeability.[67]

6

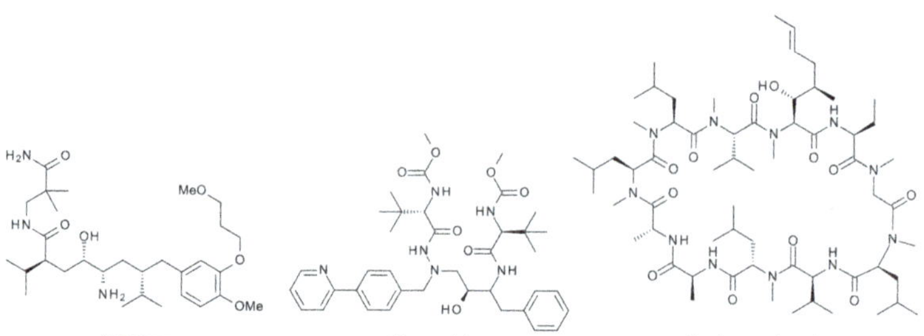

Aliskiren Atazanivir Cyclosporine A

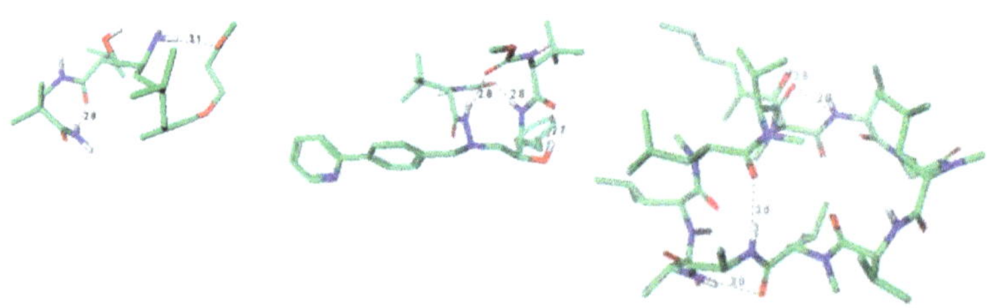

MDCK P_{app} ($\times 10^{-6}$ cm s^{-1})	1	35	n/a
MW	552	705	1202
ACD clogP/clogD	3.5/2.4	5.9/3.9	14.4/4.6

Examples

HBA/HBD (count)	9/6	13/5	23/5
TPSA (Å^2)	146	171	279
***F* (rat; %)**	0.1	10	24
Fraction absorbed [F_a (%) (rat)]	0.5	54	26
***F*% (human)**	3	60	30

Design tactics:

- Macrocyclise to restrict conformation, favour multiple IMHB formations and reduce solvent accessible surface area.

Although these compounds are not part of a related design strategy, together they serve to illustrate just how powerful IMHBs and macrocyclization strategies can be in delivering chameleonic compound properties. The example of the natural product cyclosporine A in particular is commonly used to exemplify how even high MW and highly polar peptide molecules can in theory be optimized to exhibit good levels of passive permeability.[68]

3.3.4 Reduce Numbers of HBD

Examples

1

MDCK-MDR1 AB P_{app} (×10^{-6} cm s^{-1})	0.51	8.41
MDCK-MDR1 BA P_{app} (×10^{-6} cm s^{-1})	2.94	25.7
HBA/HBD (count)	8/3	8/2
ACD clogD/clogP	0.1/0.2	1.3
TPSA (Å^2)	117	107

Design tactics:

- Remove HBD
- Increase logD
- Lower TPSA

In a systematic study of permeability, changing a heterocycle from an imidazole with a free NH to an N-linked pyrazole resulted in a reduction in HBD count, an increase in clogD and reduction in TPSA. Overall, this led to a significant improvement in permeability.[63]

(continued)

Examples

2

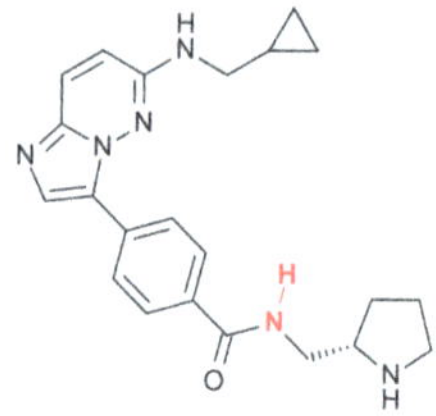 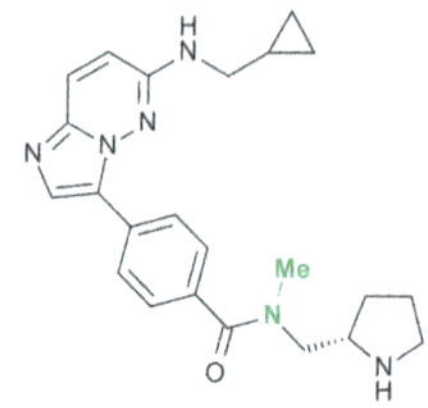

PAMPA P_{app} (×10^{-6} cm s^{-1})	<2	6.1
HBA/HBD (count)	7/3	7/2
ACD clogD/clogP	−1.53/1.29	−2.0/0.86
ACD cpK_a	10.2	10.3
TPSA (Å^2)	85	72

Design tactics:

- Reduce number of HBDs
- Lower TPSA

Removal of a HBD for a series if imidazo[1,2-*b*]pyridazines as IKK$_\beta$ inhibitors by changing from a secondary to tertiary amide led to an increase in permeability.[58]

3

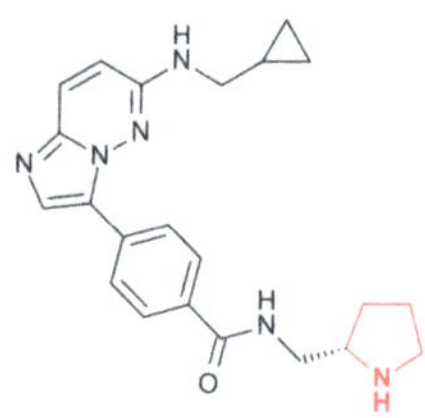 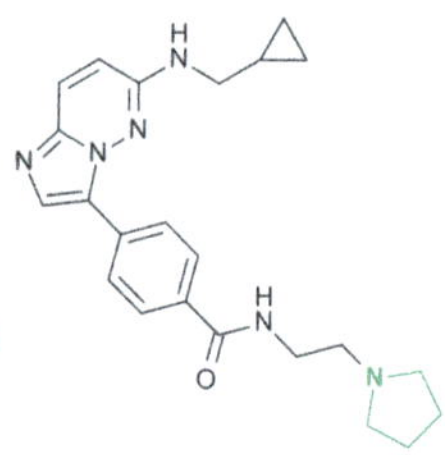 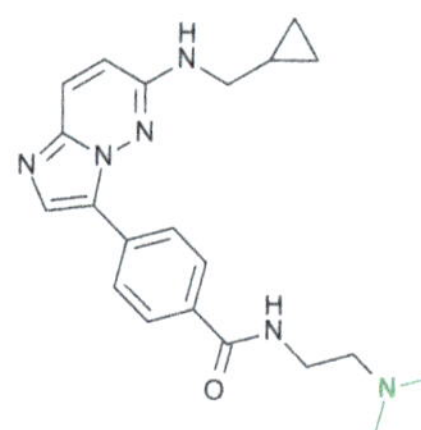

PAMPA P_{app} (×10^{-6} cm s^{-1})	0.84	>50	>50
logD (pH 7.4)	1.8	2.6	2.5
HBA/HBD (count)	7/3	7/2	7/2
ACD clogD/clogP	−1.53/1.29	−0.15/1.55	0.26/1.31
ACD cpK_a	10.2	9.0	8.9
TPSA (Å^2)	85	72	72

Design tactics:

- Reduce number of HBDs
- Increase logD
- Lower TPSA

For a series of imidazo[1,2-*b*]pyridazines as IKK$_\beta$ inhibitors moving from a pyrrolidine with a free NH to an N-linked pyrrolidine or dimethylamine resulted in a large increase in permeability probably due to the combination of fewer HBDs, reduced TPSA and increased logD due to changing to a less basic amine.[57]

Examples		

4

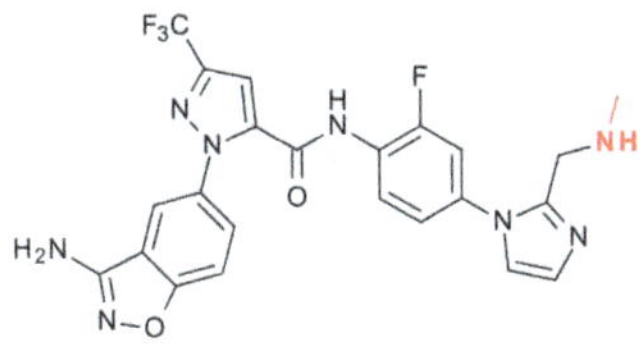

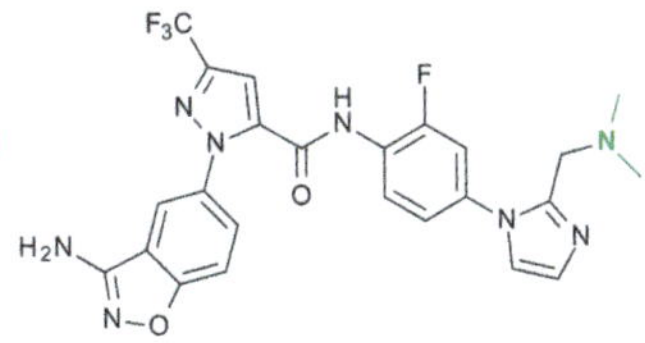

Caco-2 P_{app} ($\times 10^{-6}$ cm s^{-1})	0.2	5.6
CL (l h^{-1} kg^{-1})	1.1	1.1
%*F* (rat)	24	84
HBA/HBD (count)	10/4	10/3
ACD clogD/clogP	0.11/1.38	1.09/1.87
ACD cpK_a	8.5	7.8
TPSA (Å^2)	122	109

Design tactics:
- Reduce number of HBDs
- Increase logD
- Reduce TPSA

Replacement of a secondary amine with the equivalent *N*-dimethyl amine in a series of factor Xa inhibitors led to a reduction in the number of HBDs, reduction in TPSA and an increase in logD, factors which correlate with increased permeability.[59,61,69]

3.3.5 Examples of Prodrugs

Examples		

1

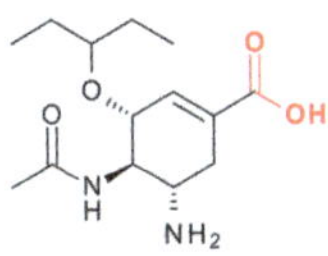

Oseltamivir carboxylate

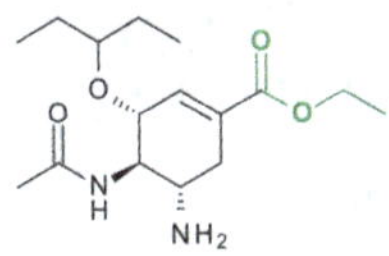

Oseltamivir

	Oseltamivir carboxylate	Oseltamivir
***F*%**	<5%	79%
***F*$_a$%**	Low	>80%
ACD clogP/clogD	0.45/−2.1	1.5/0.1
HBA/HBD (count)	6/4	6/3
TPSA (Å^2)	110	97

Design tactics:
- Remove acid (ionizable)
- Increase logD
- Reduce number of HBDs
- Reduce TPSA

The ester prodrug of the carboxylic acid drug oseltamivir led to a significant improvement in fraction absorbed and, in turn, oral bioavailability *in vivo* due to masking of a charge at physiological pH.[70,71]

(*continued*)

Examples

2

	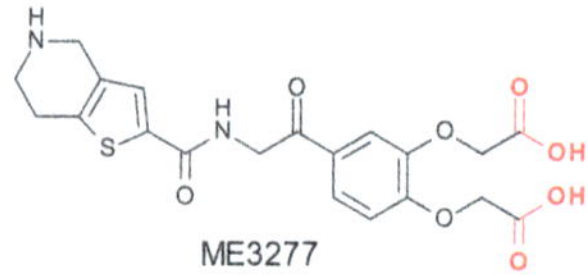 ME3277	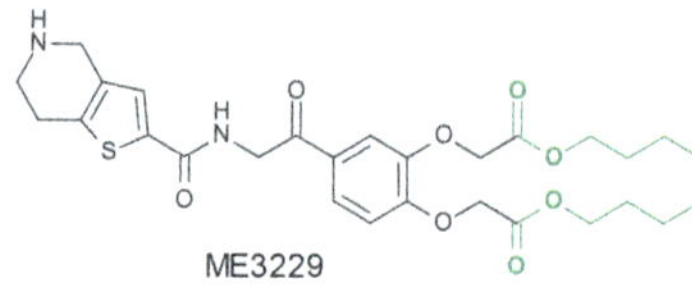 ME3229
Caco-2 P_{app} ($\times 10^{-6}$ cm s^{-1})	0.09	7.9
logD	−3.0	1.3
ACD clogP/clogD	−0.06/−6.6	3.3/1.8
HBA/HBD (count)	10/4	10/2
TPSA (Å^2)	166	138

Design tactics:

- Remove acid (ionizable)
- Reduce number of HBDs
- Increase logD
- Reduce TPSA

A double ester prodrug of diacid compound ME3277 gave a significant improvement in permeability measurements due to masking two charged groups.[70]

3

	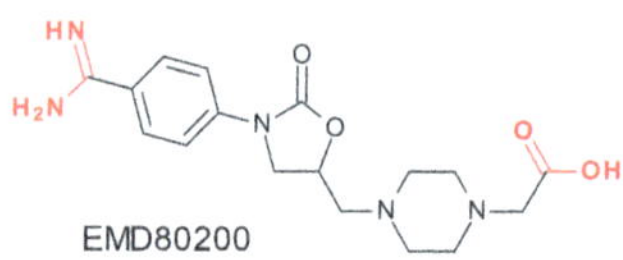 EMD80200	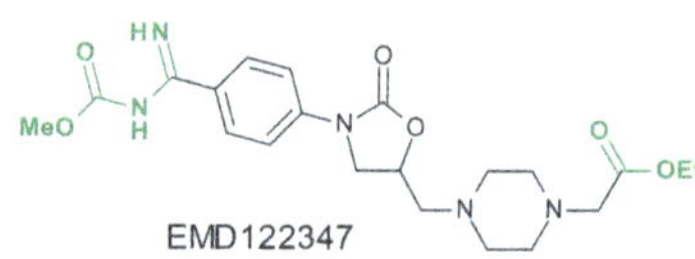 EMD122347
***F*% (cyno)**	3–10%	40%
Caco-2 P_{app} ($\times 10^{-6}$ cm s^{-1})	0.28	7.3
ACD clogP/clogD	−0.35/−3.6	0.85/0.82
HBA/HBD (count)	9/4	11/2
TPSA (Å^2)	122	122

Design tactics:

- Remove acid (ionizable)
- Remove base (ionizable)
- Reduce number of HBDs
- Increase logD

A double prodrug to mask both a basic amidine and a carboxylic acid in zwitterionic compound EMD80200 led to significant improvements in permeability (of the prodrug) and *in vivo* oral bioavailability of the parent molecule.[70]

Examples

4	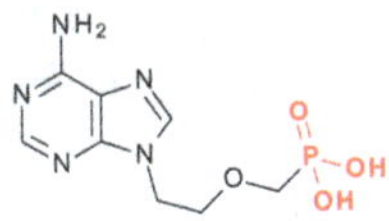 Adefovir	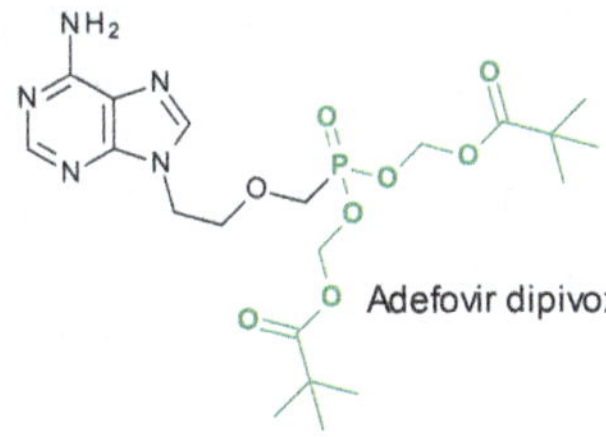 Adefovir dipivoxil
F (%)	8% (rat), 4% (cyno), 12% (human)	38% (rat), 25% (cyno), 40% (human)
Caco-2 (% h^{-1})	<0.1	9
logD	−4.11	2.48
HBA/HBD (count)	9/4	13/2
TPSA (Å^2)	132	158

Design tactics:

- Mask acid (ionizable)
- Reduce number of HBDs
- Increase logD

The phosphoric acid prodrug of Adefovir demonstrates a modest improvement in both permeability and *in vivo* oral bioavailability due to removing the charged phosphoric acid group.[70]

References

1. Y.-L. He, S. Murby, G. Warhurst, L. Gifford, D. Walker, J. Ayrton, R. Eastmond and M. Rowland, *J. Pharm. Sci.*, 1998, **87**, 626.
2. H. J. R. Lemmer and J. H. Hamman, *Expert Opin. Drug Delivery*, 2013, **10**, 103.
3. N. L. Trevaskis, L. M. Kaminskas and C. J. H. Porter, *Nat. Rev. Drug Discovery*, 2015, **14**, 781.
4. E. F. Choo, J. Boggs, C. Zhu, J. W. Lubach, N. D. Catron, G. Jenkins, A. J. Souers and R. Voorman, *Drug Metab. Dispos.*, 2014, **42**, 207.
5. C. Pidgeon, S. Ong, H. Liu, X. Qiu, M. Pidgeon, A. H. Dantzig, J. Munroe, W. J. Hornback and J. S. Kasher, *et al.*, *J. Med. Chem.*, 1995, **38**, 590.
6. G. H. Goetz, W. Farrell, M. Shalaeva, S. Sciabola, D. Anderson, J. Yan, L. Philippe and M. J. Shapiro, *J. Med. Chem.*, 2014, **57**, 2920.
7. G. H. Goetz, L. Philippe and M. J. Shapiro, *ACS Med. Chem. Lett.*, 2014, **5**, 1167.
8. M. Kansy, F. Senner and K. Gubernator, *J. Med. Chem.*, 1998, **41**, 1007.
9. F. Assmus, A. Ross, H. Fischer, J. Seelig and A. Seelig, *Mol. Pharmaceutics*, 2017, **14**, 284.

10. I. J. Hidalgo, T. J. Raub and R. T. Borchardt, *Gastroenterology*, 1989, **96**, 736.
11. E. Callegari, B. Malhotra, P. J. Bungay, R. Webster, K. S. Fenner, S. Kempshall, J. L. LaPerle, M. C. Michel and G. G. Kay, *Br. J. Clin. Pharmacol.*, 2011, **72**, 235.
12. L. Di, C. Whitney-Pickett, J. P. Umland, H. Zhang, X. Zhang, D. F. Gebhard, Y. Lai, J. J. Federico, R. E. Davidson, R. Smith, E. L. Reyner, C. Lee, B. Feng, C. Rotter, M. V. Varma, S. Kempshall, K. Fenner, A. F. El-kattan, T. E. Liston and M. D. Troutman, *J. Pharm. Sci.*, 2011, **100**, 4974.
13. J. D. Irvine, L. Takahashi, K. Lockhart, J. Cheong, J. W. Tolan, H. E. Selick and J. R. Grove, *J. Pharm. Sci.*, 1999, **88**, 28.
14. T. Takenaka, N. Harada, J. Kuze, M. Chiba, T. Iwao and T. Matsunaga, *Drug Metab. Dispos.*, 2014, **42**, 1947.
15. R. Hayeshi, C. Masimirembwa, S. Mukanganyama and A.-L. B. Ungell, *Biopharm. Drug Dispos.*, 2008, **29**, 324.
16. V. Rozehnal, D. Nakai, U. Hoepner, T. Fischer, E. Kamiyama, M. Takahashi, S. Yasuda and J. Mueller, *Eur. J. Pharm. Sci.*, 2012, **46**, 367.
17. C. A. Lipinski, *J. Pharmacol. Toxicol. Methods*, 2000, **44**, 235.
18. M. M. Hann and G. M. Keserue, *Nat. Rev. Drug Discovery*, 2012, **11**, 355.
19. M. D. Shultz, *ACS Med. Chem. Lett.*, 2014, **5**, 2.
20. M. D. Shultz, *J. Med. Chem.*, 2019, **62**, 1701.
21. P. D. Leeson, A. M. Davis and J. Steele, *Drug Discovery Today: Technol.*, 2004, **1**, 189.
22. P. D. Leeson and B. Springthorpe, *Nat. Rev. Drug Discovery*, 2007, **6**, 881.
23. P. D. Leeson and J. R. Empfield, *Annu. Rep. Med. Chem.*, 2010, **45**, 393.
24. C. W. Murray, D. A. Erlanson, A. L. Hopkins, G. M. Keseru, P. D. Leeson, D. C. Rees, C. H. Reynolds and N. J. Richmond, *ACS Med. Chem. Lett.*, 2014, **5**, 616.
25. A. L. Hopkins, G. M. Keserue, P. D. Leeson, D. C. Rees and C. H. Reynolds, *Nat. Rev. Drug Discovery*, 2014, **13**, 105.
26. M. J. Waring, J. Arrowsmith, A. R. Leach, P. D. Leeson, S. Mandrell, R. M. Owen, G. Pairaudeau, W. D. Pennie, S. D. Pickett, J. Wang, O. Wallace and A. Weir, *Nat. Rev. Drug Discovery*, 2015, **14**, 475.
27. P. D. Leeson and R. J. Young, *ACS Med. Chem. Lett.*, 2015, **6**, 722.
28. M. K. Bayliss, J. Butler, P. L. Feldman, D. V. S. Green, P. D. Leeson, M. R. Palovich and A. J. Taylor, *Drug Discovery Today*, 2016, **21**, 1719.
29. P. D. Leeson, *Adv. Drug Delivery Rev.*, 2016, **101**, 22.
30. R. J. Young and P. D. Leeson, *J. Med. Chem.*, 2018, **61**, 6421.
31. M. J. Waring, *Expert Opin. Drug Discovery*, 2010, **5**, 235.
32. T. W. Johnson, K. R. Dress and M. Edwards, *Bioorg. Med. Chem. Lett.*, 2009, **19**, 5560.
33. M. J. Waring, *Bioorg. Med. Chem. Lett.*, 2009, **19**, 2844.
34. P. W. Kenny and C. A. Montanari, *J. Comput.-Aided Mol. Des.*, 2013, **27**, 1.
35. K. D. Freeman-Cook, R. L. Hoffman and T. W. Johnson, *Future Med. Chem.*, 2013, **5**, 113.

36. L. S. Scanker, D. J. Tocco, B. B. Brodie and C. A. M. Hogben, *J. Pharmacol. Exp. Ther.*, 1958, **123**, 81.
37. M. V. S. Varma, R. S. Obach, C. Rotter, H. R. Miller, G. Chang, S. J. Steyn, A. El-Kattan and M. D. Troutman, *J. Med. Chem.*, 2010, **53**, 1098.
38. K. Palm, K. Luthman, J. Ros, J. Grasjo and P. Artursson, *J. Pharmacol. Exp. Ther.*, 1999, **291**, 435.
39. K. Palm, P. Stenberg, K. Luthman and P. Artursson, *Pharm. Res.*, 1997, **14**, 568.
40. G. H. Goetz, M. Shalaeva, G. Caron, G. Ermondi and L. Philippe, *Mol. Pharmaceutics*, 2017, **14**, 386.
41. D. F. Veber, S. R. Johnson, H.-Y. Cheng, B. R. Smith, K. W. Ward and K. D. Kopple, *J. Med. Chem.*, 2002, **45**, 2615.
42. B. C. Doak, B. Over, F. Giordanetto and J. Kihlberg, *Chem. Biol.*, 2014, **21**, 1115.
43. A. Ritzen and L. David, *Abstracts of Papers, 253rd ACS National Meeting & Exposition, San Francisco, CA, United States, 2–6, April, 2017*, MEDI, 2017.
44. V. Poongavanam, B. C. Doak and J. Kihlberg, *Curr. Opin. Chem. Biol.*, 2018, **44**, 23.
45. C. R. W. Guimaraes, A. M. Mathiowetz, M. Shalaeva, G. Goetz and S. Liras, *J. Chem. Inf. Model.*, 2012, **52**, 882.
46. D. A. DeGoey, H.-J. Chen, P. B. Cox and M. D. Wendt, *J. Med. Chem.*, 2018, **61**, 2636.
47. K. Sugano, M. Kansy, P. Artursson, A. Avdeef, S. Bendels, L. Di, G. F. Ecker, B. Faller, H. Fischer, G. Gerebtzoff, H. Lennernaes and F. Senner, *Nat. Rev. Drug Discovery*, 2010, **9**, 597.
48. M. Rossi Sebastiano, B. C. Doak, M. Backlund, V. Poongavanam, B. Over, G. Ermondi, G. Caron, P. Matsson and J. Kihlberg, *J. Med. Chem.*, 2018, **61**, 4189.
49. M. R. Naylor, A. M. Ly, M. J. Handford, D. P. Ramos, C. R. Pye, A. Furukawa, V. G. Klein, R. P. Noland, Q. Edmondson, A. C. Turmon, W. M. Hewitt, J. Schwochert, C. E. Townsend, C. N. Kelly, M.-J. Blanco and R. S. Lokey, *J. Med. Chem.*, 2018, **61**, 11169.
50. D. A. Smith, *Curr. Opin. Drug Discovery Dev.*, 2007, **10**, 550.
51. A. Pike, R. I. Storer, R. M. Owen, E. Armstrong, C. L. Benn, M. Bictash, K. F. K. Cheung, K. Costelloe, E. Dardennes, E. Impey, P. H. Milliken, E. Mortimer-Cassen and H. J. Pearce, *MedChemComm*, 2016, **7**, 1572.
52. R. I. Storer, R. M. Owen, A. Pike, C. L. Benn, E. Armstrong, D. C. Blakemore, M. Bictash, K. Costelloe, E. Impey, P. H. Milliken, E. Mortimer-Cassen, H. J. Pearce, B. Pibworth and G. Toschi, *MedChemComm*, 2016, **7**, 1587.
53. N. A. Swain, D. Batchelor, S. Beaudoin, B. M. Bechle, P. A. Bradley, A. D. Brown, B. Brown, K. J. Butcher, R. P. Butt, M. L. Chapman, S. Denton, D. Ellis, S. R. G. Galan, S. M. Gaulier, B. S. Greener, M. J. de Groot, M. S. Glossop, I. K. Gurrell, J. Hannam, M. S. Johnson, Z. Lin, C. J. Markworth, B. E. Marron, D. S. Millan, S. Nakagawa, A. Pike, D. Printzenhoff, D. J. Rawson, S. J. Ransley, S. M. Reister, K. Sasaki, R. I. Storer, P. A. Stupple and C. W. West, *J. Med. Chem.*, 2017, **60**, 7029.

54. J. T. Goodwin, R. A. Conradi, N. F. H. Ho and P. S. Burton, *J. Med. Chem.*, 2001, **44**, 3721.
55. P. V. Desai, T. J. Raub and M.-J. Blanco, *Bioorg. Med. Chem. Lett.*, 2012, **22**, 6540.
56. C. D. Cox, M. J. Breslin, D. B. Whitman, P. J. Coleman, R. M. Garbaccio, M. E. Fraley, M. M. Zrada, C. A. Buser, E. S. Walsh, K. Hamilton, R. B. Lobell, W. Tao, M. T. Abrams, V. J. South, H. E. Huber, N. E. Kohl and G. D. Hartman, *Bioorg. Med. Chem. Lett.*, 2007, **17**, 2697.
57. H. Shimizu, T. Yamasaki, Y. Yoneda, F. Muro, T. Hamada, T. Yasukochi, S. Tanaka, T. Toki, M. Yokoyama, K. Morishita and S. Iimura, *Bioorg. Med. Chem. Lett.*, 2011, **21**, 4550.
58. H. Shimizu, I. Yasumatsu, T. Hamada, Y. Yoneda, T. Yamasaki, S. Tanaka, T. Toki, M. Yokoyama, K. Morishita and S. Iimura, *Bioorg. Med. Chem. Lett.*, 2011, **21**, 904.
59. L. Di and E. Kerns, *Drug-like Properties: Concepts, Structure Design and Methods from ADME to Toxicity Optimization*, 2nd edn, 2016, p. 580.
60. H. Ellens, E. P. Eddy, C.-P. Lee, P. Dougherty, A. Lago, J.-N. Xiang, J. D. Elliott, H.-Y. Cheng, E. Ohlstein and P. L. Smith, *Adv. Drug Delivery Rev.*, 1997, **23**, 99.
61. *Optimizing the "Drug-like" Properties of Leads in Drug Discovery. [In: Biotechnol.: Pharm. Aspects, 2006, 4]*, ed. R. T. Borchardt, M. J. Hageman, J. L. Stevens, E. H. Kerns and D. R. Thakker, 2006, p. 511.
62. C. Liljebris, S. D. Larsen, D. Ogg, B. J. Palazuk and J. E. Bleasdale, *J. Med. Chem.*, 2002, **45**, 1785.
63. S. B. Rafi, B. R. Hearn, P. Vedantham, M. P. Jacobson and A. R. Renslo, *J. Med. Chem.*, 2012, **55**, 3163.
64. A. Ettorre, P. D'Andrea, S. Mauro, M. Porcelloni, C. Rossi, M. Altamura, R. M. Catalioto, S. Giuliani, C. A. Maggi and D. Fattori, *Bioorg. Med. Chem. Lett.*, 2011, **21**, 1807.
65. F. Wakenhut, T. D. Tran, C. Pickford, S. Shaw, M. Westby, C. Smith-Burchnell, L. Watson, M. Paradowski, J. Milbank, D. Stonehouse, K. Cheung, R. Wybrow, F. Daverio, S. Crook, K. Statham, D. Leese, D. Stead, F. Adam, D. Hay, L. R. Roberts, J.-Y. Chiva, C. Nichols, D. C. Blakemore, G. H. Goetz, Y. Che, I. Gardner, S. Dayal, A. Pike, R. Webster and D. C. Pryde, *ChemMedChem*, 2014, **9**, 1387.
66. T. D. Tran, F. Wakenhut, C. Pickford, S. Shaw, M. Westby, C. Smith-Burchnell, L. Watson, M. Paradowski, J. Milbank, R. A. Brimage, R. Halstead, R. Glen, C. P. Wilson, F. Adam, D. Hay, J.-Y. Chiva, C. Nichols, D. C. Blakemore, I. Gardner, S. Dayal, A. Pike, R. Webster and D. C. Pryde, *ChemMedChem*, 2014, **9**, 1378.
67. T. W. Johnson, P. F. Richardson, S. Bailey, A. Brooun, B. J. Burke, M. R. Collins, J. J. Cui, J. G. Deal, Y.-L. Deng, D. Dinh, L. D. Engstrom, M. He, J. Hoffman, R. L. Hoffman, Q. Huang, R. S. Kania, J. C. Kath, H. Lam, J. L. Lam, P. T. Le, L. Lingardo, W. Liu, M. McTigue, C. L. Palmer, N. W. Sach, T. Smeal, G. L. Smith, A. E. Stewart, S. Timofeevski, H. Zhu, J. Zhu, H. Y. Zou and M. P. Edwards, *J. Med. Chem.*, 2014, **57**, 4720.

68. A. Alex, D. S. Millan, M. Perez, F. Wakenhut and G. A. Whitlock, *MedChemComm*, 2011, **2**, 669.
69. M. L. Quan, P. Y. S. Lam, Q. Han, D. J. P. Pinto, M. Y. He, R. Li, C. D. Ellis, C. G. Clark, C. A. Teleha, J.-H. Sun, R. S. Alexander, S. Bai, J. M. Luettgen, R. M. Knabb, P. C. Wong and R. R. Wexler, *J. Med. Chem.*, 2005, **48**, 1729.
70. K. Beaumont, R. Webster, I. Gardner and K. Dack, *Curr. Drug Metab.*, 2003, **4**, 461.
71. G. He, J. Massarella and P. Ward, *Clin. Pharmacokinet.*, 1999, **37**, 471.

CHAPTER 4

Targeting Gastrointestinal Uptake Transporters

SIMONE H. STAHL[a], KATHERINE S. FENNER[a], M. RAYMOND V. FINLAY[b], RAVINDRA V. ALLURI[a], BETH WILLIAMSON[c], JOHAN X. JOHANSSON[d] AND JASON KETTLE*[b]

[a]ADME Sciences, Clinical Pharmacology & Safety Sciences, R&D, AstraZeneca, 310 Darwin Building, Cambridge Science Park, Milton Road, Cambridge, CB4 0WG, UK; [b]Medicinal Chemistry, Research and Early Development, Oncology R&D, AstraZeneca, 310 Darwin Building, Cambridge Science Park, Milton Road, Cambridge, CB4 0WG, UK; [c]Drug Metabolism and Pharmacokinetics, Research and Early Development, Oncology R&D, Hodgkin Building, Chesterford Research Campus, Little Chesterford, Saffron Walden, CB10 1XL, UK; [d]Medicinal Chemistry, Research and Early Development, Cardiovascular, Renal and Metabolism, BioPharmaceutical R&D, AstraZeneca, Pepparedsleden 1, Mölndal 431 83, Sweden
*E-mail: jason.kettle@astrazeneca.com

4.1 Introduction

The solute carrier (SLC) protein family encompasses approximately 400 transporter proteins that enable substrates, including drugs, to be transported across cell membranes. A number of SLC transporters expressed on enterocytes in the gastrointestinal (GI) tract (Figure 4.1) have a physiological role in the absorption of endogenous molecules, such as peptides, vitamins or bile acids. In theory, enterocyte influx transporters should increase the oral absorption of drug molecules that are substrates of these

Drug Discovery Series No. 79
The Medicinal Chemist's Guide to Solving ADMET Challenges
Edited by Patrick Schnider
© The Royal Society of Chemistry 2021
Published by the Royal Society of Chemistry, www.rsc.org

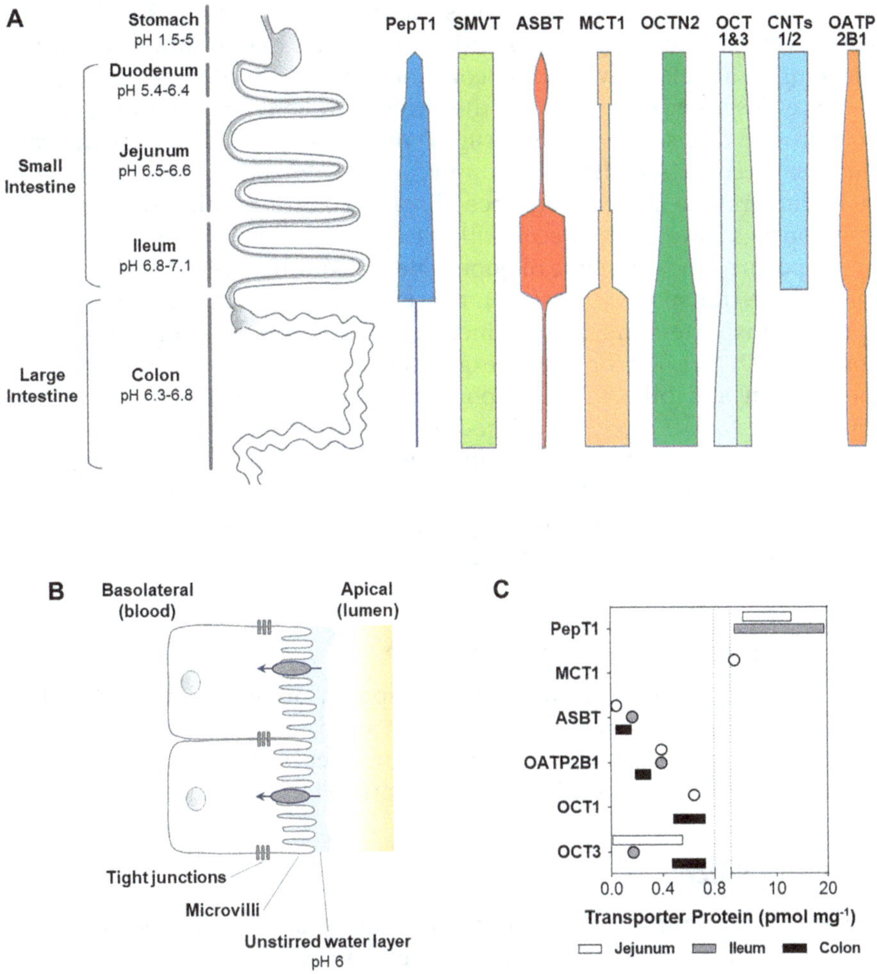

Figure 4.1 SLC transporters expressed in the GI tract. (A) Illustration of relative expression levels of selected apical SLC transporters along the intestine in relation to the expression in the jejunum.[1] The width of the symbols shows approximate expression levels for the respective transporter comparing different intestinal segments (but not different transporters) with wider symbols indicating higher expression levels. Where available data on human protein levels were referred to. More detailed information on intestinal transporter mRNA and protein expression is provided in a range of studies on the subject,[2–8] with the recent study by Harwood *et al.*, analysing data from several studies.[1] (B) Illustration of intestinal epithelial cells outlining the polarised nature with distinct apical and basolateral domains. The transporters discussed in this chapter are mostly expressed at the apical domain facing the gut lumen (with exceptions detailed in the descriptions of individual transporters in the text). Whilst luminal pH can vary (*e.g.* depending on feeding status), the unstirred water layer maintains an acidic microclimate which is important for the activity of some transporters like PepT1. (C) Summary of absolute protein expression levels (normalised to

transporters. However, many of these transporters show high affinity for substrates, thus limiting the potential dosage size due to saturation of the transporter at relatively modest concentrations (<50 μM). Several examples are described below where the drug is targeted either to the transport protein directly or as a prodrug. Typically, such prodrugs incorporate chemical recognition elements from known endogenous substrates using facile linkers that are cleaved once absorbed. For a number of transporters prodrugs have been successfully designed to target them to increase oral exposure. These include oligopeptide transporter 1 (PepT1) and monocarboxylate transporter 1 (MCT1), two of the most abundantly expressed uptake transporters in the intestine, as well as organic cation transporter novel 2 (OCTN2). Furthermore, examples exist with *in vitro* evidence for the vitamin and bile acid transporters sodium-dependent multivitamin transporter (SMVT) and apical sodium-dependent bile acid transporter (ASBT), respectively. Whilst literature evidence of successful drug targeting is limited, the potential of transporters such as organic cation transporters (OCTs), organic anion transporting polypeptides (OATPs) and nucleoside transporters (NTs) is also discussed.

4.2 Experimental Approaches

The interaction of drugs with uptake transporter proteins can be evaluated using a range of tools which are described in more detail elsewhere.[9–14] Examples of the most frequently used methods are discussed below; however, specific approaches may exist which are appropriate for individual transporters.

4.2.1 *In Silico*

If available, structural protein information can be used to understand transport mechanisms and interactions between a transporter protein and its substrate. Computational docking can evaluate the fit of the molecule in the binding pocket or how strong interactions with key amino acids are, to identify novel substrates from virtual compound libraries. However, high resolution structures for transporter proteins are scarce, and often structures of bacterial homologues or protein substructures are used instead. Examples of protein-based models include PepT1 (SLC15A1)[15,16] and ASBT (SLC10A2).[17] Alternatively, three-dimensional quantitative structure–activity relationship (QSAR) or pharmacophore modelling may be used, which does not require previous knowledge of protein structures and is based on molecular descriptors quantifying a compound's predicted or measured properties.[9] Modelling is aimed

Figure 4.1 (*continued*) amount of protein in the plasma membrane fraction) of selected uptake transporters determined using proteomic methods. Data were taken from Sawant-Basak *et al.*[6] except for MCT1 which was taken from Miyauchi *et al.*[8] Bars indicate the range of expression levels reported across studies whereas circles represent individual datapoints.

to allow improved understanding of the relationship between these molecular descriptors and a given activity so that ultimately the activity can be predicted from the descriptors. An example is the OCTN2 (SLC22A5) substrate pharmacophore model.[18] Due to the scarcity of *in silico* models for transporter proteins they are employed on a case-by-case basis and often used in conjunction with other approaches rather than being applied as a sole screening tool.

4.2.2 *In Vitro*

In vitro cell based models are a convenient and relatively simple and cost-effective way to assess compound interactions with a transporter of interest.[10,11] They employ heterologous expression systems, immortalised cell lines [*e.g.* the human colon adenocarcinoma (Caco-2) cells], or isolated primary cells and allow the easy manipulation of experimental conditions to evaluate the specificity of a given process (*e.g.* pH-dependency of PepT1, Na^+-dependency of ASBT). Commonly used cell lines in which transporters can be overexpressed include Madin–Darby canine kidney (MDCK), Chinese hamster ovary (CHO), human embryonic kidney (HEK), COS or HeLa cells and typically uptake of a substrate into transporter-expressing cells is compared with uptake into cells transfected with a vector control. In the examples discussed in this chapter it is typically the human isoform which is transfected unless indicated otherwise. Caco-2 cells, which can be grown on semi-permeable filters, also allow the assessment of *trans*-epithelial flux across the cell monolayer from the apical to the basolateral compartment, mimicking transport from the gut lumen across the intestinal cells into the bloodstream.[11–13] Less frequently used tools include *Xenopus laevis* oocytes,[19,20] where proteins of interest are expressed after injection of complementary DNA (cDNA), or brush border membrane vesicles (BBMV)[12,20,21] prepared from intestinal tissue.

4.2.3 *Ex Vivo/In Situ*

Several specialist techniques to study intestinal absorption utilise intact tissue obtained from humans as well as preclinical species. An advantage of such methods is the representation of multiple pathways (*e.g.* uptake, intestinal metabolism or apical as well as basolateral efflux) which may contribute to the overall absorption across the intestine. These techniques also allow the study of regional absorption across the length of the intestinal tract. The contribution of a specific transporter to the overall absorption of a compound can be studied using specific inhibitors or defined assay conditions (*e.g.* pH dependency for PepT1). For *ex vivo* methods, the viability of the tissue needs to be considered, and access, in particular to human tissue, may be challenging.

4.2.3.1 Ex Vivo *Methods*

In the Ussing chamber model,[12,13,22,23] samples of human or animal intestinal segments are fixed between a diffusion pool containing the test compound and a receiving pool. The drug concentrations at both sides of the membrane are measured over time to determine the drug absorption rate. Typically, the dermal layer of the tissue is removed to provide access to the active isolated intestinal mucosa. The everted gut sac method uses freshly excised intestinal tissue (usually jejunum).[13,24–27] After washing, the excised intestine is everted using a glass rod, ligated at both ends to form a closed sac and placed into an oxygenated buffer solution containing the test compound. After a given incubation time, samples are taken from both sides of the intestine to determine the drug diffusion rate. Limitations of this model are that the drug has to cross all intestinal layers, including the muscle, and the low volume in the sac potentially limiting sink conditions.

4.2.3.2 In Situ *Methods*

Perfusions can be performed *in situ* on intestinal segments.[12,14] *Via* incisions at the proximal and distal ends of intestinal sections perfusion and collection tubes can be inserted through which test drugs can be applied and collected. Test substances can either be applied in a single bolus (open loop) or recirculating (closed loop) perfusion. Determination of compound concentrations in the perfusate enables the calculation of absorption rates and effective permeability (P_{eff}) values. *In situ* methods have the advantage that blood supply is maintained and innervation is intact.

4.2.4 *In Vivo*

In vivo studies in a range of species typically measure pharmacokinetic parameters such as bioavailability and exposure [*e.g.* maximum serum concentration (C_{max}), area under the curve (AUC)] after oral administration.[12,13] These data can be used to determine the bioavailability of compounds with and without a targeting moiety or of a prodrug with its parent to assess the effect the targeting group has on absorption. However, overall properties introduced with the targeting moiety (*e.g.* permeability, lipophilicity) which may affect absorption need to be considered when interpreting results. For several transporters knockout mouse models have been established and employed to study the involvement of specific transporters in drug absorption or distribution. Examples include PepT1,[28] OATP2B1,[29] OCT1[30] and ASBT.[31] Moreover, knockout mice can be further engineered to express the human orthologue of a transporter; so-called humanised mice have so far only been generated for PepT1.[32] Other *in vivo* based approaches include drug–drug interaction studies where a transporter substrate is co-administered with an inhibitor.[33,34] Such studies can be conducted in a range of species, including humans, although careful design is required to select appropriate dose levels and inhibitors.

4.3 Oligopeptide Transporter 1 (PepT1, SLC15A1)

4.3.1 Transporter Family

PepT1 is a member of the POT family of proton-coupled oligopeptide transporters (SLC15).

4.3.2 Expression

PepT1 is expressed at the apical membrane of enterocytes in the small intestine, with little or no expression in healthy colon. Earlier studies tend to report a decrease in PepT1 mRNA expression levels along the intestinal tract[35–37] whereas more recent studies find increasing levels of PepT1 protein from duodenum to jejunum to ileum.[2,3,8] Both mRNA and protein across species (human, rat, mouse) increase along the intestinal tract from duodenum and jejunum to ileum. PepT1 expression in colon is very low to undetectable.[2,3,8,36,37] However, levels of PepT1 can be sensitive to food intake and are increased under conditions of high protein intake as well as fasting.[38–40,166]

4.3.3 Structure

Mammalian PepT1 is a protein with approximately 700 amino acids and a calculated molecular mass of approximately 79 kDa, but approximately 105 kDa when glycosylated. PepT1 is highly homologous across many mammalian species, such as human, mouse, rabbit, rat and monkey, with the latter having a higher level of homology with human transporters than other species at 92% identity.[41] PepT1 is predicted to have 12 transmembrane helices with a long extracellular loop between segments 9 and 10 and both the C- and N-termini facing the cytosol.[40]

4.3.4 Activity

Glycosylation of the conserved residue asparagine N50 has been shown to affect transport functionality; replacement of N50 increased transport rate and decreased affinity for substrates like glycylsarcosine (GlySar) or cefadroxil without affecting cell surface expression levels.[42,43] PepT1 polymorphisms appear to be limited in number and frequency, indicating a high evolutionary pressure on maintaining the transporter's function, although the number of subjects studied so far is low. Genetic variants described to date are mostly synonymous or without apparent functional consequences and only a small number have been demonstrated to either reduce expression levels (*e.g.* P586L) or affinity (*e.g.* F28Y).[44,45] The knowledge of the potential role of PepT1 and its variants in pathophysiology and disease is very limited thus far.[166] The rs2297322 single nucleotide polymorphism (SNP; Ser117Asn) has been associated with susceptibility to Crohn's disease. Swedish patients carrying the polymorphic form show an increased risk, in contrast with Finnish patients showing an increased protection;[46] the reasons for the differences in susceptibility are not understood. Whilst drug substrates are recognised across species, differences in affinities across mammalian species have been observed. For example, mouse, rat and human PepT1 expressed in yeast demonstrated threefold to fivefold differences in their Michaelis–Menten constant (K_m) values for GlySar with the highest affinity seen for rat, followed by mouse and then human.[47] In contrast, *in situ* jejunal perfusion studies with GlySar, cefadroxil and valacyclovir have consistently found lower K_m and maximal flux values in mice expressing humanised PepT1 compared with wild-type mice.[32,48,49]

4.3.5 Function and Endogenous Substrates

PepT1 mediates the cellular uptake of dipeptides and tripeptides (Figure 4.2) and hence is responsible for their absorption from dietary protein and gastro-intestinal secretions. PepT1 function is essential under nutritional conditions with high protein content; under such conditions amino acid transporters are saturated and the uptake of small peptides offers an additional supply of nutrients. Whilst the phenotype of PepT1 knockout mice is similar to that of wild-type mice when fed a carbohydrate-rich diet, the plasma amino acid pattern changes considerably when mice are fed a high protein diet.[50,51] PepT1 can also transport substances without obvious peptide-like bonds such as δ-amino-levulinic acid (Figure 4.2) or ω-amino fatty acids.[40,52,53] In the kidney, PepT1 mediates, in conjunction with its close analogue the high affinity peptide transporter PepT2, the reabsorption from the filtrate (and thereby conservation) of peptide-bound amino acids. Several PepT1 substrates are also transported by PepT2 such as valacyclovir (*vide infra*) and δ-amino-levulinic acid, albeit with differing affinities between the PepT analogues.[54–56]

4.3.6 Known Drug Substrates

PepT1 can transport a range of drugs that contain peptide-like recognition motifs (Figure 4.3). These include β-lactam antibiotics (*e.g.* the cephalosporins cefadroxil, cefaclor and cephradine and the penicillins amoxicillin and benzpenicillin), antineoplastic drugs (*e.g.* bestatin) and angiotensin-converting enzyme (ACE) inhibitors (*e.g.* lisinopril).[57–59] PepT1 has also been successfully targeted to improve the oral bioavailability of drugs employing prodrug approaches (*e.g.* antivirals, such as valacyclovir, and ACE inhibitors, such as enalapril, *vide infra*).[57,60,61]

4.3.7 Mechanism

PepT1 is a proton-coupled co-transporter, working in conjunction with the Na^+–H^+ exchanger (maintaining the H^+ gradient across the membrane) and the Na^+–K^+ ATPase (maintaining the Na^+ gradient at the cost of ATP).

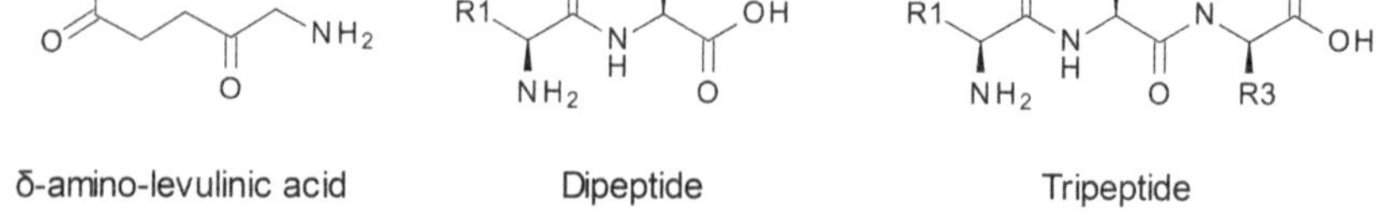

Figure 4.2 Endogenous substrates of PepT1.

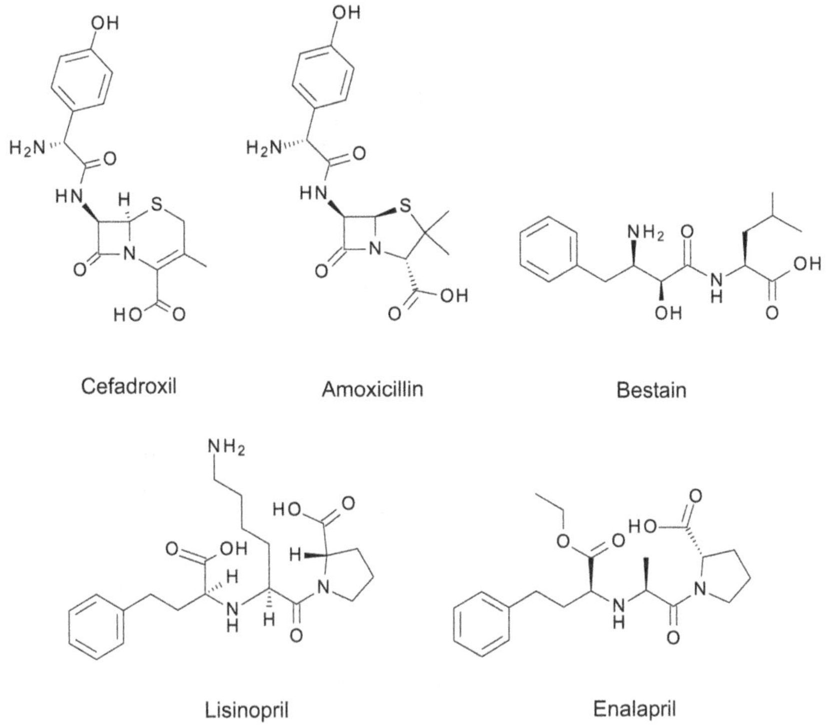

Figure 4.3 Drug substrates of PepT1.

The uphill symport of the peptide or mimetic substrate with protons into cells is electrogenic and driven by the electrochemical gradient and membrane potential.[40] Whilst the pH-dependency of PepT1 transport is not fully characterised, generally, it is the neutral species of the substrate which is transported most efficiently. The proton substrate stoichiometry of peptide transport for PepT1 is 1 : 1, 1 : 1 and 2 : 1 for neutral, anionic (*e.g.* Gly–Glu) and cationic (Gly–Lys) dipeptides, respectively.[40] Experimentally, most studies are conducted at pH 6.0 to mimic the physiological pH of the unstirred layer near the intestinal brush border membrane.

4.3.8 Pharmacophore

Several publications[62–65] in the literature have attempted to describe models for the key molecular recognition motifs required for PepT1 transport. Key aspects of these are summarised below and schematically visualised in Figure 4.4. It should be noted however, that there are some disagreements between models. Although PepT1 is primarily focused on the transport of dipeptide and tripeptide motifs, the peptide bond is not required

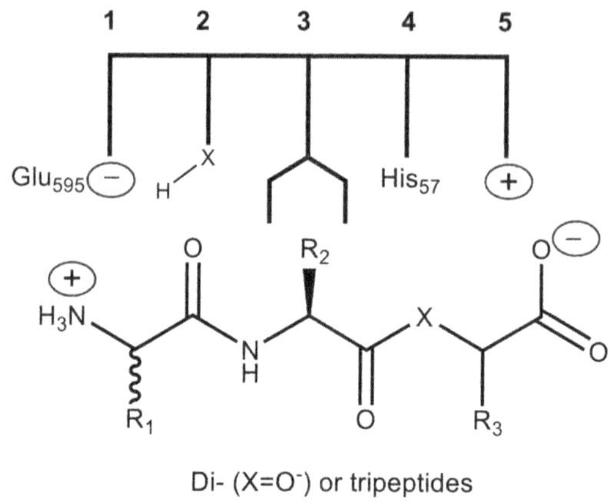

Figure 4.4 Proposed pharmacophore for recognition by PepT1. Reproduced from ref. 65 with permission from John Wiley and Sons, Copyright © 2003 Federation of European Biochemical Societies.

for substrate binding specificity of PepT1 (*e.g.* δ-amino-levulinic acid). Key proposed elements are:

- An N-terminus (NH_2) of L stereochemical configuration for binding to Glu_{595} (1).
- A suitably placed hydrogen bond acceptor in the region of the first peptide bond (2).
- The ability to orient a sidechain into a key binding pocket (3).
- A negatively charged "C-terminus" capable of binding to a positively charged site in PepT1, probably His_{57}. In the case of tripeptides, either this histidine or His_{121} provides the salt-bridge binding (4/5).

4.3.9 Relevance

PepT1 is a high-capacity, low-affinity transporter with high expression in the small intestine mediating cellular uptake of di- and tripeptides of L-amino acids, independent of sequence, as well as peptide-like drugs. A successful example of improving oral bioavailability through targeting PepT1 is valacyclovir, the L-valine ester prodrug of acyclovir; the oral bioavailability (F) in humans of acyclovir is 12–20% compared with 54% for valacyclovir (see Table 4.1 Example 1).[66] K_m values for PepT1 substrates range from 0.2 to 10 mM.[59,67,68] Therefore, PepT1 is unlikely to saturate[69,70] even at very high substrate concentrations typically present in the intestine, making it an attractive target to enable absorption of some drugs, though structure–activity relationship (SAR) data are limited. However, non-proportional increases in AUC and C_{max} were reported after increasing oral doses for some drugs in rat and human (*e.g.* cefadroxil).[71,72] Due to its high capacity, clinical drug–drug interactions are not expected (nor reported) for PepT1.[73]

Key References for PepT1

- D. E. Smith, B. Clémençon and M. A. Hediger, Proton-coupled oligopeptide transporter family SLC15: physiological, pharmacological and pathological implications, *Mol. Aspects Med.*, 2013, **34**(2–3), 323–336.
- M. Brandsch, Drug transport *via* the intestinal peptide transporter PepT1, *Curr. Opin. Pharmacol.*, 2013, **13**(6), 881–887.
- D. W. Foley, J. Rajamanickam, P. D. Bailey and D. Meredith, Bioavailability through PepT1: the role of computer modelling in intelligent drug design, *Curr. Comput.-Aided Drug Des.*, 2010, **6**(1), 68–78.
- C. Y. Wang, S. Liu, X. N. Xie and Z. R. Tan, Regulation profile of the intestinal peptide transporter 1 (PepT1), *Drug Des., Dev. Ther.*, 2017, **8**(11), 3511–3517.

Table 4.1 Examples of targeting PepT1.

Examples

1

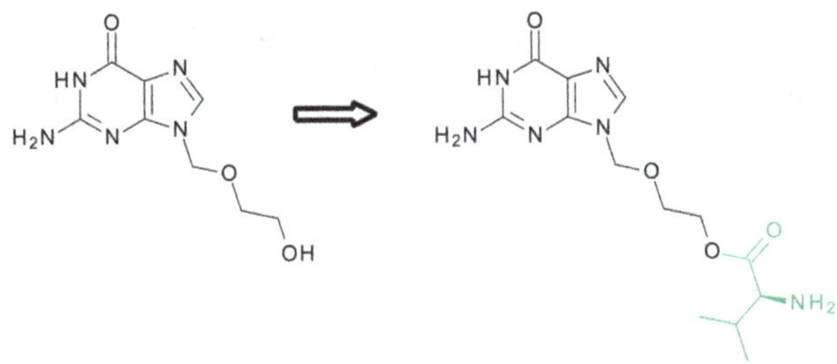

Acyclovir
F_{human} 12-20%

Valacyclovir
F_{human} 54%

In humans, the ester prodrug valacyclovir has improved oral bioavailability over acyclovir.[66] To determine the contribution of PepT1 to the intestinal permeability of valacyclovir and acyclovir intestinal perfusions in both wild-type and PepT1 knockout mice were used to assess the effective permeability (P_{eff}) of [^{3}H]-valacyclovir.[70] Also, the pharmacokinetics of acyclovir were studied following oral administration of valacyclovir. Valacyclovir showed good P_{eff} in wild-type animals (2.4×10^{-4} cm s^{-1}), while in knockout animals P_{eff} was only 10% of that seen in the wild-type cohort. The P_{eff} of valacyclovir was reduced by high concentrations of the PepT1 substrates GlySar and cefadroxil. In addition, co-incubation with non-PepT1 substrates such as L-valine, L-histidine or *para*-aminohippurate had no effect on valacyclovir P_{eff}. The rate and extent of *in vivo* valacyclovir absorption was reduced by threefold to fivefold in PepT1 knock-out compared with wild-type mice. Taken as a whole, the data indicate a key role for PepT1 in the permeability and absorption of valacyclovir.

(*continued*)

Table 4.1 Examples of targeting PepT1. (*continued*)

Examples	
2	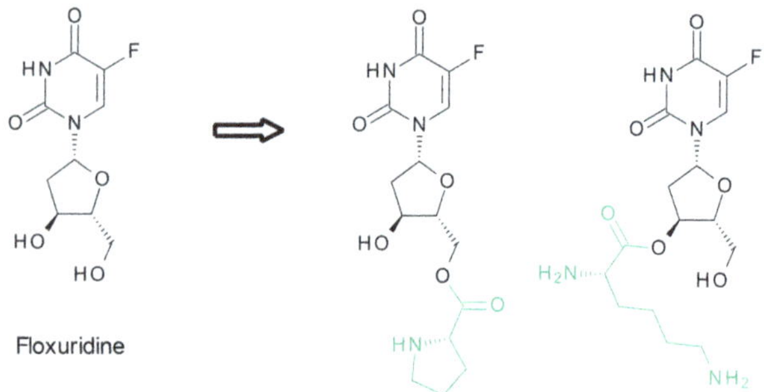 Amino acid ester prodrugs of floxuridine were tested in cell systems overexpressing PepT1, while also assessing prodrug stability in buffer, plasma and cell homogenates.[74] Prolyl and lysyl prodrugs exhibited enhanced uptake (twofold to eightfold) in HeLa-PepT1 cells compared with HeLa cells. Prodrugs selectively inhibited the growth of MDCK-PepT1 cells but not MDCK-vector control cells, which was consistent with the extent of their PepT1-mediated transport. All ester prodrugs hydrolysed to floxuridine fastest in Caco-2 and MDCK cell homogenates and more slowly in human plasma and were chemically most stable in pH 6.0 buffer. Prolyl and lysyl prodrugs were relatively less stable compared with aspartyl prodrugs in buffers and in cell homogenates. No *in vivo* data were reported.
3	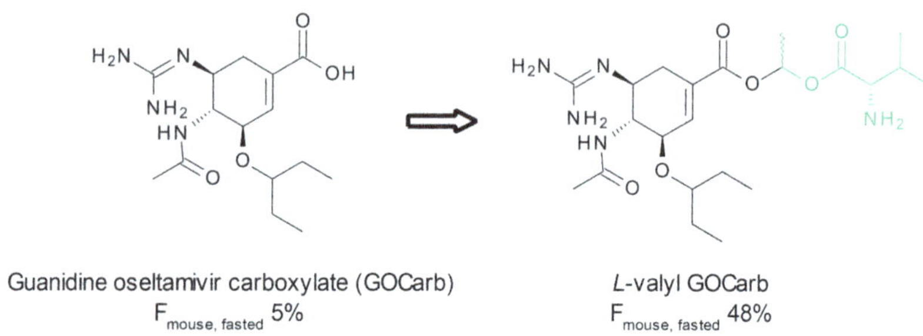

Table 4.1 Examples of targeting PepT1. (*continued*)

	Examples
	Acyloxy(alkyl) ester based amino acid linked prodrugs of guanidine oseltamivir carboxylate (GOCarb) were prepared and evaluated for stability and PepT1 transport in HeLa-PepT1 cells, Caco-2 cells and *in situ* in the rat intestinal jejunal perfusion model.[75] In competition experiments with [^{3}H]-GlySar L-valyl and L-isoleucyl prodrugs showed threefold to sixfold greater affinity for PepT1 than valacyclovir and 30-fold greater affinity compared with the parent compound. The L-valyl prodrug also showed good evidence of transport in HeLa-PepT1 cells. In further studies using Caco-2 cells, only GOCarb parent appeared in the basolateral compartment, indicating payload liberation in enterocytes. Results of studies in the rat intestinal jejunal perfusion model indicated that GOCarb-L-Val and GOC-L-Ile prodrugs were well absorbed but absorption of the parent compound was not evident. Payload liberation occurred rapidly in cell homogenates. Mouse oral bioavailability of the L-valyl GOCarb prodrug was 23% under fed conditions and 48% under fasted conditions *vs.* 5% for the parent.
4	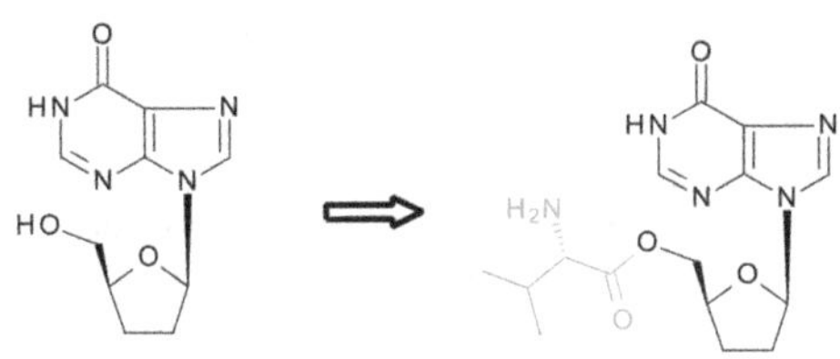 Didanosine F_{rat} 8% — *L*-valyl Didanosine F_{rat} 47% Amino acid ester prodrugs of didanosine were prepared to improve oral exposure *via* PepT1 transport.[76] The lead prodrug (valine ester) showed high Caco-2 permeability relative to the underivatised payload. Uptake of the PepT1 substrate GlySar by Caco-2 cells could be inhibited by the prodrug in a concentration-dependent manner. The uptake of the prodrug compound markedly increased in leptin-treated Caco-2 cells (leptin induces enhanced PepT1 expression in Caco-2 cells) compared with control-treated Caco-2 cells and was clearly inhibited by 20 mM GlySar in both conditions. The oral absolute bioavailability of the valine derived didanosine prodrug was 47% *vs.* 8% for the parent in rats. Prodrug bioavailability in rats was reduced to 34% when orally co-administrated with the PepT1 substrate GlySar. The authors also studied the stability of the prodrugs in simulated gastric fluid (SGF), phosphate buffers under various pH conditions, rat tissue homogenates and plasma at 37 °C. The bioactivation mechanism of the prodrugs was also investigated by comparing the *in vivo* levels of didanosine and prodrugs in the jugular and portal veins in rats. The plasma concentrations of the intact valine prodrug were very low in portal veins and could hardly be detected in the jugular vein, indicating payload liberation in enterocytes and a small amount in the liver.

(*continued*)

Table 4.1 Examples of targeting PepT1. (*continued*)

Examples	
5	 Eglumegad~(LY354740) F_{rat} 10% → LY544344 F_{rat} 85% Eglumegad (LY354740) is a potent and selective group II metabotropic glutamate (mGlu) 2/3 receptor agonist with very limited oral bioavailability. Conversion to the amidic prodrug LY544344 greatly improves oral exposure of eglumegad.[77] Varma *et al.* investigated the mechanism of absorption of LY544344 and eglumegad in Caco-2 cells.[78] The rates of LY544344 absorption and eglumegad basolateral appearance, as well as their cellular accumulation after incubation with LY544344 were concentration-dependent. In addition, inhibition of PepT1 reduced transepithelial transport of LY544344 to 22% of the level in controls. Further studies detected apical efflux and the existence of specific transporters for LY544344 and intracellularly released eglumegad on the apical and basolateral membranes. P-glycoprotein-mediated efflux was not involved in transport of either compound, as evidenced by studies in MDCK-MDR1 cells. The authors concluded that LY544344 is a PepT1 substrate and that enterocytic peptidases release eglumegad.

4.4 Sodium-dependent Multivitamin Transporter (SMVT, SLC5A6)

4.4.1 Transporter Family

Sodium-dependent multivitamin transporter (SMVT) is a member of the SLC5 family, denoted as SLC5A6.

4.4.2 Expression

SMVT is widely expressed in mammals, with high expression in many tissues, including intestine. SVMT is also expressed in placenta, brain, liver, lung, kidney, eye and heart.

4.4.3 Structure

SMVT has approximately 635 amino acids and is highly conserved across mammalian species with the human orthologue sharing amino acid sequence similarity of 92% with the rabbit and 89% with the rat equivalent.[79] SMVT is a putative 12 transmembrane domain protein, with both the N- and C-termini located intracellularly.[80]

4.4.4 Activity

Uptake of biotin has been demonstrated to be sodium-dependent in studies using BBMVs prepared from human small intestine.[81] High levels of SMVT mRNA are expressed along the entire length of the GI tract in both rats and rabbits. However, activity studies show highest transport of biotin in the jejunum with only low levels of biotin transport in the rat colon.[82] Mutations in histidine residues predicted to be in the intracellular loop and the third transmembrane domain decrease the maximal rate of transport of biotin, without changing affinity, indicating that the expression of SMVT is reduced while substrate affinity is not affected.[83]

4.4.5 Function and Endogenous Substrates

The function of SMVT is to translocate the water-soluble B vitamins (Figure 4.5), biotin (vitamin B7) and pantothenic acid (vitamin B5) across membranes. The transporter has higher affinity for pantothenic acid (1–3 μM) compared with biotin (8–20 μM).[79] Biotin is a co-factor for fatty acid biosynthesis; as humans are unable to produce it endogenously they must absorb it from their diet. Pantothenic acid is an essential nutrient, utilised in the synthesis of Coenzyme A.

4.4.6 Known Drug Substrates

Carbamazepine and primidone (Figure 4.6) have been shown to inhibit biotin uptake *in vitro*, and treatment with these anticonvulsant drugs is linked to biotin deficiency in humans.[84] Currently, there are no data available on whether the absorption of these drugs is directly due to SMVT.

4.4.7 Mechanism

SMVT is a sodium-dependent co-transporter which requires the presence of extracellular sodium (Na^+) for transport. The stoichiometry of this is not fully resolved and the mechanism may differ between species, tissues and substrates, with some suggesting that two Na^+ ions are co-transported for every

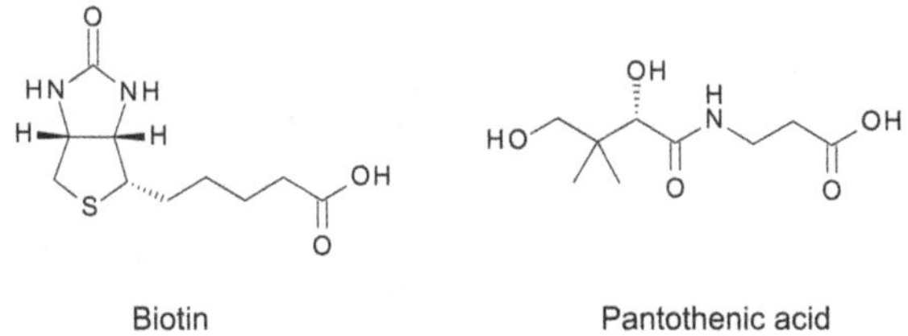

Figure 4.5 Endogenous substrates of SMVT.

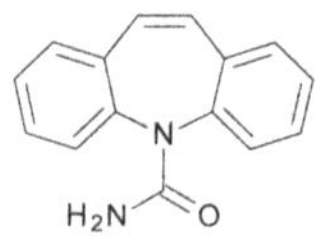

Carbamazepine

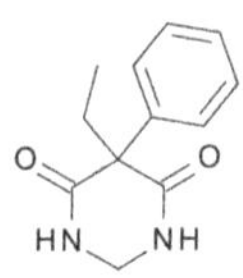

Primidone

Figure 4.6 Drug substrates of SMVT.

vitamin molecule, with a complementary transfer of charge (outward H^+). However, others have demonstrated a 1 : 1 ratio of Na^+:biotin, with an electroneutral transport.[85]

4.4.8 Relevance

SMVT has been suggested as an attractive target for drug transport across the GI tract by several groups, all employing a similar strategy of the attachment of an endogenous substrate, such as biotin or pantothenic acid, to the drug molecule.[86–88] The structure–transport relationship is not clear, with varying reports suggesting that the tetrahydrothiophene ring of biotin and/or a carboxylic acid may or may not be important for SMVT-mediated transport. Carrier-mediated uptake of biotin derivatives has been reported. However, none of these approaches have enabled a drug molecule to reach the market so far. As described below, there are some examples where *in vitro* data indicate that SMVT recognition may enable crossing of cellular membranes (Table 4.2).

Key References for SMVT

- A. D. Vadlapudi, R. K. Vadlapatla and A. K. Mitra, Sodium dependent multivitamin transporter (SMVT): a potential target for drug delivery, *Curr. Drug Targets*, 2012, **13**(7), 994–1003.
- P. D. Prasad and V. Ganapathy, Structure and function of mammalian sodium-dependent multivitamin transporter, *Curr. Opin. Clin. Nutr. Metab. Care*, 2000, **3**(4), 263–266.
- H. M. Said, R. Redha and W. Nylander, Biotin transport in the human intestine: inhibition by anticonvulsant drugs, *Am. J. Clin. Nutr.*, 1989, **49**(1), 127–131.
- S. M. Grassl, Human placental brush-border membrane Na(+)-pantothenate cotransport, *J. Biol. Chem.*, 1992, **267**(32), 22902–22906.

Table 4.2 Examples of targeting SMVT.

	Examples
1	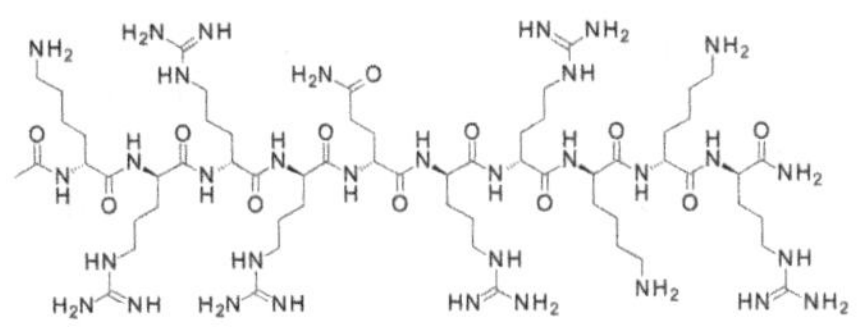 R.I._K_Tat9 peptide 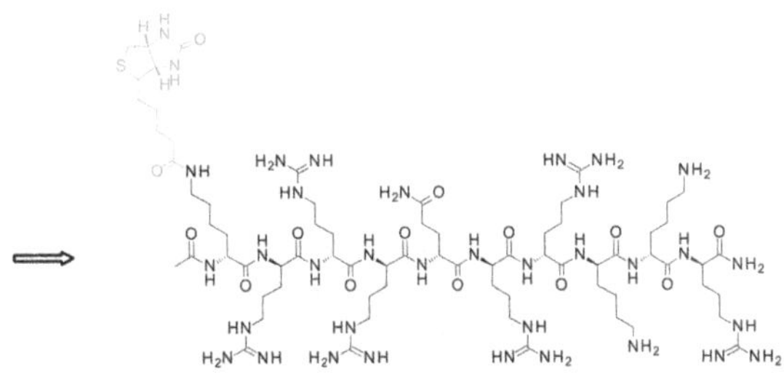 R.I._K_Biotin Tat9 conjugate The *retro-inverso* (R.I.-) K-Tat9 peptide and R.I.-K (biotin)-Tat9 conjugate have been well studied *in vitro*.[86] The peptide showed low absorptive transport across Caco-2 cell monolayers (0.8×10^{-7}–1×10^{-6} cm s^{-1}), which was not concentration dependent. The biotin-conjugate on the other hand showed a 3.2-fold increase in absorptive transport (3.2×10^{-6} cm s^{-1}) which was temperature- and concentration-dependent and displayed saturable kinetics, indicating involvement of carrier-mediated transport. Results of competitive inhibition studies with known SMVT substrates, including biotin, biocytin and desthiobiotin, support a SMVT-mediated transport mechanism. Uptake of the biotin conjugate in CHO-SMVT cells was significantly higher than in vector transfected CHO-pSPORT cells. The passive permeability of the biotin conjugate was not significantly different from that of the peptide in Caco-2 cells. However, the uptake of the biotin conjugate in CHO-pSPORT cells was significantly higher than that for the peptide, indicating that biotin conjugation led to increased uptake even in the absence of SVMT.
2	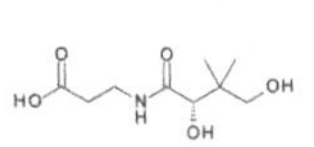 Pantothenic acid 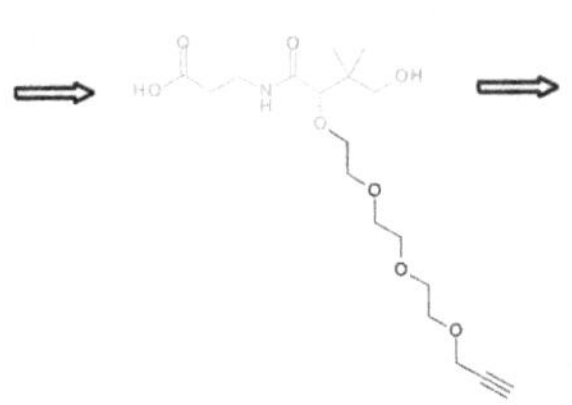 PEG linked Pantothenic acid 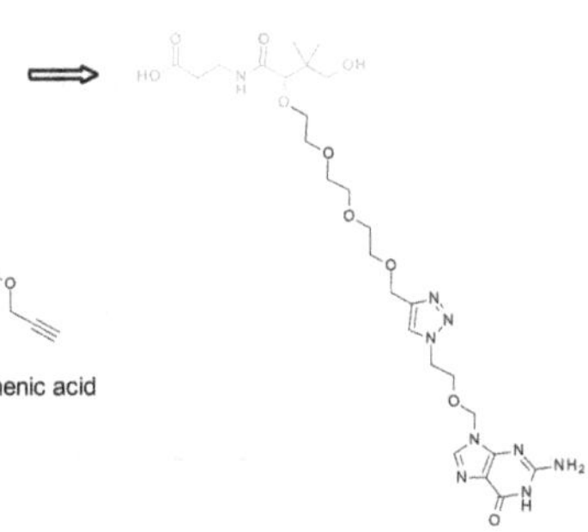 Pantothenic acid drug conjugate

(*continued*)

Table 4.2 Examples of targeting SMVT. (*continued*)

Examples	
	Pantothenic acid is a natural substrate for SMVT and may serve as a targeting ligand for SMVT in a similar way to biotin. In an extensive study, the SAR of pantothenic acid modifications and conjugates was investigated *in vitro*.[87] To assess if these conjugates interact with the SMVT transporter their ability to inhibit the uptake of [^{3}H]-biotin into HEK-SMVT cells was evaluated. Subsequently, the conjugates' own uptake was measured in the presence and absence of Na^+ ions in the buffer. Uptake ratios >tenfold verify active transport *via* SMVT, while compounds showing inhibition of [^{3}H]-biotin uptake without sodium dependence are inhibitors and not substrates for SMVT. Small extensions on the secondary alcohol on pantothenic acid, like methyl, propargyl and methoxymethyl, showed effective inhibition of [^{3}H]-biotin uptake (42–60%) and large sodium-dependent uptake ratios (46- to 107-fold). Interestingly, medium sized polyethylene glycol (PEG)-based linkers such as $-(CH_2CH_2O)_2CH_2C{\equiv}CH$, $-(CH_2CH_2O)_4CH_2C{\equiv}CH$ and $-(CH_2CH_2O)_3CH_2CH_2N_3$ are also tolerated and showed sodium-dependent uptake ratios (14- to 27-fold), but unfortunately once further conjugated to drug-like motifs *via* copper-catalysed azide/alkyne cycloaddition (CuAAC)-reaction, all active transport *via* SMVT was halted.
3	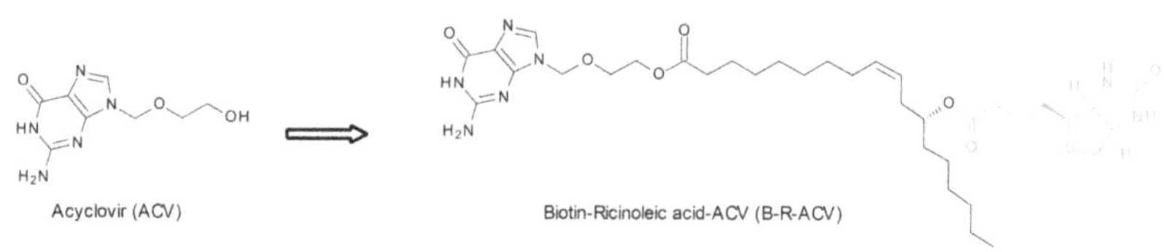 Acyclovir (ACV) suffers from poor oral absorption as a consequence of low permeability due to its high polarity. Vadlapudi *et al.* have studied the effect of biotinylation of ACV *in vitro*, as a strategy to utilise the SMVT transporter to improve oral absorption.[89,90] The biotinylated conjugates displayed concentration-, temperature- and sodium-dependent increased uptake (approximately 2–13-fold), compared with the parent compound, and they inhibited [^{3}H]-biotin uptake. All observations together indicate active uptake *via* SMVT along with improved passive diffusion due to increased lipophilicity of the biotinylated conjugates. Cell accumulation of biotin ribonucleic acid ACV (B-R-ACV) in human corneal epithelial, MDCK-MDR1 and Caco-2 cells was increased 13.6-, 9.5- and 10-fold over ACV in the respective cell systems. The corresponding lipid prodrug (R-ACV) and direct biotin conjugate (B-ACV) showed more moderate increases in uptake between 1.4- and 6-fold, demonstrating that both the lipid and biotin contribute. Despite promising *in vitro* results and claimed ongoing anti-viral efficacy studies *in vivo*, no such data have been published so far, to the best of our knowledge.

4.5 Apical Sodium-dependent Bile Acid Transporter (ASBT, SLC10A2)

4.5.1 Transporter Family

Apical sodium-dependent bile acid transporter (ASBT) is a member of the sodium-dependent SLC family 10. The gene name for ASBT is *SLC10A2*.

4.5.2 Expression

ASBT is primarily expressed on the apical membrane of enterocytes of the terminal ileum with no expression in the proximal small intestine across a range of mammals, with humans following a similar pattern. ASBT is expressed to a lesser extent in renal tubule cells, large cholangiocytes and epithelial cells of the gall bladder (*i.e.* tissues that facilitate the circulation of bile acids). Unlike rodents and monkeys, humans do express low but detectable levels of ASBT in the duodenum.

4.5.3 Structure

ASBT comprises 348 amino acids, has a molecular weight of 43 kDa and seven transmembrane domains with a cytosolic C-terminus and an extracellular N-terminus.[91] ASBT shows some homology to its liver counterpart Na^+-taurocholate co-transporting polypeptide (NTCP), exhibiting 63% agreement in amino acid sequence. Human ASBT shares 88% sequence homology with rabbit and rat ASBT.[92]

4.5.4 Activity

Polymorphisms of ASBT (G292A and G431A) are associated with impaired transport activity, but the clinical effects on transporting bile acids are unknown.[93] Additional mutations have been noted for ASBT; however, to date there is no direct link between dysfunctional ASBT and idiopathic bile acid malabsorption.

4.5.5 Function and Endogenous Substrates

ASBT mediates the transport of bile acids across the intestinal membrane. Transport of bile acids and derivatives in the intestine is the first stage of their reabsorption which, in humans, cycles six to eight times a day thereby turning over 12–18 g of bile acids daily (Figure 4.7).[94,95] Major substrates of ASBT include unconjugated bile acids, such as ursodeoxycholic acid, deoxycholic acid, cholic acid and chenodeoxycholic acid, as well as their taurine or glycine conjugates (Figure 4.8).

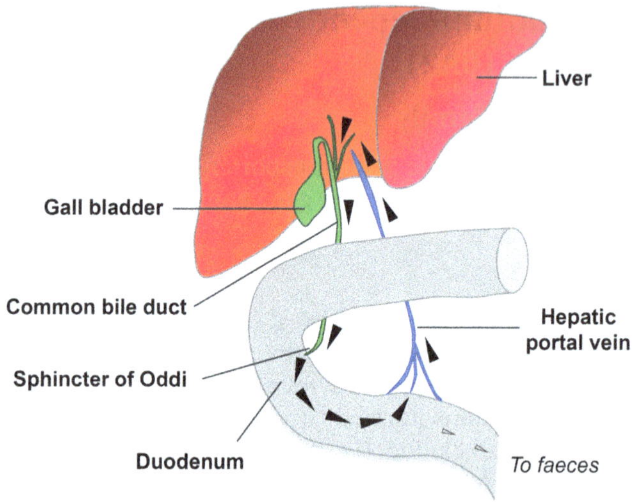

Figure 4.7 Enterohepatic circulation of bile acids. Enterohepatic circulation (or recycling) involves the biliary excretion and subsequent intestinal reabsorption of bile acids and other solutes, including drugs.[96–98] Bile acids are excreted from the liver into bile which is then secreted from the gall bladder into the duodenum *via* the Sphincter of Oddi. Bile acids are actively reabsorbed (~95%) from the lumen of the small distal intestine *via* ASBT, exported into the portal blood circulation and transported back to the liver for systemic circulation. This enterohepatic circulation ensures that bile acids are repeatedly used throughout the day to aid digestion of multiple meals. Only a small amount of bile acids (~5%) escapes reabsorption and is excreted into faeces.

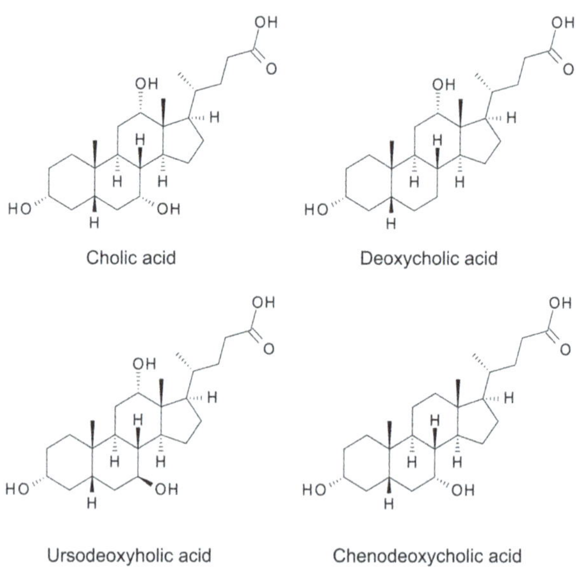

Figure 4.8 Endogenous substrates of ASBT.

4.5.6 Known Drug Substrates

While ASBT has no known parent drug substrates, it can accommodate drug-like structures containing bile acid conjugates. Manipulations at the bile acid C-24 region, including monoanions, are reported to be accommodated by the transporter.[99]

4.5.7 Mechanism

ASBT is a co-transporter, with two or more Na^+ ions transported with each ASBT substrate. This mechanism is driven by the intracellular Na^+ gradient and an electronegative intracellular potential.

4.5.8 Relevance

Bile acids are suitable building blocks for conjugation due to their rigid steroidal core containing multiple hydroxyl groups. The specificity and abundance of ASBT in addition to the stability of bile acids offers the transporter as a potential target for specific cell or tissue exposure or to improve the oral bioavailability of drugs *via* a prodrug approach. Drug conjugates of cholic acid and chenodeoxycholic acid targeting ASBT are examples of conjugates that have been applied to poorly permeable compounds (Table 4.3). The enterohepatic circulation of bile acids from the intestine to the liver provides an additional advantage and opportunity to target ASBT to increase the systemic exposure of bile acid–drug conjugates. Specifically, circulating reservoirs may result in the extended release of the drug and subsequently sustain plasma concentrations.

Key References for ASBT

- P. A. Dawson, Role of the intestinal bile acid transporters in bile acid and drug disposition, *Handb. Exp. Pharmacol.*, 2011, **201**, 169–203.
- D. A. Lionarons, J. L. Boyer and S. Y. Cai, Evolution of substrate specificity for the bile salt transporter ASBT (SLC10A2), *J. Lipid Res.*, 2012, **53**(8), 1535–1542.
- L. Lin, S. W. Yee, R. B. Kim and K. M. Giacomini, SLC transporters as therapeutic targets: emerging opportunities, *Nat. Rev. Drug Discovery*, 2015, **14**(8), 543–560.
- F. Annaba, K. Ma, P. Kumar, A. K. Dudeja, R. D. Kineman, B. L. Shneider, S. Saksena, R. K. Gill and W. A. Alrefai, Ileal apical Na+-dependent bile acid transporter ASBT is upregulated in rats with diabetes mellitus induced by low doses of streptozotocin, *Am. J. Physiol.: Gastrointest. Liver Physiol.*, 2010, **299**(4), G898–G906.
- X. Zheng, S. Ekins, J. P. Raufman and J. E. Polli, Computational models for drug inhibition of the human apical sodium-dependent bile acid transporter, *Mol. Pharm.*, 2009, **6**(5), 1591–1603.
- H. Sun, E. C. Chow, S. Liu, Y. Du and K. S. Pang, The Caco-2 cell monolayer: usefulness and limitations, *Expert Opin. Drug Metab. Toxicol.*, 2008, **4**(4), 395–411.

Table 4.3 Examples of targeting ASBT.

	Examples
1	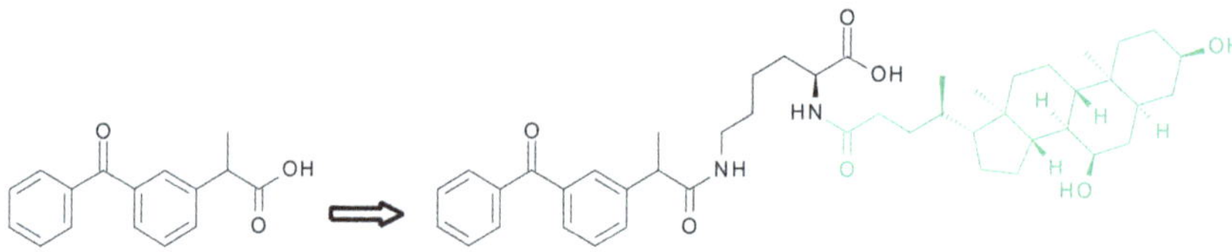 *Rac*-ketoprofen → *Rac*-ketoprofen-lysine-chenodeoxycholic acid conjugate Chenodeoxycholic acid was conjugated *via* a lysine linker to ketoprofen and separately nicotinic acid (not shown) to target uptake of these carboxylic acids *via* human ASBT.[100] Both prodrugs were found to be substrates for ASBT through inhibition of taurocholate uptake using stably transfected MDCKII-ASBT monolayers, with neither parent drug being a substrate. Both conjugates were potent inhibitors of ASBT, with affinities in a range similar to those of the natural substrates. The capacity of ASBT to transport the nicotinic acid prodrug was found to be approximately 9% of that of taurocholate while the transport capacity of the ketoprofen prodrug was 1.9-fold that of taurocholate. Enzymatic stability of the conjugates was also evaluated in Caco-2 cells and liver homogenate with nicotinic acid release from the prodrug higher in Caco-2 cells and for ketoprofen higher in the liver homogenate S9 fraction. No *in vivo* data were presented.
2	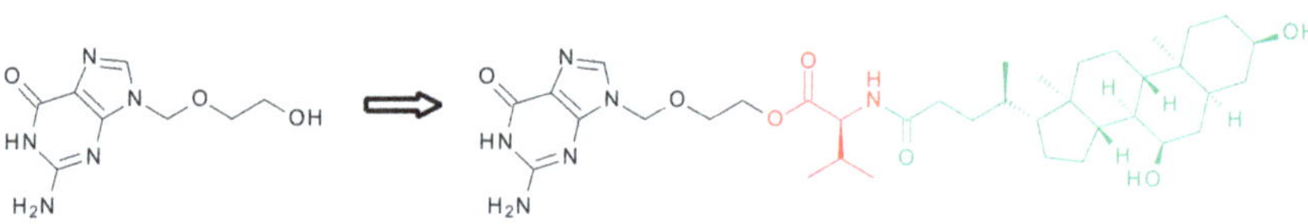 Acyclovir F_{rat} 25% → Acyclovir-lysine-chenodeoxycholate conjugate F_{rat} 48% An acyclovir conjugate of chenodeoxycholic acid *via* a valine linker showed affinity for ASBT [inhibitor constant (K_i) = 36 μM] which was similar to that of cholic acid (K_i = 25 μM).[101] Using a COS-ASBT cell line, acyclovir–lysine–chenodeoxycholate conjugate was actively translocated by ASBT. In the presence of Na^+, cellular uptake of the prodrug was 16-fold greater than uptake of acyclovir. Encouragingly, *in vivo* data were presented that showed an approximately twofold enhancement in acyclovir bioavailability from 25% to 48% in rats following oral dosing, demonstrating adequate absorption and release of active drug. However, the authors noted the prodrug has a more than tenfold higher passive permeability than acyclovir due to conjugation with the lipophilic steroid, and that this may be the major contributor to the enhanced oral bioavailability of the prodrug.

4.6 Monocarboxylate Transporter 1 (MCT1, SLC16A1)

4.6.1 Transporter Family

The monocarboxylate transporter MCT1 (SLC16A1) is one of 14 members of the SLC family 16 (SLC16).

4.6.2 Expression

MCT1 is ubiquitously expressed and located on both apical and basolateral membranes, dependent on cell type. MCT1 is expressed in the disposition organs liver, kidney, brain and GI tract, where it is the predominant MCT isoform. Furthermore, MCT1 is one of the most abundantly expressed transporters in the GI tract.[2] CD147, an accessory protein, is required for trafficking, localisation and functional expression of MCT, with an MCT1–CD147 pair dimerising with another pair.[102]

4.6.3 Structure

MCT1 is highly conserved across species with >85% sequence homology across mouse, hamster and human. It consists of approximately 495 amino acids and is around 45 kDa in size. The putative structure consists of 12 transmembrane domains (TMD), with a large intracellular loop between TMD6 and TMD7. Both the N- and C-termini are intracellularly.[103]

4.6.4 Activity

Similar affinity across species is reported for known endogenous substrates, such as lactate and pyruvate.

4.6.5 Function and Endogenous Substrates

The function of monocarboxylate transporters is – as the name suggests – to transport endogenous monocarboxylate substrates, such as lactate and pyruvate (Figure 4.9). MCTs are involved in maintaining metabolic homeostasis

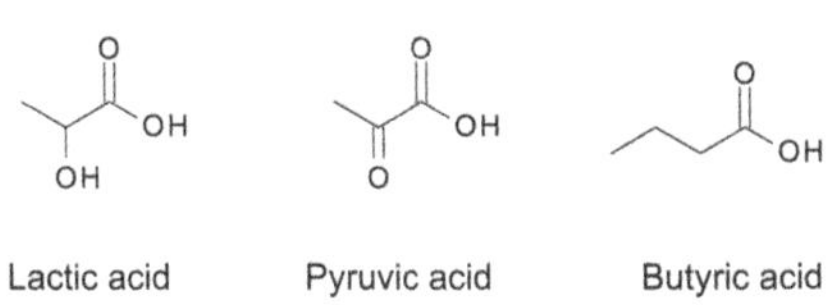

Figure 4.9 Endogenous substrates of MCT1.

via transfer of lactate not only between tissues but also between different cell types in the same tissue, *via* both influx and efflux mechanisms. In tumours, MCTs have been shown to export lactate from the cells, thereby preventing intracellular acidification. As MCT1, MCT2 and MCT4 have been shown to be highly expressed in several cancers, they have been targeted in oncology therapies, including as targets for drug delivery to tumours. They are also proposed as biomarkers of cancer outcomes.[104] In the GI tract, MCT1 has a role in the dietary absorption of monocarboxylate substrates, such as propionate and butyrate.[105]

4.6.6 Known Drug Substrates

γ-Hydroxybutyric acid (GHB; Figure 4.10) is an approved medicine in both the USA and Europe, however, it is also a drug of abuse. It shows non-linear pharmacokinetics with saturable absorption, renal clearance and metabolism being implicated. Many studies have been completed, using several *in vitro* systems derived from the GI tract and the kidney, demonstrating that GHB is actively transported in a pH- and sodium-dependent fashion.[102] Data from Caco-2 cells in the presence of the known MCT inhibitor alpha-cyano-4-hydroxycinnamate,[106] along with saturable absorption in the GI tract,[107] imply that MCT1 (the major isoform expressed in the GI tract) is responsible for the transporter-mediated absorption of GHB. XP-13512 is a prodrug of gabapentin (gabapentin enacarbil; Figure 4.10) and was

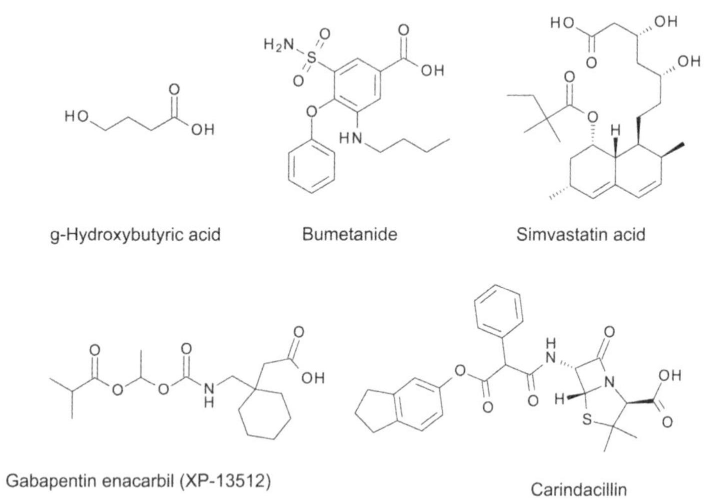

Figure 4.10 Drug substrates of MCT1.

designed to be recognised by solute carriers expressed in the GI tract. XP-13512 has been shown to be a substrate of MCT1 *in vitro*, though the exact contribution of MCT1 to its fraction absorbed has not been determined. Carindacillin (Figure 4.10), a monoester prodrug of carbenicillin, has been reported to be a substrate of a monocarboxylate transporter in rat intestinal membrane vesicles.[108]

4.6.7 Mechanism

MCT1 is a proton-coupled bidirectional co-transporter. The binding of the proton precedes the binding of the substrate *prior* to translocation; after transport the substrate is released first.[103]

4.6.8 Relevance

MCT1 has been shown to transport low molecular weight acids (<150 kDa), such as salicylic acid, along with some drugs, such as carindacillin and XP-13512. Therefore, the potential exists to target MCT1 to enable the transport of drugs across the GI tract; however, positive examples of this strategy are limited (Table 4.4).

Key References for MCT1

- P. Fisel, E. Schaeffeler and M. Schwab, Clinical and Functional Relevance of the Monocarboxylate Transporter Family in Disease Pathophysiology and Drug Therapy, *Clin. Transl. Sci.*, 2018, **11**(4), 352–364.
- B. E. Enerson and L. R. Drewes, Molecular features, regulation, and function of monocarboxylate transporters: implications for drug delivery, *J. Pharm. Sci.*, 2003, **92**(8), 1531–1544.
- M. E. Morris and M. A. Felmlee, Overview of the proton-coupled MCT (SLC16A) family of transporters: characterization, function and role in the transport of the drug of abuse gamma-hydroxybutyric acid, *AAPS J.*, 2008, **10**(2), 311–321.
- K. C. Cundy, R. Branch, T. Chernov-Rogan, T. Dias, T. Estrada, K. Hold, K. Koller, X. Liu, A. Mann, M. Panuwat, S. P. Raillard, S. Upadhyay, Q. Q. Wu, J. N. Xiang, H. Yan, N. Zerangue, C. X. Zhou, R. W. Barrett and M. A. Gallop, XP13512 [(+/−)-1-([(alpha-isobutanoyloxyethoxy)carbonyl] aminomethyl)-1-cyclohexane acetic acid], a novel gabapentin prodrug: I. Design, synthesis, enzymatic conversion to gabapentin, and transport by intestinal solute transporters, *J. Pharmacol. Exp. Ther.*, 2004, **311**(1), 315–323.

Table 4.4 Examples of targeting MCT1.

	Examples
1	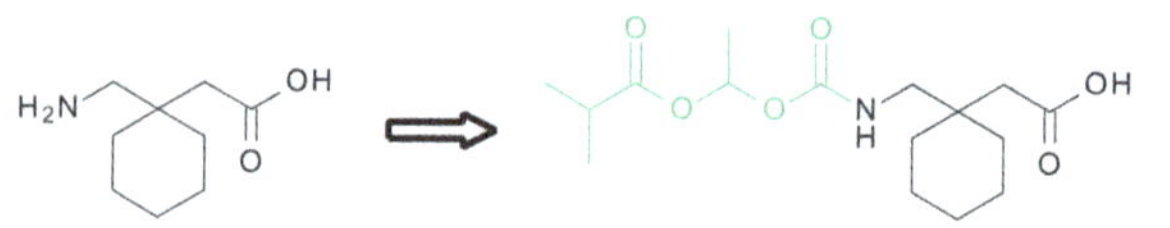 Gabapentin F_{cyno} 25% F_{human} 37% — Gabapentin enacarbil F_{cyno} 84% F_{human} 75%
	Gabapentin bioavailability is variable and exhibits a dose-dependency which is thought to be due to saturation of a low-capacity solute transporter localised in the upper small intestine.[102] Gabapentin enacarbil is an oral prodrug approved for treatment of postherpetic neuralgia and restless legs syndrome. The prodrug consists of a cleavable promoiety that masks the amine group of gabapentin, yielding a monoanionic compound at physiological pH, making it a potential substrate for monocarboxylate transporters. It has been shown to be a substrate for transport across membranes by both MCT1 and SMVT [but not large neutral amino acid transporter (LAT1)], while gabapentin itself is not transported.[109] In monkeys, bioavailability of gabapentin following prodrug administration was 84% compared with 25% for gabapentin itself.[109] Following intracolonic administration, exposure of gabapentin was 17-fold higher in rats and 34-fold higher in monkeys for the prodrug when compared with parent gabapentin administered *via* the same route.[109] In humans, an extended release formulation of the prodrug delivered higher oral bioavailability when compared with an equivalent dose of gabapentin (75% *vs.* 37%).[110] An identical prodrug strategy has been applied to the structurally related $GABA_B$ agonist baclofen, as the investigational prodrug arbaclofen placarbil, shown to possess a more favourable pharmacokinetic profile than baclofen with less variation in plasma levels.[111]
2	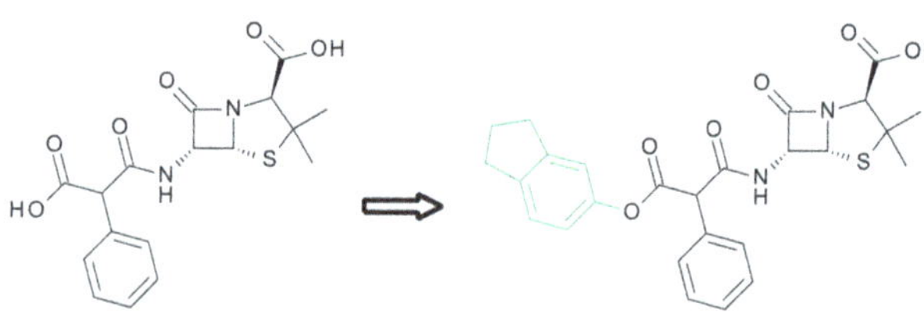 Carbenicillin — Carindacillin

Table 4.4 Examples of targeting MCT1. (*continued*)

Examples
Carindacillin is an orally active prodrug of the β-lactam antibiotic carbenicillin which is typically delivered intravenously as its disodium salt.[112] Similar to Example 1 above, prodrug modification is used to generate a mono-acid (from a di-acid) that is thus recognised by monocarboxylate transporters. In rat intestinal BBMVs, carindacillin was transported in the presence of a proton gradient, suggestive of a carrier-mediated process. Transport *via* MCT1 was demonstrated through competition with [^{14}C]-lactate. The parent drug, however, showed no affinity for MCT1.[113] Similar studies were conducted using Caco-2 cells, where transport was temperature- and pH-dependent and inhibited by substrates of MCT1 but not PepT1.[113] The increased lipophilicity of the indanyl ester prodrug may also contribute to increased passive absorption relative to carbenicillin itself.

4.7 Organic Cation Transporters (OCT, SLC22) – Isoform OCTN2 (SLC22A5)

4.7.1 Transporter Family

OCTN2 is a member of the SLC22 family of membrane transporters, denominated as SLC22A5. The SLC22 family divides into two major groups, one containing organic cation transporters and the other organic anionic transporters. The OCTN transporters form a separate subgroup within the cationic group of transporters.[114]

4.7.2 Expression

OCTN2 is expressed in several organs in humans, including the GI tract, kidney, lung, placenta, heart and cornea. OCTN2 has a similar organ expression pattern in other mammals including mice, rats and baboons, where function has been shown for the transport of L-carnitine.

4.7.3 Structure

OCTN2 consists of 557 amino acids and is 65 kDa in size. OCTN2 is a putative 12-transmembrane domain (TMD) protein, with both C- and N-termini intracellularly, a large extracellular loop between TMD1 and TMD2 and a larger intracellular loop between TMD6 and TMD7.[115]

4.7.4 Activity

Defects in OCTN2 transport can lead to muscle weakness and cardiomyopathy. Patients may be treated with dietary supplements of carnitine. Primary systemic carnitine deficiency (SCD) is a recessive inherited disorder. Both missense and nonsense mutations have been identified and linked to this disorder.[116] Mutations M352R and P478L lead to loss of transport while maintaining expression of the transporter. Mutations lead to defective OCTN2-mediated transport of L-carnitine at several sites, including the GI tract and the kidney.[117] L-Carnitine transport is mediated by OCTN2 across a range of species, including mice, rats and humans, with higher affinity for L-carnitine observed for the human homologue.

4.7.5 Function and Endogenous Substrates

The endogenous function of OCTN2 is to transport L-carnitine (Figure 4.11) across the plasma membrane of cells. L-Carnitine has an essential role in metabolism, enabling the transport of fatty acids across the mitochondrial membrane.[114] Dietary L-carnitine represents the major source of L-carnitine in animals (though some organs can synthesise L-carnitine from methionine and lysine). Influx from the GI tract is thus essential. In line with its endogenous role of transporting and maintaining L-carnitine homeostasis, evidence indicates that OCTN2 is expressed both on the apical membranes of enterocytes and on the apical membranes of proximal tubule cells to re-absorb L-carnitine.[118] Additionally, OCTN2 has a role in energy production in muscles, and it is thought that its role is to maintain 'energy' homeostasis.

4.7.6 Known Drug Substrates

Mildronate (Figure 4.12), an analogue of L-carnitine and available in some countries as a cardioprotective drug, has been demonstrated to be a substrate of OCTN2 *in vitro* and *in vivo*, mediating mildronate cellular entry and hence efficacy.[119] The nucleoside analogue entecavir and the anticholinergic bronchodilator ipratropium bromide have also been described as OCTN2 substrates *in vitro* (Figure 4.12), and transport *via* OCTN2 is suggested to have a role in renal and local lung disposition, respectively.

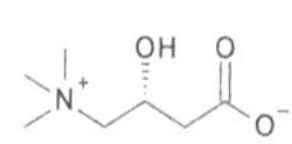

L-Carnitine

Figure 4.11 Endogenous substrate of OCTN2.

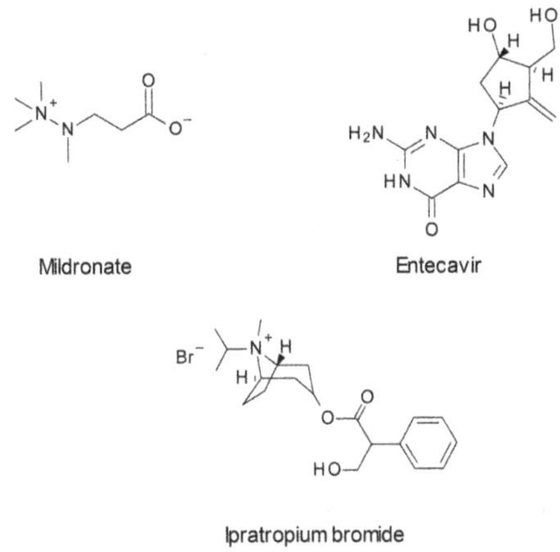

Figure 4.12 Drug substrates of OCTN2.

4.7.7 Mechanism

OCTN2 cotransports Na^+ and carnitine. L-Carnitine is a high affinity substrate of both human and rat homologues of OCTN2. While the uptake of L-carnitine is sodium-dependent, the uptake of other cationic substrates, such as tetraethylammonium ions, and some drugs, such as ipratropium bromide, is not sodium-dependent.[114]

4.7.8 Relevance

The primary effect of OCTN2 is directly linked to its transport of L-carnitine. OCTN2 has been shown to transport some drugs and drug molecules conjugated with carnitine to increase cellular uptake in the GI tract, the cardiovascular system and the lung. Examples of OCTN2 targeted uptake are limited (Table 4.5). There is only one example of a carnitine–gemcitabine conjugate where an increase in bioavailability was demonstrated. The re-uptake of both endogenous substrates (L-carnitine) and some drugs (entecavir) being facilitated by OCTN2 expressed in renal proximal tubule cells has been described.

Key References for OCTN2

- C. Zhu, K. B. Nigam, R. C. Date, K. T. Bush, S. A. Springer, M. H. Saier Jr, W. Wu and S. K. Nigam, Evolutionary Analysis and Classification of OATs, OCTs, OCTNs, and Other SLC22 Transporters: Structure-Function Implications and Analysis of Sequence Motifs, *PLoS One*, 2015, **10**(11), e0140569.
- H. Koepsell and H. Endou, The SLC22 drug transporter family, *Pflügers Arch.*, 2004, **447**(5), 666–676.

Table 4.5 Examples of targeting OCTN2.

Examples	
1	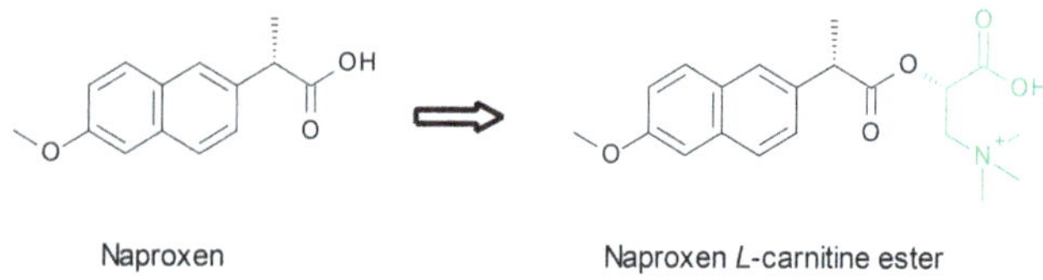 Analogues of L-carnitine were evaluated for their potential to inhibit uptake of radiolabelled L-carnitine in stably transfected MDCK-OCTN2 cells.[120] This was done to establish SAR for prodrug connectivity, and the 3′-OH was found to be dispensable for uptake. Esters of L-carnitine *via* the 3′-OH of valproic acid, naproxen (pictured) and ketoprofen were synthesised and evaluated for inhibition and uptake properties, and all were found to be substrates for OCTN2 through competition with labelled L-carnitine. Significantly enhanced uptake in the presence of Na^+ was further indicative of active transport. Naproxen–L-carnitine inhibited the uptake of L-carnitine with a K_i of approximately 6 μM. Metabolism studies were performed to assess drug release and, as analogues proved stable, a ketoprofen–glycine–L-carnitine conjugate was synthesised, which was also a substrate and inhibitor of OCTN2 (K_i approximately 14 μM). No *in vivo* data were presented, and no data showing enhanced cellular permeability compared with parent drugs were discussed.
2	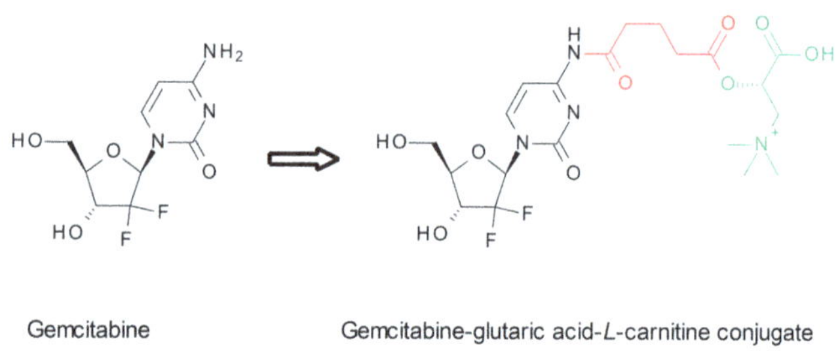 Several L-carnitine conjugates *via* di-acid linkers were evaluated for their potential to improve uptake and *in vivo* absorption of gemcitabine[121] following oral dosing of this intravenously delivered anti-cancer therapy. Transport across Caco-2 membranes was shown to be Na^+- and temperature-dependent, indicative of active transport. The rate of uptake in HEK-OCTN2 cells was threefold to eightfold higher than in mock-transfected HEK293 cells, with no difference seen for gemcitabine alone. L-Carnitine was also shown to inhibit uptake of the prodrugs and *vice versa*. Furthermore, the uptake rate was independent of LogD of the prodrug. Cytotoxicity to cancer cells was maintained, demonstrating release of payload, which was diminished by excess L-carnitine. Example prodrugs demonstrated improved stability (threefold), Caco-2 transcellular permeability (15-fold) and oral bioavailability in rats (fivefold), while causing no apparent additional cytotoxicity as compared with gemcitabine. The authors additionally demonstrated uptake of the prodrug is not *via* nucleoside transporter equilibrative nucleoside transporter 1 (ENT1), given the chemotype in question.

Table 4.5 Examples of targeting OCTN2. (*continued*)

Examples
3 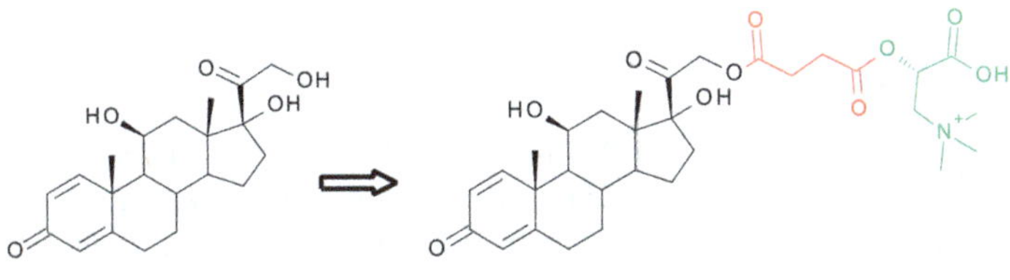 Prednisolone ⟹ Prednisolone succinate *L*-carnitine The ability of L-carnitine ester derivatives of prednisolone to increase absorption across human bronchial epithelial (BEAS-2B) cells mediated by OCTN2 was assessed.[122] The transport of prodrug was found to be superior to that of prednisolone at 37 °C (1.8-fold) and was temperature-, time- and Na^+-dependent and saturable with an apparent K_m of 330 μM, indicating involvement of carrier-mediated uptake. In HEK293T cells, which express little OCTN2, the uptake of prodrug was less than that of prednisolone itself, and addition of L-carnitine inhibited prodrug uptake in BEAS-2B but not HEK293T cells. Metabolic and chemical stabilities of the prodrug were discussed, and, although payload release was not quantified, levels of interleukin-6 (IL-6) at 16 hours were decreased 1.9-fold for prodrug relative to prednisolone. These data corresponded to the hydrolysis rate of the prodrug in BEAS-2B cell homogenates, indicative of active drug release. It appears that the aim of this study was to achieve increased drug exposure *via* inhalation; however, no *in vivo* data were reported.

4.8 Organic Cation Transporters (OCT, SLC22) – Isoforms OCT1 (SLC22A1) and OCT3 (SLC22A3)

4.8.1 Transporter Family

Organic cation transporters 1 (OCT1) and 3 (OCT3) are multi-specific SLC proteins which belong to the SLC22 transporter family. Gene names for OCT1 and OCT3 are *SLC22A1* and *SLC22A3*, respectively.[123]

4.8.2 Expression

OCT1 is expressed along the whole length of the human GI tract, albeit at low levels compared with those in liver. Localisation of OCT1 in enterocytes is unclear, with some data indicating basolateral localisation, while recent data indicate apical localisation.[124,125] Several other tissues also express low levels of OCT1 {*e.g.* kidneys, lung, heart, skeletal muscle, brain [blood–brain barrier (BBB)], spinal cord, eye, adipose tissue}.[126] OCT3 exhibits much broader tissue distribution in humans. It is expressed at low levels (compared with OCT1) along the entire length of the GI tract, with apical localisation.[3,124]

OCT3 is highly expressed in skeletal muscle and several other tissues and cell types (salivary gland, prostate, uterus, placenta, adrenal gland, liver, neurons, glial cells and epithelial cells of choroid plexus).[126]

4.8.3 Structure

All members of the SLC family including OCT1 (554 amino acids, approximately 62 kDa) and OCT3 (556 amino acids, approximately 62 kDa) are predicted to contain 12 transmembrane α-helices with intracellular N- and C-termini. Amino acid homology between OCT1 and OCT3 is approximately 50%.[127]

4.8.4 Activity

Nonsynonymous variants of OCT1 (Q97K, P117L and R206C) and OCT3 (A116S, T400I and A439V) showed reduced uptake of substrates *in vitro*.[128,129]

4.8.5 Function and Endogenous Substrates

OCT1 and OCT3 transport a broad range of structurally diverse organic cations and have extensively overlapping substrate specificities. Substrates for OC transporters are largely small hydrophilic compounds ranging from approximately 60 to 350 Da in size, with at least one positively charged amine at physiological pH. OCT1 is localised on the basolateral membrane of hepatocytes and has been shown to mediate bidirectional transport of organic cations from blood to liver and *vice versa*.[130] With results described in recent literature indicating apical localisation in enterocytes, OCT1 may play an important role in absorption of endogenous substrates and cationic drugs. OCT3 is localised on the apical membrane of enterocytes and has been shown to be involved in the absorption of drugs.[114] Examples of endogenous substrates for OCT1 and OCT3 (Figure 4.13) include 1-methyl-4-phenylpyridinium, biogenic amines (dopamine, norepinephrine) and several other endogenous compounds (choline, creatinine, serotonin, histamine and adrenalin).[130,131]

4.8.6 Known Drug Substrates

In vitro studies have shown that OCT1 and/or OCT3 transport a wide variety of drugs (Figure 4.14) such as metformin and phenformin (anti-diabetic), cimetidine and ranitidine (acid-reflux inhibitors), pentamidine (anti-infective), lamotrigine (epilepsy), acyclovir and efavirenz (antiviral), daunorubicin (anti-cancer), formoterol and salbutamol (asthma and chronic obstructive pulmonary disease).[132] However, a role of cation transporters in absorption has only be demonstrated for metformin.[133]

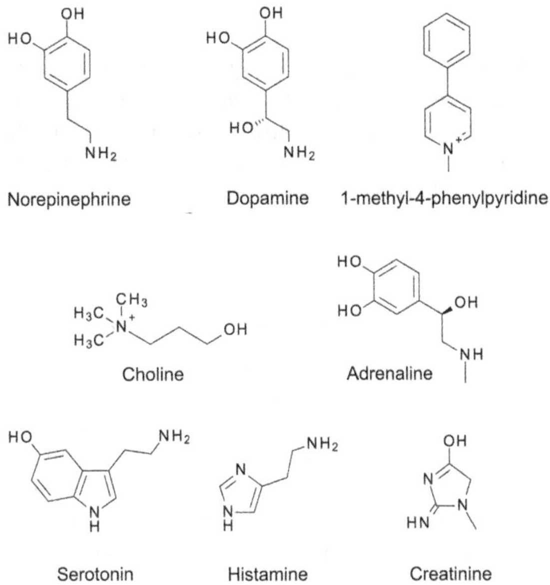

Figure 4.13 Endogenous substrates of OCTs.

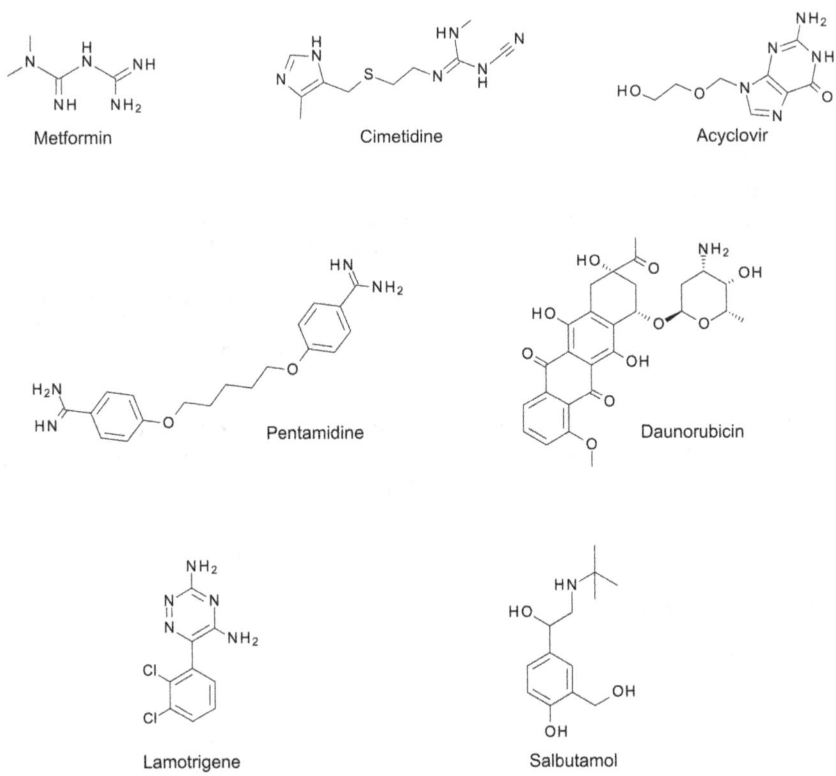

Figure 4.14 Drug substrates of OCTs.

4.8.7 Mechanism

OCT1 and OCT3 are multi-specific bidirectional transporters, driven by electrochemical gradients. Evidence indicates that OC transporters have an outward and inward conformation, allowing them to act as electrogenic uniporters (uptake) or electroneutral exchangers (uptake and efflux) based on membrane potential and concentration gradient.

4.8.8 Relevance

Using genetically engineered mice, Chen *et al.* showed that deletion of Oct3 significantly reduced oral bioavailability of metformin (from 58% in wild type to 27% in Oct3 knockout), indicating that OCT3 in humans may partly mediate oral absorption.[133,134] Prodrug strategies to increase oral absorption of payloads *via* OC transporter targeting have not been well explored. Glycosylated platinum IV prodrugs were shown to be selectively taken up into tumour tissue *via* OC transporters.[135] Similar approaches can potentially be applied to increase oral absorption for compounds with sub-optimal physicochemical properties. However, due to low expression relative to other uptake transporters, like PepT1, achieving adequate oral absorption *via* OCT1 and OCT3 may be challenging, and as such no examples are available.

Key References for OCTs

- J. Müller, K. S. Lips, L. Metzner, R. H. Neubert, H. Koepsell and M. Brandsch, Drug specificity and intestinal membrane localization of human organic cation transporters (OCT), *Biochem. Pharmacol.*, 2005, **70**(12), 1851–1860.
- T. K. Han, R. S. Everett, W. R. Proctor, C. M. Ng, C. L. Costales, K. L. Brouwer and D. R. Thakker, Organic cation transporter 1 (OCT1/mOct1) is localized in the apical membrane of Caco-2 cell monolayers and enterocytes, *Mol. Pharmacol.*, 2013, **84**(2), 182–189.
- M. Nishimura and S. Naito, Tissue-specific mRNA expression profiles of human ATP-binding cassette and solute carrier transporter superfamilies, *Drug Metab. Pharmacokinet.*, 2005, **20**(6), 452–477.
- H. Koepsell, Substrate recognition and translocation by polyspecific organic cation transporters, *Biol. Chem.*, 2011, **392**(1–2), 95–101.
- M. Roth, A. Obaidat and B. Hagenbuch, OATPs, OATs and OCTs: the organic anion and cation transporters of the SLCO and SLC22A gene superfamilies, *Br. J. Pharmacol.*, 2012, **165**(5), 1260–1287.

4.9 Organic Anion Transporting Polypeptides (OATP, SLCO)

4.9.1 Transporter Family

Members of the SLC organic anion transporter family include organic anion transporting polypeptide (OATP) 1A2 (SLCO1A2) and 2B1 (SLCO2B1) for which a role in intestinal absorption has been implicated.

4.9.2 Expression

OATP1A2 is expressed in several organs, including liver, brain and kidney, whereas OATP2B1 is expressed relatively ubiquitously, including in liver, small and large intestine, lung, spleen and other tissues.[136–139] Whilst Glaeser *et al.*[140] reported mRNA and apically located protein in the small intestine for OATP1A2, several other studies detected no or only very low OATP1A2 mRNA and/or protein levels.[3,5,8,136,141] In contrast, robust OATP2B1 expression has been reported in numerous studies.[3,5,8,36,139–142] OATP2B1 protein has been detected in the apical membrane of intestinal cells[8,142] although in some studies a basolateral location has also been reported.[143,144] The pregnane X receptor and the vitamin D receptor have been shown to increase the expression of OATPs *in vitro* although a similar effect on intestinal OATPs has not yet been demonstrated *in vivo*.[95,141,145]

4.9.3 Structure

OATP proteins contain 12 transmembrane domains and a fifth additional extracellular loop for surface expression. OATP2B1 is a 76 kDa protein, comprising 709 amino acids, and shares 77% amino acid sequence identity with rat rOatp2b1.[146]

4.9.4 Activity

The pH of the environment has been shown to alter the activity of OATPs; specifically, an acidic environment enhances activity.[146,147] Steroid hormones have been demonstrated to have a stimulatory effect on OATP2B1 for transport of sulphated steroids, but transport of drugs was not affected.[148] Several polymorphisms of OATP2B1 have been described.[141,149] Whilst the nonsynonymous SNP c.935G>A shows reduced activity *in vitro* limited effects have been observed *in vivo*.[141] For the nonsynonymous SNP c.1457C>T decreased expression and function have been described *in vitro*. Furthermore, the AUC of drugs such as fexofenadine and celiprolol was decreased by up to 50% in subjects with a c.1457CT or TT genotype.[141,150,151]

4.9.5 Function and Endogenous Substrates

OATPs in the intestine are uptake transporters that contribute to the absorption and distribution of their substrates.[146] Endogenous substrates of OATP2B1 include steroid hormones, conjugated steroids and prostaglandins (Figure 4.15).[137,149] The rat analogue, rOatp2b1, has a wider substrate specificity compared with its human orthologue.

4.9.6 Known Drug Substrates

Fexofenadine is the most documented drug example where OATPs are implicated in facilitating its oral absorption. Fexofenadine AUC is reduced up to 70% in healthy human volunteers after ingestion of grapefruit or orange juice, which have been shown *in vitro* to inhibit both OATP1A2 and OATP2B1 activity.[137] Drug substrates for OATP2B1 (Figure 4.16) include tebipenempivoxyl (ester prodrug), celiprolol, glibenclamide, atorvastatin, fluvastatin and pravastatin.[137,146,149,152]

4.9.7 Mechanism

The mechanism by which OATPs transport substrates is *via* a positively charged central core that acts through a rocker-switch mechanism.[153] In contrast to human, rat OATP-mediated uptake requires anion exchange

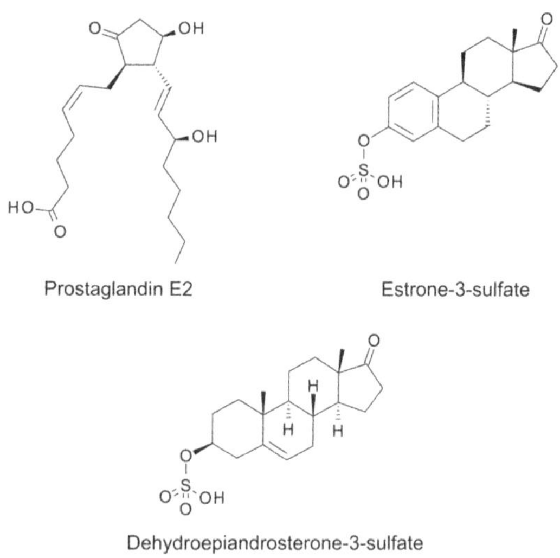

Figure 4.15 Endogenous substrates of OATP2B1.

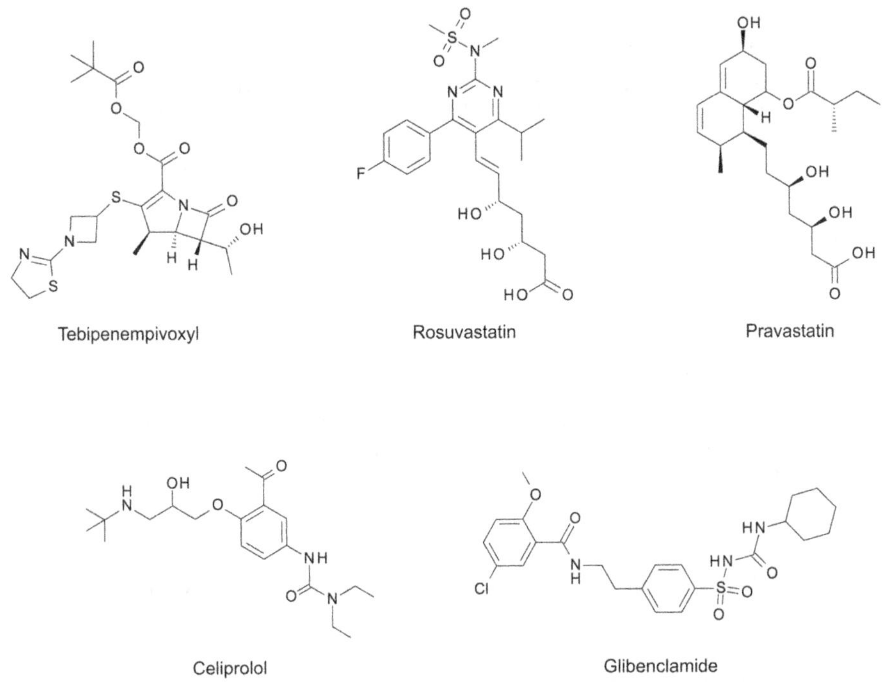

Figure 4.16 Drug substrates of OATP2B1.

(*e.g.* export of bicarbonate). This mechanism has been proposed for human OATP2B1 as well.

4.9.8 Relevance

Both OATP1A2 and OATP2B1 have been implicated in mediating the oral absorption of drugs like fexofenadine, mostly due to strong effects observed in food–drug interaction studies.[137,149] However, the role of OATP1A2 is debated due to its low expression levels and limited pharmacogenetic evidence.[141] In contrast, several pharmacogenetic, drug–drug and drug–food interaction studies in humans make a stronger case for a role of OATP2B1.[141] Nevertheless, there are currently no examples where intestinal OATP transporters have been targeted to improve permeability of poorly absorbed drugs through the gut epithelium. OATP-mediated drug uptake could theoretically be enhanced by optimising the intraluminal environment to an acidic pH. However, there are no specific examples of where this technique has been applied.[146] Ulcerative colitis and Chron's disease decrease the pH in the colon;[154] drugs targeting OATPs could thus be advantageous for treating these diseases.

Key References for OATPs

- A. Kalliokoski and M. Niemi, Impact of OATP transporters on pharmacokinetics, *Br. J. Pharmacol.*, 2009, **158**(3), 693–705.
- M. Niemi, Role of OATP transporters in the disposition of drugs, *Pharmacogenomics*, 2007, **8**(7), 787–802.
- A. Wilson and R. B. Kim, OATP Transporters: Potential Targets for Enhancing Organ and Tissue Specific Drug Delivery, *J. Pharmacol. Clin. Toxicol.*, 2014, 2(3), 1037.
- I. Tamai, Oral drug delivery utilizing intestinal OATP transporters, *Adv. Drug Delivery Rev.*, 2012, **64**(6), 508–514.
- S. Oswald, Organic Anion Transporting Polypeptide (OATP) transporter expression, localization and function in the human intestine, *Pharmacol. Ther.*, 2019, **195**, 39–53.

4.10 Nucleoside Transporters (NTs, SLC28 and SLC29)

4.10.1 Transporter Family

Human nucleoside transporters (NTs) are divided into two structurally unrelated protein families; one that contains three SLC28 genes (*SLC28A1–3*) encoding concentrative nucleoside transporter proteins (CNT1–3) and another one comprising four SLC29 genes (*SLC29A1–4*) encoding equilibrative nucleoside transporter proteins (ENT1–4).[155]

4.10.2 Expression

CNT1 and CNT2 are expressed in the plasma membrane of liver, small intestine and kidney. CNT2 is additionally expressed in heart, skeletal muscle, placenta and brain. CNT3 is more widely expressed in plasma membranes and membranes of sub-cellular organelles of various tissues, including the intestine. All CNTs are localised to the apical membrane of enterocytes. The expression of CNT2 and CNT3 is high in duodenum and small intestine, while CNT3 is expressed at low levels across the whole length of the intestine.[156,157] ENT1 and ENT2 are expressed in several tissues including intestine. In contrast to CNTs, ENT1 and ENT2 have been localised to the basolateral membrane of enterocytes (albeit being more abundant in crypt cells) and work in concert with CNTs on the apical membrane to facilitate transepithelial nucleoside flux. Both ENT1 and ENT2 are primarily located at the plasma membrane but are also expressed on nuclear membranes. ENT3 is widely expressed with intracellular localisation (*e.g.* endosomal and mitochondrial membranes), while ENT4 is primarily located at the plasma membrane with expression in heart, brain and skeletal muscle.[157,158]

4.10.3 Structure

CNTs (approximately 650–658 amino acids; approximately 70–71 kDa) consist of 13–14 transmembrane domains with the cation and substrate recognition sites in the carboxy half of the protein. ENTs (approximately 465 amino acids; approximately 52 kDa) contain 11 TMDs, a cytoplasmic N-terminus and an extracellular C-terminus.[159] Homology at the amino acid level between human, rat and mouse for CNTs is approximately 78–83%. Homology between human and rat ENT1 and ENT2 is approximately 49–78%. Tissue distribution of CNTs and ENTs can vary between species.[160]

4.10.4 Activity

In humans, variability in ribavirin serum levels has been associated with the SLC28A2 C65T SNP.[161] The mean oral bioavailability of mizoribine has been shown to be significantly lower in healthy Japanese males with a SLC28A1 G565A allele.[162]

4.10.5 Function and Endogenous Substrates

Due to apical and basolateral localisation of CNTs and ENT1 in enterocytes, respectively, it is proposed that dietary nucleosides are absorbed from the diet into blood *via* these transporters. NTs play an important role in maintaining nucleoside homeostasis through provision of nucleosides (*e.g.* guanosine, adenosine, uridine, cytosine and thymidine; Figure 4.17) and nucleobases (*e.g.* adenine) derived from the diet or produced by the liver for salvage pathways of nucleotide synthesis in tissues and cells that lack *de novo* synthesis (*e.g.* brain, muscle, erythrocytes, leukocytes and bone marrow cells). Unlike CNTs, ENTs can transport nucleobases. The nucleosides and nucleobases are converted inside the cell into nucleotides, which are important in intermediary metabolism, cell signalling pathways, phospholipid and oligosaccharide synthesis and as precursors for nucleic acid biosynthesis.[155]

4.10.6 Known Drug Substrates

Several anticancer and antiviral drugs (nucleoside and nucleobase analogues) have been shown to be substrates for CNTs and ENTs (Figure 4.18). Due to the expression of CNTs and ENTs in enterocytes and several other tissues, it is reasonable to assume that absorption and disposition of these drugs is mediated *via* NTs. Examples of drugs that interact with NTs include didanosine and zidovudine (anti-viral HIV), ribavirin (anti-viral HCV), fludarabine and trifluridine (lympho-proliferative diseases and solid tumours) and cladribine (leukaemia).[157]

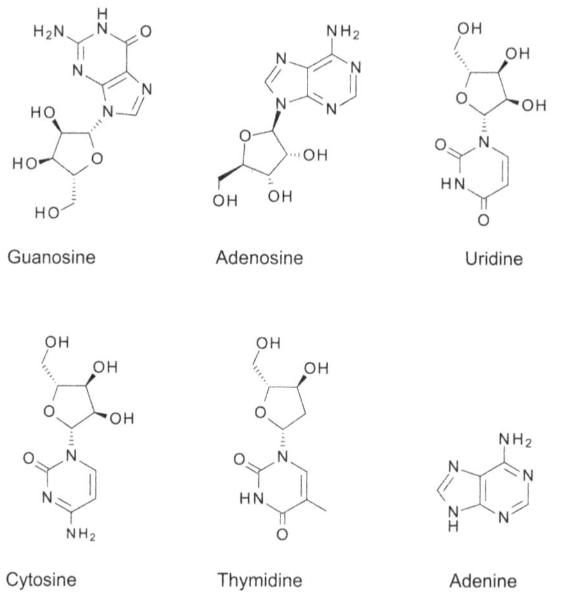

Figure 4.17 Endogenous substrates of nucleoside transporters.

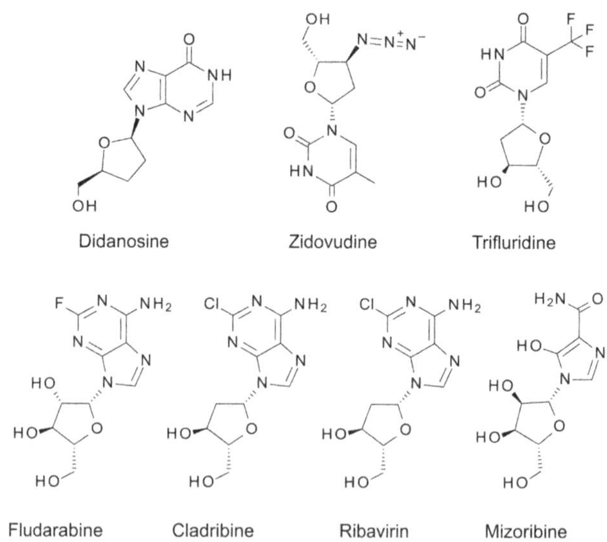

Figure 4.18 Drug substrates of nucleoside transporters.

4.10.7 Mechanism

CNTs are co-transporters that require inwardly directed Na^+- or H^+-coupling. The stoichiometry of Na^+- or H^+-coupling with nucleosides for CNT1 and CNT2 is 1 : 1, while for CNT3 the stoichiometry is largely 2 : 1 (cation:nucleoside coupling ratio of 1 : 1 for H^+, 2 : 1 for Na^+, 1 : 1 : 1 Na^+ : H^+ : nucleoside). ENTs are Na^+-independent bidirectional transporters. ENT1 and ENT2 transport substrates *via* facilitated diffusion, while transport *via* ENT3 and ENT4 appears to be proton linked. CNTs show high apparent affinities to permeants compared with ENTs.[163]

4.10.8 Relevance

Using genetically engineered Ent1 knockout mice, intestinal absorption of ribavirin was shown to be mediated by NTs.[164] The oral bioavailability of nucleoside analogues is, at best, approximately 50–60%.[165] Prodrug strategies to increase oral absorption of payloads *via* CNTs and ENTs have not been well explored.

Key References for NTs

- M. Pastor-Anglada, N. Urtasun and S. Pérez-Torras, Intestinal Nucleoside Transporters: Function, Expression, and Regulation, *Compr. Physiol.*, 2018, 8(3), 1003–1017.
- J. D. Young, S. Y. Yao, J. M. Baldwin, C. E. Cass and S. A. Baldwin, The human concentrative and equilibrative nucleoside transporter families, SLC28 and SLC29, *Mol. Aspects Med.*, 2013, **34**(2–3), 529–547.
- H. Lu, C. Chen and C. Klaassen, Tissue distribution of concentrative and equilibrative nucleoside transporters in male and female rats and mice, *Drug Metab. Dispos.*, 2004, **32**(12), 1455–1461.

References

1. M. D. Harwood, M. Zhang, S. M. Pathak and S. Neuhoff, *Drug Metab. Dispos.*, 2019, **47**, 854–864.
2. M. Drozdzik, D. Busch, J. Lapczuk, J. Müller, M. Ostrowski, M. Kurzawski and S. Oswald, *Clin. Pharmacol. Ther.*, 2019, **105**, 1204–1212.
3. M. Drozdzik, C. Groer, J. Penski, J. Lapczuk, M. Ostrowski, Y. Lai, B. Prasad, J. D. Unadkat, W. Siegmund and S. Oswald, *Mol. Pharm.*, 2014, **11**, 3547–3555.
4. G. Englund, F. Rorsman, A. Rönnblom, U. Karlbom, L. Lazorova, J. Gråsjö, A. Kindmark and P. Artursson, *Eur. J. Pharm. Sci.*, 2006, **29**, 269–277.

5. Y. Meier, J. J. Eloranta, J. Darimont, M. G. Ismair, C. Hiller, M. Fried, G. A. Kullak-Ublick and S. R. Vavricka, *Drug Metab. Dispos.*, 2007, **35**, 590–594.
6. A. Sawant-Basak, A. D. Rodrigues, M. Lech, R. Doyonnas, M. Kasaian, B. Prasad and N. Tsamandouras, *Drug Metab. Dispos.*, 2018, **46**, 1581–1587.
7. P. D. Prasad, H. Wang, W. Huang, Y. J. Fei, F. H. Leibach, L. D. Devoe and V. Ganapathy, *Arch. Biochem. Biophys.*, 1999, **366**, 95–106.
8. E. Miyauchi, M. Tachikawa, X. Decleves, Y. Uchida, J. L. Bouillot, C. Poitou, J. M. Oppert, S. Mouly, J. F. Bergmann, T. Terasaki, J. M. Schermann and C. Lloret-Linares, *Mol. Pharm.*, 2016, **13**, 2631–2640.
9. A. Schlessinger, M. A. Welch, H. van Vlijmen, K. Korzekwa, P. W. Swaan and P. Matsson, *Clin. Pharmacol. Ther.*, 2018, **104**, 818–835.
10. P. Sharma, M. I. Atari, R. Elsby, S. Thomas, S. Stahl, C. Hilgendorf and K. Fenner, in *Drug Transporters: Volume 1: Role and Importance in ADME and Drug Development*, The Royal Society of Chemistry, 2016, vol. 1, pp. 229–297.
11. K. L. Brouwer, D. Keppler, K. A. Hoffmaster, D. A. Bow, Y. Cheng, Y. Lai, J. E. Palm, B. Stieger, R. Evers and C. International Transporter, *Clin. Pharmacol. Ther.*, 2013, **94**, 95–112.
12. P. V. Balimane, S. Chong and R. A. Morrison, *J. Pharmacol. Toxicol. Methods*, 2000, **44**, 301–312.
13. S. Alqahtani, L. A. Mohamed and A. Kaddoumi, *Expert Opin. Drug Metab. Toxicol.*, 2013, **9**, 1241–1254.
14. J. Stappaerts, J. Brouwers, P. Annaert and P. Augustijns, *Int. J. Pharm.*, 2015, **478**, 665–681.
15. J. H. Beale, J. L. Parker, F. Samsudin, A. L. Barrett, A. Senan, L. E. Bird, D. Scott, R. J. Owens, M. S. P. Sansom, S. J. Tucker, D. Meredith, P. W. Fowler and S. Newstead, *Structure*, 2015, **23**, 1889–1899.
16. F. Samsudin, J. L. Parker, M. S. P. Sansom, S. Newstead and P. W. Fowler, *Cell Chem. Biol.*, 2016, **23**, 299–309.
17. N. J. Hu, S. Iwata, A. D. Cameron and D. Drew, *Nature*, 2011, **478**, 408–411.
18. S. Ekins, L. Diao and J. E. Polli, *Mol. Pharm.*, 2012, **9**, 905–913.
19. M. D. Renna, A. S. Oyadeyi, E. Bossi, G. Kottra and A. Peres, *Cell. Mol. Life Sci.*, 2011, **68**, 2961–2975.
20. I. Knutter, G. Kottra, W. Fischer, H. Daniel and M. Brandsch, *Drug Metab. Dispos.*, 2009, **37**, 143–149.
21. S. P. Shirazi-Beechey, A. G. Davies, K. Tebbutt, J. Dyer, A. Ellis, C. J. Taylor, P. Fairclough and R. B. Beechey, *Gastroenterology*, 1990, **98**, 676–685.
22. A. Sjoberg, M. Lutz, C. Tannergren, C. Wingolf, A. Borde and A. L. Ungell, *Eur. J. Pharm. Sci.*, 2013, **48**, 166–180.
23. H. H. Ussing and K. Zerahn, *Acta Physiol. Scand.*, 1951, **23**, 110–127.
24. T. H. Wilson and G. Wiseman, *J. Physiol.*, 1954, **123**, 116–125.
25. L. Barthe, J. Woodley and G. Houin, *Fundam. Clin. Pharmacol.*, 1999, **13**, 154–168.
26. D. A. Lionarons, J. L. Boyer and S. Y. Cai, *J. Lipid Res.*, 2012, **53**, 1535–1542.

27. K. Ma, Y. Hu and D. E. Smith, *J. Pharm. Sci.*, 2011, **100**, 767–774.
28. Y. Hu, D. E. Smith, K. Ma, D. Jappar, W. Thomas and K. M. Hillgren, *Mol. Pharm.*, 2008, **5**, 1122–1130.
29. S. Medwid, M. M. J. Li, M. J. Knauer, K. Lin, S. E. Mansell, C. L. Schmerk, C. Zhu, K. E. Griffin, M. D. Yousif, G. K. Dresser, U. I. Schwarz, R. B. Kim and R. G. Tirona, *Drug Metab. Dispos.*, 2019, **47**, 832–842.
30. J. W. Jonker, E. Wagenaar, S. Van Eijl and A. H. Schinkel, *Mol. Cell. Biol.*, 2003, **23**, 7902–7908.
31. P. A. Dawson, J. Haywood, A. L. Craddock, M. Wilson, M. Tietjen, K. Kluckman, N. Maeda and J. S. Parks, *J. Biol. Chem.*, 2003, **278**, 33920–33927.
32. Y. Hu, Y. Xie, Y. Wang, X. Chen and D. E. Smith, *Mol. Pharm.*, 2014, **11**, 3737–3746.
33. L. Wang, C. Wang, Q. Liu, Q. Meng, X. Huo, P. Sun, X. Yang, H. Sun, Y. Zhen, J. Peng, X. Ma and K. Liu, *Eur. J. Pharm. Sci.*, 2014, **63**, 77–86.
34. M. Johnson, D. Patel, C. Matheny, M. Ho, L. Chen and H. Ellens, *Drug Metab. Dispos.*, 2017, **45**, 27–34.
35. D. Herrera-Ruiz, Q. Wang, O. S. Gudmundsson, T. J. Cook, R. L. Smith, T. N. Faria and G. T. Knipp, *AAPS PharmSci*, 2001, **3**, E9.
36. C. Hilgendorf, G. Ahlin, A. Seithel, P. Artursson, A. L. Ungell and J. Karlsson, *Drug Metab. Dispos.*, 2007, **35**, 1333–1340.
37. D. Jappar, S. P. Wu, Y. Hu and D. E. Smith, *Drug Metab. Dispos.*, 2010, **38**, 1740–1746.
38. B. Spanier and F. Rohm, *Compr. Physiol.*, 2018, **8**, 843–869.
39. K. Naruhashi, Y. Sai, I. Tamai, N. Suzuki and A. Tsuji, *Pharm. Res.*, 2002, **19**, 1417–1423.
40. D. E. Smith, B. Clemencon and M. A. Hediger, *Mol. Aspects Med.*, 2013, **34**, 323–336.
41. E. Y. Zhang, R. M. Emerick, Y. A. Pak, S. A. Wrighton and K. M. Hillgren, *Mol. Pharm.*, 2004, **1**, 201–210.
42. T. Stelzl, T. Baranov, K. E. Geillinger, G. Kottra and H. Daniel, *Am. J. Physiol.: Gastrointest. Liver Physiol.*, 2016, **310**, G128–G141.
43. T. Stelzl, K. E. Geillinger-Kastle, J. Stolz and H. Daniel, *Am. J. Physiol.: Gastrointest. Liver Physiol.*, 2017, **312**, G580–G591.
44. E. Y. Zhang, D. J. Fu, Y. A. Pak, T. Stewart, N. Mukhopadhyay, S. A. Wrighton and K. M. Hillgren, *J. Pharmacol. Exp. Ther.*, 2004, **310**, 437–445.
45. P. Anderle, C. U. Nielsen, J. Pinsonneault, P. L. Krog, B. Brodin and W. Sadee, *J. Pharmacol. Exp. Ther.*, 2006, **316**, 636–646.
46. M. Zucchelli, L. Torkvist, F. Bresso, J. Halfvarson, A. Hellquist, F. Anedda, G. Assadi, G. B. Lindgren, M. Svanfeldt, M. Janson, C. L. Noble, S. Pettersson, M. Lappalainen, P. Paavola-Sakki, L. Halme, M. Farkkila, U. Turunen, J. Satsangi, K. Kontula, R. Lofberg, J. Kere and M. D'Amato, *Inflammatory Bowel Dis.*, 2009, **15**, 1562–1569.
47. Y. Hu, X. Chen and D. E. Smith, *Drug Metab. Dispos.*, 2012, **40**, 1328–1335.
48. Y. Hu and D. E. Smith, *Biochem. Pharmacol.*, 2016, **107**, 81–90.

49. D. Epling, Y. Hu and D. E. Smith, *Biochem. Pharmacol.*, 2018, **155**, 1–7.
50. D. Kolodziejczak, B. Spanier, R. Pais, J. Kraiczy, T. Stelzl, K. Gedrich, C. Scherling, T. Zietek and H. Daniel, *Am. J. Physiol.: Gastrointest. Liver Physiol.*, 2013, **304**, G897–G907.
51. A. M. Nassl, I. Rubio-Aliaga, H. Fenselau, M. K. Marth, G. Kottra and H. Daniel, *Am. J. Physiol.: Gastrointest. Liver Physiol.*, 2011, **301**, G128–G137.
52. F. Doring, J. Will, S. Amasheh, W. Clauss, H. Ahlbrecht and H. Daniel, *J. Biol. Chem.*, 1998, **273**, 23211–23218.
53. M. Irie, T. Terada, K. Sawada, H. Saito and K. Inui, *J. Pharmacol. Exp. Ther.*, 2001, **298**, 711–717.
54. T. Terada, K. Sawada, M. Irie, H. Saito, Y. Hashimoto and K. Inui, *Pflügers Arch.*, 2000, **440**, 679–684.
55. T. Terada, H. Saito, M. Mukai and K. Inui, *Am. J. Physiol.*, 1997, **273**, F706–F711.
56. M. E. Ganapathy, W. Huang, H. Wang, V. Ganapathy and F. H. Leibach, *Biochem. Biophys. Res. Commun.*, 1998, **246**, 470–475.
57. M. Brandsch, *Expert Opin. Drug Metab. Toxicol.*, 2009, **5**, 887–905.
58. P. V. Balimane, S. Chong, K. Patel, Y. Quan, J. Timoszyk, Y. H. Han, B. Wang, B. Vig and T. N. Faria, *Arch. Pharmacal Res.*, 2007, **30**, 507–518.
59. R. Shimizu, T. Sukegawa, Y. Tsuda and T. Itoh, *Int. J. Pharm.*, 2008, **354**, 104–110.
60. A. Guo, P. Hu, P. V. Balimane, F. H. Leibach and P. J. Sinko, *J. Pharmacol. Exp. Ther.*, 1999, **289**, 448–454.
61. P. V. Balimane, I. Tamai, A. Guo, T. Nakanishi, H. Kitada, F. H. Leibach, A. Tsuji and P. J. Sinko, *Biochem. Biophys. Res. Commun.*, 1998, **250**, 246–251.
62. Y. Zhang, J. Sun, Y. Sun, Y. Wang and Z. He, *Curr. Drug Metab.*, 2013, **14**, 675–687.
63. M. Brandsch, *Curr. Opin. Pharmacol.*, 2013, **13**, 881–887.
64. P. D. Bailey, C. A. Boyd, J. R. Bronk, I. D. Collier, D. Meredith, K. M. Morgan and C. S. Temple, *Angew. Chem., Int. Ed.*, 2000, **39**, 505–508.
65. D. Meredith, C. S. Temple, N. Guha, C. J. Sword, C. A. Boyd, I. D. Collier, K. M. Morgan and P. D. Bailey, *Eur. J. Biochem.*, 2000, **267**, 3723–3728.
66. D. Ormrod and K. Goa, *Drugs*, 2000, **59**, 1317–1340.
67. I. Rubio-Aliaga and H. Daniel, *Trends Pharmacol. Sci.*, 2002, **23**, 434–440.
68. M. Brandsch, I. Knutter and E. Bosse-Doenecke, *J. Pharm. Pharmacol.*, 2008, **60**, 543–585.
69. R. H. Barbhaiya, *Biopharm. Drug Dispos.*, 1996, **17**, 319–330.
70. B. Yang and D. E. Smith, *Drug Metab. Dispos.*, 2013, **41**, 608–614.
71. M. C. Garcia-Carbonell, L. Granero, F. Torres-Molina, J. C. Aristorena, J. Chesa-Jimenez, J. M. Pla-Delfina and J. E. Peris-Ribera, *Drug Metab. Dispos.*, 1993, **21**, 215–217.
72. T. M. Garrigues, U. Martin, J. E. Peris-Ribera and L. F. Prescott, *Eur. J. Clin. Pharmacol.*, 1991, **41**, 179–183.

73. Y. A. Pak, A. J. Long, W. F. Annes, J. W. Witcher, M. P. Knadler, M. A. Ayan-Oshodi, M. I. Mitchell, P. Leese and K. M. Hillgren, *Drug Metab. Dispos.*, 2017, **45**, 137–144.
74. C. P. Landowski, B. S. Vig, X. Song and G. L. Amidon, *Mol. Cancer Ther.*, 2005, **4**, 659–667.
75. D. Gupta, S. Varghese Gupta, A. Dahan, Y. Tsume, J. Hilfinger, K. D. Lee and G. L. Amidon, *Mol. Pharm.*, 2013, **10**, 512–522.
76. Z. Yan, J. Sun, Y. Chang, Y. Liu, Q. Fu, Y. Xu, Y. Sun, X. Pu, Y. Zhang, Y. Jing, S. Yin, M. Zhu, Y. Wang and Z. He, *Mol. Pharm.*, 2011, **8**, 319–329.
77. L. M. Rorick-Kehn, E. J. Perkins, K. M. Knitowski, J. C. Hart, B. G. Johnson, D. D. Schoepp and D. L. McKinzie, *J. Pharmacol. Exp. Ther.*, 2006, **316**, 905–913.
78. M. V. Varma, A. H. Eriksson, G. Sawada, Y. A. Pak, E. J. Perkins and C. L. Zimmerman, *Drug Metab. Dispos.*, 2009, **37**, 211–220.
79. P. D. Prasad and V. Ganapathy, *Curr. Opin. Clin. Nutr. Metab. Care*, 2000, **3**, 263–266.
80. H. Wang, W. Huang, Y. J. Fei, H. Xia, T. L. Yang-Feng, F. H. Leibach, L. D. Devoe, V. Ganapathy and P. D. Prasad, *J. Biol. Chem.*, 1999, **274**, 14875–14883.
81. H. M. Said, R. Redha and W. Nylander, *Am. J. Physiol.*, 1987, **253**, G631–G636.
82. H. M. Said and R. Redha, *Am. J. Physiol.*, 1987, **252**, G52–G55.
83. A. Ghosal and H. M. Said, *Am. J. Physiol.: Cell Physiol.*, 2011, **300**, C97–C104.
84. H. M. Said, R. Redha and W. Nylander, *Am. J. Clin. Nutr.*, 1989, **49**, 127–131.
85. A. D. Vadlapudi, R. K. Vadlapatla and A. K. Mitra, *Curr. Drug Targets*, 2012, **13**, 994–1003.
86. S. Ramanathan, S. Pooyan, S. Stein, P. D. Prasad, J. Wang, M. J. Leibowitz, V. Ganapathy and P. J. Sinko, *Pharm. Res.*, 2001, **18**, 950–956.
87. S. R. Chirapu, C. J. Rotter, E. L. Miller, M. V. Varma, R. L. Dow and M. G. Finn, *Curr. Top. Med. Chem.*, 2013, **13**, 837–842.
88. K. G. Janoria, S. H. Boddu, Z. Wang, D. K. Paturi, S. Samanta, D. Pal and A. K. Mitra, *J. Ocul. Pharmacol. Ther.*, 2009, **25**, 39–49.
89. A. D. Vadlapudi, R. K. Vadlapatla, R. Earla, S. Sirimulla, J. B. Bailey, D. Pal and A. K. Mitra, *Pharm. Res.*, 2013, **30**, 2063–2076.
90. A. D. Vadlapudi, R. K. Vadlapatla, D. Kwatra, R. Earla, S. K. Samanta, D. Pal and A. K. Mitra, *Int. J. Pharm.*, 2012, **434**, 315–324.
91. A. Balakrishnan and J. E. Polli, *Mol. Pharm.*, 2006, **3**, 223–230.
92. F. Chen, L. Ma, N. Al-Ansari and B. Shneider, *J. Biol. Chem.*, 2001, **276**, 38703–38714.
93. R. H. Ho, B. F. Leake, B. L. Urquhart, J. C. Gregor, P. A. Dawson and R. B. Kim, *J. Gastroenterol. Hepatol.*, 2011, **26**, 1740–1748.
94. J. Y. Chiang, *Compr. Physiol.*, 2013, **3**, 1191–1212.
95. W. Jia, G. Xie and W. Jia, *Nat. Rev. Gastroenterol. Hepatol.*, 2018, **15**, 111–128.

96. P. A. Dawson, *Handb. Exp. Pharmacol.*, 2011, 169–203.
97. M. S. Roberts, B. M. Magnusson, F. J. Burczynski and M. Weiss, *Clin. Pharmacokinet.*, 2002, **41**, 751–790.
98. Y. Gao, J. Shao, Z. Jiang, J. Chen, S. Gu, S. Yu, K. Zheng and L. Jia, *Drug Discovery Today*, 2014, **19**, 326–340.
99. V. Kolhatkar and J. E. Polli, *Eur. J. Pharm. Sci.*, 2012, **46**, 86–99.
100. X. Zheng and J. E. Polli, *Int. J. Pharm.*, 2010, **396**, 111–118.
101. S. Tolle-Sander, K. A. Lentz, D. Y. Maeda, A. Coop and J. E. Polli, *Mol. Pharm.*, 2004, **1**, 40–48.
102. M. E. Morris and M. A. Felmlee, *AAPS J.*, 2008, **10**, 311–321.
103. B. E. Enerson and L. R. Drewes, *J. Pharm. Sci.*, 2003, **92**, 1531–1544.
104. P. Fisel, E. Schaeffeler and M. Schwab, *Clin. Transl. Sci.*, 2018, **11**, 352–364.
105. A. P. Halestrap, *IUBMB Life*, 2012, **64**, 1–9.
106. W. K. Lam, M. A. Felmlee and M. E. Morris, *Drug Metab. Dispos.*, 2010, **38**, 441–447.
107. C. Arena and H. L. Fung, *J. Pharm. Sci.*, 1980, **69**, 356–358.
108. Y. H. Li, K. Ito, Y. Tsuda, R. Kohda, H. Yamada and T. Itoh, *J. Pharmacol. Exp. Ther.*, 1999, **290**, 958–964.
109. K. C. Cundy, R. Branch, T. Chernov-Rogan, T. Dias, T. Estrada, K. Hold, K. Koller, X. Liu, A. Mann, M. Panuwat, S. P. Raillard, S. Upadhyay, Q. Q. Wu, J. N. Xiang, H. Yan, N. Zerangue, C. X. Zhou, R. W. Barrett and M. A. Gallop, *J. Pharmacol. Exp. Ther.*, 2004, **311**, 315–323.
110. K. C. Cundy, S. Sastry, W. Luo, J. Zou, T. L. Moors and D. M. Canafax, *J. Clin. Pharmacol.*, 2008, **48**, 1378–1388.
111. R. Lal, J. Sukbuntherng, E. H. Tai, S. Upadhyay, F. Yao, M. S. Warren, W. Luo, L. Bu, S. Nguyen, J. Zamora, G. Peng, T. Dias, Y. Bao, M. Ludwikow, T. Phan, R. A. Scheuerman, H. Yan, M. Gao, Q. Q. Wu, T. Annamalai, S. P. Raillard, K. Koller, M. A. Gallop and K. C. Cundy, *J. Pharmacol. Exp. Ther.*, 2009, **330**, 911–921.
112. A. R. English, J. A. Retsema, V. A. Ray and J. E. Lynch, *Antimicrob. Agents Chemother.*, 1972, **1**, 185–191.
113. Y. H. Li, M. Tanno, T. Itoh and H. Yamada, *Int. J. Pharm.*, 1999, **191**, 151–159.
114. H. Koepsell and H. Endou, *Pflügers Arch.*, 2004, **447**, 666–676.
115. C. Zhu, K. B. Nigam, R. C. Date, K. T. Bush, S. A. Springer, M. H. Saier Jr, W. Wu and S. K. Nigam, *PLoS One*, 2015, **10**, e0140569.
116. K. Lahjouji, G. A. Mitchell and I. A. Qureshi, *Mol. Genet. Metab.*, 2001, **73**, 287–297.
117. X. Wu, W. Huang, P. D. Prasad, P. Seth, D. P. Rajan, F. H. Leibach, J. Chen, S. J. Conway and V. Ganapathy, *J. Pharmacol. Exp. Ther.*, 1999, **290**, 1482–1492.
118. I. Tamai, *Biopharm. Drug Dispos.*, 2013, **34**, 29–44.
119. S. Grigat, C. Fork, M. Bach, S. Golz, A. Geerts, E. Schomig and D. Grundemann, *Drug Metab. Dispos.*, 2009, **37**, 330–337.
120. L. Diao and J. E. Polli, *J. Pharm. Sci.*, 2011, **100**, 3802–3816.

121. G. Wang, H. Chen, D. Zhao, D. Ding, M. Sun, L. Kou, C. Luo, D. Zhang, X. Yi, J. Dong, J. Wang, X. Liu, Z. He and J. Sun, *J. Med. Chem.*, 2017, **60**, 2552–2561.
122. J. X. Mo, S. J. Shi, Q. Zhang, T. Gong, X. Sun and Z. R. Zhang, *Mol. Pharm.*, 2011, **8**, 1629–1640.
123. M. Roth, A. Obaidat and B. Hagenbuch, *Br. J. Pharmacol.*, 2012, **165**, 1260–1287.
124. J. Muller, K. S. Lips, L. Metzner, R. H. Neubert, H. Koepsell and M. Brandsch, *Biochem. Pharmacol.*, 2005, **70**, 1851–1860.
125. T. K. Han, R. S. Everett, W. R. Proctor, C. M. Ng, C. L. Costales, K. L. Brouwer and D. R. Thakker, *Mol. Pharmacol.*, 2013, **84**, 182–189.
126. M. Nishimura and S. Naito, *Drug Metab. Pharmacokinet.*, 2005, **20**, 452–477.
127. H. Koepsell, B. M. Schmitt and V. Gorboulev, *Rev. Physiol., Biochem. Pharmacol.*, 2003, **150**, 36–90.
128. L. Chen, M. Takizawa, E. Chen, A. Schlessinger, J. Segenthelar, J. H. Choi, A. Sali, M. Kubo, S. Nakamura, Y. Iwamoto, N. Iwasaki and K. M. Giacomini, *J. Pharmacol. Exp. Ther.*, 2010, **335**, 42–50.
129. T. Sakata, N. Anzai, T. Kimura, D. Miura, T. Fukutomi, M. Takeda, H. Sakurai and H. Endou, *J. Pharmacol. Sci.*, 2010, **113**, 263–266.
130. K. N. Faber, M. Muller and P. L. Jansen, *Adv. Drug Delivery Rev.*, 2003, **55**, 107–124.
131. J. W. Jonker and A. H. Schinkel, *J. Pharmacol. Exp. Ther.*, 2004, **308**, 2–9.
132. H. Koepsell, *Expert Opin. Drug Metab. Toxicol.*, 2015, **11**, 1619–1633.
133. X. Liang and K. M. Giacomini, *J. Pharm. Sci.*, 2017, **106**, 2245–2250.
134. E. C. Chen, X. Liang, S. W. Yee, E. G. Geier, S. L. Stocker, L. Chen and K. M. Giacomini, *Mol. Pharmacol.*, 2015, **88**, 75–83.
135. J. Ma, Q. Wang, Z. Huang, X. Yang, Q. Nie, W. Hao, P. G. Wang and X. Wang, *J. Med. Chem.*, 2017, **60**, 5736–5748.
136. I. Tamai, J. Nezu, H. Uchino, Y. Sai, A. Oku, M. Shimane and A. Tsuji, *Biochem. Biophys. Res. Commun.*, 2000, **273**, 251–260.
137. A. Kalliokoski and M. Niemi, *Br. J. Pharmacol.*, 2009, **158**, 693–705.
138. G. A. Kullak-Ublick, B. Hagenbuch, B. Stieger, C. D. Schteingart, A. F. Hofmann, A. W. Wolkoff and P. J. Meier, *Gastroenterology*, 1995, **109**, 1274–1282.
139. G. A. Kullak-Ublick, M. G. Ismair, B. Stieger, L. Landmann, R. Huber, F. Pizzagalli, K. Fattinger, P. J. Meier and B. Hagenbuch, *Gastroenterology*, 2001, **120**, 525–533.
140. H. Glaeser, D. G. Bailey, G. K. Dresser, J. C. Gregor, U. I. Schwarz, J. S. McGrath, E. Jolicoeur, W. Lee, B. F. Leake, R. G. Tirona and R. B. Kim, *Clin. Pharmacol. Ther.*, 2007, **81**, 362–370.
141. S. Oswald, *Pharmacol. Ther.*, 2019, **195**, 39–53.
142. D. Kobayashi, T. Nozawa, K. Imai, J. Nezu, A. Tsuji and I. Tamai, *J. Pharmacol. Exp. Ther.*, 2003, **306**, 703–708.
143. M. Keiser, L. Kaltheuner, C. Wildberg, J. Muller, M. Grube, L. I. Partecke, C. D. Heidecke and S. Oswald, *J. Pharm. Sci.*, 2017, **106**, 2657–2663.

144. M. G. Mooij, B. E. de Koning, D. J. Lindenbergh-Kortleve, Y. Simons-Oosterhuis, B. D. van Groen, D. Tibboel, J. N. Samsom and S. N. de Wildt, *Drug Metab. Dispos.*, 2016, **44**, 1014–1019.
145. G. A. Kullak-Ublick, C. Gubler, K. Spanaus, M. G. Ismair, T. Claro da Silva and A. Jetter, *Eur. J. Clin. Pharmacol.*, 2016, **72**, 797–805.
146. I. Tamai, *Adv. Drug Delivery Rev.*, 2012, **64**, 508–514.
147. P. Martinez-Becerra, O. Briz, M. R. Romero, R. I. Macias, M. J. Perez, C. Sancho-Mateo, M. P. Lostao, J. M. Fernandez-Abalos and J. J. Marin, *Mol. Pharmacol.*, 2011, **79**, 596–607.
148. A. Koenen, K. Kock, M. Keiser, W. Siegmund, H. K. Kroemer and M. Grube, *Eur. J. Pharm. Sci.*, 2012, **47**, 774–780.
149. J. Yu, Z. Zhou, J. Tay-Sontheimer, R. H. Levy and I. Ragueneau-Majlessi, *J. Pharm. Sci.*, 2017, **106**, 2312–2325.
150. W. Lee, H. Glaeser, L. H. Smith, R. L. Roberts, G. W. Moeckel, G. Gervasini, B. F. Leake and R. B. Kim, *J. Biol. Chem.*, 2005, **280**, 9610–9617.
151. F. Zhou, J. Zheng, L. Zhu, A. Jodal, P. H. Cui, M. Wong, H. Gurney, W. B. Church and M. Murray, *AAPS J.*, 2013, **15**, 1099–1108.
152. K. Kato, Y. Shirasaka, E. Kuraoka, A. Kikuchi, M. Iguchi, H. Suzuki, S. Shibasaki, T. Kurosawa and I. Tamai, *Mol. Pharm.*, 2010, **7**, 1747–1756.
153. M. Svoboda, J. Riha, K. Wlcek, W. Jaeger and T. Thalhammer, *Curr. Drug Metab.*, 2011, **12**, 139–153.
154. A. G. Press, I. A. Hauptmann, L. Hauptmann, B. Fuchs, M. Fuchs, K. Ewe and G. Ramadori, *Aliment. Pharmacol. Ther.*, 1998, **12**, 673–678.
155. J. D. Young, S. Y. Yao, J. M. Baldwin, C. E. Cass and S. A. Baldwin, *Mol. Aspects Med.*, 2013, **34**, 529–547.
156. J. H. Gray, R. P. Owen and K. M. Giacomini, *Pflügers Arch.*, 2004, **447**, 728–734.
157. M. Pastor-Anglada, N. Urtasun and S. Perez-Torras, *Compr. Physiol.*, 2018, **8**, 1003–1017.
158. R. C. Boswell-Casteel and F. A. Hays, *Nucleosides, Nucleotides Nucleic Acids*, 2017, **36**, 7–30.
159. M. F. Vickers, J. D. Young, S. A. Baldwin, J. R. Mackey and C. E. Cass, *Emerging Ther. Targets*, 2000, **4**, 515–539.
160. H. Lu, C. Chen and C. Klaassen, *Drug Metab. Dispos.*, 2004, **32**, 1455–1461.
161. M. Rau, F. Stickel, S. Russmann, C. N. Manser, P. P. Becker, M. Weisskopf, J. Schmitt, M. T. Dill, J. F. Dufour, D. Moradpour, D. Semela, B. Mullhaupt and A. Geier, *J. Hepatol.*, 2013, **58**, 669–675.
162. M. Fukao, K. Ishida, T. Sakamoto, M. Taguchi, H. Matsukura, T. Miyawaki and Y. Hashimoto, *Drug Metab. Pharmacokinet.*, 2011, **26**, 538–543.
163. J. D. Young, *Biochem. Soc. Trans.*, 2016, **44**, 869–876.
164. C. J. Endres, A. M. Moss, R. Govindarajan, D. S. Choi and J. D. Unadkat, *J. Pharmacol. Exp. Ther.*, 2009, **331**, 287–296.
165. G. D. Morse, M. J. Shelton and A. M. O'Donnell, *Clin. Pharmacokinet.*, 1993, **24**, 101–123.
166. C. Y. Wang, S. Liu, X. N. Xie and Z. R. Tan, Regulation profile of the intestinal peptide transporter 1 (PepT1), *Drug Des., Dev. Ther.*, 2017, **8**(11), 3511–3517.

CHAPTER 5

Drug Efflux Transporters: P-gp and BCRP

PETER BUNGAY*[a] AND SHARAN BAGAL[b]

[a]Sympetrus Ltd., Bishops Stortford, Hertfordshire, UK; [b]AstraZeneca R&D, Great Chesterford, Cambridge, UK
*E-mail: peterjbungay@outlook.com

5.1 Introduction

5.1.1 Enzyme Family

Efflux transporter proteins within the ATP-binding cassette (ABC) superfamily catalyze the ATP-dependent transport of substrates from within cells to their exterior. The human ABC superfamily of transmembrane proteins consists of 48 members and is further classified into seven sub-families denoted ABCA–G.[1] Two ABC members of principal interest for drug discovery are the multidrug transporters ABCB1, also known as P-glycoprotein (P-gp), and ABCG2, commonly known as breast cancer resistance protein (BCRP), which were originally identified as proteins involved in resistance to anti-cancer drugs. Thus, the ABCB sub-family has been alternatively classified as multidrug resistance (MDR) proteins, with P-gp designated MDR1.

Apart from P-gp and BCRP, several other ABC transporters have more recently been identified as important to consider during drug discovery and development. These include the multidrug resistance proteins ABCC2, ABCC3 and ABCC4, otherwise known as MRP2, MRP3 and MRP4.[2] However, the precise roles and substrate preferences of these efflux transporters are not understood to a level comparable with those for P-gp and BCRP so, presently, their

Drug Discovery Series No. 79
The Medicinal Chemist's Guide to Solving ADMET Challenges
Edited by Patrick Schnider
© The Royal Society of Chemistry 2021
Published by the Royal Society of Chemistry, www.rsc.org

importance as targets for medicinal chemistry optimization cannot be easily defined. Therefore, whilst they may require attention during drug development, they will not be considered within the scope of the present chapter.

5.1.2 Expression and Activity

P-gp and BCRP are widely expressed in mammalian tissues, including gastrointestinal tract, liver, brain capillary endothelium, kidney, testis and placenta. The expression and distribution of efflux transporters is connected to their physiological function of transporting substrates out of cells, contributing to the regulatory properties of the respective tissues. Tissues expressing P-gp and BCRP contain functionally polarized cells or cell layers within which transporters are located on a particular side of the cell, corresponding to directing transport out of cells and into an extracellular space. For example, location of P-gp and BCRP on the apical membrane of hepatocytes facing the bile canaliculus achieves the movement of substrates from hepatocytes into the biliary system. In an analogous fashion, location of P-gp and BCRP on the apical membrane of the epithelium of the intestinal mucosa effects the extrusion of ingested xenobiotic compounds from intestinal epithelial cells, a protective mechanism that counteracts the absorption of potentially harmful compounds into the circulation.

The expression and orientation of drug transporters, including ABC family members, in tissues that are particularly relevant to drug development has been summarized within a series of reviews of drug transporters by members of the International Transporter Consortium (ITC).[3,4] Several studies have examined levels of P-gp and BCRP expression across mammalian species. An example of particular interest is comparison of expression in brain microvessels, where it appears that in humans and non-human primates BCRP expression is slightly higher than that of P-gp, whereas P-gp expression is higher than that of BCRP in rats and mice.[5] The significance of the findings of such work will be discussed below when the relevance of P-gp and BCRP to drug design are discussed.

5.1.3 Structure

Human P-gp is a transmembrane protein consisting of a single polypeptide of 1280 amino acids with a molecular weight of approximately 170 kDa. P-gp consists of two homologous halves, each of which possesses a transmembrane domain (TMD) and a nucleotide-binding domain (NBD). The NBDs are located on the cytoplasmic side of the plasma membrane and both bind and hydrolyze ATP. The three-dimensional molecular structure of P-gp has been determined by X-ray diffraction[6,7] and electron microscopy,[8] and this has provided enough understanding of the substrate binding site to explain its broad specificity. The primary structure of P-gp displays high homology between mammalian species whereby sequence homology with human is 85% in rat, 81–87% in mouse, 87% in dog and 93–97% in non-human primates.[9] The structure of functionally expressed BCRP is analogous to that of P-gp, but it occurs as a homodimer of approximately 144 kDa with each monomer, in humans, consisting of 655 amino acids. Thus, it is described as a 'half transporter' as, in common with other ABCG sub-family members, it has one NBD and one TMD in a single polypeptide chain. In common with P-gp, it is located in the plasma membrane such that the NBD is oriented on the cytoplasmic side of the membrane. The high-resolution structure of BCRP has been determined by cryo-electron microscopy.[10] Like P-gp, BCRP shows high sequence homology between humans and other mammalian species. For mouse BCRP, which has been used for most structural studies, sequence identity is 81% with homology of 86%.[11]

5.1.4 Function and Substrates

P-gp and BCRP are multidrug transporters that can transport a very structurally diverse range of hydrophobic and amphipathic compounds that includes endogenous factors, such as peptides and lipid-like compounds, as well as xenobiotics, including drugs. In this way, they support secretory and excretory functions of tissues and help protect the body from potentially harmful substances arising from ingestion. The flexible structure of the substrate binding sites of P-gp and BCRP helps to explain the diversity of their substrates. It is difficult to define individual residues in these efflux transporters that interact with substrate, such that an 'induced fit' model has been proposed whereby a substrate creates its own binding site.[12] Hence, in most cases it is difficult to identify specific moieties that determine substrate affinity. Rather, it is broad physicochemical characteristics such as molecular weight (MW), lipophilicity, as indicated by calculated log of the partition coefficient (clogP), polar surface area (PSA), hydrogen bond donor (HBD) number and ionization that determine how well compounds act as efflux substrates.

5.1.5 Mechanism

In the catalytic cycle of efflux transporters, substrates are not altered structurally, but are translocated, with the hydrolysis of ATP providing the energy required for the movement. Hypotheses that explain the mechanism of transport are based on structural studies of P-gp and BCRP so that the proposed mechanism is best described by conformational changes that occur during the catalytic cycle. In the case of P-gp, it is thought that binding of ATP to the NBDs is asymmetric, *i.e.* the NBDs possess differing affinities for ATP, dependent on whether substrate is bound, and hydrolyze ATP non-simultaneously. Conformational states have been identified that are either open-inward, whereby substrates can access a binding cavity, and closed, corresponding to ejection of substrate to the cell exterior. Basal ATPase activity appears to be an intrinsic property of P-gp[13] so that it is constantly changing between open and closed conformations which are linked to the helices of the TMD. Hence, it is thought that the substrate binding pocket changes continuously and so can accept a wide range of substrates. It is presumed that binding of substrate from within the plasma membrane and/or cytoplasm when P-gp is in the open-inward conformation subsequently leads to substrate ejection outwards in the closed conformation (Figure 5.1).[6]

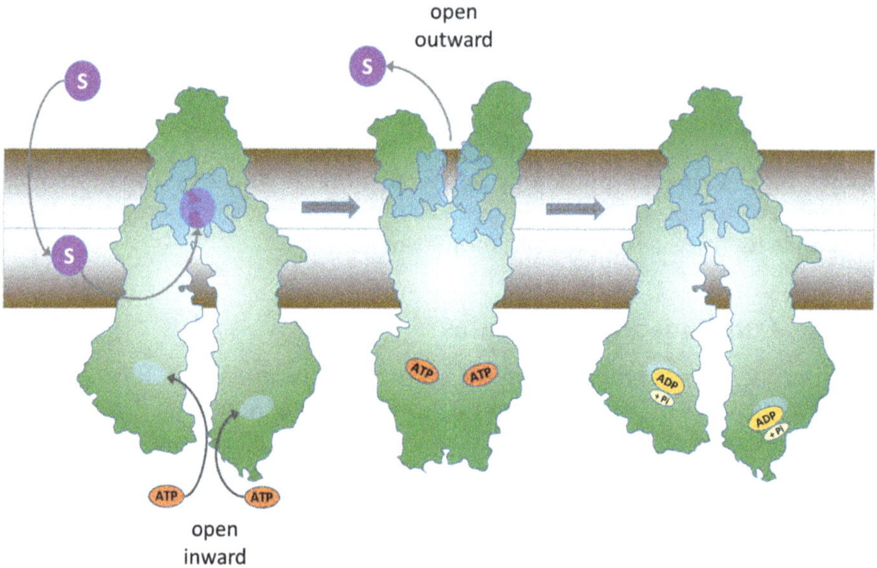

Figure 5.1 Schematic representation of the proposed catalytic mechanism of efflux transport by P-gp. Binding of substrate (S, in purple shape) takes place from within the inner leaflet of the plasma membrane when P-gp is in an open inward conformation (substrate binding pocket indicated in blue). Subsequent binding of ATP causes NBD dimerization and change to an open outward conformation and unidirectional transport of substrate. Hydrolysis of ATP returns P-gp to an open inward state.

A similar proposed transport mechanism is more explicitly described for BCRP whereby the substrate binding cavity formed between the BCRP monomers in an inward-open conformation collects substrate from the inner leaflet of the membrane or from cytoplasm.[10] Closing of the interface between NBDs upon ATP binding changes the conformation to an outward-facing state that releases substrate to the exterior.[8]

5.1.6 Relevance

The significance of the efflux transporters P-gp and BCRP in drug discovery and development is related to their expression in four major organ systems: (1) the intestinal epithelium of the gastrointestinal tract, (2) the microvascular capillaries of the brain, (3) the bile canalicular membrane of hepatocytes and (4) the proximal tubule of the kidneys.

5.1.6.1 Intestinal Epithelium

In the absorptive region of the gastrointestinal tract, the cells of the epithelium are joined together by tight junctions that prevent passage of solutes between cells (paracellular route). Absorption of substances from the lumen of the intestine therefore involves their movement across the apical epithelial cell membrane. The positioning of efflux transporters at this point provides a means of limiting uptake of potentially harmful molecules into the circulation if they act as substrates for P-gp and/or BCRP. Hence, drug molecules that are efflux transporter substrates face potential limitations of absorption, with consequent low or variable bioavailability, if administered by the oral route.

5.1.6.2 Brain Microvascular Capillaries

Unlike most blood capillaries, those that supply the brain possess an endothelium that, in common with the intestinal epithelium, consists of cells linked by tight junctions that restrict paracellular movement of solutes. The expression of P-gp and BCRP on the apical membrane of the capillary endothelium that faces the blood represents a mechanism for restricting the passage of solutes from the blood into the brain parenchyma. This system forms the blood–brain barrier (BBB) which is an important protective mechanism that can limit the access of potentially harmful substances to the brain, but that is also capable of restricting drug exposure in the brain if they are substrates of efflux transporters.

5.1.6.3 Hepatocytes

Hepatocytes are polarized such that their apical plasma membrane possessing P-gp and BCRP faces the bile canalicular intercellular space. In this way, compounds within the hepatocyte can be secreted into bile if they are efflux substrates so that P-gp and BCRP represent potential hepatobiliary clearance mechanisms.

5.1.6.4 Renal Proximal Tubule

The kidney tubule contains a wide array of transporters that regulate secretion and reabsorption of compounds from renal filtrate. P-gp located on the apical (urine side) of the tubule can play a role in preventing reabsorption of substrates into the tubular epithelium and hence contribute to net excretion of such compounds.

Considerations related to efflux transport in the liver and kidney focus on drug excretion and the possibility of drug–drug interaction (DDI) when concomitant medications are efflux substrates or inhibitors. Such excretory pathways for unchanged drug are often of relatively low magnitude and so would not usually merit attention as ways of minimizing clearance. Potential liabilities of P-gp or BCRP mediated renal and hepatobiliary drug excretion focus more on risks of DDI which may lead to changes in drug exposure. In any case, such DDI liabilities could be mitigated if design strategies already aimed to minimize efflux liabilities in line with the first two objectives above. Therefore, at the medicinal chemistry design stage, hepatobiliary and renal clearance mediated by efflux transporters is unlikely to merit great attention. If potential for DDI remains in a drug candidate by virtue of efflux transporter inhibition, this is likely to be managed most effectively during development according to regulatory guidance.[14,15]

5.1.7 Screening Strategies

The most widely used and cost-effective *in vitro* assay system for assessing P-gp and BCRP substrates measures apparent permeability (P_{app}) in polarized cell monolayers ('transwell method'). Cells expressing efflux transporters are cultured as a monolayer on a permeable membrane that separates culture medium facing the apical cell surface (A) from medium facing the basolateral side (B).[5,16] Cell lines employed include those expressing endogenous levels of transporters (*e.g.* CaCo-2) or lines derived from porcine kidney (LLC-PK1)[17] or canine kidney (MDCK) transfected and stably expressing P-gp or BCRP. Measurement of compound concentrations in medium following incubation allows calculation of P_{app} in A–B and B–A directions across the monolayer, and the ratio of P_{app} in B–A/A–B directions represents an index of efflux (efflux ratio, ER). Passive permeability can also be assessed in a transwell system using a cell line with minimal expression of efflux transporters (P_{app} A–B).[18] Recently, a variation of the transwell assay has been described whereby unidirectional transport of potential substrates is determined in the presence and absence of the P-gp inhibitor cyclosporin A. This more rapid method reduces the assay workload for early discovery screening and has been used to categorize compounds according to high, medium or low substrate potential.[19] For projects concerned with central nervous system (CNS) penetration, in each laboratory it is important to establish how ER determined in transwell assays translates to distribution of a compound in the CNS. This is usually achieved by *in vivo* studies in which plasma and CNS tissue [brain and/or cerebrospinal fluid (CSF)] are sampled over suitable time-courses [neuropharmokinetics ('neuroPK')], usually in rodents.[5] Following measurement of tissue concentrations, the fractional binding of compound to plasma and brain tissue is also determined so that the ratio of unbound concentrations in brain and plasma ($Kp_{u,u}$) can be determined. The use of CSF in such studies should be treated with caution as its origin as a secretion from the choroid plexus and exchange *via* the ependyma means that it is not equivalent to brain interstitial fluid. However, in most cases, it is the only matrix that can be sampled in humans and usually the only way that CNS penetration can be estimated clinically and translated to pre-clinical findings, so may provide some value as a laboratory measurement. Translation of results of *in vivo* studies to humans may depend on how well a compound acts as a substrate for either or both of P-gp or BCRP. The higher relative expression of BCRP in humans and non-human primates compared with rodents means that rodents may under-predict brain penetration for P-gp substrates whereas translation may be more accurate for dual P-gp and BCRP substrates.[5]

Key References

- S. G. Aller, J. Yu, A. Ward, Y. Weng, S. Chittaboina and R. Zhou, *et al.*, Structure of the P-glycoprotein reveals a molecular basis for poly-specific drug binding, *Science*, 2009;**323**, 1718–1722
- N. M. Taylor, I. Manolaridis, S. M. Jackson, J. Kowal, H. Stahlberg and K. P. Locher, Structure of the human multidrug transporter ABCG2, 2017, *Nature*, **546**, 504–509
- Z. Rankovic, CNS drug design: balancing physicochemical properties for optimal brain exposure, *J. Med. Chem.*, 2015, **58**(6), 2584–2608
- S. A. Hitchcock, Structural modifications that alter the P-glycoprotein efflux properties of compounds, *J. Med. Chem.*, 2012, **55**, 4877–4895
- B. Feng, A. C. Doran, L. Di, M. A. West, S. M. Osgood and J. Y. Mancuso, *et al.*, Prediction of Human Brain Penetration of P-glycoprotein and Breast Cancer Resistance Protein Substrates Using *In Vitro* Transporter Studies and Animal Models, *J. Pharm. Sci.*, 2018, **107**, 2225–2235
- R. Ohashi, R. Watanabe, T. Esaki, T. Taniguchi, N. Torimoto-Katori and T. Watanabe, *et al.* Development of simplified *in vitro* P-glycoprotein substrate assay and *in silico* prediction models to evaluate transport potential of P-glycoprotein, *Mol. Pharmaceutics*, 2019, **16**, 1851–1863
- P. J. Bungay, S. K. Bagal and A. Pike, Designing peripheral drugs for minimal brain exposure, in *Blood–Brain Barrier in Drug Discovery: Optimizing Brain Exposure of CNS Drugs and Minimizing Brain Side Effects for Peripheral Drugs*, ed. L. Di and E. H. Kerns, New Jersey, USA, 2015, pp. 446–462

5.2 Design Strategies Targeting Efflux Transporters

5.2.1 Consideration of Project Objectives

The strategies employed to address liabilities or opportunities associated with efflux transporters depend upon the objectives of the drug discovery project. The relevance of efflux transporters in drug disposition described earlier in this chapter focus attention at the discovery design stage on drug absorption and CNS penetration, so it is reasonable to categorize the objectives as:

- Maximizing absorption of an orally administered drug

  Efflux transport in the intestine has the potential to limit drug absorption in cases where passive permeability is insufficient to overcome rates of efflux and/or where drug concentrations in the intestinal lumen are below those required to saturate efflux transporters.

- Maximizing brain penetration of a drug aimed at a target within the CNS (all routes of administration)

As restriction of brain penetration *via* efflux at the BBB increases, the unbound systemic concentration required to maintain pharmacologically active concentrations also increases. This has the potential to raise dose requirements and to erode safety margins based on peripheral exposure [area under the curve (AUC) and maximum serum concentration (C_{max})].

- Minimizing brain penetration of an orally administered drug aimed at a peripheral (non-CNS) target

  Incorporating efflux to reduce brain penetration can offer the advantage of reducing potential for CNS-related side effects. This may have implications for drug absorption if significant efflux occurs in the intestine, depending on its balance against passive permeability and solubility.[16,20]

Thus, the effects of transporter-mediated efflux must be considered alongside other properties which influence passage of molecules across cell membranes. From the perspective of drug absorption, passive permeability and aqueous solubility remain key parameters to consider alongside efflux transport because the balance of permeability, efflux and concentration of dissolved drug in the intestinal lumen will determine the net rate and extent of absorption.

5.2.2 Broad Approaches to Altering Efflux Liabilities

The wide range of substrates accepted by P-gp and BCRP may be explained by structural studies that describe the flexibility of the substrate binding sites during the catalytic cycle of these proteins. Current knowledge indicates that physicochemical attributes including MW, hydrogen bonding potential, ionization (pK_a), lipophilicity and PSA are major factors that influence efflux transport of molecules. Retrospective assessments of the properties of CNS drugs have identified physicochemical characteristics that are compatible with successful development for their indications. However, results of such surveys do not solely reflect efflux liability, as physicochemistry influences pharmacological, permeability and other ADME and toxicological properties that are important for drug development. Other approaches have examined *in vitro* and *in vivo* laboratory datasets that indicate the rate or extent of brain permeation.[15,16,21] Once sufficient data have been assembled over a range of compounds, it may be possible to apply *in silico* predictions of efflux based on internally trained data sets and where sufficient data exist, specific computational approaches may be employed.[21] Such approaches based on physicochemical properties continue to evolve as datasets expand which may enhance the accuracy of prediction of P-gp substrate potential that can be used early in the drug discovery process.[19]

A large body of literature describes limiting or optimal values for calculated or measured physicochemical and laboratory parameters associated with optimizing permeability and efflux.[22,23] Whilst, unsurprisingly, there is some variation between publications in their emphasis and conclusions reached, it is possible to indicate broad convergence (Table 5.1).

Table 5.1 Summary of parameters associated with efflux optimization.[a]

	Minimize efflux		Incorporate efflux
	Maximize absorption of oral drug	Maximize CNS penetration	Minimize CNS penetration, retain absorption of oral drug
Physicochemical parameter			
MW (Da)	<500	<400	>400
Lipophilicity: logP	<5	2–5	
PSA (Å^2)	<120–140	<70–90	>60–80
HBD count	<5	<2	≥3
HBA count	<10		
pK_a		Reduce basicity (<8)	
***In vitro* laboratory measure**			
Passive P_{app} (cm s^{-1} × 10^{-6})	>5	>5	>5 and <20
ER[b]	Low in assays for P-gp and BCRP (*e.g.* <2)	Low in assays for P-gp and BCRP (*e.g.* <2)	High in assays for P-gp and BCRP (*e.g.* >5)

[a]Abbreviation: PSA, polar surface area, HBD/HBA, hydrogen bond donor/acceptor, P_{app}, apparent permeability, ER, efflux ratio (P_{app} B–A : P_{app} A–B).
[b]ER value will be dependent on individual laboratory outcomes and should be calibrated against *in vivo* outcomes *e.g.* $Kp_{u,u}$ in rodent.

The picture that emerges is that efflux is favoured by increasing molecular size, polarity and the ability to participate in hydrogen-bond formation. As these parameters can be regarded as interdependent, understanding which property is the most critical may be best examined by the structure–activity relationship within a series of close analogues. It is also evident that, as passive permeability is affected by the same physiochemical properties, the balance of permeability and efflux should be considered as determinants of absorption and CNS penetration. For instance, the presence of a high local concentration of a compound with high solubility and permeability in the GI tract is likely to saturate efflux and favour absorption due to a high concentration gradient.

5.3 Examples of Mitigation Strategies

Examples of medicinal chemistry approaches that address efflux are categorized according to project objectives that aim to maximize or minimize brain penetration. Excellent reviews by Rankovic[22,23] and Hitchcock[24] have collated many examples where modification of physicochemical properties *via* changes in functional groups have optimized brain exposure and P-gp efflux. Several of these examples are included here in order to illustrate tactics employed, but the reader is referred to the original reviews and references therein for greater detail. At the time of writing, there is little information about that specifically relates to optimizing efflux *via* BCRP. Given the structural and mechanistic similarity between P-gp and BCRP, it is reasonable to expect a broad alignment of physicochemical characteristics of P-gp and BCRP substrates.

5.3.1 Maximizing Brain Penetration

5.3.1.1 Reduction of HBD Capacity

Examples

1

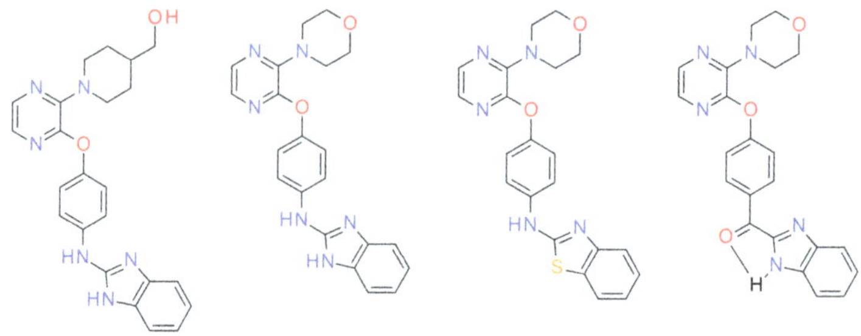

HBD count	3	2	1	1
MDCK-MDR1 ER	76.7	11.1	2.4	0.9

PDE10A inhibitors[25]

Design tactics:

- Replacement of OH by incorporation of O within ring system (12 → 13)
- Replacement of ring system NH with S (13 → 14)
- Incorporation of intramolecular hydrogen bond to replace one of the neighbouring NH groups (14 → 15)

2

MDCK-MDR1 ER	3.1	1.1

Peptidic small molecules[26]

Design tactic: Incorporation of intramolecular hydrogen-bond by shifting ring N, so that effective HBD count is reduced

Examples

3

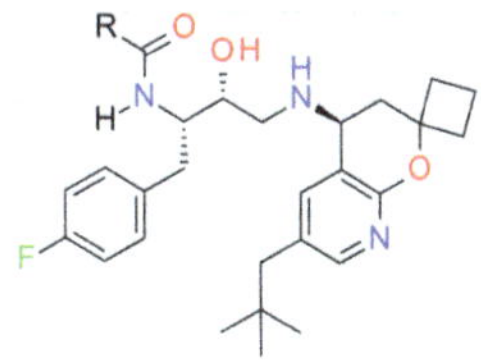

9 R=Me
12 R=

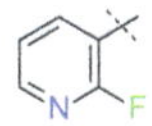

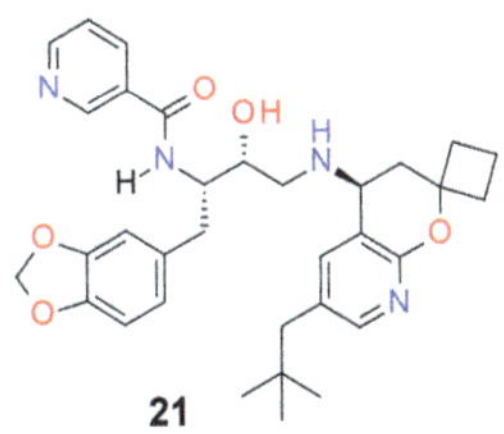

21

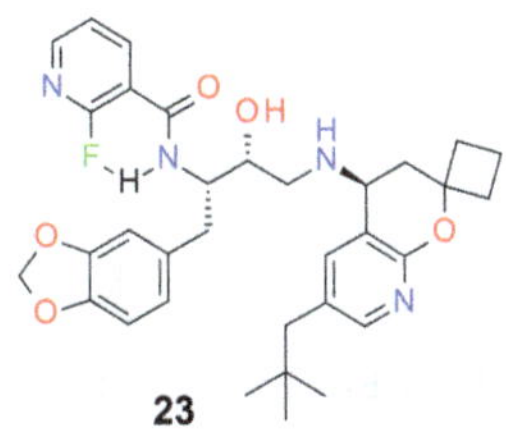

23

Compound	PSA	HBD count	Passive P_{app} (cm $s^{-1} \times 10^{-6}$)	MDCK-MDR1 ER
9	83.5	3	20	17
12	96.4	3	38	1.6
21	96.4	3	11	49
23	96.4	3	17	4

β-site amyloid precursor protein-cleaving enzyme (BACE) inhibitors[24,27]

Design tactics:

- Position HBA groups close to amide NH (9 → 12)
- Incorporation of intramolecular halogen bond, masking a HBD (21 → 23)

5.3.1.2 Reduction of PSA

Example

1

	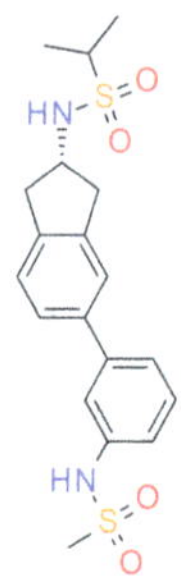	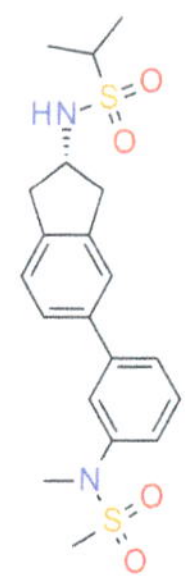	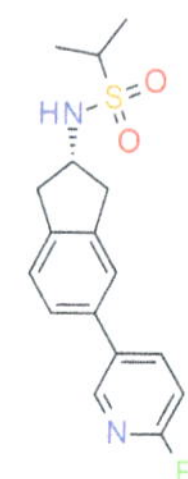
PSA (Å^2)	109.1	100.3	67.4
HBD count	2	1	1
MDR1-MDCK ER	5.8	3.2	1.1

α-Amino-3-hydroxy-5-methyl-4-isoxazolepropionic acid (AMPA) allosteric modulators[28]
Design tactics:

- Mask NH by methylation
- Remove polar group, *e.g.* remove sulphonamide

5.3.1.3 Modulation of Amine Basicity

Examples

1

	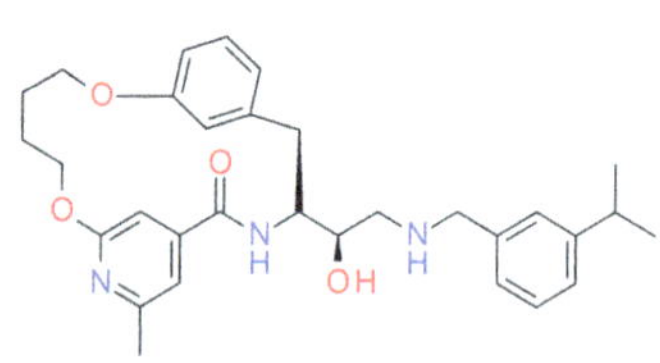	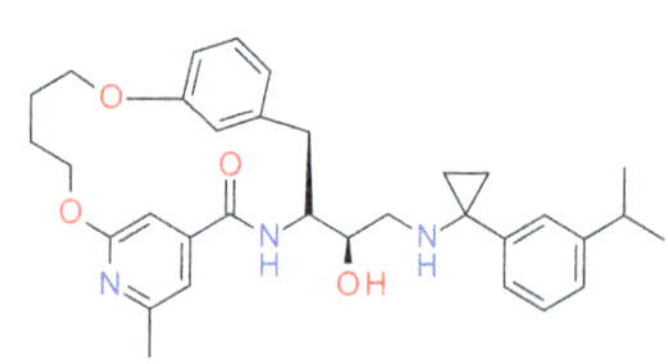
pK_a	8.5	7.3
MDR1-MDCK ER	23	3.5

BACE-1 inhibitors[29]
Design tactic: Introduction of cyclopropane proximal to amine

Examples		
2	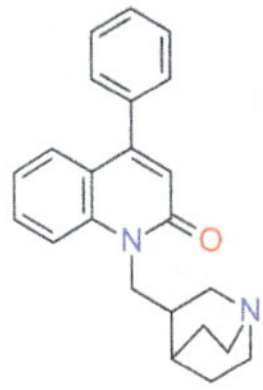	
pK_a	10.1	8.1
Caco-2 ER	6.9	1.0

α7 nicotinic acetylcholine receptor (NAcChR) agonists[30]

Design tactic: Decrease efflux through reduction of amine pK_a

5.3.1.4 Reduction of Rotatable Bond Count

Examples		
1	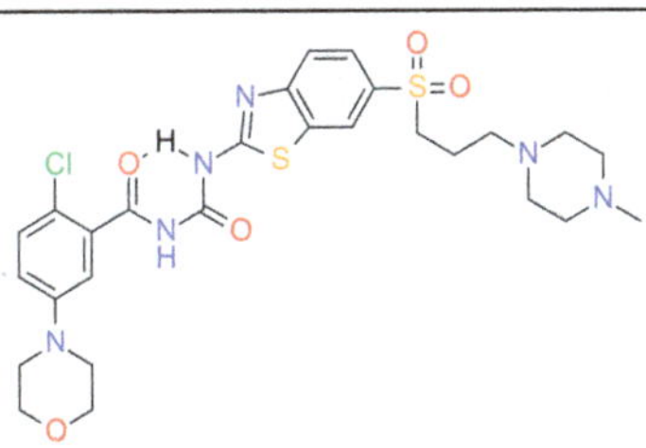	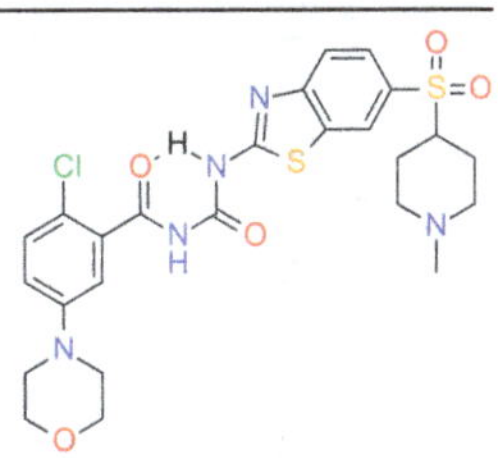
Rotatable bond count	8	5
pK_a	7.8	8.1
MDCK-MDR1 ER	69	2.6

GSH-R1a inverse agonists[31]

Design tactic: Decrease flexibility and/or rotatable bond count through reduction of carbon chain length leading to decreased efflux

2	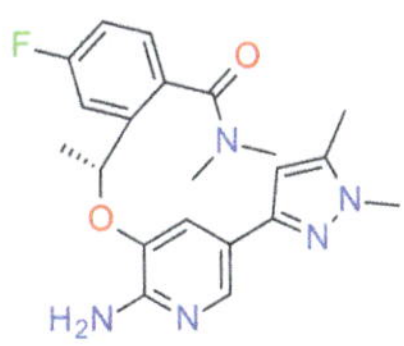	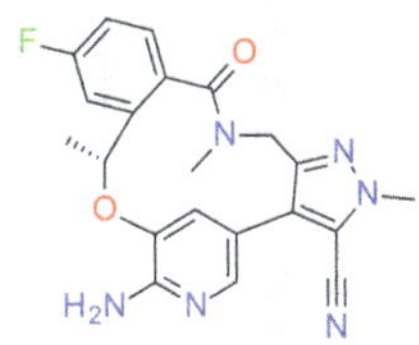
Rotatable bond count	6	0
PSA	86	110
MDCK-MDR1 ER	7.6	1.5

Macrocyclization in anaplastic lymphoma kinase (ALK) inhibitors[32]

Design tactic: Decrease bond rotation through 'locking' structure in a cyclic configuration, with decreased efflux. PSA was also increased with addition of nitrile, although this did not appear to overcome the effect of decreased flexibility.

5.3.2 Minimizing Brain Penetration of Orally Administered Drugs Aimed at Non-CNS Targets

In this instance, a balance between permeability and efflux is needed that allows sufficient absorption across the intestinal mucosa but effectively restricts CNS penetration. In addition, incorporating both P-gp and BCRP activities is an ideal approach considering the relative abundance of these transporters in the BBB, particularly in humans.[15,20]

Examples

1

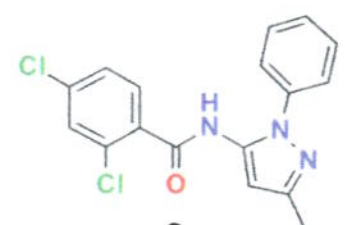

2

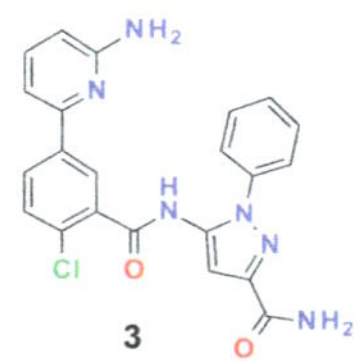

3

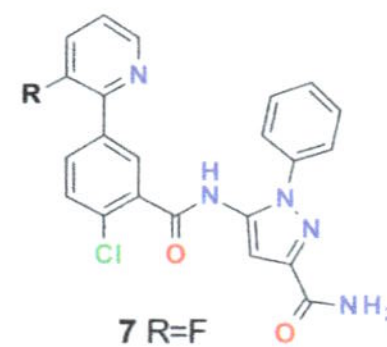

7 R=F
8 R=Cl

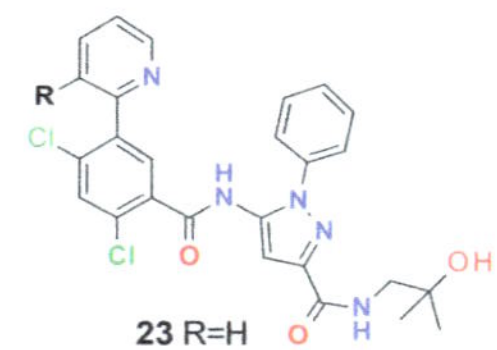

23 R=H
24 R=Cl

Compound	MW	$logD_{7.4}$	PSA (Å^2)	P-gp ER	BCRP ER	Cb,u : Cp,u[a] (in rat)	RRCK P_{app} (cm s^{-1} ×10^{-6})	p.o. F% (cm s^{-1} × 10^{-6})	est. F_a (in rat)
2	346	3.3	47	1	—	—	—	—	
3	432	2.3	129	5	7	0.041	17	6	N.D.[b]
7	436	2.6	101	2.1	4.8	0.10	19	~100	1
8	452	2.8	101	3.0	2.0	0.076	14	85	~0.9
23	524	3.3	111	28	103	0.014	16	72	~0.8
24	542	3.3	111	23	69	0.015	12	46	~0.5

Allosteric TrkA inhibitors[33]

Design tactics:

- Increasing MW and PSA from compound 2 to 3 introduces P-gp
- Decrease in P-gp from compound 3 to 7 increases brain penetration despite retaining BCRP
- Increases in MW and PSA from compounds 7 and 8 to compounds 23 and 24 increase P-gp and BCRP and increase CNS restriction whilst retaining sufficient passive permeability for adequate absorption

Examples

2

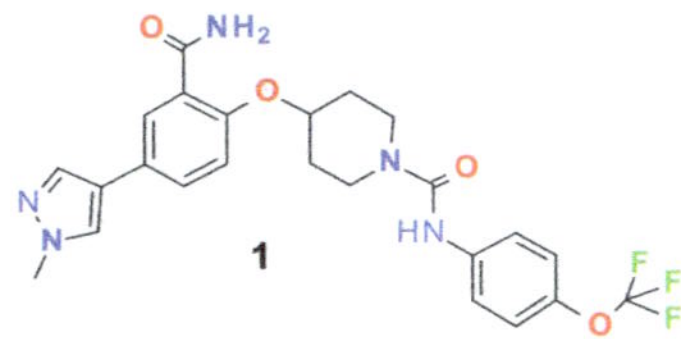

1

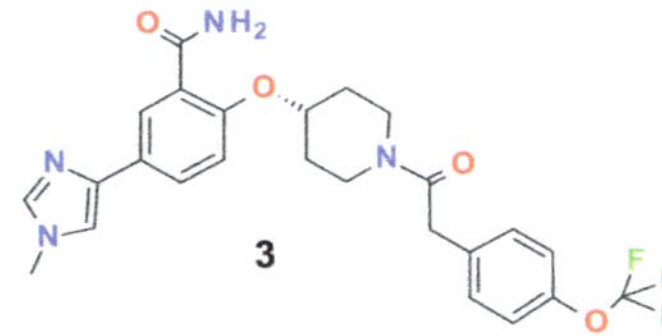

3

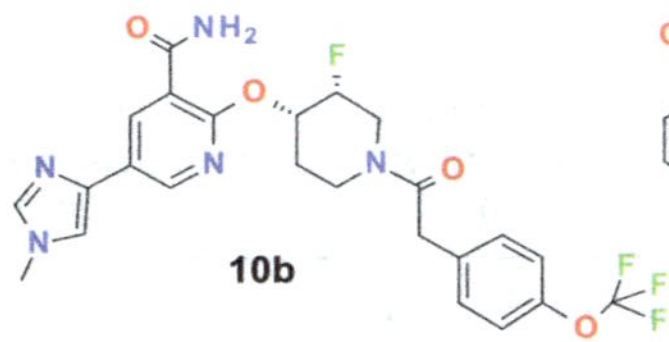

10b

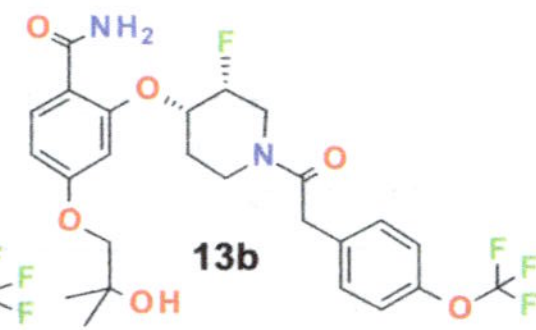

13b

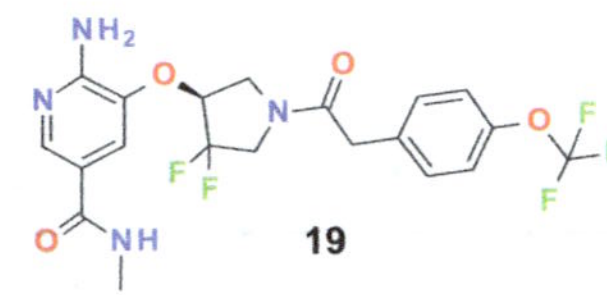

19

Compound	MW	$logD_{7.4}$	PSA ($Å^2$)	P-gp (ER)	BCRP (ER)	Cb,u : Cp,u	RRCK P_{app} (in rat)	p.o. F% (cm s^{-1} × 10^{-6})	est. F_a in rat (in rat)
1	502	3.7	112	4	N.D.	—	—	—	—
3	501	2.9	95	18	2.5	0.0065	10	45	~0.5
10b	484	3.0	105	10	4.6	0.043	17	52	~0.5
13b	520	3.1	116	72	7.0	0.026	5.2	35	~0.4
19	460	2.5	108	22	4.0	0.018	11	56	~0.6

Pan tyrosine receptor kinase (Trk) inhibitors[34]

Design tactics:

- Changing pyrazole to imidazole and substituting internal N for C (compound 1 to 3) increased P-gp ER sufficiently to cause CNS restriction
- Passive permeability of 13b is at the edge of the requirement to maintain adequate absorption in the face of high P-gp and BCRP activity, whereas high passive permeability of 10b and 19 are sufficient to achieve ≥50% absorption.

(continued)

Examples			
3		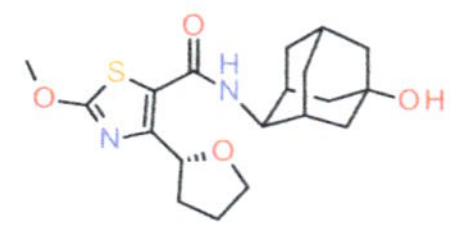	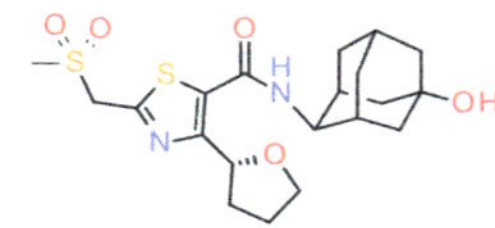
		Structure 25	Structure 27
	PSA (Å²)	84	111
	$\log D_{7.4}$	2.3	0.4
	$Kp_{u,u}$	0.7	0.03
	11β-hydroxysteroid dehydrogenase type 1 (11β-HSD1) inhibitors[35]		
	Design tactic: Increasing PSA is associated with decreased brain penetration *in vivo*		

[a]Cb,u : Cp,u, unbound brain to unbound plasma concentration ratio; est, estimated; F%, percentage bioavailability; F_a, fraction absorbed; P_{app}, apparent permeability; p.o. *per os* (oral administration) RRCK, Ralph Russ Canine Kidney.

[b]Low oral bioavailability likely to have reflected a combination of hepatic and intestinal extraction.

References

1. M. Dean, A. Rzhetsky and R. Allikmets, The human ATP-binding cassette (ABC) transporter superfamily, *Genome Res.*, 2001, **11**, 1156–1166.
2. K. M. Hillgren, D. Keppler, A. A. Zur, K. M. Giacomini, B. Steiger, C. E. Cass and L. Zhang, Emerging transporters of clinical importance: an update from the International Transporter Consortium, *Nature*, 2013, **94**, 52–63.
3. K. M. Giacomini, S.-M. Huang, D. J. Tweedie, L. Z. Benet, K. L. Brouwer and X. Chu, *et al.*, Membrane transporters in drug development, *Nat. Drug Discovery*, 2010, **9**, 215–236.
4. K. M. Giacomini and S.-M. Huang, Transporters in drug development and clinical pharmacology, *Clin. Pharm. Ther.*, 2013, **94**(1), 3–9.
5. B. Feng, A. C. Doran, L. Di, M. A. West, S. M. Osgood and J. Y. Mancuso, *et al.*, Prediction of Human Brain Penetration of P-glycoprotein and Breast Cancer Resistance Protein Substrates Using In Vitro Transporter Studies and Animal Models,*J. Pharm. Sci.*, 2018, **107**, 2225–2235.
6. S. G. Aller, J. Yu, A. Ward, Y. Weng, S. Chittaboina and R. Zhou, *et al.*, Structure of the P-glycoprotein reveals a molecular basis for poly-specific drug binding, *Science*, 2009, **323**, 1718–1722.
7. L. Esser, F. Zhou, K. M. Pluchino, J. Shiloach, J. Ma and W. Tang, *et al.*, Structures of the Multidrug Transporter P-glycoprotein reveal asymmetric ATP binding and the mechanism of polyspecificity, *J. Biol. Chem.*, 2017, **292**(2), 446–461.
8. M. F. Rosenberg, R. Callaghan, S. Modok, C. F. Higgins and R. C. Ford, Three-dimensional Structure of P-glycoprotein: the transmembrane regions adopt an asymmetric configuration in the nucleotide-bound state,*J. Biol. Chem.*, 2005, **280**, 2857–2862.
9. S. Syvanen, O. Lindhe, M. Palner, B. R. Kornum, O. Rahman and B. Langstrom, *et al.*, Species Differences in Blood-Brain Barrier Transport of Three Positron Emission Tomography Radioligands with Emphasis on P-Glycoprotein Transport, *Drug Metab. Dispos.*, 2009, **37**, 635–643.
10. N. M. Taylor, I. Manolaridis, S. M. Jackson, J. Kowal, H. Stahlberg and K. P. Locher, Structure of the human multidrug transporter ABCG2, *Nature*, 2017, **546**, 504–509.
11. J. D. Allen, R. F. Brinkhuis, J. Wijnholds and A. H. Schinkel, The mouse Bcrp1/Mxr/Abcp gene: amplification and overexpression in cell lines selected for resistance to Topotecan, Mitoxantrone, or Doxorubicin, *Cancer Res.*, 1999, **59**, 4237–4241.
12. T. W. Loo and D. M. Clarke, Identification of residues within the drug-binding domain of the human multidrug resistance P-glycoprotein by cysteine-scanning mutagenesis and reaction with dibromobimane, *J. Biol. Chem.*, 2000, **275**(50), 39272–39278.
13. M. K. Al Shawi, M. K. Polar, H. Omote and R. A. Figler, Transition state analysis of the coupling of drug transport to ATP hydrolysis by P-glycoprotein,*J. Biol. Chem.*, 2003, **278**(52), 52629–52640.

14. http://www.fda.gov/Drugs/GuidanceComplianceRegulatoryInformation/Guidances/default.htm.
15. http://www.ema.europa.eu/docs/en_GB/document_library/Scientific_guideline/2012/07/WC500129606.pdf.
16. T. T. Wager, J. L. Liras, S. Mente and P. Trapa, Strategies to minimize CNS toxicity: in vitro high-throughput assays and computational modeling, *Expert Opin. Drug Metab. Toxicol.*, 2012, **8**(5), 531–542.
17. C. L. Booth-Genthea, S. W. Louie, E. J. Carlini, B. Li, B. F. Leake and R. Eisenhandler, *et al.*, Development and characterization of LLC-PK1 cells containing Sprague–Dawley rat Abcb1a (Mdr1a): Comparison of rat P-glycoprotein transport to human and mouse, *J. Pharmacol. Toxicol. Methods*, 2006, **54**(1), 78–89.
18. L. Di, C. Whitney-Pickett, J. P. Umland, H. Zhang, X. Zhang and D. F. Gebhard, *et al.*, Development of a new permeability assay using low-efflux MDCKII cells, *J. Pharm. Sci.*, 2011, **100**(11), 4974–4985.
19. R. Ohashi, R. Watanabe, T. Esaki, T. Taniguchi, N. Torimoto-Katori and T. Watanabe, *et al.*, Development of simplified in vitro P-glycoprotein substrate assay and in silico prediction models to evaluate transport potential of P-glycoprotein, *Mol. Pharmaceutics*, 2019, **16**, 1851–1863.
20. P. J. Bungay, S. K. Bagal and A. Pike, Designing peripheral drugs for minimal brain exposure, in *Blood-Brain Barrier in Drug Discovery: Optimizing Brain Exposure of CNS Drugs and Minimizing Brain Side Effects for Peripheral Drugs*, ed. L. Di and E. H. Kerns, New Jersey, USA, 2015, pp. 446–462.
21. H. Gunaydin, M. M. Weiss and Y. Sun, De novo prediction of P-glycoprotein-mediated efflux liability for drug-like compounds, *ACS Med. Chem. Lett.*, 2013, **4**, 108–112.
22. Z. Rankovic, CNS drug design: balancing physicochemical properties for optimal brain exposure, *J. Med. Chem.*, 2015, **58**(6), 2584–2608.
23. Z. Rankovic, Designing CNS drugs for optimal brain exposure, in *Blood-Brain Barrier in Drug Discovery: Optimizing Brain Exposure of CNS Drugs and Minimizing Brain Side Effects for Peripheral Drugs*, ed. L. Di and E. H. Kerns, New Jersey, USA, 2015, pp. 387–424.
24. S. A. Hitchcock, Structural modifications that alter the P-glycoprotein efflux properties of compounds, *J. Med. Chem.*, 2012, **55**, 4877–4895.
25. E. Hu, R. K. Kunz, N. Chen, S. Rumfelt, A. Siegmund and K. Andrews, *et al.*, Design, optimization and biological evaluation of novel keto-benzimidazoles as potent and selective inhibitors of phosphodiesterase 10A (PDE10A), *J. Med. Chem.*, 2013, **56**, 8781–8792.
26. S. B. Rafi, B. R. Hearn, P. Vedantham, M. P. Jacobson and A. R. Renslo, Predicting and improving the membrane permeability of peptidic small molecules, *J. Med. Chem.*, 2012, **55**, 3163–3169.
27. M. M. Weiss, T. Williamson, S. Babu-Khan, M. D. Bartberger, J. Brown and K. Chen, *et al.*, Design and preparation of a potent series of hydroxyethylamine containing β-secretase inhibitors that demonstrate robust reduction of central β-amyloid, *J. Med. Chem.*, 2012, **55**, 9009–9024.

28. S. E. Ward, M. Harries, L. Aldegheri, D. Andreotti, S. Ballantine and B. D. Bax, *et al.*, Discovery of N-[(2S)-5-(6-fluoro-3-pyridinyl)-2,3-dihydro-1H-inden-2-yl]-2-propanesulfonamide, a novel clinical AMPA receptor positive modulator, *J. Med. Chem.*, 2010, **53**, 5801–5812.
29. A. Lerchner, R. Machauer, C. Betschart, S. Veenstra, H. Rueeger and C. McCarthy, *et al.*, Macrocyclic BACE-1 inhibitors acutely reduce Aβ in brain after po application, *Bioorg. Med. Chem. Lett.*, 2010, **20**, 603–607.
30. I. M. McDonald, R. A. Mate, F. C. Zusi, H. Huang, D. J. Post-Munson and M. A. Ferrante, *et al.*, Discovery of a novel series of quinolone α7 nicotinic acetylcholine receptor agonists, *Bioorg. Med. Chem. Lett.*, 2013, **23**, 1684–1688.
31. W. McCoull, P. Barton, A. J. Brown, S. S. Bowker, J. Cameron and D. S. Clarke, *et al.*, Identification, optimization, and pharmacology of acylurea GHS-R1a inverse agonists, *J. Med. Chem.*, 2014, **57**, 6128–6140.
32. T. W. Johnson, P. F. Richardson, S. Bailey, A. Brooun, B. J. Burke and M. R. Collins, *et al.*, Discovery of (10R)-7-Amino-12-fluoro-2,10,16-trimethyl-15-oxo-10,15,16,17-tetrahydro-2H-8,4-(metheno)pyrazolo-[4,3-h][2,5,1 1]-benzoxadiazacyclotetradecine-3-carbonitrile (PF-06463922), a macrocyclic inhibitor of anaplastic lymphoma kinase (ALK) and cros oncogene 1 (ROS1) with preclinical brain exposure and broad-spectrum potency against ALK-resistant mutations, *J. Med. Chem.*, 2014, **57**, 4720–4744.
33. S. K. Bagal, K. Omoto, D. C. Blakemore, P. J. Bungay, J. G. Bilsland and P. J. Clarke, *et al.*, Discovery of Allosteric, Potent, Subtype Selective, and Peripherally Restricted TrkA Kinase Inhibitors, *J. Med. Chem.*, 2019, **62**, 247–265.
34. S. K. Bagal, M. Andrews, B. M. Bechle, J. Bian, J. G. Bilsland and D. C. Blakemore, *et al.*, Discovery of Potent, Selective, and Peripherally Restricted Pan-Trk Kinase Inhibitors for the Treatment of Pain, *J. Med. Chem.*, 2018, **61**(15), 6779–6800.
35. F. W. Goldberg, A. G. Dossetter, J. S. Scott, R. Robb, S. Boyd and S. D. Groombridge, *et al.*, Optimization of brain penetrant 11β-hydroxysteroid dehydrogenase type I inhibitors and in vivo testing in diet-induced obese mice, *J. Med. Chem.*, 2014, **57**, 970–986.

CHAPTER 6

OATs and OCTs: The SLC22 Family of Organic Anion and Cation Transporters

PÄR MATSSON*[a] AND MARIA KARLGREN*[b]

[a]Department of Pharmacology, Institute of Neuroscience and Physiology, The Sahlgrenska Academy, University of Gothenburg, Box 431, SE-405 30 Gothenburg, Sweden; [b]Department of Pharmacy, Uppsala University, Box 580, SE-751 23 Uppsala, Sweden
*E-mail: par.matsson@gu.se, maria.karlgren@farmaci.uu.se

6.1 Introduction

6.1.1 Transporter Family

The SLC22 family of transporters belongs to the functionally and structurally diverse solute carrier (SLC) protein superfamily.[1] Currently, 29 human SLC22 transporters are known, among which the organic cation transporters 1 and 2 (OCT1–2; SLC22A1–2) and the organic anion transporters 1–3 (OAT1–3; SLC22A6–8) have the most evidence of affecting drug disposition.[2,3]

6.1.2 Expression

The different OCTs and OATs have distinctly different tissue expression patterns in humans.[2,3] OCT1 is highly expressed on the basolateral membrane of hepatocytes (facing the liver blood vasculature), but has also been detected

Drug Discovery Series No. 79
The Medicinal Chemist's Guide to Solving ADMET Challenges
Edited by Patrick Schnider
© The Royal Society of Chemistry 2021
Published by the Royal Society of Chemistry, www.rsc.org

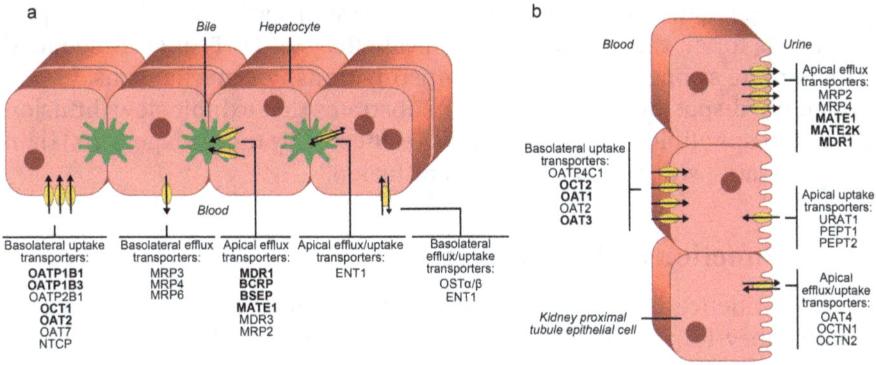

Figure 6.1 Schematic illustration of the expression of pharmacokinetically important transporters in human liver (a) and kidney (b). The selection of transporters is based on the International Transporter Consortium publications by Giacomini, *et al.* (2010) and Zamek-Gliszczynski, *et al.* (2018).[2,4] Transporters included in current Food and Drug Administration (FDA) and European Medicines Agency (EMA) drug interaction guidelines, as well as OAT2, which is of emerging clinical importance, are shown in bold type.[5,6] Adapted from ref. 7 with permission from American Chemical Society, Copyright 2017.

at low levels in multiple other tissues. OCT2, OAT1 and OAT3 are predominantly expressed on the basolateral (blood-facing) membrane of renal tubule cells. OAT2 is expressed in hepatocytes and in renal epithelium, and OCT3 is expressed at intermediate levels in several tissues, including smooth and skeletal muscle and male and female reproductive tissues. The two latter transporters are considerably less studied, but data on their effects on the disposition of certain classes of drugs is emerging. For an overview of clinically important drug transporters in human liver and kidney, see Figure 6.1.

6.1.3 Structure

The SLC22 transporters belong to the major facilitator superfamily (MFS).[8] So far, no crystal structure has been published, but biochemical evidence supports a common structure consisting of 12 transmembrane helices, a relatively large extracellular loop domain in the N-terminal end of the protein between transmembrane (TM) domains 1 and 2, containing *N*-glycosylation sites, and another extracellular loop domain between TM domains 6 and 7.[8,9] Sequence-based similarity indicates relatively close correspondence to members of the SLC2 family of sugar transporters, for which a number of structures of the human forms have been resolved.[1,8]

6.1.4 Activity

Transport rates vary depending on the substrate and the level of transporter expression in the tissue or cell type of interest. Orthologs exist in typical preclinical animal models, including rat, mouse, rabbit and dog, with species-dependent differences in tissue expression levels and substrate specificity.[10] As for all transporters, the clinical effects of OCT- or OAT-mediated transport are strongly affected by competing transport mechanisms; in particular, the disposition of OCT or OAT substrates that exhibit slow diffusion across phospholipid bilayers will be more affected by the transporter(s) than will substrates with rapid background diffusion.

6.1.5 Function and Substrates

Current evidence indicates that OCTs are facilitative transporters, meaning that they form substrate-selective channels across the cell membrane through which the substrate is translocated down its concentration gradient; OCT substrates are typically cationic at physiological pH, and the potential difference across the plasma membrane adds to the driving force.[3,9–11] Thus, the net transport of OCT substrates is typically from the cell exterior to the interior, although reversed transport can be observed *in vitro* after pre-loading of the cell with substrate.[9] The physiological relevance of such inside-to-outside concentration gradients is, however, debatable. Examples of OCT1 and 2 substrates are shown in Figure 6.2.

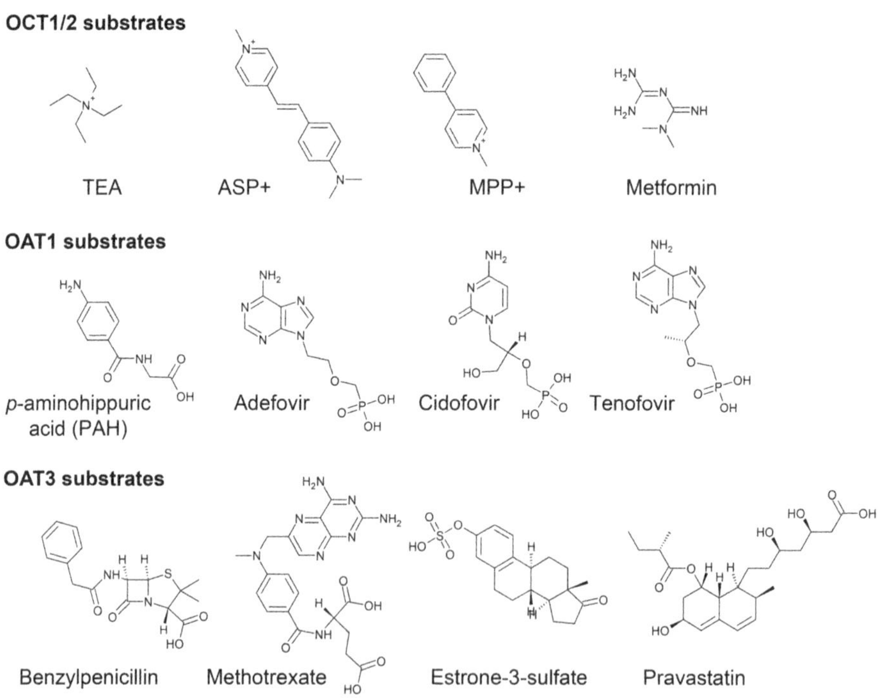

Figure 6.2 Commonly used *in vitro* probe substrates for OCT1/2 and OAT1/3.

OAT1 and OAT3 mediate the first step in the renal secretion of organic anion substrates (Figure 6.2) *via* an exchanger mechanism with endogenous dicarboxylic acids (*e.g.*, glutarate or α-ketoglutarate).[12,13]

6.1.6 Mechanism

Structural evidence for other transporters sharing the MFS-type protein fold indicate an 'alternating-access' transport cycle in which the transporter upon binding the substrate undergoes a conformational change from an outward-open to an occluded state, in which a centrally located substrate binding site is separated from both the extracellular and the intracellular space (Figure 6.3).[8] Additional conformational changes result in an inward-open state from which the substrate is released. While this general mechanism has not been proven for the OCTs and OATs, their predicted protein folding is indicative of a shared transport mechanism with other MFS transporters.[8]

6.1.7 Screening Strategies

Transporter-transfected cell lines are the most common experimental system to study transport and/or drug interactions with OCTs and OATs. The human embryonic kidney cell line HEK293 is a commonly used host.[2,14–16] The experimental setup for identifying transported substrates typically involves measurements of time-dependent cellular uptake in transporter transfected cells, using cells transfected with a transporter-deficient

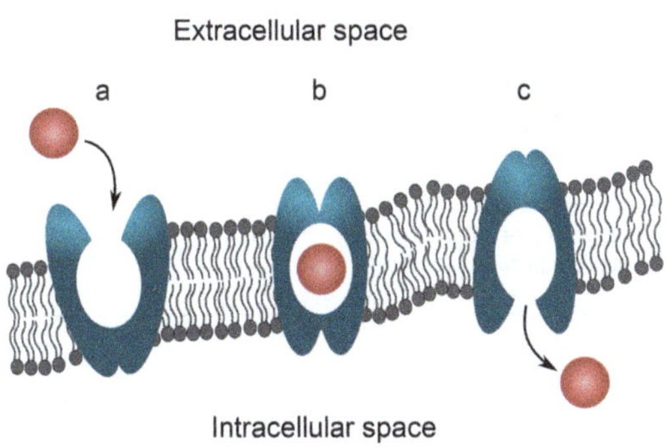

Figure 6.3 Suggested transport mechanism for SLC22 and other major facilitator Superfamily (MFS) transporters. First, the substrate binds to an exposed, centrally located binding site in the transporter. Upon binding, the transporter undergoes a conformational change from an outward-open state [binding site accessible from the extracellular space (a)] to an occluded state (b). Additional conformational changes result in an inward-open state [binding site accessible from the intracellular space (c)] and release of the substrate. Depending on substrate concentration, transport in the reverse direction can also be possible.

('empty') vector as a control, with the assumption that the uptake in the control cells represent nonspecific binding and uptake processes that are identical in the two cell types. Often, the control-cell uptake is simply subtracted from that in the transporter transfected cells to yield a transporter-mediated uptake rate (Scheme 6.1).

Drug-mediated inhibition [potentially leading to a transporter mediated drug–drug interaction (DDI)] is typically studied in an analogous setup (Scheme 6.2).[2,14–16] The cellular uptake of a defined probe substrate is detected using fluorescence, radiolabelling, or mass spectrometry. The effect of a potential inhibitor is quantified as the percentage of inhibited transport (at a single screening concentration) or as a half-maximal inhibitory concentration (IC_{50}) or inhibition constant (K_i).

For OCTs, common probe substrates include dye 4-(-4-(dimethylamino)-styryl)-*N*-methylpyridinium (ASP+), tetraethylammonium (TEA), 1-methyl-4-phenylpyridinium ion (MPP+) and the biguanide antidiabetic metformin.[2,3,9,17] For OAT1, example substrates include *p*-aminohippurate, adefovir, cidofovir and tenofovir and for OAT3 benzylpenicillin, estrone-3-sulfate, methotrexate and pravastatin are suggested *in vitro* substrates (Figure 6.2).[2,17]

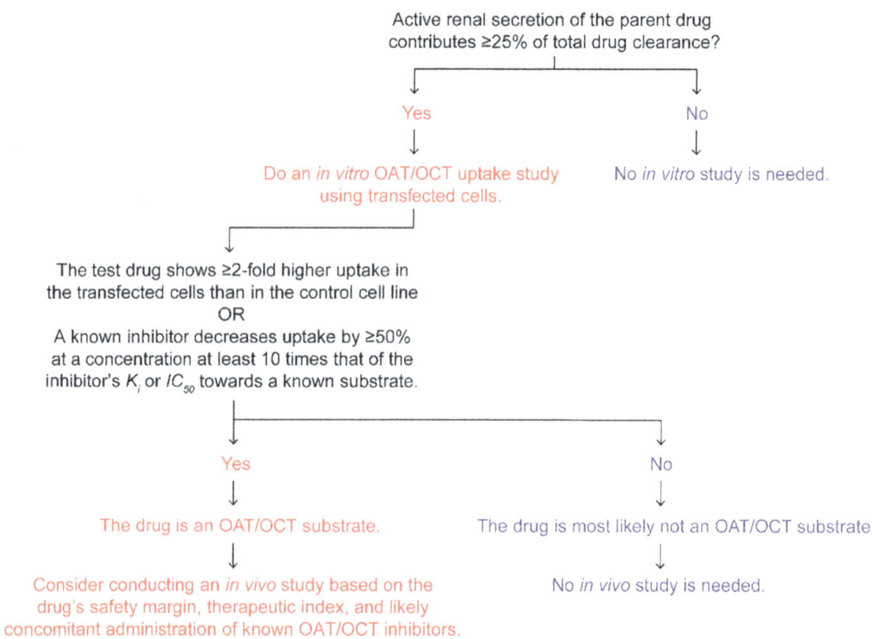

Scheme 6.1 Decision tree to guide screening of potential OCT or OAT substrates. The decision tree is based on the U.S. Food and Drug Administration (FDA) draft guidance on *in vitro* drug–drug interaction studies (2017).[6]

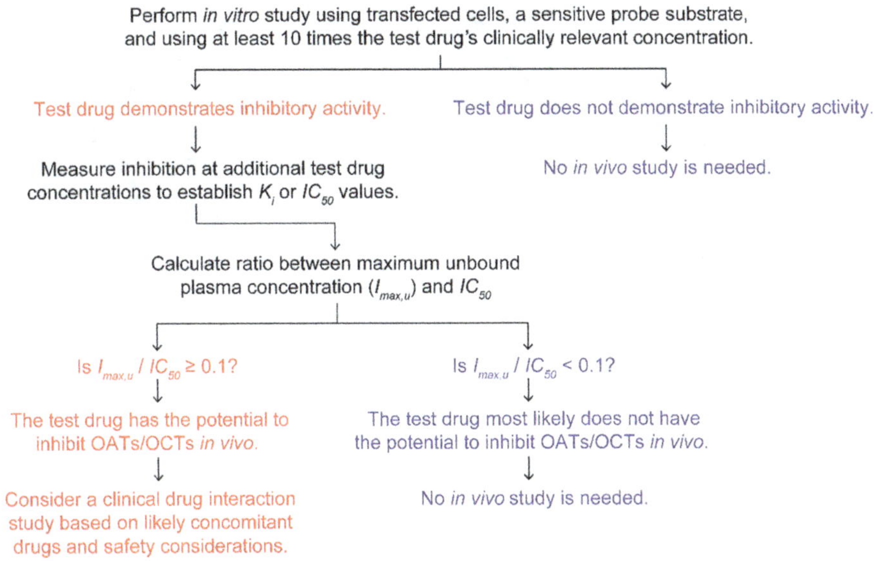

Scheme 6.2 Decision tree to guide screening of potential OCT or OAT inhibitors. The decision tree is based on the U.S. Food and Drug Administration (FDA) draft guidance on *in vitro* drug–drug interaction studies (2017).[6]

6.1.8 Relevance

The number of drugs with a clear clinical effect from OCTs and OATs is, to date, relatively low. For OCT1 and OCT2, the most prominent example is the antidiabetic metformin.[2,3,18] Clinically observed drug–drug interactions involving OCT substrates have been reported, often observed as a decreased renal excretion and a modestly increased plasma AUC. Examples include dofetilide, pilsicainide, procainamide, metformin and ranitidine (Figures 6.2 and 6.4a). However, the major effect on renal excretion of cationic small-molecule drugs has been suggested to be related to inhibition of the multidrug and toxin extrusion transporters (MATE1/*SLC47A1* and MATE2-K/*SLC47A2*), which have similar substrate selectivity to OCT2 and are expressed on the opposite (*i.e.* the apical) membrane of the renal epithelium.[19]

Examples of drug–drug interactions affecting the disposition of OAT1 and OAT3 substrate drugs include adefovir, cidofovir, cefaclor, ceftizoxime, benzylpenicillin, famotidine, furosemide, ganciclovir, methotrexate and oseltamivir carboxylate (Figures 6.2 and 6.4b).[18] In most cases, these reported clinical interactions were perpetrated by the established OAT inhibitor probenecid (Figure 6.4c). The inhibitory effect of probenecid on drug uptake in renal epithelium and secretion to urine is, in fact, exploited clinically by using it as a comedication to increase systemic exposure of antibiotics and as a nephroprotective agent to avoid the renal toxicities associated with cidofovir treatment.[20] Similar effects have been proposed for the structurally related drug tenofovir.[21]

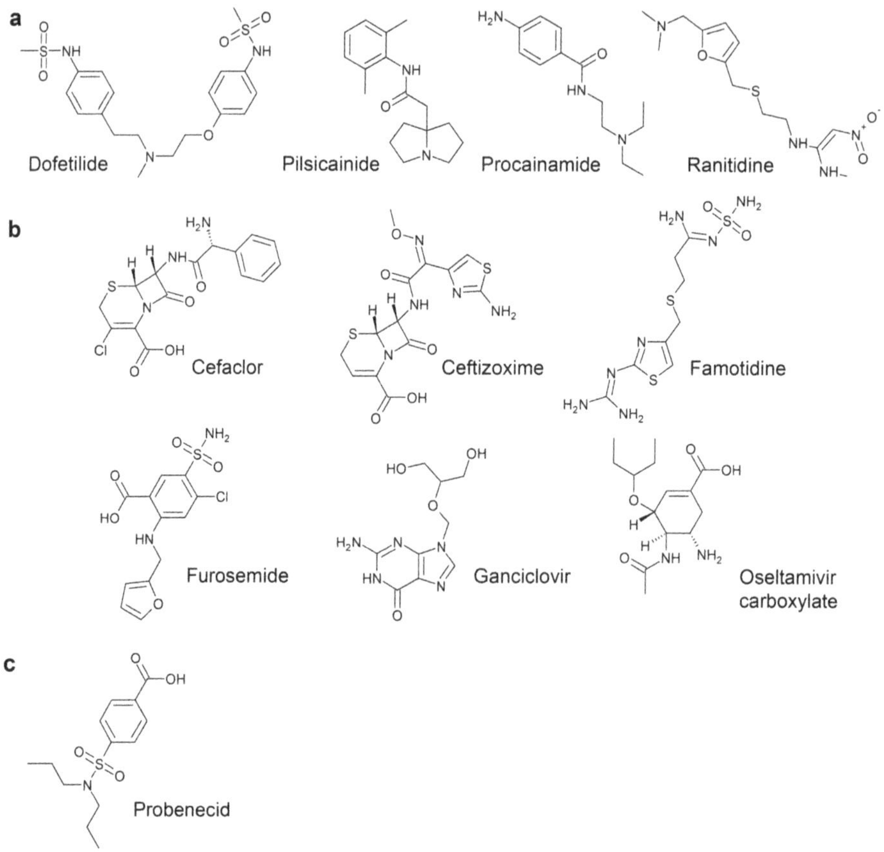

Figure 6.4 Example OCT and OAT substrates for which clinically observed drug–drug interactions have been reported.[18] Examples of affected OCT substrates and OAT substrates are shown in (a) and (b), respectively. The structure of probenecid, the most common perpetrator in OAT-mediated drug–drug interactions, is shown in (c).

Key References on OAT and OCT Transporters

- International Transporter Consortium, K. M. Giacomini, S. M. Huang, D. J. Tweedie, L. Z. Benet, K. L. Brouwer, X. Chu, A. Dahlin, R. Evers, V. Fischer, K. M. Hillgren, K. A. Hoffmaster, T. Ishikawa, D. Keppler, R. B. Kim, C. A. Lee, M. Niemi, J. W. Polli, Y. Sugiyama, P. W. Swaan, J. A. Ware, S. H. Wright, S. W. Yee, M. J. Zamek-Gliszczynski and L. Zhang, Membrane transporters in drug development, *Nat. Rev. Drug Discovery*, 2010, **9**(3), 215–236.
- H. Koepsell, Role of organic cation transporters in drug–drug interaction, *Expert Opin. Drug Metab. Toxicol.*, 2015, **11**(10), 1619–1633.
- M. J. Zamek-Gliszczynski, K. M. Giacomini and L. Zhang, Emerging Clinical Importance of Hepatic Organic Cation Transporter 1 (OCT1) in Drug Pharmacokinetics, Dynamics, Pharmacogenetic Variability, and Drug Interactions, *Clin. Pharmacol. Ther.*, 2018, **103**(5), 758–760.

- Y. Kido, P. Matsson and K. M. Giacomini, Profiling of a prescription drug library for potential renal drug–drug interactions mediated by the organic cation transporter 2, *J. Med. Chem.*, 2011, **54**(13), 4548–4558.
- R. Hendrickx, J. G. Johansson, C. Lohmann, R. M. Jenvert, A. Blomgren, L. Börjesson and L. Gustavsson, Identification of Novel Substrates and Structure–Activity Relationship of Cellular Uptake Mediated by Human Organic Cation Transporters 1 and 2, *J. Med. Chem.*, 2013, **56**, 7232–7242.
- P. J. Sandoval, K. M. Zorn, A. M. Clark, S. Ekins and S. H. Wright, Assessment of Substrate-Dependent Ligand Interactions at the Organic Cation Transporter OCT2 Using Six Model Substrates, *Mol. Pharmacol.*, 2018, **94**(3), 1057–1068.
- P. Duan, S. Li, N. Ai, L. Hu, W. J. Welsh and G. You, Potent inhibitors of human organic anion transporters 1 and 3 from clinical drug libraries: discovery and molecular characterization, *Mol. Pharm.*, 2012, **9**(11), 3340–3346.

6.2 Mitigation Strategies

Reports of using structure–activity relationships (SARs) to guide design of substrates or to avoid transport by OCT or OAT are rare. Most knowledge of drug interactions with the respective transporters is based either on anecdotal observations for substrates or on larger-scale screening assays of transporter *inhibition*. Importantly, evidence of a compound binding to the transporter (and/or inhibiting the transport of other compounds) does not necessarily indicate that the compound itself is a transported substrate. Conversely, transported substrates can competitively inhibit transport of other substrates, although often relatively high concentrations are needed for inhibition.

6.2.1 Ionization State

6.2.1.1 Substrates and Inhibitors

Molecular charge is a defining character of the vast majority of transported substrates and of a considerable fraction of inhibitors.[9,16,22–25] OCT substrates almost always carry positive charge at physiological pH, whereas OAT substrates are typically negatively charged. Modification (primarily addition or removal) of charged functional groups is thus a prime candidate for altering substrate and inhibitor affinity. Data is limited regarding the effects of tuning the pK_as of interacting groups *via*, *e.g.*, bioisosteric replacement or by affecting charge delocalization, but could provide viable options when the charged functionality is necessary to maintain pharmacological activity.

6.2.2 Molecular Size

6.2.2.1 Substrates

Reported OCT and OAT substrates tend to be small relative to the overall chemical space of registered drugs, with molecular weight rarely exceeding 400 Da (Figure 6.5).[26] In particular, bulky groups near charge centers appear detrimental for substrate translocation in some studies. Results from a larger-scale analysis of OCT1 and OCT2 substrates indicated that transporter-mediated uptake rates decreased with increasing molecular volume, with a more pronounced trend for OCT1 than for OCT2.[25]

6.2.2.2 Inhibitors

Results from some studies have indicated that molecular descriptors related to size and/or shape, including molecular weight, volume and the number of rotatable bonds are discriminating properties for inhibition.[16,22,23] Greater molecular size is associated with a higher likelihood of inhibition, although differences in size between inhibitors and non-inhibitors are typically small. Notably, the influence of size *per se* can often not be

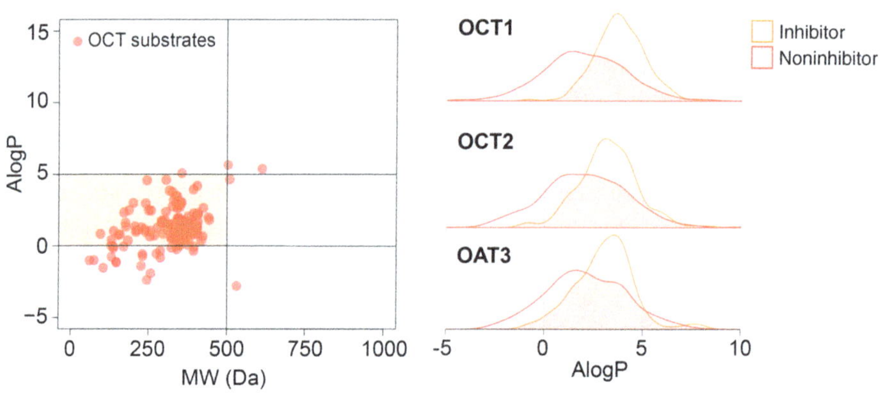

Figure 6.5 Molecular properties of OAT and OCT substrates and inhibitors. Left: distribution of reported substrates of organic cation transporters 1 and 2 (OCT1/2) and of organic anion transporters 1 and 3 (OAT1/3). Shaded area indicates property ranges compliant with the Lipinski rule-of-five. Right: distribution of calculated octanol–water partition coefficients (AlogP) for inhibitors and non-inhibitors of OCT1, OCT2 and OAT3. Similar trends are obtained using calculated charge-dependent partition coefficients ($\log D_{7.4}$) (data not shown). Inhibition data were collated from Kido *et al.*, 2011, Ahlin *et al.*, 2008, Chen *et al.*, 2017 and Duan *et al.*, 2012.[16,22–24] (Left) Adapted from ref. 26 with permission from Elsevier, Copyright 2016.

ascertained since other discriminating properties (often related to polarity and/or hydrophobicity, see later in this chapter, are typically correlated with molecular size).

6.2.3 Lipophilicity and Polarity

6.2.3.1 Substrates

Lipophilic drugs tend to be poor OCT or OAT substrates.[2,25] However, increased lipophilicity also tracks with increased diffusive membrane permeability (see, for example, the renal secretion case study below). It is therefore often not possible to discern experimentally whether a lack of OCT- or OAT-mediated transport—typically manifesting as near-identical transport rates in transporter-transfected and transporter-deficient cells—is due directly to a lower carrier-mediated transport or to the transmembrane diffusion being high in both cell models, masking the effect of the transporter. For OCT2, an increased polar surface area correlated with a decrease in transporter-mediated uptake.[25] This was accompanied by a similar tendency of higher uptake with increasing log of the distribution coefficient at pH 7.4 ($\log D_{7.4}$). In the same study, no correlation was observed with either descriptor for substrates of OCT1.

6.2.3.2 Inhibitors

A common feature in SAR studies of SLC22 transporter inhibition is compound lipophilicity.[16,22–24] Typically, a greater (calculated) octanol–water partition coefficient leads to increased likelihood of inhibition (Figure 6.5). This has been reported for inhibitors of OCT1 and OCT2, as well as for OAT1 and OAT3, using datasets on the order of a few hundred to a few thousand compounds. Additional molecular properties of importance for distinguishing inhibitors from non-inhibitors include descriptors of localized charge distribution (for OCT1) and a lower number of hydrogen bond donors and acceptors among inhibitors (for OAT1 and OAT3, as well as for OCT1 and OCT2).[16,22–24] Different trends have been observed for topological polar surface area (smaller among OCT1[23] and OCT2 inhibitors[16] than in corresponding non-inhibitors, but the reverse trend for OAT1 and 3[24]). The intercorrelation between topological polar surface area and molecular size complicates interpretations, as discussed above.

6.3 Examples of Structure–Activity Relationships for OATs/OCTs

6.3.1 Examples of Structure–Activity Relationships for OCT and OAT-mediated Cellular Uptake

Examples

1

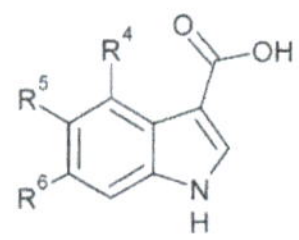

R^5	R^4	R^6	log $D_{7.4}$	P_{app} (10^{-6} cm s^{-1})	OAT3 uptake ratio	
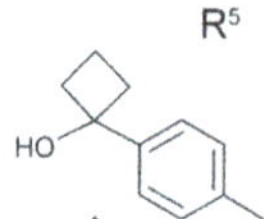	H	Cl	2.0	5.8	4.7	OAT3 substrate
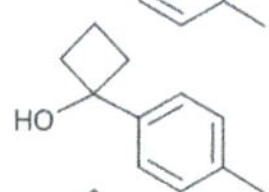	H	F	1.7	3.3	3.3	OAT3 substrate
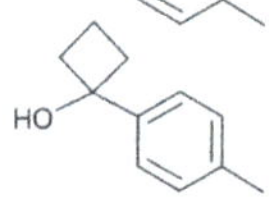	F	F	1.3	1.9	17	OAT3 substrate
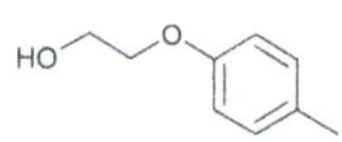	H	Cl	1.1	2.9	7.1	OAT3 substrate
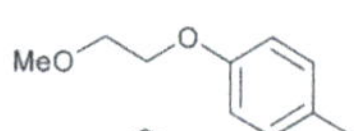	H	Cl	1.8	12	1.6	OAT3 non-substrate
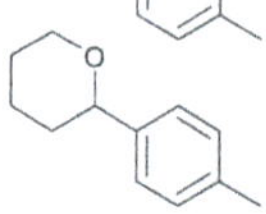	H	Cl	2.8	9.8	1.1	OAT3 non-substrate
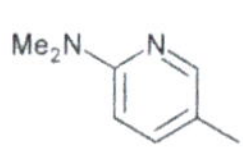	H	Cl	1.7	12	1.6	OAT3 non-substrate

Structural modifications in the indole core or in the aryl substituent of a 5′-adenosine monophosphate-activated protein kinase (AMPK) direct activator modulates OAT3-mediated cellular uptake and active tubular secretion primarily by affecting passive diffusive permeability.[27]

Examples

2

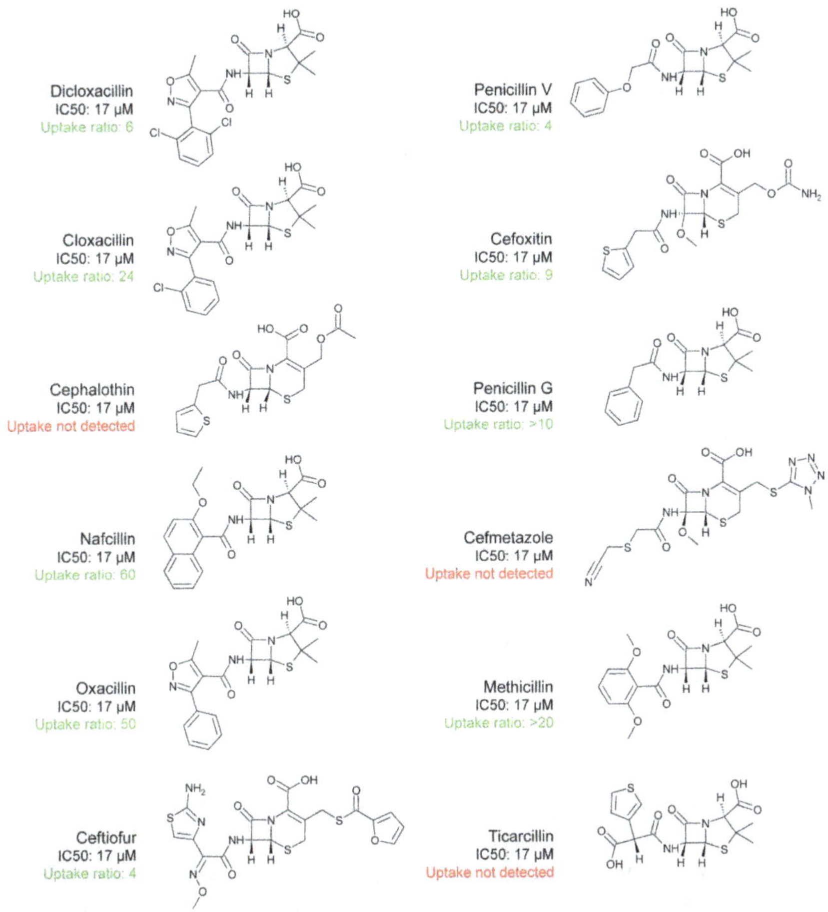

Structure–activity relationship for OAT3-mediated cellular uptake and OAT3 inhibition for beta-lactam antibiotics.[28]

(*continued*)

Examples

3

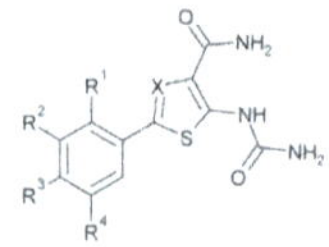

X	R¹	R², R³, R⁴	OCT1 uptake ratio	OCT2 uptake ratio
C	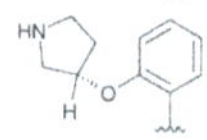	H, H, H	26 OCT1 substrate	1.9 OCT2 non-substrate
C	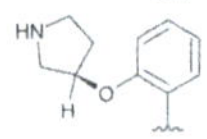	H, H, H	15 OCT1 substrate	1.1 OCT2 non-substrate
C	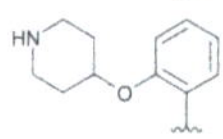	H, H, H	17 OCT1 substrate	1.8 OCT2 non-substrate
C	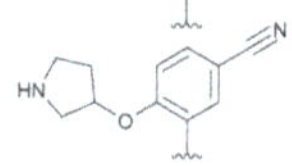	H, H, CN	39 OCT1 substrate	1.6 OCT2 non-substrate
C	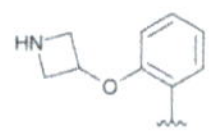	H, H, H	11 OCT1 substrate	2.2 OCT2 substrate
C	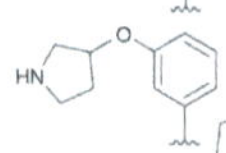	pyrrolidin-3-ol, H, H	68 OCT1 substrate	1.6 OCT2 non-substrate
C	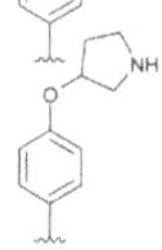	H, pyrrolidin-3-ol, H	73 OCT1 substrate	1.4 OCT2 non-substrate
C	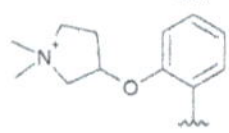	H, H, H	53 OCT1 substrate	4.3 OCT2 substrate
N	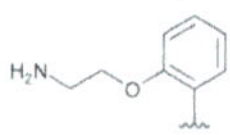	H, H, H	13 OCT1 substrate	1.3 OCT2 non-substrate
N	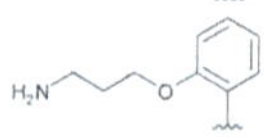	H, H, H	14 OCT1 substrate	1.5 OCT2 non-substrate

Structural modifications in the ring structure of a series of phenylthiophene-carboxamide ureas affect OCT1-mediated cellular uptake and selectivity over OCT2-mediated transport.[25]

References

1. A. Schlessinger, P. Matsson, J. E. Shima, U. Pieper, S. W. Yee, L. Kelly, L. Apeltsin, R. M. Stroud, T. E. Ferrin, K. M. Giacomini and A. Sali, *Protein Sci.*, 2010, **19**, 412–428.
2. International Transporter Consortium, K. M. Giacomini, S. M. Huang, D. J. Tweedie, L. Z. Benet, K. L. Brouwer, X. Chu, A. Dahlin, R. Evers, V. Fischer, K. M. Hillgren, K. A. Hoffmaster, T. Ishikawa, D. Keppler, R. B. Kim, C. A. Lee, M. Niemi, J. W. Polli, Y. Sugiyama, P. W. Swaan, J. A. Ware, S. H. Wright, S. W. Yee, M. J. Zamek-Gliszczynski and L. Zhang, *Nat. Rev. Drug Discovery*, 2010, **9**, 215–236.

3. M. J. Zamek-Gliszczynski, K. M. Giacomini and L. Zhang, *Clin. Pharmacol. Ther.*, 2018, **103**, 758–760.
4. M. J. Zamek-Gliszczynski, M. E. Taub, P. P. Chothe, X. Y. Chu, K. M. Giacomini, R. B. Kim, A. S. Ray, S. L. Stocker, J. D. Unadkat, M. B. Wittwer, C. Xia, S. W. Yee, L. Zhang, Y. Zhang and I. T. Consortium, *Clin. Pharmacol. Ther.*, 2018, **104**, 890–899.
5. European Medicines Agency, *Guideline on the Investigation of Drug Interactions*, 2013.
6. U.S. Department of Health and Human Services, Food and Drug Administration and Center for Drug Evaluation and Research, *In Vitro Metabolism- and Transporter-mediated Drug–Drug Interaction Studies – Guidance for Industry*, 2017.
7. C. Wegler, F. Z. Gaugaz, T. B. Andersson, J. R. Wisniewski, D. Busch, C. Groer, S. Oswald, A. Noren, F. Weiss, H. S. Hammer, T. O. Joos, O. Poetz, B. Achour, A. Rostami-Hodjegan, E. van de Steeg, H. M. Wortelboer and P. Artursson, *Mol. Pharm.*, 2017, **14**, 3142–3151.
8. C. Colas, P. M. Ung and A. Schlessinger, *MedChemComm*, 2016, **7**, 1069–1081.
9. H. Koepsell, *Expert Opin. Drug Metab. Toxicol.*, 2015, **11**, 1619–1633.
10. H. Koepsell, *Mol. Aspects Med.*, 2013, **34**, 413–435.
11. P. J. Sandoval, K. M. Zorn, A. M. Clark, S. Ekins and S. H. Wright, *Mol. Pharmacol.*, 2018, **94**, 1057–1068.
12. A. Bakhiya, A. Bahn, G. Burckhardt and N. Wolff, *Cell. Physiol. Biochem.*, 2003, **13**, 249–256.
13. R. Lu, B. S. Chan and V. L. Schuster, *Am. J. Physiol.*, 1999, **276**, F295–F303.
14. T. De Bruyn, G. J. van Westen, A. P. Ijzerman, B. Stieger, P. de Witte, P. F. Augustijns and P. P. Annaert, *Mol. Pharmacol.*, 2013, **83**, 1257–1267.
15. M. Karlgren, A. Vildhede, U. Norinder, J. R. Wisniewski, E. Kimoto, Y. Lai, U. Haglund and P. Artursson, *J. Med. Chem.*, 2012, **55**, 4740–4763.
16. Y. Kido, P. Matsson and K. M. Giacomini, *J. Med. Chem.*, 2011, **54**, 4548–4558.
17. U.S. Department of Health and Human Services, Food and Drug Administration, Center for Drug Evaluation and Research, *In Vitro Metabolism- and Transporter-mediated Drug–Drug Interaction Studies – Guidance for Industry*, 2017.
18. K. M. Morrissey, C. C. Wen, S. J. Johns, L. Zhang, S. M. Huang and K. M. Giacomini, *Clin. Pharmacol. Ther.*, 2012, **92**, 545–546.
19. M. B. Wittwer, A. A. Zur, N. Khuri, Y. Kido, A. Kosaka, X. Zhang, K. M. Morrissey, A. Sali, Y. Huang and K. M. Giacomini, *J. Med. Chem.*, 2013, **56**, 781–795.
20. J. P. Lalezari and B. D. Kuppermann, *J. Acquired Immune Defic. Syndr.*, 1997, **14**(Suppl 1), S27–S31.
21. B. Fernandez-Fernandez, A. Montoya-Ferrer, A. B. Sanz, M. D. Sanchez-Nino, M. C. Izquierdo, J. Poveda, V. Sainz-Prestel, N. Ortiz-Martin, A. Parra-Rodriguez, R. Selgas, M. Ruiz-Ortega, J. Egido and A. Ortiz, *AIDS Res. Treat.*, 2011, **2011**, 354908.

22. G. Ahlin, J. Karlsson, J. M. Pedersen, L. Gustavsson, R. Larsson, P. Matsson, U. Norinder, C. A. Bergstrom and P. Artursson, *J. Med. Chem.*, 2008, **51**, 5932–5942.
23. E. C. Chen, N. Khuri, X. Liang, A. Stecula, H. C. Chien, S. W. Yee, Y. Huang, A. Sali and K. M. Giacomini, *J. Med. Chem.*, 2017, **60**, 2685–2696.
24. P. Duan, S. Li, N. Ai, L. Hu, W. J. Welsh and G. You, *Mol. Pharm.*, 2012, **9**, 3340–3346.
25. R. Hendrickx, J. G. Johansson, C. Lohmann, R. M. Jenvert, A. Blomgren, L. Borjesson and L. Gustavsson, *J. Med. Chem.*, 2013, **56**, 7232–7242.
26. P. Matsson, B. C. Doak, B. Over and J. Kihlberg, *Adv. Drug Delivery Rev.*, 2016, **101**, 42–61.
27. D. J. Edmonds, D. W. Kung, A. S. Kalgutkar, K. J. Filipski, D. C. Ebner, S. Cabral, A. C. Smith, G. E. Aspnes, S. K. Bhattacharya, K. A. Borzilleri, J. A. Brown, M. F. Calabrese, N. L. Caspers, E. C. Cokorinos, E. L. Conn, M. S. Dowling, H. Eng, B. Feng, D. P. Fernando, N. E. Genung, M. Herr, R. G. Kurumbail, S. Y. Lavergne, E. C. Lee, Q. Li, S. Mathialagan, R. A. Miller, J. Panteleev, J. Polivkova, F. Rajamohan, A. R. Reyes, C. T. Salatto, A. Shavnya, B. A. Thuma, M. Tu, J. Ward, J. M. Withka, J. Xiao and K. O. Cameron, *J. Med. Chem.*, 2018, **61**, 2372–2383.
28. A. T. Wolman, M. R. Gionfriddo, G. A. Heindel, P. Mukhija, S. Witkowski, A. Bommareddy and A. L. VanWert, *Drug Metab. Dispos.*, 2013, **41**, 791–800.

CHAPTER 7

OATPs: The SLCO Family of Organic Anion Transporting Polypeptide Transporters

MARIA KARLGREN*[a] AND PÄR MATSSON*[b]

[a]Department of Pharmacy, Uppsala University, Box 580, SE-751 23 Uppsala, Sweden; [b]Department of Pharmacology, Institute of Neuroscience and Physiology, The Sahlgrenska Academy, University of Gothenburg, Box 431, SE-405 30 Gothenburg, Sweden
*E-mail: maria.karlgren@farmaci.uu.se, par.matsson@gu.se

7.1 Introduction

7.1.1 Transporter Family

The OATP/*SLCO* family (previously called *SLC21*) of transporters belongs to the functionally and structurally diverse solute carrier (SLC) protein superfamily. Currently, 11 human organic anion transporting polypeptide (OATP) transporters are known, of which OATP1B1 (*SLCO1B1*) and OATP1B3 (*SLCO1B3*) have been shortlisted as transporters of considerable importance for drug disposition.[1–3] However, other OATP family members, including OATP1A2 (*SLCO1A2*), OATP2B1 (*SLCO2B1*) and OATP4C1 (*SLCO4C1*), are also involved in drug transport.[2]

Drug Discovery Series No. 79
The Medicinal Chemist's Guide to Solving ADMET Challenges
Edited by Patrick Schnider
© The Royal Society of Chemistry 2021
Published by the Royal Society of Chemistry, www.rsc.org

7.1.2 Expression

OATP1B1 and OATP1B3 are both highly and specifically expressed in the basolateral membrane of hepatocytes (facing the liver blood vasculature).[2] OATP1B1 is the most highly expressed drug uptake transporter in the liver, followed by OATP1B3, as revealed by liquid chromatography–tandem mass spectrometry (LC–MS/MS) based protein quantification.[4] In contrast, many of the other human OATP transporters show a less discriminatory tissue expression pattern.[5] For example, OATP2B1 is the predominant OATP isoform in intestinal enterocytes, but is also expressed in, for example, muscle and pancreatic tissue, and proteomic data also indicate a significant contribution from OATP2B1 to the total OATP abundance in hepatocytes.[4,6]

7.1.3 Structure

The OATP/*SLCO* transporters belong to the major facilitator superfamily (MFS) protein fold family.[7] So far, no crystal structure has been published, but biochemical evidence supports a common structure consisting of 12 transmembrane domains and a large extracellular loop between transmembrane domains 9 and 10 containing several conserved cysteine residues.[1] Furthermore, the second and fifth extracellular loops contain *N*-glycosylation sites, and an "OATP superfamily signature" amino acid sequence (D-X-RWI/V-GAWW-X-G-F/L-L) is found in extracellular loop 3 close to the border to transmembrane domain 6.[1]

7.1.4 Activity

Transport rates vary depending on the substrate and the level of transporter expression in the tissue or cell type of interest. In contrast to other SLC families, the OATPs are less conserved across species, and orthologs may hence not exist in typical preclinical animal models.[8,9] Humanized rodent models have therefore been explored as alternatives for *in vitro–in vivo* correlations studies.[10–12] As for all transporters, the clinical effects of OATP mediated transport are strongly affected by competing transport mechanisms; in particular, the disposition of OATP substrates that exhibit slow diffusion across phospholipid bilayers will be more affected by the transporter(s) than will substrates with rapid background diffusion. The specific expression of OATP1B1 and OATP1B3 in liver strongly contributes to the hepatic uptake of substrate drugs and has been exploited for tissue targeting (see Zhou *et al.*[13]).

7.1.5 Function and Substrates

Current evidence indicates that OATPs/*SLCOs* are facilitative transporters, meaning that they form substrate-selective channels across the cell membrane through which the substrate is translocated down its concentration gradient. A pH dependence has been reported for OATP2B1, with more efficient transport at pH 5.5 than 7.4; however, whether this indicates a proton-coupled mechanism or an effect on the ionization of the substrate has not been concluded.[14,15] The amino acids lining the transport channel are predominantly positively charged, which is consistent with the anionic charge of the majority the OATP/*SLCO* substrates. Positively charged amino acids in several of the OATP/SLCO transmembrane spanning regions or in adjacent parts have been identified as important for recognition and/or transport of different substrates.[16,17] Multiple binding sites have been reported for OATP1B1 and OATP2B1, based on biphasic uptake kinetics, and different inhibitor susceptibility depending on substrates.[15,18–20] This could reflect the presence of a large central ligand-binding cavity, capable of simultaneously binding multiple smaller ligands.

Common *in vitro* and *in vivo* probe substrates for OATPs include estradiol-17β-glucuronide (E17βG), estrone-3-sulfate (E3S), pitavastatin, rosuvastatin, pravastatin and bromosulfophthalein (BSP) (Figure 7.1).[2] Inhibitors include cyclosporin A, rifampicin and antiretroviral drugs like ritonavir and lopinavir (Figure 7.2).[2]

OATPs accept substrates that are relatively large compared with other human MFS-type protein fold drug transporters,[25] such as the OCTs and OATs (see Chapter 6. OATs and OCTs: The SLC22 Family of Organic Anion and Cation Transporters). While the latter appear to have an upper limit on substrate size well within the limits of traditional rule-of-five compliant chemical space, several OATP substrates in the 750–900 Da range have been reported, for example the hepatitis C virus (HCV) NS3/4A protease inhibitors simeprevir, faldaprevir, danoprevir and grazoprevir (Figure 7.3).[25]

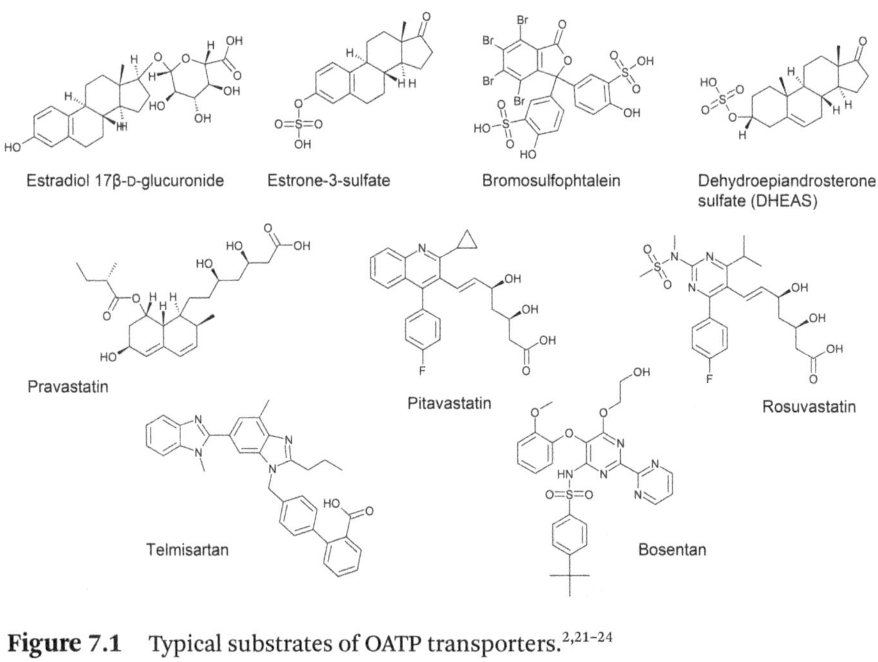

Figure 7.1 Typical substrates of OATP transporters.[2,21–24]

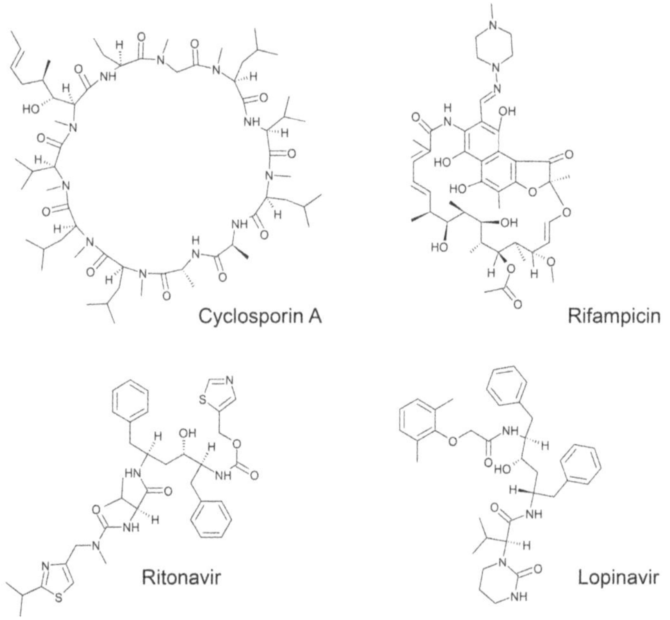

Figure 7.2 Typical inhibitors of OATP transporters.[2,21–24]

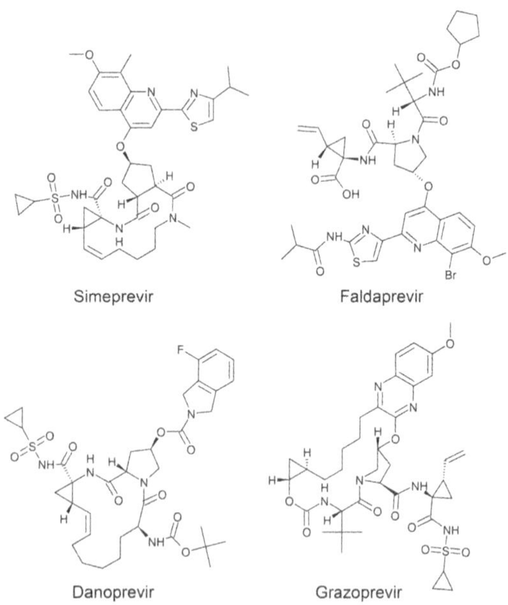

Figure 7.3 HCV NS3/4 protease inhibitors as examples of larger OATP substrates.

7.1.6 Mechanism

Structural evidence for other transporters sharing the MFS-type protein fold indicates an 'alternating-access' transport cycle in which the transporter upon binding the substrate undergoes a conformational change from an outward-open to an occluded state, in which a centrally located substrate binding site is separated from both the extracellular and the intracellular space. Additional conformational changes result in an inward-open state from which the substrate is released (for a schematic illustration see Figure 7.4). While this mechanism has not been proven for OATPs, their predicted protein folding indicates a shared transport mechanism with other MFS transporters.[7]

7.1.7 Screening Strategies

Transporter-transfected cell lines (*e.g.* HEK293 or HeLa cells) or oocytes are the most common experimental system to study transport and/or drug interactions with OATPs, although, during recent years an increasing number of OATP transport or interaction studies has been conducted in primary human hepatocytes.[26] The experimental setup for identifying transported substrates typically involves measurements of time-dependent cellular uptake in transporter-transfected cells, using cells transfected with a transporter-deficient ('empty') vector as a control, with the assumption that the uptake in the control cells represents nonspecific binding and uptake processes that are identical in the two cell types.[2] Often, the control-cell uptake is simply subtracted from that in the transporter transfected cells to yield a transporter-mediated uptake rate.[26]

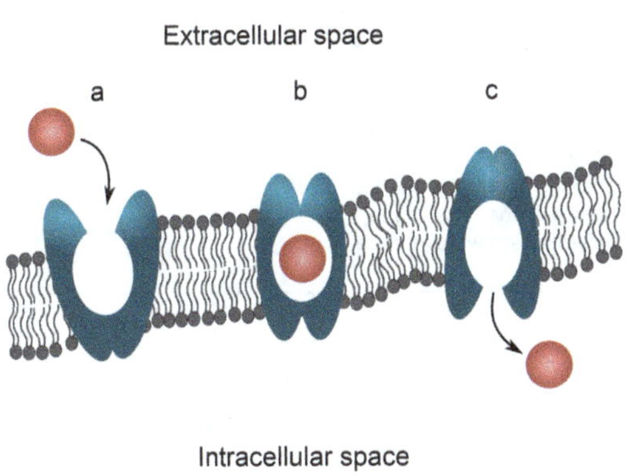

Figure 7.4 Suggested transport mechanism for MFS transporters. First, the substrate binds to an exposed, centrally located binding site in the transporter. Upon binding, the transporter undergoes a conformational change from an outward-open state [binding site accessible from the extracellular space (a)] to an occluded state (b). Additional conformational changes result in an inward-open state [binding site accessible from the intracellular space (c)] and release of the substrate. Depending on substrate concentration, transport in the reverse direction can also be possible.

Drug-mediated inhibition (potentially leading to a transporter-mediated drug–drug interaction; DDI) is typically studied in an analogous setup. The cellular uptake of a defined probe substrate is detected using fluorescence, radiolabelling or mass spectrometry.[27–30] The effect of a potential inhibitor is quantified as the percentage of inhibited transport (at a single screening concentration) or as a half-maximal inhibitory concentration (IC_{50}) or inhibition constant (K_i).[26]

In recent years, the importance of OATPs in drug disposition and drug–drug interactions has gained more attention from regulatory agencies. In current guidelines, OATP uptake must be studied for drug candidates where the hepatic metabolism and/or the biliary secretion contribute significantly to the total drug clearance or if uptake to the liver is clinically important (see Scheme 7.1). In addition, the potential for OATP inhibition needs to be investigated for all drug candidates (Scheme 7.2).

Studies in fresh or cryopreserved primary human hepatocytes rely on prototypical OATP probe substrates and/or inhibitors. However, as there is a large substrate and inhibitor overlap between the different OATPs and also an overlap with other hepatic transport proteins, such studies are often more difficult to interpret.[30] Furthermore, as OATP1B1 is genetically polymorphic

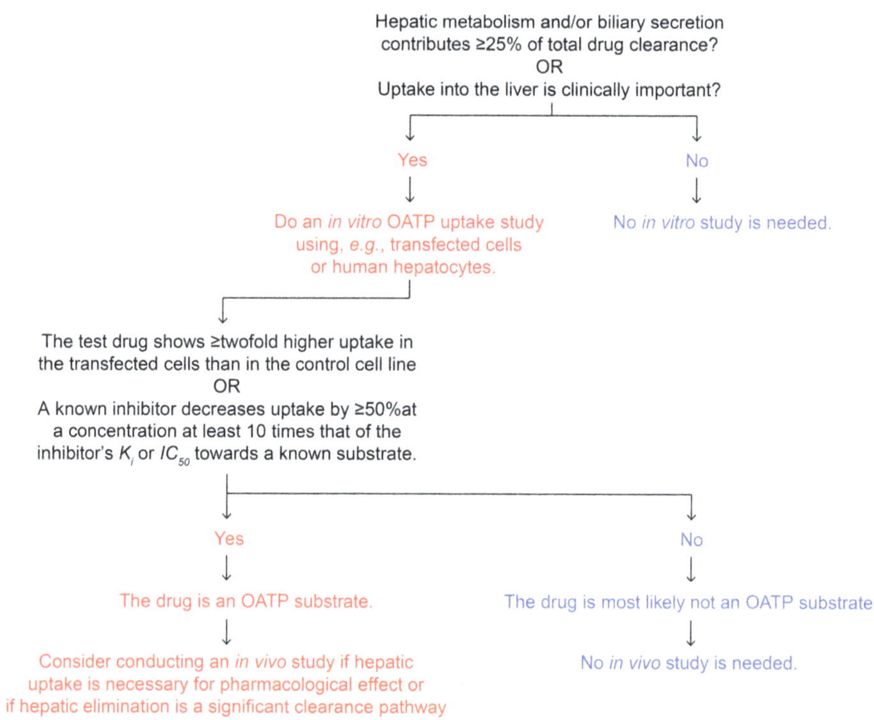

Scheme 7.1 Decision tree to guide screening of potential OATP substrates. The decision tree is based on the U.S. Food and Drug Administration (FDA) draft guidance on *in vitro* drug–drug interaction studies (2017).[31]

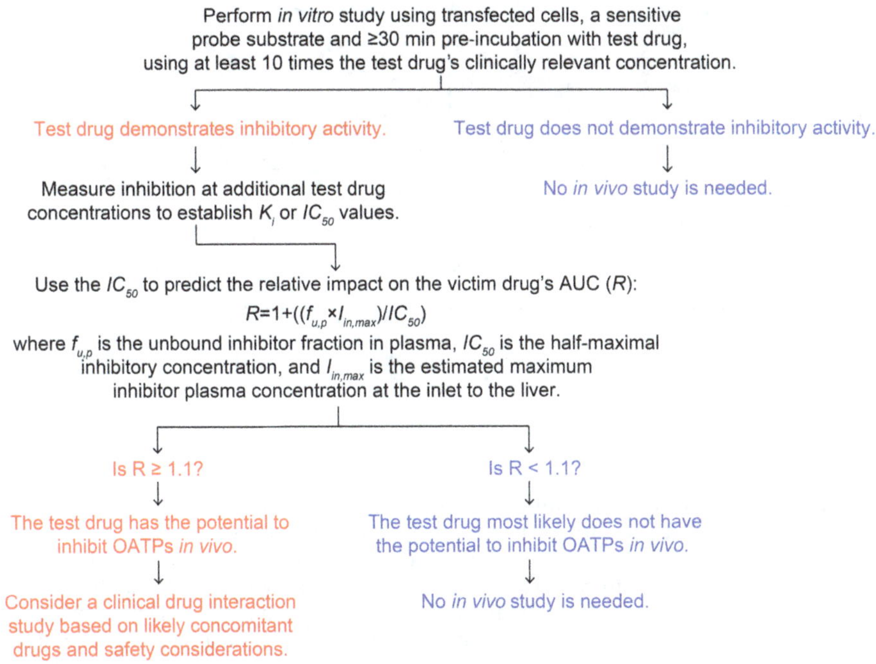

Scheme 7.2 Decision tree to guide screening of potential OATP inhibitors. The decision tree is based on the U.S. Food and Drug Administration (FDA) draft guidance on *in vitro* drug–drug interaction studies (2017).[31]

and has been identified as an important pharmacogenomic biomarker for statin-induced adverse effects,[32] the donor genotype needs to be taken into consideration in studies using primary tissues or cells.

7.1.8 Relevance

OATP1B1 and OATP1B3 are highly expressed in the liver and are important for the hepatocellular uptake of many drugs. However, many substrates are transported by multiple OATP isoforms and also by additional hepatic drug transporters. Hence, many of the substrates and inhibitors of hepatic OATPs are also, to varying extents, interacting with other hepatic uptake and efflux transporters or with hepatic drug metabolizing enzymes.[30] (For an overview of the most important hepatic drug transporters and drug-metabolizing enzymes see Figure 7.5). One example of such transporter interactions is the compound MK571, which has historically been used as a specific inhibitor of drug efflux transporters belonging to the multidrug resistance associated protein or ATP binding cassette subfamily C (MRP/ABCC) family. However, it was demonstrated that MK571 inhibited P-glycoprotein (P-gp/ABCB1) and breast cancer resistance protein (BCRP/ABCG2) with similar potency.[33] Furthermore, OATP1B1, OATP1B3 and OATP2B1 were also inhibited by MK571, with equal or greater potency to MRPs.[30]

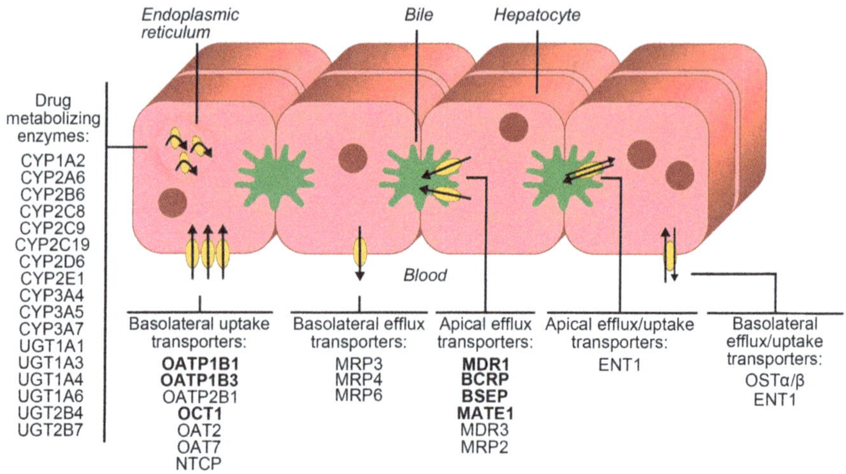

Figure 7.5 Localization of important drug transporters and drug metabolizing enzymes in human hepatocytes. The selection of included transporters is based on the International Transporter Consortium publications by Giacomini *et al.* (2010) and Zamek-Gliszczynski *et al.* (2018) and the selection of drug metabolizing enzymes is based on Evans and Relling (1999) and Williams *et al.* (2004).[2,3,39,40] Transporters included in current FDA and European Medicines Agency (EMA) drug interaction guidelines are shown in bold type.[31,41] Adapted from ref. 42 with permission from American Chemical Society, Copyright 2017.

In addition to such substrate and/or inhibitor overlap, expression of many hepatic drug transporters and drug-metabolizing enzymes are regulated *via* the same nuclear receptors,[34] emphasizing the significant interplay between different transport and metabolism processes, and the difficulties that often arise in delineating these.

Methods to predict transporter-mediated *in vivo* drug disposition and the potential for transporter-mediated drug–drug interactions include both static models and dynamic ones, such as physiologically-based pharmacokinetic (PBPK) models. For an overview of different modeling approaches, as well as of their advantages and disadvantages, see Zamek-Gliszczynski *et al.* 2013.[35] Progress has been made in assessing the contribution of OATP transporters to *in vivo* hepatic uptake clearance using methods that take *in vitro* kinetics and *in vitro* and *in vivo* expression of the transporter(s) into account.[4,30,36,37] So far, such methods have only been applied to a limited number of compounds, but show a good correlation with *in vivo* data. Recent modeling approaches and regulatory applications that include the use of PBPK transporter modeling are discussed in more detail in Guo *et al.*, 2018.[38]

OATPs significantly influence the hepatic disposition of several substrate drugs. For example, OATP-mediated hepatic uptake is the rate-limiting step in the elimination of the endothelin receptor antagonist bosentan.[21] Among the most well studied examples of OATP-mediated disposition are the 3-hydroxy-3-methylglutaryl coenzyme A (HMG-CoA) reductase inhibitors ('statins').[43] For several statins, OATPs are the dominant mechanism for uptake into hepatocytes, which are also the target cells for these drugs.[2,44] Cerivastatin, the first statin on the market, was withdrawn in 2001 after adverse drug reactions and DDIs resulting in 52 lethalities.[45] Many of these cases were in patients concomitantly taking gemfibrozil.[46] Cerivastatin is a substrate of the OATP1B1 transporter and is metabolized by CYP2C8 and CYP3A4. However, gemfibrozil and one of its metabolites, gemfibrozil 1-*O*-β-glucuronide, are inhibitors of both OATP1B1 and CYP2C8, with gemfibrozil 1-*O*-β-glucuronide being the more potent inhibitor.[47] Hence the drug–drug interaction between cerivastatin and gemfibrozil was found to be a combination of interactions at the metabolism level and hepatic uptake, leading to increased cerivastatin plasma concentrations and an increased risk of adverse effects and toxicity in peripheral tissues.[44,47] Also for other, still marketed statins, OATPs—in particular OATP1B1—have a clear clinical effect.[48] In 2013, the FDA updated the simvastatin product label with dosing recommendations based on OATP1B1/*SLCO1B1* genotype, and in 2014 the Clinical Pharmacogenetics Implementation Consortium updated their guidelines regarding OATP1B1/*SLCO1B1* and simvastatin-induced myopathy.[49,50]

Key References on OATP Transporters

- International Transporter Consortium, K. M. Giacomini, S. M. Huang, D. J. Tweedie, L. Z. Benet, K. L. Brouwer, X. Chu , A. Dahlin , R. Evers, V. Fischer, K. M. Hillgren, K. A. Hoffmaster, T. Ishikawa, D. Keppler, R. B. Kim, C. A. Lee, M. Niemi, J. W. Polli, Y. Sugiyama, P. W. Swaan, J. A. Ware, S. H. Wright, S. W. Yee, M. J. Zamek-Gliszczynski and L. Zhang, Membrane transporters in drug development, *Nat. Rev. Drug Discov.*, 2010, **9**(3), 215–336.

- M. J.Zamek-Gliszczynski, M. E. Taub, P. P. Chothe, X. Chu, K. M. Giacomini, R. B. Kim, A. S. Ray, S. L. Stocker, J. D. Unadkat, M. B. Wittwer, C. Xia, S. W. Yee, L. Zhang and Y. Zhang, International Transporter Consortium. Transporters in Drug Development: 2018 ITC Recommendations for Transporters of Emerging Clinical Importance, *Clin. Pharmacol Ther.*, 2018, **104**(5), 890–899.

- B. Hagenbuch and P. J. Meier, Organic anion transporting polypeptides of the OATP/SLC21 family: phylogenetic classification as OATP/SLCO superfamily, new nomenclature and molecular/functional properties, *Pflugers Arch.*, 2004, **447**(5), 653–665.

- A. Kalliokoski and M. Niemi, Impact of OATP transporters on pharmacokinetics, *Br. J. Pharmacol.*, 2009, **158**(3), 693–705.

- M. Karlgren, A. Vildhede, U. Norinder, J. R. Wisniewski, E. Kimoto, Y. Lai, U. Haglund and P. Artursson, Classification of inhibitors of hepatic organic anion transporting polypeptides (OATPs): influence of protein expression on drug–drug interactions, *J. Med. Chem.*, 2012, **55**(10), 4740–4763.

- T. De Bruyn, G. J. van Westen, A. P. Ijzerman, B. Stieger, P. de Witte, P. F. Augustijns and P. P. Annaert, Structure-based identification of OATP1B1/3 inhibitors, *Mol. Pharmacol.*, 2013;**83**(6):1257–1267.

- J. Zhou, J. Xu, Z. Huang and M. Wang, Transporter-mediated tissue targeting of therapeutic molecules in drug discovery, *Bioorg. Med. Chem. Lett.*, 2015, **25**(5), 993–997.

7.2 Mitigation Strategies

Typically, structure–activity relationship (SAR) information for OATP-mediated transport has been derived from indirect data on liver selectivity, or from anecdotal observations in direct substrate transport assays. Larger-scale screening assays have been used to determine molecular properties of importance for OATP *inhibition*. Importantly, evidence of compound binding (and/or transport inhibition) does not necessarily indicate that the compound is transported. Conversely, transported substrates can competitively inhibit transport of other substrates, depending on their relative affinities for the transporter binding site(s) and the applied concentration.

Optimization of OATP mediated uptake has been used to achieve hepatocellular selectivity for therapies where the pharmacological effect is in the liver, including statins, antidiabetics and HCV inhibitors. In a summary of Pfizer projects aiming for liver selectivity, optimization of OATP-mediated cellular uptake was found to be an important strategy.[51,52] Incorporation of polar acidic functionality was found to be successful for achieving OATP activity and simultaneously limiting off-target activity by decreasing passive transmembrane diffusion into other tissues than the intended. Limiting passive membrane permeability was found to be critical for tissue selectivity, and it was suggested that a logarithm of the distribution coefficient (logD) in the range between 0.5 and 2 would be optimal for achieving liver selectivity in combination with oral bioavailability. A number of different carboxylic acid bioisosters have been reported to maintain hepatic selectivity and/or OATP mediated hepatocyte uptake activity.

Most SARs reported for series of liver targeted drugs have not directly used transporter assays in the optimization, but instead inferred tissue selectivity from activity differences between cell types. For example, in multiple series of statins—all maintaining a carboxylic acid moiety necessary for activity—the hepatocyte : myocyte IC_{50} ratio for inhibition of cholesterol synthesis indicated a varying, but consistently highly significant liver-specific activity.[53] OATP mediated hepatocyte uptake was confirmed for the optimized compounds only. Thus, in these series the SAR for the OATP interaction appeared relatively flat.

7.2.1 Ionization State

7.2.1.1 Substrates

The majority of OATP1B1/1B3 substrates are negatively charged. Some neutral substrates have been reported; however, such compounds tend to be weaker substrates than the acidic ones.[51]

7.2.1.2 Inhibitors

Screening studies aiming to identify inhibitors of OATP1B1, OATP1B3 or OATP2B1 have revealed that 60–75% of identified OATP inhibitors were negatively charged at physiological pH, whereas the remaining inhibitors were predominantly neutral. SAR models based on these inhibition datasets also indicate charge-related molecular descriptors among the most important ones.[28–30] Similarly, pharmacophore and 3D-quantitative SAR (QSAR) models based on smaller datasets of OATP inhibitors indicate the importance of negatively ionizable features.[54,55]

Modifications that affect negatively ionizable functional groups are thus prime candidates for altering substrate and inhibitor affinity.

7.2.2 Molecular Size

7.2.2.1 Substrates

As compared with other SLC drug transporters, like the OATs and the OCTs, OATP substrates often have a relatively high molecular weight. Approximately 75% of reported OATP substrates have a molecular weight of ≥400 Da and, as mentioned previously, substrates of up to approximately 900 Da have been reported.[25] Given the shared transport mechanism within the MFS family there should exist an upper physical boundary to substrate size compatible with transport, but this boundary is clearly higher than for most other of the known SLC drug transporters.

7.2.2.2 Inhibitors

Molecular weight is also an important property for OATP inhibitors. Results of screening studies have indicated that OATP inhibitors have significantly higher molecular weight as compared with non-inhibitors,[29,30] with the median molecular weight for OATP1B1/1B3/2B1 inhibitors ranging from 481 to 514 g mol^{-1} and for structurally diverse non-inhibitors from 325 to 336 g mol^{-1}.[30] Similarly, molecular volume has been identified as an important discriminating property between OATP inhibitors and non-inhibitors, with greater volumes indicative of OATP inhibition.[28–30]

7.2.3 Lipophilicity and Polarity

7.2.3.1 Substrates

For substrate translocation, increased lipophilicity tracks with increased diffusive membrane permeability. Lipophilic (and, thus, highly permeable) drugs tend to be less efficient OATP substrates. However, as a high transmembrane diffusion is likely to mask the carrier-mediated transport in the *in vitro* setting this is typically not possible to differentiate experimentally. Many reported OATP substrates are still relatively lipophilic, and it is thus likely that lipophilicity does not limit the transporter-mediated flux itself. Rather, the detrimental effects on hepatic selectivity observed at higher lipophilicities (logD >2) probably result from increased transmembrane diffusion into off-target tissues.

7.2.3.2 Inhibitors

A common feature in SAR studies of OATP transporter inhibition is compound lipophilicity. Typically, a greater (calculated) octanol–water partition coefficient leads to an increased likelihood of inhibition.[28–30] Additional molecular properties of importance for distinguishing inhibitors from non-inhibitors include the number of hydrogen bond acceptors and the topological polar surface area—both having greater values in OATP inhibitors than non-inhibitors in multiple studies of drug-like datasets, ranging from 200 to 2000 compounds in size.[28–30] The seemingly conflicting results of increased lipophilicity and polar surface both increasing the likelihood of inhibition may be related to the fact that polar surface typically correlates with molecular size; however, the binding mechanisms resulting in these observations have not yet been fully clarified.

7.3 Examples of Structure–Activity Relationships for OATP-mediated Cellular Uptake and Hepatic Targeting

Examples

1

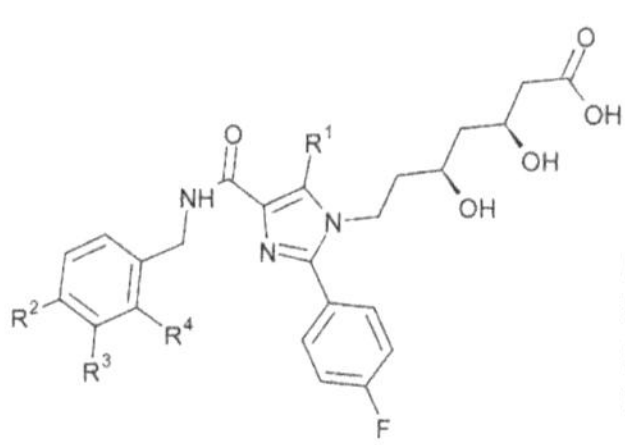

R^1: Me, Et, *n*Pr, *i*Pr, *c*Pr
R^2: H, Me, OMe, F, C(O)NH_2, SO_2NMe_2
R^3: H, Me, OMe, F
R^4: H, Me, OMe, F

Hepatocyte:myocyte IC_{50} ratio: 1,800-250,000 (rosuvastatin: 800)

Structural modifications in regions of an HMG-CoA reductase inhibitor distal to the carboxylic acid moiety resulted in altered, but consistently high, enrichment in hepatocytes, indicating modest effects on OATP-mediated uptake.[53]

Examples

2

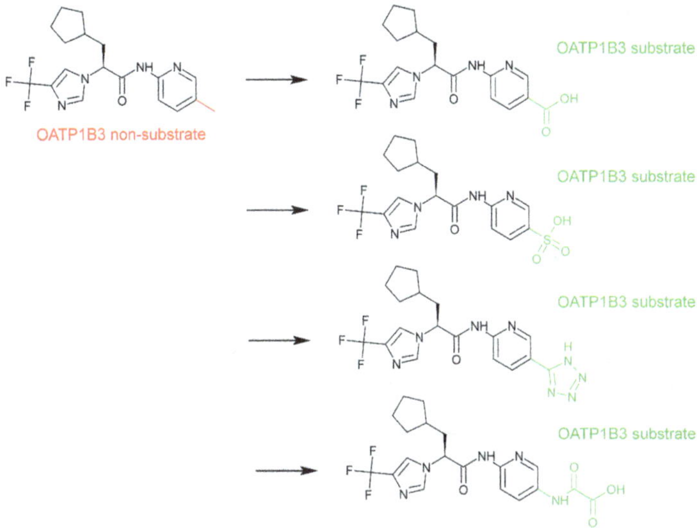

Incorporation of a carboxylic acid in a glucokinase activator resulted in hepatoselectivity *via* OATP1B3 mediated cellular uptake. Multiple carboxylic acid bioisosters maintained OATP1B3 substrate activity.[56]

3

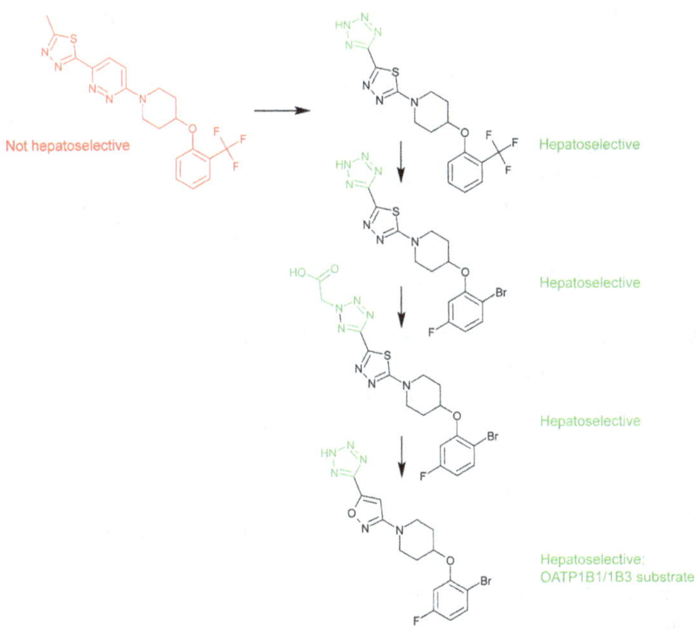

Optimization of hepatoselectivity *via* OATP1B1/1B3 mediated cellular uptake through incorporation of a tetrazole and/or a tetrazole-carboxylic acid moiety in a stearoyl-coenzyme A desaturase inhibitor.[57]

(*continued*)

Examples

4

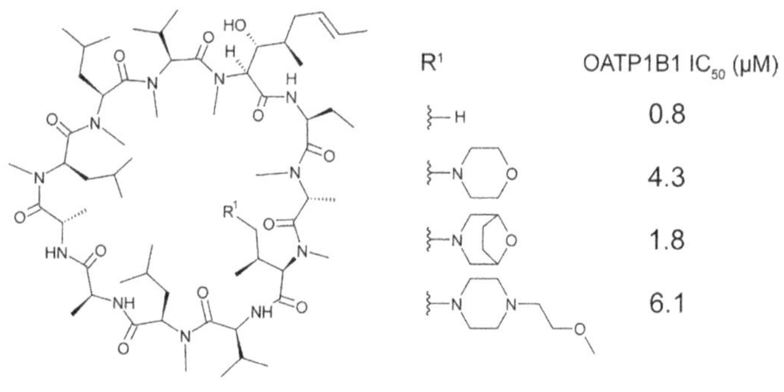

Decreasing inhibitory potency towards OATP1B1 in derivatives of the cyclophilin inhibitor alisporivir (R^1 = H).[58]

References

1. B. Hagenbuch and P. J. Meier, *Pflügers Arch.*, 2004, **447**, 653–665.
2. Consortium International Transporter, K. M. Giacomini, S. M. Huang, D. J. Tweedie, L. Z. Benet, K. L. Brouwer, X. Chu, A. Dahlin, R. Evers, V. Fischer, K. M. Hillgren, K. A. Hoffmaster, T. Ishikawa, D. Keppler, R. B. Kim, C. A. Lee, M. Niemi, J. W. Polli, Y. Sugiyama, P. W. Swaan, J. A. Ware, S. H. Wright, S. W. Yee, M. J. Zamek-Gliszczynski and L. Zhang, *Nat. Rev. Drug Discovery*, 2010, **9**, 215–236.
3. M. J. Zamek-Gliszczynski, M. E. Taub, P. P. Chothe, X. Y. Chu, K. M. Giacomini, R. B. Kim, A. S. Ray, S. L. Stocker, J. D. Unadkat, M. B. Wittwer, C. Xia, S. W. Yee, L. Zhang, Y. Zhang and I. T. Consortium, *Clin. Pharmacol. Ther.*, 2018, **104**, 890–899.
4. A. Vildhede, J. R. Wisniewski, A. Noren, M. Karlgren and P. Artursson, *J. Proteome Res.*, 2015, **14**, 3305–3314.
5. I. Tamai, J. Nezu, H. Uchino, Y. Sai, A. Oku, M. Shimane and A. Tsuji, *Biochem. Biophys. Res. Commun.*, 2000, **273**, 251–260.
6. E. Kimoto, K. Yoshida, L. M. Balogh, Y. A. Bi, K. Maeda, A. El-Kattan, Y. Sugiyama and Y. Lai, *Mol. Pharm.*, 2012, **9**, 3535–3542.
7. C. Colas, P. M. Ung and A. Schlessinger, *MedChemComm*, 2016, **7**, 1069–1081.
8. B. Hagenbuch and P. J. Meier, *Biochim. Biophys. Acta, Biomembr.*, 2003, **1609**, 1–18.
9. K. Bleasby, J. C. Castle, C. J. Roberts, C. Cheng, W. J. Bailey, J. F. Sina, A. V. Kulkarni, M. J. Hafey, R. Evers, J. M. Johnson, R. G. Ulrich and J. G. Slatter, *Xenobiotica*, 2006, **36**, 963–988.

10. S. Durmus, S. van Hoppe and A. H. Schinkel, *Drug Resist. Updates*, 2016, **27**, 72–88.
11. J. W. Higgins, J. Q. Bao, A. B. Ke, J. R. Manro, J. K. Fallon, P. C. Smith and M. J. Zamek-Gliszczynski, *Drug Metab. Dispos.*, 2014, **42**, 182–192.
12. M. Uchida, Y. Tajima, M. Kakuni, Y. Kageyama, T. Okada, E. Sakurada, C. Tateno and R. Hayashi, *Drug Metab. Dispos.*, 2018, **46**, 11–19.
13. J. Y. Zhou, J. F. Xu, Z. Huang and M. M. Wang, *Bioorg. Med. Chem. Lett.*, 2015, **25**, 993–997.
14. O. Kis, J. A. Zastre, M. Ramaswamy and R. Bendayan, *J. Pharmacol. Exp. Ther.*, 2010, **334**, 1009–1022.
15. Y. Shirasaka, T. Mori, M. Shichiri, T. Nakanishi and I. Tamai, *Drug Metab. Pharmacokinet.*, 2012, **27**, 360–364.
16. M. Gruetz, H. Sticht, H. Glaeser, M. F. Fromm and J. Konig, *Biochim. Biophys. Acta*, 2016, **1858**, 2894–2902.
17. F. Meier-Abt, Y. Mokrab and K. Mizuguchi, *J. Membr. Biol.*, 2005, **208**, 213–227.
18. Y. Hoshino, D. Fujita, T. Nakanishi and I. Tamai, *MedChemComm*, 2016, **7**, 1775–1782.
19. J. Noe, R. Portmann, M. E. Brun and C. Funk, *Drug Metab. Dispos.*, 2007, **35**, 1308–1314.
20. I. Tamai, T. Nozawa, M. Koshida, J. Nezu, Y. Sai and A. Tsuji, *Pharm. Res.*, 2001, **18**, 1262–1269.
21. A. Treiber, R. Schneiter, S. Hausler and B. Stieger, *Drug Metab. Dispos.*, 2007, **35**, 1400–1407.
22. T. Abe, M. Kakyo, T. Tokui, R. Nakagomi, T. Nishio, D. Nakai, H. Nomura, M. Unno, M. Suzuki, T. Naitoh, S. Matsuno and H. Yawo, *J. Biol. Chem.*, 1999, **274**, 17159–17163.
23. B. N. Hsiang, Y. J. Zhu, Z. Q. Wang, Y. L. Wu, V. Sasseville, W. P. Yang and T. G. Kirchgessner, *J. Biol. Chem.*, 1999, **274**, 37161–37168.
24. J. Konig, Y. H. Cui, A. T. Nies and D. Keppler, *Am. J. Physiol.*, 2000, **278**, G156–G164.
25. P. Matsson, B. C. Doak, B. Over and J. Kihlberg, *Adv. Drug Delivery Rev.*, 2016, **101**, 42–61.
26. P. Artursson, P. Matsson and M. Karlgren, in *Transporters in Drug Development*, ed. T. Sugiyama and B. Steffensen, Springer, 2014.
27. T. De Bruyn, S. Fattah, B. Stieger, P. Augustijns and P. Annaert, *J. Pharm. Sci.*, 2011, **100**, 5018–5030.
28. T. De Bruyn, G. J. van Westen, A. P. Ijzerman, B. Stieger, P. de Witte, P. F. Augustijns and P. P. Annaert, *Mol. Pharmacol.*, 2013, **83**, 1257–1267.
29. M. Karlgren, G. Ahlin, C. A. Bergstrom, R. Svensson, J. Palm and P. Artursson, *Pharm. Res.*, 2012, **29**, 411–426.
30. M. Karlgren, A. Vildhede, U. Norinder, J. R. Wisniewski, E. Kimoto, Y. Lai, U. Haglund and P. Artursson, *J. Med. Chem.*, 2012, **55**, 4740–4763.
31. US Department of Health and Human Services, Food and Drug Administration, Center for Drug Evaluation and Research, *In Vitro Metabolism- and Transporter-Mediated Drug-Drug Interaction Studies - Guidance for Industry*, 2017.

32. S. C. Sim and M. Ingelman-Sundberg, *Trends Pharmacol. Sci.*, 2011, **32**, 72–81.
33. P. Matsson, J. M. Pedersen, U. Norinder, C. A. Bergstrom and P. Artursson, *Pharm. Res.*, 2009, **26**, 1816–1831.
34. B. L. Urquhart, R. G. Tirona and R. B. Kim, *J. Clin. Pharmacol.*, 2007, **47**, 566–578.
35. M. J. Zamek-Gliszczynski, C. A. Lee, A. Poirier, J. Bentz, X. Chu, H. Ellens, T. Ishikawa, M. Jamei, J. C. Kalvass, S. Nagar, K. S. Pang, K. Korzekwa, P. W. Swaan, M. E. Taub, P. Zhao, A. Galetin and C. International Transporter, *Clin. Pharmacol. Ther.*, 2013, **94**, 64–79.
36. A. Kunze, J. Huwyler, G. Camenisch and B. Poller, *Drug Metab. Dispos.*, 2014, **42**, 1514–1521.
37. A. Vildhede, A. Mateus, E. K. Khan, Y. Lai, M. Karlgren, P. Artursson and M. C. Kjellsson, *Drug Metab. Dispos.*, 2016, **44**, 505–516.
38. Y. Guo, X. Chu, N. J. Parrott, K. L. R. Brouwer, V. Hsu, S. Nagar, P. Matsson, P. Sharma, J. Snoeys, Y. Sugiyama, D. Tatosian, J. D. Unadkat, S. M. Huang, A. Galetin and C. International Transporter, *Clin. Pharmacol. Ther.*, 2018, **104**, 865–889.
39. W. E. Evans and M. V. Relling, *Science*, 1999, **286**, 487–491.
40. J. A. Williams, R. Hyland, B. C. Jones, D. A. Smith, S. Hurst, T. C. Goosen, V. Peterkin, J. R. Koup and S. E. Ball, *Drug Metab. Dispos.*, 2004, **32**, 1201–1208.
41. European Medicines Agency, *Guideline on the Investigation of Drug Interactions*, 2013.
42. C. Wegler, F. Z. Gaugaz, T. B. Andersson, J. R. Wisniewski, D. Busch, C. Groer, S. Oswald, A. Noren, F. Weiss, H. S. Hammer, T. O. Joos, O. Poetz, B. Achour, A. Rostami-Hodjegan, E. van de Steeg, H. M. Wortelboer and P. Artursson, *Mol. Pharm.*, 2017, **14**, 3142–3151.
43. A. Kalliokoski and M. Niemi, *Br. J. Pharmacol.*, 2009, **158**, 693–705.
44. P. J. Neuvonen, M. Niemi and J. T. Backman, *Clin. Pharmacol. Ther.*, 2006, **80**, 565–581.
45. C. D. Furberg and B. Pitt, *Curr. Controlled Trials Cardiovasc. Med.*, 2001, **2**, 205–207.
46. J. A. Staffa, J. Chang and L. Green, *N. Engl. J. Med.*, 2002, **346**, 539–540.
47. Y. Shitara, M. Hirano, H. Sato and Y. Sugiyama, *J. Pharmacol. Exp. Ther.*, 2004, **311**, 228–236.
48. S. C. Group, E. Link, S. Parish, J. Armitage, L. Bowman, S. Heath, F. Matsuda, I. Gut, M. Lathrop and R. Collins, *N. Engl. J. Med.*, 2008, **359**, 789–799.
49. R. A. Wilke, L. B. Ramsey, S. G. Johnson, W. D. Maxwell, H. L. McLeod, D. Voora, R. M. Krauss, D. M. Roden, Q. Feng, R. M. Cooper-DeHoff, L. Gong, T. E. Klein, M. Wadelius and M. Niemi, *Clin. Pharmacol. Ther.*, 2012, **92**, 112–117.
50. L. B. Ramsey, S. G. Johnson, K. E. Caudle, C. E. Haidar, D. Voora, R. A. Wilke, W. D. Maxwell, H. L. McLeod, R. M. Krauss, D. M. Roden, Q. Feng, R. M. Cooper-DeHoff, L. Gong, T. E. Klein, M. Wadelius and M. Niemi, *Clin. Pharmacol. Ther.*, 2014, **96**, 423–428.

51. M. Tu, A. M. Mathiowetz, J. A. Pfefferkorn, K. O. Cameron, R. L. Dow, J. Litchfield, L. Di, B. Feng and S. Liras, *Curr. Top. Med. Chem.*, 2013, **13**, 857–866.
52. M. V. Varma, G. Chang, Y. Lai, B. Feng, A. F. El-Kattan, J. Litchfield and T. C. Goosen, *Drug Metab. Dispos.*, 2012, **40**, 1527–1537.
53. J. A. Pfefferkorn, J. Litchfield, R. Hutchings, X. M. Cheng, S. D. Larsen, B. Auerbach, M. R. Bush, C. Lee, N. Erasga, D. M. Bowles, D. C. Boyles, G. Lu, C. Sekerke, V. Askew, J. C. Hanselman, L. Dillon, Z. Lin, A. Robertson, K. Olsen, C. Boustany, K. Atkinson, T. C. Goosen, V. Sahasrabudhe, J. Chupka, D. B. Duignan, B. Feng, R. Scialis, E. Kimoto, Y. A. Bi, Y. Lai, A. El-Kattan, R. Bakker-Arkema, P. Barclay, E. Kindt, V. Le, J. W. Mandema, M. Milad, B. D. Tait, R. Kennedy, B. K. Trivedi and M. Kowala, *Bioorg. Med. Chem. Lett.*, 2011, **21**, 2725–2731.
54. C. Gui, B. Wahlgren, G. H. Lushington and B. Hagenbuch, *Pharmacol. Res.*, 2009, **60**, 50–56.
55. C. Chang, K. S. Pang, P. W. Swaan and S. Ekins, *J. Pharmacol. Exp. Ther.*, 2005, **314**, 533–541.
56. J. A. Pfefferkorn, A. Guzman-Perez, J. Litchfield, R. Aiello, J. L. Treadway, J. Pettersen, M. L. Minich, K. J. Filipski, C. S. Jones, M. Tu, G. Aspnes, H. Risley, J. Bian, B. D. Stevens, P. Bourassa, T. D'Aquila, L. Baker, N. Barucci, A. S. Robertson, F. Bourbonais, D. R. Derksen, M. Macdougall, O. Cabrera, J. Chen, A. L. Lapworth, J. A. Landro, W. J. Zavadoski, K. Atkinson, N. Haddish-Berhane, B. Tan, L. Yao, R. E. Kosa, M. V. Varma, B. Feng, D. B. Duignan, A. El-Kattan, S. Murdande, S. Liu, M. Ammirati, J. Knafels, P. Dasilva-Jardine, L. Sweet, S. Liras and T. P. Rolph, *J. Med. Chem.*, 2012, **55**, 1318–1333.
57. R. M. Oballa, L. Belair, W. C. Black, K. Bleasby, C. C. Chan, C. Desroches, X. Du, R. Gordon, J. Guay, S. Guiral, M. J. Hafey, E. Hamelin, Z. Huang, B. Kennedy, N. Lachance, F. Landry, C. S. Li, J. Mancini, D. Normandin, A. Pocai, D. A. Powell, Y. K. Ramtohul, K. Skorey, D. Sorensen, W. Sturkenboom, A. Styhler, D. M. Waddleton, H. Wang, S. Wong, L. Xu and L. Zhang, *J. Med. Chem.*, 2011, **54**, 5082–5096.
58. J. Fu, M. Tjandra, C. Becker, D. Bednarczyk, M. Capparelli, R. Elling, I. Hanna, R. Fujimoto, M. Furegati, S. Karur, T. Kasprzyk, M. Knapp, K. Leung, X. Li, P. Lu, W. Mergo, C. Miault, S. Ng, D. Parker, Y. Peng, S. Roggo, A. Rivkin, R. L. Simmons, M. Wang, B. Wiedmann, A. H. Weiss, L. Xiao, L. Xie, W. Xu, A. Yifru, S. Yang, B. Zhou and Z. K. Sweeney, *J. Med. Chem.*, 2014, **57**, 8503–8516.

CHAPTER 8

Bile Salt Export Pump (BSEP) Inhibition

ALEXANDER TREIBER[a] AND MARTIN H. BOLLI*[a]

[a]Departments of Nonclinical Drug Metabolism and Pharmacokinetics (AT) and Drug Discovery Chemistry (MB), Idorsia Pharmaceuticals Ltd, Hegenheimermattweg 91, 4123 Allschwil, Switzerland
*E-mail: martin.bolli@idorsia.com

8.1 Introduction

8.1.1 Enzyme Family

The bile salt export pump BSEP (in humans)/Bsep (in animals) is a member of the ATP-binding cassette (ABC) superfamily of transport proteins. Its systematic name is ABCB11. The human BSEP gene maps to chromosome 2q24 and codes for a protein with a mass of 150–170 kDa.[1]

8.1.2 Expression

BSEP is selectively located in the apical (canalicular) domain of hepatocytes and catalyzes the rate-limiting step in bile salt secretion from hepatocytes into the bile duct.[2]

8.1.3 Structure

BSEP is a membrane-bound protein. Like most other ABC transporters, BSEP is a homodimer with 12 transmembrane domains and two intracellular binding domains with Walker A and B motifs for binding and hydrolysis of ATP.[1]

Drug Discovery Series No. 79
The Medicinal Chemist's Guide to Solving ADMET Challenges
Edited by Patrick Schnider
© The Royal Society of Chemistry 2021
Published by the Royal Society of Chemistry, www.rsc.org

8.1.4 Activity

BSEP has been cloned and expressed from all species of interest for preclinical safety testing and humans.[1,3–5] On the basis of results from enzyme kinetic analysis using taurocholate as a prototypical bile salt, species differences in BSEP function are small – as expected in light of the fundamental role of the transport protein to maintain bile flow. BSEP function has been evolutionarily conserved, as shown by studies with Bsep of the marine skate *Raja erinacea*, a 200 million years old vertebrate.[6]

8.2 BSEP Function, Substrates and Inhibition

BSEP mediates the rate-limiting step in bile salt secretion from hepatocytes into the bile duct (Figure 8.1). High affinity transport has been demonstrated for conjugated monovalent bile acids like taurochenodeoxycholic acid, taurocholic acid and tauroursodeoxycholic acid (Figure 8.2).[4,5] Divalent bile acids are mainly transported by the multi-drug resistance-associated

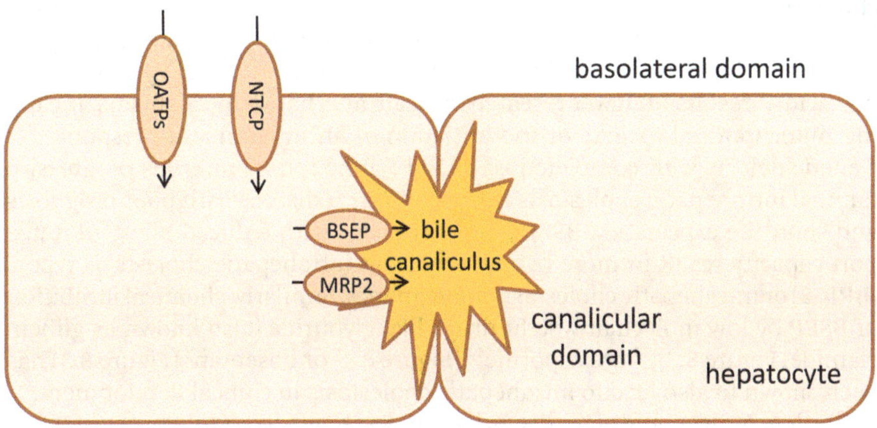

Figure 8.1 Localization of transport proteins involved in hepatic bile acid trafficking.

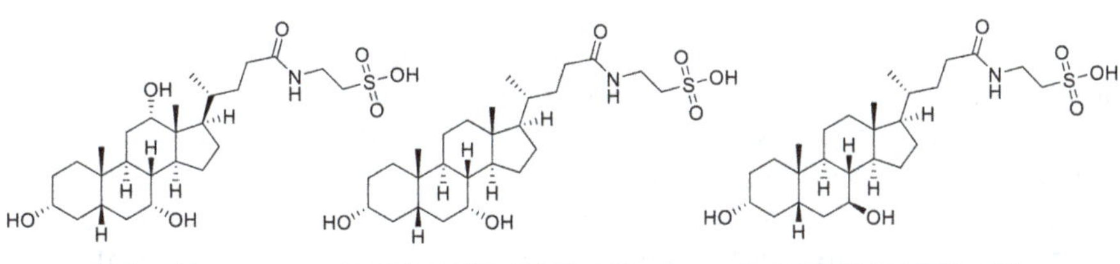

Figure 8.2 Structures of the monovalent bile acids taurocholic, taurochenodeoxycholic and tauroursodeoxycholic acid.

protein 2 (MRP2, ABCC2). The driving force for bile acid transport is the hydrolysis of ATP to ADP. The major transport protein for the import of bile acids from blood into hepatocytes is the sodium taurocholate co-transporting polypeptide (NTCP/SLC10A1). The organic anion transporting polypeptides (OATPs) also import bile salts from blood into hepatocytes, but their overall role in bile salt trafficking is limited.

Bile is a complex biological fluid mainly consisting of bile acids, cholesterol, phospholipids, conjugated bile pigments, inorganic electrolytes and water.[7] Bile acids are initially produced in the liver by cytochrome P450-mediated oxidation of cholesterol and then excreted into the gut lumen after conjugation with taurine or glycine. There they assist in the absorption of lipids *via* formation of micelles. Intestinal uptake of bile acids is mediated by the apical sodium-dependent bile salt transporter ASBT (SLC10A2) while the re-uptake into hepatocytes is mainly driven by the sodium taurocholate co-transporting polypeptide NTCP (SLC10A1). Enterohepatic recirculation of bile acids is highly efficient and a reflection of the expensive process of bile acid biosynthesis.

Disturbances in bile acid trafficking at the level of liver export is defined as cholestasis. Reduced BSEP activity, either by disease or chemical inhibition, leads to hepatocellular accumulation of bile acids and eventually liver injury. While liver injury was initially believed to result from direct cytotoxicity of bile acids,[8] results of newer research indicate alternative mechanisms, including mitochondrial toxicity or the initiation of an inflammatory response.[9,10] Genetic defects with complete loss in BSEP function manifest as progressive familial intrahepatic cholestasis type 2 (PFIC2), a disease with poor prognosis and short life expectancy. BSEP polymorphisms with reduced bile acid transport capacity result in more benign recurrent intrahepatic cholestasis type 2 (BRIC2) or intrahepatic cholestasis of pregnancy. Similarly, chemical inhibition of BSEP by low molecular weight drugs like glyburide (also known as glibenclamide, Figure 8.3),[11] cyclosporine[12] (Figure 8.4) or bosentan[8] (Figure 8.3) has been shown to also lead to intrahepatic cholestasis in clinical development.

Few low molecular weight drugs have been shown to be substrates of BSEP. The contribution of BSEP-mediated drug efflux to total drug clearance is currently unclear.[13]

8.3 Relevance

Inhibition of BSEP function by low molecular weight drugs may lead to intrahepatic cholestasis and eventually drug-induced liver injury (DILI).[8,14–17]

Though the mechanism and consequences of BSEP inhibition are well understood, a risk assessment on the basis of *in vitro* BSEP data alone is not straightforward. The magnitude of BSEP inhibition *in vivo* can only be estimated in conjunction with drug concentration data in the liver. Unbound plasma exposure can be used as a surrogate for drugs whose hepatic disposition is driven by passive diffusion. For drugs with active hepatic uptake, the degree of drug accumulation in the liver needs to be taken into account.

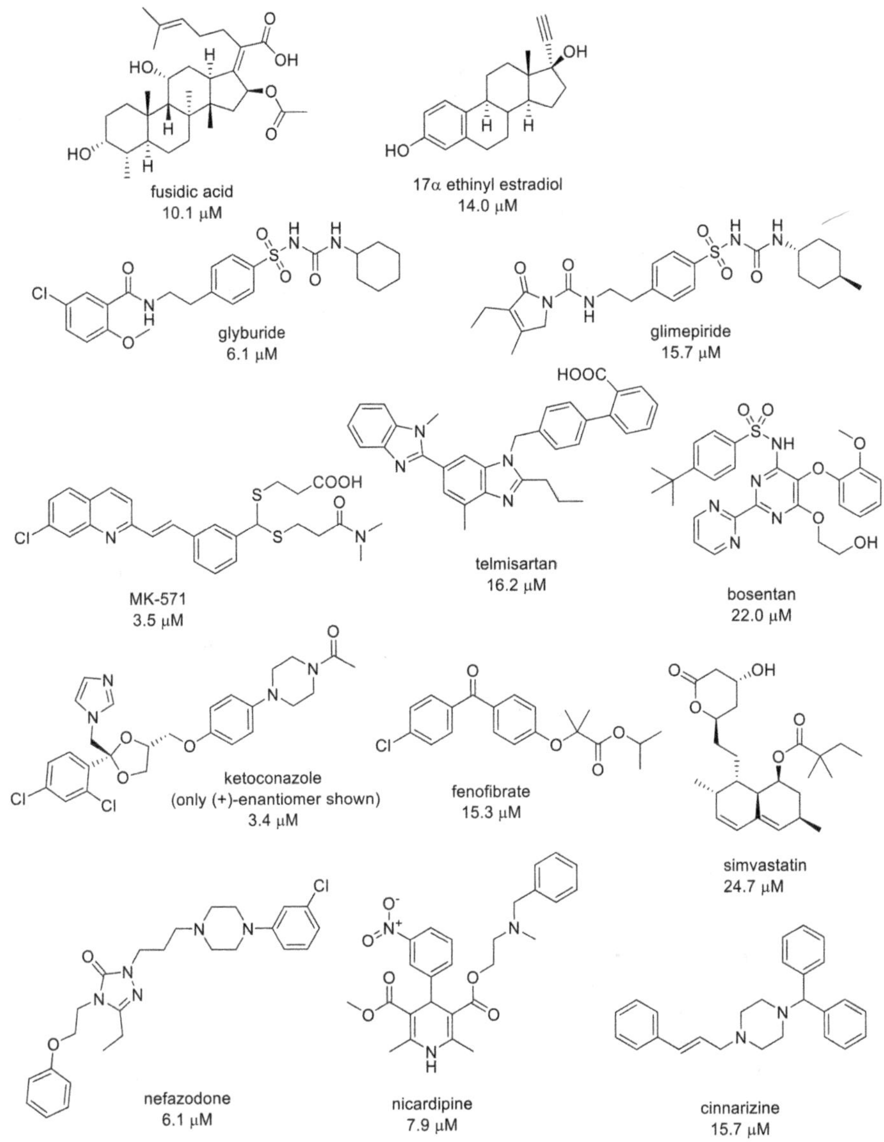

Figure 8.3 Structures of selected drugs with significant *in vitro* BSEP inhibition. IC_{50} values were measured using membrane vesicles from transiently transfected Sf9 insect cells.[15]

An illustrative example of the complexity of predicting the effects of *in vitro* BSEP inhibition on human safety are the two endothelin receptor antagonists bosentan and macitentan, both approved for the treatment of pulmonary arterial hypertension. Bosentan and macitentan inhibit BSEP *in vitro* to a similar extent, with concentrations giving 50% of maximum inhibition (IC_{50}) values of 25–77 μM and 18 μM, respectively, and also show similar plasma exposure in the low micromolar range. While macitentan does not

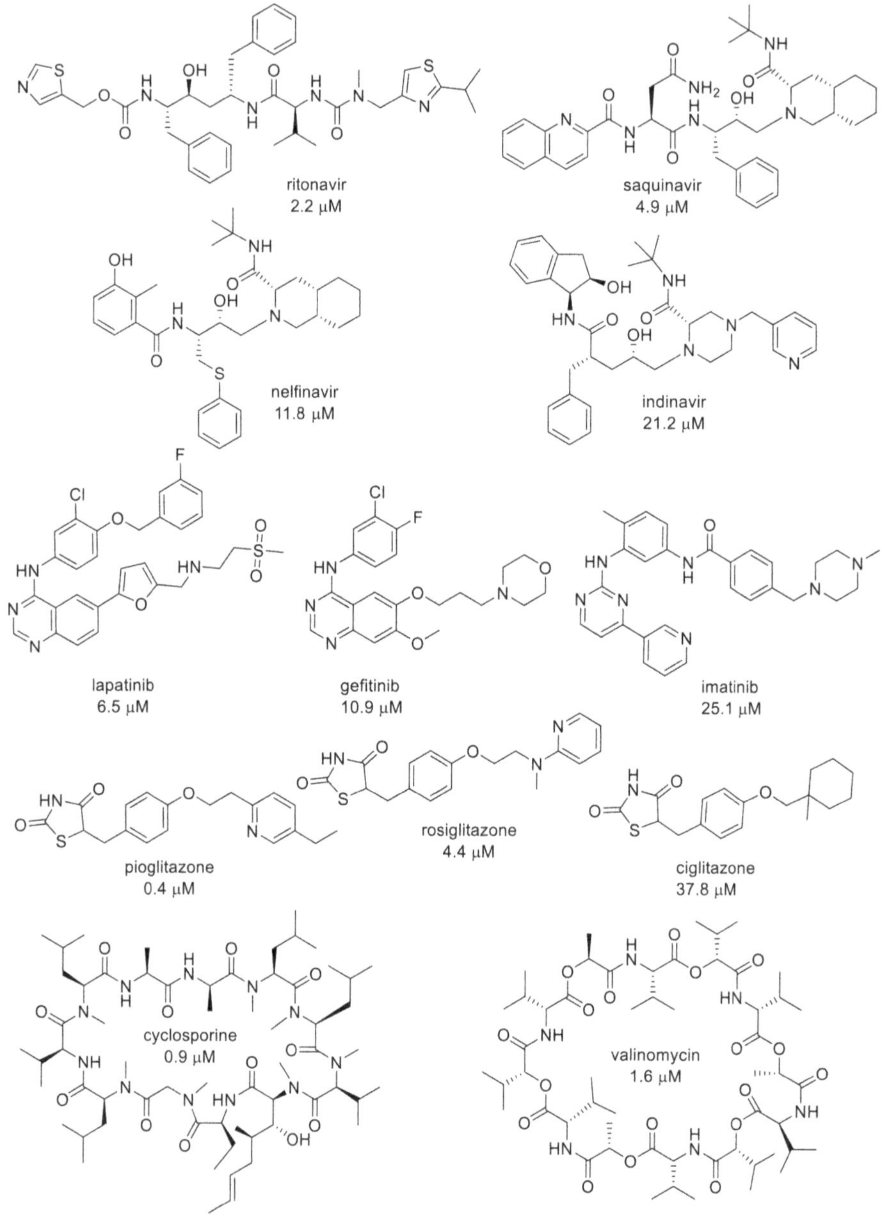

Figure 8.4 Structures of selected drugs with significant *in vitro* BSEP inhibition. IC_{50} values were measured using membrane vesicles from transiently transfected Sf9 insect cells.[15]

cause DILI,[18] bosentan has a proven track record for cholestasis.[8] The difference between the two endothelin receptor antagonists is their mechanism of drug distribution into liver. While the more lipophilic macitentan enters liver cells through passive diffusion,[18] bosentan is a substrate of OATP transporters and consequently accumulates in liver cells.[19]

As human pharmacokinetic (PK) data are often not available at the stage of drug discovery, the use of animal models looks appealing, in particular in light of the functional similarity of the human and animal transporters. For most drugs, differences in BSEP inhibition between the human and rat transport protein are within twofold, with few notable exceptions, such as pioglitazone and rosiglitazone.[15,16] However, rat and dog safety studies have been shown to be poorly predictive for a compound's potential to induce cholestasis or DILI in humans. Two main explanations have been provided to rationalize this species difference. Humans have a more lipophilic bile salt pool compared with rats and dogs,[20] and bile salt lipophilicity has been shown to positively correlate with hepatotoxicity.[21–23] Moreover, rodents, but not man, are known to metabolize bile salts by extensive cytochrome P450-mediated hydroxylation[24] which might serve as a compensatory mechanism for BSEP-mediated intrahepatic bile salt accumulation.

8.4 Screening Strategies

From a regulatory perspective there is no consistent guidance on the generation of BSEP data during drug development. While the current Food and Drug Administration (FDA) guidelines[25] on drug–drug interactions do not mention BSEP at all, the corresponding European Medicines Agency (EMA) guidelines[26] recommend the assessment of BSEP inhibition. The International Transporter Consortium recommends a proactive evaluation of BSEP inhibition during drug discovery and development to mechanistically assist liver safety assessment.[7]

From an experimental perspective, two complementary screening strategies have been established over the last two decades. A first, more holistic approach aims at assessing a compound's potential to interfere with bile salt transport *in vivo*, while in a more classical mechanistic approach the inhibition of human BSEP or rat Bsep is studied by means of dedicated *in vitro* assays using primary or transfected cells, or cell fragments containing the transport protein (Scheme 8.1).

In 2001, Fattinger *et al.*[8] performed rat studies to identify the potential mechanism underlying bosentan's hepatic adverse reactions. In the corresponding experiment, the cholestatic potential of bosentan was assessed by measuring serum bile salts after intravenous administration of the compound. The observed increase in serum bile salt concentrations led to the conclusion that bosentan's liver injury in humans can, at least in part, be explained by BSEP inhibition.

Similarly, Kostrubsky *et al.*[14] characterized the cholestatic effect of nefazodone (Figure 8.3), an antidepressant associated with clinical idiosyncratic hepatotoxicity and strong *in vitro* BSEP inhibition, by measuring serum bile salts after oral administration to rats.

During one of our drug discovery programs, the above approach was successfully used to identify the novel endothelin receptor antagonist macitentan devoid of bile salt transport interference in preclinical safety assessment,[18,27,28] and clinical trials.[29,30]

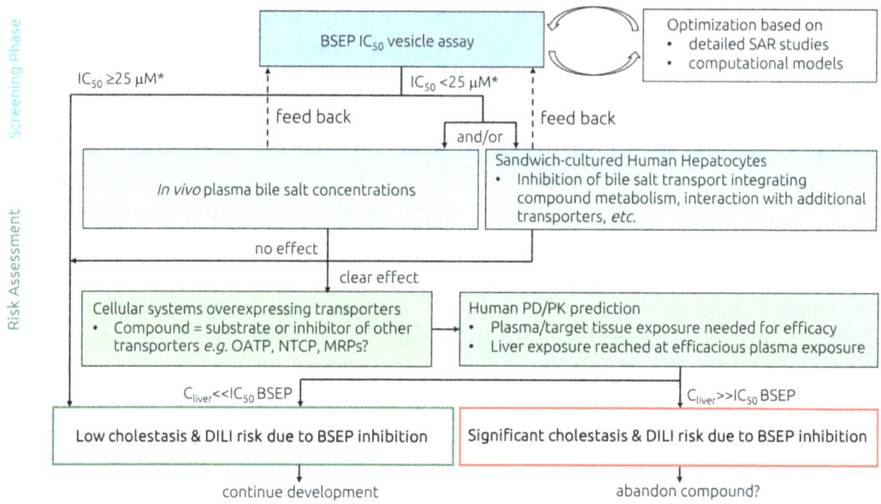

Scheme 8.1 Proposed workflow for the screening and subsequent risk assessment for cholestasis and DILI due to BSEP inhibition. Adapted from Kenna *et al.*[7] and based on our own experience.[18,19,28] †Although an IC_{50} value of 25 µM has been proposed by several groups[7,15] as a threshold for *in vitro* screening approaches, the *in vitro* IC_{50} values, as illustrated with the examples of bosentan and macitentan,[8,18,19] need to be put into perspective with liver exposure data.[7,33]

Thus, while the above *in vivo* approach may represent an attractive screening strategy as it allows for an assessment beyond a mere IC_{50} value on the BSEP transporter, mechanistic conclusions need to be drawn with care in light of the fact that all individual steps in bile salt trafficking are dependent on transport proteins (see Section 8.3 Relevance).

In parallel to the above *in vivo* studies, several *in vitro* assays have been established to assess a compound's potential to interfere with rat Bsep or human BSEP.[7,15,17,31] BSEP inhibition of a compound can be assessed using either membrane vesicles from Sf9 or Sf21 insect cells expressing recombinant BSEP, or mammalian cell lines overexpressing BSEP. Today, the corresponding assays have become the method of choice for the assessment of BSEP interaction, as they are relatively easy to perform and allow for high-throughput measurements.

Primary hepatocytes cultured in a sandwich configuration,[32] *i.e.* in between two layers of collagen, foster the development of hepatobiliary canaliculi in between two hepatocytes. This set-up enables experiments on bile salt excretion to be performed in a physiologically more relevant setting, taking into account hepatic drug uptake, passive diffusion in and out of liver cells, drug metabolism and drug efflux.

8.5 Examples of BSEP Inhibition

Morgan *et al.*[15] used insect cell membrane vesicles containing BSEP to assess the inhibitory potency of more than 200 approved drugs. While about 16% of these compounds showed potent BSEP inhibition ($IC_{50} \leq 25$ µM), 9% showed moderate (25 μM < IC_{50} ≤ 100 μM) and 75% showed no interaction with BSEP (IC_{50} > 100 μM).

A selection of compounds with significant BSEP inhibition is shown in Figures 8.3 and 8.4. The structures compiled in these two figures illustrate that there is no obvious common structural feature that is responsible for high affinity to the transporter. The steroid derivatives fusidic acid and 17α-ethinyl estradiol (Figure 8.3) both bear structural similarity to the natural substrates of BSEP (see Figure 8.2), and their ability to block this transporter may therefore not come as a surprise. Hence, one is tempted to speculate that an acidic function attached to a rather lipophilic moiety represents a pharmacophore that is recognized by the transporter. This would then explain the affinity of (weak) lipophilic acids like glyburide, glimepiride, MK-571, telmisartan, and possibly bosentan (Figure 8.3). On the other hand, ketoconazole, fenofibrate[†] and simvastatin[†], all lacking an acidic function, exhibit similar IC_{50} values for BSEP. Furthermore, nefazodone, nicardipine[†] and cinnarizine are basic amines and still bind to BSEP with low micro-molar affinity (Figure 8.3).

The structures compiled in Figure 8.4 further illustrate the structural diversity of drugs that show significant BSEP inhibition *in vitro*. The four HIV protease inhibitors selected in Figure 8.4 demonstrate that rather large structural variations of a given scaffold may not remove the BSEP liability completely. Even though the structural similarity of, for instance, ritonavir and saquinavir is only 0.62 (Tanimoto coefficient based on SkelSpheres fingerprint calculation using DataWarrior[34]), the difference in their affinity to BSEP is only about twofold. Similarly, the three kinase inhibitors lapatinib, gefitinib and imatinib show significant structural differences, yet their inhibitory potency on BSEP differs only by a factor of four. Conversely, with their thiazolidine-dione heads attached to rather lipophilic phenyl-ether tails, the three glitazones in Figure 8.4 appear rather similar in structure (similarity 0.78–0.82). However, their affinity for BSEP is quite different and ciglitazone is almost 100-fold less potent than pioglitazone. Finally, the two macrocyclic peptides cyclosporine and valinomycin share no similarity at all with the natural substrates, yet they are the most potent BSEP inhibitors in the above study.

In summary, many pharmacologically active and structurally very diverse molecules show inhibitory affinities to BSEP in the low micromolar range. In view of the physiological function of BSEP, namely the transport of bile salts from liver cells into bile, the large diversity of structures that interact with BSEP is surprising. Therefore, guidance on how to remove this undesired affinity to BSEP by a rational design approach is most desired by the medicinal chemist.

[†]Although Morgan *et al.* do not comment on compound stability during the BSEP membrane vesicle assay, we assume that the esters present in these molecules remained intact during the 25 min assay time.

8.6 Mitigation Strategies

In the previous section, we have seen that it is very difficult to pinpoint obvious structural elements that lead to BSEP inhibition. To give the medicinal chemist guidance on how to design compounds with minimal BSEP inhibition, Warner *et al.*[35] built an *in silico* model for the prediction of BSEP inhibition. To this end, BSEP inhibition was quantified for a total of 624 compounds using vesicles of Sf21 insect cells containing the human transporter. While 115 compounds showed an IC_{50} value <10 μM, 269 compounds showed 10 μM ≤ IC_{50} ≤ 1000 μM, and 240 compounds were inactive (IC_{50} > 1000 μM). To build their *in silico* BSEP inhibition model, Warner *et al.* categorized the compounds as active (IC_{50} < 300 μM) or inactive (IC_{50} ≥ 300 μM). The data set was randomly split into training (437 compounds) and test (187 compounds) sets. A first model was based on molecular weight and calculated lipophilicity of the compounds. To obtain more refined models, the structure–activity relationships (SARs) were analyzed using two- and three-dimensional descriptors for molecular size, lipophilicity, hydrogen bonding, electrostatics and topology calculated with a proprietary program, or by applying descriptor sets comprising combinations of finger-print based descriptors encoding *e.g.* functional groups and molecular bulk properties. Partial least-square and two non-linear machine learning methods were used to build these classification models.

From these analyses, the authors concluded that BSEP inhibition is significantly driven by the molecular weight and the lipophilicity of a given compound. In particular, for compounds with a molecular weight >309 the lipophilicity (clogP) is the most important parameter influencing BSEP inhibitory potency, and a clogP >2 increases the likelihood for a compound to inhibit BSEP significantly (>90%). The authors also observed that more basic compounds and compounds with more hydrogen bond donors tended to be less potent inhibitors. In addition, a reduced incidence of cationic and zwitterionic compounds in the category of molecules classified as 'BSEP active' (*i.e.* IC_{50} < 300 μM) was noted.

More recently, two additional descriptions of computational approaches enabling *in silico* screening for BSEP inhibition have been published.[36,37] Based on the structure of the mouse P-glycoprotein, Jain *et al.*[36] built a homology model for BSEP. A subset of known inhibitors (IC_{50} ≤ 10 μM) and non-inhibitors (IC_{50} ≥ 300 μM) of BSEP was selected from the compounds compiled by Warner *et al.*[35] and used as a training set for docking studies. This allowed for the classification of a test set into inhibitors and non-inhibitors with an accuracy of 73%. On the basis of results from a statistical analysis, certain structural features, like halides, ether, carbonyl and vinyl carbons, as well as amide groups, appeared to be associated with BSEP inhibition, indicating hydrophobicity-driven protein–inhibitor interactions.

Xi *et al.*[37] followed a ligand-based approach to construct a model for the prediction of BSEP inhibition. The authors employed a random forest algorithm ranking structural descriptors and complemented this approach with a pharmacophore model that was produced with the so-called HipHop method. As in the work by Jain *et al.*,[36] the data set published by Warner *et al.*[35] served as a basis to train the model, however, with a different categorization. Compounds with IC_{50} values ≥300 µM were classified as inactive while compounds with IC_{50} values <300 µM were considered as positive samples. The structures were minimized in energy before they entered the model generation process. Rigorous validation tests revealed that both the random forest and the pharmacophore approach delivered reliable models. Hydrophobicity and molecular size emerged as key factors influencing the BSEP inhibition potential of a compound.

Unfortunately, literature on more detailed SAR studies that would enable the medicinal chemist to reduce the BSEP inhibition potential of a compound by specific, targeted structural changes is very scarce. As part of their computational approach to analyse the SAR of BSEP inhibition, Warner *et al.* included a matched molecular pair analysis. The authors reported a few molecular pairs for which the measured IC_{50} value did not match with the prediction of their machine learning approach. These 'rule breakers' are interesting as they may hint at specific structural features that are recognized by BSEP. Two examples are shown Figure 8.5. For proxicromil, removal of the 5-hydroxy group reduces the affinity for BSEP about 20-fold, indicating that this group is important for the compound's interaction with BSEP. Similarly, replacing the indole moiety in mepindolol by a naphthyl group to give propanolol, significantly attenuated the compound's

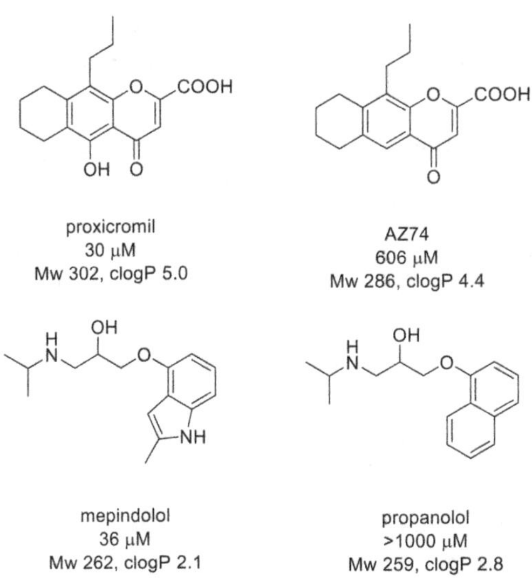

Figure 8.5 Matched molecular pairs with IC_{50} values for BSEP illustrating potential recognition motifs, selected from Warner *et al.*[35]

affinity for the transporter, suggesting that the indole is recognized by the transporter.

In conclusion, results from computational SAR analyses indicate that molecular and physicochemical properties, such as molecular weight, lipophilicity and ionic state, influence the inhibitory potency of a compound on BSEP. Unfortunately, however, key structural elements or specific functional groups that are common motifs of BSEP inhibitors have not been identified and remain elusive.

Key References

- J. G. Kenna, K. S. Taskar, C. Battista, D. L. Bourdet, K. L. R. Brouwer, K. R. Brouwer, D. Dai, C. Funk, M. J. Hafey, Y. Lai, J. Maher, Y. A. Pak, J. M. Pedersen, J. W. Polli, A. D. Rodrigues, P. B. Watkins, K. Yang, R. W. Yucha and International Transporter Consortium, *Clin. Pharmacol. Ther.*, 2018, **104**, 916–932.
    - *Describes a detailed workflow of how to generate in vitro BSEP inhibition data, interpret and mitigate BSEP inhibition, and discusses cut-off values for decision making in drug discovery.*
- K. Fattinger, C. Funk, M. Pantze, C. Weber, J. Reichen, B. Stieger and P. J. Meier, *Clin. Pharmacol. Ther.*, 2001, **69**, 223–231.
    - *Describes a possible mechanistic explanation for the reversible liver injury observed in patients treated with bosentan using in vitro BSEP inhibition data as well as a rat model of cholestasis.*
- M. Trauner and J. L. Boyer, *Physiol. Rev.*, 2003, **83**, 633–671.
    - *A comprehensive summary of the molecular characterization, function, and regulation of bile salt transporters in normal physiology, cholestatic liver disease and liver regeneration.*
- R. E. Morgan, M. Trauner, C. J. van Staden, P. H. Lee, B. Ramachandran, M. Eschenberg, C. A. Afshari, C. W. Qualls Jr, R. Lightfoot-Dunn and H. K. Hamadeh, *Toxicol. Sci.*, 2010, **118**, 485–500.
    - *Describes the in vitro BSEP inhibition and effects on the liver in humans of a large number of drugs studied in clinical trials, and comments on the clinical relevance of in vitro data.*
- S. Dawson, S. Stahl, N. Paul, J. Barber and J. G. Kenna, *Drug Metab. Dispos.*, 2012, **40**, 130–138.
    - *An analysis with 85 drugs on human and rat BSEP to investigate whether inhibition of bile salt trafficking correlates with drug-induced liver toxicity (DILI).*
- D. J. Warner, H. Chen, L. D. Cantin, J. G. Kenna, S. Stahl, C. L. Walker and T. Noeske, *Drug Metab. Dispos.*, 2012, **40**, 2332–2341.
    - *Describes physicochemical properties and structural features associated with BSEP inhibition.*

References

1. M. Arrese and M. Ananthanarayanan, *Pflügers Arch.*, 2004, **449**, 123–131.
2. T. Gerloff, B. Stieger, B. Hagenbuch, J. Madon, L. Landmann, J. Roth, A. F. Hofmann and P. J. Meier, *J. Biol. Chem.*, 1998, **273**, 10046–10050.
3. R. M. Green, F. Hoda and K. L. Ward, *Gene*, 2000, **241**, 117–123.
4. J. A. Byrne, S. S. Strautnieks, G. Mieli-Vergani, C. F. Higgins, K. J. Linton and R. J. Thompson, *Gastroenterology*, 2002, **123**, 1649–1658.
5. J. Noe, B. Stieger and P. J. Meier, *Gastroenterology*, 2002, **123**, 1659–1666.
6. S. Y. Cai, L. Wang, N. Ballatori and J. L. Boyer, *Am. J. Physiol.: Gastrointest. Liver Physiol.*, 2001, **281**, G316–G322.
7. J. G. Kenna, K. S. Taskar, C. Battista, D. L. Bourdet, K. L. R. Brouwer, K. R. Brouwer, D. Dai, C. Funk, M. J. Hafey, Y. Lai, J. Maher, Y. A. Pak, J. M. Pedersen, J. W. Polli, A. D. Rodrigues, P. B. Watkins, K. Yang, R. W. Yucha and C. International Transporter, *Clin. Pharmacol. Ther.*, 2018, **104**, 916–932.
8. K. Fattinger, C. Funk, M. Pantze, C. Weber, J. Reichen, B. Stieger and P. J. Meier, *Clin. Pharmacol. Ther.*, 2001, **69**, 223–231.
9. K. Yang, C. Guo, J. L. Woodhead, R. L. St Claire 3rd, P. B. Watkins, S. Q. Siler, B. A. Howell and K. L. R. Brouwer, *J. Pharm. Sci.*, 2016, **105**, 443–459.
10. S. Y. Cai, X. Ouyang, Y. Chen, C. J. Soroka, J. Wang, A. Mennone, Y. Wang, W. Z. Mehal, D. Jain and J. L. Boyer, *JCI Insight*, 2017, **2**, e90780.
11. V. N. Tholakanahalli, A. Potti and M. F. Heyworth, *West. J. Med.*, 1998, **168**, 274–277.
12. B. Le Thai, M. Dumont, A. Michel, S. Erlinger and D. Houssin, *Transplant. Proc.*, 1987, **19**, 4149–4151.
13. M. Trauner and J. L. Boyer, *Physiol. Rev.*, 2003, **83**, 633–671.
14. S. E. Kostrubsky, S. C. Strom, A. S. Kalgutkar, S. Kulkarni, J. Atherton, R. Mireles, B. Feng, R. Kubik, J. Hanson, E. Urda and A. E. Mutlib, *Toxicol. Sci.*, 2006, **90**, 451–459.
15. R. E. Morgan, M. Trauner, C. J. van Staden, P. H. Lee, B. Ramachandran, M. Eschenberg, C. A. Afshari, C. W. Qualls Jr., R. Lightfoot-Dunn and H. K. Hamadeh, *Toxicol. Sci.*, 2010, **118**, 485–500.
16. S. Dawson, S. Stahl, N. Paul, J. Barber and J. G. Kenna, *Drug Metab. Dispos.*, 2012, **40**, 130–138.
17. S. Chatterjee and P. Annaert, *Curr. Drug Metab.*, 2018, **19**, 808–818.
18. A. Treiber, P. Aanismaa, R. de Kanter, S. Delahaye, M. Treher, P. Hess and P. Sidharta, *J. Pharmacol. Exp. Ther.*, 2014, **350**, 130–143.
19. A. Treiber, R. Schneiter, S. Hausler and B. Stieger, *Drug Metab. Dispos.*, 2007, **35**, 1400–1407.
20. D. M. Heuman, *J. Lipid Res.*, 1989, **30**, 719–730.
21. A. F. Attili, M. Angelico, A. Cantafora, D. Alvaro and L. Capocaccia, *Med. Hypotheses*, 1986, **19**, 57–69.
22. N. M. Delzenne, P. B. Calderon, H. S. Taper and M. B. Roberfroid, *Toxicol. Lett.*, 1992, **61**, 291–304.
23. M. J. Perez and O. Briz, *World J. Gastroenterol.*, 2009, **15**, 1677–1689.

24. S. Takahashi, T. Fukami, Y. Masuo, C. N. Brocker, C. Xie, K. W. Krausz, C. R. Wolf, C. J. Henderson and F. J. Gonzalez, *J. Lipid Res.*, 2016, **57**, 2130–2137.
25. FDA Food and Drug Administration, *US Department of Health and Human Services Food and Drug Administration Center for Drug Evaluation and Research (CDER)*, January 2020.
26. EMA European Medicines Agency, *Committee for Human Medicinal Products (CHMP)*, 2012, CPMP/EWP/560/95/Rev. 1.
27. M. H. Bolli, C. Boss, C. Binkert, S. Buchmann, D. Bur, P. Hess, M. Iglarz, S. Meyer, J. Rein, M. Rey, A. Treiber, M. Clozel, W. Fischli and T. Weller, *J. Med. Chem.*, 2012, **55**, 7849–7861.
28. M. H. Bolli, C. Boss, J. Gatfield, M. Iglarz and A. Treiber, in *Comprehensive Medicinal Chemistry III, Vol. 8*, ed. S. Chackalamannil, D. P. Rotella and S. E. Ward, Elsevier, Oxford, 2017, vol. 8, pp. 252–283.
29. P. N. Sidharta, P. L. M. van Giersbergen, A. Halabi and J. Dingemanse, *Eur. J. Clin. Pharmacol.*, 2011, **67**, 977–984.
30. P. N. Sidharta, P. L. van Giersbergen and J. Dingemanse, *J. Clin. Pharmacol.*, 2013, **53**, 1131–1138.
31. Y. Cheng, T. F. Woolf, J. Gan and K. He, *Chem.-Biol. Interact.*, 2016, **255**, 23–30.
32. O. Fardel, A. Moreau, M. Le Vee, C. Denizot and Y. Parmentier, *Eur. J. Drug Metab. Pharmacokinet.*, 2019, **44**, 13–30.
33. M. D. Aleo, D. Davie and A. S. Kalgutkar, in *Medicinal Chemistry Reviews*, 2017, vol. 52, ch. 24, pp. 465–481.
34. T. Sander, *DataWarrior*, 2019, http://www.openmolecules.org/datawarrior/features.html.
35. D. J. Warner, H. Chen, L. D. Cantin, J. G. Kenna, S. Stahl, C. L. Walker and T. Noeske, *Drug Metab. Dispos.*, 2012, **40**, 2332–2341.
36. S. Jain, M. Grandits, L. Richter and G. F. Ecker, *J. Comput.-Aided Mol. Des.*, 2017, **31**, 507–521.
37. L. Xi, J. Yao, Y. Wei, X. Wu, X. Yao, H. Liu and S. Li, *Mol. BioSyst.*, 2017, **13**, 417–424.

CHAPTER 9

Cytochrome P450 Metabolism

ANTONIA F. STEPAN*[a] AND R. SCOTT OBACH[b]

[a]Pharma Research and Early Development, F. Hoffmann-La Roche Ltd, Grenzacherstrasse 124, 4070 Basel, Switzerland; [b]Pfizer Worldwide Research and Development, Eastern Point Road, Groton, Connecticut 06340, USA
*E-mail: Antonia.Stepan@roche.com, r.scott.obach@pfizer.com

9.1 Introduction

9.1.1 Enzyme Family

Cytochrome P450 (CYP) comprises a large family of enzymes, some of which are responsible for specific metabolic transformations of endogenous substrates and others of which are involved in the metabolism of xenobiotics. Cytochrome P450 is a heme containing protein. The name derives from a characteristic spectrum wherein the enzyme is reduced and bound to carbon monoxide, yielding a wavelength of maximum absorbance (λ_{max}) of 450 nm[1]; "P" refers to pigment. This spectrum is unique among the heme containing proteins and is caused by the fact that the fifth ligand axial to the heme iron is a cysteine thiol. Among the more than 40 human CYP enzymes, the ones most important in drug metabolism include CYP1A2, CYP2A6, CYP2B6, CYP2C8, CYP2C9, CYP2C19, CYP2D6, CYP2E1, CYP3A4 and CYP3A5.[2] Laboratory animal species also possess CYP enzymes of varying degrees of structural relatedness to the human enzymes, but they are not identical and can exhibit different drug metabolizing activities (substrate selectivity, rates of reaction and regiochemical sites of biotransformation).

Drug Discovery Series No. 79
The Medicinal Chemist's Guide to Solving ADMET Challenges
Edited by Patrick Schnider
© The Royal Society of Chemistry 2021
Published by the Royal Society of Chemistry, www.rsc.org

9.1.2 Expression

Xenobiotic metabolizing CYP enzymes are expressed in many tissues but are found at the highest amounts in the liver. CYP enzymes in some other tissues, such as intestine,[3] kidney,[4] and lung,[5] can make meaningful contributions to drug metabolism. Mammalian drug-metabolizing CYP enzymes are membrane bound and present in the endoplasmic reticulum (ER). Upon homogenization, the ER forms spherical vesicles called microsomes, and these are commonly utilized in drug metabolism experiments to study CYP activities. In human liver microsomes (HLM), the level of the sum of CYP enzymes is approximately 0.3 nmoles per mg protein.[6] Of that, CYP3A4 and CYP2C9 represent at least half. Several of the individual human CYP enzymes are subject to genetic polymorphisms, which can render enzymes inactive or partially active, alter substrate specificities or cause ultrarapid metabolism. Enzymes with genetic polymorphisms that have sizeable subpopulations, many of which are somewhat associated with ethnicity, include CYP2D6, CYP2C9, CYP2C19 and CYP3A5.[7]

9.1.3 Structure

CYP enzymes have molecular weights (MWs) of approximately 50 kDa. All contain a heme wherein the two axial ligands for the heme iron are a cysteine residue on one side and a water molecule on the other (Figure 9.1). X-ray crystal structures of several of the human CYP enzymes have been solved.[8–12] They have domains that bind to CYP oxidoreductase (OR), another membrane-bound enzyme that is responsible for feeding reducing equivalents to CYP.

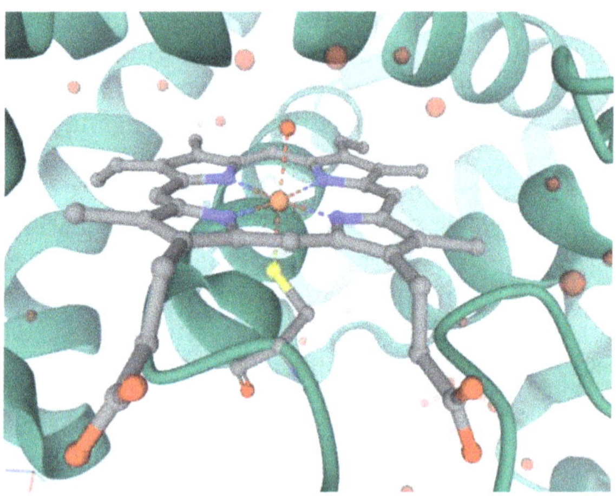

Figure 9.1 X-ray crystallography structure of human microsomal cytochrome P450 3A4 (carbons in grey, nitrogens in blue, iron in orange, sulfur in yellow, oxygens in red, water molecules in light red, heme, backbone shown in green).[9,13–17]

9.1.4 Activity

CYP activities are highly substrate-dependent. In general, laboratory animal species, such as rodents, have greater quantities of CYP enzymes and activities than humans.[18]

9.1.5 Function and Substrates

CYP catalyzes several different reaction types to a wide variety of substituents on small organic molecules. The mechanism is common (see below) and the ultimate reaction catalyzed depends on the chemical substituent being acted upon. It can catalyze aliphatic and aromatic hydroxylations, *N*-, *O*- and *S*-dealkylations, *N*- and *S*-oxygenations, epoxidation of alkene and arenes, among others (Table 9.1). Substrate specificities are extremely promiscuous, but some substrate types are more associated with some of the specific CYPs.[19,20] CYP3A4 has the widest substrate specificity and can metabolize organic molecules ranging in MW from 200 to 1500 Daltons of moderate

Table 9.1 Common drug metabolism reactions catalyzed by cytochrome P450 enzymes.[21]

Drug substituent	Reaction	Classic examples in drug metabolism
Alkane	Alkane → Alcohol	Testosterone 6β-hydroxylation Midazolam 1′-hydroxylation Tolbutamide 4-hydroxylation
Alkane	Alkane → Alkene	Valproic acid 4-dehydrogenation
Alkene	Alkene → Epoxide	Secobarbital epoxidation
Ether	Alkyl Ether → Alcohol + Alkanal	Dextromethorphan *O*-demethylation
	Aryl Ether → Phenol + Alkanal	Phenacetin *O*-deethylation
Amine	1° Amine → Ammonia + Alkanal	Sertraline *N*-demethylation
	2° Amine → 1° Amine + Alkanal	Diltiazem *N*-demethylation
	3° Amine → 2° Amine + Alkanal	Amodiaquine *N*-deethylation
Amine	1° or 2° Amine → Hydroxylamine	Dapsone *N*-hydroxylation
	3° Amine → *N*-Oxide	Olanzapine *N*-oxidation
Amine	2° Alicyclic Amine → Imine	Nifedipine dehydrogenation
	3° Alicyclic Amine → Iminium Ion	Nicotine 5′-dehydrogenation
Alcohol	Alcohol → Alkanal	Ethanol dehydrogenation
Thioether	Thioether → *S*-Oxide	Thioridazine *S*-oxidation
S-Oxide	*S*-Oxide → Sulfone	Omeprazole *S*-oxidation
Aromatic	Phenyl → Phenol	Propranolol 4-hydroxylation Diclofenac 4′-hydroxylation Mephenytoin 4-hydroxylation
Aromatic	Phenyl → Epoxide → Dihydrodiol	Phenytoin oxidation
Azaaromatic	Azaaromatic→ Azaaromatic *N*-Oxide	Lamotrigine *N*-oxidation
Azaaromatic	Azaaromatic→ Lactam	Pindolol 2-oxidation

to high lipophilicity. CYP2C8 and 2C9 are most frequently associated with the metabolism of large and small anionic organic molecules, respectively, and CYP2D6 is commonly associated with the metabolism of cationic drugs. CYP1A2 is associated with possessing a substrate selectivity for relatively planar aromatic compounds. CYP2A6 and CYP2E1 are known to metabolize small organic molecules, generally of lower MW than 200 Daltons. Substrate specificities of CYP2B6 and 2C19 are less descript.

9.1.6 Mechanism

The catalytic cycle of CYP enzymes is complex.[21] The enzyme starts from a basal low spin state of low redox potential where the iron is in the +3 oxidation state and is bound by a water molecule. The catalytic cycle commences with the displacement of the water with a substrate, which converts the Fe^{3+} to a high spin state of a redox potential of about −160 mV. This binding can be observed spectrally and is referred to as a Type I binding spectrum. With the increased redox potential, the heme iron can now accept an electron, derived from reduced nicotinamide adenine dinucleotide phosphate (NADPH) and delivered by CYP OR. The Fe^{2+} center can now bind molecular oxygen to form an iron–peroxo species. A second electron is introduced (from either OR per cytochrome b_5) which leads to scission of the oxygen peroxo bond to yield water and the highly active iron–oxo species (represented by either Fe^{4+}–O• or Fe^{5+}=O). This is the species that will react with the substrate. In aliphatic hydroxylation and *O*-dealkylation reactions, a hydrogen atom is abstracted, bonds to the iron–oxygen and the hydroxyl radical reacts with the carbon-centered radical from where the H atom was abstracted ("radical rebound") to yield the hydroxyl product. For *O*-dealkylations, the hemiacetal product decomposes to the alcohol and aldehyde. For *N*-dealkylations, the Fe^{4+}–O• abstracts an electron from the lone pair of the amine to form an aminium cation that decomposes to the *N*-dealkylated products either through deprotonation to an iminium ion or carbinolamine intermediate. For arene oxidations, an electron is abstracted from the ring and the iron bound oxygen radical adds across the C–C bond to yield an initial epoxide. A 1,2-hydride shift of the arene oxide results in the phenol product.

In some cases, one of the iron-peroxo intermediate species can decompose back to the resting enzyme, releasing unreacted substrate and a reactive oxygen species (peroxide, peroxide anion or hydroxyl radical, Scheme 9.1).[21] This is referred to as uncoupling. In other cases, an organic molecule can bind to the resting low spin Fe^{3+} state but a lone pair of electrons on the substrate (like an amine or an imidazole) directly binds to the iron and keeps it in the low spin, low redox state. This is an inactive complex (referred to as type II binding) and results in enzyme inhibition.[22]

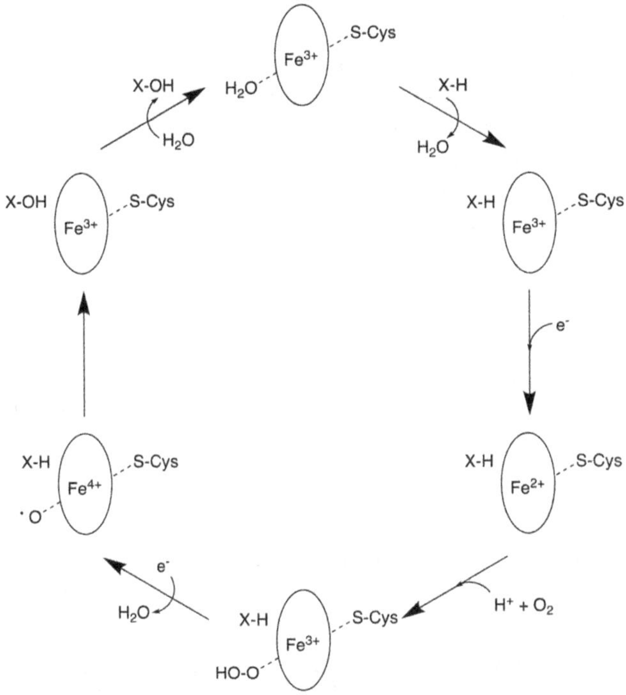

Scheme 9.1 Proposed catalytic cycle of CYP activity. The circles represent the active site porphyrin.

9.1.7 Screening Strategies

Newly synthesized drug candidates are analyzed for their potential as substrates and inhibitors for CYP enzymes. Most commonly, lability is measured in HLM supplemented with NADPH or an NADPH generation system to support activity. The loss of substrate over time is measured by high performance liquid chromatography–mass spectrometry (HPLC–MS). In some cases, heterologously expressed individual CYP enzymes are used to measure lability, to aid in the identification of which specific enzymes are most important in metabolism of a compound. Lability values can be converted into intrinsic clearance values, which, in turn, can be scaled to the whole organism. In humans, it is assumed that there are approximately 45 mg liver microsomes per gram liver and that there is 20 grams of liver per kg body weight. Intrinsic clearance values can be combined with plasma free fraction and liver blood flow to make estimates of clearance.[23] While metabolic lability assays that use human liver microsomes have been used for decades, more recently commercially available quality pooled human hepatocytes have become available. This reagent offers a broader complement of drug clearance mechanisms [*e.g.* other drug metabolizing enzyme families like UDP-glucuronosyltransferases (UGTs) as well as uptake transporters] and, thus, can accommodate compounds that may have other clearance pathways besides P450 metabolism.[24–26]

9.1.8 Relevance

Among all drug metabolizing enzymes, the CYP family is by far the most important.[27] They are highly relevant for metabolic clearance, drug–drug interactions,[28] and interpatient variability in pharmacokinetics.[29,30] An understanding of the sites of metabolism and the bioorganic mechanisms that give rise to these metabolites permit the design of analogues with reduced metabolic clearance. Strategies to slow or block CYP catalyzed metabolism are described later in this chapter.

Key References

- M. B. Fisher, K. R. Henne and J. Boer, The complexities inherent in attempts to decrease drug clearance by blocking sites of CYP-mediated metabolism, *Curr. Opin. Drug Discovery Dev.*, 2006, **9**, 101–109.
- D. F. Lewis and M. Dickins, Substrate SARs in human P450s, *Drug Discovery Today*, 2002, 7, 918–925.
- F. P. Guengerich, Common and uncommon cytochrome P450 reactions related to metabolism and chemical toxicity, *Chem. Res. Toxicol.*, 2001, **14**, 611–650.
- U. Zanelli, N. P. Caradonna, D. Hallifax, E. Turlizzi and J. B. Houston, Comparison of cryopreserved HepaRG cells with cryopreserved human hepatocytes for prediction of clearance for 26 drugs, *Drug Metab. Dispos.*, 2012, **40**, 104–110.
- L. Di, B. Feng, T. C. Goosen, Y. Lai, S. J. Steyn, M. V. Varma and R. S. Obach, A perspective on the prediction of drug pharmacokinetics and disposition in drug research and development, *Drug Metab. Dispos.*, 2013, **41**, 1975–1993.
- S. F. Zhou, J. P. Liu and B. Chowbay, Polymorphism of human cytochrome P450 enzymes and its clinical impact, *Drug Metab. Rev.*, 2009, **41**, 89–295.

9.2 Mitigation Strategies

9.2.1 Reduction of Lipophilicity

CYP3A4 is the major constitutively expressed human CYP isoform and has a large and lipophilic binding pocket due to the presence of seven phenylalanine residues above the heme prosthetic group.[9,31–35] Typically, lipophilic [logarithm of the distribution coefficient (logD) > 3][36] and highly permeable drugs are therefore being metabolized by CYP3A4, with the goal of introducing polarity for subsequent excretion through the bile or kidney.[37] Reduction of overall compound lipophilicity [represented by measured log of the partition coefficient (logP)/logD values or the calculated 1-octanol–water partition coefficient clogP][38] is therefore a key strategy to increase microsomal stability and ultimately lead to an improved pharmacokinetic profile.[39,40] The effects of changes in lipophilicity upon the metabolic stability of aromatics have been analysed through a matched molecular pair analysis,[41] and the medicinal chemistry design parameter lipophilic metabolism efficiency (LipMetE) has been proposed to discern the contribution of lipophilicity from that of other factors (*e.g.*, chemical stability, metabolic hot spot) to metabolic stability.[42]

9.2.2 Modification of the Site of Metabolism

Compound optimization through lowering of lipophilicity alone does not always lead to an improved pharmacokinetic profile, as this strategy may also lead to a reduced volume of distribution and thus not necessarily improve the half-life ($T_{1/2}$) *in vivo*.[43] Alternative strategies that directly remove the metabolic soft spot are therefore also of importance. The site of metabolism can be altered such that the oxidized carbon–hydrogen bonds are either being removed completely (*e.g.* through blocking with methyl groups and fluorines, see next section for a discussion on the use of fluorine) or their binding to the CYP enzyme is made unfavourable by significantly altering the molecular structure. Examples of these structural modifications are compound cyclization, reduction in planarity and changing configuration. Deuterium can also be incorporated at sites of metabolism if hydrogen abstraction is the rate-determining step.[44] Due to the kinetic isotope effect, this hydrogen to deuterium substitution may lead to an up to tenfold reduced rate of metabolism. However, alternate metabolites may be formed *via* compensatory pathways and therefore prevent an increased microsomal stability.[45]

9.2.3 Fluorine Addition

Due to its electron withdrawing nature, fluorines (*e.g.* F, CF_2H, CF_3) can be added to an aromatic ring in order to reduce its electrophilicity and thus propensity towards oxidative metabolism. At 108 kcal mol^{-1}, the fluorine–carbon bond is the strongest bond between carbon and any other element and is inert under most biological conditions.[46–48] Substituting an aliphatic hydrogen with a fluorine can therefore attenuate metabolism of saturated carbons. While this strategy is often used to block oxidation at a specific metabolic soft spot, introduction of one or two aliphatic fluorines also reduces the overall compound lipophilicity in the case of neutral molecules and can therefore increase metabolic stability through increased compound polarity.[49,50] Conversely, fluorine addition to aromatic groups typically increases the lipophilicity of compounds.[49,51] Fluorine addition to bases can decrease the basicity and lead to increases in logD relative to the parent compounds).[52]

9.3 Examples of Mitigation Strategies

9.3.1 Reduction of Compound Lipophilicity

Examples

1 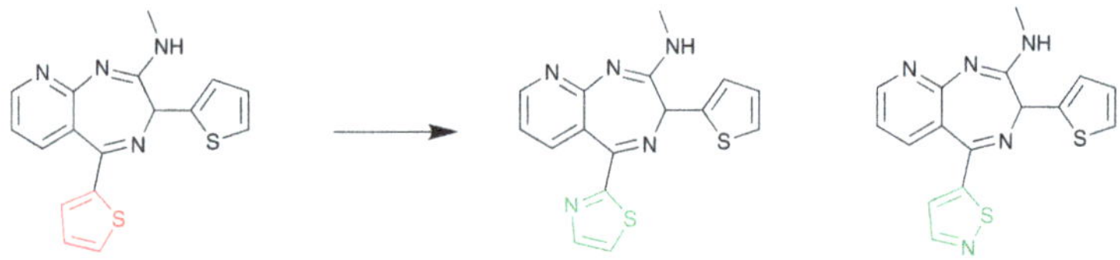

	1	2	3
IC_{50} (μM) =	0.7	1.7	1.1
HLM CL (μL/min/kg) =	30	<14	18
MLM CL (μL/min/kg) =	>100	33	37
$clogD_{7.4}$ =	1.4	0.2	0.3

Thiophene-containing glutamate racemase inhibitor **1** had a high intrinsic clearance in both human and mouse liver microsomes (HLM CL = 30 μL min^{-1} kg^{-1} and MLM CL > 100 μL min^{-1} kg^{-1}, respectively). Introduction of heteroatoms on the thiophene ring led to thiazole **2** and isothiazole **3** with significantly increased polarity as indicates by more than one log unit reduced $clogD_{7.4}$ values. This lower lipophilicity translated into a more than twofold reduced turnover in both HLM and MLM with **2** and **3**, relative to **1**.[53]

2 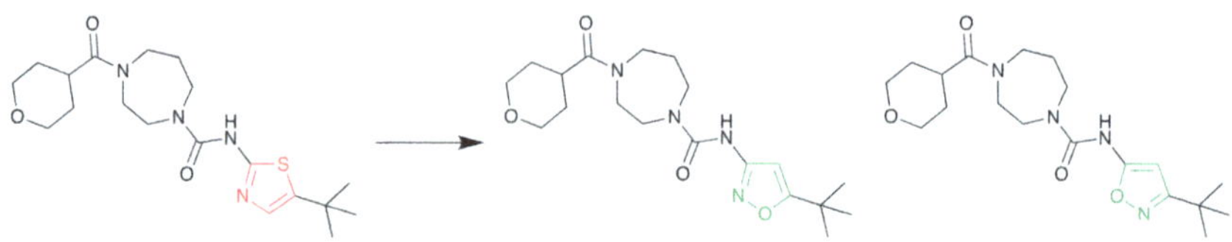

	4	5	6
EC_{50} (nM) =	150	67	26
HLM $T_{1/2}$ (min) =	27	>120	>120
clogP =	0.8	0.1	0.1

Thiazole **4** is a cannabinoid receptor type 2 (CB2) agonist lead with a short half-life in HLM ($T_{1/2}$ = 27 min). Replacing the thiazole with more polar isoxazole heterocycles gave compounds **5** and **6** that both have half-lives of greater than 120 minutes in HLM.[54]

Examples

3

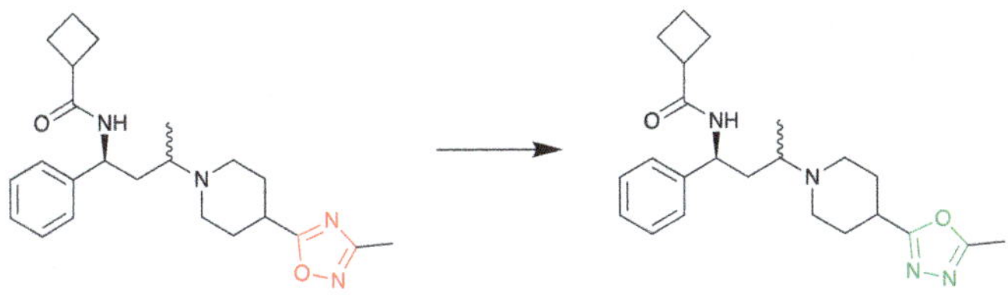

	7	8
IC_{50} (nM)	21.4	1849
HLM CL (μL/min/mg)	56	<7
$logD_{7.4}$	2.1	1.3

On the C–C chemokine receptor type 5 (CCR5) program for HIV, replacement of the 1,2,4-oxadiazole in **7** with the regioisomeric 1,3,4-oxadiazole led to analogue **8**, a compound with reduced lipophilicity and a more than eightfold improved metabolic stability in HLM. The reduction in compound polarity probably contributed significantly to the lower turnover in HLM. It may also be possible that the switch to the 1,3,4-oxadiazole prevented a metabolic N–O-ring opening possible in the 1,2,4-oxadiazole, although a metabolite identification study has not been conducted with **8** to confirm this hypothesis.[55]

4

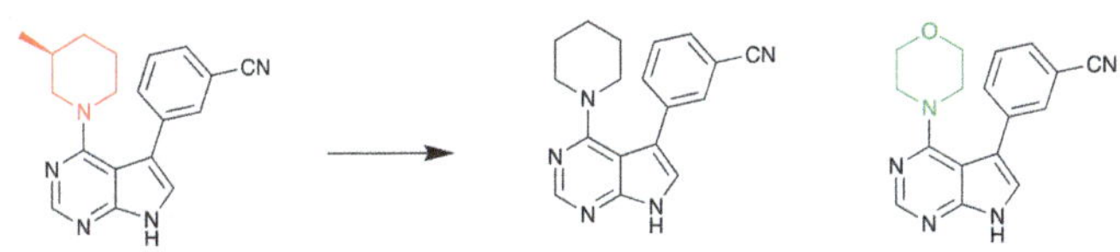

	9	10	11
IC_{50} (nM)	16	6	3
HLM CL (mL/min/kg)	220	131	36
clogP	4.0	3.6	2.4

Reproduced from ref. 56 with permission from American Chemical Society, Copyright 2015.

Leucine-rich repeat kinase 2 (LRRK2) inhibitor **9** was identified as a virtual screen follow-up to a hit from a high-throughput screen. Successive reduction in lipophilicity first led to inhibitor **10** and then to analogue **11**. Relative to compound **9**, and due to its much reduced lipophilicty, compound **11** had a greater than sixfold improved stability in HLM.[56]

(*continued*)

Examples

5

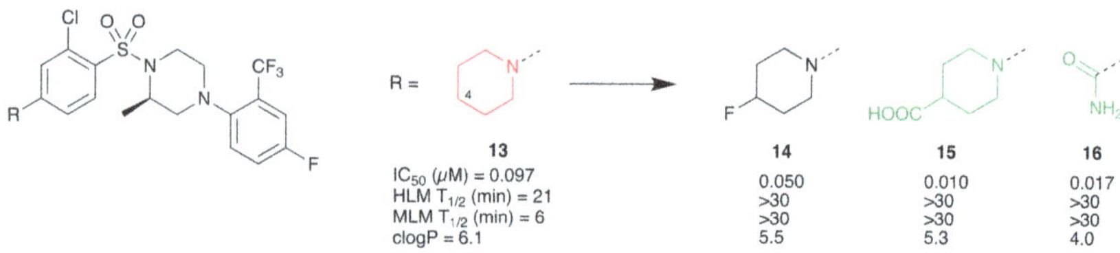

On the 11β-hydroxysteroid dehydrogenase type I program, piperidine analogue **13** had a low to moderate metabolic stability in mouse and human liver microsomes ($T_{1/2}$ = 6 min in MLM and $T_{1/2}$ = 21 min in HLM). Fluorine substitution at position four gave analogue **14** with much improved metabolic stability, indicating that this four-piperidine position may be a possible site of metabolism that is being blocked by the fluorine. Thus, further modification or replacement of the piperidine and a reduction in overall compound lipophilicity led to acid **15** and amide **16**, analogues that have half-lives of greater than 30 min in both human and mouse liver microsomes.[57]

6

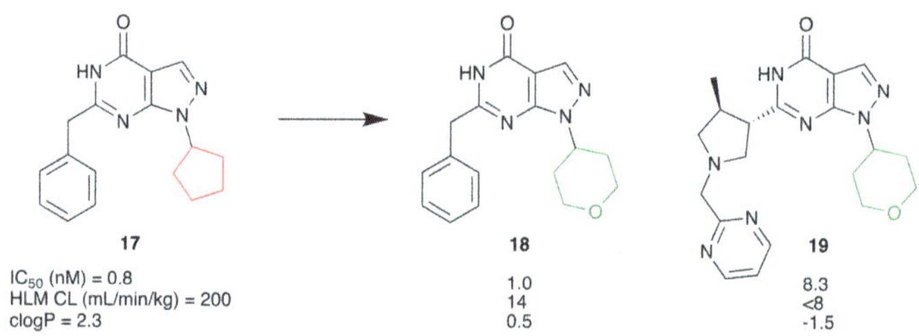

Adapted from ref. 58 with permission from American Chemical Society, Copyright 2012.

Lead compound **17** was identified in a high throughput screen and displayed good inhibitory potency at phosphodiesterase (PDE) 9A, but insufficient metabolic stability in HLM. To reduce microsomal turnover, the lipophilic cyclopentyl moiety was replaced with the more polar tetrahydropyran ring to give compound **18**. This replacement lowered the lipophilicity represented by clogP by almost 2 log units, which led to a greater than 14-fold improved stability in HLM, while maintaining the inhibitory potency at the primary target. Further optimization using both structure- and property-based drug design furnished clinical candidate **19** with an even further reduced lipophilicity and turnover in HLM, relative to compound **18**.[58]

Examples

7

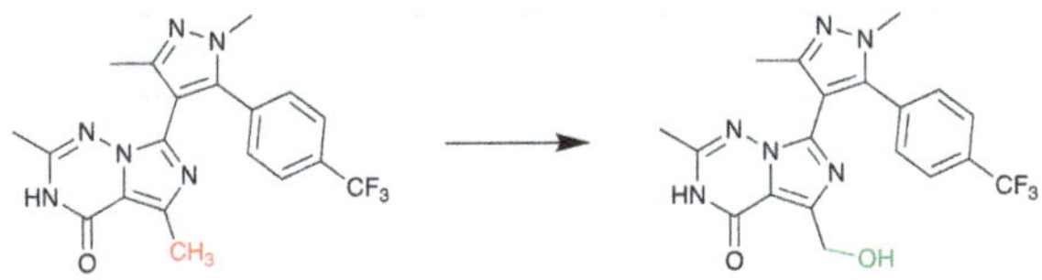

20	**21**
IC_{50} (nM) = 0.6	0.4
HLM CL (mL/min/kg) = 16	<8.0
clogP = 3.3	2.0

Adapted from ref. 59 with permission from American Chemical Society, https://pubs.acs.org/doi/10.1021/acsmedchemlett.7b00343, Copyright 2018.

Analogue **20** was characterized by a favorable inhibitory potency at PDE2 and moderate metabolic stability in human liver microsomes and hepatocytes. In addition, compound **20** was at risk for a clinical drug–drug interaction (DDI) due to its exclusive clearance by CYP3A4 in a panel of human recombinant CYPs. To improve metabolic stability and lower its DDI risk, oxidative metabolites of compound **20** were evaluated due to their increased polarity that, in addition to reducing turnover in HLM, may also introduce a renal clearance component and thus minimize the DDI risk. From this effort compound **21** emerged with reduced metabolic turnover in HLM and, importantly, the desired renal clearance component in three preclinical species (22, 48 and 26% renal clearance *in vivo* in rats, dogs and non-human primates, respectively). The favorable preclinical pharmacokinetics translated into a human projection of low plasma clearance (2.9 mL min^{-1} kg^{-1}), moderate volume of distribution (0.9 L kg^{-1}) and a moderate $T_{1/2}$ of 3.7 h. It is interesting to note that compound **21** was originally prepared biosynthetically at the microgram scale to gain the initial insight into its *in vitro* potency and improved metabolic stability.[59]

8

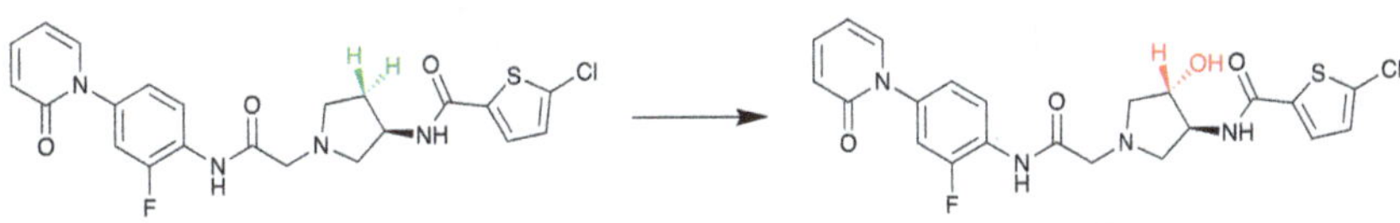

22	**23**
K_i (nM) = 8	10
RLM CL (µL/min/mg) = 79.2	15.8
$clogD_{7.4}$ = 1.6	1.0

Despite its good primary pharmacology and low lipophilicity (clogD = 1.6), Factor Xa inhibitor **22** only had moderate stability in rat liver microsomes. Hydroxy substitution of the central pyrollidine ring further reduced lipophilicity and led to the desired reduction in rat liver microsomal turnover (compound **23**). However, this added polarity also introduced a significant amount of renal clearance in rats, leading to a compound with an undesirable high rat *in vivo* clearance.[60]

9.3.2 Modification of the Site of Metabolism

Examples

1

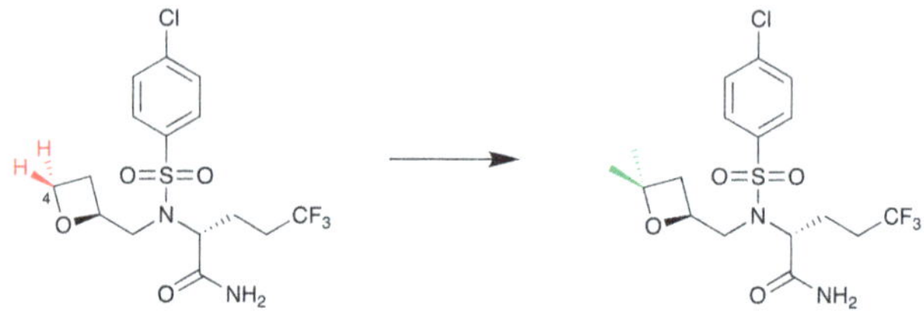

24 → **25**

IC$_{50}$ (nM) = 8.0 → 16.5
HLM CL (mL/min/kg) = 62.1 → 25.9
ElogD = 2.6 → 3.8

Reproduced from ref. 61 with permission from American Chemical Society, Copyright 2011.

The 2-oxetane containing gamma-secretase inhibitor (GSI) **24** had a single digit nanomolar potency (IC$_{50}$ = 8.0 nM), but only moderate metabolic stability in HLM. Metabolite identification studies revealed that the major metabolite of compound **24** came from methylene oxidation at position four of the oxetane. Blocking this position with a gem-dimethyl group increased metabolic stability by more than two-fold (compound **25**), despite an increase in the experimental logD (ElogD) value by more than 1 log unit.[61]

2

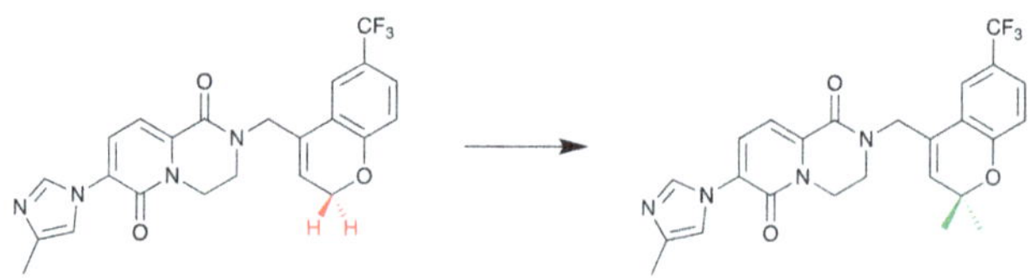

26 → **27**

IC$_{50}$ (nM) = 30 → 13
HLM CL (μL/min/mg) = 108 → 47.8
clogP = 2.4 → 3.4

Adapted from ref. 62 with permission from the Royal Society of Chemistry.

Gamma secretase modulator (GSM) lead **26**, a compound in good lipophilicity space (clogP = 2.4), was characterized by moderate potency (Abeta$_{42}$ IC$_{50}$ = 30 nM), but high turnover in human liver microsomes (HLM CL = 108 μL min^{-1} mg^{-1}). Results from metabolite identification studies indicate that CYP-mediated oxidation of the methylene group between the olefin and the chromene oxygen afforded the corresponding chromen-2-one and was the main culprit for the instability of **26** in liver microsomes. Introduction of a gem-dimethyl moiety blocked the metabolism at this position and furnished gem-dimethyl chromene **27** with improved metabolic stability in HLM, despite an increase in lipophilicity relative to compound **26**. Importantly, this transformation also increased the potency relative to the original lead **26**.[62]

Examples

3

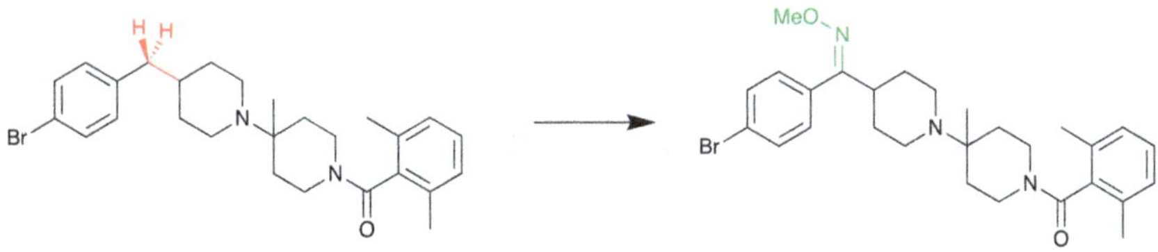

28	29
K_i (nM) = 66	2
rat PK (10 mg/kg, po)	
$AUC_{0\text{-}6h}$ (h μg/mL): 0.04	1.4
clogP = 6.1	5.6

Adapted from ref. 63 with permission from American Chemical Society, Copyright 2002.

CCR5 antagonist **28** was a promising lead due to its potency, but poor oral bioavailability in rats, probably due to extensive oxidation at its benzylic position, hampered its further development. While conventional strategies, such as adding substituents to the benzylic site, did not improve potency at the primary target, it was found that changing its hybridisation to sp^2 improved potency for compound **29** by more than 30-fold. This transformation also blocked oxidation at the benzylic position, thereby improving metabolic stability significantly, as indicated by plasma levels that are 35-fold higher relative to **28**.[63]

4

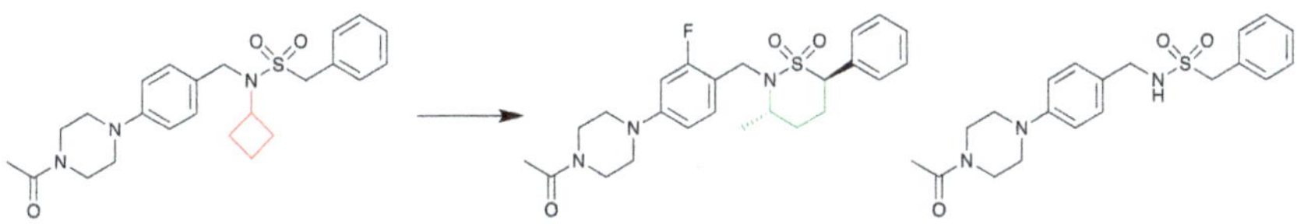

30	32	31
EC_{50} (μM) = 0.031	0.012	
HLM CL_{hep} (mL/min/kg) = 19	10	
RLM CL_{hep} (mL/min/kg) = 47	21	
clogP = 3.2	3.8	

Adapted from ref. 64 with permission from American Chemical Society, Copyright 2015.

Tertiary sulfonamide **30** was a potent retinoic acid receptor-related orphan receptor (ROR) C inverse agonist with only moderate stability in human and rat liver microsomes. Oxidation of the *N*-alkyl group in **30** followed by cleavage of the unstable hemiaminal formed the major metabolite **31**. Lowering the lipophilicity of compound **30** by introduction of polar groups on the *N*-alkyl substituent removed the activity at the primary target and the team therefore devised a cyclisation approach to the sulphonamide to increase the intrinsic metabolic stability of compounds such as **30**. This led to cyclized, potent and orally bioavailable sulphonamide **32** with appropriate microsomal stability in human and rat and was therefore a tool compound to be further evaluated in preclinical pharmacokinetic and pharmacodynamic studies.[64]

(*continued*)

Examples

5 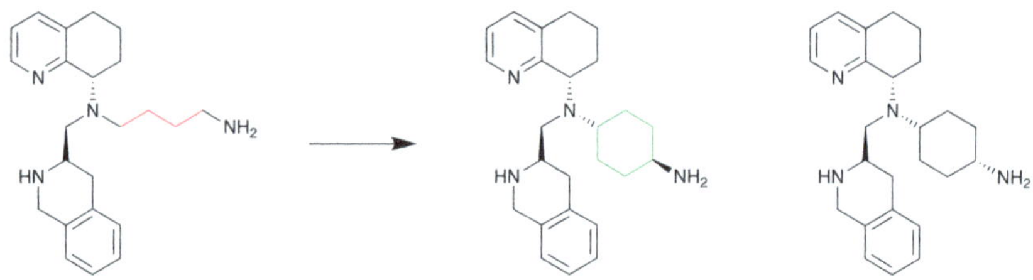

33	34	35
IC_{50} (nM) = 6.3	34	1800
H / R / M LM, % remaining = 77 / 37 / 17	95 / 71 / 77	86 / 24 / 56
clogP = 2.8	3.1	3.1

Reproduced from ref. 65 with permission from American Chemical Society, Copyright 2018.

Compound **33** was a potent C–X–C chemokine receptor type 4 (CXCR4) antagonist with good metabolic stability in HLM. However, its microsomal stability in rodent (mouse and rat) liver microsomes was only poor, which meant the compound could not be used as a preclinical *in vivo* tool to study efficacy. To reduce microsomal turnover in rodents, cyclic side chain congeners were investigated and cyclohexyl analogue **34** was found to largely maintain the potency at the primary target, while also significantly improving the metabolic stability in mouse and rat. It is worth noting, that the diastereomer **35** did not show a similar degree of metabolic stability, highlighting the important influence that chirality has on CYP-mediated oxidation processes.[65]

6 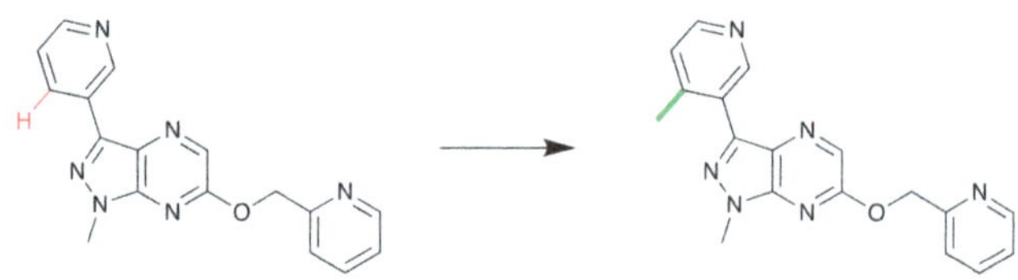

36	37
IC_{50} (nM) = 6.5	2.5
HLM CL (mL/min/kg) = 94	23.7
clogP = 2.3	2.8

The advanced metabotropic glutamate receptor 5 negative allosteric modulator (mGluR5 NAM) **36** suffered from high turnover in human liver microsomes (HLM CL = 94 mL min^{-1} kg^{-1}), despite being in a good lipophilicity space. Introduction of an *ortho*-methyl group on the pyridyl ring maintained the good mGluR5 potency and gave analogue **37** with an almost four-fold better stability in HLM (HLM CL = 23.7 mL min^{-1} kg^{-1}). Quantum mechanics based dihedral torsion scans (pyridyl–central pyrazolopyrazine angle) indicated that the pyridyl and pyrazolopyrazine core of compound **36** are in plane, while the substituted pyridyl moiety was at an angle of 30° for **37**. It was believed that this conformational change and the concomitant increase in three-dimensionality positively affected microsomal stability. Compound **37** was a key compound from this series and was efficacious in a non-human primate model of Levadopa-induced dyskinesia in Parkinson's disease.[66]

Examples

7

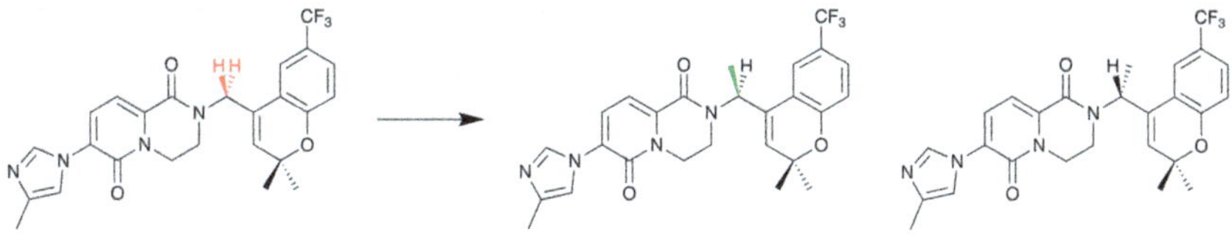

	38	39	40
IC_{50} (nM) =	13	4.9	801
HLM CL (μL/min/mg) =	47.8	14.3	242
clogP =	3.4	3.7	3.7

Chromene **38** was an advanced GSM lead with a good balance of potency and metabolic stability. A (*S*)-methyl group was introduced on the methylene linker to enforce the bioactive, turned conformation and improve potency by almost three-fold as seen with compound **39.** Despite an increase in lipophilicity, this methyl substitution also improved metabolic stability, probably due to the increased three-dimensionality of the turned conformation. Interestingly, the (*R*)-methyl substituent negatively affected metabolic stability (compound **40**), highlighting the importance of chirality to CYP-mediated turnover.[67]

8

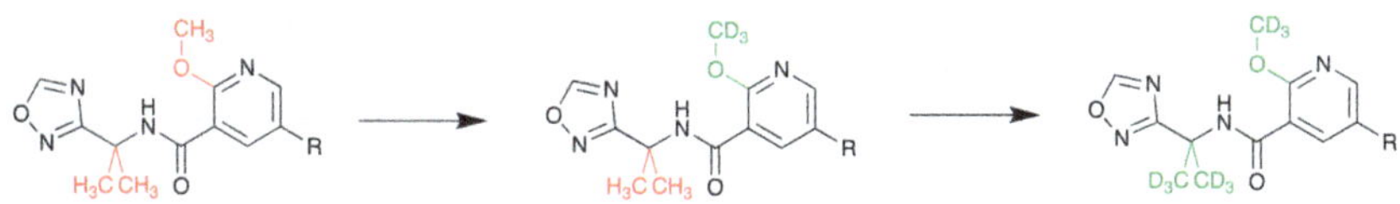

	41	42	43
EC_{50} (nM) =	3	5	4
H / cyno LM $T_{1/2}$ (min) =	14 / 5	24 / 14	59 / 48
clogP =	4.0	4.0	4.0

R =

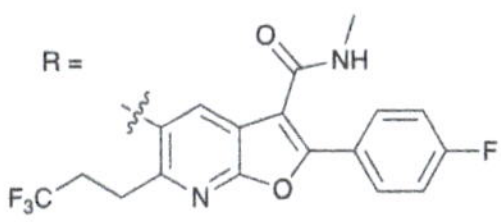

Compound **41** was a potent pan-genotypic hepatitis C virus nonstructural protein 5B NS5B (HCV NS5B) polymerase primer grip inhibitor lead, suffering from a short half-life of 14 and 5 minutes in human and cynomolgus monkey (cyno) liver microsomes, respectively. Step-wise deuterium introduction generated an analogue with reduced metabolic turnover. First, replacing the methoxy group with its deuterated version furnished analogue **42** with a slightly increased stability in human and cyno liver microsomes ($T_{1/2}$ = 24 and 14 min, respectively). Incorporation of deuterium into the gem-dimethyl group then gave the potential clinical candidate **43** with half-lives of 59 and 48 minutes in human and cyno liver microsomes, respectively. The favorable *in vitro* properties of **43** translated *in vivo* and gave a compound characterized by a low clearance (CL = 4.7 mL min^{-1} kg^{-1}), moderate volume of distribution (V_{DSS} = 2.3 L kg^{-1}) and good oral bioavailability (%F = 85) in cyno.[68]

9.3.3 Fluorine Addition

Examples

1

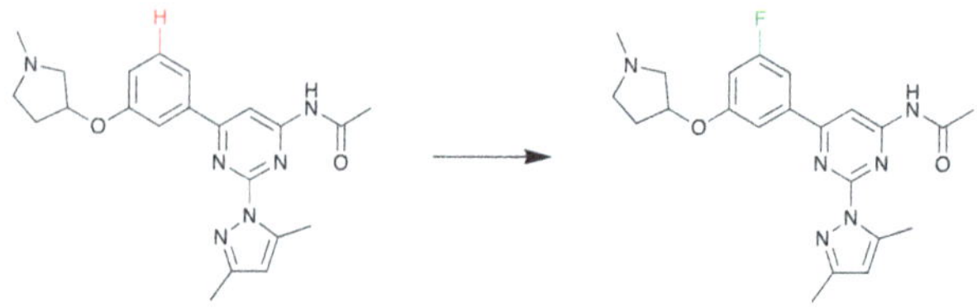

	44	45
K_i (nM)	2	1
HLM CL (mL/min/kg)	54	3
$clogD_{7.4}$	-1.1	-1.0

Lead compound **44** demonstrated good non-xanthine selective A2A antagonist activity but was rapidly cleared in HLM with an intrinsic clearance of 54 mL min^{-1} kg^{-1}. Addition of a single fluorine atom to the central phenyl ring led to analogue **45**, which had an 18-fold improved metabolic stability in HLM (CL = 3 mL min^{-1} kg^{-1}) and retained the single digit nanomolar potency of compound **44**. It is believed that the electron-withdrawing fluorine rendered the phenyl ring more resistant to CYP oxidation, without changing the overall polarity of the parent molecule.[69]

2

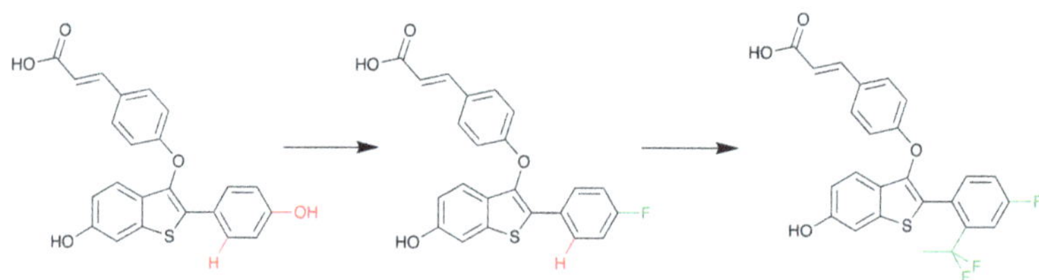

	46	47	48
IC_{50} (nM)	748	1823	6
CL_p (mL/min/kg)	54 (rat)	24 (rat) / 22 (mouse)	28 (mouse)

Adapted from ref. 70 with permission from American Chemical Society, Copyright 2018.

Compound **46** is a selective estrogen receptor degrader (SERD) lead for the treatment of estrogen receptor positive breast cancer. Its moderate transcriptional potency was combined with insufficient bioavailability (6%) and peak plasma concentrations [C_{max} (orally) = 80 nM] due to its high clearance (54 mL min^{-1} kg^{-1}) in rats. Substituents in place of the electron-donating hydroxyl group were therefore investigated and it was found that the *para*-fluoro substituted analogue **47** maintained the original potency (ERα transcription IC_{50} = 1823 nM), but improved the bioavailability and clearance in both mice and rats (oral bioavailability = 19 and 33% and clearance = 22 and 24 mL min^{-1} kg^{-1}, respectively). Addition of an *ortho*-difluoroethyl group then improved potency by more than 150-fold, while maintaining an acceptable pharmacokinetic (PK) profile (bioavailability: 12%, clearance: 28 mL min^{-1} kg^{-1} in mouse) and furnished analogue **48**, a compound in clinical development for the treatment of estrogen receptor alpha-positive breast cancer.[70]

Examples

3

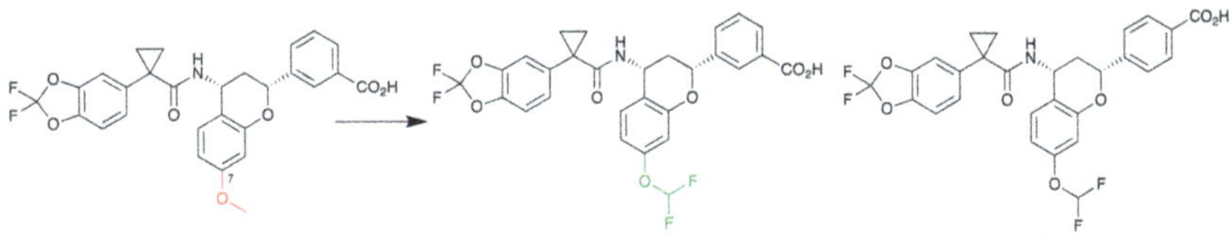

49 — EC_{50} (nM) = 240; H / R Hepatocyte $CL_{int,u}$ (L/h/kg) = 25 / 17; ElogD = 3.8

50 — 75; 14 / 7; 4.0

51 — 5; 7 / 3

Adapted from ref. 71 with permission from American Chemical Society, Copyright 2018.

Compound **49** is an advanced cystic fibrosis transmembrane conductance regulator (CFTR) lead for the treatment of cystic fibrosis with a favorable combination of biological activity and properties. In order to reduce its unbound turnover in human and rat hepatocytes, various fluorinated and non-fluorinated substituents were explored on the chromane ring, mostly on position seven. As seen with compound **50**, it was found that replacing the methoxy group with a difluoromethoxy group improved both potency and metabolic stability in human and rat hepatocytes. Further optimization of potency gave *para*-substituted acid analogue **51** that is mainly cleared *via* glucuronidation and has good bioavailability in rats (74%) and dogs (40%). Compound **51** is currently being investigated in clinical trials.[71]

4

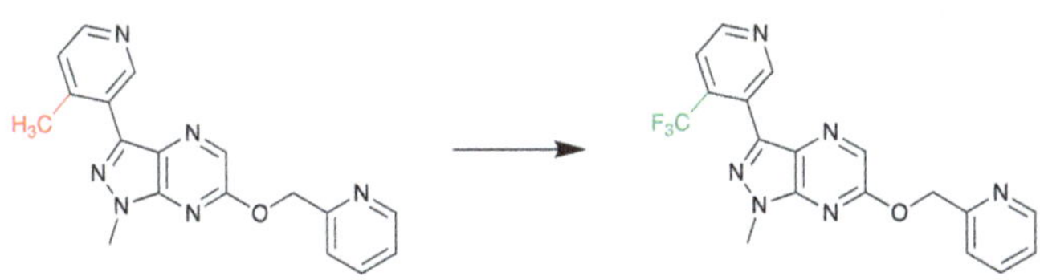

52 — IC_{50} (nM) = 2.5; HLM CL (mL/min/kg) = 23.7; clogP = 2.1

53 — 4.8; <8.0; 2.5

As described in one of the previous examples, key mGluR5 NAM **52** was in a good space both with regards to its potency as well as PK profile.[66] As seen by its *in vitro* microsomal clearance number (CL = 23.7 mL min^{-1} kg^{-1}), compound **52** had some turnover in HLM, presumably due to oxidation at the pyridyl methyl group. The oxidation at this site could be fully blocked by replacing the methyl group with a trifluoromethyl group. As seen with analogue **53**, this change led to an undetectable turnover in the standard HLM assay (HLM CL < 8 mL min^{-1} kg^{-1}). It should be noted that while placement of electron-withdrawing groups on azaaromatic moieties can decrease P450 catalyzed metabolism, it can make the molecule more susceptible to aldehyde oxidase catalyzed oxidation.[72] Aldehyde oxidase is a cytosolic enzyme that is not present in liver microsomes and therefore not detected in HLM metabolic lability experiments.

(*continued*)

Examples

5

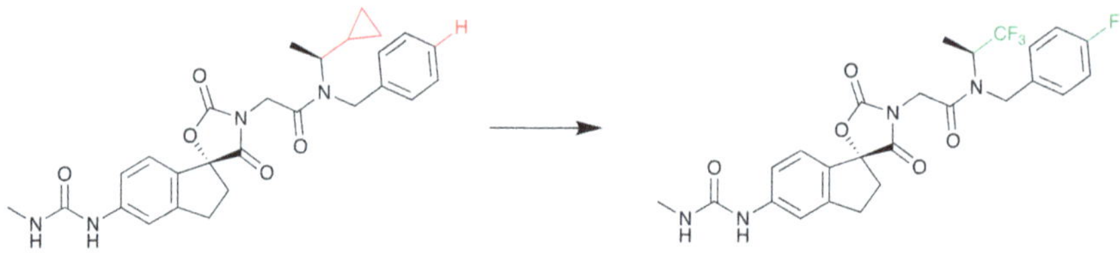

54
IC_{50} (nM) = 25
MLM $CL_{int,u}$ (L/h/kg) = 188
M $CL_{p,iv}$ (L/h/kg) = 1.62
clogP = 2.4

55
60
37.6
0.56
2.8

Adapted from ref. 73 with permission from American Chemical Society, https://doi.org/10.1021/acsmedchemlett.7b00395, Copyright 2017.

Oxazolidinedione **54** is a potent histone acetyltransferase (HAT) domain inhibitor with high clearance in MLM, translating into a mouse PK with moderate plasma clearance and bioavailability. It was believed that the microsomal instability of **54** was due to oxidation of the terminal phenyl group and *N*-substituents on the amide. Therefore, fluorinated analogues were prepared that mitigate metabolism at these two sites. As seen with compound **55**, introduction of a fluorine atom on the phenyl ring as well as replacing the cyclopropyl group with a CF_3 group on the amide led to a significantly lower microsomal turnover, translating into a low unbound plasma clearance and high oral exposure (>100%) in mouse. The pharmacokinetic behaviour of compound **55** in rats and dogs was also favourable with modest clearance in rats (1.2 L h^{-1} kg^{-1}), very low clearance in dogs (0.1 L h^{-1} kg^{-1}) and good oral bioavailability in both species (rat: %F = 63, dog: %F = 74).[73]

6

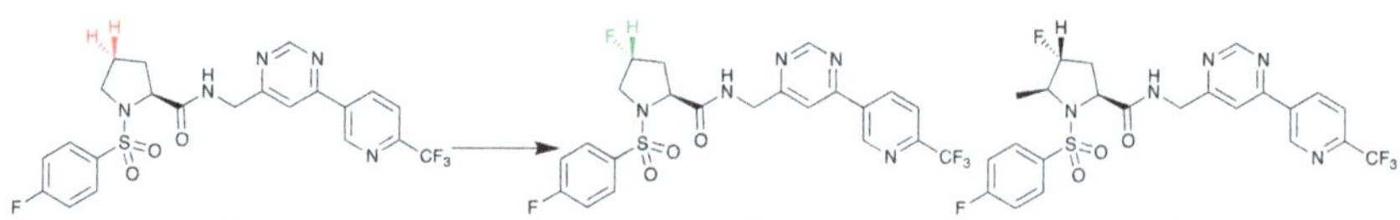

56
IC_{50} (nM) = 154
H / R LM CL (mL/min/kg) = 16 / 43
$logD_{7.4}$ = 2.7

57
1200
5 / 10
2.4

59
179
10 / 19
2.7

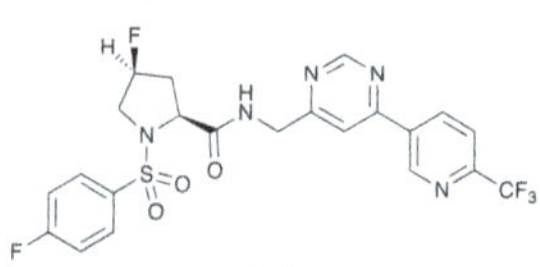

58
H TRP1 IC_{50} (nM) = 4400
H / R LM CL (mL/min/kg) = 7 / 40
$logD_{7.4}$ = 1.8

Adapted from ref. 74 with permission from American Chemical Society, Copyright 2018.

Examples

Compound **56** was a potent transient receptor potential ankyrin 1 (TRPA1) antagonist with moderate metabolic stability in human and rat liver microsomes. Results of metabolite identification studies indicated that pyrrolidine oxidation was the main reason for the microsomal instability. Thus, fluorinated pyrrolidine analogues were explored to introduce polarity and potentially block CYP metabolism. As can be seen with analogue **57**, this strategy reduced microsomal turnover in human and rat liver microsomes (RLM) by threefold and fourfold, respectively. The low value in the *in vitro* liver microsome assay (RLM CL = 10 mL min^{-1} kg^{-1}) was consistent with a very low clearance *in vivo* (rat iv CL_p = 15 mL min^{-1} kg^{-1}). The fluorine stereochemistry was critical to achieve this degree of microsomal stability and was not obtained with diastereomer **58**. Introduction of a methyl group in position five on the pyrrolidine (compound **59**) then improved potency by about tenfold while maintaining the favourable *in vivo* biopharmaceutical attributes and was a key compound *en route* to a novel TRPA1 antagonist tool.[74]

7

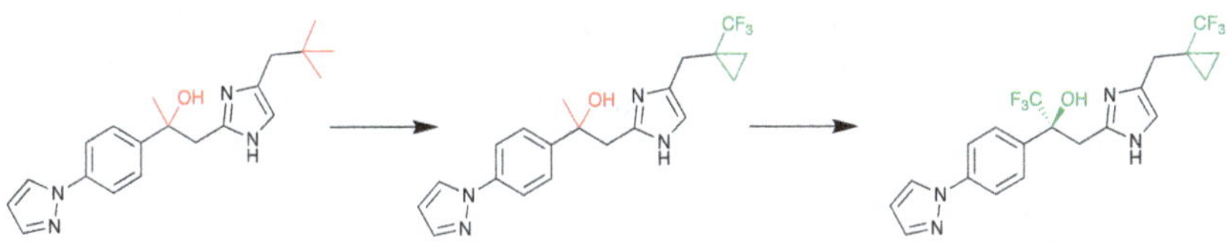

	60	61	62
IC_{50} (nM)	5	11	27
CL_p (mL/min/kg)	110	52	20
clogP	2.5	2.4	3.1

Reproduced from ref. 75 with permission from American Chemical Society, Copyright 2011.

Compound **60** was a potent human bombesin receptor subtype-3 (hBRS-3) agonist with high rat plasma clearance (rat CL_p = 110 mL min^{-1} kg^{-1}). To improve stability *in vivo*, the methyl groups available for oxidation were replaced in a stepwise manner. First, rat plasma clearance of agonist **60** was reduced twofold by replacing the *tert*-butyl group with the cyclo-propyl-CF_3 moiety and therefore removing several hydrogen atoms available for oxidation (compound **61**: CL_p = 52 mL min^{-1} kg^{-1}). Another twofold improvement in the plasma clearance was then achieved by replacing the central methyl group with a CF_3 group to give tool compound **62** that was further evaluated in pharmacological studies.[75]

(continued)

Examples

8

63 H / R LM $T_{1/2}$ (min) = 51 / 24; clogP = 3.7

64 H / R LM $T_{1/2}$ (min) = 88 / 98; clogP = 3.6

65 H / R LM $T_{1/2}$ (min) = 29 / 12; clogP = 5.5

66 H / R LM $T_{1/2}$ (min) = >405 / 40; clogP = 5.4

A thorough investigation into the metabolic stability of the cyclopropyl-CF_3 group relative to the *tert*-butyl group was disclosed by Barnes-Seeman *et al.* As seen with the representative model compounds **63**/**64** and **65**/**66**, the cyclopropyl-CF_3-containing analogues had consistently higher metabolic stability, ranging from an about twofold (compounds **63** and **64**) to 13-fold (compounds **65** and **66**) reduced turnover in HLM. RLM stabilities were also consistently improved. The increased metabolic stability of the cyclopropyl-CF_3 group relative to the *tert*-butyl group is attributed to the removal of all fully sp^3-hybridised C–H bonds from the *tert*-butyl group.[76]

9

67
EC$_{50}$ (nM) = 44 nM
HLM CL (mL/min/kg) > 300
ElogD = 4.3

68
80
42.4
3.4

Reproduced from ref. 77 with permission from Elsevier, Copyright 2011.

Potent G protein-coupled receptor 119 (GPR119) agonist **67** displayed low human liver microsomal stability (CL_{int} > 300 mL min^{-1} kg^{-1}), probably due to its high lipophilicity (ELogD = 3.9). Substituting the central methyl group with a fluorine gave compound **68** with a reduced lipophilicity (ElogD = 3.4) and microsomal turnover (CL = 42.4 mL min^{-1} kg^{-1}). Importantly, this substitution did not significantly affect potency at the primary target.[77]

Examples

10 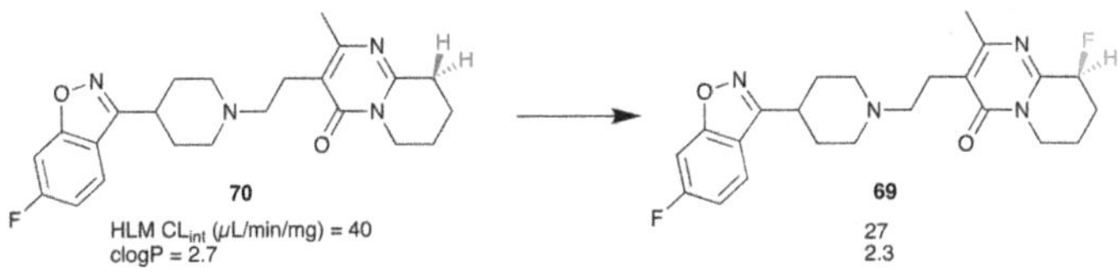

In a different approach to introduce fluorine specifically at metabolized positions, four known drugs (risperidone, midazolam, ramelteon and celecoxib) were subjected to CYP mediated hydroxylation by liver microsomes followed by deoxyfluorination using (diethylamino)sulfur trifluoride. Exemplified above is the generation of 9-fluororisperidone (**69**) from risperidone (**70**) wherein the scaled intrinsic clearance decreased from 40 to 27 mL min^{-1} kg^{-1}. It should be noted that in those four examples, not every fluorinated analogue had decreased metabolic lability relative to its hydrogen-containing counterpart and that deoxyfluorination was still a metabolic pathway for the fluorinated analogues.[78]

References

1. T. Omura and R. Sato, *J. Biol. Chem.*, 1964, **239**, 2379–2385.
2. S. Rendic and F. P. Guengerich, *Chem. Res. Toxicol.*, 2015, **28**, 38–42.
3. M. F. Paine, H. L. Hart, S. S. Ludington, R. L. Haining, A. E. Rettie and D. C. Zeldin, *Drug Metab. Dispos.*, 2006, **34**, 880–886.
4. J. Hukkanen, O. Pelkonen, J. Hakkola and H. Raunio, *Crit. Rev. Toxicol.*, 2002, **32**, 391–411.
5. K. M. Knights, A. Rowland and J. O. Miners, *Br. J. Clin. Pharmacol.*, 2013, **76**, 587–602.
6. H. Yamazaki, K. Inoue, C. G. Turvy, F. P. Guengerich and T. Shimada, *Drug Metab. Dispos.*, 1997, **25**, 168–174.
7. https://www.pharmvar.org/genes, accessed 24 March 2020.
8. R. L. Reynald, S. Sansen, C. D. Stout and E. F. Johnson, *J. Biol. Chem.*, 2012, **287**, 44581–44591.
9. J. K. Yano, M. R. Wester, G. A. Schoch, K. J. Griffin, C. D. Stout and E. F. Johnson, *J. Biol. Chem.*, 2004, **279**, 38091–38094.
10. M. R. Wester, J. K. Yano, G. A. Schoch, C. Yang, K. J. Griffin, C. D. Stout and E. F. Johnson, *J. Biol. Chem.*, 2004, **279**, 35630–35637.
11. G. A. Schoch, J. K. Yano, M. R. Wester, K. J. Griffin, C. D. Stout and E. F. Johnson, *J. Biol. Chem.*, 2004, **279**, 9497–9503.
12. S. Sansen, J. K. Yano, R. L. Reynald, G. A. Schoch, K. J. Griffin, C. D. Stout and E. F. Johnson, *J. Biol. Chem.*, 2007, **282**, 14348–14355.
13. M. R. Wester, J. K. Yano, G. A. Schoch, K. J. Griffin, C. D. Stout and E. F. Johnson, *PDB ID: 1TQN*, 2004.

14. A. S. Rose, D. Sehnal, J. Koca, S. K. Burley and S. Velankar, *Workshop on Molecular Graphics and Visual Analysis of Molecular Data*, 2018, pp. 29–33, DOI: 10.2312/molva.20181103.
15. https://www.rcsb.org/, accessed 19 August 2020.
16. H. M. Berman, J. Westbrook, Z. Feng, G. Gilliland, T. N. Bhat, H. Weissig, I. N. Shindyalov and P. E. Bourne, *Nucleic Acids Res.*, 2000, **28**, 235–242.
17. S. K. Burley, H. M. Berman, C. Bhikadiya, C. Bi, L. Chen, L. Di Costanzo, C. Christie, K. Dalenberg, J. M. Duarte, S. Dutta, Z. Feng, S. Ghosh, D. S. Goodsell, R. K. Green, V. Guranovic, D. Guzenko, B. P. Hudson, T. Kalro, Y. Liang, R. Lowe, H. Namkoong, E. Peisach, I. Periskova, A. Prlic, C. Randle, A. Rose, P. Rose, R. Sala, M. Sekharan, C. Shao, L. Tan, Y. P. Tao, Y. Valasatava, M. Voigt, J. Westbrook, J. Woo, H. Yang, J. Young, M. Zhuravleva and C. Zardecki, *Nucleic Acids Res.*, 2019, **47**, D464–D474.
18. M. Spatzenegger, S. Born and J. Halpert, *Cytochrome P450 in Drug Discovery and Development*, ed. M. B. Fisher, R. S. Obach and J. S. Lee, 2003, pp. 179–209.
19. D. F. Lewis, *Biochem. Pharmacol.*, 2000, **60**, 293–306.
20. D. A. Smith and B. C. Jones, *Biochem. Pharmacol.*, 1992, **44**, 2089–2098.
21. F. P. Guengerich, *Chem. Res. Toxicol.*, 2001, **14**, 611–650.
22. M. M. Ahlstrom and I. Zamora, *J. Med. Chem.*, 2008, **51**, 1755–1763.
23. R. S. Obach, *Drug Metab. Dispos.*, 1999, **27**, 1350–1359.
24. H. S. Brown, M. Griffin and J. B. Houston, *Drug Metab. Dispos.*, 2007, **35**, 293–301.
25. U. Zanelli, N. P. Caradonna, D. Hallifax, E. Turlizzi and J. B. Houston, *Drug Metab. Dispos.*, 2012, **40**, 104–110.
26. L. Di, B. Feng, T. C. Goosen, Y. Lai, S. J. Steyn, M. V. Varma and R. S. Obach, *Drug Metab. Dispos.*, 2013, **41**, 1975–1993.
27. M. A. Cerny, *Drug Metab. Dispos.*, 2016, **44**, 1246–1252.
28. S. Fowler and H. J. Zhang, *AAPS J.*, 2008, **10**, 410–424.
29. S. F. Zhou, J. P. Liu and B. Chowbay, *Drug Metab. Rev.*, 2009, **41**, 89–295.
30. J. McGraw and D. Waller, *Expert Opin. Drug Metab.*, 2012, **8**, 371–382.
31. P. A. Williams, J. Cosme, D. M. Vinkovic, A. Ward, H. C. Angove, P. J. Day, C. Vonrhein, I. J. Tickle and H. Jhoti, *Science*, 2004, **305**, 683–686.
32. M. Ekroos and T. Sjogren, *Proc. Natl. Acad. Sci. U. S. A.*, 2006, **103**, 13682–13687.
33. H. Z. Bu, *Curr. Drug Metab.*, 2006, **7**, 231–249.
34. D. F. Lewis, M. N. Jacobs and M. Dickins, *Drug Discovery Today*, 2004, **9**, 530–537.
35. D. F. Lewis and M. Dickins, *Drug Discovery Today*, 2002, **7**, 918–925.
36. M. V. S. Varma, R. S. Obach, C. Rotter, H. R. Miller, G. Chang, S. J. Steyn, A. El-Kattan and M. D. Troutman, *J. Med. Chem.*, 2010, **53**, 1098–1108.
37. M. V. Varma, S. J. Steyn, C. Allerton and A. F. El-Kattan, *Pharm. Res.*, 2015, **32**, 3785–3802.
38. L. K. Di and E. H. Kerns, in *Drug-Like Properties*, Elsevier, 2nd edn, 2016, pp. 299–306, DOI: 10.1016/C2013-0-18378-X.

39. R. J. Riley, *Curr. Opin. Drug Discovery Dev.*, 2001, **4**, 45–54.
40. C. A. Lipinski, F. Lombardo, B. W. Dominy and P. J. Feeney, *Adv. Drug Delivery Rev.*, 2001, **46**, 3–26.
41. G. Chang, K. Huard, G. W. Kauffman, A. F. Stepan and C. E. Keefer, *Bioorg. Med. Chem.*, 2017, **25**, 381–388.
42. A. F. Stepan, G. W. Kauffman, C. E. Keefer, P. R. Verhoest and M. Edwards, *J. Med. Chem.*, 2013, **56**, 6985–6990.
43. F. Broccatelli, I. Aliagas and H. Zheng, *ACS Med. Chem. Lett.*, 2018, **9**, 522–527.
44. T. G. Gant, *J. Med. Chem.*, 2014, **57**, 3595–3611.
45. M. B. Fisher, K. R. Henne and J. Boer, *Curr. Opin. Drug Discovery Dev.*, 2006, **9**, 101–109.
46. N. A. Meanwell, *J. Med. Chem.*, 2011, **54**, 2529–2591.
47. W. K. Hagmann, *J. Med. Chem.*, 2008, **51**, 4359–4369.
48. H. J. Bohm, D. Banner, S. Bendels, M. Kansy, B. Kuhn, K. Muller, U. Obst-Sander and M. Stahl, *Chembiochem*, 2004, **5**, 637–643.
49. Q. A. Huchet, B. Kuhn, B. Wagner, N. A. Kratochwil, H. Fischer, M. Kansy, D. Zimmerli, E. M. Carreira and K. Muller, *J. Med. Chem.*, 2015, **58**, 9041–9060.
50. E. P. Gillis, K. J. Eastman, M. D. Hill, D. J. Donnelly and N. A. Meanwell, *J. Med. Chem.*, 2015, **58**, 8315–8359.
51. A. B. Shtarov, P. J. Krusic, B. E. Smart and W. R. Dolbier Jr, *J. Am. Chem. Soc.*, 2001, **123**, 9956–9962.
52. M. Morgenthaler, E. Schweizer, A. Hoffmann-Roder, F. Benini, R. E. Martin, G. Jaeschke, B. Wagner, H. Fischer, S. Bendels, D. Zimmerli, J. Schneider, F. Diederich, M. Kansy and K. Muller, *ChemMedChem*, 2007, **2**, 1100–1115.
53. B. Geng, G. Basarab, J. Comita-Prevoir, M. Gowravaram, P. Hill, A. Kiely, J. Loch, L. MacPherson, M. Morningstar, G. Mullen, E. Osimboni, A. Satz, C. Eyermann and T. Lundqvist, *Bioorg. Med. Chem. Lett.*, 2009, **19**, 930–936.
54. D. Riether, L. Wu, P. F. Cirillo, A. Berry, E. R. Walker, M. Ermann, B. Noya-Marino, J. E. Jenkins, D. Albaugh, C. Albrecht, M. Fisher, M. J. Gemkow, H. Grbic, S. Lobbe, C. Moller, K. O'Shea, A. Sauer, D. T. Shih and D. S. Thomson, *Bioorg. Med. Chem. Lett.*, 2011, **21**, 2011–2016.
55. C. G. Barber, D. C. Blakemore, J. Y. Chiva, R. L. Eastwood, D. S. Middleton and K. A. Paradowski, *Bioorg. Med. Chem. Lett.*, 2009, **19**, 1075–1079.
56. J. L. Henderson, B. L. Kormos, M. M. Hayward, K. J. Coffman, J. Jasti, R. G. Kurumbail, T. T. Wager, P. R. Verhoest, G. S. Noell, Y. Chen, E. Needle, Z. Berger, S. J. Steyn, C. Houle, W. D. Hirst and P. Galatsis, *J. Med. Chem.*, 2015, **58**, 419–432.
57. Z. K. Wan, E. Chenail, J. Xiang, H. Q. Li, M. Ipek, J. Bard, K. Svenson, T. S. Mansour, X. Xu, X. Tian, V. Suri, S. Hahm, Y. Xing, C. E. Johnson, X. Li, A. Qadri, D. Panza, M. Perreault, J. F. Tobin and E. Saiah, *J. Med. Chem.*, 2009, **52**, 5449–5461.

58. P. R. Verhoest, K. R. Fonseca, X. Hou, C. Proulx-Lafrance, M. Corman, C. J. Helal, M. M. Claffey, J. B. Tuttle, K. J. Coffman, S. Liu, F. Nelson, R. J. Kleiman, F. S. Menniti, C. J. Schmidt, M. Vanase-Frawley and S. Liras, *J. Med. Chem.*, 2012, **55**, 9045–9054.
59. A. F. Stepan, T. P. Tran, C. J. Helal, M. S. Brown, C. Chang, R. E. O'Connor, M. De Vivo, S. D. Doran, E. L. Fisher, S. Jenkinson, D. Karanian, B. L. Kormos, R. Sharma, G. S. Walker, A. S. Wright, E. X. Yang, M. A. Brodney, T. T. Wager, P. R. Verhoest and R. S. Obach, *ACS Med. Chem. Lett.*, 2018, **9**, 68–72.
60. K. G. Zbinden, L. Anselm, D. W. Banner, J. Benz, F. Blasco, G. Decoret, J. Himber, B. Kuhn, N. Panday, F. Ricklin, P. Risch, D. Schlatter, M. Stahl, S. Thomi, R. Unger and W. Haap, *Eur. J. Med. Chem.*, 2009, **44**, 2787–2795.
61. A. F. Stepan, K. Karki, W. S. McDonald, P. H. Dorff, J. K. Dutra, K. J. Dirico, A. Won, C. Subramanyam, I. V. Efremov, C. J. O'Donnell, C. E. Nolan, S. L. Becker, L. R. Pustilnik, B. Sneed, H. Sun, Y. Lu, A. E. Robshaw, D. Riddell, T. J. O'Sullivan, E. Sibley, S. Capetta, K. Atchison, A. J. Hallgren, E. Miller, A. Wood and R. S. Obach, *J. Med. Chem.*, 2011, **54**, 7772–7783.
62. M. Pettersson, D. S. Johnson, D. A. Rankic, G. W. Kauffman, C. W. Am Ende, T. W. Butler, B. Boscoe, E. Evrard, C. J. Helal, J. M. Humphrey, A. F. Stepan, C. M. Stiff, E. Yang, L. Xie, K. R. Bales, E. Hajos-Korcsok, S. Jenkinson, B. Pettersen, L. R. Pustilnik, D. S. Ramirez, S. J. Steyn, K. M. Wood and P. R. Verhoest, *Medchemcomm*, 2017, **8**, 730–743.
63. A. Palani, S. Shapiro, H. Josien, T. Bara, J. W. Clader, W. J. Greenlee, K. Cox, J. M. Strizki and B. M. Baroudy, *J. Med. Chem.*, 2002, **45**, 3143–3160.
64. B. P. Fauber, O. Rene, Y. Deng, J. DeVoss, C. Eidenschenk, C. Everett, A. Ganguli, A. Gobbi, J. Hawkins, A. R. Johnson, H. La, J. Lesch, P. Lockey, M. Norman, W. Ouyang, S. Summerhill and H. Wong, *J. Med. Chem.*, 2015, **58**, 5308–5322.
65. E. Jecs, E. J. Miller, R. J. Wilson, H. H. Nguyen, Y. A. Tahirovic, B. M. Katzman, V. M. Truax, M. B. Kim, K. M. Kuo, T. Wang, C. S. Sum, M. E. Cvijic, G. M. Schroeder, L. J. Wilson and D. C. Liotta, *ACS Med. Chem. Lett.*, 2018, **9**, 89–93.
66. L. Zhang, G. Balan, G. Barreiro, B. P. Boscoe, L. K. Chenard, J. Cianfrogna, M. M. Claffey, L. G. Chen, K. J. Coffman, S. E. Drozda, J. R. Dunetz, K. R. Fonseca, P. Galatsis, S. Grimwood, J. T. Lazzaro, J. Y. Mancuso, E. L. Miller, M. R. Reese, B. N. Rogers, I. Sakurada, M. Skaddan, D. L. Smith, A. F. Stepan, P. Trapa, J. B. Tuttle, P. R. Verhoest, D. P. Walker, A. S. Wright, M. M. Zaleska, K. Zasadny and C. L. Shaffer, *J. Med. Chem.*, 2014, **57**, 861–877.
67. M. Pettersson, D. S. Johnson, D. A. Rankic, G. W. Kauffman, C. W. A. Ende, T. W. Butler, B. Boscoe, E. Evrard, C. J. Helal, J. M. Humphrey, A. F. Stepan, C. M. Stiff, E. Yang, L. F. Xie, K. R. Bales, E. Hajos-Korcsok, S. Jenkinson, B. Pettersen, L. R. Pustilnik, D. S. Ramirez, S. J. Steyn, K. M. Wood and P. R. Verhoest, *Medchemcomm*, 2017, **8**, 730–743.

68. K. Parcella, K. Eastman, K. S. Yeung, K. A. Grant-Young, J. L. Zhu, T. Wang, Z. X. Zhang, Z. W. Yin, D. Parker, K. Mosure, H. Fang, Y. K. Wang, J. Lemm, X. L. Zhuo, U. Hanumegowda, M. P. Liu, K. Rigat, M. Donoso, M. Tuttle, T. Zvyaga, Z. Haarhoff, N. A. Meanwell, M. G. Soars, S. B. Roberts and J. F. Kadow, *ACS Med. Chem. Lett.*, 2017, **8**, 771–774.
69. M. Moorjani, Z. Y. Luo, E. Lin, B. G. Vong, Y. S. Chen, X. H. Zhang, J. K. Rueter, R. S. Gross, M. C. Lanier, J. E. Tellew, J. P. Williams, S. M. Lechner, S. Malany, M. Santos, M. I. Crespo, J. L. Diaz, J. Saunders and D. H. Slee, *Bioorg. Med. Chem. Lett.*, 2008, **18**, 5402–5405.
70. G. S. Tria, T. Abrams, J. Baird, H. E. Burks, B. Firestone, L. A. Gaither, L. G. Hamann, G. He, C. A. Kirby, S. Kim, F. Lombardo, K. J. Macchi, D. P. McDonnell, Y. Mishina, J. D. Norris, J. Nunez, C. Springer, Y. C. Sun, N. M. Thomsen, C. R. Wang, J. L. Wang, B. Yu, C. L. Tiong-Yip and S. Peukert, *J. Med. Chem.*, 2018, **61**, 2837–2864.
71. X. Q. Wang, B. Liu, X. Searle, C. Yeung, A. Bogdan, S. Greszler, A. Singh, Y. H. Fan, A. M. Swensen, T. Vortherms, C. Balut, Y. Jia, K. Desino, W. Q. Gao, H. Yong, C. Tse and P. Kym, *J. Med. Chem.*, 2018, **61**, 1436–1449.
72. D. C. Pryde, D. Dalvie, Q. Y. Hu, P. Jones, R. S. Obach and T. D. Tran, *J. Med. Chem.*, 2010, **53**, 8441–8460.
73. M. R. Michaelides, A. Kluge, M. Patane, J. H. Van Drie, C. Wang, T. M. Hansen, R. M. Risi, R. Mantei, C. Hertel, K. Karukurichi, A. Nesterov, D. McElligott, P. de Vries, J. W. Langston, P. A. Cole, R. Marmorstein, H. Liu, L. Lasko, K. D. Bromberg, A. Lai and E. A. Kesicki, *ACS Med. Chem. Lett.*, 2018, **9**, 28–33.
74. H. F. Chen, M. Volgraf, S. Do, A. Kolesnikov, D. G. Shore, V. A. Verma, E. Villemure, L. Wang, Y. Chen, B. H. Hu, A. J. Lu, G. S. Wu, X. F. Xu, P. W. Yuen, Y. M. Zhang, S. D. Erickson, M. Dahl, C. Brotherton-Pleiss, S. Tay, J. Q. Ly, L. J. Murray, J. Chen, D. Amm, W. Lange, D. H. Hackos, R. M. Reese, S. D. Shields, J. P. Lyssikatos, B. S. Safina and A. A. Estrada, *J. Med. Chem.*, 2018, **61**, 3641–3659.
75. I. K. Sebhat, C. Franklin, M. M. C. Lo, D. Chen, J. P. Jewell, R. Miller, J. M. Pang, O. Palyha, Y. Q. Kan, T. M. Kelly, X. M. Guan, D. J. Marsh, J. A. Kosinski, J. M. Metzger, K. Lyons, J. Dragovic, P. R. Guzzo, A. J. Henderson, M. L. Reitman, R. P. Nargund, M. J. Wyvratt and L. S. Lin, *ACS Med. Chem. Lett.*, 2011, **2**, 43–47.
76. D. Barnes-Seeman, M. Jain, L. Bell, S. Ferreira, S. Cohen, X. H. Chen, J. Amin, B. Snodgrass and P. Hatsis, *ACS Med. Chem. Lett.*, 2013, **4**, 514–516.
77. V. Mascitti, B. D. Stevens, C. Choi, K. F. McClure, C. R. W. Guimaraes, K. A. Farley, M. J. Munchhof, R. P. Robinson, K. Futatsugi, S. Y. Lavergne, B. A. Lefker, P. Cornelius, P. D. Bonin, A. S. Kalgutkar, R. Sharma and Y. Chen, *Bioorg. Med. Chem. Lett.*, 2011, **21**, 1306–1309.
78. R. S. Obach, G. S. Walker and M. A. Brodney, *Drug Metab. Dispos.*, 2016, **44**, 634–646.

CHAPTER 10

Cytochrome P450 Induction

HUA LV[a], WEI ZHU*[a] AND HONG SHEN*[a]

[a]Roche Innovation Center Shanghai, Roche Pharma Research and Early Development, Bld 5, 371 Lishizhen Road, Shanghai, 201203, China
*E-mail: wei.zhu.wz1@roche.com, hong.shen.hs1@roche.com

10.1 Introduction

10.1.1 Definition

The induction of drug metabolism enzymes refers to the increased expression of drug-metabolizing enzymes, which is associated with enhanced drug metabolism. Drug metabolism enzyme induction is usually the result of increased transcription of the associated gene. The most well studied area for metabolism enzyme induction is related to one or several forms of cytochrome P450 (CYP).

While drugs may also induce their own metabolism, a phenomenon referred to as "auto-induction", induction is of particular importance in drug–drug interactions when one drug increases the clearance of a second, by increasing its rate of metabolism. This type of drug–drug interaction may lead to significant clinical consequences, such as lower plasma concentration levels, decreased half-life and possible therapeutic failure. Thus, addressing such a liability could be highly critical in the early stages of drug discovery programs. Human CYP1A, CYP2A, CYP3A, CYP2B, CYP2C and CYP2E are known to be inducible. Although CYP induction in humans and in animals is mechanistically analogous, the quantitative and qualitative inductive responses to inducers are, however, significantly different among different species.

Drug Discovery Series No. 79
The Medicinal Chemist's Guide to Solving ADMET Challenges
Edited by Patrick Schnider
© The Royal Society of Chemistry 2021
Published by the Royal Society of Chemistry, www.rsc.org

10.1.2 Mechanism

The most common mechanism of CYP enzyme induction is transcriptional gene activation (Figure 10.1).[1,54,55] Transcriptional activation is mediated *via* nuclear receptors (NRs) that function as transcription factors, such as constitutive androstane receptor (CAR)[2] and pregnane X receptor (PXR).[3] In addition, the aryl hydrocarbon receptor (AhR) regulates CYP1A genes.[4] These NRs are activated either through ligand (drug) binding or other mechanisms, such as regulation of coactivators or corepressors. In the absence of a ligand, the NR is associated with NR corepressor complexes resulting in only a baseline level of transcription. When a ligand binds to the ligand binding domain (LBD) of the NR, it induces conformational changes which subsequently result in the release of corepressors and recruitment of coactivators. The LBDs of many NRs are different among species, especially between animals and humans. Therefore, *in vitro* and *in vivo* animal models of CYP induction can be misleading, and these models are not recommended to be employed to predict the potential induction effect in humans.

10.1.3 Consequences

CYP induction can have profound and extensive effects on pharmacology and clinical practices. Specifically, CYP induction may lead to therapeutic efficacy reduction of a co-medication or the CYP inducer itself in the case of auto-induction. In addition, the augmented production of metabolites could induce toxicity.[55]

The ramifications of CYP induction on clinical and toxicological consequences depend on multiple factors, such as (1) the induction potency and magnitude of the compound, (2) the threshold concentration and exposure duration of the inducer for the induction to proceed, (3) the metabolic properties of the inducer, (4) the duration of treatment of the inducer, (5) the duration of the induction after the inducer is withdrawn, (6) the route of administration, (7) the tissue location of the CYP induction, (8) the CYP inhibition potential of the inducer in addition to induction, (9) the CYP inhibiting nature of the co-medication, (10) the safety margin of the co-medication, (11) the exposure variability in the target patient population and (12) the translatability from pre-clinical to clinical observations and from animals to humans.

10.1.4 Screening Strategies

As a strong metabolic liability, CYP induction has prompted the pharmaceutical industry to establish and routinely apply *in vitro* methods to screen compounds in order to establish structure–activity relationships (SARs), which may help to eventually eliminate the liability.[56]

Several models to evaluate the induction of CYP enzymes have been established over the past decade (Figure 10.2). Since animal models are not

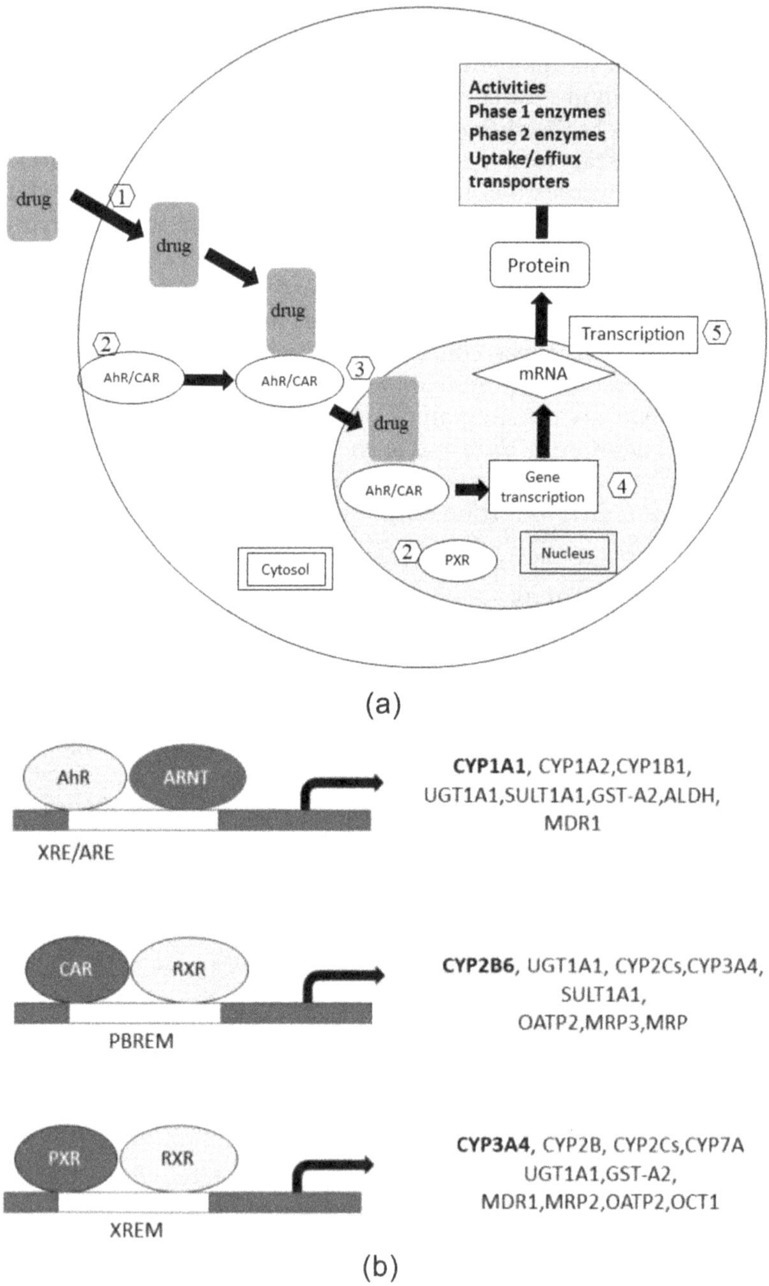

Figure 10.1 (a) Overview of the receptor-mediated mechanism of enzyme induction. (1) The drug enters the cell through passive diffusion and/or active uptake. (2) The nuclear receptors, aryl hydrocarbon receptor (AhR) and constitutive androstane receptor (CAR), are both located

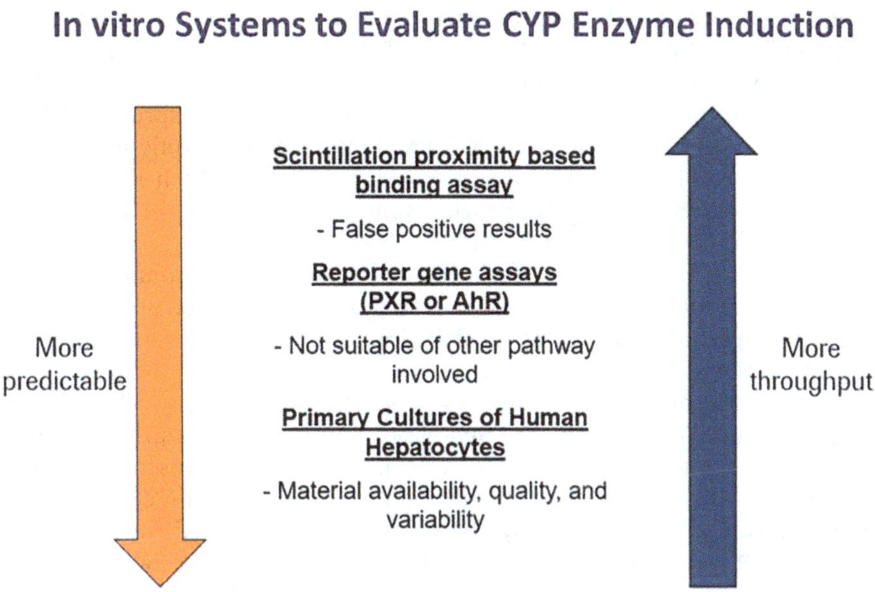

Figure 10.2 *In vitro* systems to evaluate CYP enzyme induction.

---

in the cytoplasm, and pregnane X receptor (PXR) is mainly located in the nucleus. This scheme describes AhR and CAR-mediated pathways only; however, PXR is activated in the same manner in the nucleus. (3) Once the drug binds to receptors they are translocated to the nucleus. (4) The activated receptors form heterodimers with other factors [shown in (b)] which bind to response elements leading to the transcription of the respective CYP isoform. (5) mRNA translocates to the cytoplasm where the translation into CYP and other active proteins occurs. Activities regulated are phase 1 and 2 drug-metabolizing enzymes, such as CYPs and conjugating enzymes, as well as uptake and efflux transporters. (b) Nuclear receptors and their regulation of drug-metabolizing enzymes and transporter proteins. Activated nuclear receptors – AhR, CAR and PXR – form heterodimers with AhR nuclear translocator (ARNT, for AhR) and retinoid X receptor (RXR, for both PXR and CAR). The heterodimers bind to target xenobiotic response elements (XRE) located in both the proximal and distal P450 gene promoters, resulting in the transcription of the respective CYP isoform. Proteins other than CYPs, such as Phase 2 enzymes and transporters, are regulated *via* these pathways. ARE, aryl hydrocarbon response element; PBREM, phenobarbital-responsive enhancer module; XREM, xenobiotic-responsive enhancer module. Reproduced from ref. 1 with permission from Taylor & Francis, http://www.tandfonline.com/, Copyright 2008.

predictable for the induction potential in humans due to the species differences in the LBD of most NRs, most of the induction models are based on human reagents. There are several cell-free high-throughput models which involve measurement of the ligand binding to the expressed NR in scintillation proximity type assays (SPA).[5] Another commonly used high-throughput cell-based model is a reporter gene assay which involves the coexpression of an NR together with a gene promoter region coupled to a reporter gene.[6]

Since the induction of CYP3A4 is considered most relevant and common for drug–drug interactions, the high throughput gene reporter assay (PXR) is used for screening compounds for their induction potential with respect to CYP3A4 in the discovery stage.[7]

However, the disadvantages of these high-throughput assays should also be considered. For example, the SPA-based binding assay is prone to false positive results, while the PXR reporter gene assay only represents the PXR-dependent induction pathway, neglecting other pathways of CYP3A4 induction that exist *in vivo*. Ultimately, the present "gold standard" for assessing human CYP induction is to employ primary cultures of fresh or cryopreserved human hepatocytes. In this system, all of the NRs, corepressors and activators, response elements and target genes are present in their natural environment. The Food and Drug Administration (FDA) considers data from a cell-based system more relevant to the *in vivo* situation. An FDA accepted method is to evaluate the enzyme induction potential of a compound in primary human hepatocytes from three individual donors after treatment with the drug of interest for three days. The extent of CYP expression and induction could be quantitatively assessed by measuring CYP enzyme mRNA levels when incubated with the investigational drug. A cut-off is determined from known positive and negative controls to calibrate the system. For example, a twofold or greater increase in mRNA and a response change of 20% or more than that of the positive control in the presence of an investigational drug are considered as a positive finding.[8] Although there are drawbacks associated with the acquisition of human hepatocytes from multiple donors and big variability among different donors, a primary hepatocyte assay remains the most representative of the *in vivo* situation in humans.

10.1.5 Relevance

P450 enzyme induction, especially of CYP3A4, 1A2 and 2B6, can have profound clinical consequences as they represent the most important CYP enzymes due to their high hepatic expression. As shown in Figure 10.1b, CYP1A2 can be induced *via* AhR activation. It has been also reported that CYP1A2 can be induced *via* PXR activation or a combined CAR–PXR activation indirectly with an unknown mechanism. The direct PXR activation and the combined CAR–PXR activation can induce both CYP2B6 and CYP3A4 enzymes. Examples of representative inducers of CYP isoforms 3A4, 1A2 and 2B6 are shown in Figure 10.3. A more comprehensive list of CYP inducers can be found in a review by Hukkanen.[33]

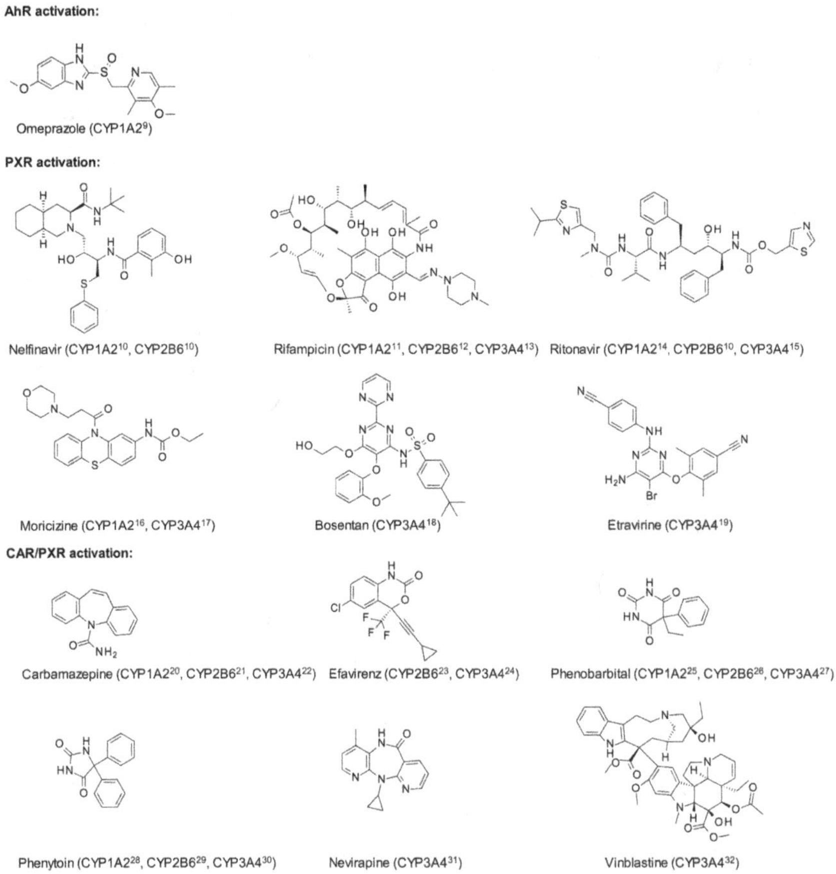

Figure 10.3 Examples of representative CYP inducers.

Key References

- O. Pelkonen, *et al.*, Inhibition and induction of human cytochrome P450 enzymes: current status, *Arch. Toxicol.*, 2008, **82**, 667–715.
- N. J. Hewitt, *et al.*, Induction of hepatic cytochrome P450 enzymes: methods, mechanisms, recommendations, and *in vitro–in vivo* correlations, *Xenobiotica*, 2007, **37**(10–11), 1196–1224.
- J. H. Lin, CYP induction-mediated drug interactions: *in vitro* assessment and clinical implications, *Pharm. Res.*, 2006, **23**(6), 1089–1115.
- V. Chu, *et al.*, *In vitro* and *in vivo* induction of cytochrome P450: a survey of the current practices and recommendations: a Pharmaceutical Research and Manufacturers of America perspective, *Drug Metab. Dispos.*, 2009, **37**(7), 1339–1354.
- J. Hukkanen, Induction of cytochrome P450 enzymes: a view on human *in vivo* findings, *Expert Rev. Clin. Pharmacol.*, 2012, **5**(5), 569–585.
- Y. D. Gao, *et al.*, Attenuating pregnane X receptor (PXR) activation: a molecular modeling approach, *Xenobiotica*, 2007, **37**(2), 124–138.

10.2 Medicinal Chemistry Strategies to Mitigate CYP3A4 Induction Mediated by PXR Activation

Among the induction of several CYP enzymes, CYP3A4 induction represents the most significant clinical issue because CYP3A4 is involved in the metabolism of more than 50% of marketed drugs. As aforementioned, the predominant mechanism leading to CYP3A4 induction is through the activation of PXR, a member of the NR superfamily. A ligand-regulated DNA-binding transcription factor, human PXR contains a LBD, which is large, hydrophobic and flexible. Thus PXR is able to accommodate structurally diverse ligands. It is generally accepted that the pharmacophore of hPXR agonists comprises an essential hydrogen-bond acceptor interacting with residue H407, and at least two flanking hydrophobic groups (preferably aromatic groups) forming critical hydrophobic interactions with the hPXR LBD. A third hydrophobic group would potentially increase potency, but is not necessary (Figure 10.4).[34,35]

Based on this putative model, several strategies to mitigate hPXR activation have been adopted by different groups, *e.g.* Gao, *et. al.* as follows:[34]

1. Introducing a polar substituent to the hydrophobic group.
2. Removing or replacing the key hydrophobic group with a less hydrophobic group.
3. Introducing steric hindrance or rigidifying the structure.

However, these structural elements are often responsible for the binding with the targeted protein to exert the desired pharmacology too. As such, it is very important to compare pharmacophore models of the desired target *vs.* hPXR in order to identify the precise scaffold optimization direction, which enables medicinal chemists to dial out the hPXR activity while maintaining the desirable on-target affinity.

In the following sections, we exemplify the hPXR mitigation strategies with literature examples which led to reduced off-target PXR activation liability without sacrificing the desired on-target activity. For clarity, examples which

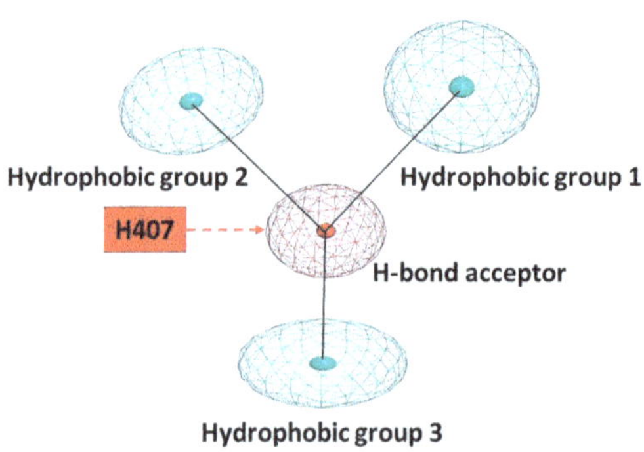

Figure 10.4 Pharmacophore of hPXR agonists.

could not sufficiently separate PXR activation from the on-target activity, or are difficult to rationalize, are omitted.

10.2.1 Introducing a Polar Substituent to the Hydrophobic Group

Introducing a polar substituent to the hydrophobic group in order to disrupt its interaction with hPXR is the most successful strategy to mitigate CYP3A4 induction. However, this polar substituent has to be tolerated in the binding pocket of target protein.

Examples

1

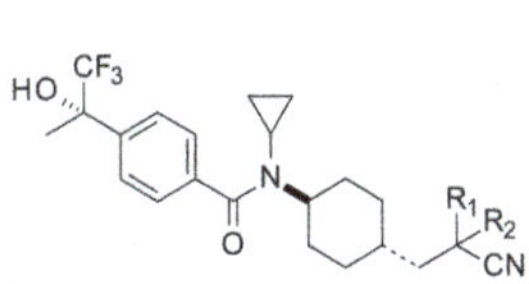

1.1-1 – 1.1-3

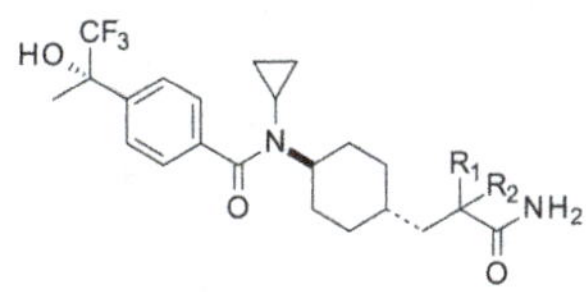

1.1-4 – 1.1-6

Cpd	R_1/R_2	11β-HSD1 (K_i, nM)	PXR activation (%)[a]	Cpd	R_1/R_2	11β-HSD1 (K_i, nM)	PXR activation (%)[a]
1.1-1	H/H	0.7	30	**1.1-4**	H/H	2.4	5
1.1-2	Me/Me	0.88	60	**1.1-5**	Me/Me	3.8	4
1.1-3	Cyclopropyl	0.6	60	**1.1-6**	Cyclopropyl	1.7	6.5

[a]PXR activity with a 2 μM concentration of individual compounds relative to the activity achieved with 10 μM rifampin in a luciferase reporter gene assay.

Rew *et al.* from Amgen reported a novel cyclohexyl benzamide series of 11β-HSD1 inhibitors. They found that conversion of the nitrile group to the more polar primary amide significantly reduced PXR activation consistently for a series of analogues. As rationalized, the key to successfully diverting SAR of PXR *vs.* 11β-HSD1 was largely that this polar carboxamide group was tolerated in the 11β-HSD1 binding pocket, where it formed a weak water-mediated hydrogen bond with a tyrosine residue as shown in a co-crystal structure.[36]

(*continued*)

Examples

2

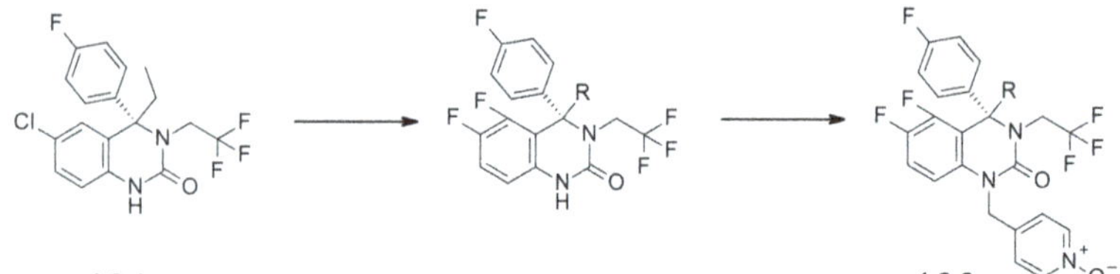

1.2-1 1.2-2a-c 1.2-3a-c

Cpd	R	α1I FLIPR IP depolarized (nM)	PXR (% Rif at 10 μM)[a]
1.2-1	—	61	63
1.2-2a	Ethyl	16 ± 6	52
1.2-2b	*n*-propyl	9 ± 4	73
1.2-2c	CH_2CH_2F	17 ± 4	53
1.2-3a	Ethyl	12 ± 4	26
1.2-3b	*n*-propyl	6 ± 1	11
1.2-3c	CH_2CH_2F	38 ± 5	56

[a]PXR activity with a 10 μM concentration of individual compounds relative to the activity achieved with 10 μM rifampin in a luciferase reporter gene assay.

Schlegel *et al.* from Merck disclosed a chemical series of 4,4-disubstituted quinazolinones as T-type calcium channel antagonists for the treatment of CNS disorders. The lead compound **1.2-1** showed good potency, selectivity and *in vivo* efficacy, but it was a PXR activator. The fine-tuning of the halogen substitution pattern on the quinazolinone core was explored with basically no impact on PXR activation mitigation as seen for compounds **1.2-2a-c**. A more drastic approach was then explored by introducing the polar methyl 4-pyridyl *N*-oxide moiety at the 1-position of the quinazolinone core, which led to decreased PXR activation while maintaining potency for compounds **1.2-3a-b** bearing an ethyl and *n*-propyl substituent at the 4-position, respectively, but surprisingly not for monofluoroethyl derivative **1.2-3c**.[37]

3

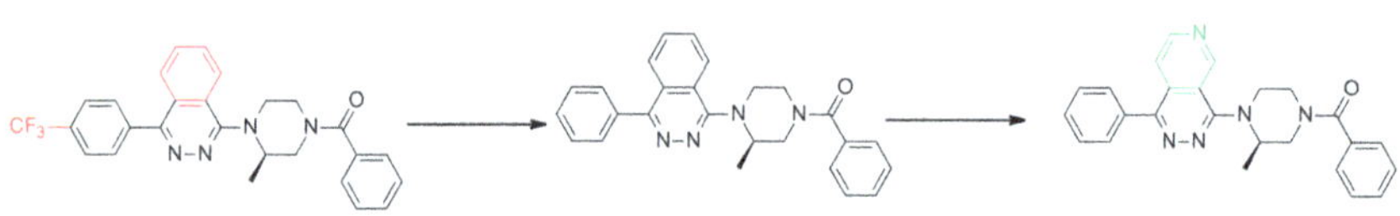

1.3-1

hSMO IC_{50}: 2.8 nM
PXR, % cntl @ 2 μM: 81
LogD (pH 7.4): 3.74

1.3-2

hSMO IC_{50}: 2.2 nM
PXR, % cntl @ 2 μM: 30
LogD (pH 7.4): 2.29

1.3-3

hSMO IC_{50}: 5.8 nM
PXR, % cntl @ 2 μM: 4.9

Examples

Kaizerman *et al.* from Amgen reported their efforts in addressing the PXR liabilities of compound **1.3-1**, a potent hedgehog/smoothened (SMO) antagonist for the treatment of cancer. The initial efforts to mitigate the PXR activation focused on the removal of the lipophilic CF_3 group on the left-hand phenyl ring or replacement with polar substituents, *e.g.* carboxamide, carboxyl, or hydroxymethyl. Indeed, there was a general trend that reduced lipophilicity (reflected by LogD values) led to decreased PXR activation. The same trend was also observed when the phenyl group was replaced with more polar heteroaryl groups such as pyridyl, oxazolyl, imidazolyl, or pyrimidinyl. However, both approaches led to either reduced SMO potency or suboptimal drug metabolism and pharmacokinetics (DMPK) properties. Subsequently, simple removal of the CF_3 gave analogue **1.3-2** possessing the desired combination of SMO activity and reduced PXR liability. Encouraged by the effect of reducing lipophilicity on PXR mitigation, the authors introduced an additional *N* atom to the central phthalazine core, which led to the discovery of analogue **1.3-3** with significantly reduced PXR activation while maintaining good SMO activity.[38]

4

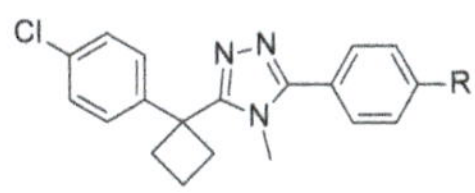

Cpd	R	11β-HSD1 (IC_{50}, nM)	PXR activation (%)
1.4-1	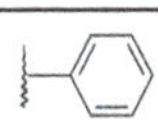	11	92
1.4-2	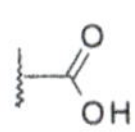	76	−1
1.4-3	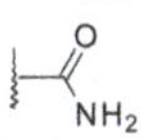	1	21
1.4-4	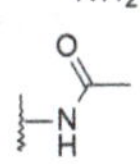	5.8	21
1.4-5	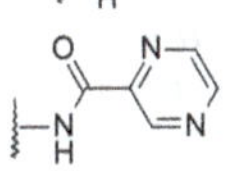	69	−2.5

Sun *et al.* from Merck identified a series of triazole analogs as potent and selective 11β-HSD1 inhibitors; however, the early lead compound **1.4-1** was found to be a strong PXR activator. This team focused on the strategy of adding polar groups on the distal phenyl ring to mitigate this off-target liability based on a PXR docking model. Indeed, replacement of the terminal phenyl with a series of polar groups led to decreased PXR activation, especially for compounds **1.4-2** and **1.4-5**, but at the cost of reduced HSD1 activity. Compounds **1.4-3** and **1.4-4** demonstrated a good balance of reduced PXR activation and potent HSD1 inhibition, though their ADMET properties had to be improved.[39]

(*continued*)

Examples

5

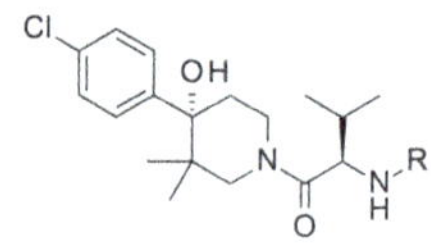

Cpd	R	CCR1 IC_{50} (nM)	PXR EC_{20}/EC_{60} (μM)
1.5-1	Ph	18	1.1/2.9
1.5-2	3-(SO_2NH_2)Ph	2.3 ± 1.1	>50/>50
1.5-3	3-(CO_2H)Ph	7.2 ± 1.1	>50/>50
1.5-4	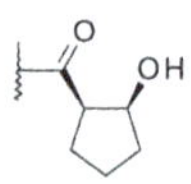	2.3 ± 0.6	1.4 (EC_{50})
1.5-5	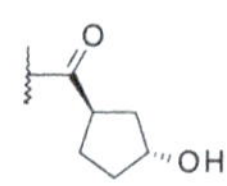	0.8 ± 0	41/>50
1.5-6	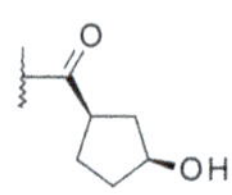	0.8 ± 0.2	5.3/>50
1.5-7	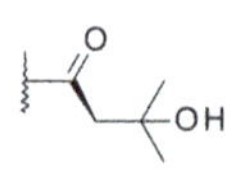	5.2	1.8 (EC_{50})

SAR exploration to mitigate PXR activation was reported by Gardner *et al.* from BMS in a chemical series of CCR1 antagonists as shown below. The lead compound **1.5-1** was a potent CCR1 antagonist, but also a strong PXR activator as revealed by its low EC_{20} and EC_{60}. The introduction of polar substituents, *e.g.* sulfonamide and carboxyl, could eliminate PXR off-target liability. However, both compounds, **1.5-2** and **1.5-3**, suffered from poor pharmacokinetic properties. The attempts were then to replace the phenyl with polar group-substituted cyclopentyls. An interesting SAR study showed that the effect of a polar hydroxyl group on PXR activation was both site- and stereochemistry-sensitive. The *trans*-3-hydroxycyclopentane analogue **1.5-5** showed greatly reduced PXR activation. Later on, the co-crystal structure of analogue **1.5-7** with the PXR LBD was solved, which revealed that the intramolecular hydrogen bonding between the terminal hydroxyl and the central valine NH facilitated the *gem*-dimethyl group to occupy the hydrophobic pocket of the PXR LBD. The same hydrogen bonding effect has been proposed to explain the increased PXR activation of the *cis* analogue **1.5-6** with an EC_{20} of 5.3 μM. Though it was not discussed in the paper, a similar H-bonding between the hydroxyl and the carbonyl could account for the increased PXR activation of analogue **1.5-4**. In contrast, it was not feasible to form such an intramolecular H-bond interaction for the *trans* isomer **1.5-5**, and the polarity of the hydroxyl group could lead to reduced PXR binding in this hydrophobic pocket. Interestingly, the CCR1 activity was not affected along the PXR SAR exploration, which indicated the larger tolerability of the CCR1 binding pocket to modifications in the right-hand amide region.[40]

Examples

6

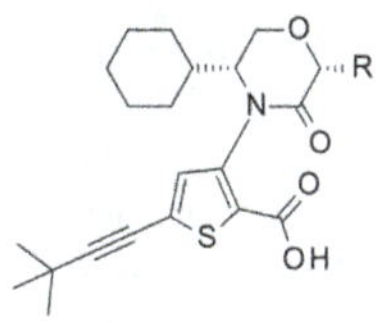

Cpd	R	NS5B IC_{50} (μM)	Replicon 1b EC_{50} (μM)	PXR activation at 10 μM[a]
1.6-1	H	0.003	0.017	87%
1.6-2	CH_2CH_2OH	0.002	0.007	3%
1.6-3	$CH_2CH_2CH_2OH$	0.002	0.014	5%
1.6-4	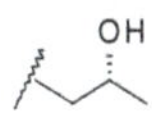	0.003	0.008	9%

[a]PXR activity with a 10 μM concentration of individual compounds relative to the activity achieved with 10 μM rifampin in a luciferase reporter gene assay.

Fu *et al.* from Novartis reported a chemical class of thiophene carboxylic acids as Hepatitis C Virus (HCV) RNA-dependent RNA polymerase (NS5B) inhibitors. The lead compound **1.6-1** was a strong CYP3A4 inducer despite its excellent anti-HCV potency in both enzymatic and replicon assays. Based on the co-crystal structure with NS5B, the substituent R in the *α*-position to carbonyl group was identified to be exposed to solvent, which allowed the use of a polar group at this position in order to destabilize the interaction of the molecule with PXR while maintaining potent NS5B inhibition. As expected, this strategy successfully led to the discovery of new hydroxyalkyl substituted analogues with significantly reduced CYP3A4 induction, as determined in both the PXR reporter gene assay and the human hepatocyte assay.[41]

10.2.2 Removing or Replacing the Key Hydrophobic Group with a Less Hydrophobic Group

Examples

1

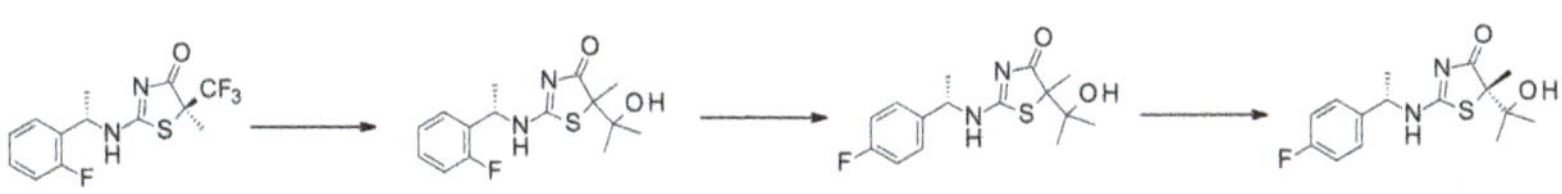

2.1-1	2.1-2	2.1-3	(s)-2.1-3
11β-HSD1 Ki = 22 nM	11β-HSD1 Ki = 34 ± 11 nM	11β-HSD1 Ki = 65 ± 10 nM	11β-HSD1 Ki = 35 ± 1 nM
PXR = 45 ± 5% of control[a]	PXR = 19 ± 3% of control[a]	PXR = 10 ± 2% of control[a]	PXR = 6 ± 2% of control[a]

[a]PXR activity with a 20 μM concentration of individual compounds relative to the activity achieved with 12.5 μM rifampin.

(*continued*)

Examples

Fotsch *et al.* from Amgen described thiazolone derivatives as selective inhibitors of 11β-HSD1 for potential treatment of diabetes. An early lead compound **2.1-1** showed potent and selective 11β-HSD1 activity. However, this compound was associated with the liability of PXR activation, determined in the human PXR reporter gene assay. The medicinal chemistry strategy of mitigating PXR activation was to replace the lipophilic substituent at the 5-position of **2.1-2** with a polar group. Indeed, with the introduction of a polar tertiary alcohol, the PXR activation could be reduced. The subtle change on the left phenyl ring also had an impact on the PXR activation. The *para*-fluoro substituted analogue **2.1-3** demonstrated slightly reduced PXR activation. The (*S*)-isomer of **2.1-3** gave further improved 11β-HSD1 activity and reduced PXR activation. The reduced PXR activation was nicely translated into the decreased CYP3A4 mRNA expression and enzyme activity in human hepatocytes. Interestingly, during the PXR optimization, the 11β-HSD1 activity was basically not affected. The co-crystal structure of (*S*)-**2.1-3** with 11β-HSD1 elucidated this diverted SAR for PXR and 11β-HSD1. The *gem*-dimethyl of the tertiary alcohol formed favorable hydrophobic interactions with the residues of 11β-HSD1, while the hydroxyl group pointed to the solvent exposing area.[42]

2

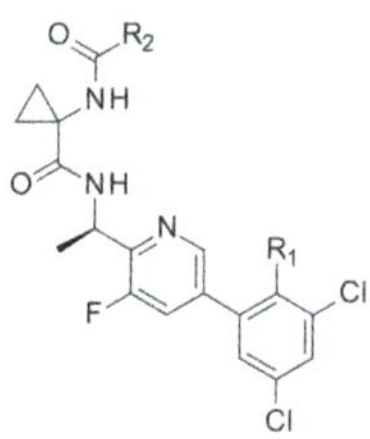

Cpd	R_1	R_2	hBK_1 (K_i, nM)	PXR (%)[a]
2.2-1	CH_3CH_2O	CF_3	1.5	89
2.2-2	CH_3CH_2O	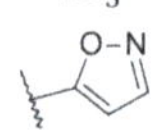	0.6	15
2.2-3	CHF_2CH_2O	CF_3	1.5	99
2.2-4	CHF_2CH_2O	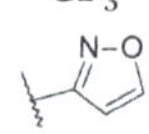	1.2	23
2.2-5	CHF_2CH_2O	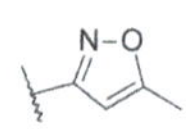	1.4	28
2.2-6	CHF_2CH_2O	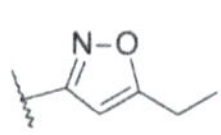	1.9	37
2.2-7	CHF_2CH_2O	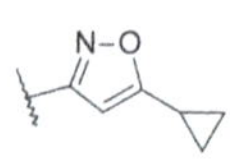	2.6	58

Examples				
2.2-8	CHF_2CH_2O	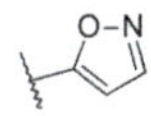	0.4	21
2.2-9	CHF_2CH_2O	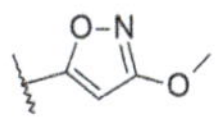	0.5	8
2.2-10	CHF_2CH_2O	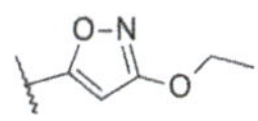	1.1	37
2.2-11	CHF_2CH_2O	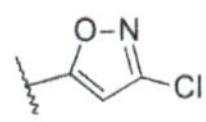	0.2	68

[a]PXR activity with a 10 μM concentration of individual compounds relative to the activity achieved with 10 μM rifampin in a luciferase reporter gene assay.

In an attempt of solving PXR off-target liability associated with the trifluoroacetamide biaryl class of bradykinin B_1 receptor antagonists, Kuduk *et al.* at Merck discovered that the replacement of the trifluoroacetamide with the isoxazole carboxamide was essential to address this issue. Compared to the trifluoromethyl group, the introduction of the relatively polar isoxazolyl likely destabilized the hydrophobic interaction with the PXR LBD at this position. Also, this hypothesis could be supported by the PXR SAR trend that the more lipophilic the substituent on the isoxazole ring, the higher the PXR activation. The polar methoxyl substituted isoxazole analogue showed the least PXR activation.[43]

3

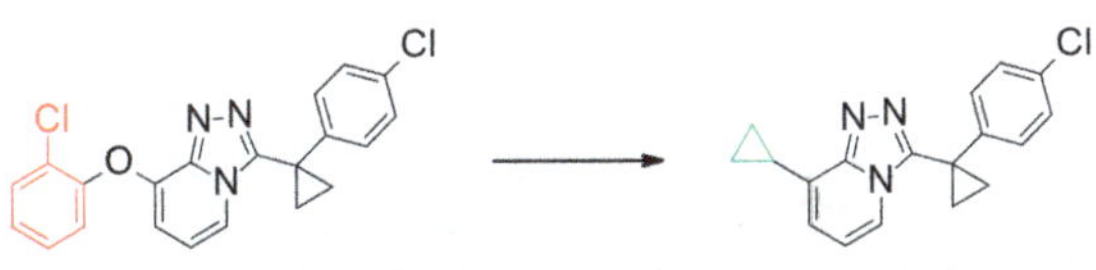

2.3-1

11β-HSD1 IC_{50} (nM): 1.0
PXR EC_{50} (μM) / Ymax(%): 0.97 / 100

2.3-2

11β-HSD1 IC_{50} (nM): 1.6
PXR EC_{50} (μM) / Ymax(%): 50 / 14

BMS scientists reported a novel triazolopyridine class of human 11β-HSD1 inhibitors, *e.g.* compound **2.3-1**, which possessed excellent HSD1 inhibitory activity. However, this class of compounds was typically associated with a strong PXR activation liability. A systematic SAR study was carried out to separate the PXR transactivation from the desired HSD1 inhibition, which eventually led to the discovery of a new class of shortened HSD1 inhibitors with much reduced PXR activation as exemplified by compound **2.3-2**. No explanation for reduced PXR activation was given, but very likely it could be ascribed to the removal of the lipophilic phenyl ring essential for the hydrophobic interactions with PXR. Fortunately, the absence of the left-hand phenyl ring did not affect HSD1 activity.[44]

10.2.3 Introducing Steric Hindrance or Rigidifying the Structure

Examples

1

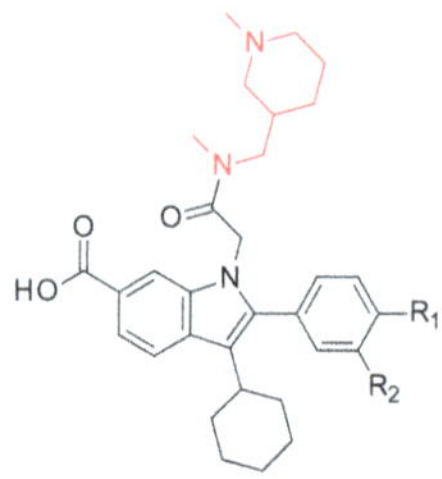

3.1-1 – 3.1-3

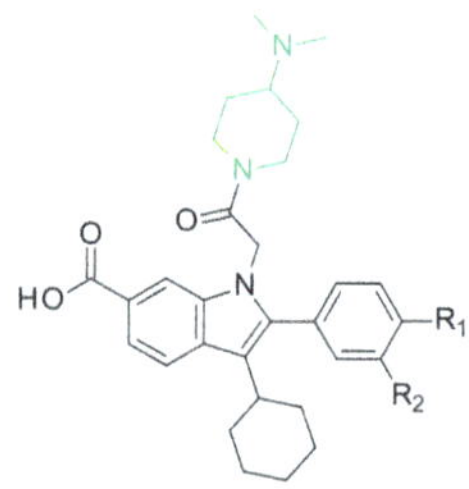

3.1-4 – 3.1-6

Cpd	R_1/R_2	HCV EC_{50} (μM)	PXR activation (%)[a]	Cpd	R_1/R_2	HCV EC_{50} (μM)	PXR activation (%)[a]
3.1-1	H/H	0.7	82	3.1-4	H/H	0.3	14
3.1-2	H/F	—	72	3.1-5	H/F	—	30
3.1-3	Cl/H	0.3	92	3.1-6	Cl/H	—	15

[a]Results are expressed as % PXR mediated induction of the SEAP reporter gene with respect to positive control (10 μM rifampicin).

Harper *et al.* from Merck reported a series of zwitterionic indole-*N*-acetamides as allosteric inhibitors of HCV NS5B polymerase, which demonstrated not only potent anti-HCV activity, but also unwanted strong PXR activation as exemplified by compounds **3.1-1-3**. Through systematic PXR SAR study, the authors postulated that this off-target liability was associated with the acetamide side chain. To solve this issue, they pursued a strategy of increasing rigidity to presumably destabilize PXR binding. As a result, compounds **3.1-4-6** showed consistently decreased PXR activation as determined in a reporter gene assay.[45]

Examples

2

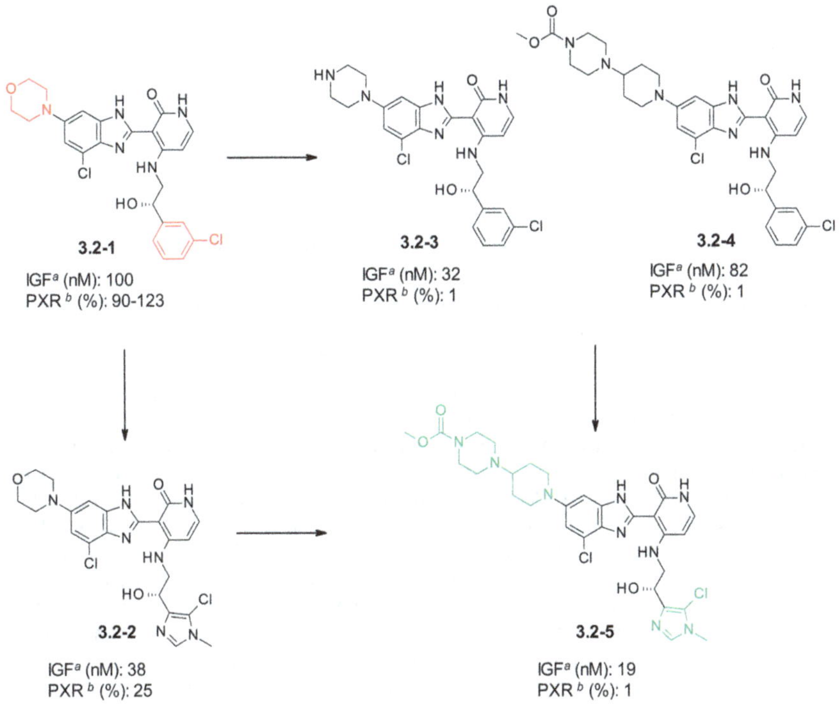

[a]IGF-1R kinase IC_{50}.
[b]% Transactivation of PXR receptor, compared to Rifampicin at 10 μM concentration.

Zimmermann *et al.* from BMS systematically studied the PXR SAR of a benzimidazole class of IGF-1R inhibitors. Their representative first generation IGF-1R inhibitor compound **3.2-1** was a strong PXR activator. Its aromatic 3-chlorophenyl substituent was likely an essential element for the PXR binding. As their first attempt to mitigate PXR activation, the lipophilic 3-chlorophenyl ring of **3.2-1** was replaced with substituted imidazoles, which led to the discovery of a representative compound **3.2-2** with partially reduced PXR activation. As an alternative approach, the replacement of the left-hand side morpholine with larger and more polar groups was explored, which turned out to be a win–win solution for both IGF-1R activity improvement and PXR mitigation. Compounds **3.2-3** and **3.2-4** showed essentially no PXR activation and improved IGF-1R activity, which might indicate that the left hand side was exposed to the solvent area in the IGF-1R binding pocket. Finally, combining the polar and rigid left-hand piperidine and the right-hand imidazole moiety gave PD study candidate compound **3.2-5** with optimal DMPK properties. However, the story did not end here as compound **3.2-5** showed strong CYP3A4 induction in human hepatocytes despite no PXR transactivation. This unexpected finding indicated the possible existence of non-PXR meditated CYP3A4 induction pathways and also highlighted the translational importance of the human hepatocyte assay for CYP3A4 induction assessment.[46]

Examples

3

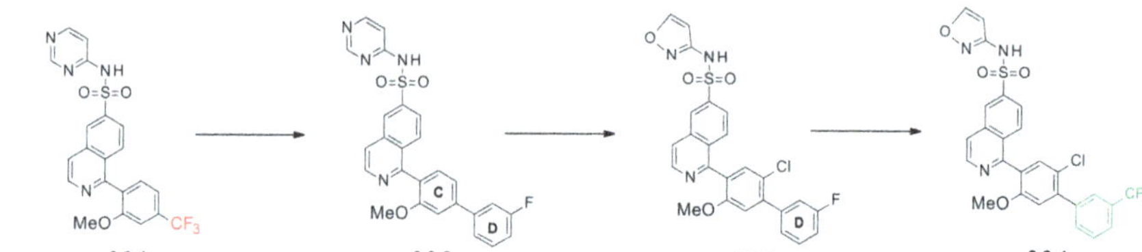

3.3-1 → 3.3-2 → 3.3-3 → 3.3-4

3.3-1	3.3-2	3.3-3	3.3-4
$Na_v1.7$ IC_{50} (μM): 0.73	$Na_v1.7$ IC_{50} (μM): 0.15	$Na_v1.7$ IC_{50} (μM): 0.049	$Na_v1.7$ IC_{50} (μM): 0.036
CYP3A4 induction (%)[a]: 177	CYP3A4 induction (%)[a]: 12	PXR activation (%)[a]: 68	PXR activation (%)[a]: 1

[a]Expressed as a percentage of the induction seen with rifampicin when both compounds are incubated with human hepatocytes at a concentration of 10 μM.

Weiss *et al.* from Amgen explored the bicyclic heteroarylsulfonamides as $Na_V1.7$ inhibitors for the treatment of pain. The prototype compound **3.3-1** demonstrated strong CYP3A4 induction in human hepatocytes. Replacing the CF_3 substituent on ring C with a meta-fluoro substituted aryl ring provided compound **3.3-2** with significantly reduced CYP3A4 induction and improved $Na_V1.7$ activity, but unfortunately suboptimal pharmacokinetics. The reduced CYP3A4 induction was likely due to the introduction of steric hindrance which led to a disfavored interaction with PXR. Next, substituting the 4-aminopyrimidine head group with 4-amino isoxazole afforded compound **3.3-3** with increased $Na_V1.7$ activity and desirable pharmacokinetics at the cost of elevated CYP3A4 induction, as reflected by the high PXR activation. Replacing the small fluoro-substituent on D ring with a large CF_3 group (and the chloro-substituent on ring C by fluoro) to give compound **3.3-4** eventually dialed out PXR activation while maintaining $Na_V1.7$ activity and the favorable pharmacokinetic profile of compound **3.3-3**.[47]

4

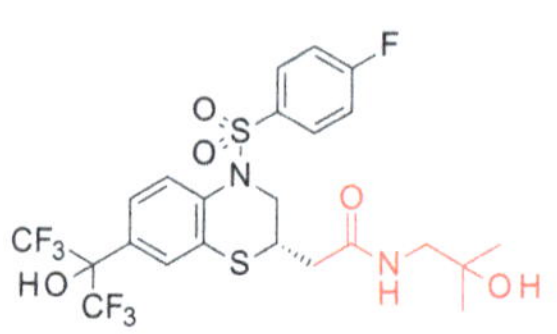

3.4-1

RORγτ EC_{50} nM: 31
PXR EC_{50} nM (%max.): 175 (90)

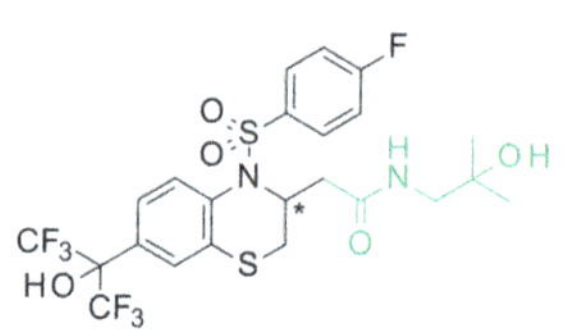

3.4-2

RORγτ EC_{50} nM: 32
PXR EC_{50} nM (%max.): 1100 (90)
* Absolute stereochemistry not established

The chemical class of benzothiazine sulfonamides represented by compound **3.4-1** was described to show potent RORγτ inverse agonism. However, most of the analogues were also strong PXR activators assessed using a transactivation assay in HepG2 cells. Based on the X-ray co-crystal structures of **3.4-1** with the LBDs of both RORγτ and PXR, Murali Dhar *et al.* from BMS moved the hydroxypropyl acetamide side chain from position 2 of the benzothiazine ring to position 3 in order to improve selectivity of RORγτ *vs.* PXR by creating a steric clash with the corresponding residues in the PXR LBD. This rational design successfully led to the discovery of compound **3.4-2** with similar RORγτ potency and much decreased PXR activity.[48]

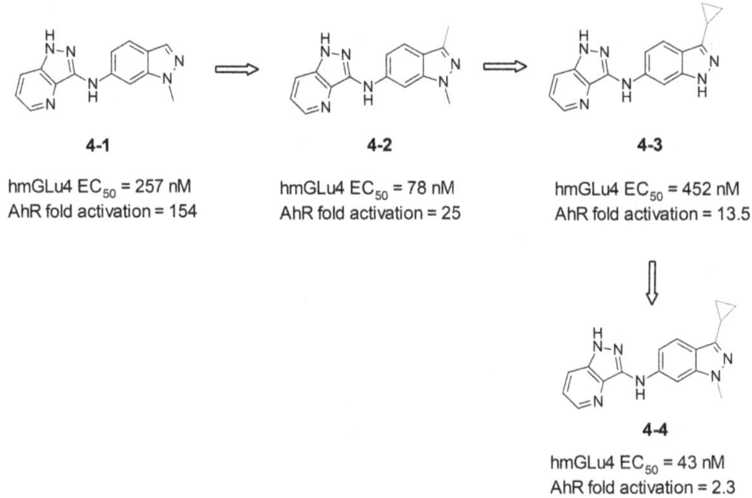

Scheme 10.1 mGlu4 examples from Vanderbilt University.

10.3 Medicinal Chemistry Strategies to Mitigate Induction of Other P450 Enzymes

To the best of our knowledge, there are no literature reports on the mitigation of CYP2B6 induction and only very few examples of CYP1A2 induction mitigation. However, considering that both CYP2B6 and CYP1A2 enzymes can be induced by PXR activation, the general strategies of mitigating CYP3A4 induction can be potentially applied to the mitigation of CYP2B6 and CYP1A2 inductions also. There is one example of mitigating CYP1A2 induction reported by Engers *et al.* from Vanderbilt University in their discovery of selective metabotropic glutamate receptor 4 (mGlu4) positive allosteric modulators (PAMs) (Scheme 10.1). Early compound **4-1** suffered from strong CYP1A2 induction as measured by AhR activation studies. Optimization of the substituents at the 1- and 3-position of the 6-indazolyl group led to compound **4-4,** which was not only devoid of CYP1A2 activation liability but also a more potent mGlu4 PAM as compared with **4-1.**[49]

10.4 *In Silico* Models for Predicting P450 Enzyme Induction

To date, a number of computational models based on the three-dimensional structures of the PXR-ligand binding domain (protein based) as well as the pharmacophores generated from known PXR activators (ligand based) have been developed for quantitative structure–activity relationship (QSAR) prediction of new drug candidates to varying degrees of accuracy.[50–52]

The correlation of P450 induction, *e.g.* CYP3A4 and CYP2B6 induction, and the molecular properties of structurally diverse activators and non-activators have been extensively studied. In particular, molecular weight and lipophilicity are reported to be correlated to a certain level with CYP3A4 induction.[53]

In general, the *in silico* models are of interest to medicinal chemists for designing new compounds of low P450 induction potential at the early stage of drug discovery and prioritizing compounds for *in vitro* testing if there is limited assay capacity. However, considering the existence of different mechanisms to induce P450 enzymes, the overall large and flexible ligand binding pockets of nuclear receptors, and the limitation of the size of the training data sets used for the development of *in silico* models, predicting P450 enzyme induction, especially CYP2B6 induction, will be associated with significant uncertainty.

References

1. N. J. Hewitt, E. L. Lecluyse and S. S. Ferguson, *Xenobiotica*, 2007, **37**, 1196–1224.
2. M. Qatanani and D. D. Moore, *Curr. Drug Metab.*, 2005, **6**, 329–339.
3. P. Honkakoski, T. Sueyoshi and M. Negishi, *Ann. Med.*, 2003, **35**, 172–182.
4. P. K. Mandal, *J. Comp. Physiol., B*, 2005, **175**, 221–230.
5. Z. Zhu, S. Kim, T. Chen, J. H. Lin, A. Bell, J. Bryson, Y. Dubaquie, N. Yan, J. Yanchunas, D. Xie, R. Stoffel, M. Sinz and K. Dickinson, *J. Biomol. Screening*, 2004, **9**, 533–540.
6. G. Luo, M. Cunningham, S. Kim, T. Burn, J. Lin, M. Sinz, G. Hamilton, C. Rizzo, S. Jolley, D. Gilbert, A. Downey, D. Mudra, R. Graham, K. Carroll, J. Xie, A. Madan, A. Parkinson, D. Christ, B. Selling, E. Lecluyse and L. S. Gan, *Drug Metab. Dispos.*, 2002, **30**, 795–804.
7. M. Sinz, S. Kim, Z. Zhu, T. Chen, M. Anthony, K. Dickinson and A. D. Rodrigues, *Curr. Drug Metab.*, 2006, **7**, 375–388.
8. *In Vitro Metabolism and Transporter Mediated Drug–Drug Interaction Studies Guidance for Industry*, https://www.fda.gov/downloads/Drugs/Guidances/UCM581965.pdf, October, 2017.
9. D. Diaz, I. Fabre, M. Daujat, B. Saint Aubert, P. Bories, H. Michel and P. Maurel, *Gastroenterology*, 1990, **99**, 737–747.
10. B. J. Kirby, A. C. Collier, E. D. Kharasch, V. Dixit, P. Desai, D. Whittington, K. E. Thummel and J. D. Unadkat, *Drug Metab. Dispos.*, 2011, **39**, 2329–2337.
11. J. T. Backman, M. T. Granfors and P. J. Neuvonen, *Eur. J. Clin. Pharmacol.*, 2006, **62**, 451–461.
12. K. K. Loboz, A. S. Gross, K. M. Williams, W. S. Liauw, R. O. Day, J. K. Blievernicht, U. M. Zanger and A. J. McLachlan, *Clin. Pharmacol. Ther.*, 2006, **80**, 75–84.

13. E. E. Ohnhaus and B. K. Park, *Eur. J. Clin. Pharmacol.*, 1979, **15**, 139–145.
14. A. Hsu, G. R. Granneman and R. J. Bertz, *Clin. Pharmacokinet.*, 1998, **35**, 275–291.
15. D. Ouellet, A. Hsu, J. Qian, C. S. Locke, C. J. Eason, J. H. Cavanaugh, J. M. Leonard and G. R. Granneman, *Br. J. Clin. Pharmacol.*, 1998, **46**, 111–116.
16. H. J. Pieniaszek Jr, A. F. Davidson and I. H. Benedek, *Ther. Drug Monit.*, 1993, **15**, 199–203.
17. L. Shum, H. J. Pieniaszek Jr, C. A. Robinson, A. F. Davidson, P. J. Widner, I. H. Benedek and W. Flamenbaum, *J. Clin. Pharmacol.*, 1996, **36**, 1161–1168.
18. C. Weber, R. Schmitt, H. Birnboeck, G. Hopfgartner, H. Eggers, J. Meyer, S. Van Marle, H. W. Viischer and J. H. G. Jonkman, *J. Clin. Pharmacol.*, 1999, **39**, 703–714.
19. M. Schöller–Gyüre, T. N. Kakuda, A. Raoof, G. De Smedt and R. M. Hoetelmans, *Clin. Pharmacokinet.*, 2009, **48**, 561–574.
20. R. A. Lucas, D. J. Gilfillan and R. F. Bergstrom, *Eur. J. Clin. Pharmacol.*, 1998, **54**, 639–643.
21. M. Oscarson, U. M. Zanger, O. F. Rifki, K. Klein, M. Eichelbaum and U. A. Meyer, *Clin. Pharmacol. Ther.*, 2006, **80**, 440–456.
22. T. A. Moreland, B. K. Park and G. W. Rylance, *Br. J. Clin. Pharmacol.*, 1982, **14**, 861–865.
23. S. M. Robertson, F. Maldarelli, V. Natarajan, E. Formentini, R. M. Alfaro and S. R. Penzak, *J. Acquired Immune Defic. Syndr.*, 2008, **49**, 513–519.
24. S. Mouly, K. S. Lown, D. Kornhauser, J. L. Joseph, W. D. Fiske, I. H. Benedek and P. B. Watkins, *Clin. Pharmacol. Ther.*, 2002, **72**, 1–9.
25. R. A. Landay, M. A. Gonzalez and J. C. Taylor, *J. Allergy Clin. Immunol.*, 1978, **62**, 27–29.
26. J. Y. Jao, W. J. Jusko and J. L. Cohen, *Cancer Res.*, 1972, **32**, 2761–2764.
27. D. J. Back, M. Bates, A. Bowden, A. M. Breckenridge, M. J. Hall, H. Jones, M. MacIver, M. Orme, E. Perucca, A. Richens, P. H. Rowe and E. Smith, *Contraception*, 1980, **22**, 495–503.
28. H. Wietholtz, T. Zysset, K. Kreiten, D. Kohl, R. Büchsel and S. Matern, *Eur. J. Clin. Pharmacol.*, 1989, **36**, 401–406.
29. M. L. Williams, I. W. Wainer, L. Embree, M. Barnett, C. L. Granvil and M. P. Ducharme, *Chirality*, 1999, **11**, 569–574.
30. P. Crawford, D. J. Chadwick, C. Martin, J. Tjia, D. J. Back and M. Orme, *Br. J. Clin. Pharmacol.*, 1990, **30**, 892–896.
31. D. Mildvan, R. Yarrish, A. Marshak, H. W. Hutman, M. McDonough, M. Lamson and P. Robinson, *J. Acquired Immune Defic. Syndr.*, 2002, **29**, 471–477.
32. N. F. Smith, S. Mani, E. G. Schuetz, K. Yasuda, T. M. Sissung, S. E. Bates, W. D. Figg and A. Sparreboom, *Ann. Pharmacother.*, 2010, **44**, 1709–1717.
33. J. Hukkanen, *Expert Rev. Clin. Pharmacol.*, 2012, **5**, 569–585.
34. Y. D. Gao, S. H. Olson, J. M. Balkovec, Y. Zhu, I. Royo, J. Yabut, R. Evers, E. Y. Tan, W. Tang, D. P. Hartley and R. T. Mosley, *Xenobiotica*, 2007, **37**, 124–138.

35. S. Ekins and J. A. Erickson, *Drug Metab. Dispos.*, 2002, **30**, 96–99.
36. Y. Rew, D. L. McMinn, Z. Wang, X. He, X. R. W. Hungate, J. C. Jaen, A. Sudom, D. Sun, H. Tu, S. Ursu, E. Villemure, N. P. C. Walker, X. Yan, Q. Ye and J. P. Powers, *Bioorg. Med. Chem. Lett.*, 2009, **19**, 1797–1801.
37. K. A. S. Schlegel, Z. Q. Yang, T. S. Reger, Y. Shu, R. Cube, K. E. Rittle, P. Bondiskey, M. G. Bock, G. D. Hartman, C. Tang, J. Ballard, Y. Kuo, T. Prueksaritanont, C. E. Nuss, S. M. Doran, S. V. Fox, S. L. Garson, R. L. Kraus, Y. Li, V. N. Uebele, J. J. Renger and J. C. Barrow, *Bioorg. Med. Chem. Lett.*, 2010, **20**, 5147–5152.
38. J. A. Kaizerman, W. Aaron, S. An, R. Austin, M. Brown, A. Chong, T. Huang, R. Hungate, B. Jiang, M. G. Johnson, G. Lee, B. S. Lucas, J. Orf, M. Rong, M. M. Toteva, D. Wickramasinghe, G. Xu, Q. Ye, W. Zhong and D. L. McMinn, *Bioorg. Med. Chem. Lett.*, 2010, **20**, 4607–4610.
39. W. Sun, M. Maletic, S. S. Mundt, K. Shah, H. Zokian, K. Lyons, S. T. Waddell and J. Balkovec, *Bioorg. Med. Chem. Lett.*, 2011, **21**, 2141–2145.
40. D. S. Gardner, J. B. Santella, J. V. Duncia, P. H. Carter, T. G. M. Dhar, H. Wu, W. Guo, C. Cavallaro, K. Van Kirk, M. Yarde, S. W. Briceno, R. Grafstrom, R. Liu, S. R. Patel, A. J. Tebben, D. Camac, J. Khan, A. Watson, G. Yang, A. Rose, W. R. Foster, M. E. Cvijic, P. Davies and J. Hynes, *Bioorg. Med. Chem. Lett.*, 2013, **23**, 3833–3840.
41. D. Barnes-Seeman, C. Boiselle, C. Capacci-Daniel, R. Chopra, K. Hoffmaster, C. T. Jones, M. Kato, K. Lin, S. Ma, G. Pan, L. Shu, J. Wang, L. Whiteman, M. Xu, R. Zheng and J. Fu, *Bioorg. Med. Chem. Lett.*, 2014, **24**, 3979–3985.
42. C. Fotsch, M. D. Bartberger, E. A. Bercot, M. Chen, R. Cupples, M. Emery, J. Fretland, A. Guram, C. Hale, N. Han, D. Hickman, R. W. Hungate, M. Hayashi, R. Komorowski, Q. Liu, G. Matsumoto, D. J. St. Jean, S. Ursu, M. Veniant, G. Xu, Q. Ye, C. Yuan, J. Zhang, X. Zhang, H. Tu and M. Wang, *J. Med. Chem.*, 2008, **51**, 7953–7967.
43. D. M. Feng, R. M. DiPardo, J. M. Wai, R. K. Chang, C. N. Di Marco, K. L. Murphy, R. W. Ransom, D. R. Reiss, C. Tang, T. Prueksaritanont, D. J. Pettibone, M. G. Bock and S. D. Kuduk, *Bioorg. Med. Chem. Lett.*, 2008, **18**, 682–687.
44. J. Li, L. J. Kennedy, H. Wang, J. J. Li, S. J. Walker, Z. Hong, S. P. O'Connor, A. Nayeem, D. M. Camac, P. E. Morin, S. Sheriff, M. Wang, T. Harper, R. Golla, R. Seethala, T. Harrity, R. P. Ponticiello, N. N. Morgan, J. R. Taylor, R. Zebo, D. A. Gordon and J. A. Robl, *ACS Med. Chem. Lett.*, 2014, **5**, 803–808.
45. S. Harper, S. Avolio, B. Pacini, M. Di Filippo, S. Altamura, L. Tomei, G. Paonessa, S. Di Marco, A. Carfi, C. Giuliano, J. Padron, F. Bonelli, G. Migliaccio, R. De Francesco, R. Laufer, M. Rowley and F. Narjes, *J. Med. Chem.*, 2005, **48**, 4547–4557.
46. K. Zimmermann, M. D. Wittman, M. G. Saulnier, U. Velaparthi, X. Sang, D. B. Frennesson, C. Struzynski, S. P. Seitz, L. He, J. M. Carboni, A. Li, A. F. Greer, M. Gottardis, R. M. Attar, Z. Yang, P. Balimane, L. N. Discenza, F. Y. Lee, M. Sinz, S. Kim and D. Vyas, *Bioorg. Med. Chem. Lett.*, 2010, **20**, 1744–1748.

47. M. M. Weiss, T. A. Dineen, I. E. Marx, S. Altmann, A. Boezio, H. Bregman, M. Chu-Moyer, E. F. DiMauro, E. F. Bojic, R. S. Foti, H. Gao, R. Graceffa, H. Gunaydin, A. Guzman-Perez, H. Huang, L. Huang, M. Jarosh, T. Kornecook, C. R. Kreiman, J. Ligutti, D. S. La, M. J. Lin, D. Liu, B. D. Moyer, H. N. Nguyen, E. A. Peterson, P. E. Rose, K. Taborn, B. D. Youngblood, V. Yu and R. T. Fremeau Jr., *J. Med. Chem.*, 2017, **60**, 5969–5989.
48. H. Gong, D. S. Weinstein, Z. Lu, J. J. W. Duan, S. Stachura, L. Haque, A. Karmakar, H. Hemagiri, D. K. Raut, A. K. Gupta, J. Khan, D. Camac, J. S. Sack, A. Pudzianowski, D. R. Wu, M. Yarde, D. R. Shen, V. Borowski, J. H. Xie, H. Sun, C. D'Arienzo, M. Dabros, M. A. Galella, F. Wang, C. A. Weigelt, Q. Zhao, W. Foster, J. E. Somerville, L. M. Salter–Cid, J. C. Barrish, P. H. Carter and T. G. M. Dhar, *Bioorg. Med. Chem. Lett.*, 2018, **28**, 85–93.
49. D. W. Engers, S. R. Bollinger, J. L. Engers, J. D. Panarese, M. M. Breiner, A. Gregro, A. L. Blobaum, J. J. Bronson, Y. J. Wu, J. E. Macor, A. L. Rodriguez, R. Zamorano, P. J. Conn, C. W. Lindsley, C. M. Niswender and C. R. Hopkins, *Bioorg. Med. Chem. Lett.*, 2018, **28**, 2641–2646.
50. S. A. Rosenberg, M. Xia, R. Huang, N. G. Nikolov, E. B. Wedebye and M. Dybdahl, *Comput. Toxicol.*, 2017, **1**, 39–48.
51. N. Torimoto–Katori, R. Huang, H. Kato, R. Ohashi and M. Xia, *J. Pharm. Sci.*, 2017, **106**, 1752–1759.
52. L. Xiao, E. Nickbarg, W. Wang, A. Thomas, M. Ziebell, W. W. Prosise, C. A. Lesburg, S. S. Taremi, V. L. Gerlach, H. V. Le and K. C. Cheng, *Biochem. Pharmacol.*, 2011, **81**, 669–679.
53. M. Nagai, Y. Konno, M. Satsukawa, S. Yamashita and K. Yoshinari, *Drug Metab. Dispos.*, 2016, **44**, 1390–1398.
54. O. Pelkonen, M. Turpeinen, J. Hakkola, P. Honkakoski, J. Hukkanen and H. Raunio, *Arch. Toxicol.*, 2008, **82**, 667–715.
55. J. H. Lin, *Pharm. Res.*, 2006, **23**(6), 1089–1115.
56. V. Chu, H. J. Einolf, R. Evers, G. Kumar, D. Moore, S. Ripp, J. Silva, V. Sinha, M. Sinz and A. Skerjanec, *Drug Metab. Dispos.*, 2009, **37**, 1339.

CHAPTER 11

Strategies to Mitigate CYP450 Inhibition

ALEXANDER G. DOSSETTER[a], MARCEL J. DE GROOT[b] AND SARAH E. SKERRATT*[c]

[a]MedChemica Limited, Ebenezer House, Ryecroft, Newcastle-Under-Lyme, Staffordshire, ST5 2BE, UK; [b]LEO Pharma A/S, Industriparken 55, 2750 Ballerup, Denmark; [c]MSD, The Francis Crick Institute, 1 Midland Road, London NW1 1AT, UK
*E-mail: sarah.skeratt@msd.com

11.1 Cytochrome-P450 Inhibition

11.1.1 Introduction

Cytochrome P450s (CYPs) comprise a superfamily of haem-containing proteins that are predominantly expressed in the liver and catalyse the metabolism of a broad range of exogenous and endogenous molecules.[1–3] The inhibition of drug metabolism may result in undesirable elevations in plasma drug concentrations; therefore an understanding of the potential for CYP inhibition is important from both a therapeutic efficacy and a safety stand-point.[4] Most drug-like compounds whose clearance is dominated by cytochrome P450-mediated oxidation are metabolized by the following isoforms: 1A2, 2C9, 2C19, 2D6 and 3A4 (see Table 11.1).[5,6] Inhibition of any of these CYP isoforms may lead to concerns about co-medication with other compounds. In general, significant drug–drug interactions occur only when

Drug Discovery Series No. 79
The Medicinal Chemist's Guide to Solving ADMET Challenges
Edited by Patrick Schnider
© The Royal Society of Chemistry 2021
Published by the Royal Society of Chemistry, www.rsc.org

Table 11.1 Comparison between enzyme levels in human hepatic P450 complement and percentage involvement in drug metabolism. Adapted from ref. 5 with permission from Bentham Science Publishers Ltd.

CYP	Percentage P450 complement	Percentage drugs metabolized
1A2	13	10
2A6	4	3
2B6	1	4
2C	19	35
2D6	3	15
2E1	7	3
3A4	28	36

two or more drugs compete for the same enzyme, and when the metabolic reaction catalysed by this enzyme is a major elimination pathway.[7] The mechanisms of CYP inhibition can be broadly categorized into reversible and irreversible inhibition. Reversible inhibition is probably the most common mechanism responsible for documented drug interactions and will be the focus of this chapter. Drug–drug interactions (DDI) can also occur as a result of the induction of CYPs following prolonged drug treatment;[3] this topic is covered in Chapter 10. In terms of DDI, compounds are often described as 'victims' or 'perpetrators'. Perpetrators are compounds that inhibit or induce drug-metabolizing enzymes and/or drug-transporting proteins. Victim compounds are those which are metabolized by drug metabolizing enzymes and/or transported by drug-transporter proteins.

11.1.2 Structure

CYPs have molecular weights of approximately 50 kDa and a single haem–iron centre in the active site.[8] The iron is tethered to the protein *via* an axial cysteine thiolate ligand. This cysteine and several flanking residues are highly conserved in known CYPs. A water molecule resides at the additional Fe-axial site, which is displaced upon binding of a substrate. X-ray crystal structures of several human CYP enzymes have been solved, enabling structure-based ligand design to overcome CYP inhibition in several drug discovery projects.[9] Representative co-crystal structures of ligands bound to the active site of CYP 1A2, CYP 2C9, CYP 2C19, CYP 2D6 and CYP 3A4 are shown in Figure 11.1. Key protein residues are highlighted.

11.1.3 Mechanisms

The mechanisms of CYP inhibition can be broadly categorized into reversible and irreversible inhibition.[10] Reversible inhibition may be competitive or non-competitive. Competitive reversible inhibition occurs when two drugs compete for the same CYP enzyme, irrespective of whether they are substrates for that enzyme.[11–13] In non-competitive reversible inhibition,

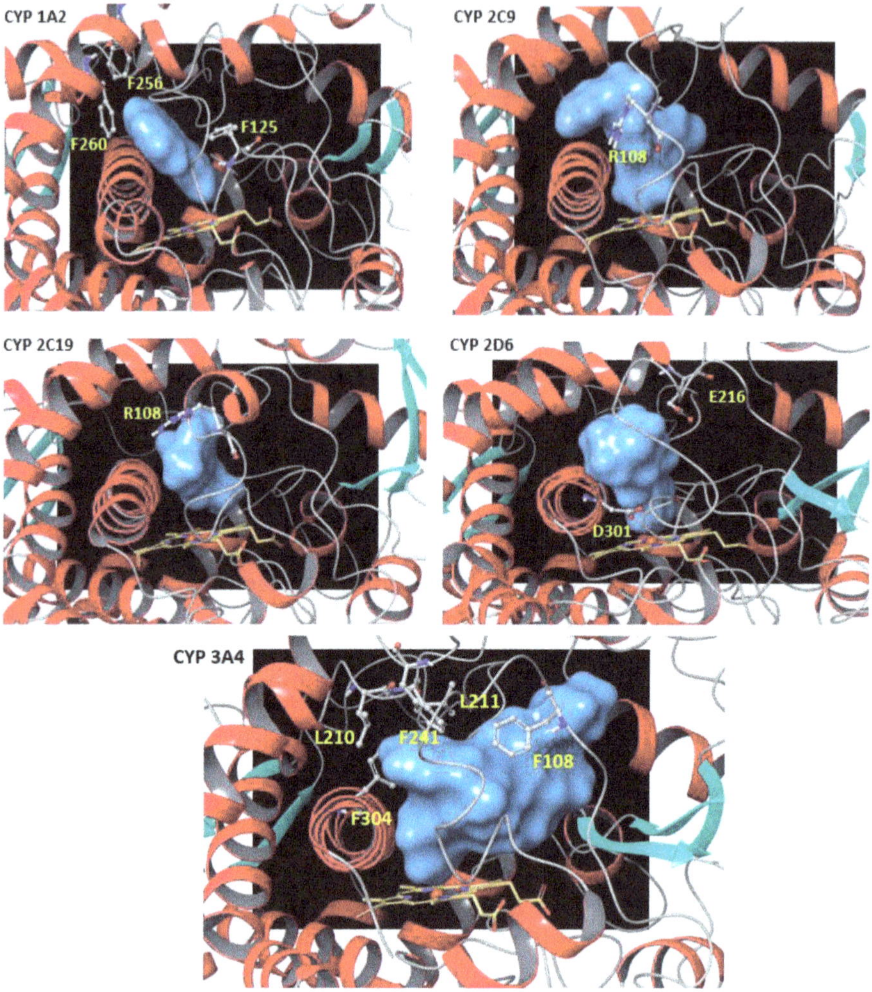

Figure 11.1 Representative co-crystal structures of ligands bound to the active site of CYP 1A2 (PDB ID 2HI4), CYP 2C9 (PDB ID 5K7K), CYP 2C19 (PDB ID 2GQS), CYP 2D6 (PDB ID 3TDA) and CYP 3A4 (PDB ID 3NUX). The haem is depicted as a stick model and coloured yellow. The solvent-accessible molecular surfaces of bound ligands are coloured blue. Key protein residues are highlighted.

drugs bind to a site other than the active site (an allosteric site).[14] Irreversible inhibition of CYP enzymes occurs when compounds are transformed into reactive species in the CYP active site. The reactive species can then covalently bind to residues in the active site, leading to enzyme inactivation. This process is often termed "time-dependent" inhibition.[15–17] The CYP catalytic cycle consists of at least seven discrete steps that include binding of the substrate to the ferric form of the enzyme, reduction of the haem group from the

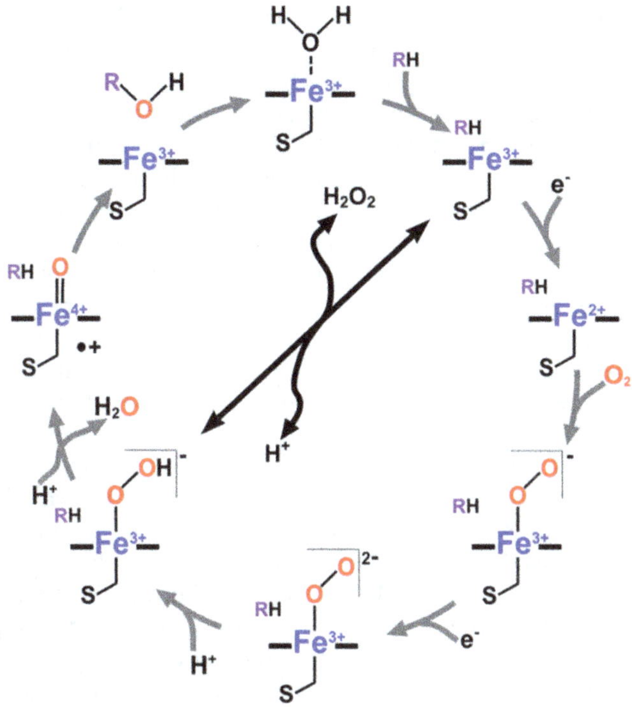

Figure 11.2 The catalytic cycle of cytochrome P450. Reproduced from ref. 18 with permission from The American Society for Biochemistry and Molecular Biology, Inc.

ferric to the ferrous state by an electron provided by reduced nicotinamide adenine dinucleotide phosphate (NADPH) *via* CYP reductase, binding of molecular oxygen, transfer of a second electron from CYP reductase and/or cytochrome b5, cleavage of the O–O bond, substrate oxygenation and product release.[8] A representative schematic of the catalytic cycle of cytochrome P450 is highlighted in Figure 11.2.[18] In mechanistic terms, reversible interactions arise as a result of competition at the CYP active site, which probably involves only the first step of the CYP catalytic cycle. Compounds that act to inhibit during or subsequent to the oxygen transfer step are generally irreversible inhibitors.

11.1.4 Screening Strategies

It is highly desirable to avoid developing drug candidates with the potential for drug–drug interactions. Hence, the assessment of potential drug interactions of compounds during the early stages of drug discovery is essential.[10,19–22] Since human liver samples and recombinant human CYPs are now readily available, they are widely used as *in vitro* screening tools to predict the potential for *in vivo* drug interactions in drug discovery programs. Isoform-specific substrates are incubated individually with a range of test compound concentrations in the presence of human liver microsomes (or expressed enzyme) and cofactor. At the end of the incubation, the formation of metabolite is monitored by liquid–chromatography tandem mass spectrometry (LC–MS/MS) or fluorescence, at each of the test compound concentrations. Typically, percentage inhibition data (percentage inhibition at a single concentration of test compound at a single time point) are used to initially assess/rank order the reversible DDI potential of compounds. For compounds of significant interest, CYP inhibition data expressed as the concentration giving 505 of maximum inhibition (IC_{50}) or inhibition constant (K_i) data may be generated (using multiple substrate concentrations, single time point and multiple inhibitor concentrations) and used in conjunction with (predicted) clinical pharmacokinetic data to determine if an *in vivo* clinical DDI study is required. This analysis can use a number of different models, including basic, mechanistic static or dynamic (*e.g.*, physiologically based pharmacokinetic) models.[19,21] A study by Obach *et al.* evaluated the utility of *in vitro* data for prediction of drug–drug interactions for a variety of drugs in the clinical setting. The authors concluded that *in vitro* inhibition data could be reliably used for predictions of CYP 1A2, CYP 2C9, CYP 2C19 and CYP 2D6 DDI, while for CYP 3A4, the effects of both hepatic and intestinal metabolism should be considered.[22] *In silico* tools are available for predicting CYP inhibition and are used to guide medicinal chemistry design towards compounds with a lower DDI risk.[20] For example, Gleeson *et. al.* generated *in silico* quantitative structure–activity relationship (QSAR) models to predict the extent of inhibition of a number of cytochrome P450 isoenzymes using both linear and non-linear statistical methods, and a set of easily interpretable descriptors. The models were statistically significant, agreed with known SAR, and could be used as a guide to assess the P450 liability of molecules for a particular isoform. The authors noted that hybrid models using bulk and fragmental descriptors performed better than bulk property QSAR descriptors alone.[23] A report by Leach and Kidley described how quantum mechanical calculations of binding energies for aromatic heterocycles to the haem iron can provide a quantitative link between observed co-crystal structures and measured biochemical inhibition. Indeed, the compounds predicted to be the tightest binders corresponded to the most potent inhibitors in experimental assays of cytochrome P450 inhibition, especially when weighted for lipophilicity-based expectation.[24]

11.1.5 Relevance

Drug–drug interactions (DDI) are a significant safety concern as substantial changes in blood and tissue concentrations can alter the safety and efficacy profile of a drug and/or its metabolites, especially for drugs with a narrow therapeutic index or when patients take multiple medications concurrently. It is reported that DDIs from polypharmacy are responsible for approximately 26% of adverse events leading to hospitalization.[25] DDIs are a major regulatory hurdle which can lead to early termination of development, refusal of approval, dosage adjustments, prescribing restrictions or withdrawal of drugs from the market. Guidance documents are available from regulatory agencies that detail a number of *in vitro* assays designed to detect potential DDIs and identify whether additional clinical studies are required prior to launch of the drug to market.[26]

Key References

- U.S. FDA, *Draft Guidance*, October, 2017,https://www.fda.gov/regulatory-information/search-fda-guidance-documents/vitro-metabolism-and-transporter-mediated-drug-drug-interaction-studies-guidance-industry.
- T. D. Bjornsson, J. T. Callaghan, H. J. Einolf, V. Fischer, L. Gan, S. Grimm, J. Kao, S. P. King, G. Miwa, L. Ni, G. Kumar, J. McLeod, R. S. Obach, S. Roberts, A. Roe, A. Shah, F. Snikeris, J. T. Sullivan, D. Tweedie, J. M. Vega, J. Walsh, S. A. Wrighton, The conduct of *in vitro* and *in vivo* drug–drug interaction studies: A Pharmaceutical Research and Manufacturers of America (PhRMA) perspective. Pharmaceutical, G. Manufacturers of America Drug Metabolism/ Clinical Pharmacology Technical Working, F. D. A. C. f. D. Evaluation and Research, *Drug Metab. Dispos.*, 2003, **31**, 815–832.
- O. Pelkonen, M. Turpeinen, J. Hakkola, P. Honkakoshi, J. Hukkanen and H. Raunio, Inhibition and induction of human cytochrome P450 enzymes: current status, *Arch. Toxicol.*, 2008, **82**, 667–715.
- M. P. Gleeson, Generation of a set of simple, interpretable ADMET rules of thumb, *J. Med. Chem.*, 2008, **51**, 817–834.
- M. P. Gleeson, A. M. Davis, K. K. Chohan, S. W. Paine, S. Boyer, C. L. Gavaghan, C. H. Arnby, C. Kankkonen and N. Albertson, Generation of *in silico* cytochrome P450 1A2, 2C9, 2C19, 2D6, and 3A4 inhibition QSAR models, *J. Comput.-Aided Mol. Des.*, 2007, 21, 559–573.
- P. A. Williams, J. Cosme, A. Ward, H. C. Angove, D. M. Vinkovic and H. Jhoti, Crystal structure of human cytochrome P450 2C9 with bound warfarin, *Nature*, 2003, **424**, 464–468.
- G. Braenden, T. Sjoegren, V. Schnecke and Y. Xue, Structure-based ligand design to overcome CYP inhibition in drug discovery projects, *Drug Discovery Today*, 2014, **19**, 905–911.

11.2 Medicinal Chemistry Strategies to Modulate CYP Inhibitory Activity

11.2.1 The Effects of Molecular Properties on CYP Inhibitory Activity

In 2008, Gleeson analysed data from multiple ADMET assays run within GSK, including cytochrome P450 1A2/2C9/2C19/2D6/3A4 inhibition assays, and concluded molecular weight (MW) and lipophilicity [log of the partition coefficient (logP)] to be key characteristics that determine ADMET liabilities, with almost all ADMET parameters deteriorating with either increasing MW, logP or both, and with ionization state having either a beneficial or detrimental effect depending on the parameter in question.[27] In a follow-up study, Gleeson *et al.* reviewed the effects of common substituents on ADMET parameters, again including data from GSK cytochrome P450 1A2/2C9/2C19/2D6/3A4 inhibition assays, using a matched molecular pair analysis.[28] Key trends relating to specific CYP isoforms from these analyses, alongside data from the matched molecular pair analysis of the ChEMBL database, will be highlighted in the CYP isoform-specific sections of this chapter. In a later report from Young *et al.*,[29] the effects of chromatographic hydrophobicity measurements and aromaticity on ADMET parameters that include CYP inhibition was investigated. The authors showed compound hydrophobicity (either measured or predicted Chrom log $D_{pH7.4}$) and the number of aromatic rings (#Ar) to be key molecular descriptors in determining likely levels of CYP inhibitory activity. In this analysis, data were interrogated across the five P450 isoforms regularly screened at GSK in bactosome assays with 50 000–70 000 data points available. Table 11.2 summarises wherein particular descriptors showed effects. The 1A2 isoform only interacted with smaller, flatter, molecules (those with a higher proportion of aromatic rings rather than #Ar *per se*). The effect of increasing activity with increased size was apparent for the 2D6, 2C9, 2C19 and 3A4 isoforms; evidence was also observed for an activity increase with particular charge states, as expected. The effect of measured Chrom log $D_{pH7.4}$ on 2D6, 2C9, 2C19 and 3A4 was demonstrated, with bi-linear responses for 2D6, 2C9, 2C19 and 3A4. Aromatic ring count also had a clear influence on the activity of 2C9, 2C19 and 3A4.

In addition, over the last approximately 15 years, the pharmaceutical industry and academia have produced a wealth of CYP structural information (liganded and *apo*), enabling structure-based ligand design to overcome CYP inhibition. Representative examples are highlighted in later sections.[9]

Table 11.2 Influence of descriptors on P450 binding activity. +, ++, +++ represent increasing effects of the parameter; minus signs at either end (*e.g.* −+++−) are indicative of a bi-linear relationship. Adapted from ref. 29 with permission from Elsevier, Copyright 2011.

CYP isoform	Chrom log $D_{pH7.4}$	Size (CMR)	#Ar	Recognition factors
1A2	−	− ++ −	%Ar not #Ar	Highly aromatic/flat structures, smaller/ hydrophobic
2D6	− +++ −	− ++ −	+++	Hydrophobic, optimum size, basic
2C9	− +++ −	− ++ −	+++	Hydrophobic, optimum size, aromatic, acidic
2C19	− +++ −	− ++ −	+++	Hydrophobic, optimum size aromatic, basic
3A4	− +++ −	+++	+++	Hydrophobic, large, aromatic, basic

11.2.2 Matched Molecular Pair Changes to Modulate CYP Inhibitory Activity

A matched molecular pair analysis of the cytochrome P450 inhibition assay data in the ChEMBL database,[30] was performed using MedChemica proprietary software.[31] In the matched molecular pair method, the average effect of a substituent on a biological parameter is estimated by analysing all the molecular pairs obtained where the only change in structure involves a single, localized substituent.[32,33] A number of MedChem design strategies to reduce CYP inhibition are highlighted from this analysis, that are applicable across CYP isoforms:

11.2.2.1 Blocking Aromatic Nitrogen With a Flanking Methyl Group

A general approach for all CYPs (3A4, 2D6, 2C9, 2C19, 1A2) is to add a flanking group (*e.g.* a methyl group) next to the aromatic nitrogen to reduce the capability of the nitrogen lone pair to effectively coordinate to the haem group. Examples are highlighted in Table 11.3.

Table 11.3 Blocking aromatic nitrogen with a flanking methyl group can minimise CYP inhibitory activity.

Examples	Matched Pair[a]	Notes	Reference
1	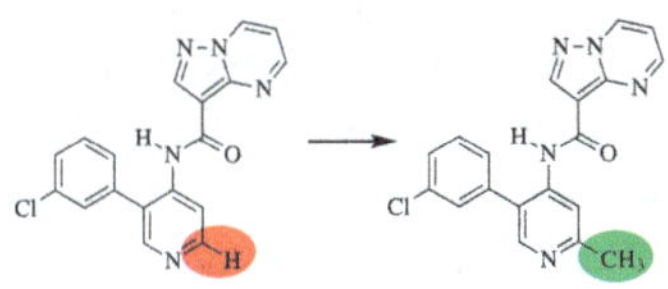 0.20 μM → >10 μM 3A4 inhibition	Janus kinase 2 (JAK2) inhibitors. Pyridine nitrogen was suggested to bind the iron haem group. Adding a flanking methyl group reduced binding to 3A4, without affecting JAK2 inhibition.	34
2	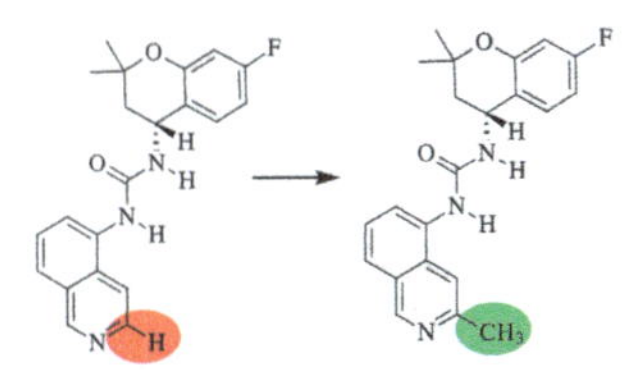 0.66 μM, <0.08 μM → >20 μM 3A4 inhibition, 17 μM 2D6 inhibition	Human transient receptor potential vanilloid type 1 (hTRPV1) binders. The flanking methyl group reduced binding to 3A4 and 2D6, without affecting hTRPV1 inhibition.	35

(*continued*)

Table 11.3 Blocking aromatic nitrogen with a flanking methyl group can minimise CYP inhibitory activity. (*continued*)

Examples	Matched Pair[a]	Notes	Reference
3	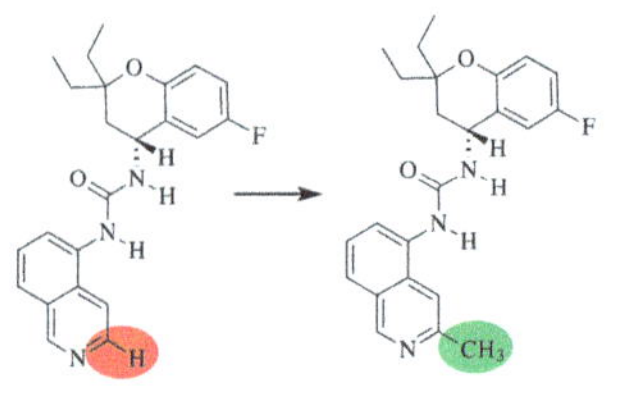 2.8 μM → >20 μM 3A4 inhibition <0.08 μM → >20 μM 2D6 inhibition	hTRPV1 binders from the same series as Example 2. The flanking methyl group reduced binding to 3A4 and 2D6, however this matched pair lost sevenfold in hTRPV1 binding.	35
4	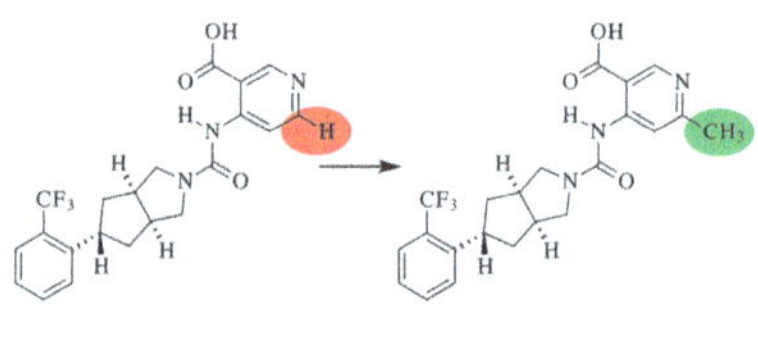 <0.05 μM → 1.1 μM 3A4 inhibition 5.6 μM → 1.0 μM 2C9 inhibition	Retinol binding protein 4 (RBP4) binder. The flanking methyl group reduced binding to 3A4 without effect on RBP4. However, inhibition of 2C9 increased approximately fivefold.	36
5	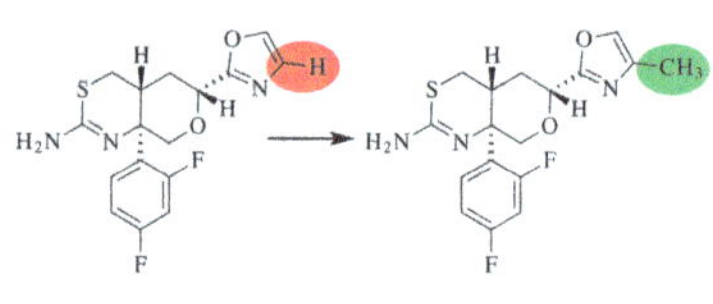 0.33 μM. → 5.5 μM 2D6 inhibition	Beta-site amyloid precursor protein cleaving enzyme 1 (BACE1) inhibitors. The flanking methyl group reduced binding to 2D6 without affecting BACE1 activity. The program used structure-based design.	37
6	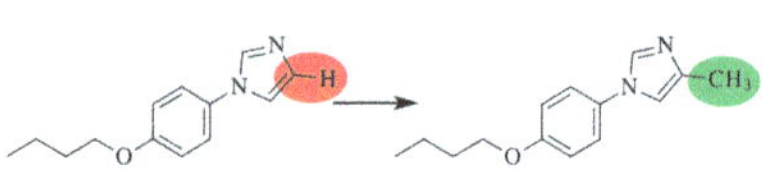 0.35 μM → 9.2 μM 3A4 inhibition 0.46 μM → 7.8 μM 2D6 inhibition 0.07 μM → 0.36 μM 2C19 inhibition 0.100 μM → 6.9 μM 2C9 inhibition 0.07 μM → 0.39 μM 1A2 inhibition	20-Hydroxyeicosatetraenoic acid (20-HETE) inhibitors had pan CYP inhibition. The flanking methyl group reduced binding to all CYPs but also caused a 30-fold drop in target potency.	38
7	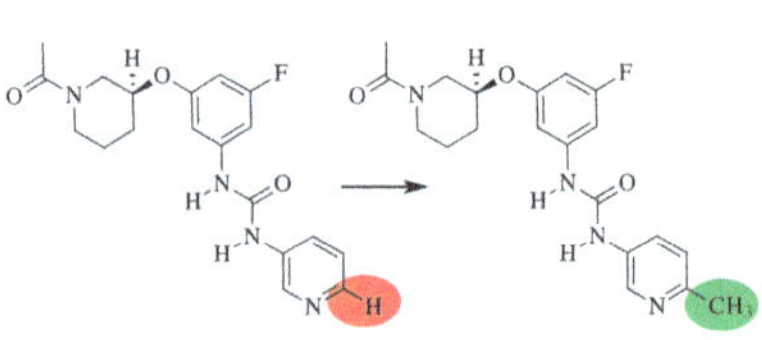 0.08 μM → >10 μM 1A2 inhibition	Activators of cardiac myosin.	39

[a]Cytochrome P450 inhibition data is taken directly from the publication and converted to IC_{50} in μM concentration.

11.2.2.2 Changing the Heterocycle

A general approach for all CYPs, examples for 3A4 and 2C9 are highlighted in Table 11.4.

Table 11.4 Changing the heterocycle can minimise CYP inhibitory activity.

Examples	Matched Pair[a]	Notes	Reference
8	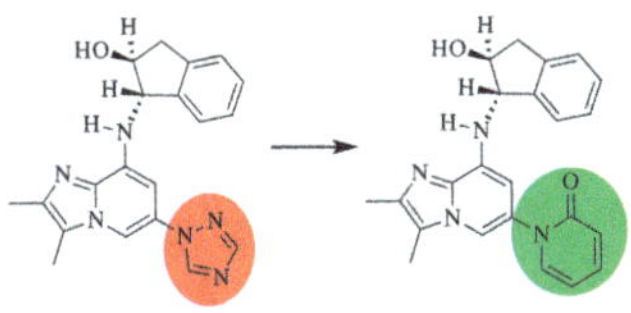 6.8 μM → 79 μM 3A4 inhibition	Acid pump antagonists (APAs), several matched pairs are represented in this publication where triazole is replaced with pyrimidone.	40
9	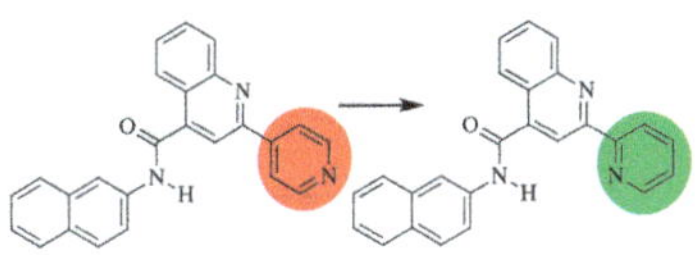 0.033 μM → 11 μM 2C9 inhibition	The inhibition of 2C9 was studied; there are many matched pairs in this paper.	41

[a]Cytochrome P450 inhibition data is taken directly from the publication and converted to IC_{50} in μM concentration.

11.2.3 Strategies to Mitigate CYP 1A2 Inhibition

CYP 1A2 is implicated in the metabolism of drug molecules, such as paracetamol and phenacetin, as well as caffeine.[27,42] As highlighted by Gleeson, (see Figure 11.3a–c), molecular weight is an important parameter in determining the extent of 1A2 inhibition, with 1A2 activity generally decreasing with increasing molecular weight.[27] This finding is consistent with the reported crystal structure of 1A2, which displays a narrow cavity of limited volume compared with other isoforms,[42] and is also consistent with the results of Burton *et al.*[43] who found that size and aromaticity were the key features discriminating inhibitors *vs.* non-inhibitors. Collectively, these results indicate that steric factors associated with the 1A2 active site restrict all but the smallest molecules from binding effectively.[44] Ionization state plays a relatively small role in determining 1A2 activity, and this can be rationalized by the lack of ionizable residues capable of making interactions with ligands within the active site.[42] The influence of lipophilicity on 1A2 inhibitory activity is also weak, relative to the influence of MW.[27,45]

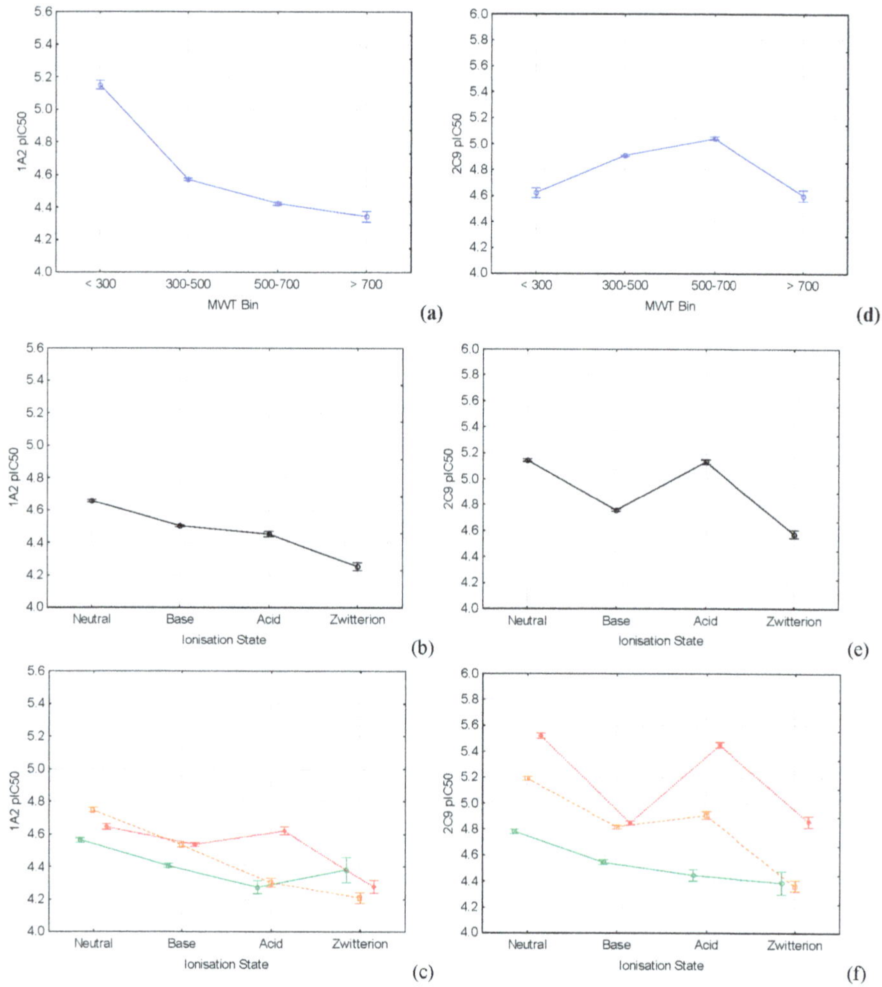

Figure 11.3 Relationship between CYP-1A2 inhibition (a–c) and CYP-2C9 inhibition (d–f) with respect to MW (top), ionization state (middle), and a combination of calculated log partition coefficient (clogP) and ionization state (bottom) for 49 837 and 51 097 molecules measured in the respective assays. The green solid lines denote clogPs < 3, the orange dashed lines denote clogPs 3–5, and the red dotted lines denote clogPs > 5. The error bars denote the 95% confidence limits of the mean. Reproduced from ref. 27 with permission from American Chemical Society, Copyright 2008.

The reported co-crystal structure highlights the relatively narrow, planar substrate binding cavity that is highly adapted for the size and shape of substrates that are oxidized by the enzyme.[42] Knowledge of the key recognition features driving 1A2 activity may be useful to enable structure-based ligand design to overcome CYP 1A2 inhibition in drug discovery projects, although no examples are to-date available in the publicly accessible scientific literature.

11.2.3.1 Matched Molecular Pair Transformations Specific to Reducing CYP 1A2 Inhibition

A matched molecular pair analysis of the cytochrome P450 inhibition assay data in the ChEMBL database,[30] performed using MedChemica proprietary software[31] highlighted a number of examples specific to 1A2, Table 11.5.

Table 11.5 Matched molecular pair examples specific to CYP 1A2, derived from analysis of the cytochrome P450 inhibition assays in the ChEMBL database,[30] performed using MedChemica proprietary software.[31]

Examples	Matched Pair[a]	Notes	Reference
10	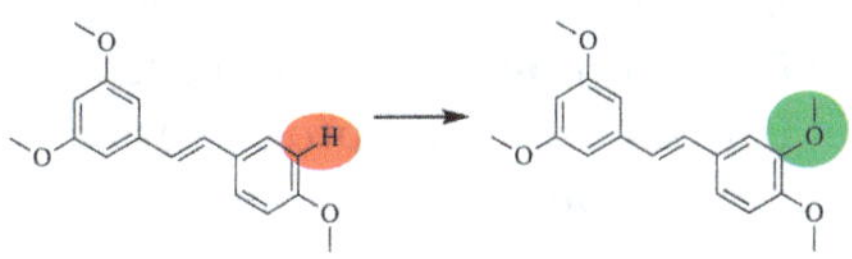 6.2 μM → 570 μM 1A2 inhibition	CYP 1B1 inhibitors. A small increase in size can be enough to reduce inhibition in CYP 1A2.	46
11	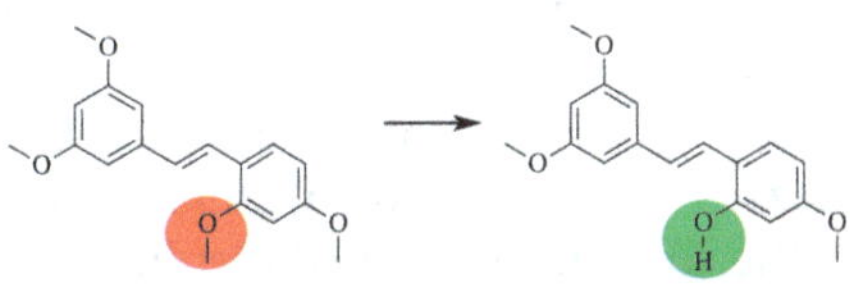 3.1 μM → 31 μM 1A2 inhibition	In the same publication as Example 10, the change from non-polar to polar group yields a large change in inhibition	46
12	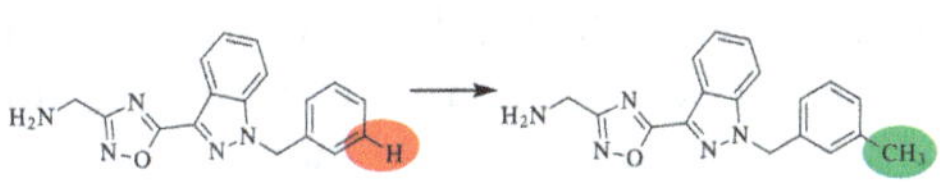 0.009 μM 0.09 μM → 0.84 μM 1A2 inhibition 4.0 μM 2D6 inhibition	Sodium ion-channel inhibitors. In this case the addition of a methyl group gave a dramatic decrease in CYP 1A2 activity.	47

[a]Cytochrome P450 inhibition data is taken directly from the publication and converted to IC_{50} in μM concentration.

11.2.4 Strategies to Mitigate CYP 2C9 Inhibition

CYP 2C9 is polymorphically expressed and is implicated in the metabolism of drug molecules, such as phenytoin, tolbutamide and warfarin.[48] Molecular weight plays an important role in determining the extent of 2C9 inhibition with molecules with MW's between 300 and 700 being on average the most potent, indicating that steric factors within the 2C9 active size generally prevent very large or very small molecules from binding optimally. Ionization state is also an important factor, with acidic molecules having the highest 2C9 affinity on average.[27,48] Indeed, Arg108 plays an important role in binding the acidic molecule flurbiprofen to 2C9.[49] In terms of lipophilicity, Gleeson's 2008 analysis revealed that molecules with clogP < 3 had a mean 2C9 negative log of the IC_{50} value when converted to molar (pIC_{50}) value of 4.6, those with clogP between 3 and 5 had a mean pIC_{50} value of 4.9, and those with clogP > 5 had a mean pIC_{50} value of 5.2 (see Figure 11.3d–f).[27] A number of reports have demonstrated how structure-based ligand design has been used to overcome CYP 2C9 inhibition issues in drug discovery projects.[9,49–53] An example of this is from the Pfizer progesterone receptor (PR) antagonist program in which trifluoromethanesulphonamide compounds such as **1**, although antagonists of PR, were found to strongly inhibit CYP 2C9. In order to design analogues with reduced CYP 2C9 activity, a CYP 2C9 homology model was used to predict the potential binding mode of compounds within 2C9. The model predicted a key binding determinant to be a hydrogen bond network between the two sulphonamide oxygen atoms of the ligands and Arg108 and Asn204 of CYP 2C9. Electrostatic mapping showed a significant increase in negative electrostatic potential on the sulphonamide oxygen atoms in the ionized *versus* the neutral form, and it was reasoned the interaction of CYP 2C9 with the ionized form of the sulphonamide group would lead to increased CYP 2CP binding through an enhanced hydrogen bonding interaction. The less acidic methanesulphonamide analogue (**2**) was synthesised, with the hypothesis that at physiological pH, the propensity of the sulphonamide motif to exist in an ionized form would be much reduced. Indeed, compound **2** was found to retain potency at the progesterone receptor but lost inhibitory activity at CYP 2C9. A subsequently determined X-ray co-crystal structure of **1** bound to 2C9 confirmed the binding interactions predicted by the homology model, with the nitrile motif of **1** binding to the haem iron (Figure 11.4).[52]

11.2.4.1 Matched Molecular Pair Transformations Specific to Reducing CYP 2C9 Inhibition

A matched molecular pair analysis of the cytochrome P450 inhibition assay data in the ChEMBL database,[30] performed using MedChemica proprietary software[31] highlighted a number of examples specific to 2C9, see Table 11.6.

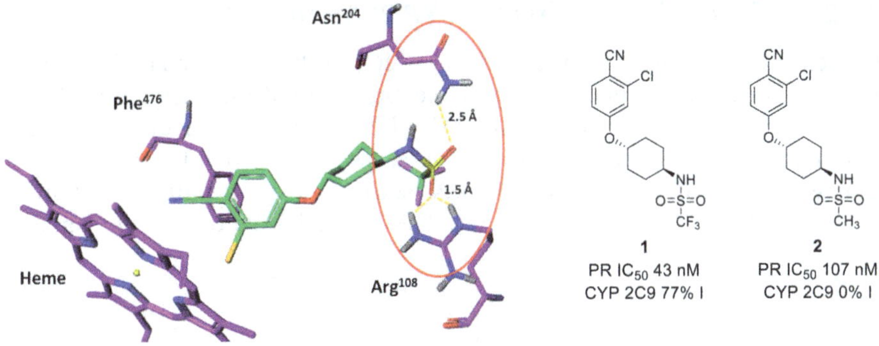

Figure 11.4 Predicted hydrogen bonding interactions between sulphonamide oxygens of compound **1** (green) and Arg108 and Asn204 in the CYP 2C9 homology model (purple). Structure and properties of compounds **1** and **2**; the CYP 2C9 percentage ionization (% I) data were generated at a 3 μM concentration. Reproduced from ref. 52 with permission from the Royal Society of Chemistry.

Table 11.6 Matched molecular pair examples specific to CYP 2C9, derived from analysis of the cytochrome P450 inhibition assays in the ChEMBL database,[30] performed using MedChemica proprietary software.[31]

Examples	Matched Pair[a]	Notes	Reference
13	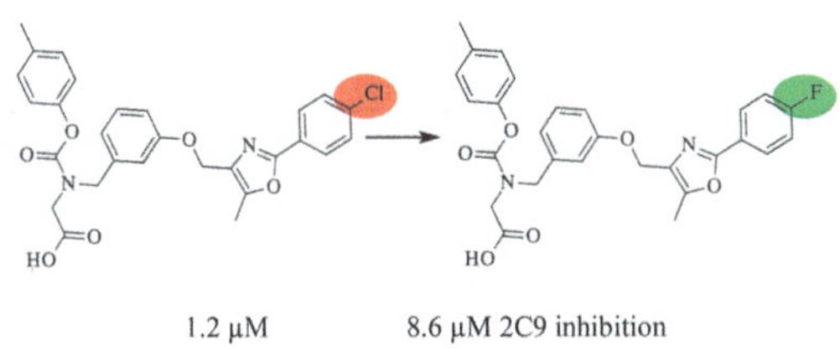	Peroxisome proliferator activated receptor α selective agonists. A sixfold reduction in 2C9 activity was achieved by a chloro to fluoro switch, with only a twofold loss in primary potency.	54
14	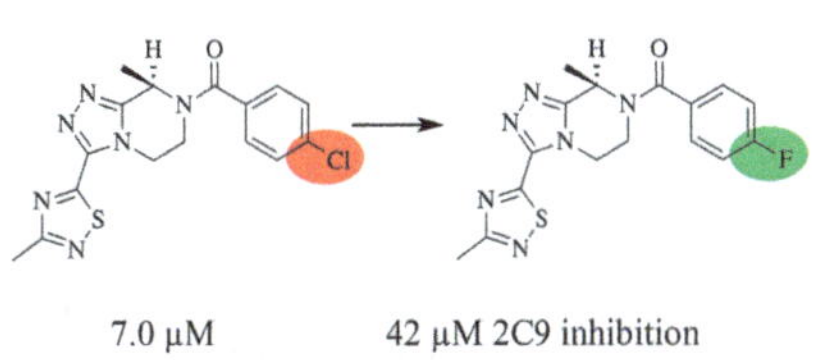	Neurokinin-3 Receptor antagonists (NK_3R).	55

[a]Cytochrome P450 inhibition data is taken directly from the publication and converted to IC_{50} in μM concentration.

11.2.5 Strategies to Mitigate CYP 2C19 Inhibition

Cytochrome P450 2C19 is a member of the 2C family implicated in the metabolism of compounds such as omeprazole, propranolol and diazepam.[48] From Gleeson's 2008 analysis of the SAR derived from 48 464 molecules with measured 2C19 pIC_{50}s, CYP 2C19 inhibition shows little dependence on molecular weight (MWT), even though this isoform is structurally very similar (91%) to CYP 2C9 (Figure 11.5a–c).[27] The ionization state of a molecule is

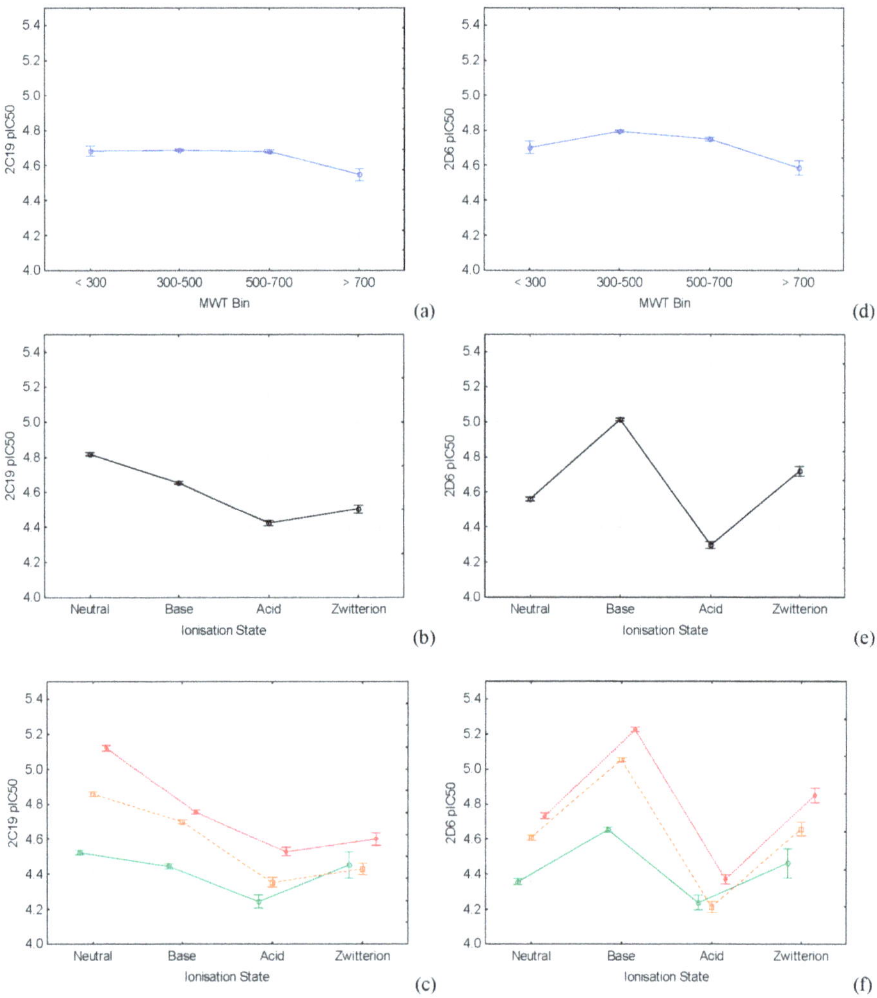

Figure 11.5 Relationship between CYP 2C19 inhibition (a–c) and CYP 2D6 inhibition (d–f) with respect to MW (top), ionization state (middle), and a combination of clogP and ionization state (bottom) for 48 464 and 50 886 molecules measured in the respective assays. The green solid lines denote clogPs < 3, orange dashed lines denote clogPs 3–5, and red dotted lines denote clogPs > 5. The error bars denote the 95% confidence limits of the mean. Reproduced from ref. 27 with permission from American Chemical Society, Copyright 2008.

of minor importance in determining the extent of inhibition of 2C19, with neutral and basic molecules having only slightly higher mean 2C19 pIC_{50}s than zwitterions and acids. The key observation being that 2C19 does not have the same affinity for acids as that found for 2C9. CYP 2C19 inhibition does show dependence on compound lipophilicity, with 2C19 inhibition on average increasing with increasing clogP.[27]

The first structure of human P450 2C19 was reported in 2012, providing insights into substrate and inhibitor binding to the enzyme. Comparison of the structure of the P450 2C19 0XV complex with structures of the closely related P450 2C9 isoform indicate the 2C19 active site cavity to be similar to that of the 2C9 flurbiprofen complex, but with more divergency in the more distal regions that include the helix B–C loop and the turn in the C-terminal loop.[56] Knowledge of the key structural features of 2C19 should enable structure-based ligand design to reduce 2C19 inhibition issues in drug discovery programs.

11.2.5.1 *Matched Molecular Pair Transformations to Reduce CYP 2C19 Inhibition*

A matched molecular pair analysis of the cytochrome P450 inhibition assay data in the ChEMBL database,[30] was performed using MedChemica proprietary software.[31] From this analysis, the design strategies that were identified to minimise CYP 2C19 inhibition were those included in the general strategies highlighted in Tables 11.3 and 11.4, *i.e.* blocking aromatic nitrogen with a flanking methyl group and changing the heterocycle (*e.g.* Table 11.3, example 6).

11.2.6 Strategies to Mitigate CYP 2D6 Inhibition

CYP 2D6 is polymorphically expressed and is implicated in the metabolism of drug molecules, such as codeine, ondansetron, and quinidine.[48,57] An analysis by Gleeson revealed that molecular weight displays a weak but clearly evident parabolic relationship with extent of 2D6 inhibition, and (similar to the trends observed for 2C9) this indicates that steric factors associated with the 2D6 active generally prevent very large or very small molecules from binding optimally [see Figure 11.5d–f].[27] Ionization state is also an important factor, with basic molecules having the highest 2D6 inhibitory activity, followed by zwitterions, neutral molecules, and finally acidic molecules.[58,59] This is consistent with X-ray crystallography data, in which an acidic residue in the active site is shown to be capable of hydrogen bonding to bound inhibitors or substrates.[58,60,61]

Gleeson's analysis also revealed that as lipophilicity increases 2D6 inhibition on average increases, although the dependency between 2D6 inhibition and clogP differs when broken down by ionization state, with an increase in clogP having a larger effect on neutral, basic and zwitterionic molecules compared with that of acids. This would be expected due to the acidic nature of the active site.[27] A number of reports have described how structure-based ligand design has been used to overcome CYP 2D6 inhibition issues in drug discovery projects.[37,61–63] An example of this is from the Pfizer β-secretase (BACE1) inhibitor program, in which CYP profiling of the lead thioamidine series revealed exclusive metabolism *via* CYP 2D6 which precluded advancement of these molecules due to the potential for clinical DDI. Co-crystal structures of CYP 2D6 complexes with substrate **3** and its corresponding metabolic product pyrazole **4**, a CYP 2D6 inhibitor, indicated that substituents adjacent to the thioamidine sulphur could reduce binding of these compounds to CYP 2D6.[37] The combination of S2′ heterocycles with a thioamidine methyl group resulted in compounds (*e.g.* **5**) with lower clearance and a more balanced CYP metabolism profile, significantly reducing the potential for DDIs (Figures 11.6 and 11.7).

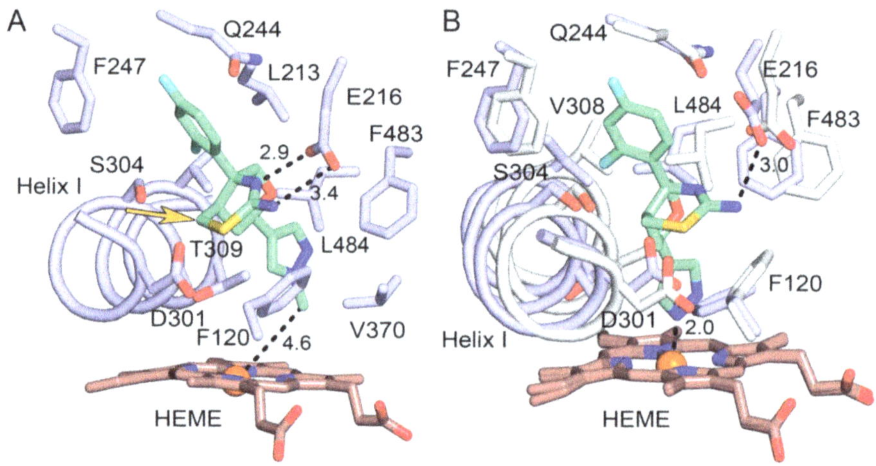

Figure 11.6 Compound **3** bound in the active site of CYP 2D6. Notable distance measurements are shown by dashed lines with the distance indicated in Å. Protein, haem and substrate carbons are coloured light-blue, brown, and pale-green, respectively. Oxygen, nitrogen sulphur, fluorine, potassium and iron atoms are coloured red, blue, yellow, light-blue, grey and orange, respectively. The carbon between the sulphur and the THP ring of **3** is indicated by the arrow. (B) Compound **4** bound in the active site of CYP 2D6, where it forms a coordinate covalent bond to the haem iron. The difference in position of **4** relative to **3** is associated with significant changes for helix I, the helix F–G region and surrounding residues. For reference, the protein structure of the complex with **3** is shown with grey carbons (B). Reproduced from ref. 37 with permission from American Chemical Society, Copyright 2015.

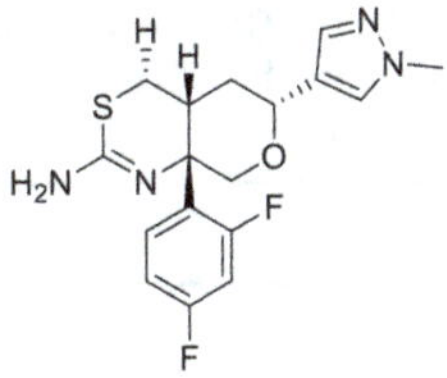

3

BACE1 IC_{50} 52 nM
HLM Cl_{int} = 9.1 mL min Kg^{-1}
CYP2D6 Inhib. IC_{50} =13.9 μM
Metabolism = 100% 2D6

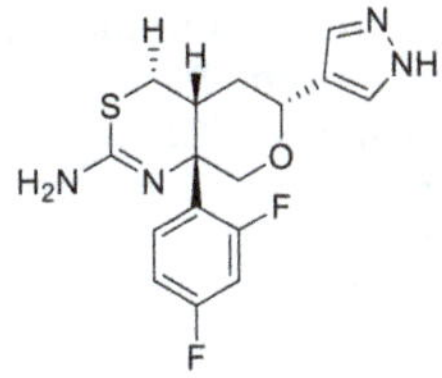

4

BACE1 IC_{50} 599 nM
HLM Cl_{int} < 8 mL min Kg^{-1}
CYP2D6 Inhib. IC_{50} = 0.37 μM

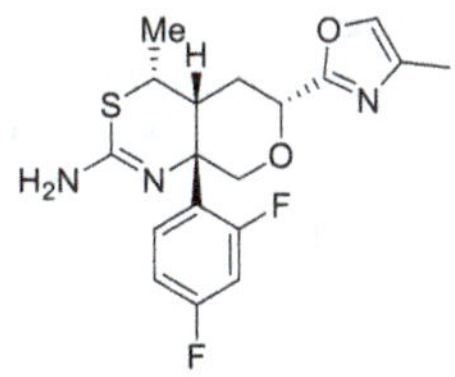

5

BACE1 IC_{50} 103 nM
HLM Cl_{int} < 8 mL min Kg^{-1}
CYP2D6 Inhib. IC_{50} > 30 μM
Metabolism = 38% 3A4
34% 2C19
28% 2D6

Figure 11.7 Structure and properties of compounds **3**, **4** and **5** and effects of a methyl adjacent to sulphur on metabolism.[37]

11.2.6.1 Matched Molecular Pair Transformations Specific to Reducing CYP 2D6 Inhibition

A matched molecular pair analysis of the cytochrome P450 inhibition assay data in the ChEMBL database,[30] performed using MedChemica proprietary software,[31] highlighted a number of examples specific to 2D6, see Table 11.7.

Table 11.7 Matched molecular pair examples specific to CYP 2D6, derived from analysis of the cytochrome P450 inhibition assays in the ChEMBL database,[30] performed using MedChemica proprietary software.[31]

Examples	Matched Pair[a]	Notes	Reference
15	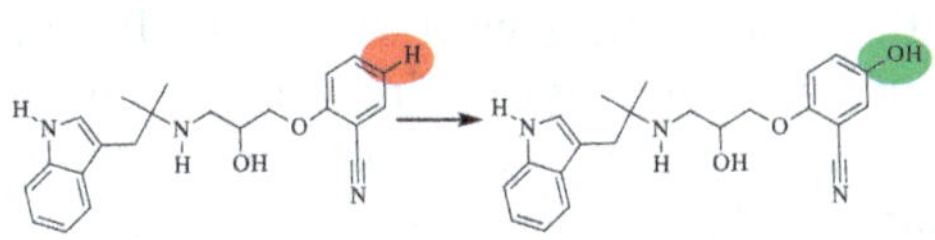 0.05 μM → 0.31 μM 2D6 inhibition	A study of 2D6 inhibition revealed that the addition of a phenol group reduced binding. A phenol may have detrimental effects on other ADMET properties.	64
16	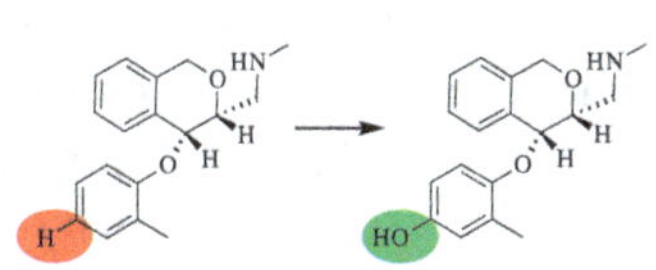 0.80 μM → 9.0 μM 2D6 inhibition	Selective norepinephrine reuptake inhibitors (sNRIs). In a similar fashion addition of the polar OH group decreased 2D6 inhibition.	65

(*continued*)

Table 11.7 Matched molecular pair examples specific to CYP 2D6, derived from analysis of the cytochrome P450 inhibition assays in the ChEMBL database,[30] performed using MedChemica proprietary software.[31] (*continued*)

Examples	Matched Pair[a]	Notes	Reference
17	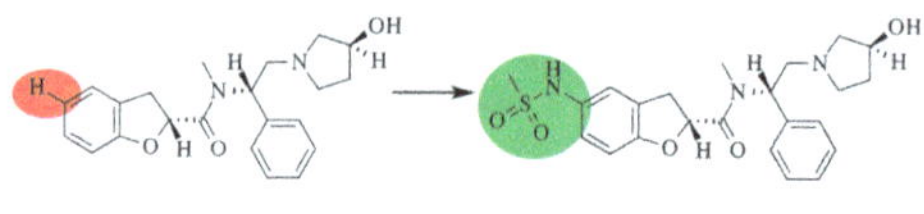 0.25 μM → 5.4 μM 2D6 inhibition	Kappa opioid receptor agonists could be improved by the addition of a sulphonamide side chain.	66
18	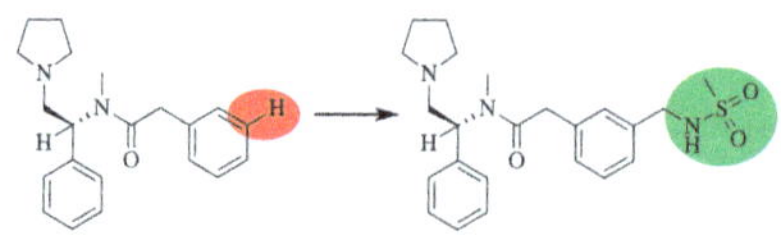 0.05 μM → 1.82 μM 2D6 inhibition	Similarly, addition of a sulphonamide group improved the 2D6 DDI profile of another series of Kappa opioid receptor agonists.	67

[a]Cytochrome P450 inhibition data is taken directly from the publication and converted to IC_{50} in μM concentration.

11.2.7 Strategies to Mitigate CYP 3A4 Inhibition

CYP 3A4 is the most abundant CYP3 isoform[68] and is implicated in the metabolism of drug molecules, such as ketoconazole, lidocaine and erythromycin.[48,57] Molecular weight is an important parameter, with 3A4 inhibition activity generally increasing with molecular weight.[27] Indeed, the results of Gleeson's analysis indicate that compounds with MW > 700 have the same mean pIC_{50} value as in the 500–700 range, indicating that the 3A4 active site is considerably larger than those of the 2C9, 2D6 and 1A2 isoforms (Figure 11.8).[27] In a publication by Yano *et al.*, the authors highlighted that whilst the active site of 3A4 may be the same size as other isoforms, the cavity toward the haem is considerably larger than that of other isoforms.[69] Furthermore, results of studies by Ekroos and Sjoegren indicate that the 3A4 cavity has the potential to expand considerably on substrate or inhibitor binding.[70] Ionization state plays a role in determining the extent of 3A4 inhibition, with neutral molecules generally being more active at 3A4 than bases or zwitterions and with acids being the least active. CYP 3A4 shows the characteristic dependency on lipophilicity, with the mean 3A4 pIC_{50} generally increasing as with clogP.[27] Compounds with a larger *f*MF [a descriptor that describes the structural complexity of a compound based on the size of its molecular framework (MF) in relation to its overall size, with MF defined as the fraction of the size of the molecular framework *versus* the size of the whole molecule (*f*MF)[71]] may also have higher CYP 3A4 inhibition. For cases in lead optimization where it is difficult to decrease clogP and molecular size, it might be a viable alternative to lower the *f*MF to avoid CYP 3A4 inhibition.[72]

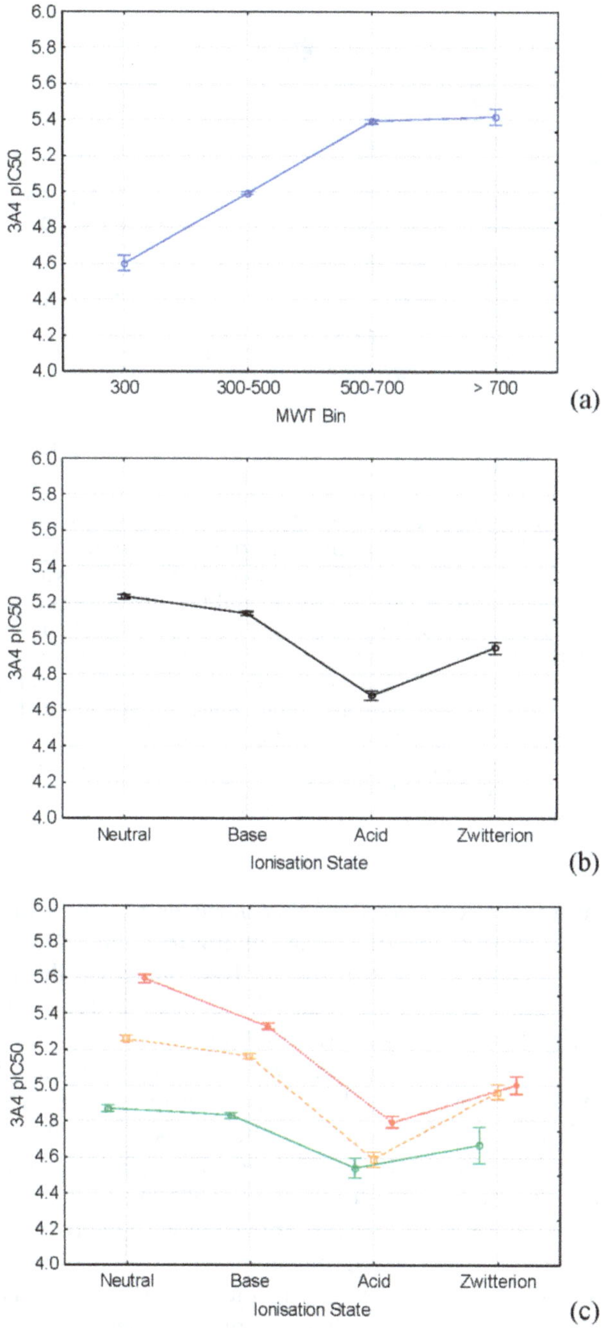

Figure 11.8 Relationship between CYP 3A4 inhibition with respect to MW (a), ionization state (b), and a combination of clogP and ionization state (c) for 42 987 molecules measured in the 3A4 assay. The green solid line denotes clogPs < 3, the orange dashed line denotes clogPs 3–5, and the red dotted line denotes clogPs > 5. The error bars denote the 95% confidence limits of the mean. Reproduced from ref. 27 with permission from American Chemical Society, Copyright 2008.

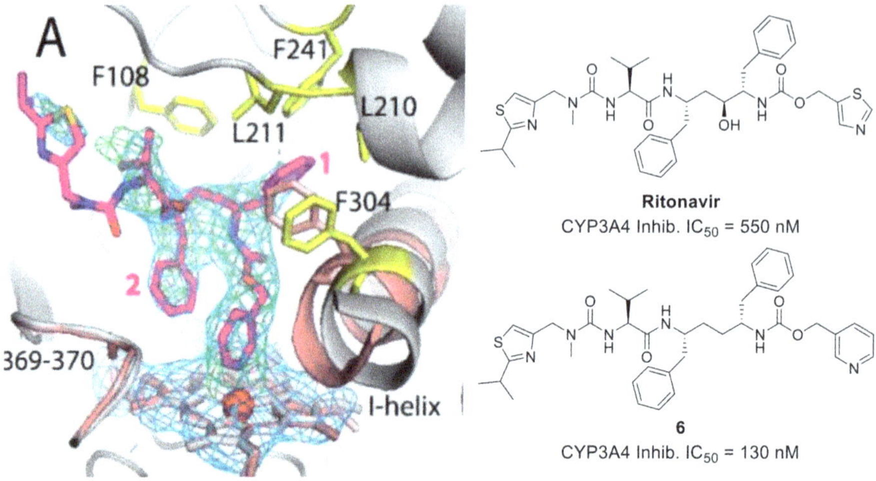

Figure 11.9 Crystal structure of the CYP 3A4–**6** complex shown in panel A. Compound **6** (in magenta) binds to the haem *via* the pyridine nitrogen. Most of the ligand–protein interactions are provided by the phenyl side groups, labelled as (1) and (2). Phenyl-1 is imbedded into a hydrophobic pocket lined with Phe108, Leu210, Leu211, Phe241 and Phe304 (shown in yellow sticks). Steric clashing between phenyl-1 and Phe304, which is also observed in the CYP 3A4–ritonavir complex, leads to the I-helix displacement. For comparison, the haem, 369–370 peptide, and I-helix of the superimposed ligand-free 1TQN structure are shown in pink. The structure and properties of ritonavir and **6** are shown. Reproduced from ref. 76 with permission from American Chemical Society, Copyright 2013.

CYP 3A4 has a large degree of flexibility of the active site, which can make CYP 3A4 X-ray co-crystal structures challenging to obtain. However, a number of CYP 3A4-liganded structures have been successfully obtained and these have been used to generate design hypotheses to reduce the CYP 3A4 inhibitory activity or conversely for rational design of potent CYP 3A4 inhibitors.[9,30,31,73–79] An example of the latter case is highlighted (Figure 11.9). In this work, Sevrioukova and Poulos sought to improve the CYP 3A4 inhibitory potency of ritonavir (often used as a pharmacoenhancer) to develop new CYP 3A4 inhibitors for anti-HIV combination therapy.[76] Ritonavir (CYP 3A4 IC_{50} = 550 nM) is known to coordinate to the CYP 3A4 haem *via* the thiazole motif, and the effect on CYP 3A4 activity was investigated for a number of desoxyritonavir analogues in which the haem-ligating thiazole in ritonazir was replaced by an imidazole, oxazole, or pyridine groups. Pyridyl **6** is a representative example from this work and is superior to ritonavir in terms of inhibitory potency (CYP 3A4 IC_{50} = 130 nM). This increase in 3A4 activity is likely to be due to the favourable stereoelectronic properties of the pyridine nitrogen and the increased structural flexibility of the desoxyritonavir backbone, which facilitates a stronger Fe–N bond and a binding conformation that optimizes protein–ligand interactions, in particular with Ser119 (Figure 11.9).

11.2.7.1 Matched Molecular Pair Transformations Specific to Reducing CYP 3A4 Inhibition

A matched molecular pair analysis of the cytochrome P450 inhibition assay data in the ChEMBL database,[30] performed using MedChemica proprietary software[31] highlighted a number of examples specific to 3A4, see Table 11.8.

Table 11.8 Matched molecular pair examples specific to CYP 3A4, derived from analysis of the cytochrome P450 inhibition assays in the ChEMBL database,[30] performed using MedChemica proprietary software.[31]

Examples	Matched Pair[a]	Notes	Reference
19	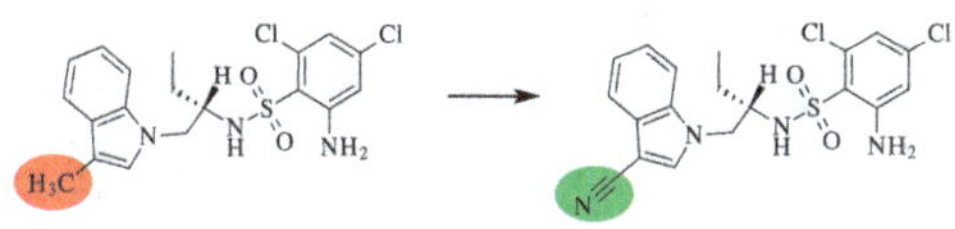 0.94 µM → 5.8 µM 3A4 inhibition	Glucocorticoid receptor (GR) modulators optimized for agonist activity.	80
20	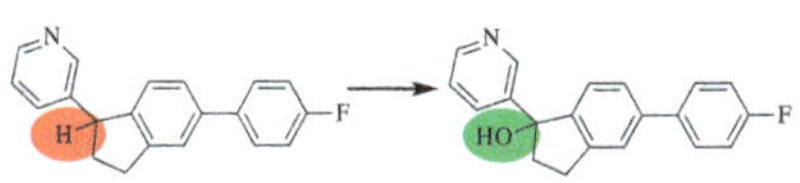 1.6 µM → >20 µM 3A4 inhibition	Compounds were targeting CYP 17 inhibition. Addition of a polar group to a potential site of metabolism reduced inhibition of 3A4. Several other matched pairs are described in the publication	81
21	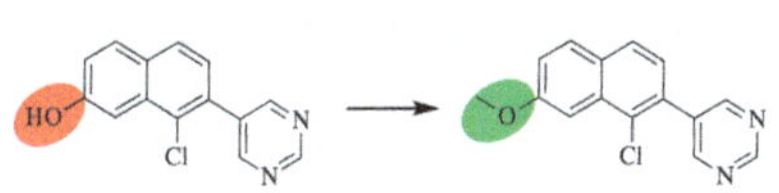 2.2 µM → >10 µM 3A4 inhibition	Compounds were designed as CYP 17/CYP 11B2 inhibitors; the change from OH to OCH_3 reduced 3A4 inhibition but without affecting target inhibition.	82

(continued)

Table 11.8 Matched molecular pair examples specific to CYP 3A4, derived from analysis of the cytochrome P450 inhibition assays in the ChEMBL database,[30] performed using MedChemica proprietary software.[31] *(continued)*

Examples	Matched Pair[a]	Notes	Reference
22	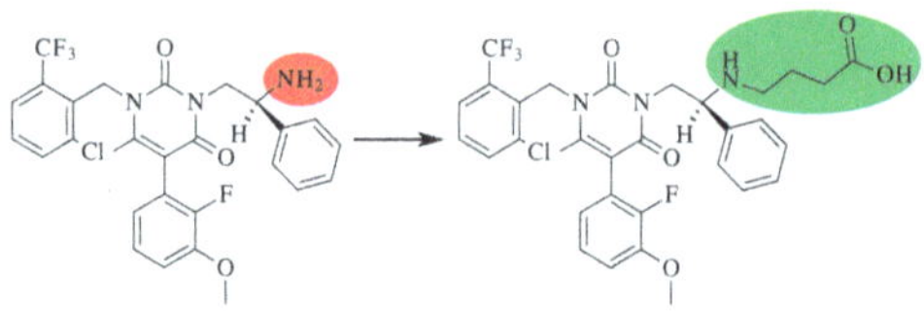 0.1 μM → 36 μM 3A4 inhibition	Human gonadotropin-releasing hormone-receptor (hGnRH-R) antagonists. Addition of the acidic side chain significantly reduced 3A4 inhibition.	83
23	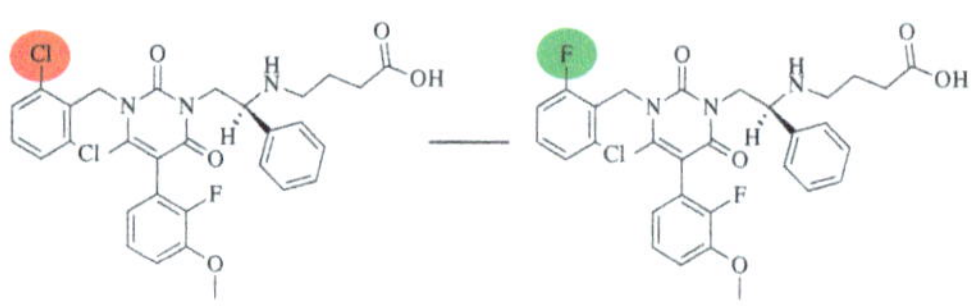 11 μM — >135 μM 3A4 inhibition	hGnRH-R antagonists. A small change of chloro to fluoro effected a dramatic improvement. Such examples are rare.	83

[a]Cytochrome P450 inhibition data is taken directly from the publication and converted to IC_{50} in μM concentration.

11.3 Summary

Reducing or eliminating CYP inhibition during a drug design program is now possible with the wealth of specific knowledge for the five major isoforms. Knowledge of the mechanisms, function, shape and electrostatics of CYP binding pockets, combined with analysis of inhibitory activity data of known CYP inhibitors and their physicochemical properties has led to clear tactics to design new molecules with reduced CYP liabilities.

References

1. S. Rendic and F. J. Di Carlo, *Drug Metab. Rev.*, 1997, **29**, 413–580.
2. F. P. Guengerich, *Drug Metab. Rev.*, 2002, **34**, 7–15.
3. J. H. Lin and A. Y. H. Lu, *Clin. Pharmacokinet.*, 1998, **35**, 361–390.
4. O. Pelkonen, M. Turpeinen, J. Hakkola, P. Honkakoshi, J. Hukkanen and H. Raunio, *Arch. Toxicol.*, 2008, **82**, 667–715.
5. D. F. V. Lewis, *Curr. Med. Chem.*, 2003, **10**, 1955–1972.
6. F. P. Guengerich, in *Oxidative, Reductive, and Hydrolytic Metabolism of Drugs*, John Wiley & Sons, Inc., 2008, pp. 15–35.

7. J. H. Lin, *Pharm. Res.*, 2006, **23**, 1089–1116.
8. P. Manikandan and S. Nagini, *Curr. Drug Targets*, 2018, **19**, 38–54.
9. G. Braenden, T. Sjoegren, V. Schnecke and Y. Xue, *Drug Discovery Today*, 2014, **19**, 905–911.
10. D. F. McGinnity and R. J. Riley, *Biochem. Soc. Trans.*, 2001, **29**, 135–139.
11. K. T. Olkkola, K. Aranko, H. Luurila, A. Hiller, L. Saarnivaara, J. J. Himberg and P. J. Neuvonen, *Clin. Pharmacol. Ther.*, 1993, **53**, 298–305.
12. R. Janknegt, *J. Antimicrob. Chemother.*, 1990, **26**, 7–29.
13. L. L. Von Moltke, D. J. Greenblatt, J. Schmider, S. X. Duan, C. E. Wright, J. S. Harmatz and R. I. Shader, *J. Clin. Pharmacol.*, 1996, **36**, 783–791.
14. H. E. Kang, S. K. Bae, M. Yoo, D. C. Lee, Y. G. Kim and M. G. Lee, *Br. J. Pharmacol.*, 2009, **156**, 1009–1018.
15. M. Murray, *Clin. Pharmacokinet.*, 1992, **23**, 132–146.
16. U. M. Kent, M. I. Jushchyshyn and P. F. Hollenberg, *Curr. Drug Metab.*, 2001, **2**, 215–243.
17. S.-I. Kanamitsu, K. Ito, C. E. Green, C. A. Tyson, N. Shimada and Y. Sugiyama, *Pharm. Res.*, 2000, **17**, 419–426.
18. J. Belcher, K. J. McLean, S. Matthews, L. S. Woodward, K. Fisher, S. E. J. Rigby, D. R. Nelson, D. Potts, M. T. Baynham, D. A. Parker, D. Leys and A. W. Munro, *J. Biol. Chem.*, 2014, **289**, 6535–6550.
19. T. D. Bjornsson, J. T. Callaghan, H. J. Einolf, V. Fischer, L. Gan, S. Grimm, J. Kao, S. P. King, G. Miwa, L. Ni, G. Kumar, J. McLeod, R. S. Obach, S. Roberts, A. Roe, A. Shah, F. Snikeris, J. T. Sullivan, D. Tweedie, J. M. Vega, J. Walsh and S. A. Wrighton, *Drug Metab. Dispos.*, 2003, **31**, 815–832.
20. R. J. Dimelow, P. D. Metcalfe and S. Thomas, *Encycl. Drug Metab. Interact.*, 2012, 1–55.
21. O. A. Fahmi, S. Hurst, D. Plowchalk, J. Cook, F. Guo, K. Youdim, M. Dickins, A. Phipps, A. Darekar, R. Hyland and R. S. Obach, *Drug Metab. Dispos.*, 2009, **37**, 1658–1666.
22. R. S. Obach, R. L. Walsky, K. Venkatakrishnan, E. A. Gaman, J. B. Houston and L. M. Tremaine, *J. Pharmacol. Exp. Ther.*, 2006, **316**, 336–348.
23. M. P. Gleeson, A. M. Davis, K. K. Chohan, S. W. Paine, S. Boyer, C. L. Gavaghan, C. H. Arnby, C. Kankkonen and N. Albertson, *J. Comput.-Aided Mol. Des.*, 2007, **21**, 559–573.
24. A. G. Leach and N. J. Kidley, *J. Chem. Inf. Model.*, 2011, **51**, 1048–1063.
25. P. J. McDonnell and M. R. Jacobs, *Ann. Pharmacother.*, 2002, **36**, 1331–1336.
26. https://www.fda.gov/regulatory-information/search-fda-guidance-documents/vitro-metabolism-and-transporter-mediated-drug-drug-interaction-studies-guidance-industry.
27. M. P. Gleeson, *J. Med. Chem.*, 2008, **51**, 817–834.
28. P. Gleeson, G. Bravi, S. Modi and D. Lowe, *Bioorg. Med. Chem.*, 2009, **17**, 5906–5919.
29. R. J. Young, D. V. S. Green, C. N. Luscombe and A. P. Hill, *Drug Discovery Today*, 2011, **16**, 822–830.
30. I. F. Sevrioukova and T. L. Poulos, *J. Biol. Chem.*, 2012, **287**, 3510–3517.

31. P. Kaur, A. R. Chamberlin, T. L. Poulos and I. F. Sevrioukova, *J. Med. Chem.*, 2016, **59**, 4210–4220.
32. C. Kramer, A. Ting, H. Zheng, J. Hert, T. Schindler, M. Stahl, G. Robb, J. J. Crawford, J. Blaney, S. Montague, A. G. Leach, A. G. Dossetter and E. J. Griffen, *J. Med. Chem.*, 2018, **61**, 3277–3292.
33. I. Lukac, J. Zarnecka, E. J. Griffen, A. G. Dossetter, S. A. St-Gallay, S. J. Enoch, J. C. Madden and A. G. Leach, *J. Chem. Inf. Model.*, 2017, **57**, 2424–2436.
34. E. J. Hanan, A. van Abbema, K. Barrett, W. S. Blair, J. Blaney, C. Chang, C. Eigenbrot, S. Flynn, P. Gibbons, C. A. Hurley, J. R. Kenny, J. Kulagowski, L. Lee, S. R. Magnuson, C. Morris, J. Murray, R. M. Pastor, T. Rawson, M. Siu, M. Ultsch, A. Zhou, D. Sampath and J. P. Lyssikatos, *J. Med. Chem.*, 2012, **55**, 10090–10107.
35. E. A. Voight, A. R. Gomtsyan, J. F. Daanen, R. J. Perner, R. G. Schmidt, E. K. Bayburt, S. DiDomenico, H. A. McDonald, P. S. Puttfarcken, J. Chen, T. R. Neelands, B. R. Bianchi, P. Han, R. M. Reilly, P. H. Franklin, J. A. Segreti, R. A. Nelson, Z. Su, A. J. King, J. S. Polakowski, S. J. Baker, D. M. Gauvin, L. R. Lewis, J. P. Mikusa, S. K. Joshi, C. R. Faltynek, P. R. Kym and M. E. Kort, *J. Med. Chem.*, 2014, **57**, 7412–7424.
36. C. L. Cioffi, N. Dobri, E. E. Freeman, M. P. Conlon, P. Chen, D. G. Stafford, D. M. C. Schwarz, K. C. Golden, L. Zhu, D. B. Kitchen, K. D. Barnes, B. Racz, Q. Qin, E. Michelotti, C. L. Cywin, W. H. Martin, P. G. Pearson, G. Johnson and K. Petrukhin, *J. Med. Chem.*, 2014, **57**, 7731–7757.
37. M. A. Brodney, E. M. Beck, C. R. Butler, G. Barreiro, E. F. Johnson, D. Riddell, K. Parris, C. E. Nolan, Y. Fan, K. Atchison, C. Gonzales, A. E. Robshaw, S. D. Doran, M. W. Bundesmann, L. Buzon, J. Dutra, K. Henegar, E. LaChapelle, X. Hou, B. N. Rogers, J. Pandit, R. Lira, L. Martinez-Alsina, P. Mikochik, J. C. Murray, K. Ogilvie, L. Price, S. M. Sakya, A. Yu, Y. Zhang and B. T. O'Neill, *J. Med. Chem.*, 2015, **58**, 3223–3252.
38. T. Nakamura, M. Sato, H. Kakinuma, N. Miyata, K. Taniguchi, K. Bando, A. Koda and K. Kameo, *J. Med. Chem.*, 2003, **46**, 5416–5427.
39. B. P. Morgan, A. Muci, P.-P. Lu, X. Qian, T. Tochimoto, W. W. Smith, M. Garard, E. Kraynack, S. Collibee, I. Suehiro, A. Tomasi, S. C. Valdez, W. Wang, H. Jiang, J. Hartman, H. M. Rodriguez, R. Kawas, S. Sylvester, K. A. Elias, G. Godinez, K. Lee, R. Anderson, S. Sueoka, D. Xu, Z. Wang, N. Djordjevic, F. I. Malik and D. J. Morgans, *ACS Med. Chem. Lett.*, 2010, **1**, 472–477.
40. N. Bailey, M. J. Bamford, D. Brissy, J. Brookfield, E. Demont, R. Elliott, N. Garton, I. Farre-Gutierrez, T. Hayhow, G. Hutley, A. Naylor, T. A. Panchal, H.-X. Seow, D. Spalding and A. K. Takle, *Bioorg. Med. Chem. Lett.*, 2009, **19**, 3602–3606.
41. C.-C. Peng, J. L. Cape, T. Rushmore, G. J. Crouch and J. P. Jones, *J. Med. Chem.*, 2008, **51**, 8000–8011.
42. S. Sansen, J. K. Yano, R. L. Reynald, G. A. Schoch, K. J. Griffin, C. D. Stout and E. F. Johnson, *J. Biol. Chem.*, 2007, **282**, 14348–14355.
43. J. Burton, I. Ijjaali, O. Barberan, F. Petitet, D. P. Vercauteren and A. Michel, *J. Med. Chem.*, 2006, **49**, 6231–6240.

44. K. K. Chohan, S. W. Paine, J. Mistry, P. Barton and A. M. Davis, *J. Med. Chem.*, 2005, **48**, 5154–5161.
45. L. E. Korhonen, M. Rahnasto, N. J. Maehoenen, C. Wittekindt, A. Poso, R. O. Juvonen and H. Raunio, *J. Med. Chem.*, 2005, **48**, 3808–3815.
46. S. Kim, H. Ko, J. E. Park, S. Jung, S. K. Lee and Y.-J. Chun, *J. Med. Chem.*, 2002, **45**, 160–164.
47. L. A. Clutterbuck, C. G. Posada, C. Visintin, D. R. Riddall, B. Lancaster, P. J. Gane, J. Garthwaite and D. L. Selwood, *J. Med. Chem.*, 2009, **52**, 2694–2707.
48. D. F. V. Lewis, S. Modi and M. Dickins, *Drug Metab. Drug Interact.*, 2001, **18**, 221–242.
49. M. R. Wester, J. K. Yano, G. A. Schoch, C. Yang, K. J. Griffin, C. D. Stout and E. F. Johnson, *J. Biol. Chem.*, 2004, **279**, 35630–35637.
50. R. Liu, X. Lyu, S. M. Batt, M.-H. Hsu, M. B. Harbut, C. Vilcheze, B. Cheng, K. Ajayi, B. Yang, Y. Yang, H. Guo, C. Lin, F. Gan, C. Wang, S. G. Franzblau, W. R. Jacobs Jr, G. S. Besra, E. F. Johnson, M. Petrassi, A. K. Chatterjee, K. Fuetterer and F. Wang, *Angew. Chem., Int. Ed.*, 2017, **56**, 13011–13015.
51. P. A. Williams, J. Cosme, A. Ward, H. C. Angove, D. M. Vinkovic and H. Jhoti, *Nature*, 2003, **424**, 464–468.
52. S. E. Skerratt, M. J. de Groot and C. Phillips, *MedChemComm*, 2016, **7**, 813–819.
53. N. A. Swain, D. Batchelor, S. Beaudoin, B. M. Bechle, P. A. Bradley, A. D. Brown, B. Brown, K. J. Butcher, R. P. Butt, M. L. Chapman, S. Denton, D. Ellis, S. R. G. Galan, S. M. Gaulier, B. S. Greener, M. J. de Groot, M. S. Glossop, I. K. Gurrell, J. Hannam, M. S. Johnson, Z. Lin, C. J. Markworth, B. E. Marron, D. S. Millan, S. Nakagawa, A. Pike, D. Printzenhoff, D. J. Rawson, S. J. Ransley, S. M. Reister, K. Sasaki, R. I. Storer, P. A. Stupple and C. W. West, *J. Med. Chem.*, 2017, **60**, 7029–7042.
54. J. Li, L. J. Kennedy, Y. Shi, S. Tao, X.-Y. Ye, S. Y. Chen, Y. Wang, A. S. Hernández, W. Wang, P. V. Devasthale, S. Chen, Z. Lai, H. Zhang, S. Wu, R. A. Smirk, S. A. Bolton, D. E. Ryono, H. Zhang, N.-K. Lim, B.-C. Chen, K. T. Locke, K. M. O'Malley, L. Zhang, R. A. Srivastava, B. Miao, D. S. Meyers, H. Monshizadegan, D. Search, D. Grimm, R. Zhang, T. Harrity, L. K. Kunselman, M. Cap, P. Kadiyala, V. Hosagrahara, L. Zhang, C. Xu, Y.-X. Li, J. K. Muckelbauer, C. Chang, Y. An, S. R. Krystek, M. A. Blanar, R. Zahler, R. Mukherjee, P. T. W. Cheng and J. A. Tino, *J. Med. Chem.*, 2010, **53**, 2854–2864.
55. H. R. Hoveyda, G. L. Fraser, G. Dutheuil, M. El Bousmaqui, J. Korac, F. Lenoir, A. Lapin and S. Noël, *ACS Med. Chem. Lett.*, 2015, **6**, 736–740.
56. R. L. Reynald, S. Sansen, C. D. Stout and E. F. Johnson, *J. Biol. Chem.*, 2012, **287**, 44581–44591.
57. D. F. V. Lewis, *Toxicology*, 2000, **144**, 197–203.
58. R. Snyder, R. Sangar, J. Wang and S. Ekins, *Quant. Struct.-Act. Relat.*, 2002, **21**, 357–368.
59. S. Ekins, G. Bravi, S. Binkley, J. S. Gillespie, B. J. Ring, J. H. Wikel and S. A. Wrighton, *Pharmacogenetics*, 1999, **9**, 477–489.

60. P. Rowland, F. E. Blaney, M. G. Smyth, J. J. Jones, V. R. Leydon, A. K. Oxbrow, C. J. Lewis, M. G. Tennant, S. Modi, D. S. Eggleston, R. J. Chenery and A. M. Bridges, *J. Biol. Chem.*, 2006, **281**, 7614–7622.
61. A. Wang, C. D. Stout, Q. Zhang and E. F. Johnson, *J. Biol. Chem.*, 2015, **290**, 5092–5104.
62. A. Wang, U. Savas, M.-H. Hsu, C. D. Stout and E. F. Johnson, *J. Biol. Chem.*, 2012, **287**, 10834–10843.
63. C. R. Butler, K. Ogilvie, L. Martinez-Alsina, G. Barreiro, E. M. Beck, C. E. Nolan, K. Atchison, E. Benvenuti, L. Buzon, S. Doran, C. Gonzales, C. J. Helal, X. Hou, M.-H. Hsu, E. F. Johnson, K. Lapham, L. Lanyon, K. Parris, B. T. O'Neill, D. Riddell, A. Robshaw, F. Vajdos and M. A. Brodney, *J. Med. Chem.*, 2017, **60**, 386–402.
64. R. J. Vaz, A. Nayeem, K. Santone, G. Chandrasena and A. V. Gavai, *Bioorg. Med. Chem. Lett.*, 2005, **15**, 3816–3820.
65. S. Hudson, M. Kiankarimi, W. Eccles, W. Dwight, Y. S. Mostofi, M. J. Genicot, B. A. Fleck, K. Gogas, A. Aparicio, H. Wang, J. Wen and W. S. Wade, *Bioorg. Med. Chem. Lett.*, 2008, **18**, 4491–4494.
66. G.-H. Chu, M. Gu, J. A. Cassel, S. Belanger, T. M. Graczyk, R. N. DeHaven, N. Conway-James, M. Koblish, P. J. Little, D. L. DeHaven-Hudkins and R. E. Dolle, *Bioorg. Med. Chem. Lett.*, 2005, **15**, 5114–5119.
67. B. Le Bourdonnec, C. W. Ajello, P. R. Seida, R. G. Susnow, J. A. Cassel, S. Belanger, G. J. Stabley, R. N. DeHaven, D. L. DeHaven-Hudkins and R. E. Dolle, *Bioorg. Med. Chem. Lett.*, 2005, **15**, 2647–2652.
68. T. Shimada, H. Yamazaki, M. Mimura, Y. Inui and F. P. Guengerich, *J. Pharmacol. Exp. Ther.*, 1994, **270**, 414–423.
69. J. K. Yano, M. R. Wester, G. A. Schoch, K. J. Griffin, C. D. Stout and E. F. Johnson, *J. Biol. Chem.*, 2004, **279**, 38091–38094.
70. M. Ekroos and T. Sjoegren, *Proc. Natl. Acad. Sci. U. S. A.*, 2006, **103**, 13682–13687.
71. Y. Yang, H. Chen, I. Nilsson, S. Muresan and O. Engkvist, *J. Med. Chem.*, 2010, **53**, 7709–7714.
72. Y. Yang, O. Engkvist, A. Llinas and H. Chen, *J. Med. Chem.*, 2012, **55**, 3667–3677.
73. M. Ekroos and T. Sjogren, *Proc. Natl. Acad. Sci. U. S. A.*, 2006, **103**, 13682–13687.
74. I. F. Sevrioukova and T. L. Poulos, *Proc. Natl. Acad. Sci. U. S. A.*, 2010, **107**, 18422–18427.
75. I. F. Sevrioukova and T. L. Poulos, *Arch. Biochem. Biophys.*, 2012, **520**, 108–116.
76. I. F. Sevrioukova and T. L. Poulos, *J. Med. Chem.*, 2013, **56**, 3733–3741.
77. I. F. Sevrioukova and T. L. Poulos, *Biochemistry*, 2013, **52**, 4474–4481.
78. I. F. Sevrioukova, T. L. Poulos, T. L. Poulos and T. L. Poulos, *Proc. Natl. Acad. Sci. U. S. A.*, 2017, **114**, 486–491.
79. E. R. Samuels and I. Sevrioukova, *Mol. Pharmaceutics*, 2018, **15**, 279–288.

80. J. P. Taygerly, L. R. McGee, S. M. Rubenstein, J. B. Houze, T. D. Cushing, Y. Li, A. Motani, J.-L. Chen, W. Frankmoelle, G. Ye, M. R. Learned, J. Jaen, S. Miao, P. B. Timmermans, M. Thoolen, P. Kearney, J. Flygare, H. Beckmann, J. Weiszmann, M. Lindstrom, N. Walker, J. Liu, D. Biermann, Z. Wang, A. Hagiwara, T. Iida, H. Aramaki, Y. Kitao, H. Shinkai, N. Furukawa, J. Nishiu and M. Nakamura, *Bioorg. Med. Chem.*, 2013, **21**, 979–992.
81. M. A. E. Pinto-Bazurco Mendieta, M. Negri, C. Jagusch, U. Müller-Vieira, T. Lauterbach and R. W. Hartmann, *J. Med. Chem.*, 2008, **51**, 5009–5018.
82. M. A. E. Pinto-Bazurco Mendieta, Q. Hu, M. Engel and R. W. Hartmann, *J. Med. Chem.*, 2013, **56**, 6101–6107.
83. C. Chen, D. Wu, Z. Guo, Q. Xie, G. J. Reinhart, A. Madan, J. Wen, T. Chen, C. Q. Huang, M. Chen, Y. Chen, F. C. Tucci, M. Rowbottom, J. Pontillo, Y.-F. Zhu, W. Wade, J. Saunders, H. Bozigian and R. S. Struthers, *J. Med. Chem.*, 2008, **51**, 7478–7485.

CHAPTER 12

Aldehyde and Xanthine Oxidase Metabolism

DAVID C. PRYDE*[a], DHARMENDRA B. YADAV[b] AND RAJIB GHOSH[b]

[a]Curadev Pharma Ltd., Sandwich, Kent, CT13 9ND, UK; [b]Curadev Pharma Pvt. Ltd., B-87, Sector 83, Noida 201305, India
*E-mail: david@curadev.co.uk

12.1 Introduction

Aldehyde oxidase (AO, EC 1.2.3.1) and xanthine oxidase (XO, EC 1.2.3.2) are metabolizing enzymes contained within the cytosolic compartment of many tissues in many species. The first literature reference to AO dates from the 1930s, but citations to this enzyme have been steadily growing ever since, especially in the last decade. As drug discovery research has moved towards different target gene families, synthetic methodology has allowed facile access to heteroaromatic systems, and the industry has developed a greater understanding of how to reduce P450-mediated metabolism, the role of non-P450 metabolism, such as by AO and XO, has been brought into sharp focus. In this chapter, we highlight the main features of this family of metabolising enzymes and support our view that this is an enzyme family of increasing importance in xenobiotic metabolism with several real-life examples.

Drug Discovery Series No. 79
The Medicinal Chemist's Guide to Solving ADMET Challenges
Edited by Patrick Schnider
© The Royal Society of Chemistry 2021
Published by the Royal Society of Chemistry, www.rsc.org

12.1.1 Enzyme Family

AO and XO belong to the xanthine oxidase family of enzymes,[1] which are cytosolic complex molybdoflavoproteins containing two [2Fe–2S] clusters, flavin adenine dinucleotide (FAD) and the molybdenum cofactor MoCo that are essential for enzyme activity.[2–4] These enzymes catalyse the metabolism of a variety of aldehydes and azaheterocyclic-containing drug molecules.[5]

12.1.2 Expression

AO and XO are widely distributed across the animal kingdom from insects to fish to humans[6] and expressed in many tissues.[7] Humans express a single functional protein AOX1, while AO in rodents and other animals exists in up to four isoforms (AOX1, AOX3, AOX3L1, AOX4). It is notable that human AOX1 is highly homologous with rodent AOX1, but significantly less homologous to other isotypes. XO exists in two isoforms (oxidase and dehydrogenase forms). Expression of both enzymes is by far the highest in liver in all species, including humans, and drug clearance mediated by both enzymes is dominated by liver-based activity. AO distribution in other tissues is highly species dependent. In humans, aside from the liver, AO activity has also been detected in excretory organs, such as lung, the gastrointestinal tract and kidney, as well as in endocrine tissue and brain.[8,9] Despite similarities in amino acid sequence there are marked species differences in protein expression.[7,10]

Genetic deficits in human XO have been described as resulting in xanthinuria and the appearance of xanthine urinary stones.[11] but no genetic deficiency in AOX1 has been reported to date.

12.1.3 Activity

The catalytic activity of AO differs widely between species and across individuals within a single species. Understanding enzyme variations between species is crucial to being able to accurately predict human pharmacokinetic parameters. In the case of AO in particular this has proven to be lacking, leading to several clinical failures. Although the order of AO activity among animal species may vary depending on the substrate, it generally seems to be high in monkeys and humans and low in guinea pigs, rats and mice, whereas dogs are to a large extent deficient in activity. A recent study[12] using several known AO substrates has demonstrated that minipig *in vivo* data is suitable to predict human *in vivo* clearance, with other reports also indicating that minipig is a good model for predicting human AO clearance.[13] The active site may be sized differently in different species, contributing to this variability.[8,14] Large activity differences have been observed among various strains in rats and mice[15–17] and inter-individual variability has been noted within the human population.[18,19] Indeed, Hutzler *et al.* have reported large variations in AO activity between different batches of human hepatocytes.[20] While the exact causes of the observed variability in activity have yet to be fully elucidated, the presence of allelic variants of the human AOX1 gene has been noted.[21] While there have been some reports[22] of variability in species activity of XO, variability in this enzyme appears to be considerably lower than in AO.

12.1.4 Structure

AO and XO show an amino acid sequence identity of approximately 40% in various animal species[6,23] and both enzymes are active as a homodimer composed of two identical subunits of approximately 150 kDa. Each subunit is subdivided into three distinct domains; a 20 kDa N-terminal domain, which binds two non-equivalent iron-containing clusters, a 40 kDa central domain, that contains a flavin adenine dinucleotide (FAD) binding site, and an 85 kDa C-terminal domain, that accommodates the molybdenum cofactor in close proximity to the substrate pocket.[6] Molybdenum (Mo) is an essential component of the enzyme and is required for enzyme catalysis along with FAD. It is biologically inactive until it becomes complexed to form the tetracyclic pyranopteridine complex[24] termed "MoCo" (Figure 12.1).

For many years, the crystal structure of bovine xanthine dehydrogenase was used as the model for mammalian XO and AO. The first crystal structure of a mammalian AO, that of mouse AOX3 (PDB 3ZYV), was published in 2012.[25] This was subsequently followed by the first crystal structure of human AOX1 in 2015[26] of the free enzyme (PDB 4UHW) and of the enzyme bound with both an inhibitor (thioridazine) and a substrate (phthalazine) simultaneously (PDB 4UHX).

These studies enabled several hypotheses to be proposed about how AO functions and why AO and XO enzymes behave differently. The authors proposed that in mammalian AOs, substrate access is *via* a wide, deep funnel that leads into the active site. Entrance to the funnel is determined by the flexibility of two loops, which they termed gates 1 and 2. These gates are longer and more flexible than the corresponding sites in XO, thereby allowing a much wider tolerance of substrate size and structure in AOs. Gate 2 in mammalian AOs features two acidic residues which can be envisaged as directing polar and/or positively charged groups toward the catalytic centre of the enzyme. Given the non-competitive nature of the inhibitor binding site, they also proposed this remote site as a general site for inhibition of AOs.

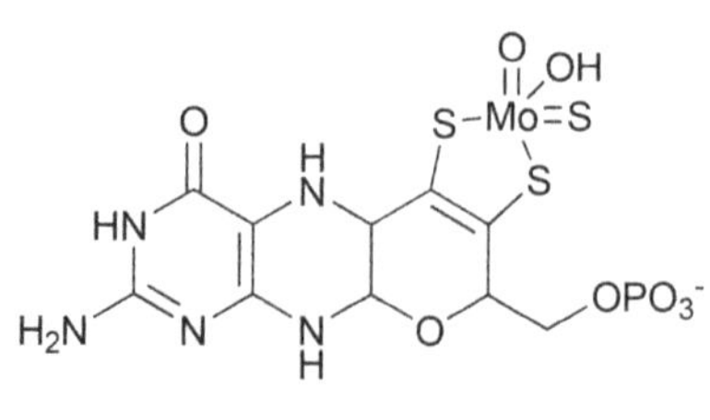

Figure 12.1 Structure of the Mo–pterin cofactor MoCo.

12.1.5 Mechanism

In a typical catalytic cycle, substrates are oxidized to products at the Mo center.[27] The reducing equivalents are then passed to FAD, which is reoxidized by molecular oxygen, the final acceptor of the reducing equivalents produced, with H_2O_2 as the main by-product. The iron-containing centres function as mediators of electron transfer between MoCo and the flavin cofactor and serve as electron sinks, storing reducing equivalents during catalysis.

The oxidative hydroxylation of substrates by AO and XO is complementary to that mediated by CYP450. Even though both enzyme families utilize molecular oxygen as the ultimate electron acceptor, the oxygen atom that is incorporated into the product during AO-mediated oxidative hydroxylation comes from water and not oxygen. The AO-catalyzed oxidation of azaheteroaromatic rings involves an initial nucleophilic attack at the carbon atom adjacent to the heteroatom. The susceptibility of a heterocycle to this nucleophilic attack therefore defines whether or not that heterocycle is a substrate for AO. A number of plausible mechanisms have been proposed for this initial nucleophilic attack; a recent report provides evidence favouring a concerted reaction mechanism (Figure 12.2).[28]

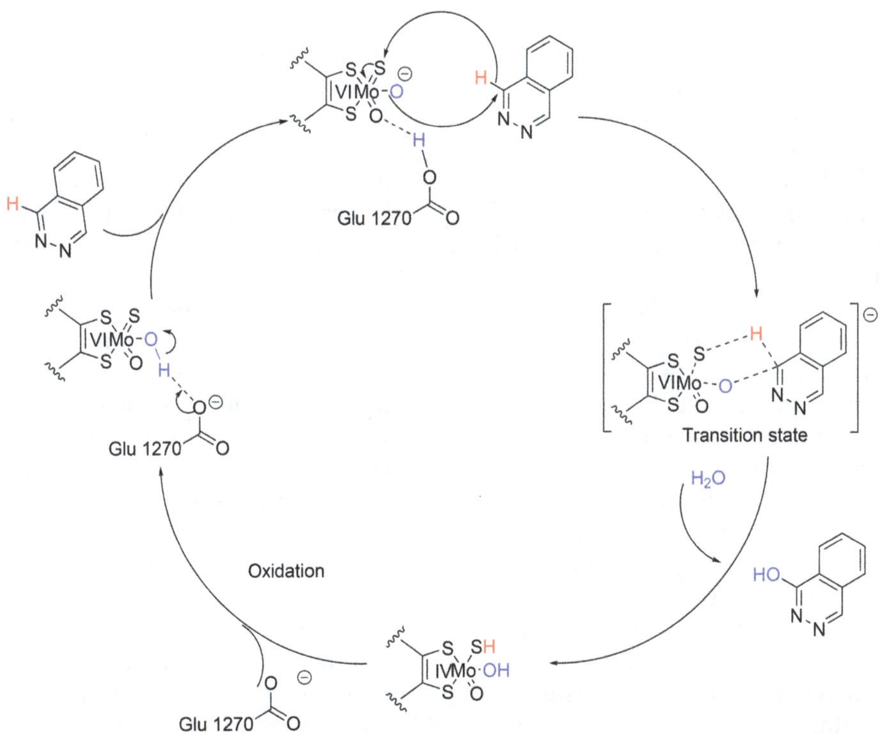

Figure 12.2 AO catalyzed oxidation of heteroaromatic rings. Reproduced from ref. 27 with permission from American Chemical Society, Copyright 2013.

12.1.6 Function and Substrates

The homeostatic function of XO is to oxidise hypoxanthine and xanthine into the terminal catabolite uric acid, but the physiological role of AO is as yet unknown. AO substrates include indole-3-acetate, retinaldehyde and the two vitamins, nicotinamide and pyridoxal. Both enzymes catalyse an oxidation reaction by a nucleophilic mechanism and using water as the source of oxygen that is incorporated into the metabolite. The primary difference in the two enzymes however is that XO can exist in two interconvertible forms, xanthine oxidase and xanthine dehydrogenase, while AO exists only in the oxidase form. AO utilizes only molecular oxygen as an electron acceptor, in contrast to XO, which can transfer electrons to both oxygen and nicotinamide adenine dinucleotide (NAD^+). Both enzymes catalyze oxidation and reduction reactions of a large array of substrates although oxidation reactions are far more common.[29] XO oxidizes primarily pyrimidines and purines. In general, AO has the ability to oxidize a broader range of substrates than XO, which is reflected in the AO literature being much more extensive in the context of drug discovery.[4] Typical substrates of AO are compounds containing either an aldehyde function, nitro/nitroso compounds and nitrogen-containing aromatic heterocycles.[5] Since drugs themselves rarely possess aldehyde moieties, the oxidation of aromatic azaheterocyclic groups to oxo-heterocycles is of most importance in drug discovery as these substituents are present in many drug molecules. Nitrogen is often introduced into aromatic systems to reduce global lipophilicity and decrease P450-mediated clearance. Azaheteroaromatic systems have also lent themselves to particular gene families in recent years, for example as hinge-binding groups against kinase targets.[30] Finally, the increased use of high-throughput and highly efficient aromatic cross-coupling reaction methods has greatly increased the number of azaheteroaromatic drug candidates being made in discovery programmes across the industry.[31,32]

While XO recognises primarily purines and pyrimidine rings, quinolines, quinazolines, pyridines, pyrimidines, pyrazines and/or isomers and fused-ring analogues are all typical targets for oxidation by AO. In cases where a common substrate is recognised by both enzymes, the rate of metabolism by either in the species of interest will dictate which is the primary metabolizing enzyme.[33] We have surveyed the literature for azaheterocycle structures that are substrates for AO and XO and these are summarised in Figure 12.3. This is not intended to be exhaustive, but indicative of the chemotypes that may be turned over by these enzymes and to highlight that any drug molecule that contains one of these ring systems *could* be a substrate for AO or XO. AO in particular is of most concern, given the much greater substrate recognition and species variability; indeed, in several examples described later, when compound metabolism was surveyed for both XO and AO liabilities, it is almost invariably AO that is responsible for unexpected outcomes.

It is important to recognise that substrates for AO do not correlate with any specific area of molecular space, and there is a wide spread of properties for

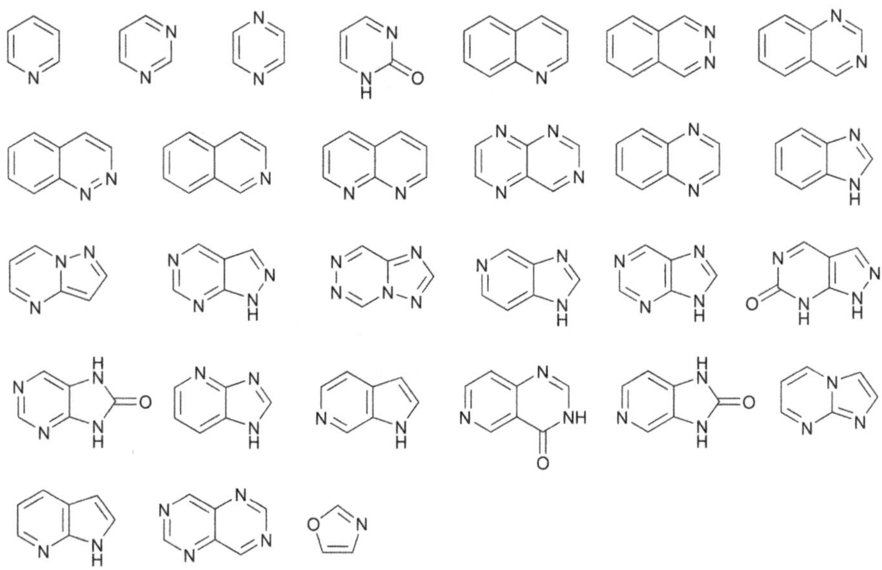

Figure 12.3 Literature-precedented chemotypes that are substrates for oxidation by AO or XO.

AO substrates across structure, polarity and lipophilicity ranges, unlike substrates that are recognised by CYP450s. It is clear that specific structural features alone are a good indicator of AO turnover, which all without exception feature an aromatic carbon–hydrogen bond adjacent to an aromatic nitrogen atom. *For any structural feature of this type, there is a reasonable chance of AO oxidation across all physicochemical space.*

12.2 Screening Strategies

Metabolic screens that use simple *in vitro* systems, such as liver microsomes (LMs) and hepatocytes, are commonly employed early in drug discovery programmes to inform compound design. LMs do not provide information on the contribution of cytosolic enzymes to compound metabolism. Since AO and XO are cytosolic enzymes, normal LM assays are not useful. Hepatocytes, on the other hand, contain all human phase I and phase II enzymes. Thus, the parent depletion profile following incubation of a compound with hepatocytes can help with assessment of its metabolism by enzymes other than CYP450 when the data is used in conjunction with that generated from LMs. Faster turnover in hepatocytes is a good indicator that a compound is a substrate for a non-CYP450 enzyme and possibly a cytosolic enzyme such as AO and XO. We propose a simple decision tree in Figure 12.4 to elucidate a possible involvement of AO/XO-catalyzed oxidation in compound metabolism.[33]

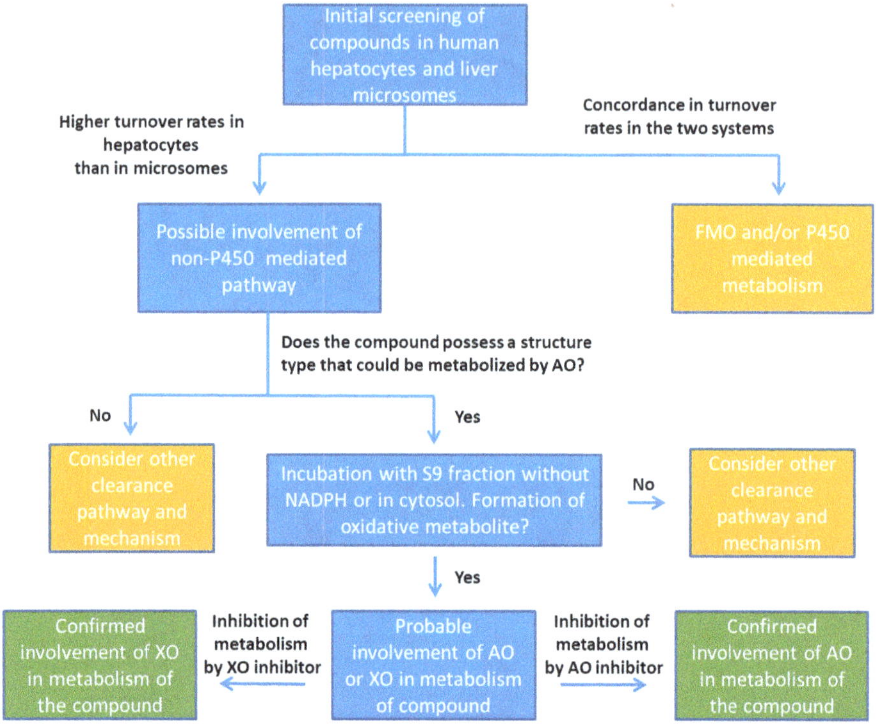

Figure 12.4 Decision tree for exploring metabolism by the cytosolic AO and XO enzymes. Adapted from ref. 33 with permission from Taylor and Francis, Copyright 2013.

The above process should be followed up with experiments in which the lead compound is incubated with a cytosolic or an S9 fraction (the supernatant fraction obtained from liver homogenate by centrifuging at 9000 *g* in a suitable medium; this fraction contains both cytosol and microsomes) in the absence of a co-factor (such as NADPH) which is generally included in incubations with LMs. Formation of an oxidative metabolite will reveal contribution of a non-CYP450 enzyme. The involvement of AO or XO can then be confirmed by carrying out the same incubation in the presence of a specific inhibitor and observing a reduction in metabolite production. The most common selective inhibitors used for research purposes are allopurinol (XO), raloxifene (AO) and hydralazine (AO). The reader is referred to several other sources for more detailed descriptions of inhibitors.[34,35]

Zientek *et al.* have developed a 'yardstick approach' to classify compounds as being 'low', 'medium' and 'high' intrinsic clearance by comparison to marker compounds in cytosol or S9.[36] Baran and co-workers have developed a simple and practical 'litmus test' for assessing AO liabilities in heteroaromatic substrates[37] using bis-(((difluoromethyl)sulfinyl)oxy)zinc (DFMS) and liquid chromatography–mass spectrometry (LC-MS) detection. The principle of the method is that known AO substrates are closely correlated with substrates that would be predicted to undergo functionalization by nucleophilic radicals. Indeed, the mechanism of nucleophilic attack on an aromatic C atom, followed by C–H bond cleavage to re-establish aromaticity is common to both processes. The empirical reactivity of a heteroaromatic substrate towards an alkylsulfinyl radical correlates with susceptibility to AO metabolism. While the DFMS litmus test is somewhat less discriminating than enzymatic AO oxidation, the correlation for a range of different substrates is strong and provides useful information that is informative about where further more detailed testing is appropriate. An interesting consequence of the DFMS radical substitution method is that the products of the reaction install a blocking CF_2H group at the vulnerable position on the heteroaromatic system, which themselves can be tested for AO liability. A similar method uses ecoAO,[38] a cell paste of MoCo-producing *E. coli* expressing human AO for metabolite generation.

In the decision tree in Figure 12.4, an important step involves the review of compound structure as a prospective indicator of possible AO metabolism in the context of the substrates depicted above. Given the clear correlation between chemotype and AO recognition, a number of researchers have developed methods to predict computationally the susceptibility of substrates to AO metabolism. Dastmalchi *et al.* reported the construction of a three-dimensional model of human AO using the crystal structure of bovine xanthine dehydrogenase (XDH) as a template to study the mode of interaction between the enzyme and its substrates.[39] Torres *et al.* have reported the use of density functional theory methods[40] and geometry optimization of tetrahedral intermediates resulting from the nucleophilic attack of a hydroxyl nucleophile to predict the regioselectivity of oxidation of a substrate by AO[41] with more recent optimisation of the method reported.[42,43]

Docking methods have been reported to accurately predict an unusual AO-mediated hydrolysis reaction of an amide.[44]

These methods only allow an assessment of the likely position and extent of AO catalysed oxidation of a heterocycle substrate, but not whether an oxidation by AO will actually take place.

12.3 Effects on Drug Discovery

Due to species differences in enzyme activity, a number of recent examples have illustrated the effects of AO metabolism on clinical failures, due to higher than expected clearance in humans or a safety-related finding attributed to a human metabolite that was not foreseen preclinically.[45–47] Further examples have been included from other advanced drug discovery programmes that have run across AO issues. The following examples illustrate clearly that predictions of human clearance built from LM data alone and/or pre-clinical species pharmacokinetic studies do not account for higher AO activity in humans and are an inherently risky translational strategy.

12.3.1 Carbazeran

The potent ionotropic agent carbazeran **1** is predominantly metabolized by AO (Figure 12.5) in humans to a phthalazinone **2**.[48] Functional efficacy was initially demonstrated in dogs but not reproduced in humans at similar doses. It was subsequently found that while oral bioavailability in dogs was approximately 68%, the oral bioavailability in humans was too low to be measurable and compound development was stopped. Comparison of the metabolic rates of this compound in baboons, humans and dog cytosolic fractions revealed that it was rapidly inactivated in baboons and humans but not in dogs, indicating minimal AO activity in the latter species.

12.3.2 RO1

Routine pre-clinical evaluation of the p38 inhibitor RO1 **3** in monkeys, rats and dogs indicated a projected pharmacokinetic half-life in humans of approximately 6 h. Unexpectedly, in a Phase 1 clinical study, RO1 was found to have very low exposure levels and a half-life of just 0.7 h, leading to compound termination.[49] A computational genetic analysis, followed by experiments using specific inhibitors confirmed that AOX1 catalyzed the formation of the major 4-hydroxy metabolite **4** in humans (Figure 12.6), which was not produced in preclinical model species.

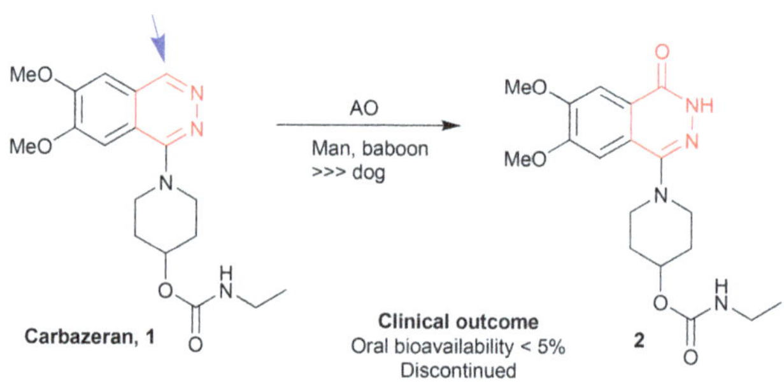

Figure 12.5 AO-mediated metabolism of Carbazeran.

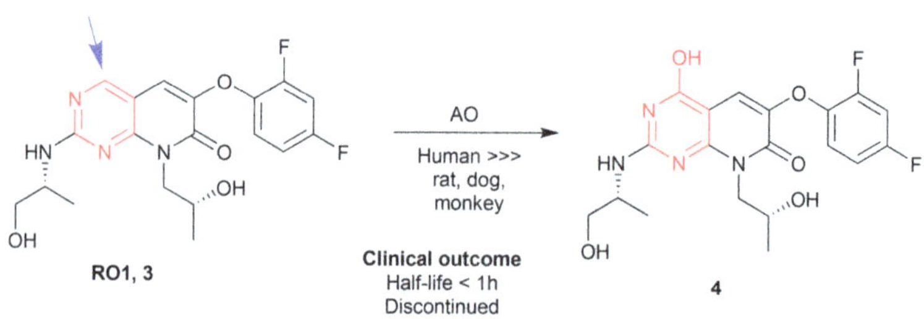

Figure 12.6 AO-mediated metabolism of RO1 *via* a pathway greatly favoured in humans.

12.3.3 FK3453

A report from Astellas highlighted a novel adenosine A1/2 dual inhibitor for the treatment of Parkinson's Disease, for which pre-clinical data indicated a favourable pharmacokinetic profile with oral bioavailability of up to 90% in rats and dogs and good metabolic stability in liver microsomes. The development of FK3453 **5** was stopped prematurely due to extremely low plasma concentrations of unchanged drug in a Phase 1 study.[50] Results of radio-chromatography analysis of [^{3}H]-FK3453 incubated with either human liver microsomes or S9 fraction revealed the production of a major oxidised metabolite **6** (Figure 12.7) and incubations with and without AO and XO inhibitors confirmed that AO was responsible for its production.

12.3.4 BIBX1382

The epidermal growth factor receptor (EGFR) inhibitor BIBX1382 **7** failed in clinical studies due to poor oral exposure, despite having an attractive ADME profile in preclinical model species. In man, BIBX1382 undergoes extensive oxidative metabolism into the metabolite BIBU1476 **8** by AO (Figure 12.8).[51] It was subsequently found that the pharmacokinetic profile of cynomolgus monkeys compared favourably with the human clinical data.

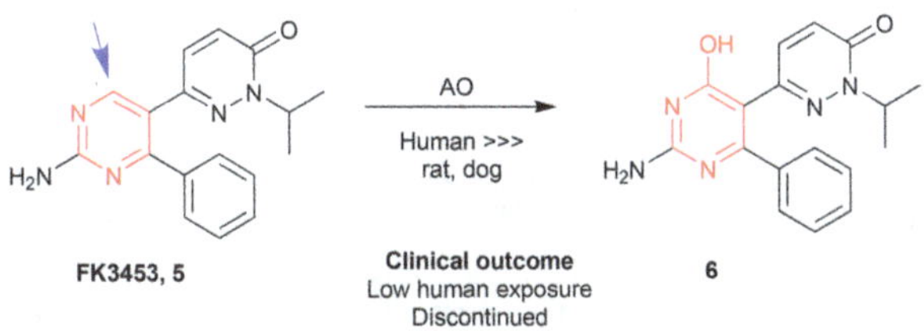

Figure 12.7 AO-mediated metabolism of FK3453.

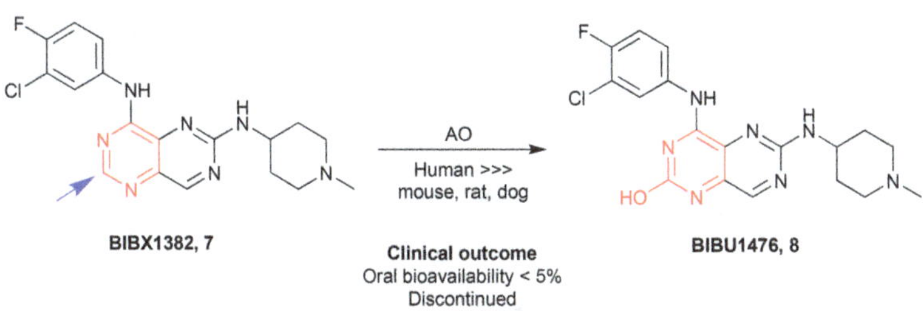

Figure 12.8 AO-mediated metabolism of BIBX1382.

12.3.5 SGX523

The c-met inhibitor SGX523 **9** entered clinical development for the treatment of solid tumors on the basis of a good pharmacokinetic profiles in rats and dogs. In humans, the compound produced obstructive renal failure in patients and was withdrawn from further evaluation. In an elegant series of investigations, it was demonstrated that SGX523 is extensively metabolized in a species-specific manner to a quinolone species **10** by human and monkey AO, but to a much lesser extent in rat and not at all in dog S9 incubations (Figure 12.9).[52]

The parent SGX523 had modest aqueous solubility at neutral pH of 4 μg mL^{-1}, but notably the quinolone metabolite had markedly lower solubility at the same pH of just 0.1 μg mL^{-1}. The authors of the report proposed that the metabolite forms insoluble crystals in renal tubules, causing the kidney toxicity.

12.3.6 Zoniporide

Zoniporide **11** is a potent sodium–hydrogen exchanger isoform-1 (NHE-1) inhibitor which was being developed for myocardial ischemic injury. When dosed in humans, an AO-mediated quinolone metabolite **12** was the major excretory species, which was also found in rats but not in dogs (Figure 12.10).[53]

Through cytosol and S9 fraction experiments, the role of AO was confirmed and zoniporide was predicted to be a high-clearance compound in human.

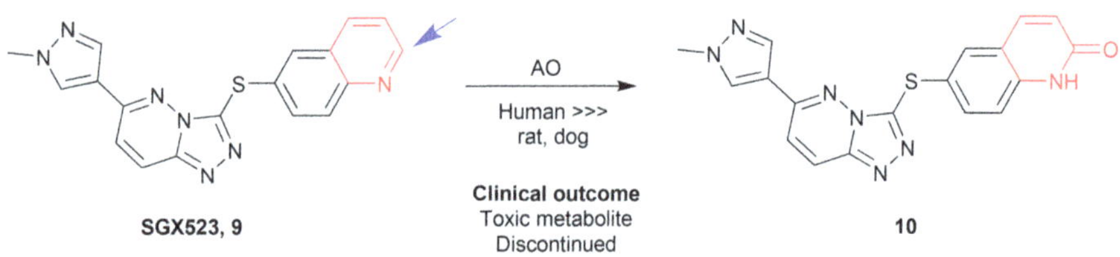

Figure 12.9 AO-mediated metabolism of SGX523.

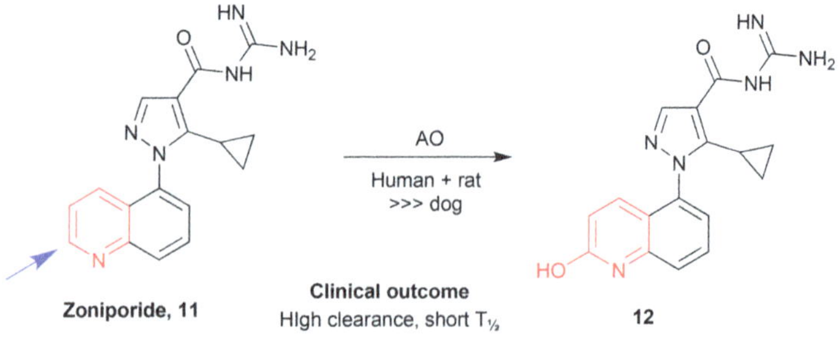

Figure 12.10 AO-mediated metabolism of Zoniporide.

12.3.7 Auglurant

Auglurant **13** is a potent negative allosteric modulator of metabotropic glutamate receptor 5 (mGluR5), which has shown quite marked species differences in its metabolism. A major metabolite **14** is the result of AO-mediated oxidation on the 6-position of the pyrimidine ring, which is seen across multiple species and in humans (Figure 12.11). A second oxidation at the 2-position of the pyrimidine ring **15** was also seen, but, using specific enzyme inhibitors, the authors showed this was mediated by AO in monkeys and XO in rats, whilst in humans very little of this metabolite was observed.[54]

The same authors had reported previously on a similar finding within the series and have proposed that rat AO activity at the 2-position of pyrimidines may be generally very low.

12.3.8 VX-509

VX-509 (decernotinib) **16** is an oral Janus kinase 3 (JAK3) inhibitor which has been studied in patients with rheumatoid arthritis. The reactive hydroxyl metabolite **17** was formed in human liver cytosol by AO (Figure 12.12) and then binds irreversibly to inactivate drug metabolizing enzymes, especially CYP3A4. It is highly probable that this hydroxyl metabolite is the major perpetrator of the time-dependent inhibition (TDI) based drug–drug interactions (DDIs) seen in clinical settings with VX-509.[55]

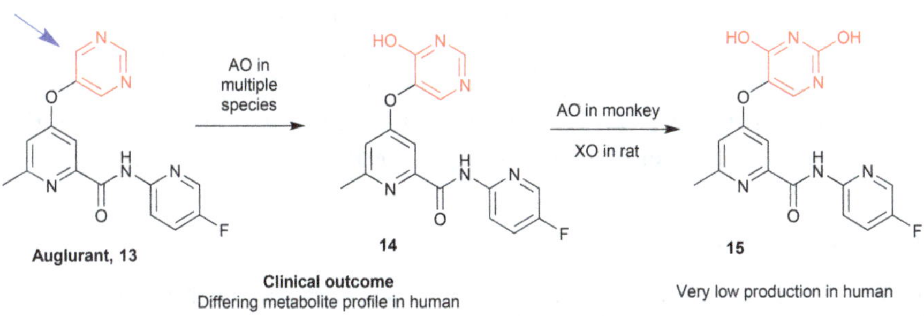

Figure 12.11 AO-mediated metabolism of Auglurant.

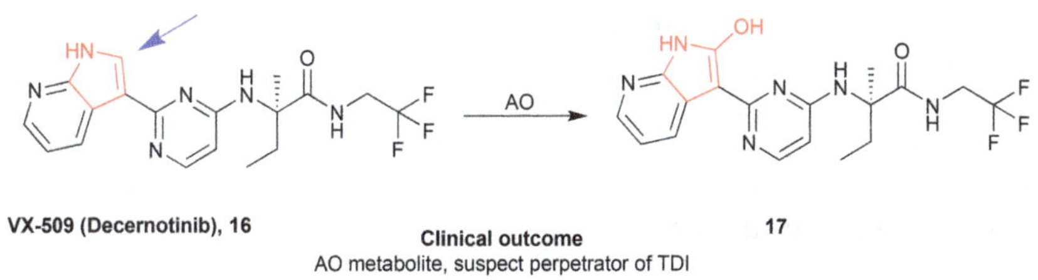

Figure 12.12 AO-mediated metabolism of VX-509.

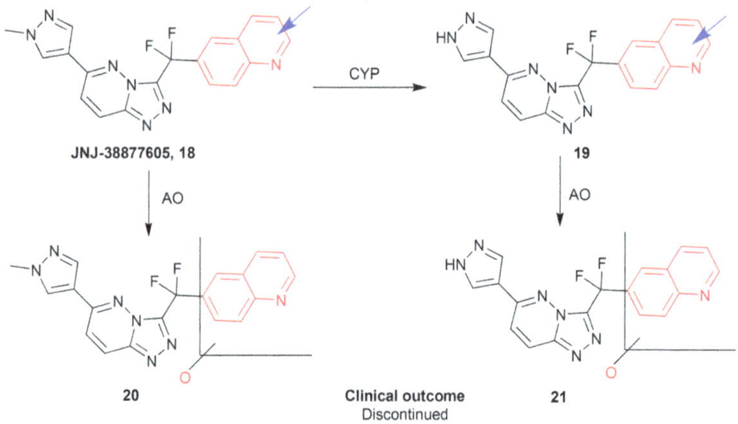

Figure 12.13 AO-mediated metabolism of JNJ-38877605.

12.3.9 JNJ-38877605

A potent and highly selective c-met tyrosine kinase inhibitor JNJ-38877605 **18** showed renal toxicity in all patients during first-in-human Phase 1 clinical trials.[56] Renal toxicity had not been observed in pre-clinical studies in rats and dogs. Subsequently, the same findings were also observed in rabbits, indicating that **18** induced species-specific renal toxicity. Histopathological studies in rabbits confirmed the presence of crystals of insoluble metabolites **20** and **21** formed from oxidative metabolism of **18** (and its demethylated derivative **19**) by AO (Figure 12.13). These observations resulted in the cessation of any further clinical development of JNJ-38877605.

12.3.10 BILR355

BILR355 **22** is an HIV-1 non-nucleoside reverse transcriptase inhibitor but was found to have a short half-life and low exposure when administered orally in humans.[57] The results of *in vitro* metabolic studies revealed that CYP3A4 was mainly responsible for limiting the systemic exposure of **22**, and co-dosing with ritonavir was investigated. However, this led to an unexpected metabolite BILR516 **24** being formed which was not detected earlier in humans when **22** was given alone. It was established that **24** was formed *via* reduction of the N-oxide **22** to quinolone BILR402 **23** followed by AO oxidation (Figure 12.14). The metabolism of the parent to the reduced species **23** was further investigated and it was suggested that gut bacteria were responsible for this reductive biotransformation.

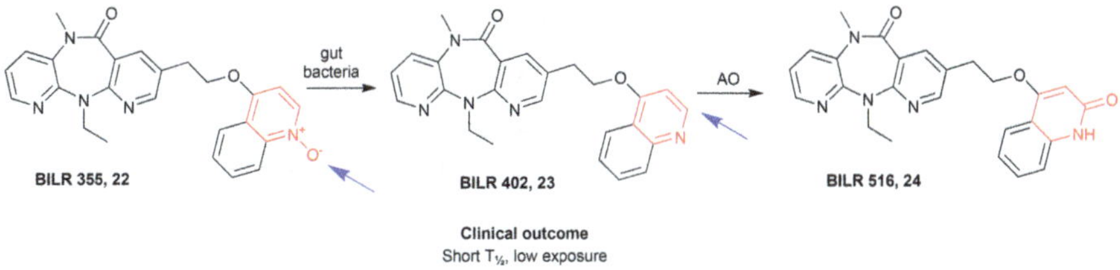

Figure 12.14 AO-mediated metabolism of BILR355.

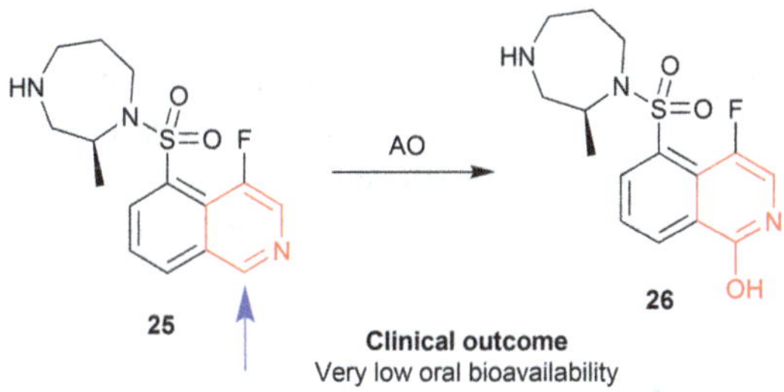

Figure 12.15 AO-mediated metabolism of Ripasudil.

12.3.11 Ripasudil

Ripasudil (K-115) **25** is a potent inhibitor of Rho-associated protein kinase (ROCK) that has been shown to reduce intraocular pressure in glaucoma and ocular hypertension patients. The development of ripasudil as an oral agent was stopped due to very low oral bioavailability in humans.[58] Incubating radiolabelled ripasudil in S9 liver fractions from various species indicated a major metabolite **26** in human S9 which was confirmed in cryopreserved human hepatocytes (Figure 12.15). Menadione and raloxifene inhibited the formation of this major metabolite, but allopurinol did not, indicating that AO, but not XO, was responsible for its formation.

12.3.12 SB-277011

The selective dopamine D3 receptor antagonist SB-277011 **27**, was initially found to be quite stable to liver microsomes from rats, dogs, cynomolgus monkeys and humans, while it was metabolised much more rapidly in cynomolgus monkey and human total liver homogenates.[59] The human *in vitro* data was especially striking, with clearance values some 35-fold higher in homogenates than in microsomes. SB-277011 **27** was shown to have a high clearance and low (2%) bioavailability in cynomolgus monkeys compared with moderate to

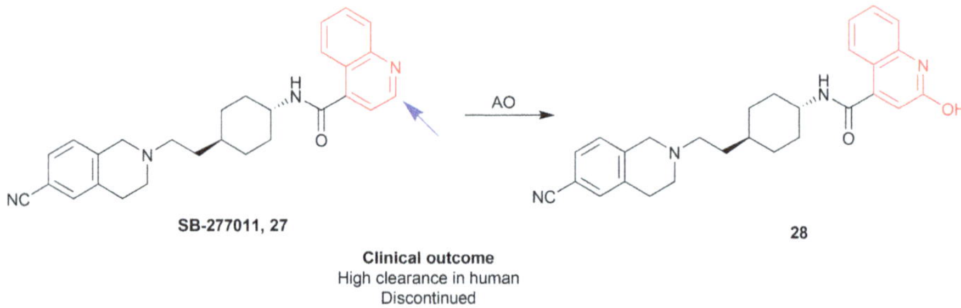

Figure 12.16 AO-mediated metabolism of SB-277011.

good bioavailability (35–43%) in rats and dogs with the major metabolic route confirmed as AO oxidation to the quinolone **28** (Figure 12.16).

It was predicted on the basis of these data that the bioavailability of the compound would be low in human, and its progression as a drug candidate was terminated.

12.3.13 Lenvatinib

Lenvatinib **29** is a multi-tyrosine kinase inhibitor that inhibits vascular endothelial growth factor receptors and is being developed as an anticancer drug.[60] The oxidative metabolic pathways of **29** in animals and human liver S9 fractions were investigated and established as being dominated by AO metabolism, but only in monkeys and humans. The major metabolites were identified by high-resolution mass spectrometry as the quinolone **31** and hydroxy-quinolone **32** form of *O*-desmethylated lenvatinib **30**, formed by an intermediate CYP450 demethylation (Figure 12.17).

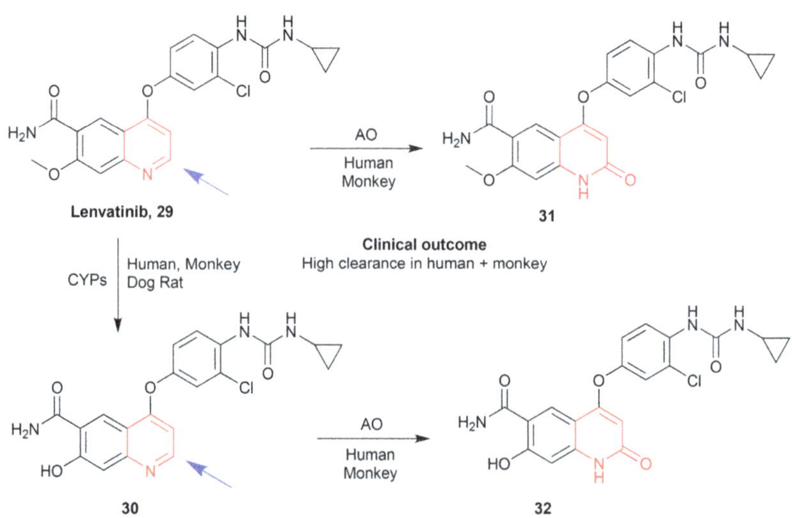

Figure 12.17 AO-mediated metabolism of Lenvatinib.

The involvement of AO in oxidative metabolism of **29** was examined by studying the effects of AO or XO inhibitors on the formation of metabolites **31** and **32**. In addition, ^{18}O-water and recombinant human AO (rhAO) were used to confirm the involvement of AO.

12.3.14 BIIB021

BIIB021 **33** is a potent inhibitor of heat shock protein 90 (HSP90) which exhibited promising antitumor activity in preclinical models and was being developed for the treatment of breast cancer. BIIB021 **33** was profiled extensively in rat, dog and human liver cytosolic fractions which revealed a major metabolite **34**, formed by AO- catalyzed purine ring oxygenation, in human preparations but not in those from rats and dogs (Figure 12.18).[61]

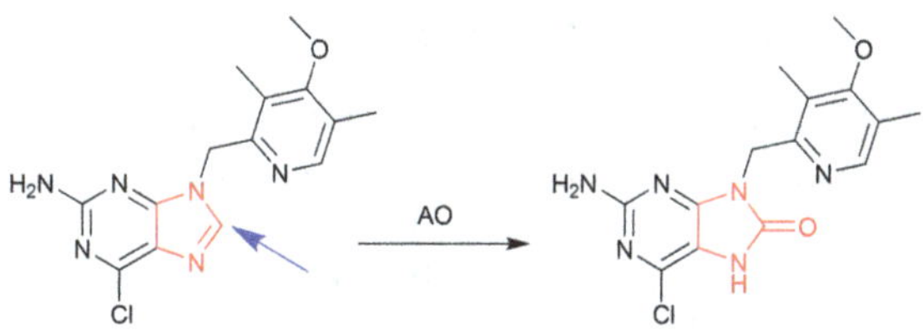

Figure 12.18 AO-mediated metabolism of BIIB021.

Key References

- E. Garattini and M. Terao, in *Comprehensive Toxicology*, ed. C. McQueen, Elsevier, 3rd edn, 2018, ch. 10.10, vol. 10, pp. 208–232.
  - *A contemporary review of the AO/XO family.*
- D. C. Pryde, D. Dalvie, Q. Hu, P. Jones, R. S. Obach and T. Tran, *J. Med. Chem.*, 2010, **53**, 8441.
  - *A general review of AO metabolism, and the impact AO has had on drug discovery.*
- J. F. Alfaro and J. P. Jones, *J. Org. Chem.*, 2008, **73**, 9469.
  - *A landmark study of modelling for AO susceptibility.*
- J. M. Hutzler, R. S. Obach, D. Dalvie and M. A. Zientek, *Exp. Opin. Drug Metab. Toxicol.*, 2013, **9**, 153.
  - *An excellent review of screening strategies for AO and XO metabolism.*

(*continued*)

- C. Coelho, A. Foti, T. Hartmann, T. Santos-Silva, S. Leimkuhler and M. J. Ramao, *Nat. Chem. Biol.*, 2015, **11**, 779.
  - *The first crystal structure of human AO.*
- A. Mallinger, K. Schiemann, C. Rink, J. Sejberg, M. A. Honey, P. Czodrowski, M. Stubbs, O. Poeschke, M. Busch, R. Schneider, D. Schwarz, D. Musil, R. Burke, K. Urbahns, P. Workman, D. Wienke, P. A. Clarke, F. I. Raynaud, S. A. Eccles, C. Esdar, F. Rohdich and J. Blagg, *ACS Med. Chem. Lett.*, 2016, 7, 573.
  - *An excellent recent example of structure-metabolism relationships for an AO substrate and optimizing away from AO metabolism.*

12.4 Key Mitigation Strategies

12.4.1 Remote Functionalization

A change in substitution, chirality or relative position of a functional group at a site distant from the susceptible centre can cause a dramatic change in reactivity and in some cases a complete abrogation of AO metabolism, related to the ability of a compound to fit into the AO active site. This is currently very difficult to predict.

12.4.2 Alternative Heterocycles

Heterocycles that are prone to AO oxidation feature a C–H bond adjacent to an aromatic N atom. A simple way to mitigate AO metabolism is to swap the heterocycle for alternative systems with decreased reactivity towards an initial nucleophilic attack, or with no available C–H bond, provided that the primary target SAR will tolerate such a change.

12.4.3 Blocking Group Adjacent to Aromatic N Atom

The most common and straightforward approach to mitigate AO metabolism of a heterocycle is to block or substitute the C–H group adjacent to the aromatic N, without affecting the potency or other major properties of the molecule. This is the most successfully applied strategy to date.

12.5 Examples of Successful Mitigation Strategies

Having established that structure is an important indicator of potential for a compound to undergo AO- or XO-mediated oxidation, identified a range of 'at-risk' substructures and proposed with the above illustrative examples that AO related metabolism issues in particular are likely to increase in drug discovery programmes, we now turn to how best to address AO metabolism.

Given the clear substrate preferences outlined in Figure 12.3, it does appear that avoiding AO metabolism should be quite straightforward, provided that the primary pharmacological target is tolerant, and the literature offers several real examples where drug discoverers have successfully moved away from AO oxidation.

Please refer to the examples illustrated in sections 12.5.1–12.5.3.

12.5.1 Remote Functionalization

Examples

1

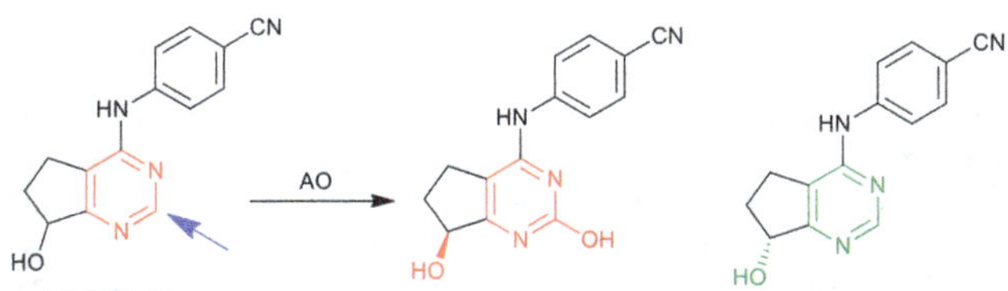

RS-8359, 35 **36** **37**

For a series of monoamine oxidase (MAO) inhibitors, remote alcohol stereochemistry was found to control AO-mediated oxidation of a pyrimidine.[62] RS-8359 **35** is a racemate, but only the *S*-enantiomer **36** at a remote centre undergoes AO-mediated oxidation, while the *R*-enantiomer **37** is stable.

2

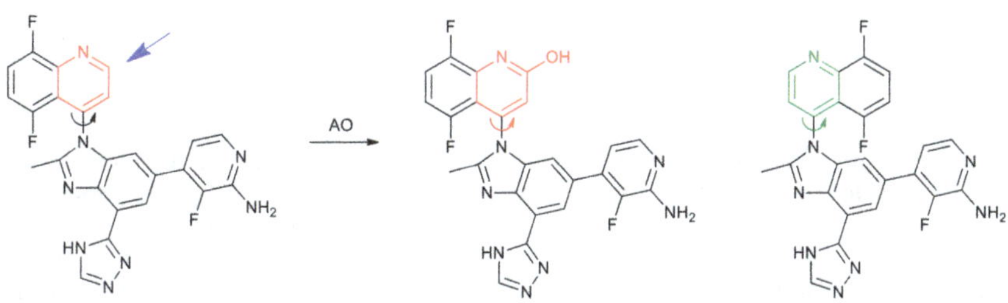

38 **39** **40**

In a similar recent example, a quinoline phosphatidylinositol-3-kinase delta (PI3Kδ) inhibitor **38** was shown to display atropisomerism through steric clash of a benzimidazole with the quinoline ring system, with the atropisomers **39** and **40** readily separable by chromatography.[63] One of the atropisomers **40** possessed almost all of the observed potency seen in the racemate, while the other atropisomer **39** was substantially weaker for all PI3K isoforms. The weaker **39** was also much more rapidly cleared in human hepatocytes than its isomer This was confirmed to be *via* AO-mediated metabolism to a quinolone.

(*continued*)

Examples

3

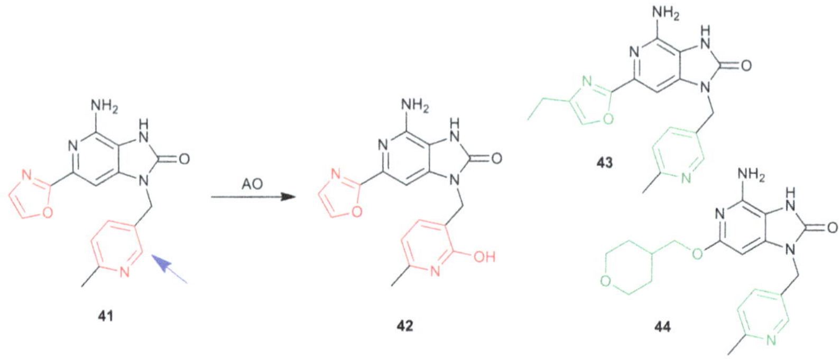

41 42 43 44

Remote functionalization of a deazapurine template was found to affect AO oxidation of a remote pyridine ring. Lead compounds from an oxazole-substituted Toll-like receptor 7 (TLR7)-active series *e.g.* **41** were rapidly metabolized by rat AO, *but not human by AO*, at the pyridine ring to a hydroxy-pyridine **42**. Remote heterocycle functionalization or swapping of the oxazole (**43** and **44**) was found to ablate AO activity for both species.[64]

4

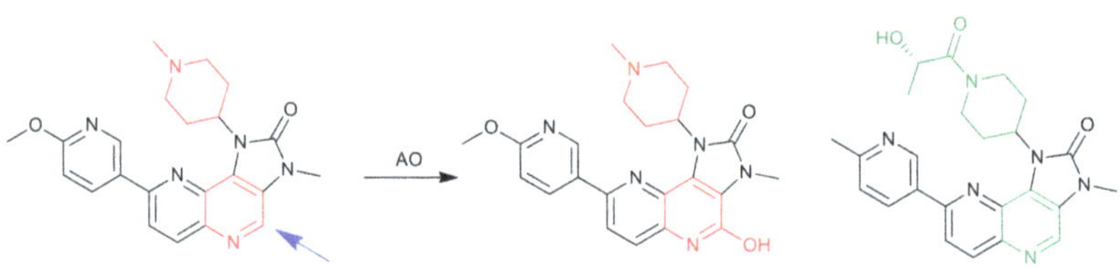

45 46 47

Large amide groups can retard AO oxidation at a remote site on a tricyclic core. A lead series of PI3K/mammalian target of rapamycin (mTOR) inhibitors **45** showed good rat pharmacokinetics but rapid turnover in human S9 to a pyridine **46**. Systematic increase in molecular size at the remote piperidine centre, *e.g.* the extended amide **47**, could ablate AO recognition at the remote pyridine ring.[65]

12.5.2 Alternative Heterocycles

Examples

1

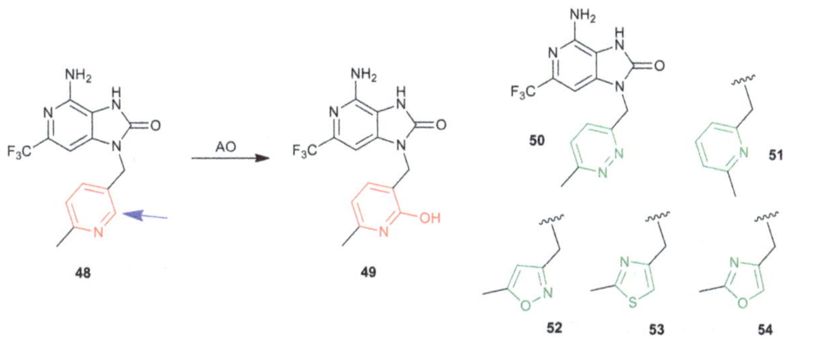

Related to the TLR7-active series above, the same Pfizer group observed AO-mediated oxidation of a pyridine ring **48** to a hydroxy-pyridine metabolite **49**. It was found that simple heterocycle swaps of the pyridine ring to systems which either did not have an available position adjacent to the aromatic N atom, such as the isomeric pyridine **51**, or removed it completely as in the pyridazine **50**, isoxazole **52**, thiazole **53** and oxazole **54** all successfully prevented AO oxidation.[66]

2

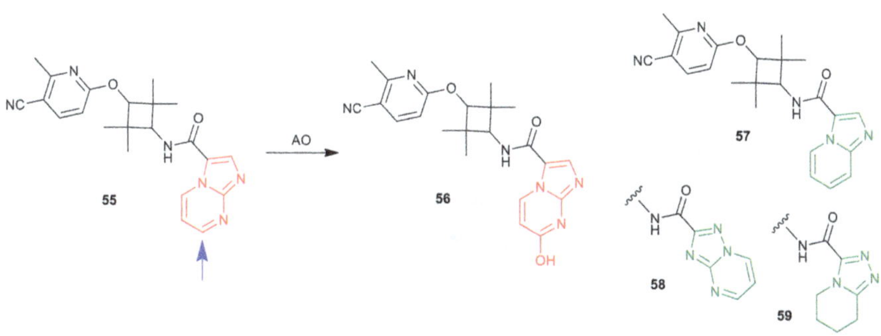

A potent series of androgen receptor antagonists was systematically evaluated for structural changes that were effective in preventing AO-mediated oxidation of the imidazo-pyrimidine **55** to a hydroxyl metabolite **56**.[67] Initial attempts to make remote changes to the aryl ether portion of the molecule did not reduce the AO liability. Similarly, early attempts at methylation of the imidazole ring of the susceptible imidazo-pyrimidine ring or moving the connection point of the amide functional group to the ring system were also unsuccessful. It was only when different heterocyclic systems were explored, for example the reduced system **59**, the related triazolo-pyrimidine **58** and the imidazo-pyridine **57** that recognition by AO was lost. Unfortunately, none of these ring systems retained high potency at the androgen receptor.

(*continued*)

Examples

3

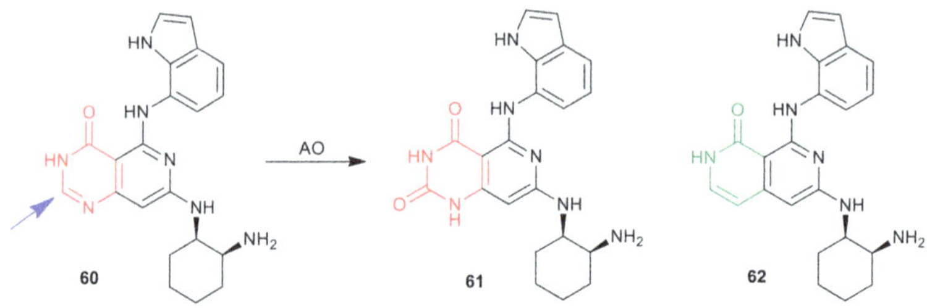

60 61 62

A series of potent spleen tyrosine kinase (Syk) inhibitors have been described by a group at Novartis based on analogues of the clinically-precedented agent BIIB-057.[68] Several novel scaffold investigations led to the discovery of a potent series of pyrido pyrimidinones **60**. These were found to be rapidly eliminated in rat pharmacokinetic studies, with subsequent human and rat liver S9 experiments indicating AO involvement in the metabolic pathway to a uracil derivative **61** confirmed with *Escherichia coli* overexpressing human AO. Attempts to avoid this pathway included blocking with a methyl group at the vulnerable centre, which eliminated AO metabolism, but also ablated much of the Syk activity. Removing a heteroatom from the core **62** and increasing its electron density in the corresponding naphthyridinones did result in complete loss of AO metabolism. However, this series of compounds suffered from very high *in vivo* clearance through an unknown pathway which did not appear to be related to Phase II or CYP-mediated metabolism.

4

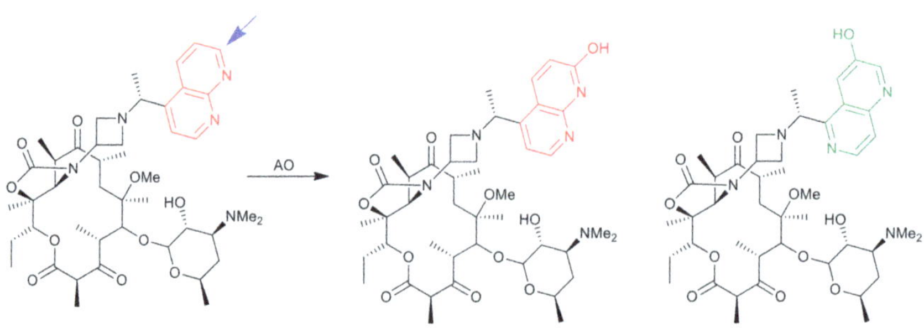

63 64 65

Magee *et al.* have reported on a ketolide series of antibiotics based on the clarithromycin skeleton with a heterocycle-substituted azetidine functional group.[69] Early heterocycle analogues featured a 1,8-naphthyridine system **63** which was found to undergo rapid AO-oxidation upon single-dose exposure in humans to a hydroxylated derivative **64** with concomitant low plasma exposures of parent. A combination of a 3-hydroxy group and an isomeric 1,5-naphthyidine ring system **65** provided the best combination of ADME properties with no AO turnover observed.

12.5.3 Blocking Group Adjacent to Aromatic N Atom

Examples

1

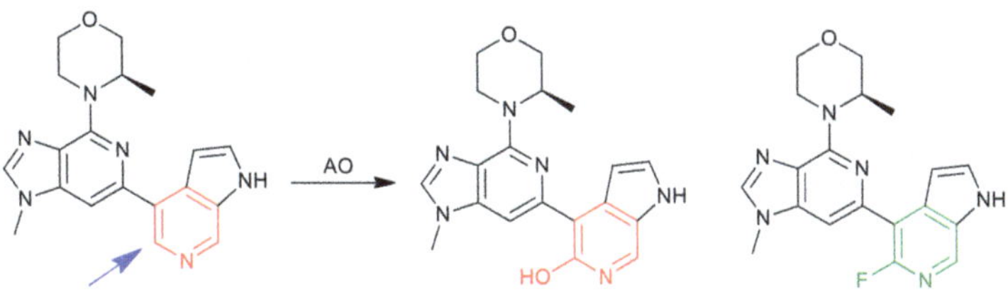

66 67 68

Novartis scientists synthesised a series of potent angiotensin receptor (ATR) antagonists based on an aza-benzimidazole core structure. The lead compound **66** displayed high clearance in rats, and upon further profiling in cytosolic incubations using ^{18}O-labelled water it was discovered that AO-mediated metabolism was a significant contributor to the compound's clearance *via* **67**.[70] A simple fluorinated derivative **68** blocked AO oxidation of the pyridine ring and displayed much improved oral bioavailability in rats.

2

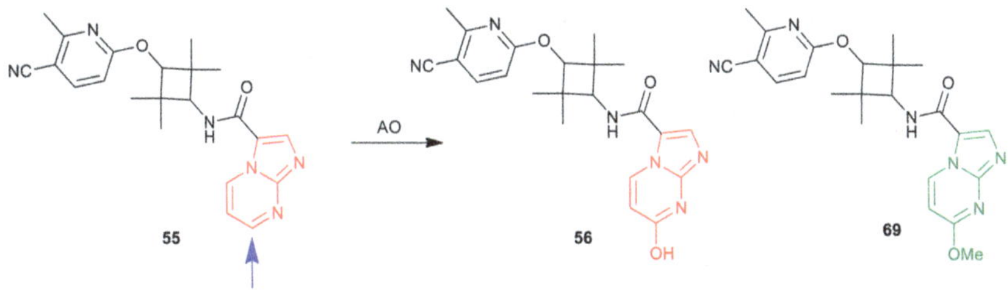

55 56 69

In the Pfizer example from above, ultimately the most successful strategy to avoid AO metabolism of the imidazo-pyrimidine **55** was to block the vulnerable position of the pyrimidine ring.[67] The methoxy derivative **69** was completely devoid of any AO activity, as was a morpholino analogue, which was rationalised using a computational docking study.

(*continued*)

Examples

3

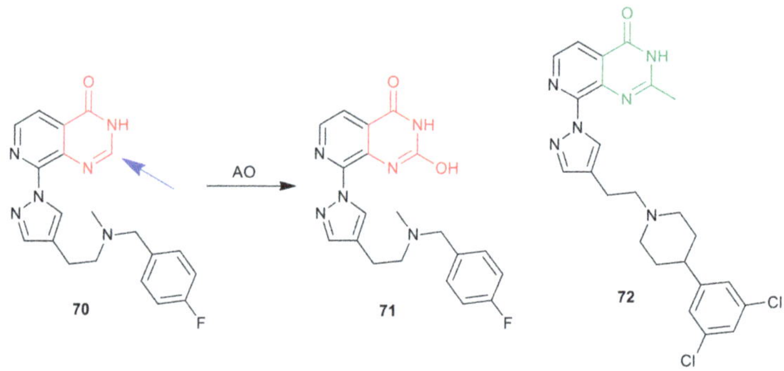

Pyrido[3,4-*d*]pyrimidin-4(3*H*)-one-based small molecules are dual inhibitors of histone lysine demethylase 4 and 5 (KDM4 and KDM5) subfamilies of Jumonji C (JmjC) histone lysine demethylase that are involved in oncogenesis and drug resistance. Hayes *et al.* have reported[71] that a C8-substituted example **70** exhibited moderate clearance in mouse and human liver microsomes but very high *in vivo* clearance. The involvement of AO in the metabolism of **70** was confirmed with an AO inhibitor, while the location of the oxidised site **71** was confirmed by mass spectrometric profiling and ^{1}H nuclear magnetic resonance (NMR) spectroscopy. A C2-substituted derivative **72** which blocked this location was stable in cytosolic incubations. The authors also computationally modelled **70** which docked into the hAOX1 crystal structure to the Mo cofactor *via* the NH and C2 positions. A methyl group placed at the C2-position disrupted this interaction.

4

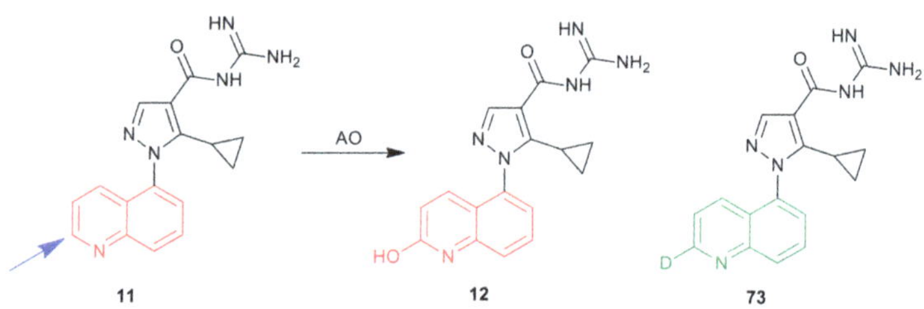

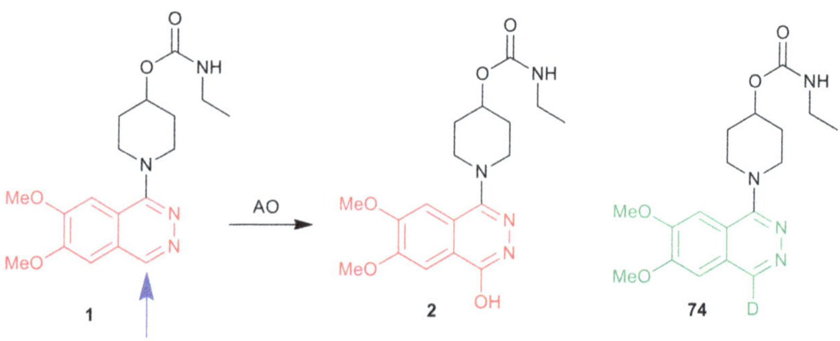

Examples

One blocking group which should be carefully considered is deuterium. Vaz and colleagues. have shown that a full kinetic deuterium isotope effect (KDIE) was expressed with a deuterated version of carbazeran **74** and zoniporide **73**, but that this only translated into modest KDIE's for both area under the curve (AUC) and maximum serum concentration (C_{max}) in rats, indicating alternative clearance mechanisms at work across the species and test systems used.[72] Deuteration does appear to be a viable option, provided this strategy does not simply redirect metabolism to a non-AO dominated pathway.

5

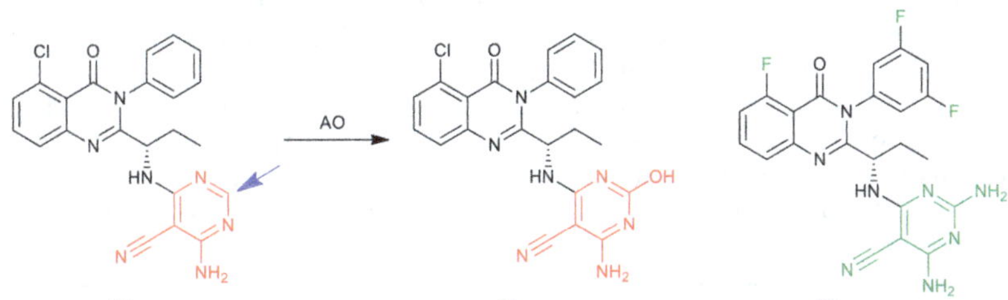

75 **76** **77**

A series of substituted quinazolinones **75** were developed as selective inhibitors of PI3Kδ but were found to be AO-susceptible at a diamino-pyrimidine substituent to form a 2-oxo-pyrimidine **76.**[73] The oxidation could be blocked with an amine applied at this 2-position, a change which caused a loss of isoform selectivity. Further modifications were then made around the quinazoline core to reinstate the PI3Kδ selectivity profile **77**.

6

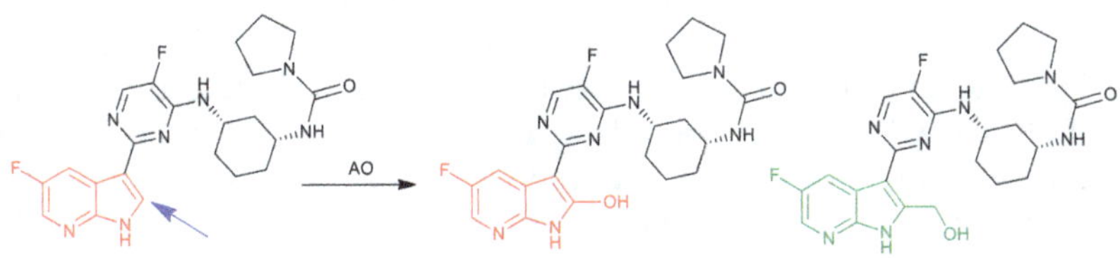

78 **79** **80**

JNJ-63623872 (VX-787) **78** is a potent influenza viral polymerase inhibitor. The 2-position of the indole core was found to be subject to oxidation by AO **79.**[74] Incorporating a heteroatom to generate azaindazole core structures surprisingly did not significantly alter stability in cytosol, possibly due to metabolism being switched to another location on the molecule. However, substitutions at the 2-position greatly increased cytosolic stability. After screening different substituents, hydroxymethyl **80** was found to be optimal, which retained all other properties and introduced stability towards AO.

(*continued*)

Examples

7

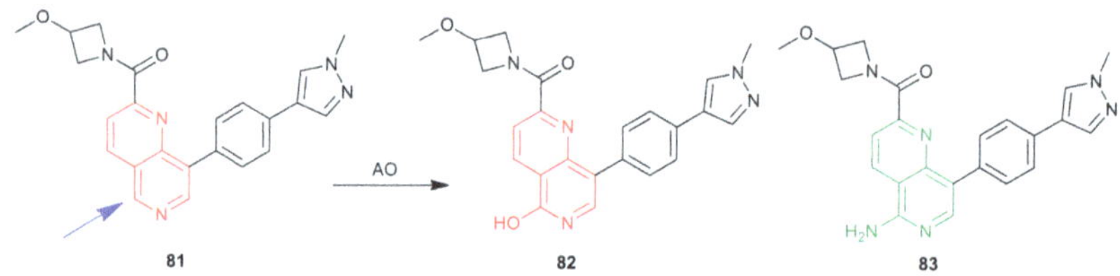

Several naphthyridine derivatives were shown to be potent and selective dual inhibitors of cyclin-dependent kinases 8 and 19 (CDK8/CDK19). Compounds within the series, *e.g.* **81**, showed good bioavailability but high clearance.[75] Analyzing the major metabolite **82** indicated AO to be the responsible enzyme for oxidizing the 5-position of the naphthyridine core, resulting in rapid clearance. Stability of the lead towards AO was substantially improved by introducing an amino group at this 5-position **83**; the resulting amino derivative retained similar potency to the original lead but with much improved oral bioavailability.

12.6 Conclusion

AO and XO have been known as cytosolic metabolizing enzymes for decades, but it is relatively recently that they have been recognised as playing an important role in the metabolism of drugs. Moreover, the difference in activity of AO in different species has contributed to clinical and preclinical drug failures, with several examples having come to light recently in which human metabolism by AO was not predicted from preclinical model species. It is therefore critical that a thorough understanding of the role of AO in compound metabolism is obtained early in drug discovery programmes. In this perspective, we have proposed a basis for an increasing role of cytosolic enzymes, particularly AO, in the metabolism of new chemical entities emerging from modern drug discovery programmes, primarily through nitrogen-containing heterocycle oxidation, and identified particular substructures which are 'at-risk' of oxidation mediated by AO/XO.

Our analysis has highlighted a straightforward process which can be followed to elucidate the role of AO/XO in the metabolism of drugs and provided illustrative examples in which structures have been modified to eliminate metabolism by AO.

We anticipate that further advances in modelling and screening methods should make the prospective design of AO-stable compounds more commonplace in the future.

Abbreviations

AO	Aldehyde oxidase
ATR	Angiotensin receptor
AUC	Area under the curve
CDK	Cyclin-dependent kinase
$C_{\max}$	Maximum serum concentration
CYP450	Cytochrome P450
EGFR	Epidermal growth factor receptor
FAD	Flavin adenine dinucleotide
FMO	Flavin-containing monooxygenase
GPCR	G-protein coupled receptor
HLM	Human liver microsomes
HSP90	Heat shock protein 90
JAK3	Janus kinase 3
JmjC	Jumonji C
KDIE	Kinetic deuterium isotope effect
KDM	Histone lysine demethylase
MAO	Monoamine oxidase
mGluR5	Metabotropic glutamate receptor 5
MoCo	Molybdenum pyranopterin cofactor
mTOR	Mammalian target of rapamycin
NADPH	Nicotinamide adenine dinucleotide phosphate
NHE-1	Sodium–hydrogen exchanger isoform-1
NMR	Magnetic resonance
PSA	Polar surface area
ROCK	Rho-associated protein kinase
S9	The supernatant fraction obtained from liver homogenate by centrifuging at 9000 *g* in a suitable medium
Syk	Spleen tyrosine kinase
TLR7	Toll-like receptor 7
UGT	Uridine 5′-diphosphoglucuronosyltransferase
XDH	Xanthine dehydrogenase
XO	Xanthine oxidase

References

1. E. Garattini and M. Terao, in *Comprehensive Toxicology*, ed. C. McQueen, Elsevier, 3rd edn, 2018, vol. 10, ch. 10.10, pp. 208–232.
2. R. R. Mendel, *BioFactors*, 2009, **35**, 429.
3. C. Beedham, *Prog. Med. Chem.*, 1987, **24**, 85.
4. S. Kitamura, K. Sugihara and S. Ohta, *Drug Metab. Pharmacokinet.*, 2006, **21**, 83.
5. D. C. Pryde, D. Dalvie, Q. Hu, P. Jones, R. S. Obach and T. Tran, *J. Med. Chem.*, 2010, **53**, 8441.

6. E. Garattini, M. Fratelli and M. Terao, *Cell. Mol. Life Sci.*, 2008, **65**, 1019.
7. C. Beedham, *Molybdenum Hydroxylases in Enzyme Systems that Metabolise Drugs and Other Xenobiotics,* ed. C. Ioannides, John Wiley and Sons, London, 2002, ch. 5, pp. 147–187.
8. M. Strolin-Benedetti, R. Whomsley and E. Baltes, *Expert Opin. Drug Metab. Toxicol.*, 2006, **2**, 895.
9. M. Nishimura and S. Naito, *Drug Metab. Pharmacokinet.*, 2006, **21**, 357.
10. Y. Moriwaki, T. Yamamoto and K. Higashino, *Histol. Histopathol.*, 1999, **14**, 1321.
11. F. Gok, K. Ichida and R. Topaloglu, *Nephrol., Dial., Transplant.*, 2003, **18**, 2278.
12. D. J. Wilkinson, R. L. Southall, M. Li, L. M. Wright, L. J. Corfield, T. A. Heeley, B. Bratby, R. Mannu, S. L. Johnson, V. Shaw, H. L. Friett, L. A. Blakeburn, J. S. Kendrick and M. B. Otteneder, *AAPS J.*, 2017, **19**, 1163.
13. R. D. Crouch, A. L. Blobaum, A. S. Felts, P. J. Conn and C. W. Lindsley, *Drug Metab. Dispos.*, 2017, **45**, 1245.
14. S. Kitamura, K. Sugihara, K. Nakatani, S. Ohta, T. Oh-Hara, S. Ninomiya, C. E. Green and C. A. Tyson, *IUBMB Life*, 1999, **48**, 607.
15. H. S. Al-Salmy, *Comp. Biochem. Physiol., C: Toxicol. Pharmacol.*, 2002, **132**, 341.
16. K. Sugihara, S. Kitamura and K. Tatsumi, *Biochem. Mol. Biol. Int.*, 1995, **37**, 861.
17. K. Itoh, H. Maruyama, M. Adachi, K. Hoshino, N. Watanabe and Y. Tanaka, *Xenobiotica*, 2007, **37**, 709.
18. H. S. Al-Salmy, *IUBMB Life*, 2001, **51**, 249.
19. J. Sahi, K. K. Khan and C. B. Black, *Drug Metab. Lett.*, 2008, **2**, 176.
20. J. M. Hutzler, Y. S. Yang, C. Brown, S. Heyward and T. Moeller, *Drug Metab. Dispos.*, 2014, **42**, 1090.
21. E. Garattini and M. Terao, *Expert Opin. Drug Metab. Toxicol.*, 2012, **8**, 487.
22. D. E. Decker and D. J. Levinson, *Arthritis Rheum.*, 1982, **25**, 326.
23. E. Garattini, M. Fratelli and M. Terao, *Hum. Genomics*, 2009, **4**, 119.
24. K. V. Rajagopalan and J. L. Johnson, *J. Biol. Chem.*, 1992, **267**, 10199.
25. C. Coelho, M. Mahro, J. Trincao, A. T. P. Carvalho, M. J. Ramos, M. Terao, E. Garattini, S. Leimkuhler and M. J. Ramao, *J. Biol. Chem.*, 2012, **287**, 40690.
26. C. Coelho, A. Foti, T. Hartmann, T. Santos-Silva, S. Leimkuhler and M. J. Ramao, *Nat. Chem. Biol.*, 2015, **11**, 779.
27. J. F. Alfaro and J. P. Jones, *J. Org. Chem.*, 2008, **73**, 9469.
28. M. Montefirori, F. S. Jorgensen and L. Olsen, *ACS Omega*, 2017, **2**, 4237.
29. T. A. Krenitsky, S. M. Neil, G. B. Elion and G. H. Hitchings, *Arch. Biochem. Biophys.*, 1972, **150**, 585.
30. Z. Knight and K. Shokat, *Chem. Biol.*, 2005, **12**, 621.
31. F. Lovering, J. Bikker and C. Humblet, *J. Med. Chem.*, 2009, **52**, 6752.
32. D. G. Brown and J. Bostrom, *J. Med. Chem.*, 2016, **59**, 4443.
33. J. M. Hutzler, R. S. Obach, D. Dalvie and M. A. Zientek, *Expert Opin. Drug Metab. Toxicol.*, 2013, **9**, 153.

34. R. S. Obach, P. Huynh, M. C. Allen and C. Beedham, *J. Clin. Pharmacol.*, 2004, **44**, 7.
35. R. Ojha, J. Singh, A. Ojha, H. Singh, S. Sharma and K. Nepali, *Expert Opin. Ther. Pat.*, 2017, **3**, 311.
36. M. Zientek, Y. Jiang, K. Youdim and R. S. Obach, *Drug Metab. Dispos.*, 2010, **38**, 1322.
37. F. O'Hara, A. C. Burns, M. R. Collins, D. Dalvie, M. A. Ornelas, A. D. N. Vaz, Y. Fujiwara and P. S. Baran, *J. Med. Chem.*, 2014, **57**, 1616.
38. E. M. Paragas, S. C. Humphreys, J. Min, C. A. Joswig-Jones, S. Leimkuhler and J. P. Jones, *ACS Omega*, 2017, **2**, 4820.
39. S. Dastmalchi and M. Hamzeh-Mivehrod, *Daru, J. Fac. Pharm., Tehran Univ. Med. Sci.*, 2005, **13**, 82.
40. R. A. Torres, K. R. Korzekwa, D. R. McMasters, C. M. Fandozzi and J. P. Jones, *J. Med. Chem.*, 2007, **50**, 4642.
41. J. P. Jones and K. R. Korzekwa, *Mol. Pharmaceutics*, 2013, **10**, 1262.
42. Y. Xu, L. Li, Y. Wang, J. Xing, L. Zhou, D. Zhong, L. Luo, H. Jiang, K. Chen, M. Zheng, P. Deng and X. Chen, *J. Med. Chem.*, 2017, **60**, 2973.
43. (a) S. Lepri, M. Ceccarelli, N. Milani, S. Tortorella, A. Cucco, A. Valeri, L. Goracci, A. Brink and G. Cruciani, *Proc. Natl. Acad. Sci. U. S. A.*, 2017, **114**, E3178; (b) G. Cruciani, N. Milani, P. Benedetti, S. Lepri, L. Cesarini, M. Baroni, F. Spyrakis, S. Tortorella and E. Mosconi, *J. Med. Chem.*, 2018, **61**, 360.
44. J. K. Sodhi, S. Wong, D. S. Kirkpatrick, L. Lichuan Liu, S. C. Khojasteh, C. E. C. A. Hop, J. T. Barr, J. P. Jones and J. S. Halladay, *Drug Metab. Dispos.*, 2015, **43**, 908.
45. R. Mohammed-Reza and S. Soltani, *Expert Opin. Drug Discovery*, 2017, **12**, 305.
46. S. Anoh, Y. Tayama, K. Sugihara, S. Kitamura and S. Ohta, *Drug Metab. Pharmacokinet.*, 2015, **30**, 52.
47. E. Garattini and M. Terao, *Drug Metab. Rev.*, 2011, **43**, 374.
48. B. Kaye, D. J. Rance and L. Waring, *Xenobiotica*, 1985, **15**, 237.
49. X. Zhang, H. Liu, P. Weller, M. Zheng, W. Tao, J. Wang, G. Liao, M. Monshouwer and G. Peltz, *Pharmacogenomics J.*, 2011, **1**, 15.
50. T. Akabane, K. Tanaka, M. Irie, S. Terashita and T. Teramura, *Xenobiotica*, 2011, **41**, 372.
51. J. M. Hutzler, M. A. Cerny, Y. Yang, C. Asher, D. Wong, K. Frederick and K. Gilpin, *Drug Metab. Dispos.*, 2014, **42**, 1751.
52. S. Diamond, J. Boer, T. P. Maduskuie, N. Falahatpisheh, Y. Li and S. Yeleswaram, *Drug Metab. Dispos.*, 2010, **38**, 1277.
53. D. Dalvie, C. Zhang, W. Chen, T. Smolarek, R. S. Obach and C. Loi, *Drug Metab. Dispos.*, 2010, **38**, 641.
54. R. D. Morrison, A. L. Blobaum, F. W. Byers, T. S. Santomango, T. M. Bridges, D. Stec, K. A. Brewer, R. Sanchez-Ponce, M. M. Corlew, R. Rush, A. S. Felts, J. Manka, B. S. Bates, D. F. Venable, A. L. Rodriguez, C. K. Jones, C. M. Niswender, P. J. Conn, C. W. Lindsley, K. A. Emmitte and J. S. Daniels, *Drug Metab. Dispos.*, 2012, **40**, 1834.

55. C. Zetterberg, C. Maltais, L. Laitinen, S. Liao, H. Tsao, A. Chakilam and N. Hariparsad, *Drug Metab. Dispos.*, 2016, **44**, 1286.
56. P. Martijn, M. P. Lolkema, H. H. Bohets, H. Hendrik-Tobias Arkenau, A. Lampo, E. Barale, J. A. Maja, M. J. A. de Jonge, L. van Doorn, P. Hellemans, J. S. de Bono and F. A. L. M. Eskens, *Clin. Cancer Res.*, 2015, **21**, 2297.
57. Y. Li, J. Xu, W. G. Lai, A. Whitcher-Johnstone and D. J. Tweedie, *Drug Metab. Dispos.*, 2012, **40**, 1130.
58. T. Isobe, M. Ohta, Y. Kaneko and H. Kawai, *Xenobiotica*, 2015, **46**, 579.
59. N. E. Austin, S. J. Baldwin, L. Cutler, N. Deeks, P. J. Kelly, M. Nash, C. E. Shardlow, G. Stemp, K. Thewlis, A. Ayrton and P. Jeffrey, *Xenobiotica*, 2001, **31**, 677.
60. K. Inoue, H. Mizuo, S. Kawaguchi, K. Fukuda, K. Kusano and T. Yoshimura, *Drug Metab. Dispos.*, 2014, **42**, 1326.
61. L. Xu, C. Woodward, J. Dai and C. Prakash, *Drug Metab. Dispos.*, 2013, **41**, 2133.
62. T. Sasaki, A. Masubuchi, M. Yamamura, N. Watanabe, M. Hiratsuka, M. Mizugaki, K. Itoh and Y. Tanaka, *Biopharm. Drug Dispos.*, 2006, **5**, 247.
63. J. Chandrasekhar, R. Dick, J. V. Veldhuizen, D. Koditek, E. Lepist, M. E. McGrath, L. Patel, G. Phillips, K. Sedillo, J. R. Somoza, J. Thierren, N. A. Till, J. Treiberg, A. G. Villasenor, Y. Zherebina and S. Perreault, *J. Med. Chem.*, 2018, **61**, 6858.
64. D. C. Pryde, T. Tran, P. Jones, J. Duckworth, M. Howard, I. Gardner, R. Hyland, R. Webster, T. Wenham, S. Bagal, K. Omoto, R. P. Schneider and J. Lin, *Bioorg. Med. Chem. Lett.*, 2012, **22**, 2856.
65. H. Cheng, C. Li, S. Bailey, S. M. Baxi, L. Goulet, L. Guo, J. Haffman, Y. Jiang, T. O. Johnson, T. W. Johnson, D. R. Knighton, J. Li, K. Liu, Z. Liu, M. A. Marx, M. Walls, P. A. Wells, M. Yin, J. Zhu and M. Zientek, *ACS Med. Chem. Lett.*, 2013, **4**, 91.
66. P. Jones, D. C. Pryde, T. Tran, F. M. Adam, G. Bish, F. Calo, G. Ciaramella, R. Dixon, J. Duckworth, D. N. A. Fox, D. A. Hay, J. Hitchin, N. Horscroft, M. Howard, C. Laxton, T. Parkinson, G. Parsons, K. Proctor, M. C. Smith, N. Smith and A. Thomas, *Bioorg. Med. Chem. Lett.*, 2012, **22**, 2856.
67. A. Linton, P. Kang, M. Ornelas, S. Kephart, Q. Hu, M. Parish, Y. Jiang and C. Guo, *J. Med. Chem.*, 2011, **54**, 7705.
68. G. Thoma, J. Blanz, P. Buhlmayer, P. Druckes, M. Kittelmann, A. B. Smith, M. van Eis, E. Vangrevelinghe, H. Zerwes, J. Che, X. He, Y. Jin, C. C. Lee, P. Michellys, T. Uno and H. Liu, *Bioorg. Med. Chem. Lett.*, 2014, **24**, 2278.
69. T. V. Magee, S. L. Ripp, B. Li, R. A. Buzon, L. Chupak, T. J. Dougherty, S. M. Finegan, D. Girard, A. E. Hagen, M. J. Falcone, K. A. Farley, K. Granskog, J. R. Hardink, M. D. Huband, B. J. Kamicker, T. Kaneko, M. J. Knickerbocker, J. L. Liras, A. Marra, I. Medina, T.-T. Nguyen, M. Noe, R. S. Obach, J. P. O'Donnell, J. B. Penzien, U. D. Reilly, J. R. Schafer, Y. Shen, G. G. Stone, T. J. Strelevitz, J. Sun, A. Tait-Kamradt, A. D. N. Vaz, D. A. Whipple, D. W. Widlicka, D. G. Wishka, J. P. Wolkowski and M. E. Flanagan, *J. Med. Chem.*, 2009, **52**, 7446.

70. P. A. Barsanti, Y. Pan, Y. Lu, R. Jain, M. Cox, R. J. Aversa, M. P. Dillon, R. Elling, C. Hu, X. Jin, M. Knapp, J. Lan, S. Ramurthy, P. Rudewicz, L. Setti, S. Subramanian, M. Mathur, L. Taricani, G. Thomas, L. Xiao and Q. Yue, *ACS Med. Chem. Lett.*, 2015, **6**, 42.
71. A. Hayes, N. Yi Mok, M. Liu, C. Thai, T. H. Henley, B. Atrash, R. M. Lanigan, J. Sejberg, Y.-V. Le Bihan, V. Bavetsias, J. Blagg, I. Florence and F. I. Raynaud, *Xenobiotica*, 2017, **47**, 771.
72. R. Sharma, T. J. Strelevitz, H. Gao, A. J. Clark, K. Schildknegt, R. S. Obach, S. L. Ripp, L. M. Tremaine, D. K. Spracklin and A. D. N. Vaz, *Drug Metab. Dispos.*, 2012, **40**, 625.
73. L. Patel, J. Chandrasekhar, J. Evarts, A. C. Haran, C. Ip, J. A. Kaplan, M. Kin, D. Koditek, L. Lad, E. Lepist, M. E. McGrath, N. Novikov, S. Perreault, K. D. Puri, J. R. Somoza, B. H. Steiner, K. L. Stevens, J. Therrien, J. Treiberg, A. G. Villasenor, A. Yeung and G. Phillips, *J. Med. Chem.*, 2016, **59**, 3532.
74. U. K. Bandarage, M. P. Clark, E. Perola, H. Gao, M. D. Jacobs, A. Tsai, J. Gillespie, J. M. Kennedy, F. Maltais, M. W. Ledeboer, I. Davies, W. Gu, R. A. Byrn, K. N. Addae, H. Bennett, J. R. Leeman, S. M. Jones, C. O'Brien, C. Memmott, Y. Bennani and P. Charifson, *ACS Med. Chem. Lett.*, 2017, **8**, 261.
75. A. Mallinger, K. Schiemann, C. Rink, J. Sejberg, M. A. Honey, P. Czodrowski, M. Stubbs, O. Poeschke, M. Busch, R. Schneider, D. Schwarz, D. Musil, R. Burke, K. Urbahns, P. Workman, D. Wienke, P. A. Clarke, F. I. Raynaud, S. A. Eccles, C. Esdar, F. Rohdich and J. Blagg, *ACS Med. Chem. Lett.*, 2016, 7, 573.

CHAPTER 13

Glucuronidation

YUE PAN*

Global Discovery Chemistry, Novartis Institutes for BioMedical Research, 22 Windsor Street, Cambridge, Massachusetts 02139, USA
*E-mail: ypan@relaytx.com

13.1 Introduction

13.1.1 Enzyme Family

Uridine 5′-diphospho-glucuronosyltransferase (UGT) is a glycosyltransferase that catalyzes the transfer of glucuronic acid from its cofactor uridine diphosphate glucuronic acid (UDPGA, Figure 13.1) to the substrate.

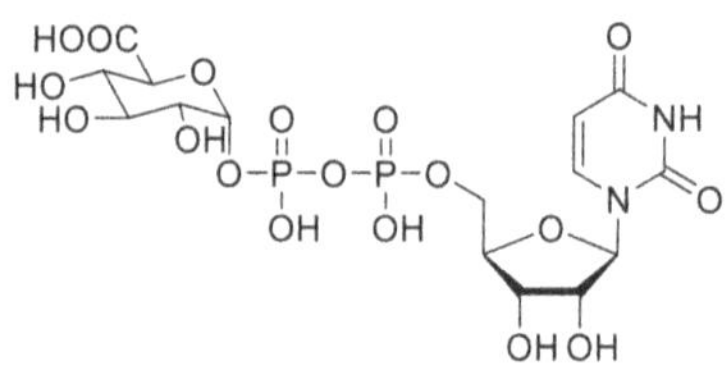

Figure 13.1 Structure of uridine diphosphate glucuronic acid (UDPGA).

13.1.2 Expression

UGT is present in all preclinical model species and humans, and there are very significant differences in specific UGT isoforms' expression between the species.[1,2] Human UGTs have over 20 isoforms from three functional families (UGT1, 2A/B and 3A).[3] There is also variation in their expression among the

Drug Discovery Series No. 79
The Medicinal Chemist's Guide to Solving ADMET Challenges
Edited by Patrick Schnider
© The Royal Society of Chemistry 2021
Published by the Royal Society of Chemistry, www.rsc.org

tissues, with the expression level high in the liver, gastrointestinal tract and kidney. For drugs that have glucuronidation as a major metabolic pathway, UGT polymorphism can lead to variable exposure in patients.[4]

13.1.3 Structure

UGTs are membrane-bound glycoproteins that have two functional domains. The N-terminal domain (NTD) contains the substrate binding site and the C-terminal domain (CTD) is where the cofactor UDPGA binds.[5] While structurally the CTD is similar between UGT isoforms, the NTD is highly variable and explains this enzyme family's broad substrate scope. The NTD of UGT is bound to the membrane of the endoplasmic reticulum or the nucleus and as a result has been elusive to crystallography. To date, among all human UGTs, only the crystal structure of the C-terminal domain of UGT2B7 has been solved (PDB 2O6L).[6]

13.1.4 Activity

UGTs' activity varies based on the substrate and the enzyme isoforms involved. Given the large number of UGT isoforms, it is not surprising that there is significant overlap in substrate specificity. UGTs' expression levels and activities also vary across species. For example, while UGT1A2 is expressed in rodents, the UGT1A2 gene is a pseudogene in humans.[1] For the same species, a good *in vitro–in vivo* correlation in clearance is still possible, so an *in vitro* UGT assay can be used to predict *in vivo* clearance through the glucuronidation pathway.[7]

13.1.5 Function and Substrates

UGTs catalyze the conjugation of alcohol, phenol, carboxylic acid, amine (primary, secondary and tertiary[8,9]), amide,[10] acidic carbon,[11] thiol, selenol (-SeH) and certain heterocycles (imidazole,[12] pyrazole,[13] triazole,[14] tetrazole,[15] pyridine,[16] indole,[17] indazole,[18] pyridone,[19] triazine[20] *etc.*, Figure 13.2) with glucuronic acid, a highly polar and solubilizing moiety. It is worth noting that a free N–H is not a prerequisite for the conjugation reaction, as *N*-glucuronidation is known to happen with tertiary amine, pyridine, N-substituted imidazole and triazole. The conjugation with glucuronic acid facilitates excretion of the metabolite into bile and urine because of its polar nature and the fact that it can be a substrate of transporters, such as multidrug resistance protein (MRP), leading to the elimination of the glucuronide from the body. Interestingly, certain MRP isoforms, such as MRP3, are also known to efflux glucuronide into the systemic circulation.[21] Glucuronides also tend to have a smaller volume of distribution due to their polarity and high albumin binding, therefore they can be major circulating metabolites. Other transporters may also transport glucuronides, leading to a complex interplay between metabolism and transport. In general, glucuronidation catalyzed by UGTs and sulfation catalyzed by sulfotransferases (Chapter 14) are important detoxification pathways. Given the large number of UGT and sulfotransferase isoforms, and the substantial overlap of substrate scope between the two enzyme classes, versatility and redundancy of these metabolic pathways protect us from chemical toxicity.[22]

13.1.6 Mechanism

In a typical UGT-mediated reaction, the glucuronic acid is transferred from the cofactor UDPGA to the substrate in an S_N2 reaction (Figure 13.3). The crystal structure of UGT2B7 indicates that the enzyme mechanism is similar to a serine protease where a serine–histidine–aspartate catalytic triad (the accepting moiety in the UGT substrate takes the role of the serine) expedites the reaction.[6] The substrate's acidic proton is removed by the histidine, rendering it a better nucleophile to react with UDPGA. The cofactor UDPGA is synthesized from UDP-glucose by UDP-glucose 6-dehydrogenase using NAD^+.

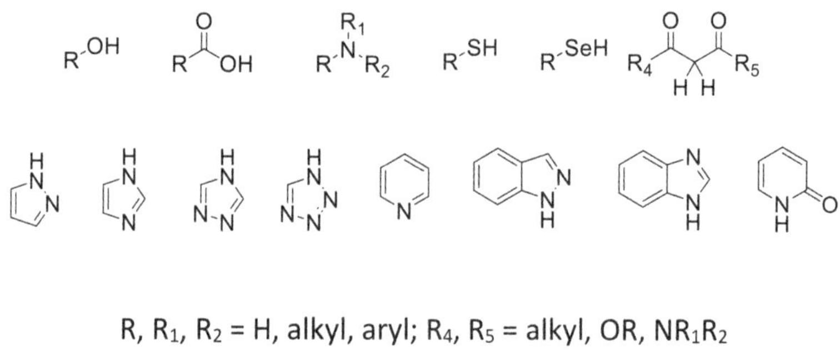

Figure 13.2 Structural features of some UGT substrates.

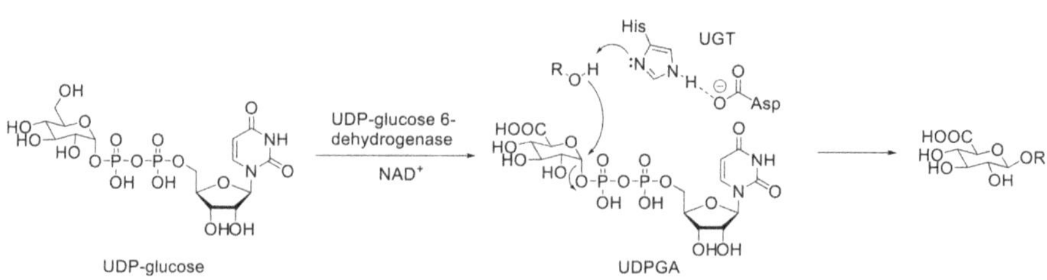

Figure 13.3 Mechanism of alcohol glucuronidation catalyzed by UGT.

13.1.7 Screening Strategies

Since UGTs are mostly present in the lumen of the endoplasmic reticulum (ER), metabolism mediated by UGTs can, in principle, be captured by the liver microsomal assay. However, the ER membrane is a barrier to the diffusion of UGT substrates and the cofactor UDPGA. As a result, UGT-mediated metabolic clearance can be underestimated in the liver microsomal assay, even with added UDPGA.[7] Membrane-disrupting detergents, such as Brij-58, and, more recently, alamethicin, a pore-forming peptide, are used with liver microsomes so the substrate and cofactor can access the UGTs. Primary hepatocytes are the model system of choice to assess both phase I and phase II metabolism, including glucuronidation, although the substrate's cell permeability, which may be affected by transporter-mediated uptake or efflux, can be complicating factors. Higher intrinsic clearance in hepatocytes or liver microsomes fortified with UDPGA than in microsomes without added UDPGA, along with a moiety that can serve as the acceptor of glucuronic acid, is indicative of UGT-mediated metabolism. Intrinsic clearance from these *in vitro* assays can be used to project *in vivo* clearance. Alternatively, recombinant UGTs can be used to confirm glucuronidation and for isoform phenotyping, although they are not typically used to predict the *in vivo* glucuronidation rate. In metabolite identification studies mass spectrometry can help to identify the glucuronide metabolite [+162 Daltons]. Due to the large number of UGT isoforms and their overlapping substrate specificities, isoform-specific UGT inhibitors are uncommon and not routinely used to help diagnose UGT-mediated metabolism.

13.1.8 Relevance

Glucuronidation is a major metabolic pathway for both endogenous chemicals and drugs. About 10% of the top 200 prescribed drugs are metabolized by UGTs.[23] UGTs are expressed in the liver, intestine and other tissues and are a major component of first pass metabolism. UGTs can work in concert with CYP450s and other oxidative enzymes, as oxidation often installs a hydroxyl group or reveals a more nucleophilic amine for efficient conjugation. The resulting glucuronides are much more polar and water-soluble, which often minimizes their passive reabsorption and facilitates excretion. UGT-mediated metabolism can be the reason why a compound has low metabolic stability and oral bioavailability, and unexpected metabolites can be generated following glucuronidation, complicating metabolite identification. For example, intramolecular nucleophilic attack of an acyl glucuronide can lead to an unanticipated cyclic metabolite (Figure 13.4).[24] On the other hand, functional groups that can undergo UGT-mediated metabolism, either directly or after being unmasked by oxidative metabolism, can be intentionally introduced to enable more metabolic pathways, which can reduce the risk of drug–drug interaction. Certain UGT isoforms, such as UGT1A1 and 2B10, are polymorphic,[25] and drugs such as irinotecan, which

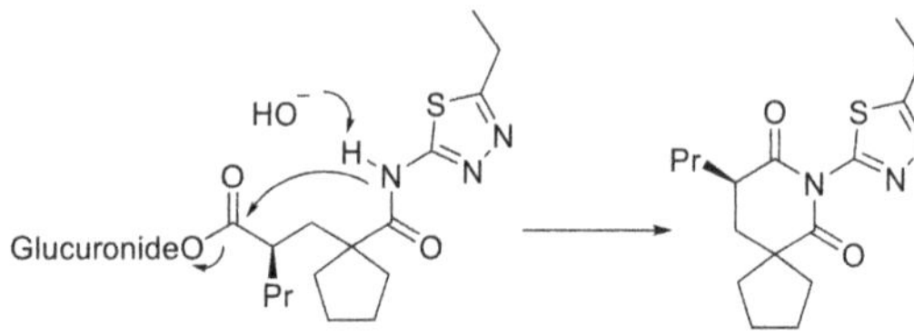

Figure 13.4 Unexpected metabolites can form following glucuronidation.

is mainly metabolized by UGT1A1, can have variable exposure in patients.[26] Some glucuronides can be hydrolyzed back to the parent molecule either due to chemical instability or catalysis by β-glucuronidase in the gut after biliary excretion, leading to enterohepatic recirculation. If the hydrolysis of a glucuronide is catalyzed by bacterial β-glucuronidase, reduction of the bacterial glucuronidase activity by concomitant administration of oral antibiotics may reduce enterohepatic recirculation.[27] Additionally, acyl glucuronides can be hydrolyzed by esterases in the plasma.[1]

While glucuronidation usually leads to an inactive metabolite, there are exceptions. For example, it is well known that morphine 6-glucuronide (M6G) is at least as potent as morphine itself towards the μ opioid receptor, and despite M6G's high polarity and poor brain penetration, it can achieve sufficient brain exposure to produce analgesic effects.[28,29] Other examples include ezetimibe, dabigatran and certain adenosine monophosphate-activated protein kinase (AMPK) activators.[30] Glucuronidation is also an important pathway leading to reactive metabolites. For example, glucuronidation of carboxylic acids is sometimes associated with toxicity as the corresponding acyl glucuronide can act as an acylating agent, which may lead to covalent modification of biological targets. Another mechanism leading to acyl glucuronide's toxicity is through the Amadori rearrangement, leading to a stable covalent adduct (see Chapter 15 on reactive metabolites). Besides toxicity associated with acyl glucuronides, glucuronidation of a hydroxylamine or hydroxamic acid can lead to an N–O-glucuronide, which may also react with a nucleophile by breaking the N–O bond.[31]

Key References

- H. Komura and M. Iwaki, *Drug Metab. Rev.*, 2011, **43**, 476–498.
- A. Rowland, J. O. Miners and P. I. Mackenzie, *Int. J. Biochem. Cell Biol.*, 2013, **45**, 1121–1132.
- M. J. Miley, A. K. Zielinska, J. E. Keenan S. M. Bratton, A. Radominska-Pandya and M. R. Redinbo, *J. Mol. Biol.*, 2007, **369**, 498–511.
- J. H. Lin and B. K. Wong, *Curr. Drug Metab*, 2002, **3**, 623–646.
- U. A. Argikar, P. M. Potter, J. M. Hutzler and P. H. Marathe, *AAPS J.*, 2016, **18**, 1391–1405.

13.2 Mitigation Strategies

13.2.1 Remove or Block the Glucuronidation Site

The most direct approach to lower UGT-mediated clearance is to remove or block the metabolic soft spot, provided that the on-target potency can be maintained. Sometimes blocking with an alkyl group may not be sufficient as it is still subject to CYP-mediated dealkylation, after which the functionality subject to glucuronidation will be revealed. Sometimes such glucuronidation sites cannot be simply removed or blocked due to large potency loss, and other strategies need to be applied.

13.2.2 Use Bioisosteres to Replace the Susceptible Moiety

Bioisosteric replacement is an effective approach to address glucuronidation. A functional group susceptible to glucuronidation, such as phenol, can be replaced with a bioisosteric replacement[32] that is less prone (azaheterocycles such as indazole and benzimidazole) or unlikely (benzolactam, for example) to be glucuronidated, while maintaining the on-target potency. Similarly, a carboxylic acid can be replaced with tetrazole, which can form a non-reactive glucuronide, or acyl sulfonamide, which does not typically form a glucuronide.

13.2.3 Sterically or Electronically Decrease Glucuronidation Rate

As glucuronidation is a nucleophilic reaction between the substrate (nucleophile) and the cofactor UDPGA (electrophile), steric and electronic factors have a strong influence on the reaction rate. Substitution close to the reactive site can increase the steric hindrance around it, and electron withdrawing groups can reduce the electron density of the soft spot, both of which can slow down glucuronidation. This is an often-used strategy to reduce glucuronidation while maintaining the on-target potency.

For phenol and alcohol, while steric hindrance close to the hydroxyl group can reduce glucuronidation, incorporation of an electron-withdrawing group and the resulting lower pK_a can have different effects on their glucuronidation rate. Although studies correlating the glucuronidation rate of various substituted phenols with the substitutions' Hammett parameter are available,[33,34] systematic studies of more drug-like molecules are rare, and it can be difficult to predict the effect of pK_a on UGT-mediated clearance. It can be imagined that with every other parameter fixed, the relationship between glucuronidation rate and pK_a is an inverted U-shaped curve with the clearance being the fastest at a certain pK_a. For example, significantly lowering a regular phenol's pK_a from 10 could increase the glucuronidation rate as more deprotonation occurs and the deprotonated phenoxide is more nucleophilic. However, once a significant amount of deprotonation is achieved, further pK_a reduction may lead to lower electron density on the phenoxide and slow down glucuronidation. One publication on the glucuronidation of a series of catechols described that the fastest glucuronidation occurred when the phenol's pK_a was between 8 and 9.[35] The reader is encouraged to experimentally test key compounds in their own chemical series to establish the pK_a's effect on glucuronidation. For other moieties susceptible to glucuronidation, such as carboxylic acids, amines and heterocycles, the rule of thumb is that nearby electron-withdrawing groups should decrease the glucuronidation rate as the electron density on the nucleophile is reduced.

13.2.4 Decrease Lipophilicity

On the basis of results reported in several papers, UGTs are similar to CYP450s in that they prefer lipophilic substrates, and incorporation of polarity, *e.g.* by means of adding heteroatoms or a basic amine, can significantly reduce the glucuronidation rate, even if their position is distal to the soft spot.

13.2.5 Sterically Disrupt the Substrate's Binding to UGT

This strategy is empirical without detailed information from a crystal structure of the involved UGT. An example described later in this chapter indicates, at least in that particular scaffold, that making the molecule more three dimensional and escaping the molecular flatland has a beneficial effect on reducing glucuronidation. Enantiomers have also been shown to have different glucuronidation rates.

13.2.6 Protection of the Soft Spot as a Prodrug

A prodrug, if sufficiently stable when passing through the gut wall and liver, can not only help to enhance absorption but also protect the glucuronidation soft spot. This can be difficult to engineer as the hydrolytic enzymes required to release the parent drug are often highly expressed in the liver, where UGTs' expression level is also high.

13.3 Examples of Mitigation Strategies

13.3.1 Remove or Block the Glucuronidation Site

Examples

1

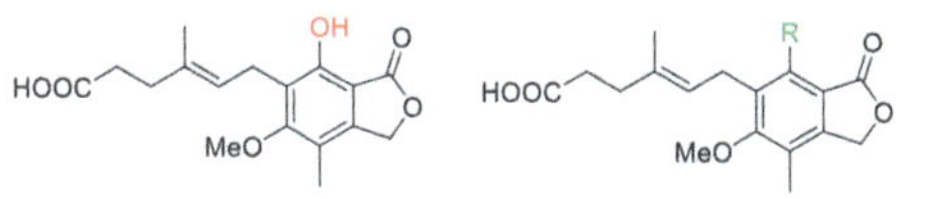

1 R = H, NH_2, CN, F

Mycophenolic acid (**1**), an inhibitor of inosine-5′-monophosphate dehydrogenase (IMPDH), suffers from rapid glucuronidation of the phenol. Removal of the hydroxyl group as well as its replacement by an amine, nitrile or fluorine completely abolished glucuronidation.[36]

2

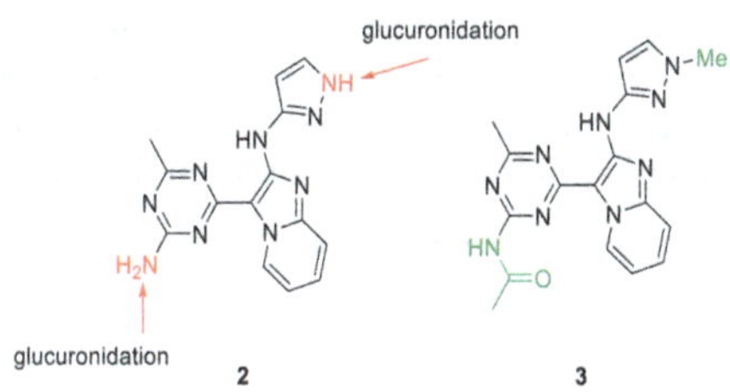

2 **3**

Compound	Rat *in vivo* clearance ($L\ h^{-1}\ kg^{-1}$)
2	1.8
3	0.6

In a series of mammalian target of rapamycin (mTOR) inhibitors represented by **2**, results from an *in vivo* rat pharmacokinetic (PK) study indicated that glucuronidation was the major metabolic pathway and both the pyrazole and the primary amine were soft spots. Upon blocking the pyrazole with a methyl group and acetylation of the primary amine, compound **3** had much lower *in vivo* clearance.[13]

13.3.2 Use Bioisosteres to Replace the Susceptible Moiety

Examples

1

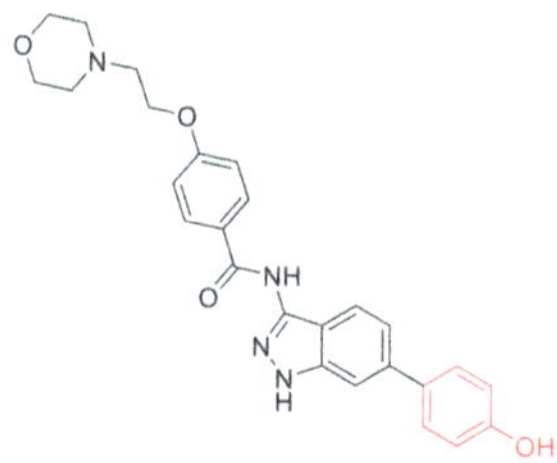

4

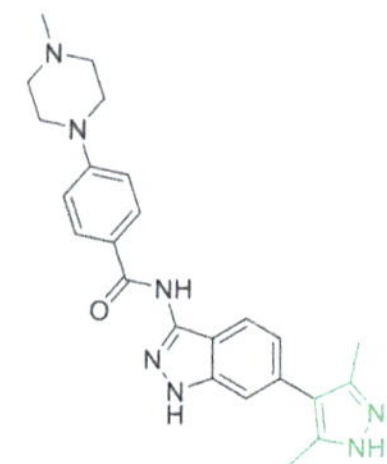

5

Compound	Percentage remaining in incubation with UGT after 30 min	AUC in mouse, 30 mg kg^{-1}, orally (h μg mL^{-1})
4	76%	0.89
5	94%	2.55

Hepcidin production inhibitor **4** showed low oral exposure in mice. A search for bioisosteres of the phenol, along with optimization of the other end of the molecule, led to compound **5**, which was more stable towards UGT-mediated metabolism and showed much higher oral exposure when dosed at 30 mg kg^{-1} in mice.[37]

2

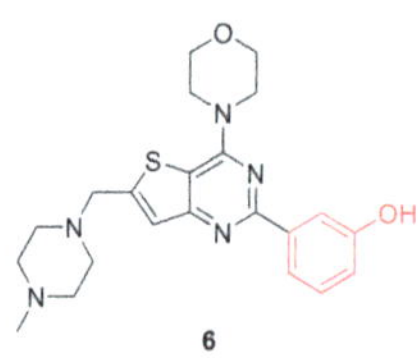

6

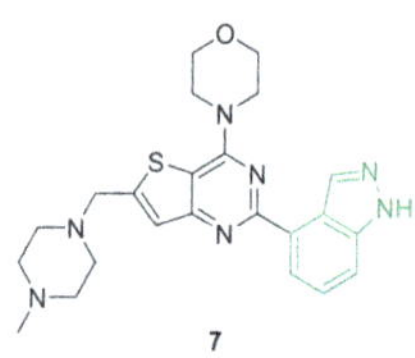

7

Compound	Oral bioavailability (mouse, rat)	Rat *in vivo* glucuronide exposure
6	10%, 2%	Higher than parent molecule
7	31%, 37%	Lower than parent molecule

For a series of PI3K inhibitors represented by the phenol **6**, results from *in vitro* microsomal studies indicated the glucuronide was the common metabolite, indicating that the phenol was the soft spot. In an *in vivo* study in mice with **6**, the phenol glucuronide was the dominant metabolite with exposure much higher than the parent molecule. As a result of the first pass effect the compound has poor oral bioavailability (10% in mouse and 2% in rat). Indazole was identified as a bioisosteric replacement for the phenol. Compound 7, albeit somewhat less potent against PI3K, was found to have significantly higher oral bioavailability (31% in mice and 37% in rats). Formation of a glucuronide (probably on the indazole) was still observed, but its exposure was now significantly lower than that of the parent. Interestingly, no significant difference in metabolic stability was observed between the two compounds in mouse and human microsomes with added NADPH and UDPGA, with 85% of **6** and 81% of **7** remaining after a 30 minute incubation.[38] There could be multiple reasons for such an *in vitro–in vivo* disconnect, such as *in vitro* assay latency without pore-forming agents, *in vitro* inhibition of UGTs by long chain unsaturated fatty acids released from microsomal membranes or extrahepatic expression of UGTs.

Examples

3

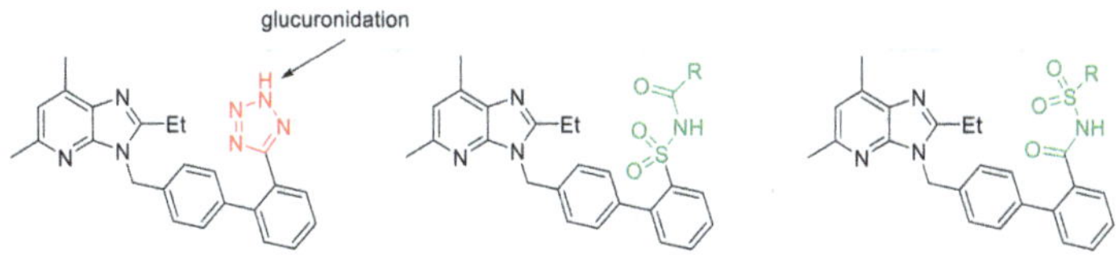

L-158,809 R = alkyl, phenyl R = alkyl, phenyl

In a series of angiotensin II receptor antagonists, L-158,809 was rapidly cleared *via* glucuronidation at the indicated tetrazole nitrogen in dogs and monkeys. Either isomer of the acyl sulfonamide as a replacement of the tetrazole was found to be resistant to glucuronidation.[39,40]

4

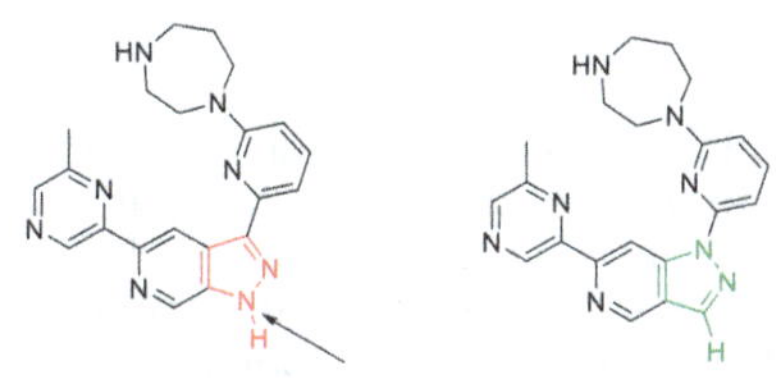

8 **9**

Compound	Percentage remaining in rat intestinal microsomes + UDPGA after 60 min	Oral bioavailability in rats
8	40	1.3%
9	85	42%

In a series of proviral integration site for Moloney murine leukemia virus (PIM; a proto-oncogene family) inhibitors it was found that the 6-azaindazole was metabolized as the glucuronide. As the azaindazole N–H is the kinase hinge binder and forms a key interaction with the protein, an extensive search for its replacement was carried out, including calculating the electrostatic potential at the molecular surface ($V_{s,max}$) of various bioisosteres to identify the most positive N–H or C–H as the hydrogen bond donor. The group with the highest $V_{s,max}$ at that position, the 5-azaindazole as represented by compound **9**, also turned out to be the best isostere as its potency across the three PIM isoforms was similar to that of **8**. The N-substituted 5-azaindazole was not subject to glucuronidation, and as a result the *in vitro* metabolic stability of **9** in UDPGA-supplemented rat intestinal microsomes was much higher, which translated nicely *in vivo* and led to much higher oral bioavailability.[41]

13.3.3 Sterically or Electronically Decrease Glucuronidation Rate

Examples

1

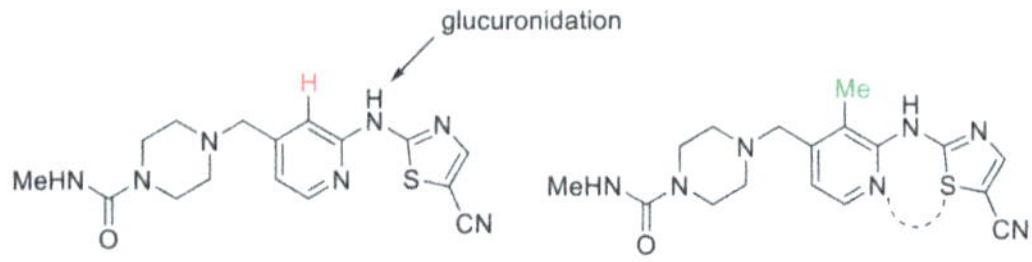

10 11

Compound	Clearance (mL min^{-1} kg^{-1}, dog, rat, rhesus monkey)	Oral bioavailability (dog, rat, rhesus monkey)
10	0.51, 13.3 ± 1.0, 17.2	99%, 49%, 43%
11	2.0, 3.0 ± 0.45, 1.7	75%, 73%, 92%

In a series of vascular endothelial growth factor (VEGF) inhibitors two compounds varying by only a methyl group, **10** and **11**, were studied in great detail. Methyl substitution on the central pyridine has a dramatic effect of lowering the *in vivo* clearance in rats and rhesus monkeys while the effect in dogs was in the opposite direction. Compared with **10**, compound **11**'s oral bioavailability was also higher in rat sand rhesus monkeys (73% *vs.* 49% and 92% *vs.* 43%, respectively), but lower in dogs (75% *vs.* 99%). The discrepancy between the species was followed up. Using ^{14}C-labeled **10** it was shown that the major route of compound elimination was metabolism and the major metabolite in rats and rhesus monkeys was the *N*-glucuronide formed at the bridging N–H group, while in dogs **10** was mostly eliminated by oxidative metabolism. This can explain the higher clearance and lower oral bioavailability of **11** in dogs, possibly due to its higher lipophilicity. The conformation of **11** is worth noting. In the low-energy conformation the pyridine and thiazole should be very close to being coplanar as the pyridine nitrogen and the thiazole sulfur can form a chalcogen bond, thereby putting **11**'s methyl group close to the central N–H and effectively blocking the nucleophilic attack towards UDPGA.[42]

2

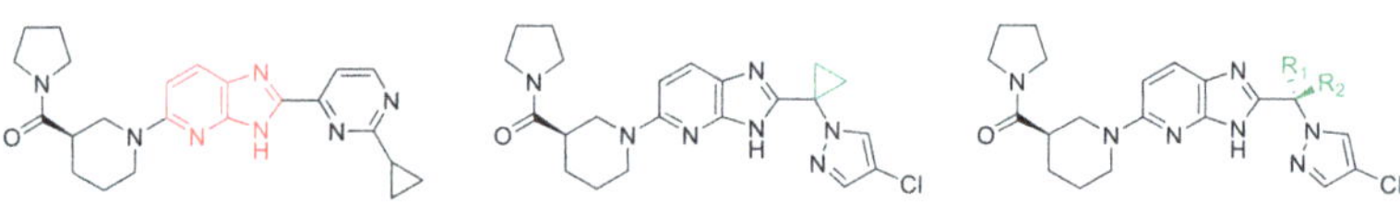

12 13 14

Compound	R_1, R_2	HLM UGT Cl_{int} (μL min^{-1} mg^{-1})
12		316
13		<1.9
14a	Me, H	10.8
14b	H, Me	7.8

Examples

The diacylglycerol *O*-acyltransferase 2 (DGAT2) inhibitor **12** was rapidly cleared in human liver microsomes (HLM) with added alamethicin and UDPGA [intrinsic clearance (Cl_{int}) 316 μL min^{-1} mg^{-1}], much faster than in the absence of UDPGA (Cl_{int} 50 μL min^{-1} mg^{-1}). Glucuronidation at the imidazole was suspected to be the cause. To address the phase II metabolism issue an sp^3 spacer was introduced. A cyclopropyl group (**13**) proved to be highly effective in reducing glucuronidation, while both enantiomers with an ethyl spacer (**14a** and **14b**) also showed much lower UGT-mediated clearance rates than **12**.[43,44]

3

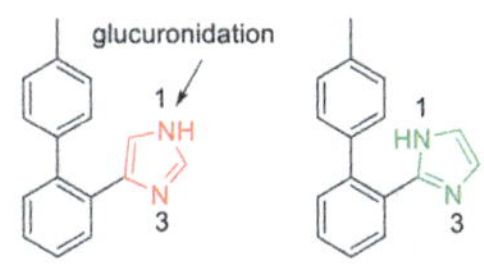

15 **16**

The 4-substituted imidazole **15** underwent glucuronidation at the indicated N1 position as it is exposed, while the 2-substituted imidazole **16** did not undergo glucuronidation.[45] It is important to note that the three rings are not coplanar, but for **16** the steric hindrance around the potential glucuronidation site is still significantly larger than that for **15**.

4

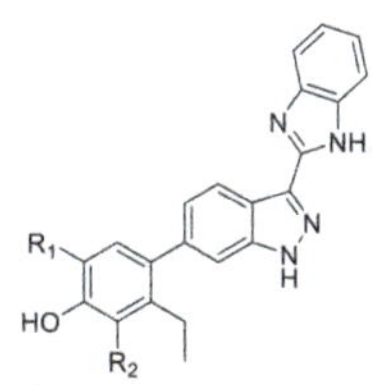

Compound	R_1, R_2	HLM UGT Cl_{int} (μL min^{-1} mg^{-1})
17	H, H	49
18	H, F	46
19	F, H	>360

The difficulty in predicting the effect of the pK_a of a phenol on its glucuronidation is well exemplified by the inhaled Janus kinase 2 (JAK2) inhibitors **17–19**. To introduce a glucuronidation soft spot and minimize systemic exposure, the researchers studied the effect of *ortho*-fluorine on the phenol's glucuronidation. The pK_a and lipophilicity of compounds **18** and **19** should be very similar, yet **19** displayed much higher UGT-mediated clearance than **18**, the glucuronidation rate of which was similar to that of the unsubstituted **17**.[46]

(*continued*)

Examples

5

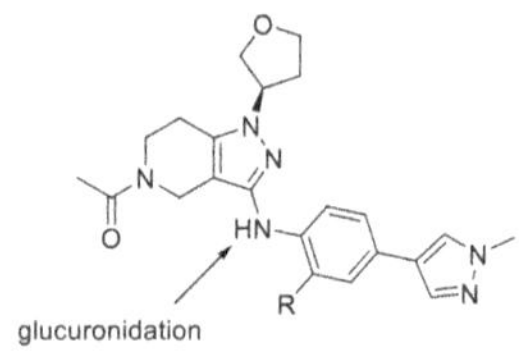

Compound	MLM Cl_{hep}, probably no UDPGA (mL min^{-1} kg^{-1})	Mouse hepatocyte Cl_{hep} (mL min^{-1} kg^{-1})
20, R = H	25	59
21, R = F	42	6.2

For a series of CREB binding protein (CBP) inhibitors represented by **20** and **21**, a disconnect in predicted hepatic clearance (Cl_{hep}) between mouse liver microsomes (MLM) and hepatocytes was identified, indicating phase 2 metabolism not captured by the microsomal assay. Results from a metabolite identification (MetID) study of a representative compound from this series in mouse hepatocytes indicated that the sole metabolite was a glucuronide. It was proposed that the N–H between the pyrazole and phenyl groups was the glucuronidation soft spot. An ortho-fluorine substituent was added to help reduce the electron density on the nitrogen, making it less nucleophilic. Compound **21** was indeed found to have much reduced clearance in mouse hepatocytes compared with **20** and was confirmed to have low clearance and high oral bioavailability (F) in a mouse PK study [Clearance (Cl) 9.6 mL min^{-1} kg^{-1}, F 100%].[47] However, it is likely that this was not entirely due to the reduced pK_a, since the reported difference in the calculated pK_a between compounds **20** and **21** is small (approximately 0.2). Steric and stereoelectronic factors, such as the intramolecular hydrogen bond/antiparallel dipole interaction between the N–H and C–F, may also play a role.

6

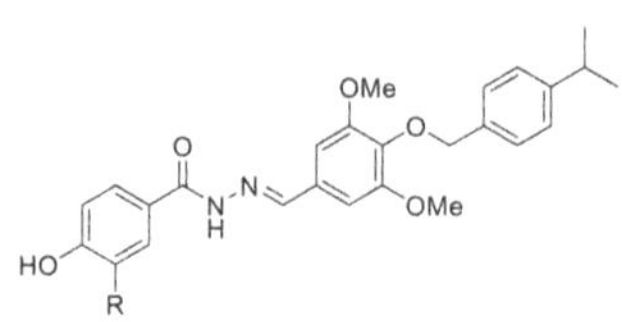

Compound	R	Glucuronidation rate (pmol min^{-1} mg^{-1})
22	Cl	169
23	CN	27

The 2-chlorophenol in a series of glucagon receptor antagonists represented by **22** was rapidly glucuronidated in rat liver microsomes. The replacement of Cl with CN led to dramatically lower UGT-mediated clearance, which translated to a longer *in vivo* half-life (60 min *vs.* 18 min). In rat liver microsomes compound **23**'s major metabolic pathway has switched from phase 2 to oxidative metabolism.[48] Compound **23** has a significantly lower pK_a than **22** (reported pK_a values of 2-chlorophenol and 2-cyanophenol are 8.5 and 6.9, respectively). Steric hindrance due to the larger size of CN compared with Cl may have contributed as well.

Examples

7

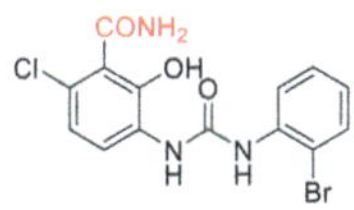

24

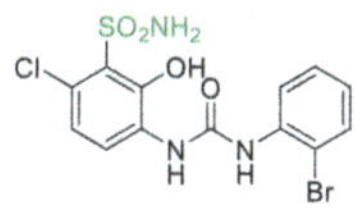

25

Compound	HLM (no UDPGA) Cl_{int} (mL min^{-1} g^{-1} of liver)	HLM + UDPGA Cl_{int} (mL min^{-1} g^{-1} of liver)
24	0.74	15
25	Stable	2.7

In a series of CXC chemokine receptor 2 (CXCR2) antagonists substitutions *ortho* to the phenol were explored to address rapid clearance and low oral bioavailability, which are probably the result of phenol glucuronidation. While the difference in HLM Cl_{int} without UDPGA was small, the sulfonamide **25** was found to be significantly more stable than the amide **24** in the presence of UDPGA, indicating reduced glucuronidation. This was rationalized by the sulfonamide's increased bulk and stronger electron-withdrawing character.[49]

8

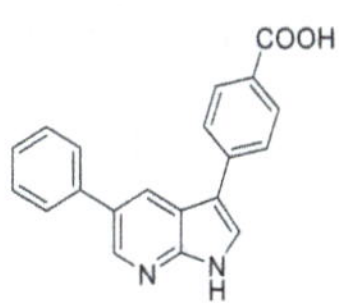

26

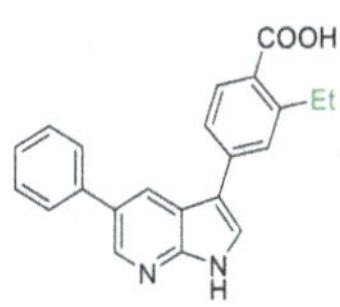

27

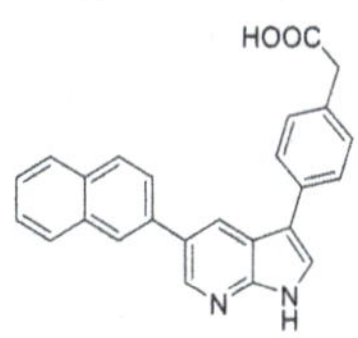

28

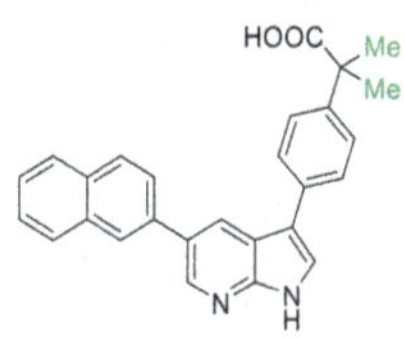

29

Compound	Rat *in vivo* Cl (mL min^{-1} kg^{-1})	Oral AUC (dose normalized, ng h^{-1} mL^{-1}/[mg kg^{-1}])
26	64.4	154 ± 28
27	9.4 ± 1.2	600 ± 80
28	21.3	225 ± 143
29	4.32 ± 0.57	2100 ± 550

In rat hepatocytes the serum/glucocorticoid regulated kinase 1 (SGK1) inhibitor **26** showed formation of a glucuronide metabolite, and the rat PK curves of **26** and **28** both showed a second peak, indicating acyl glucuronide formation and enterohepatic recirculation. For compound **26**, ethyl substitution *ortho* to the acid (**27**) led to much lower clearance and higher area under the curve (AUC). The pharmacokinetic profile of **27** still showed a second peak, albeit smaller than that of **26**, indicating that acyl glucuronide was still formed. When the phenyl acetic acid **28** was α-disubstituted (**29**), clearance was also lower, and no second peak was observed in rat pharmacokinetics.[50] This example shows that steric hindrance can be an effective approach to reduce or eliminate acyl glucuronide formation.

(*continued*)

Examples

9

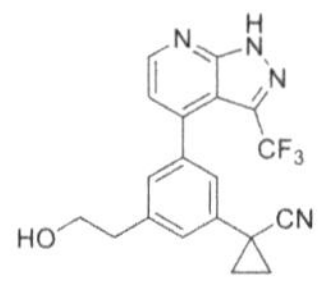

30

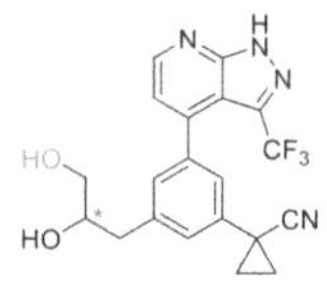

R-31/S-31

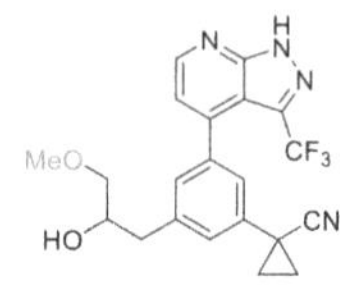

(±)32

Compound	% Remaining in rat liver S9 after 60 minutes (−NADPH, +UDPGA)
30	17
***R*-31/*S*-31**	65/67
(±) 32	51

The protein kinase c epsilon (PKCε) inhibitor **30** had low metabolic stability towards glucuronidation, as indicated by the percentage remaining after incubation with rat liver S9 in the presence of UDPGA for an hour. The hydroxy group was identified as the glucuronidation site. Polar groups such as $-CH_2OH$ (**31**) and $-CH_2OMe$ (**32**) have a beneficial effect in reducing the rate of glucuronidation, and in this case the stereochemistry did not seem to have a significant effect, as the two enantiomers of **31** are very similar in this regard. The *R*- enantiomer of **31** was chosen for further *in vivo* studies and demonstrated reduced *in vivo* clearance in rats (55 mL min^{-1} kg^{-1}) and increased AUC in dogs (4.4 h μg mL^{-1}) compared with **30** (87 mL min^{-1} kg^{-1} and 0.13 h μg mL^{-1}, respectively).[51] In the absence of a UGT co-crystal structure there could be multiple explanations for such effects, but it is conceivable that introducing such electron-withdrawing groups sterically (gauche effect) and electronically deactivates hydroxyl groups towards glucuronidation. Alternatively, such polar moieties may disrupt the molecule's interaction with certain hydrophobic amino acid residues in UGT's active site.

10

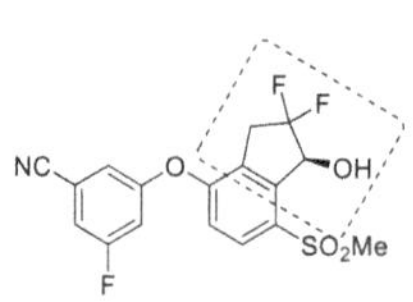

33

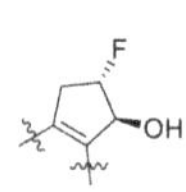

34

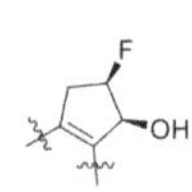

35

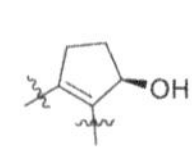

36

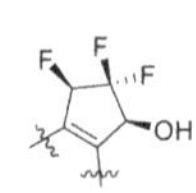

37

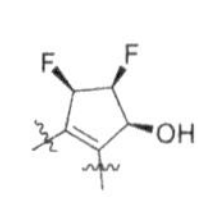

38

Compound	UGT2B17 V_{max}/K_m (μL min^{-1} mg^{-1})
33	25.5
34	3.3
35	0.93
36	0.32
37	80
38	0.47

Examples

For the hypoxia inducible factor 2 alpha (HIF-2α) inhibitor **33**, the only detectable metabolite in human hepatocytes is the glucuronide of the hydroxyl group. Removal of either fluorine led to reduced UGT2B17-mediated glucuronidation, indicating in this case that increased pK_a of the hydroxyl can retard glucuronidation. The difference between **34** and **35** can be rationalized by the intramolecular hydrogen bond between OH and F for **35** but not **34**, which gives **35** a higher pK_a and improved metabolic stability. The non-fluorinated **36** has the lowest glucuronidation rate. A similar electronic effect on glucuronidation was also observed for the matched pair of **37** and **38**.[52]

13.3.4 Decrease Lipophilicity

Examples

1

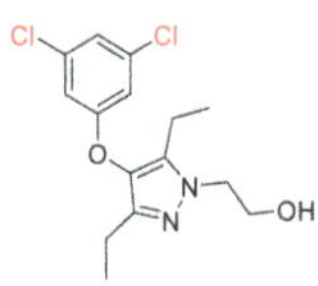

39

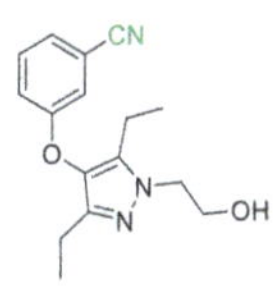

40

Compound	cLog *P*	$t_{1/2}$ (min) in HLM + UDPGA	Microsomal fraction unbound	Cl_u predicted from human hepatocytes (mL min^{-1} kg^{-1})
39	4.7	89	0.13	1000
40	2.7	27	0.80	50

The hydroxyl group in the HIV non-nucleoside reverse-transcriptase inhibitor (NNRTI) **39** was found to undergo rapid glucuronidation. As it is part of the pharmacophore for on-target potency, modification close to the hydroxyl group was not tolerated. Replacement of the 3,5-dichlorophenyl with 3-cyanophenyl led to a two-unit decrease in calculated log of the partition coefficient (cLog *P*). Although **40** appeared less stable in HLM compared with **39**, this was attributed to much lower binding to microsomal protein; conversely the predicted unbound clearance based on the data from human hepatocytes was much lower for the more polar compound **40**.[53,54]

(*continued*)

Examples

2

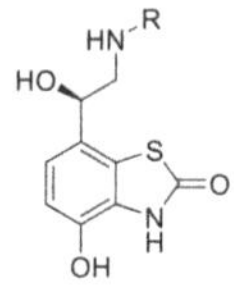

R = 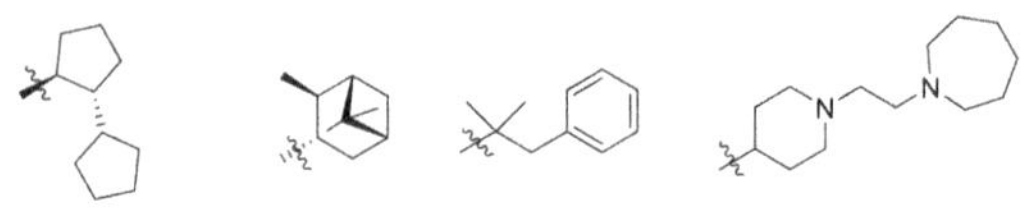

41 42 43 44

Compound	cLog *P*	RLM + UDPGA Cl_{int} (μL min^{-1} mg^{-1})
41	3.4	781
42	3.0	589
43	2.3	291
44	0.7	Resistant to metabolism

Another example supporting the correlation between lipophilicity and UGT-mediated clearance comes from a series of β2-adrenoceptor agonists. The R group distal to the phenol has a large effect on the rate of its glucuronidation, which in this study correlates well with the calculated lipophilicity of the molecules, with the least lipophilic **44** being resistant to glucuronidation.[55]

3

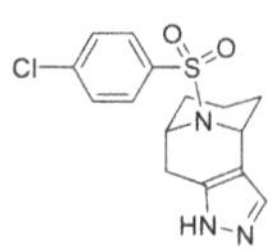

45

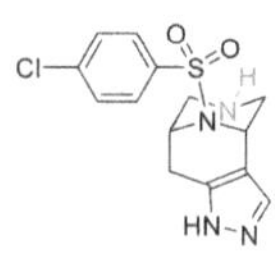

46

Compound	cLog D^a	Percentage remaining in liver microsomes + UDPGA after 30 minutes (mouse, human)
45	2.6	5, 0
46	0.9	89, 76

For the γ-secretase inhibitor **45**, the *in vitro* UGT-mediated clearance was high, and after 30 minutes of incubation in liver microsomes supplemented with UDPGA and alamethicin, little to no compound remained. Incorporation of a basic nitrogen as in **46**, which led to lower Log *D*, dramatically improved the *in vitro* metabolic stability of the compound.[56]

Examples

4

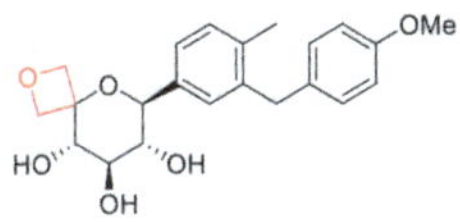

47

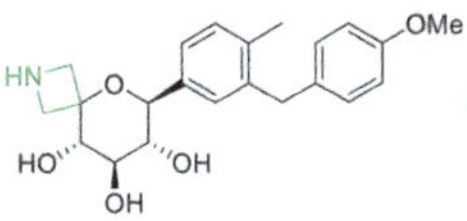

48

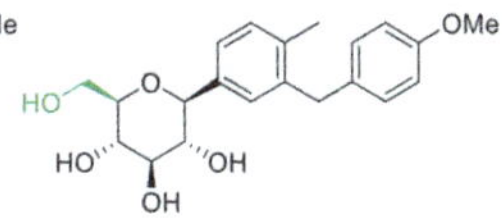

49

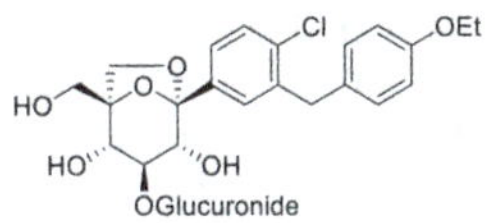

Major metabolite of ertugliflozin (PF-04971729)

Compound	HLM (no UDPGA) Cl_{int} (µL min^{-1} mg^{-1})	Human hepatocytes Cl_{int} (µL min^{-1} $million^{-1}$)	ELog *D*
47	12	44	3.0
48	32	<2.0	3.2
49	<8.0	3.4	2.4

The SGLT2 inhibitor **47** had low oxidative clearance in human liver microsomes but the clearance was significantly higher in human hepatocytes. Although the metabolite was not identified for **47**, the metabolic pathways of the later identified clinical candidate ertugliflozin that eventually received FDA approval were elucidated, and the metabolism was indeed mostly mediated by glucuronidation with three glucuronides identified. When an amine (**48**) or hydroxyl (**49**) was introduced to replace the oxetane in compound **47**, the hepatocyte clearance was much lower, leading to the hypothesis that an amine or a hydrogen bond donor close to the glucuronidation soft spot can reduce clearance.[57,58] Lower lipophilicity may also account for the reduction in glucuronidation. Although the azetidine **48** was reported to have a slightly higher ELog D than its oxetane analog **47**, its intrinsic lipophilicity (Log *P*) would usually be expected to be about 0.3 units lower.[a] The basic pK_a of the azetidine is expected to be 1–2 units above 7.4[b], leading to a high degree of protonation at physiological pH, which would result in further reduction of the effective lipophilicity of **48** by up to two units as compared with oxtane **47**.

5

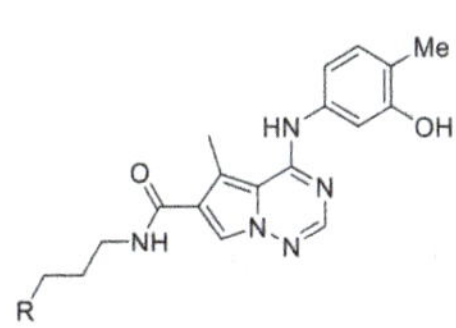

Compound	R	CLog D^b	Glucuronidation rate (nmol min^{-1} mg^{-1})
50	OMe	2.8	0.16
51	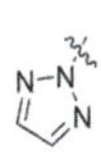	2.9	0.21

(*continued*)

Examples

52	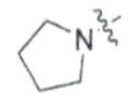	1.6	0.01
53	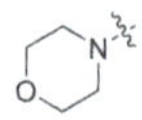	2.1	0.01

In a series of VEGF inhibitors distinct structure–activity relationships in glucuronidation rate were observed between neutral and basic groups distal to the phenol. When R was neutral (**50** and **51**), the *in vitro* UGT-mediated clearance was much higher than when R contained a basic amine (**52** and **53**), despite the fact that this group is at the other end of the molecule from the phenol.[59] This can be rationalized by the lower cLog *D* of compounds **52** and **53**.

[a]Calculated by the author using a Novartis model and not in the original publication.
[b]Calculated using a machine learning Log *D* model and not in original publication.

13.3.5 Sterically Disrupt the Substrate's Binding to UGT

Examples

1

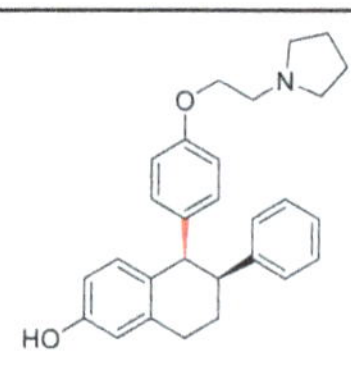

54

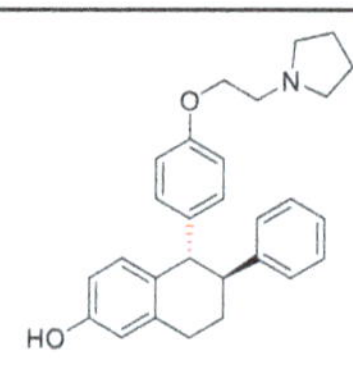

55

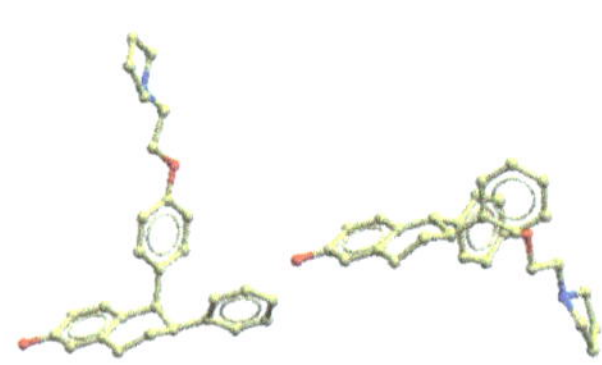

Compound	Oral bioavailability in rat
54	60%
55	16%

Examples

This example demonstrates that the shape of the molecule can have an effect on the glucuronide formation. The estrogen receptor ligand **54** had much higher oral bioavailability than **55**, which was attributed to lower glucuronidation at the gut wall in a portal vein cannulation study. The researchers proposed that the distal steric bulk was important in lowering UGT-mediated metabolism, probably due to unfavorable binding to the UGT.[60] One caveat is that both compounds were tested as racemates, which may also have an effect on the clearance.

2

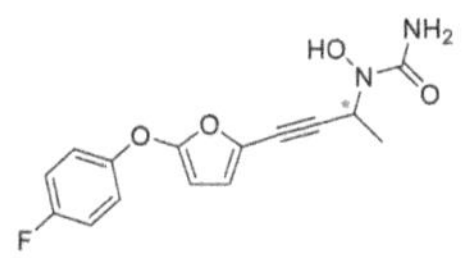

56

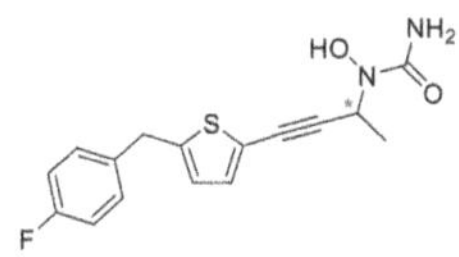

57

Compound	Glucuronidation rate (nmol min^{-1} mg^{-1})	Cynomolgus monkey *in vivo* half-life (hours)
***R*-56**	0.01	9.0
***S*-56**	0.05	1.9
***R*-57**	0.004	16.0
***S*-57**	0.02	3.7

A series of 5-lipoxygenase inhibitors nicely demonstrated the influence of stereochemistry on glucuronidation. While both enantiomers of **56** and **57** have similar potency on the target, differences in the *in vitro* glucuronidation rate were observed. *In vivo* studies in cynomolgus monkeys also revealed a large difference in the half-lives of the enantiomers, probably driven by the different *in vivo* clearance assuming similar volumes of distribution.[61]

3

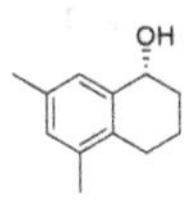

58

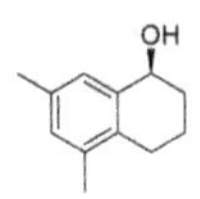

59

In the study of the glucuronidation rate of some secondary benzylic alcohols catalyzed by UGT2B7 and UGT2B17, a significant difference was found between the *R*- and *S*- enantiomers. Both UGT isoforms favor the *R*-enantiomer; *e.g.* the UGT2B17-mediated glucuronidation rate of the *R*-enantiomer **58** was 256 times that of the *S*-enantiomer **59**.[62]

13.3.6 Protection of the Soft Spot as a Prodrug

Examples

1

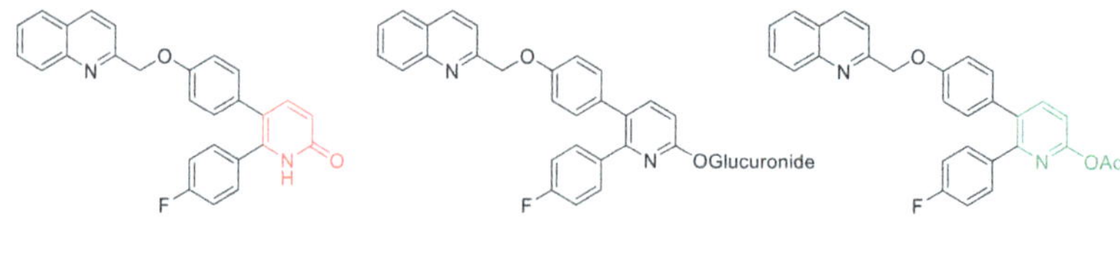

60 61 62

Compound	C_{max} (ng mL^{-1})	AUC (ng h mL^{-1})
60	44 ± 12	359 ± 272
62	372 ± 128 (compound **60**)	2436 ± 1025 (compound **60**)

A series of PDE10A inhibitors, such as **60**, suffered from low exposure when dosed orally in rats. The low exposure was determined to be due to extensive first-pass glucuronidation in the intestine and liver, as **60** dosed orally gave much higher glucuronide exposure than intravenous dosing, and the metabolite's structure (**61**) matched an authentic sample. Protection of the pyridone from intestinal and hepatic metabolism as the acetate prodrug **62** provided a large boost in both C_{max} and AUC of **60**.[19]

Acknowledgements

The author is grateful to Dr. Upendra Argikar, PK Sciences, Novartis Institutes for BioMedical Research, for proofreading this chapter and offering his suggestions. The author also thanks Dr. David J. Edmonds of Pfizer for his help.

References

1. U. A. Argikar, P. M. Potter, J. M. Hutzler and P. H. Marathe, *AAPS J.*, 2016, **18**, 1391–1405.
2. H. Komura and M. Iwaki, *Drug Metab. Rev.*, 2011, **43**, 476–498.
3. A. Rowland, J. O. Miners and P. I. Mackenzie, *Int. J. Biochem. Cell Biol.*, 2013, **45**, 1121–1132.
4. Y. Maruo, M. Iwai, A. Mori, H. Sato and Y. Takeuchi, *Curr. Drug Metab.*, 2005, **6**, 91–99.
5. A. Radominska-Pandya, S. M. Bratton, M. R. Redinbo and M. J. Miley, *Drug Metab. Rev.*, 2010, **42**, 133–144.
6. M. J. Miley, A. K. Zielinska, J. E. Keenan, S. M. Bratton, A. Radominska-Pandya and M. R. Redinbo, *J. Mol. Biol.*, 2007, **369**, 498–511.
7. J. H. Lin and B. K. Wong, *Curr. Drug Metab.*, 2002, **3**, 623–646.

8. U. Breyer-Pfaff, B. Becher, E. Nusser, K. Nill, B. Baier-Weber, D. Zaunbrecher, H. Wachsmuth and A. Prox, *Xenobiotica*, 1990, **20**, 727–738.
9. W. J. Moree, B. F. Li, S. Zamani-Kord, J. Yu, T. Coon, C. Huang, D. Marinkovic, F. C. Tucci, S. Malany, M. J. Bradbury, L. M. Hernandez, J. Wen, H. Wang, S. R. Hoare, R. E. Petroski, K. Jalali, C. Yang, A. Sacaan, A. Madan, P. D. Crowe and G. Beaton, *Bioorg. Med. Chem. Lett.*, 2010, **20**, 5874–5878.
10. D. Zhang, W. Zhao, V. A. Roongta, J. G. Mitroka, L. J. Klunk and M. Zhu, *Drug Metab. Dispos.*, 2004, **32**, 545–551.
11. T. Nishiyama, T. Kobori, K. Arai, K. Ogura, T. Ohnuma, K. Ishii, K. Hayashi and A. Hiratsuka, *Arch. Biochem. Biophys.*, 2006, **454**, 72–79.
12. S. Klieber, S. Hugla, R. Ngo, C. Arabeyre-Fabre, V. Meunier, F. Sadoun, O. Fedeli, M. Rival, M. Bourrie, F. Guillou, P. Maurel and G. Fabre, *Drug Metab. Dispos.*, 2008, **36**, 851–862.
13. E. A. Peterson, A. A. Boezio, P. S. Andrews, C. M. Boezio, T. L. Bush, A. C. Cheng, D. Choquette, J. R. Coats, A. E. Colletti, K. W. Copeland, M. DuPont, R. Graceffa, B. Grubinska, J. L. Kim, R. T. Lewis, J. Liu, E. L. Mullady, M. H. Potashman, K. Romero, P. L. Shaffer, M. K. Stanton, J. C. Stellwagen, Y. Teffera, S. Yi, T. Cai and D. S. La, *Bioorg. Med. Chem. Lett.*, 2012, **22**, 4967–4974.
14. K. Bourcier, R. Hyland, S. Kempshall, R. Jones, J. Maximilien, N. Irvine and B. Jones, *Drug Metab. Dispos.*, 2010, **38**, 923–929.
15. S. E. Huskey, J. Magdalou, M. Ouzzine, G. Siest and S. H. Chiu, *Drug Metab. Dispos.*, 1994, **22**, 659–662.
16. A. Kodimuthali, S. S. Jabaris and M. Pal, *J. Med. Chem.*, 2008, **51**, 5471–5489.
17. M. Chang, V. K. Sood, D. A. Kloosterman, M. J. Hauer, P. E. Fagerness, P. E. Sanders and J. J. Vrbanac, *Drug Metab. Dispos.*, 1997, **25**, 814–827.
18. K. Rose, Y. S. Yang, R. Sciotti and H. Cai, *Drug Metab. Lett.*, 2009, **3**, 28–34.
19. V. S. Lingam, D. H. Dahale, V. E. Rathi, Y. B. Shingote, R. R. Thakur, A. S. Mindhe, S. Kummari, N. Khairatkar-Joshi, M. Bajpai, D. M. Shah, R. S. Sapalya, S. Gullapalli, P. K. Gupta, G. S. Gudi, S. B. Jadhav, R. Pattem and A. Thomas, *J. Med. Chem.*, 2015, **58**, 8292–8308.
20. M. W. Sinz and R. P. Remmel, *Drug Metab. Dispos.*, 1991, **19**, 149–153.
21. D. A. Smith and D. Dalvie, *Xenobiotica*, 2012, **42**, 107–126.
22. R. Meech, J. O. Miners, B. C. Lewis and P. I. Mackenzie, *Pharmacol. Ther.*, 2012, **134**, 200–218.
23. L. C. Wienkers and T. G. Heath, *Nat. Rev. Drug Discovery*, 2005, **4**, 825–833.
24. X. Meng, J. L. Maggs, D. C. Pryde, S. Planken, R. E. Jenkins, T. M. Peakman, K. Beaumont, C. Kohl, B. K. Park and A. V. Stachulski, *J. Med. Chem.*, 2007, **50**, 6165–6176.
25. S. Fowler, H. Kletzl, M. Finel, N. Manevski, P. Schmid, D. Tuerck, R. D. Norcross, M. C. Hoener, O. Spleiss and V. A. Iglesias, *J. Pharmacol. Exp. Ther.*, 2015, **352**, 358–367.
26. A. Di Paolo, G. Bocci, M. Polillo, M. Del Re, T. Di Desidero, M. Lastella and R. Danesi, *Curr. Drug Metab.*, 2011, **12**, 932–943.
27. T. Sousa, R. Paterson, V. Moore, A. Carlsson, B. Abrahamsson and A. W. Basit, *Int. J. Pharm.*, 2008, **363**, 1–25.

28. D. Paul, K. M. Standifer, C. E. Inturrisi and G. W. Pasternak, *J. Pharmacol. Exp. Ther.*, 1989, **251**, 477–483.
29. D. Wu, Y. S. Kang, U. Bickel and W. M. Pardridge, *Drug Metab. Dispos.*, 1997, **25**, 768–771.
30. T. F. Ryder, M. F. Calabrese, G. S. Walker, K. O. Cameron, A. R. Reyes, K. A. Borzilleri, J. Delmore, R. Miller, R. G. Kurumbail, J. Ward, D. W. Kung, J. A. Brown, D. J. Edmonds, H. Eng, A. C. Wolford and A. S. Kalgutkar, *J. Med. Chem.*, 2018, **61**, 7273–7288.
31. T. Cai, L. Yao and R. J. Turesky, *Chem. Res. Toxicol.*, 2016, **29**, 879–891.
32. N. A. Meanwell, *J. Med. Chem.*, 2011, **54**, 2529–2591.
33. J. Magdalou, Y. Hochman and D. Zakim, *J. Biol. Chem.*, 1982, **257**, 13624–13629.
34. H. Yin, G. Bennett and J. P. Jones, *Chem.-Biol. Interact.*, 1994, **90**, 47–58.
35. L. Antonio, J. P. Grillasca, J. Taskinen, E. Elovaara, B. Burchell, M. H. Piet, B. Ethell, M. Ouzzine, S. Fournel-Gigleux and J. Magdalou, *Drug Metab. Dispos.*, 2002, **30**, 199–207.
36. T. J. Franklin, V. N. Jacobs, G. Jones and P. Ple, *Drug Metab. Dispos.*, 1997, **25**, 367–370.
37. T. Fukuda, R. Goto, T. Kiho, K. Ueda, S. Muramatsu, M. Hashimoto, A. Aki, K. Watanabe and N. Tanaka, *Bioorg. Med. Chem. Lett.*, 2017, **27**, 3716–3722.
38. A. J. Folkes, K. Ahmadi, W. K. Alderton, S. Alix, S. J. Baker, G. Box, I. S. Chuckowree, P. A. Clarke, P. Depledge, S. A. Eccles, L. S. Friedman, A. Hayes, T. C. Hancox, A. Kugendradas, L. Lensun, P. Moore, A. G. Olivero, J. Pang, S. Patel, G. H. Pergl-Wilson, F. I. Raynaud, A. Robson, N. Saghir, L. Salphati, S. Sohal, M. H. Ultsch, M. Valenti, H. J. Wallweber, N. C. Wan, C. Wiesmann, P. Workman, A. Zhyvoloup, M. J. Zvelebil and S. J. Shuttleworth, *J. Med. Chem.*, 2008, **51**, 5522–5532.
39. P. K. Chakravarty, E. M. Naylor, A. Chen, R. S. Chang, T. B. Chen, K. A. Faust, V. J. Lotti, S. D. Kivlighn, R. A. Gable and G. J. Zingaro, *et al.*, *J. Med. Chem.*, 1994, **37**, 4068–4072.
40. A. E. Colletti and P. A. Krieter, *Drug Metab. Dispos.*, 1994, **22**, 183–188.
41. X. Wang, A. Kolesnikov, S. Tay, G. Chan, Q. Chao, S. Do, J. Drummond, A. J. Ebens, N. Liu, J. Ly, E. Harstad, H. Hu, J. Moffat, V. Munugalavadla, J. Murray, D. Slaga, V. Tsui, M. Volgraf, H. Wallweber and J. H. Chang, *J. Med. Chem.*, 2017, **60**, 4458–4473.
42. M. T. Bilodeau, A. E. Balitza, T. J. Koester, P. J. Manley, L. D. Rodman, C. Buser-Doepner, K. E. Coll, C. Fernandes, J. B. Gibbs, D. C. Heimbrook, W. R. Huckle, N. Kohl, J. J. Lynch, X. Mao, R. C. McFall, D. McLoughlin, C. M. Miller-Stein, K. W. Rickert, L. Sepp-Lorenzino, J. M. Shipman, R. Subramanian, K. A. Thomas, B. K. Wong, S. Yu and G. D. Hartman, *J. Med. Chem.*, 2004, **47**, 6363–6372.
43. K. Futatsugi, D. W. Kung, S. T. Orr, S. Cabral, D. Hepworth, G. Aspnes, S. Bader, J. Bian, M. Boehm, P. A. Carpino, S. B. Coffey, M. S. Dowling, M. Herr, W. Jiao, S. Y. Lavergne, Q. Li, R. W. Clark, D. M. Erion, K. Kou, K. Lee, B. A. Pabst, S. M. Perez, J. Purkal, C. C. Jorgensen, T. C. Goosen, J. R. Gosset, M. Niosi, J. C. Pettersen, J. A. Pfefferkorn, K. Ahn and B. Goodwin, *J. Med. Chem.*, 2015, **58**, 7173–7185.

44. K. Futatsugi, K. Huard, D. W. Kung, J. C. Pettersen, D. A. Flynn, J. R. Gosset, G. E. Aspnes, R. J. Barnes, S. Cabral, M. S. Dowling, D. P. Fernando, T. C. Goosen, W. P. Gorczyca, D. Hepworth, M. Herr, S. Lavergne, Q. Li, M. Niosi, S. T. M. Orr, I. D. Pardo, S. M. Perez, J. Purkal, T. J. Schmahai, N. Shirai, A. M. Shoieb, J. Zhou and B. Goodwin, *Medchemcomm*, 2017, **8**, 771–779.
45. I. G. Robertson, B. D. Palmer, M. Officer, D. J. Siegers, J. W. Paxton and G. J. Shaw, *Biochem. Pharmacol.*, 1991, **42**, 1879–1884.
46. P. Jones, R. I. Storer, Y. A. Sabnis, F. M. Wakenhut, G. A. Whitlock, K. S. England, T. Mukaiyama, C. M. Dehnhardt, J. W. Coe, S. W. Kortum, J. E. Chrencik, D. G. Brown, R. M. Jones, J. R. Murphy, T. Yeoh, P. Morgan and I. Kilty, *J. Med. Chem.*, 2017, **60**, 767–786.
47. T. D. Crawford, F. A. Romero, K. W. Lai, V. Tsui, A. M. Taylor, G. de Leon Boenig, C. L. Noland, J. Murray, J. Ly, E. F. Choo, T. L. Hunsaker, E. W. Chan, M. Merchant, S. Kharbanda, K. E. Gascoigne, S. Kaufman, M. H. Beresini, J. Liao, W. Liu, K. X. Chen, Z. Chen, A. R. Conery, A. Cote, H. Jayaram, Y. Jiang, J. R. Kiefer, T. Kleinheinz, Y. Li, J. Maher, E. Pardo, F. Poy, K. L. Spillane, F. Wang, J. Wang, X. Wei, Z. Xu, Z. Xu, I. Yen, L. Zawadzke, X. Zhu, S. Bellon, R. Cummings, A. G. Cochran, B. K. Albrecht and S. Magnuson, *J. Med. Chem.*, 2016, **59**, 10549–10563.
48. A. Ling, M. Plewe, J. Gonzalez, P. Madsen, C. K. Sams, J. Lau, V. Gregor, D. Murphy, K. Teston, A. Kuki, S. Shi, L. Truesdale, D. Kiel, J. May, J. Lakis, K. Anderes, E. Iatsimirskaia, U. G. Sidelmann, L. B. Knudsen, C. L. Brand and A. Polinsky, *Bioorg. Med. Chem. Lett.*, 2002, **12**, 663–666.
49. Q. Jin, H. Nie, B. W. McCleland, K. L. Widdowson, M. R. Palovich, J. D. Elliott, R. M. Goodman, M. Burman, H. M. Sarau, K. W. Ward, M. Nord, B. M. Orr, P. D. Gorycki and J. Busch-Petersen, *Bioorg. Med. Chem. Lett.*, 2004, **14**, 4375–4378.
50. M. Hammond, D. G. Washburn, H. T. Hoang, S. Manns, J. S. Frazee, H. Nakamura, J. R. Patterson, W. Trizna, C. Wu, L. M. Azzarano, R. Nagilla, M. Nord, R. Trejo, M. S. Head, B. Zhao, A. M. Smallwood, K. Hightower, N. J. Laping, C. G. Schnackenberg and S. K. Thompson, *Bioorg. Med. Chem. Lett.*, 2009, **19**, 4441–4445.
51. J. J. Clemens, T. Coon, B. B. Busch, J. L. Asgian, S. Hudson, A. Termin, T. B. Flores, D. Tran, P. Chiang, S. Sperry, R. Gross, J. Abt, R. Heim, S. Lechner, H. Twin, J. Studley, G. Brenchley, P. N. Collier, F. Pierard, A. Miller, C. Mak, V. Dvornikovs, J. M. Jimenez and D. Stamos, *Bioorg. Med. Chem. Lett.*, 2014, **24**, 3398–3402.
52. R. Xu, K. Wang, J. P. Rizzi, H. Huang, J. A. Grina, S. T. Schlachter, B. Wang, P. M. Wehn, H. Yang, D. D. Dixon, R. M. Czerwinski, X. Du, E. L. Ged, G. Han, H. Tan, T. Wong, S. Xie, J. A. Josey and E. M. Wallace, *J. Med. Chem.*, 2019, **62**, 6876–6893.
53. C. E. Mowbray, R. Corbau, M. Hawes, L. H. Jones, J. E. Mills, M. Perros, M. D. Selby, P. A. Stupple, R. Webster and A. Wood, *Bioorg. Med. Chem. Lett.*, 2009, **19**, 5603–5606.

54. G. Allan, J. Davis, M. Dickins, I. Gardner, T. Jenkins, H. Jones, R. Webster and H. Westgate, *Xenobiotica*, 2008, **38**, 620–640.
55. D. Beattie, M. Bradley, A. Brearley, S. J. Charlton, B. M. Cuenoud, R. A. Fairhurst, P. Gedeck, M. Gosling, D. Janus, D. Jones, C. Lewis, C. McCarthy, H. Oakman, R. Stringer, R. J. Taylor and A. Tuffnell, *Bioorg. Med. Chem. Lett.*, 2010, **20**, 5302–5307.
56. S. Bowers, G. D. Probst, A. P. Truong, R. K. Hom, A. W. Konradi, H. L. Sham, A. W. Garofalo, K. Wong, E. Goldbach, K. P. Quinn, J. M. Sauer, W. Wallace, L. Nguyen, S. S. Hemphill, M. P. Bova and G. S. Basi, *Bioorg. Med. Chem. Lett.*, 2009, **19**, 6952–6956.
57. V. Mascitti, T. S. Maurer, R. P. Robinson, J. Bian, C. M. Boustany-Kari, T. Brandt, B. M. Collman, A. S. Kalgutkar, M. K. Klenotic, M. T. Leininger, A. Lowe, R. J. Maguire, V. M. Masterson, Z. Miao, E. Mukaiyama, J. D. Patel, J. C. Pettersen, C. Preville, B. Samas, L. She, Z. Sobol, C. M. Steppan, B. D. Stevens, B. A. Thuma, M. Tugnait, D. Zeng and T. Zhu, *J. Med. Chem.*, 2011, **54**, 2952–2960.
58. Z. Miao, G. Nucci, N. Amin, R. Sharma, V. Mascitti, M. Tugnait, A. D. Vaz, E. Callegari and A. S. Kalgutkar, *Drug Metab. Dispos.*, 2013, **41**, 445–456.
59. R. M. Borzilleri, Z. W. Cai, C. Ellis, J. Fargnoli, A. Fura, T. Gerhardt, B. Goyal, J. T. Hunt, S. Mortillo, L. Qian, J. Tokarski, V. Vyas, B. Wautlet, X. Zheng and R. S. Bhide, *Bioorg. Med. Chem. Lett.*, 2005, **15**, 1429–1433.
60. R. L. Rosati, P. Da Silva Jardine, K. O. Cameron, D. D. Thompson, H. Z. Ke, S. M. Toler, T. A. Brown, L. C. Pan, C. F. Ebbinghaus, A. R. Reinhold, N. C. Elliott, B. N. Newhouse, C. M. Tjoa, P. M. Sweetnam, M. J. Cole, M. W. Arriola, J. W. Gauthier, D. T. Crawford, D. F. Nickerson, C. M. Pirie, H. Qi, H. A. Simmons and G. T. Tkalcevic, *J. Med. Chem.*, 1998, **41**, 2928–2931.
61. J. J. Bouska, R. L. Bell, C. L. Goodfellow, A. O. Stewart, C. D. Brooks and G. W. Carter, *Drug Metab. Dispos.*, 1997, **25**, 1032–1038.
62. I. Bichlmaier, A. Siiskonen, M. Finel and J. Yli-Kauhaluoma, *J. Med. Chem.*, 2006, **49**, 1818–1827.

CHAPTER 14

Sulfation

YUE PAN*

Global Discovery Chemistry, Novartis Institutes for BioMedical Research, 22 Windsor St., Cambridge, Massachusetts 02139, USA
*E-mail: ypan@relaytx.com

14.1 Introduction

14.1.1 Enzyme Family

Sulfotransferases (SULTs) are a family of enzymes that catalyze the transfer of a sulfo group ($^{-}O\text{-}SO_{2^-}$) from its cofactor 3′-phosphoadenosine-5′-phosphosulfate (PAPS, Figure 14.1) to an alcohol to form a sulfate or to an amine to give a sulfamate. After the reaction the cofactor turns into 3′-phosphoadenosine-5′-phosphate (PAP).

14.1.2 Expression

SULT's expression level is the highest in the liver but it is found in most extrahepatic organs too, such as intestine, lung, kidney and brain.[1] To date 15 isoforms of human SULTs from four families (SULT 1, 2, 4 and 6) have been reported.[2]

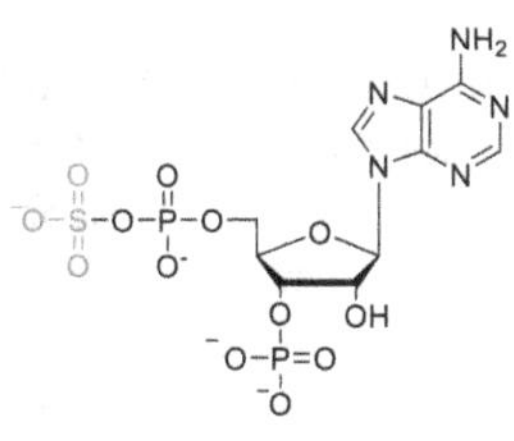

Figure 14.1 Structure of 3′-hosphoadenosine 5′-phosphosulfate (PAPS).

Drug Discovery Series No. 79
The Medicinal Chemist's Guide to Solving ADMET Challenges
Edited by Patrick Schnider
© The Royal Society of Chemistry 2021
Published by the Royal Society of Chemistry, www.rsc.org

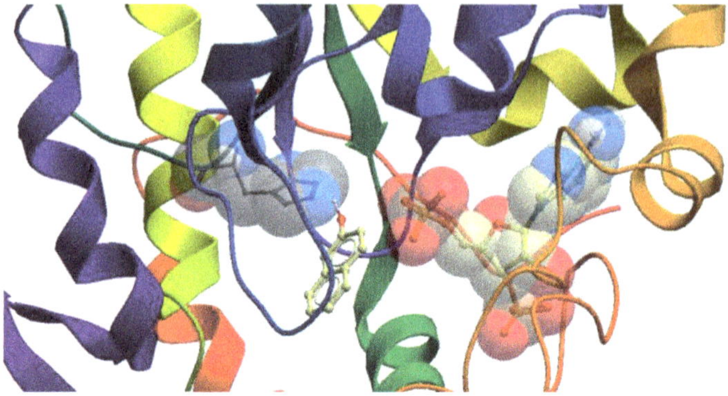

Figure 14.2 Close-up view of the substrate and PAP binding sites of human SULT1A1 complexed with PAP and 2-naphthol. The catalytic histidine and PAP are shown in a space-filling model (PDB code 3U3K).

14.1.3 Structure

SULT has a substrate binding site and a cofactor binding site. The PAPS binding site is highly conserved among all the isoforms.[3] In a cocrystal structure of 2-naphthol bound to SULT1A1 (Figure 14.2), the hydroxyl group is placed near the deprotonating histidine (*vide infra*) and PAP. Interestingly, at least for SULT1A1, one of the major SULT isoforms, there is a second substrate binding site that inhibits the enzyme's activity, which indicates that there may be auto-inhibition at high substrate concentrations (Figure 14.3).[4] There are a number of SULT crystal structures available (46 homo sapiens structures based on the latest search of the Protein Data Bank), which usually consist of the enzyme and the catalytically inactive cofactor PAP, with or without a substrate.

14.1.4 Activity

SULT accounts for a third of human phase II metabolism.[5] While the Michaelis–Menten constant (K_m) depends on both the SULT isoform and its substrate, it is typically lower compared with 5′-diphospho-glucuronosyltransferase (UGT)-catalyzed glucuronidation. The conjugation catalyzed by SULT is generally characterized as high-affinity and low-capacity, as the reaction rate is often limited by the concentration of PAPS and the speed at which it can be replenished from inorganic sulfate.[6] At low concentrations a compound's phase II metabolism can go through the sulfation pathway while at higher concentrations the glucuronidation pathway can take over. As a trade-off between affinity and capacity is difficult to avoid, the two seemingly redundant metabolic pathways provide better protection from various toxins at low and high levels.[1] Within preclinical model species PAPS concentration in

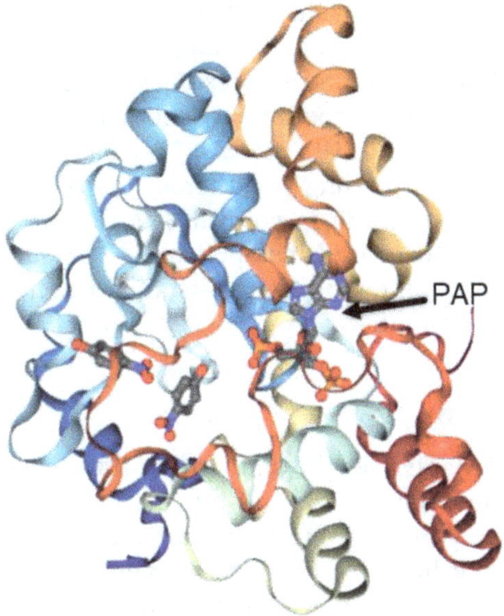

Figure 14.3 Crystal structure of human SULT1A1 complexed with PAP and two 4-nitrophenol molecules (PDB code 1LS6).

liver is in the order of rat > mouse > dog, and differences in sulfotransferase activity between male and female animals have also been reported.[7,8] Similar to glucuronidation and glucuronide hydrolysis, mediated respectively by UGT and β-glucuronidase, the amount of circulating sulfate or sulfamate is determined by both SULT and sulfatase, with the latter catalyzing the hydrolysis of sulfates or sulfamates back to the parent molecule.[9]

14.1.5 Function and Substrates

SULT catalyzes the conjugation of alcohol, phenol, amine (primary and secondary), hydroxylamine and *N*-oxide to form either a sulfate or a sulfamate, both of which are highly polar and can be excreted more easily. While conjugation with tertiary amines and heterocycles is conceivable, no such examples have been found in the literature. Similar to glucuronidation, sulfation can couple to phase I metabolism to more efficiently eliminate drugs from the body. There are two classes of SULTs: Golgi membrane-bound and cytosolic.[1] The membrane-bound SULTs are typically involved in the metabolism of endogenous peptides, lipids and glycosaminoglycans while the cytosolic SULTs tend to metabolize smaller molecules, such as drugs and biogenic amines. Similar to UGTs, there is a substantial overlap of substrate scope between the SULT isoforms. The enzyme can also adopt more than one conformation with variable sizes of the substrate binding site, which provides a broader substrate scope.[3] For major SULT isoforms such as 1A1 and 2A1, the access channel to the substrate-binding site and the binding site itself consist mostly of lipophilic amino acid residues (Val, Phe, Met and Tyr), indicating that more lipophilic compounds are better substrates of these SULTs.[4] Compared with SULT1A1 and SULT2A1, SULT1A3 has more acidic amino acid residues in the substrate binding site, thereby being able to accommodate amine-containing substrates. It is worth noting that the resulting sulfates/sulfamates from SULT-catalyzed metabolism can be substrates of organic anion transporter (OAT) and multidrug resistance-associated protein (MRP), which affects their distribution.[10,11]

14.1.6 Mechanism

The postulated mechanism of the sulfo group transfer is shown in Figure 14.4.[12] Upon substrate binding to SULT a histidine acts as the general base to deprotonate the hydroxyl group to be sulfated. For ammonium groups, deprotonation by the histidine can also happen. The enzyme also places the substrate moiety to be sulfated close to the reactive PAPS sulfur atom, thereby increasing the substrate's effective concentration and expediting the reaction.

The first step of PAPS synthesis is between inorganic sulfate and ATP, catalyzed by ATP sulfurylase (Figure 14.5). ATP sulfurylase produces adenosine 5′-phosphosulfate (APS), which is then phosphorylated by APS kinase to produce PAPS.

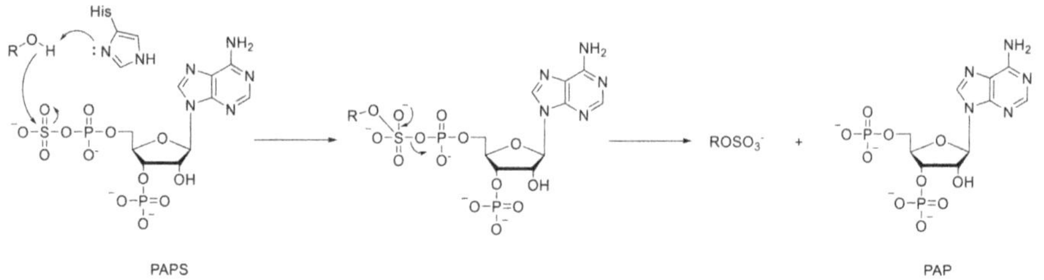

Figure 14.4 Mechanism of alcohol sulfation catalyzed by SULT.

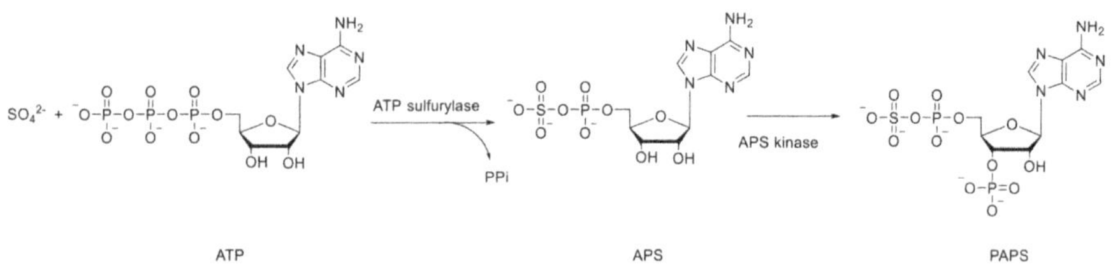

Figure 14.5 Synthesis of the cofactor PAPS.

14.1.7 Screening Strategies

Incubation of putative SULT substrates with liver cytosol and PAPS can produce the sulfation metabolite. A compound's half-life in this assay can be used to calculate its intrinsic clearance driven by SULT-mediated metabolism. As SULT and UGT can metabolize the same substrates (alcohol, phenol and primary or secondary amine), for such compounds it may be more efficient to use liver S9 fortified with PAPS and uridine-diphosphate-glucuronic acid (UDPGA) to capture both sulfation and glucuronidation. Alternatively, hepatocytes can be used, which captures even more metabolic pathways and may better predict *in vivo* clearance. SULT inhibitors can also be used to confirm SULT involvement and diagnose the isoform contributing to the metabolism, such as quercetin (SULT1A1/1E1), estrone (SULT1A3/1E1) and 2,6-dichloro-4-nitrophenol (SULT1A1).[13]

14.1.8 Relevance

Sulfation is an important phase II metabolic pathway and shares a lot of similarity with glucuronidation. As SULTs are expressed in the liver and intestine, it can limit a drug's exposure. Perhaps more importantly, as the sulfate moiety is a leaving group, a reactive species can be formed from the sulfate of benzylic or allylic alcohol and *N*-hydroxylamine, which may lead to covalent modification of biological targets and result in toxicity (see Chapter 15 on Reactive Metabolites and Chapter 16 on Genotoxicity).[14] For example, it is known that the benzylic alcohol of polycyclic aromatic hydrocarbons (PAH) can undergo metabolic activation *via* sulfation and covalently modify DNA (Figure 14.6). Pronethalol, a beta blocker with a benzylic alcohol, forms a highly reactive and mutagenic aziridine due to intramolecular displacement of the sulfate (Figure 14.7).[15] Sometimes the site of sulfation is introduced by phase I metabolism and is not present in the parent molecule,

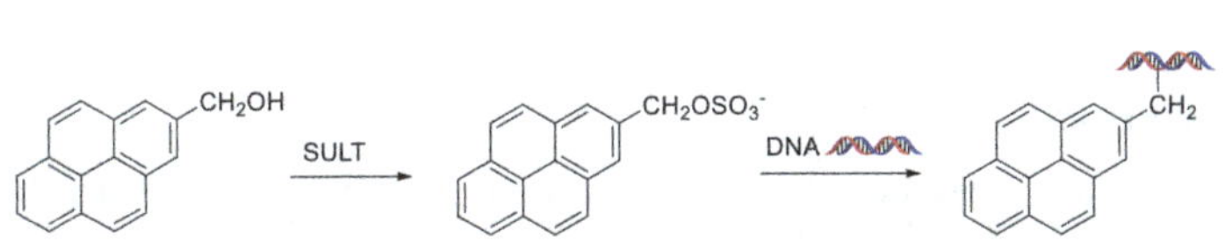

Figure 14.6 Toxicity associated with benzylic alcohol of polyaromatic hydrocarbons (PAH).

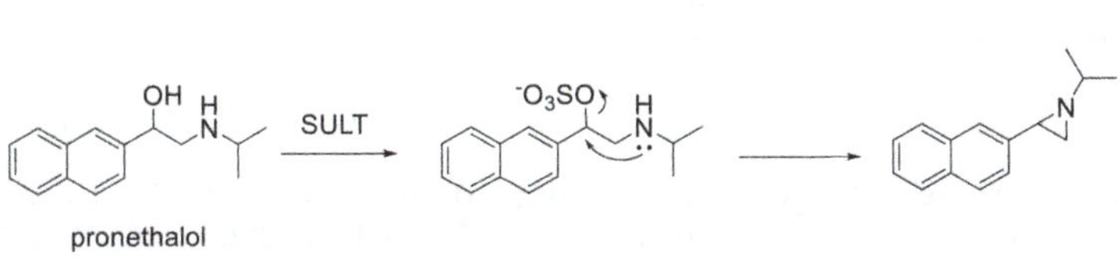

Figure 14.7 Reactive metabolite associated with pronethalol.

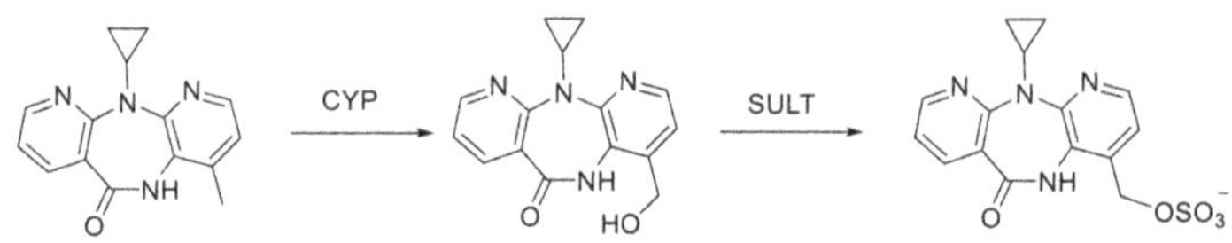

Figure 14.8 Sulfation following phase I metabolism of nevirapine generates a reactive metabolite.

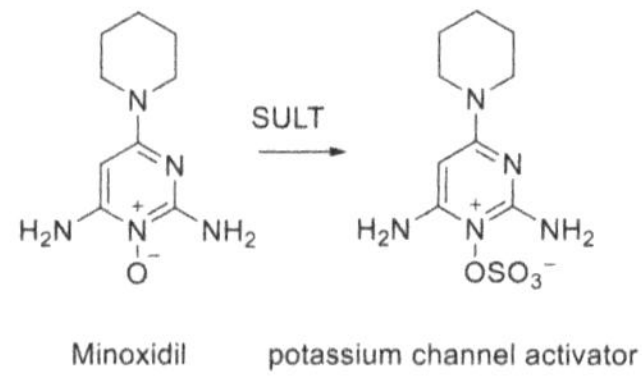

Figure 14.9 Activation of minoxidil by sulfation.

which deserves the medicinal chemist's attention. For example, the HIV non-nucleoside reverse-transcriptase inhibitor (NNRTI) nevirapine (Figure 14.8) first undergoes CYP-mediated oxidation to generate the benzylic alcohol, followed by sulfation, which then forms covalent protein adducts.[16]

Besides reactivity of the resulting sulfate metabolite, the sulfate itself may also have pharmacological activity, despite the fact that sulfation is typically a detoxification pathway and the resulting sulfate is much more polar than the parent molecule. This may warrant further profiling of the sulfate metabolite. For example, minoxidil, a drug for the treatment of hair loss, is a prodrug that requires activation by SULT to the active sulfate (Figure 14.9), which is reported to be a potassium channel activator. Sulfates of neurosteroids, such as pregnenolone, also have activities on gamma-aminobutyric acid (GABA) and *N*-methyl-D-aspartate (NMDA) receptors.[17,18]

Key References

- E. Chapman, M. D. Best, S. R. Hanson and C. H. Wong, *Angew. Chem. Int. Ed.*, 2004, **43**, 3526–3548.
- N. Gamage, A. Barnett, N. Hempel, R. G. Duggleby, K. F. Windmill, J. L. Martin and M. E. McManus, *Toxicol. Sci.*, 2006, **90**, 5–22.
- M. Suiko, K. Kurogi, T. Hashiguchi, Y. Sakakibara and M. C. Liu, *Biosci. Biotechnol. Biochem.*, 2017, **81**, 63–72.
- D. Dong, R. Ako and B. Wu, *Expert. Opin. Drug Metab. Toxicol*, 2012, **8**, 635–646
- N. U. Gamage, R. G. Duggleby, A. C. Barnett, M. Tresillian, C. F. Latham, N. E. Liyou, M. E. McManus and J. L. Martin, *J. Biol. Chem.*, 2003, **278**, 7655–7662.
- T. Wang, I. Cook and T. S. Leyh, *Drug Metab. Dispos.*, 2016, **44**, 481–484.

14.2 Mitigation Strategies

Since there is a large overlap between the substrate scope of UGTs and SULTs the strategies to improve metabolic stability are often similar.

14.2.1 Remove or Block the Sulfation Site

Such compounds are often made to assess their effects on the target potency. If the sulfation soft spot is not required for potency a solution can often be quickly identified.

14.2.2 Use Bioisosteres to Replace the Susceptible Moiety

Bioisosteres are also effective tools to maintain target potency while abrogating SULT-mediated metabolism.

14.2.3 Sterically or Electronically Decrease the Sulfation Rate

As sulfation is a reaction between a nucleophile and a mixed anhydride, sterically or electronically reducing the nucleophile's reactivity could improve its stability towards sulfation.

14.2.4 Lower Lipophilicity

It is known that the substrate binding site of major SULT isoforms (SULT1A1/1A2/1A3) contains multiple lipophilic amino acid residues. To reduce binding to such SULT isoforms, lowering lipophilicity of the substrate can be an effective approach.

14.2.5 Increase the Size of the Molecule to Disrupt Binding to SULT

While in general the medicinal chemist's aim is to keep molecules small, limited precedents indicate that a modest increase in size, while still within the drug-like space, may be sufficient to effect a clash with SULT and thus significantly decrease its catalytic efficiency. Moreover, with the crystal structures of SULT available, it is possible to rationally remove such a metabolic liability using structure-based design.

14.3 Examples of Mitigation Strategies

14.3.1 Remove or Block the Sulfation Site

Example

1

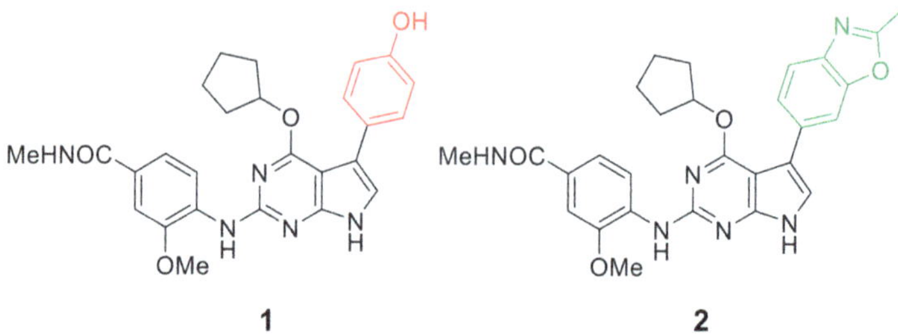

Compound	Cellular IC_{50} (nM)	Percentage remaining in rat S9 after 60 minutes
1	13	1 ± 1
2	60	98 ± 6

In a series of dual tyrosine–threonine kinase/cell division cycle-like kinase 2 (TTK/CLK2) inhibitors the phenol **1** suffered from poor metabolic stability. The major metabolites were the glucuronide and the sulfate of the phenol. The 2-methyl benzoxazole **2** was found to be a good substitute. Although it did not have a hydrogen bond donor, the loss in on-target potency was small, and the metabolic stability was dramatically improved.[19]

14.3.2 Use Bioisosteres to Replace the Susceptible Moiety

Example

1

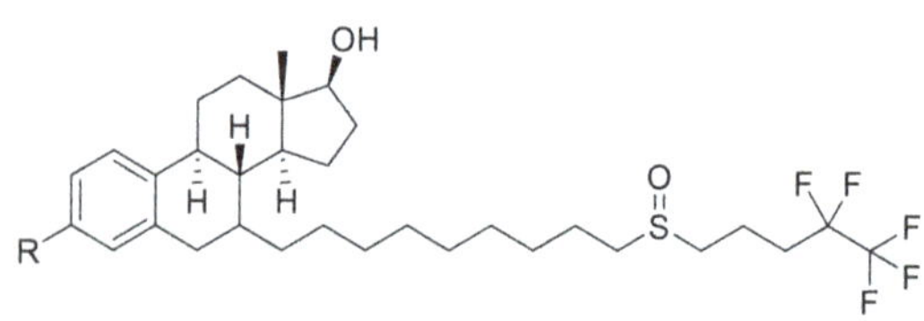

Compound	R	MCF-7 cellular IC_{50} (nM)
3	OH	1.5
4	$B(OH)_2$	3.2

Fulvestrant (**3**), a selective estrogen receptor degrader (SERD), had low oral bioavailability due to rapid glucuronidation and sulfation of the phenol. A boronic acid isostere (**4**) was identified, which maintained the antiproliferative potency and showed much improved oral exposure in a mouse pharmacokinetic (PK) study.[20]

14.3.3 Sterically or Electronically Decrease the Sulfation Rate

Example

1

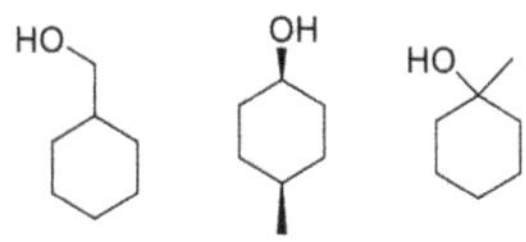

5 6 7

Compound	SULT specificity constant (k_{cat}/K_{m}) (mM^{-1} min^{-1})
5	~1.7
6	~0.4
7	~0.05

The rate of sulfation in a series of alcohols containing seven carbon atoms was investigated. The primary alcohol **5** had a catalytic efficiency about four times that of the secondary alcohol **6**, which in turn was about eight times that of the tertiary alcohol **7**.[21] Due to the small differences in other physico-chemical properties, it is reasonable to conclude that the degree of steric hindrance is the main factor accounting for the differences in sulfation rate between such compounds. The same trend was also observed for an analogous series of non-cyclic isomeric alcohols.

14.3.4 Lower Lipophilicity

Example

1

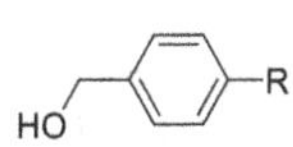

Compound	R
8	H
9	Me
10	Et
11	*n*-Pr
12	*n*-Bu
13	*n*-pentyl

In a series of 4-substituted benzylic alcohols, it was observed that the catalytic efficiency of SULT increases with the size of the substitution and peaks at the *n*-pentyl analogue **13**. For **8–13** k_{cat}/K_m was found to correlate with Log P.[21]

14.3.5 Increase the Size of the Molecule to Disrupt Binding to SULT

Example

1

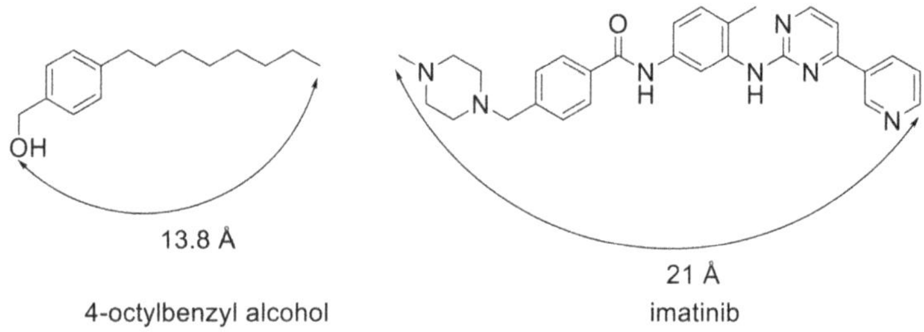

The SULT catalytic efficiency of a series of benzyl alcohols substituted in the 4-position with *n*-alkyl chains up to *n*-octyl was found to steadily increase up to *n*-pentyl and decrease again from *n*-hexyl to *n*-octyl,[21] which can be rationalized by the size and shape of the enzyme's active site being less accommodating to long and linear chains beyond *n*-pentyl. A similar observation was made on the SULT1A1/1A2-mediated sulfation of a series of 4-alkylated phenols: metabolism was significant when the chain was smaller than *n*-hexyl but dropped again when it was *n*-heptyl or larger.[22] While this indicates that increasing a molecule's size may decrease sulfation, such a strategy offers the advantage of not having to modify the soft spot itself, which sometimes is critical for potency. The length of 4-octylbenzyl alcohol is 13.8 Å, which is still well within the drug-like space; for comparison imatinib is significantly larger at 21 Å.

References

1. N. Gamage, A. Barnett, N. Hempel, R. G. Duggleby, K. F. Windmill, J. L. Martin and M. E. McManus, *Toxicol. Sci.*, 2006, **90**, 5–22.
2. M. Suiko, K. Kurogi, T. Hashiguchi, Y. Sakakibara and M. C. Liu, *Biosci., Biotechnol., Biochem.*, 2017, **81**, 63–72.
3. D. Dong, R. Ako and B. Wu, *Expert Opin. Drug Metab. Toxicol.*, 2012, **8**, 635–646.
4. N. U. Gamage, R. G. Duggleby, A. C. Barnett, M. Tresillian, C. F. Latham, N. E. Liyou, M. E. McManus and J. L. Martin, *J. Biol. Chem.*, 2003, **278**, 7655–7662.
5. T. Wang, I. Cook and T. S. Leyh, *Drug Metab. Dispos.*, 2016, **44**, 481–484.
6. C. D. Klaassen and J. W. Boles, *FASEB J.*, 1997, **11**, 404–418.
7. T. Shiraga, K. Iwasaki, K. Takeshita, H. Matsuda, T. Niwa, Z. Tozuka, T. Hata and F. P. Guengerich, *Xenobiotica*, 1995, **25**, 1063–1071.

8. W. Z. Zhong, J. Zhan, P. Kang and S. Yamazaki, *Curr. Drug Metab.*, 2010, **11**, 296–306.
9. S. R. Hanson, M. D. Best and C. H. Wong, *Angew. Chem., Int. Ed.*, 2004, **43**, 5736–5763.
10. A. Kalliokoski and M. Niemi, *Br. J. Pharmacol.*, 2009, **158**, 693–705.
11. G. Jedlitschky, I. Leier, U. Buchholz, K. Barnouin, G. Kurz and D. Keppler, *Cancer Res.*, 1996, **56**, 988–994.
12. E. Chapman, M. D. Best, S. R. Hanson and C. H. Wong, *Angew. Chem., Int. Ed.*, 2004, **43**, 3526–3548.
13. L. Wang, N. Raghavan, K. He, J. M. Luettgen, W. G. Humphreys, R. M. Knabb, D. J. Pinto and D. Zhang, *Drug Metab. Dispos.*, 2009, **37**, 802–808.
14. Y. J. Surh, *Chem.-Biol. Interact.*, 1998, **109**, 221–235.
15. U. Bicker and W. Fischer, *Nature*, 1974, **249**, 344–345.
16. A. M. Sharma, M. Novalen, T. Tanino and J. P. Uetrecht, *Chem. Res. Toxicol.*, 2013, **26**, 817–827.
17. G. Akk, J. Bracamontes and J. H. Steinbach, *J. Physiol.*, 2001, **532**, 673–684.
18. A. Malayev, T. T. Gibbs and D. H. Farb, *Br. J. Pharmacol.*, 2002, **135**, 901–909.
19. J. R. Riggs, M. Nagy, J. Elsner, P. Erdman, D. Cashion, D. Robinson, R. Harris, D. Huang, L. Tehrani, G. Deyanat-Yazdi, R. K. Narla, X. Peng, T. Tran, L. Barnes, T. Miller, J. Katz, Y. Tang, M. Chen, M. F. Moghaddam, S. Bahmanyar, B. Pagarigan, S. Delker, L. LeBrun, P. P. Chamberlain, A. Calabrese, S. S. Canan, K. Leftheris, D. Zhu and J. F. Boylan, *J. Med. Chem.*, 2017, **60**, 8989–9002.
20. J. Liu, S. Zheng, V. L. Akerstrom, C. Yuan, Y. Ma, Q. Zhong, C. Zhang, Q. Zhang, S. Guo, P. Ma, E. V. Skripnikova, M. R. Bratton, A. Pannuti, L. Miele, T. E. Wiese and G. Wang, *J. Med. Chem.*, 2016, **59**, 8134–8140.
21. G. Chen, E. Banoglu and M. W. Duffel, *Chem. Res. Toxicol.*, 1996, **9**, 67–74.
22. R. M. Harris, R. H. Waring, C. J. Kirk and P. J. Hughes, *J. Biol. Chem.*, 2000, **275**, 159–166.

CHAPTER 15

Reactive Metabolites

AMIT S. KALGUTKAR*

Medicine Design, Pfizer Worldwide Research and Development, Cambridge, MA 02139, USA
*E-mail: amit.kalgutkar@pfizer.com

15.1 Introduction

The process of bioactivation resulting in the formation of electrophilic reactive metabolites (RMs) is an unattractive feature in investigational drug substances, considering their known association with genotoxicity (*via* covalent modification of DNA),[1] clinical drug–drug interactions [*via* inactivation of human cytochrome P450 (CYP) enzymes],[2] and certain idiosyncratic adverse drug reactions (IADRs).[3–5] IADRs manifest as rare and sometimes life-threatening reactions [*e.g.*, drug-induced liver injury (DILI), skin rashes and agranulocytosis] that cannot be explained by the primary pharmacology of the drug.[3–6] The underlying mechanisms of IADRs remain unclear; however, it is believed that the vast majority are caused by immunogenic conjugates formed *via* the covalent interaction of a RM with cellular proteins resulting in direct cellular dysfunction or an immune response *via* the formation of a hapten.[6]

15.2 Screening for RMs in Preclinical Drug Discovery

IADRs are difficult to reproduce in humans, and there are few, if any, generally applicable animal models for addressing these liabilities during the drug discovery and development phase.[7] As a consequence, predicting

Drug Discovery Series No. 79
The Medicinal Chemist's Guide to Solving ADMET Challenges
Edited by Patrick Schnider
© The Royal Society of Chemistry 2021
Published by the Royal Society of Chemistry, www.rsc.org

the IADR potential of new chemical entities (NCEs) is practically difficult, if not impossible. Under the basic notion that eliminating RM formation could potentially mitigate IADRs risks, high-throughput screens for assessing CYP-catalyzed oxidations of NCEs to RMs have evolved, wherein compounds are incubated with nicotinamide adenine dinucleotide phosphate (NADPH)-supplemented human liver microsomes (HLM) in the presence of nucleophilic trapping reagents [*e.g.*, glutathione (GSH) and its derivatives, methoxylamine, semicarbazide and/or cyanide], and the trapped electrophiles (RM-nucleophile adducts) are detected using liquid chromatography tandem mass spectrometry.[8–11] RMs resulting from CYP-mediated oxidations can also be studied upon incubating radiolabeled (^{14}C or ^{3}H-labeled) NCEs with NADPH-supplemented human liver microsomes (HLMs) and/or human hepatocytes followed by quantification of the amount of unextractable radioactivity (presumably due to covalent binding of the electrophilic RM to hepatic proteins).[12]

15.3 Structural Alerts and Drug Design

Exclusion of certain functional groups [referred to as structural alerts (SAs) or toxicophores][13] that are intrinsically electrophilic or are known to undergo enzyme-catalyzed bioactivation to RMs (Figure 15.1) is a standard *modus operandi* in modern medicinal chemistry. Such a notion is backed by the observations that out of 68 drugs recalled or associated with a black box warning (BBW) for IADRs, 55 (80.8%) contained one or more SA, and evidence for RM formation is available for 36 out of the 55 drugs (65%).[14] Several examples of structure–toxicity relationship studies have emerged, wherein absence of RM liability is consistent with the improved safety profile of successor drugs (see Key References). Although anecdotal for the most part, the results of such studies indicate that avoiding SAs in drug design would potentially mitigate IADR risks due to RM formation.

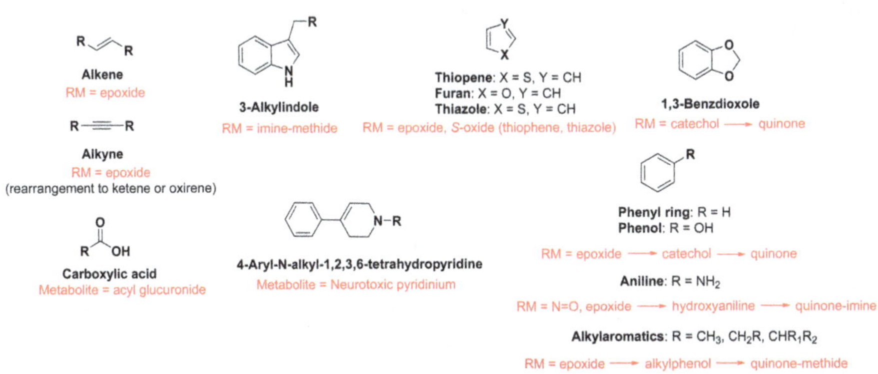

Figure 15.1 Representative examples of functional groups categorized as SAs.

15.3.1 Shortcomings of the SA Concept

There is no clear distinction as to when a particular functional motif is viewed as a SA. SAs are divided into a simple binary categorization; ones that form RMs *versus* other functional groups. The vast majority of marketed drugs possess a phenyl ring, which is a SA (Figure 15.1).[13] As such, the generation of a RM in the course of metabolism will depend on the binding pose of the NCE in the catalytic active site of the metabolizing enzyme (*e.g.*, CYP), and subsequent positioning of the SA towards the catalytic center to yield a RM. It is entirely possible that metabolism could occur in a region distinct from the SA and lead to non-reactive metabolites as described in Section 15.7. As depicted in Figure 15.2, examination of the structural trends for recently approved drugs (2009-present) also reveals several examples containing SAs that are not metabolized to RMs.[14–17]

15.3.2 Critical Evaluation of the SA Concept

Eliminating RM formation is a viable starting point in drug design, but there is a growing concern that the perceived safety risks associated with RM-positive compounds may be over accentuated. Several marketed drugs contain SAs and form RMs, but are rarely associated with IADRs. A recent meta-analysis revealed that out of the 108 most prescribed drugs in 2009, 58 (53%) contained SAs and 24 out of the 58 (41%) examples formed RMs.[14] Likewise, 13 out of the 15 most sold drugs in 2009 were found to contain SAs and experimental evidence for RM formation has been presented for 10 out of the 13 drugs.[14] Overall, the percentage of SA- and/or RM-positive compounds in the most prescribed or highest total sales categories is largely

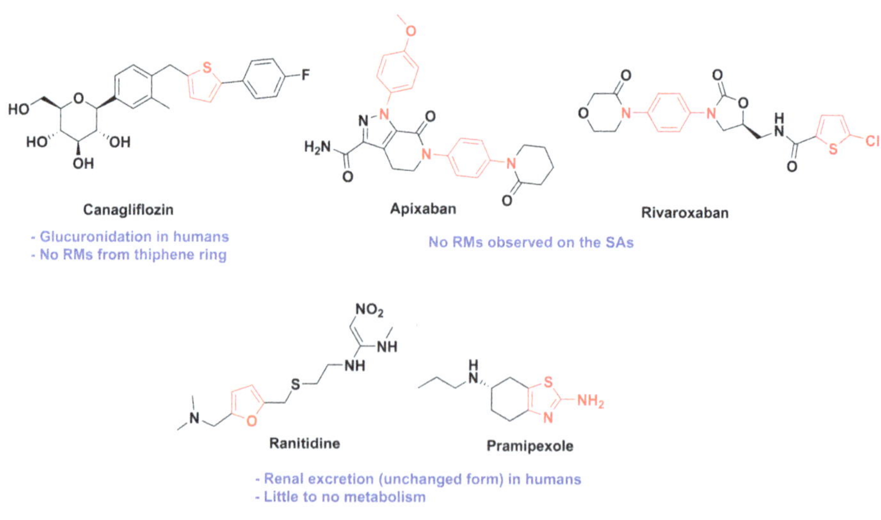

Figure 15.2 Marketed drugs that contain SAs, which are not metabolized to RMs.

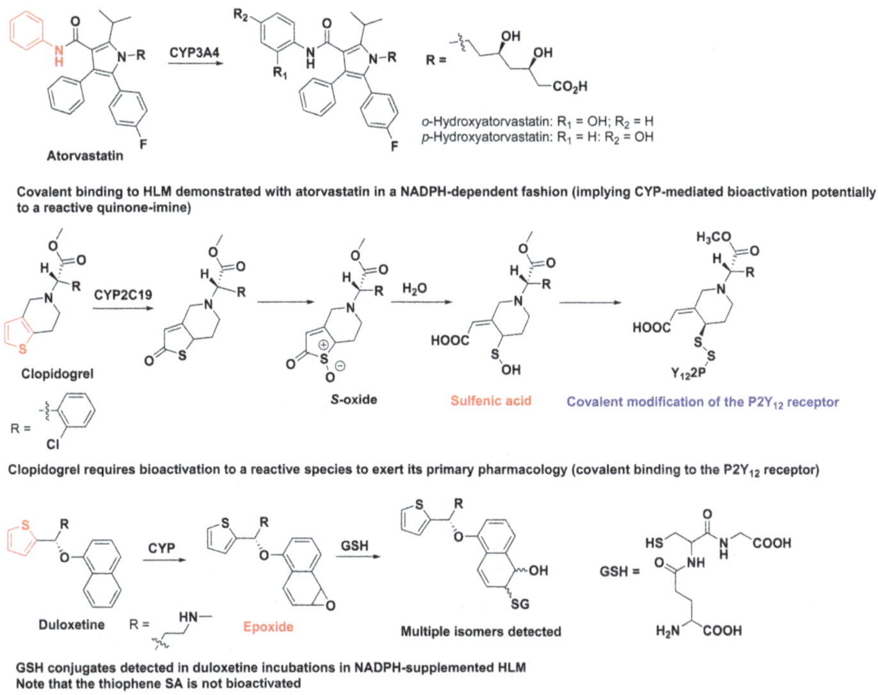

Figure 15.3 Marketed drugs with good safety records that form RMs.

similar to that noted for drugs recalled or associated with a BBW for IADRs, indicating that the SA concept and RM screening tools may be too rigorous and in its current form could halt the progression of novel and much-needed medicines. Atorvastatin, clopidogrel and duloxetine (Figure 15.3) are perhaps the most provocative illustrations of marketed agents that contain SAs and are bioactivated to RMs.[14]

15.4 Bioactivation *Versus* Detoxification

In vitro RM screens (conducted in GSH- and NADPH-supplemented HLM) are only capable of inspecting oxidative bioactivation by CYP enzymes. In several instances, CYP-dependent RM formation may be observed in HLM, but *in vivo*, the compound may undergo a distinctly different metabolic fate that bypasses and/or competes with RM formation (Figure 15.4). The antidepressant paroxetine is metabolized by CYP2D6 on the 1,3-benzdioxole SA to a catechol intermediate in humans,[18] which also leads to the mechanism-based inactivation of CYP2D6.[19] The results of studies with [^{3}H]-paroxetine demonstrated NADPH-dependent covalent binding to HLM and human liver S9 proteins and the characterization of GSH conjugates of reactive quinone metabolites.[20] Likewise, the selective estrogen receptor modulator raloxifene is metabolized by CYP3A4 on the phenolic SAs to yield reactive

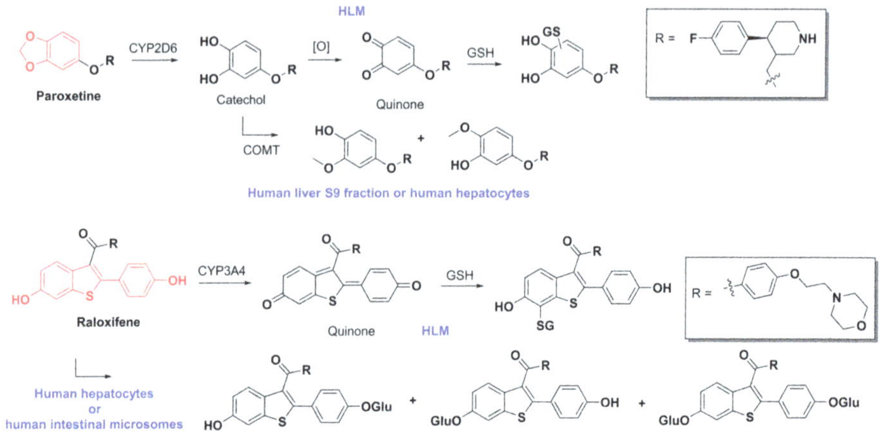

Figure 15.4 Effects of competing detoxification pathways on RM formation.

quinone species that are trapped with GSH *in vitro*.[21] However *in vivo*, the quinone precursors of these two drugs are principally metabolized *via* competing *O*-methylation and/or glucuronidation pathways, respectively.[20,22] In fact, the principal metabolites of paroxetine in humans are the corresponding guaiacol derivatives, obtained *via* catechol-*O*-methyl transferase (COMT)-catalyzed methylation of the catechol intermediate.[18] In the case of raloxifene, preincubation of raloxifene with uridine 5-diphosphoglucuronic acid-fortified human intestinal microsomes reduces the amount of [^{14}C]-raloxifene that is covalently bound to HLM, which is consistent with the efficient and extensive raloxifene glucuronidation by human intestinal-specific uridine glucuronosyl transferase (UGT) 1A10 and 1A8 that limits the amount of drug undergoing bioactivation in the liver.[22] These observations indicate the importance of detailed follow-up studies in fully integrated *in vitro* biological matrices (*e.g.* human hepatocytes and/or liver S-9 fractions) to minimize false positives generated in preliminary RM screens. Knowledge about the *in vivo* clearance mechanisms in animals and the ones anticipated in humans and how that relates to RM formation in *in vitro* matrices will lead to data-driven decision making with regards to compound selection.

15.5 Total Daily Dose as a Principal Mitigating Factor for IADRs

High daily dose drugs (>100 mg) tend to be the ones, which most frequently lead to IADRs, while low dose drugs (<50 mg) rarely are problematic in this regard (irrespective of the potential of these compounds to form RMs).[14] A statistically significant relationship has been noted between daily dose and reports of liver failure ($p = 0.009$), liver transplantation ($p < 0.001$) and death caused by DILI ($p = 0.004$) amongst prescription medicines in the USA. Of 598 Swedish DILI cases, 9% belonged to the ≤ 10 mg per day group, 14.2%

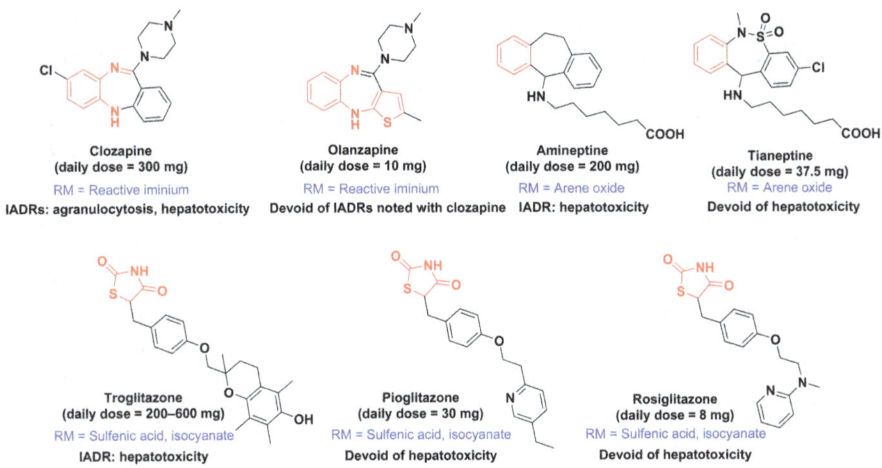

Figure 15.5 Influence of low total daily dose on IADR potential of drugs that form RMs.

to the 11–49 mg per day group, and 77% of cases were caused by medications given at doses ≥50 mg per day.[23] In the meta-analysis by Stepan *et al.*[14] the vast majority of SA- and/or RM-positive drugs, which are not associated with IADRs, are low total daily dose drugs (Figure 15.5). Likewise, in the case of clopidogrel, the majority (>70%) of its daily dose of 75 mg undergoes rapid ester hydrolysis by carboxylesterases to yield an inactive carboxylic acid metabolite (approximately 80–85% of circulating metabolites).[24] This observation implies that only a small percentage of the parent drug (20 mg or less) is theoretically available for conversion to the pharmacologically active RM. Indeed, covalent binding to platelets accounts for only 2% of radiolabeled clopidogrel in human mass balance studies (Plavix® package insert). These combined observations (discussed in Sections 15.4 and 15.5) indicate that the improved safety of low daily dose drugs arises from a marked reduction in the total body burden to RM exposure *via* efficient detoxification processes involving scavenging by the endogenous GSH pool and/or competing metabolic pathways, such that the RMs are unlikely to exceed the safety threshold needed for toxicity.

15.6 Managing RM Liability of Drug Candidates in Preclinical Discovery

The mere presence of SAs cannot in itself predict the type, severity or incidence of IADRs. Likewise, RM screening tools are not intended to predict toxicity but rather to detect the formation of RMs. Selection of drug candidates should not rely solely on SA or RM information, as overall metabolic fate and other considerations (*e.g.* toxicity arising from the parent compound itself *via* inhibition of critical hepatobiliary transport proteins such as the bile-salt export pump)[25,26] provide additional valuable information that can be used in a 'weight of evidence' approach for risk assessment and management.

Recent advances in risk assessment methodologies, such as the estimate of total daily body burden of covalent binding in hepatocytes or by zone classification taking the clinical dose into consideration, are positive steps towards quantitative prediction of IADR risks with drug candidates.[27–31] Optimization of lead compounds should focus on improving intrinsic pharmacological potency and optimizing pharmacokinetics (*e.g.* reducing metabolic clearance) as a means of decreasing the total daily dose and the associated 'body burden' of the parent drug and its metabolites. Against this backdrop, it is necessary to emphasize that a correlation has been established between DILI and drugs undergoing high hepatic metabolism.[32] Out of the approximately 207 most widely prescribed oral medications in the USA, 12 drugs with no reported hepatic metabolism had no reports of DILI. In contrast, drugs significantly metabolized in the liver (>50% hepatic metabolism, $n = 149$) had significantly higher frequency of alanine aminotransferase more than three times the upper limit of normal (35% *versus* 11%, $p = 0.001$), liver failure (28% *versus* 9%, $p = 0.004$) and fatal DILI (23% *versus* 4%, $p = 0.001$). When the relationship between DILI and combination of hepatic metabolism and daily dose was examined, compounds with both substantial hepatic metabolism and daily dose >50 mg ($n = 50$) were significantly more hepatotoxic than compounds belonging to other groups. Extension of these findings[33] led to the observation that drugs ($n = 254$) metabolized by CYP enzymes have a higher likelihood of causing DILI [odds ratio, 3.99; 95% confidence interval, 2.07–7.67; $p < 0.0001$] in a dose-dependent manner.

It is also important to note that several marketed drugs (*e.g.* aniline-based tyrosine kinase inhibitors) intended to treat life-threatening ailments generate RMs, cause IADRs and carry a BBW for adverse reactions.[34] Such drugs remain on the market and are widely prescribed because of favorable benefit–risk considerations. What these observations indicate is that the level of risk (*e.g.* idiosyncratic toxicity, pharmacokinetic interactions due to CYP inactivation, *etc.*) that would be deemed acceptable for drug candidates intended to treat major unmet medical needs, life-threatening diseases and/or orphan diseases is significantly higher relative to treatment of chronic non-debilitating conditions where alternate treatment options are already available.

Key References

- S. T. Orr, S. L. Ripp, T. E. Ballard, J. L. Henderson, D. O. Scott, R. S. Obach, H. Sun and A. S. Kalgutkar, *J. Med. Chem.*, 2012, **55**, 4896.
  - *Exhaustive review on the mechanisms and strategies to eliminate mechanism-based inactivation of human CYP enzymes.*
- A. S. Kalgutkar, I. Gardner, R. S. Obach, C. L. Shaffer, E. Callegari, K. R. Henne, A. E. Mutlib, D. K. Dalvie, J. S. Lee, Y. Nakai, J. P. O'Donnell, J. Boer and S. P. Harriman, *Curr. Drug Metab.*, 2005, **3**, 161.
  - *Categorization and mechanisms of bioactivation of structural alerts.*

- A. F. Stepan, D. P. Walker, J. Bauman, D. A. Price, T. A. Baillie, A. S. Kalgutkar and M. D. Aleo, *Chem. Res. Toxicol.*, 2011, **24**, 1345.
  - *Comprehensive analysis of bioactivation potential of marketed drugs (2009) based on most prescribed and total sales.*
- J. G. Kenna, K. S. Taskar, C. Battista, D. L. Bourdet, K. L. R. Brouwer, D. Dai, C. Funk, M. J. Hafey, Y. Lai, J. Maher, Y. A. Pak, J. M. Pedersen, J. W. Polli, A. D. Rodrigues, P. B. Watkins, K. Yang, R. W. Yucha and International Transporter Consortium, *Clin. Pharmacol. Ther.*, 2018, **104**, 916.
  - *Assessment of bile salt export pump inhibition in relation to DILI.*
- S. Nakayama, R. Atsumi, H. Takakusa, Y. Kobayashi, A. Kurihara, Y. Nagai, D. Nakai and O. Okazaki, *Drug Metab. Dispos.*, 2009, **37**, 1970.
  - *Zone classification and the potential of drugs to cause DILI.*
- C. Lammert, E. Bjornsson, A. Niklasson and N. Chalasani, *Hepatology*, 2010, **51**, 615.
  - *Relationship between hepatic metabolism and DILI.*
- K. Yu, X. Geng, M. Chen, J. Zhang, B. Wang, K. Illic and W. Tong, *Drug Metab. Dispos.*, 2014, **42**, 744.
  - *Retrospective analysis of the relationship between CYP metabolism and DILI.*

15.7 Elimination of RM Liability in Preclinical Drug Discovery

For RM-positive compounds, the structure of the reactive species (usually inferred from the characterization of stable adducts with nucleophiles), the biochemical pathway(s) and the enzyme(s) responsible for their generation have to be determined. The information can then be used, as appropriate, for structural modifications aimed at eliminating the liability. In practice, eliminating or reducing RM formation in a lead chemical series is not trivial; medicinal chemistry solutions to eliminate RM formation could result in a detrimental effect on primary pharmacology (*e.g.* changes in subtype selectivity for target receptor or enzyme, agonist or antagonist behavior) and/or pharmacokinetic attributes (*e.g.* attenuation of aqueous solubility or passive cell permeability). If the structural alert can be readily replaced with an alternate functional group without a significant loss of primary pharmacology or pharmacokinetic properties, then it is advisable to do so (examples are noted later in this chapter) and thereby avoid the need for additional risk assessment beyond standard drug safety packages and further internal debate on this topic for the remainder of the development program. On the basis of available examples in the toxicological and medicinal chemistry literature, there appear to be approximately three major groups of reactive structural entities derived from small molecule bioactivation, including:

1. Electron-deficient double (and triple) bonds, including quinones, quinone-methides, quinone-imines, imine-methides, diimines, classical Michael-acceptors and – especially electronically stabilized – iminium ions.
2. Epoxides derived from CYP-mediated oxidation of aryl rings as well as double and triple bond-containing compounds.
3. Acyl glucuronides.

Strategies towards mitigating their formation in a preclinical drug discovery setting are described in Sections 15.7.1–15.7.10 and Figures 15.6–15.15.

15.7.1 Mitigation of Aromatic (and Heteroaromatic) Ring Epoxidation *via* Introduction of Metabolic Soft Spots

Life-threatening hepatotoxicity observed in patients treated with sudoxicam during clinical trials has been attributed to thiazole ring scission yielding a reactive acylthiourea metabolite capable of oxidizing GSH and proteins.[35] Although introduction of a methyl group at the C5 position on the thiazole ring in meloxicam is the only structural difference, the change dramatically alters the metabolic profile such that oxidation of the C5 methyl group to the alcohol (and carboxylic acid) metabolite(s) constitutes the principal metabolic fate of meloxicam in humans (Figure 15.6).[36] In the case of the anxiolytic agents alpidem (hepatotoxic, withdrawn from commercial use) and zolpidem (Ambien®, non-hepatotoxic),[37] a key structural difference is the replacement of the two chlorine atoms in alpidem with two methyl groups in zolpidem. The chloro-imidazopyridine ring in alpidem undergoes CYP-mediated oxidation to a reactive epoxide intermediate that adducts with GSH; the detection of thiol conjugates in human excreta provides evidence for the pathway *in vivo*.[38] In contrast, zolpidem is metabolized *via* oxidation of both methyl groups to the corresponding alcohol and carboxylic acid metabolites and is not subject to RM formation.[38]

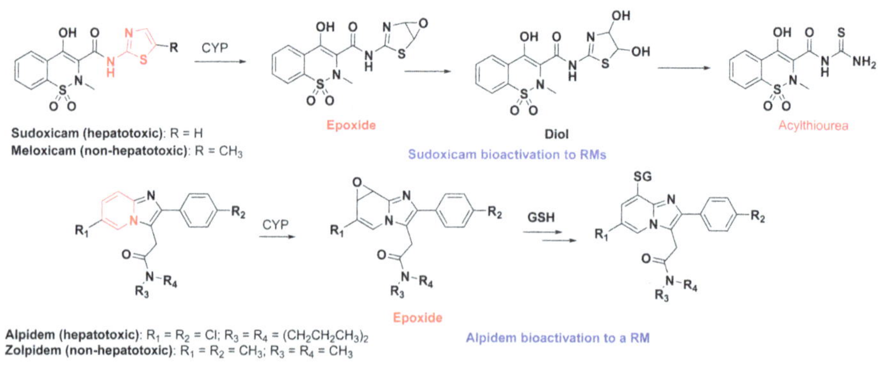

Figure 15.6 Examples of mitigation of (hetero)aromatic ring epoxidation *via* introduction of metabolic soft spots.

15.7.2 Mitigation of Heteroaromatic Ring Epoxidation *via* SA Replacement

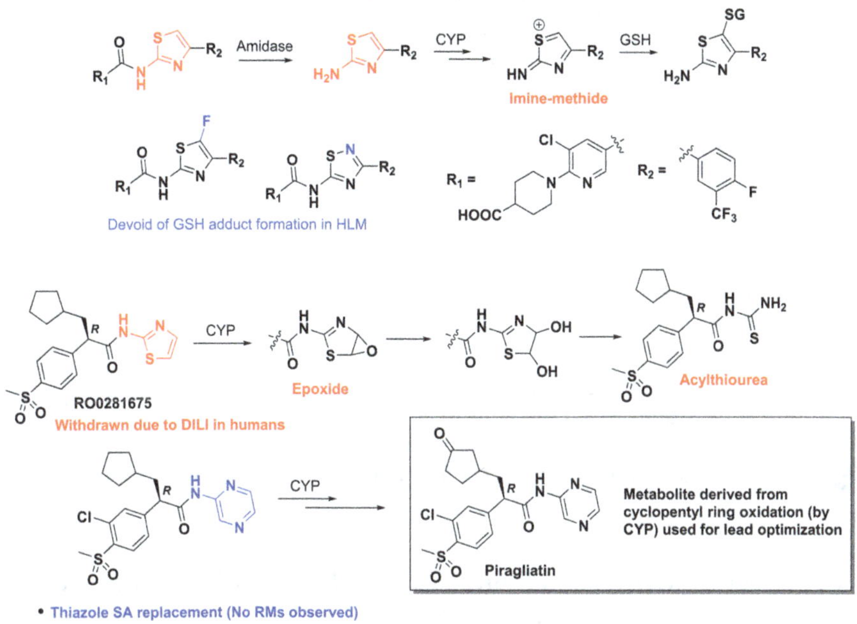

Figure 15.7 Examples of mitigation of (hetero)aromatic ring epoxidation *via* SA replacement.[39,40]

15.7.3 Mitigation of Heteroaromatic Ring Epoxidation *via* Introduction of Electron-deficient Ring Substituents

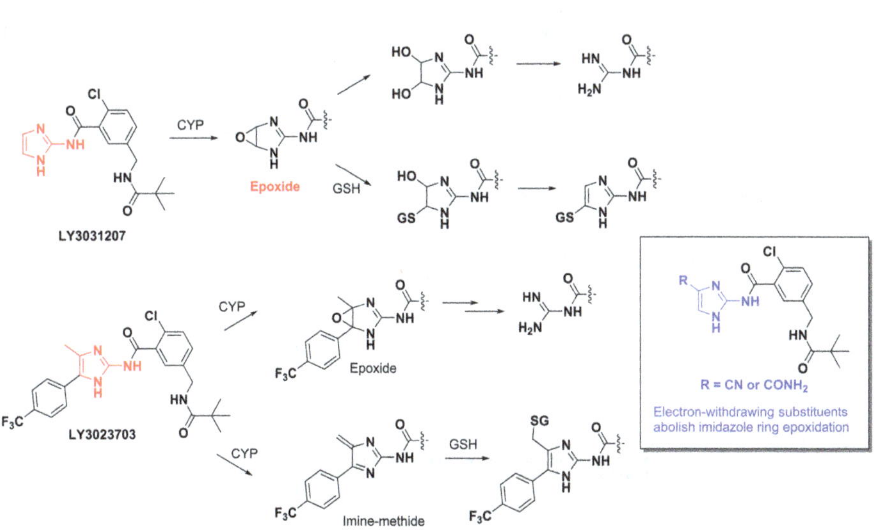

Figure 15.8 Examples of mitigation of (hetero)aromatic ring epoxidation *via* introduction of electron-deficient ring substituents.[41]

15.7.4 Mitigating the Formation of Electrophilic Iminium Ion: Structure–Toxicity Relationship Studies on Marketed Dibenzodiazepine Anxiolytics

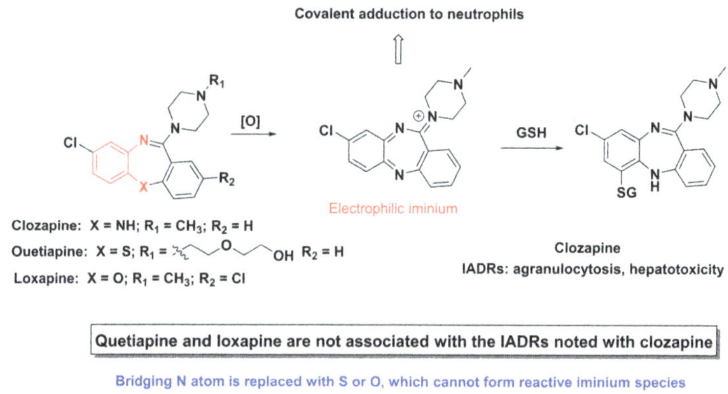

Figure 15.9 Example of mitigating the formation of an electrophilic iminium ion.[42–44]

15.7.5 Mitigating the Formation of Electrophilic Quinone-Imine and Diimines[45–49]

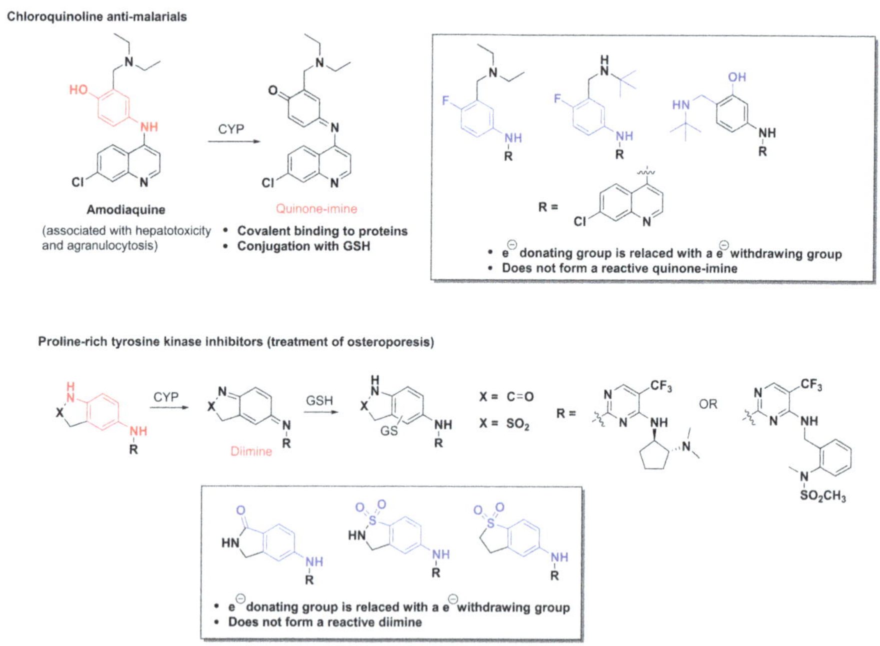

Figure 15.10 Examples of mitigating the formation of electrophilic quinone-imine and diimines.[45–49]

15.7.6 Reducing or Eliminating Bioactivation of Phenyl and Phenol Rings to Epoxides and Quinones *via* Introduction of Electron-withdrawing Substituents

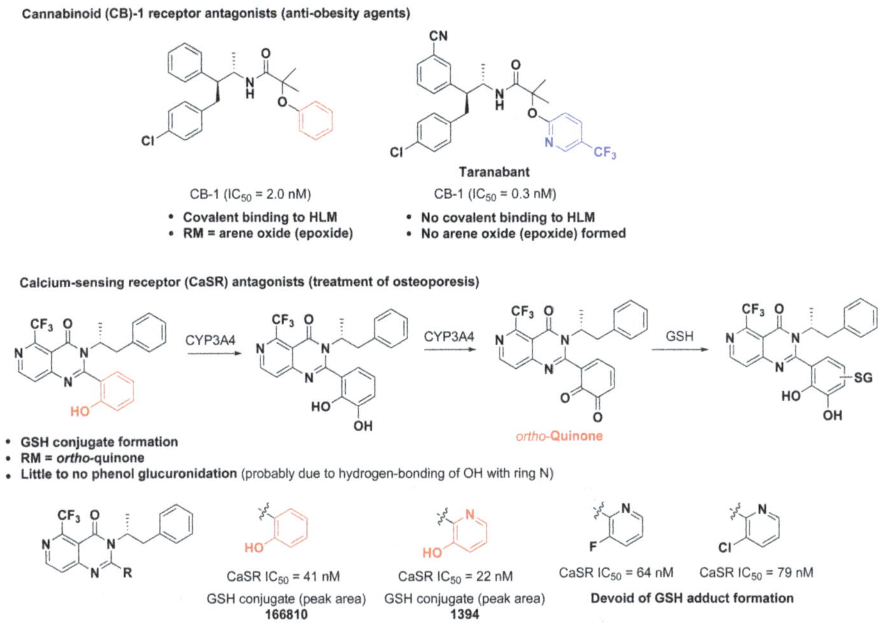

Figure 15.11 Examples of reducing or eliminating bioactivation of phenyl and phenol rings to epoxides and quinones *via* introduction of electron-withdrawing substituents.[50–52]

15.7.7 Eliminating Multiple Oxidative Bioactivation Pathways (Epoxidation and Quinone-imine Formation) in Pyrazinone-Based Corticotrophin-Releasing Factor-1 Receptor Antagonists

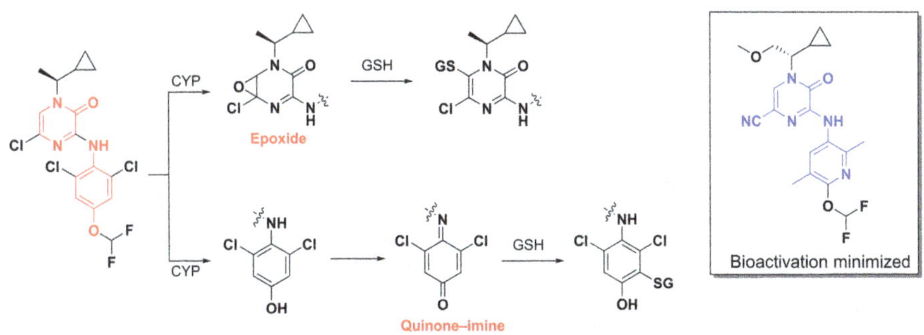

Figure 15.12 Example of eliminating multiple oxidative bioactivation pathways (epoxidation and quinone-imine formation) in pyrazinone-based corticotrophin-releasing factor-1 receptor antagonists.[53–56]

15.7.8 Eliminating S9/NADPH-dependent Genotoxicity of a 5-Hydroxytryptamine Receptor Family 2C (5-HT$_{2C}$) Receptor Agonist in the *Salmonella* Ames Assay[57,58]

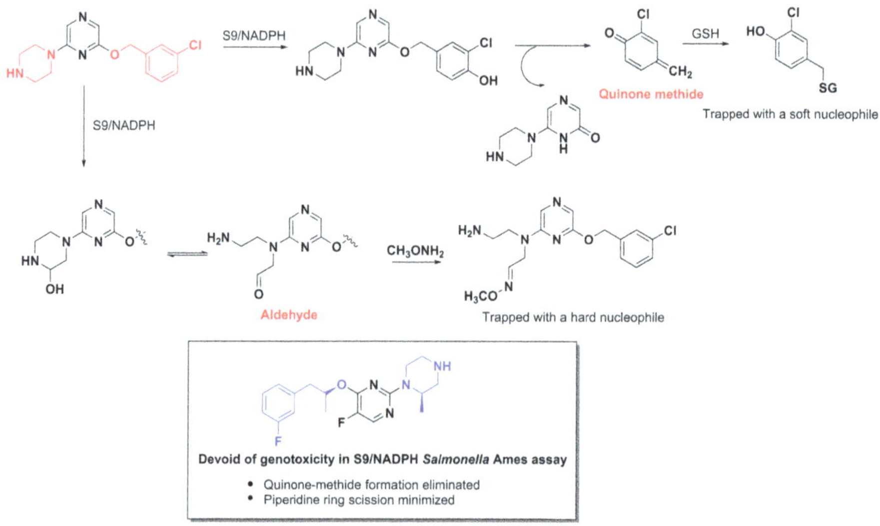

Figure 15.13 Example of eliminating S9/NADPH-dependent genotoxicity of a 5-HT$_{2C}$ receptor agonist in the *Salmonella* Ames assay.[57,58]

15.7.9 Eliminating the Formation of the GSH-Reactive α,β-Unsaturated Aldehyde Noted in the Metabolism of the Anti-convulsant Felbamate (Associated with Aplastic Anemia and Hepatotoxicity)

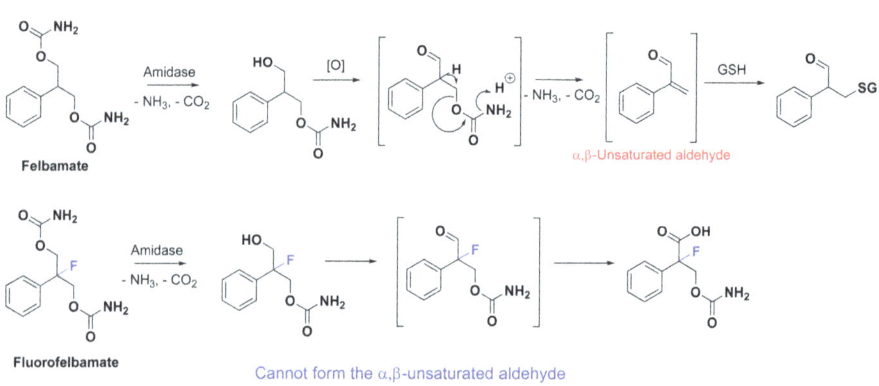

Figure 15.14 Example of eliminating the formation of the GSH-reactive α,β-unsaturated aldehyde noted in the metabolism of the anti-convulsant felbamate (associated with aplastic anemia and hepatotoxicity).[59–61]

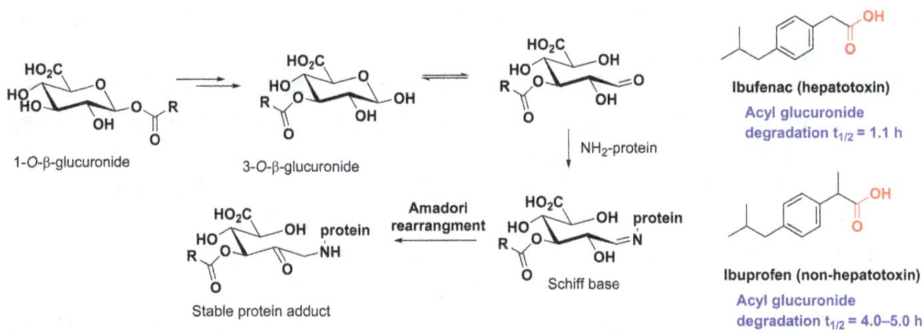

Figure 15.15 Structure–toxicity relationship analysis of carboxylic acid containing drugs.

15.7.10 Structure–Toxicity Relationships for Carboxylic Acid-based Drugs Prone to Acyl Glucuronidation

A rather dramatic illustration of a structure–toxicity relationship is evident with the hepatotoxicant ibufenac and ibuprofen, one of the safest over-the-counter nonsteroidal anti-inflammatory drugs on the market. While both drugs are extensively glucuronidated in animals and humans, the presence of the extra α-methyl substituent in ibuprofen slows acyl glucuronide formation and migration relative to ibufenac, thus providing a possible explanation for the toxicological differences (Figure 15.15). Structure–activity relationships between acyl glucuronide migration rates and protein covalent binding have revealed that carboxylic acids with a higher degree of alkyl substitution at the α-carbon exhibit lower reactivity with nucleophiles, indicating that inherent electronic and steric effects probably modulate the overall rate of acyl glucuronide rearrangement.[62–64]

References

1. K. L. Dobo, R. S. Obach, D. Luffer-Atlas and J. P. Bercu, *Chem. Res. Toxicol.*, 2009, **22**, 348.
2. S. T. Orr, S. L. Ripp, T. E. Ballard, J. L. Henderson, D. O. Scott, R. S. Obach, H. Sun and A. S. Kalgutkar, *J. Med. Chem.*, 2012, **55**, 4896.
3. F. P. Guengerich and J. S. MacDonald, *Chem. Res. Toxicol.*, 2007, **20**, 344.
4. J. Li and J. P. Uetrecht, *Handb. Exp. Pharmacol.*, 2010, **196**, 493.
5. J. Uetrecht and D. J. Naisbitt, *Pharmacol. Rev.*, 2013, **65**, 779.
6. A. S. Kalgutkar, A. D. Vaz, M. E. Lame, K. R. Henne, J. Soglia, S. X. Zhao, Y. A. Abramov, F. Lombardo, C. Collin, Z. S. Hendsch and C. E. Hop, *Drug Metab. Dispos.*, 2005, **33**, 243.
7. J. Uetrecht, *Drug Metab. Rev.*, 2006, **38**, 745.
8. D. Argoti, L. Liang, A. Conteh, L. Chen, D. Bershas, C. P. Yu, P. Vouros and E. Yang, *Chem. Res. Toxicol.*, 2005, **18**, 1537.
9. S. Ma and R. Subramanian, *J. Mass Spectrom.*, 2006, **41**, 1121.
10. A. S. Kalgutkar, *Chem. Res. Toxicol.*, 2017, **30**, 220.
11. K. Inoue, K. Fukuda, T. Yoshimura and K. Kusano, *Chem. Res. Toxicol.*, 2015, **28**, 1546.

12. D. C. Evans, A. P. Watt, D. A. Nicoll-Griffith and T. A. Baillie, *Chem. Res. Toxicol.*, 2004, **17**, 3.
13. A. S. Kalgutkar, I. Gardner, R. S. Obach, C. L. Shaffer, E. Callegari, K. R. Henne, A. E. Mutlib, D. K. Dalvie, J. S. Lee, Y. Nakai, J. P. O'Donnell, J. Boer and S. P. Harriman, *Curr. Drug Metab.*, 2005, **3**, 161.
14. A. F. Stepan, D. P. Walker, J. Bauman, D. A. Price, T. A. Baillie, A. S. Kalgutkar and M. D. Aleo, *Chem. Res. Toxicol.*, 2011, **24**, 1345.
15. D. Devineni, C. R. Curtin, D. Polidori, M. J. Gutierrez, J. Murphy, S. Rusch and P. L. Rothenberg, *J. Clin. Pharmacol.*, 2013, **53**, 601.
16. D. Zhang, K. He, N. Raghavan, L. Wang, J. Mitroka, B. D. Maxwell, R. M. Knabb, C. Frost, A. Schuster, F. Hao, Z. Gu, W. G. Humphreys and S. J. Grossman, *Drug Metab. Dispos.*, 2009, **37**, 1738.
17. C. Weinz, T. Schwarz, D. Kubiza, W. Mueck and D. Lang, *Drug Metab. Dispos.*, 2009, **37**, 1056.
18. R. E. Haddock, A. M. Johnson, P. F. Langley, D. R. Nelson, J. A. Pope, D. R. Thomas and F. R. Wood, *Acta Psychiatr. Scand., Suppl.*, 1989, **350**, 24.
19. K. Venkatakrishnan and R. S. Obach, *Drug Metab. Dispos.*, 2005, **33**, 845.
20. S. X. Zhao, D. K. Dalvie, J. M. Kelly, J. R. Soglia, K. S. Frederick, E. B. Smith, R. S. Obach and A. S. Kalgutkar, *Chem. Res. Toxicol.*, 2007, **20**, 1649.
21. Q. Chen, J. S. Ngui, G. A. Doss, R. W. Wang, X. Cai, F. P. DiNinno, T. A. Blizzard, M. L. Hammond, R. A. Stearns, D. C. Evans, T. A. Baillie and W. Tang, *Chem. Res. Toxicol.*, 2002, **15**, 907.
22. D. Dalvie, P. Kang, M. Zientek, C. Xiang, S. Zhou and R. S. Obach, *Chem. Res. Toxicol.*, 2008, **21**, 2260.
23. C. Lammert, S. Einarsson, C. Saha, A. Niklasson, E. Bjornsson and N. Chalasani, *Hepatology*, 2008, **47**, 2003.
24. N. A. Farid, A. Kurihara and S. A. Wrighton, *J. Clin. Pharmacol.*, 2010, **50**, 126.
25. S. Dawson, S. Stahl, N. Paul, J. Barber and J. G. Kenna, *Drug Metab. Dispos.*, 2012, **40**, 130.
26. J. G. Kenna, K. S. Taskar, C. Battista, D. L. Bourdet, K. L. R. Brouwer, D. Dai, C. Funk, M. J. Hafey, Y. Lai, J. Maher, Y. A. Pak, J. M. Pedersen, J. W. Polli, A. D. Rodrigues, P. B. Watkins, K. Yang, R. W. Yucha and International Transporter Consortium, *Clin. Pharmacol. Ther.*, 2018, **104**, 916.
27. J. N. Bauman, J. M. Kelly, S. Tripathy, S. X. Zhao, W. W. Lam, A. S. Kalgutkar and R. S. Obach, *Chem. Res. Toxicol.*, 2009, **22**, 332.
28. R. A. Thompson, E. M. Isin, Y. Li, L. Weidolf, K. Page, I. Wilson, S. Swallow, B. Middleton, S. Stahl, A. J. Foster, H. Dolgos, K. Weaver and J. G. Kenna, *Chem. Res. Toxicol.*, 2012, **25**, 1616.
29. S. Schadt, K. Simon, S. Kustermann, F. Boess, C. McGinnis, A. Brink, R. Lieven, S. Fowler, K. Youdim, M. Ullah, M. Marschmann, C. Zihlmann, Y. M. Siegrist, A. C. Cascais, E. Di Lenarda, E. Durr, N. Schaub, X. Ang, V. Starke, T. Singer, R. Alvarez-Sanchez, A. B. Roth, F. Schuler and C. Funk, *Toxicol. In Vitro*, 2015, **30**, 429.
30. R. A. Thompson, E. M. Isin, M. O. Ogese, J. T. Mettetal and D. P. Williams, *Chem. Res. Toxicol.*, 2016, **29**, 505.
31. S. Nakayama, R. Atsumi, H. Takakusa, Y. Kobayashi, A. Kurihara, Y. Nagai, D. Nakai and O. Okazaki, *Drug Metab. Dispos.*, 2009, **37**, 1970.

32. C. Lammert, E. Bjornsson, A. Niklasson and N. Chalasani, *Hepatology*, 2010, **51**, 615.
33. K. Yu, X. Geng, M. Chen, J. Zhang, B. Wang, K. Illic and W. Tong, *Drug Metab. Dispos.*, 2014, **42**, 744.
34. A. S. Kalgutkar and D. Dalvie, *Annu. Rev. Pharmacol. Toxicol.*, 2015, **55**, 35.
35. R. S. Obach, A. S. Kalgutkar, T. F. Ryder and G. S. Walker, *Chem. Res. Toxicol.*, 2008, **21**, 1890.
36. N. M. Davies and N. M. Skjodt, *Clin. Pharmacokinet.*, 1999, **36**, 115.
37. V. Baty, B. Denis, C. Goudot, V. Bas, P. Renkes, M. A. Bigard, P. Boissel and P. Gaucher, *Gastroenterol. Clin. Biol.*, 1994, **18**, 1129.
38. A. Durand, J. P. Thenot, G. Bianchetti and P. L. Morselli, *Drug Metab. Rev.*, 1992, **24**, 239.
39. A. S. Kalgutkar, J. Driscoll, S. X. Zhao, G. S. Walker, R. M. Shepard, J. R. Soglia, J. Atherton, L. Yu, A. E. Mutlib, M. J. Munchhof, L. A. Reiter, C. S. Jones, J. L. Doty, K. A. Trevena, C. L. Shaffer and S. L. Ripp, *Chem. Res. Toxicol.*, 2007, **20**, 1954.
40. R. Sarabu, F. T. Bizzarro, W. L. Corbett, M. T. Dvorozniak, W. Geng, J. F. Grippo, N. E. Haynes, S. Hutchings, L. Garofalo, K. R. Guertin, D. W. Hilliard, M. Kabat, R. F. Kester, W. Ka, Z. Liang, P. E. Mahaney, L. Marcus, F. M. Matshinsky, D. Moore, J. Racha, R. Radinov, Y. Ren, L. Qi, M. Pignatello, C. L. Spence, T. Steele, J. Tengi and J. Grimsby, *J. Med. Chem.*, 2012, **55**, 7021.
41. B. H. Norman, M. J. Fisher, M. A. Schiffler, S. L. Kuklish, N. E. Hughes, B. A. Czeskis, K. C. Cassidy, T. L. Abraham, J. J. Alberts and D. Luffer-Atlas, *J. Med. Chem.*, 2018, **61**, 2041.
42. I. Gardner, J. S. Leeder, T. Chin, N. Zahid and J. P. Uetrecht, *Mol. Pharmacol.*, 1998, **53**, 999.
43. Z. C. Liu and J. P. Uetrecht, *J. Pharmacol. Exp. Ther.*, 1995, **275**, 1476.
44. J. Uetrecht, N. Zahid, A. Tehim, J. M. Fu and S. Rakhit, *Chem.-Biol. Interact.*, 1997, **104**, 117.
45. P. M. O'Neill, A. C. Harrison, R. C. Storr, S. R. Hawley, S. A. Ward and B. K. Park, *J. Med. Chem.*, 1994, **37**, 1362.
46. P. M. O'Neill, A. Mukhtar, P. A. Stocks, L. E. Randle, S. Hindley, S. A. Ward, R. C. Storr, J. F. Bickley, I. A. O'Neil, J. L. Maggs, R. H. Hughes, P. A. Winstanley, P. G. Bray and B. K. Park, *J. Med. Chem.*, 2003, **46**, 4933.
47. P. M. O'Neill, B. K. Park, A. E. Shone, J. L. Maggs, P. Roberts, P. A. Stocks, G. A. Biagini, P. G. Bray, P. Gibbons, N. Berry, P. A. Winstanley, A. Mukhtar, R. Bonar-Law, S. Hindley, R. B. Bambal, C. B. Davis, M. Bates, T. K. Hart, S. L. Gresham, R. M. Lawrence, R. A. Brigandi, F. M. Gomez-delas-Heras, D. V. Gargallo and S. A. Ward, *J. Med. Chem.*, 2009, **52**, 1408.
48. P. M. O'Neill, A. E. Shone, D. Stanford, G. Nixon, E. Asadollahy, B. K. Park, J. L. Maggs, P. Roberts, P. A. Stocks, G. Biagini, P. G. Bray, J. Davies, N. Berry, C. Hall, K. Rimmer, P. A. Winstanley, S. Hindley, R. B. Bambal, C. B. Davis, M. Bates, S. L. Gresham, R. A. Brigandi, F. M. Gomez-de-Las-Heras, D. V. Gargallo, S. Parapini, L. Vivas, H. Lander, D. Taramelli and S. A. Ward, *J. Med. Chem.*, 2009, **52**, 1828.
49. D. P. Walker, F. C. Bi, A. S. Kalgutkar, J. N. Bauman, S. X. Zhao, J. R. Soglia, G. E. Aspnes, D. W. Kung, J. Klug-McLeod, M. P. Zawistoski, M. A.

McGlynn, R. Oliver, M. Dunn, J. C. Li, D. T. Richter, B. A. Cooper, J. C. Kath, C. A. Hulford, C. L. Autry, M. J. Luzzio, E. J. Ung, W. G. Roberts, P. C. Bonnette, L. Buckbinder, A. Mistry, M. C. Griffor, S. Han and A. Guzman-Perez, *Bioorg. Med. Chem. Lett.*, 2008, **18**, 6071.

50. W. K. Hagman, *Arch. Pharm.*, 2008, **341**, 405.
51. M. T. Didiuk, D. A. Griffith, J. W. Benbow, K. K. Liu, D. P. Walker, F. C. Bi, J. Morris, A. Guzman-Perez, H. Gao, B. M. Bechle, R. M. Kelley, X. Yang, K. Dirico, S. Ahmed, W. Hungerford, J. DiBrinno, M. P. Zawistoski, S. W. Bagley, J. Li, Y. Zeng, S. Santucci, R. Oliver, M. Corbett, T. Olson, C. Chen, M. Li, V. M. Paralkar, K. A. Riccardi, D. R. Healy, A. S. Kalgutkar, T. S. Maurer, H. T. Nguyen and K. S. Frederick, *Bioorg. Med. Chem. Lett.*, 2009, **19**, 4555.
52. A. S. Kalgutkar, D. A. Griffith, T. Ryder, H. Sun, Z. Miao, J. N. Bauman, M. T. Didiuk, K. S. Frederick, S. X. Zhao, C. Prakash, J. R. Soglia, S. W. Bagley, B. M. Bechle, R. M. Kelley, K. Dirico, M. Zawistoski, J. Li, R. Oliver, A. Guzman-Perez, K. K. Liu, D. P. Walker, J. W. Benbow and J. Morris, *Chem. Res. Toxicol.*, 2010, **23**, 1115.
53. R. A. Hartz, V. T. Ahuja, M. Rafalski, W. D. Schmitz, A. B. Brenner, D. J. Denhart, J. L. Ditta, J. A. Deskus, E. W. Yue, A. G. Arvanitis, S. Lelas, Y. W. Li, T. F. Molski, H. Wong, J. E. Grace, K. A. Lentz, J. Li, N. J. Lodge, R. Zaczek, A. P. Combs, R. E. Olson, R. J. Mattson, J. J. Bronson and J. E. Macor, *J. Med. Chem.*, 2009, **52**, 4161.
54. R. A. Hartz, V. T. Ahuja, X. Zhuo, R. J. Mattson, D. J. Denhart, J. A. Deskus, V. M. Vrudhula, S. Pan, J. L. Ditta, Y. Z. Shu, J. E. Grace, K. A. Lentz, S. Lelas, Y. W. Li, T. F. Molski, S. Krishnananthan, H. Wong, J. Qian-Cutrone, R. Schartman, R. Denton, N. J. Lodge, R. Zaczek, J. E. Macor and J. J. Bronson, *J. Med. Chem.*, 2009, **52**, 7653.
55. R. A. Hartz, V. T. Ahuja, W. D. Schmitz, T. F. Molski, G. K. Mattson, N. J. Lodge, J. J. Bronson and J. E. Macor, *Bioorg. Med. Chem. Lett.*, 2010, **20**, 1890.
56. X. Zhuo, R. A. Hartz, J. J. Bronson, H. Wong, V. T. Ahuja, V. M. Vrudhula, J. E. Leet, S. Huang, J. E. Macor and Y. Z. Shu, *Drug Metab. Dispos.*, 2010, **38**, 5.
57. A. S. Kalgutkar, D. K. Dalvie, J. Aubrecht, E. B. Smith, S. L. Coffing, J. R. Cheung, C. Vage, M. E. Lame, P. Chiang, K. F. McClure, T. S. Maurer, R. V. Coelho Jr, V. F. Soliman and K. Schildknegt, *Drug Metab. Dispos.*, 2007, **35**, 848.
58. A. S. Kalgutkar, J. N. Bauman, K. F. McClure, J. Aubrecht, S. R. Cortina and J. Paralkar, *Bioorg. Med. Chem. Lett.*, 2009, **19**, 1559.
59. C. M. Dieckhaus, C. D. Thompson, S. G. Roller and T. L. Macdonald, *Chem.-Biol. Interact.*, 2002, **142**, 99.
60. C. D. Thompson, M. T. Barthen, D. W. Hopper, T. A. Miller, M. Quigg, C. Hudspeth, G. Montouris, L. Marsh, J. L. Perhach, R. D. Sofia and T. L. Macdonald, *Epilepsia*, 1999, **40**, 769.
61. B. A. Roecklein, H. J. Sacks, H. Mortko and J. Stables, *Neurotherapeutics*, 2007, **4**, 97.
62. J. Wang, M. Davis, F. Li, F. Azam, J. Scatina and R. Talaat, *Chem. Res. Toxicol.*, 2004, **17**, 1206.
63. G. S. Walker, J. Atherton, J. Bauman, C. Kohl, W. Lam, M. Reily, Z. Lou and A. Mutlib, *Chem. Res. Toxicol.*, 2007, **20**, 876–886.
64. A. Baba and T. Yoshioka, *Chem. Res. Toxicol.*, 2009, **22**, 1998.

CHAPTER 16

Genotoxicity

STEPHAN KIRCHNER*[a] AND PATRICK SCHNIDER*[b]

[a]F. Hoffmann-La Roche AG, Grenzacherstrasse, 4070 Basel, Switzerland;
[b]Rejuveron Life Sciences AG, Wagistrasse 18, 8952 Schlieren, Switzerland
*E-mail: stephan.kirchner@roche.com, patrick.schnider@alumnibasel.ch

16.1 Introduction

16.1.1 Genotoxicity Assessment in Small Molecule Drug Discovery and Development

Assessing genotoxic liabilities of potential new drug candidates is mandatory for preclinical safety testing and comprises a battery of regulatory tests covering different genotoxic mechanisms. As genotoxic activity is very critical for the development of new drug candidates (and in general not acceptable despite few exceptions, such as late-stage cancer indications), reliable information on this endpoint is needed early on. Therefore, miniaturized versions and indicator assays are applied already during the lead optimization phase and to predict the regulatory tests (see Section 16.1.3).

16.1.2 Mechanisms of Genotoxicity

Small molecules may induce genotoxicity either by direct DNA-reactive mechanisms or by indirect, non-DNA-reactive mechanisms (Figure 16.1).

Direct DNA-reactive interactions (*e.g.* by alkylating agents or reactive oxygen species) may lead to DNA adduct formation and base mispairing during replication, ultimately resulting in gene mutations and/or DNA strand breaks if not repaired correctly.[1,2]

Drug Discovery Series No. 79
The Medicinal Chemist's Guide to Solving ADMET Challenges
Edited by Patrick Schnider
© The Royal Society of Chemistry 2021
Published by the Royal Society of Chemistry, www.rsc.org

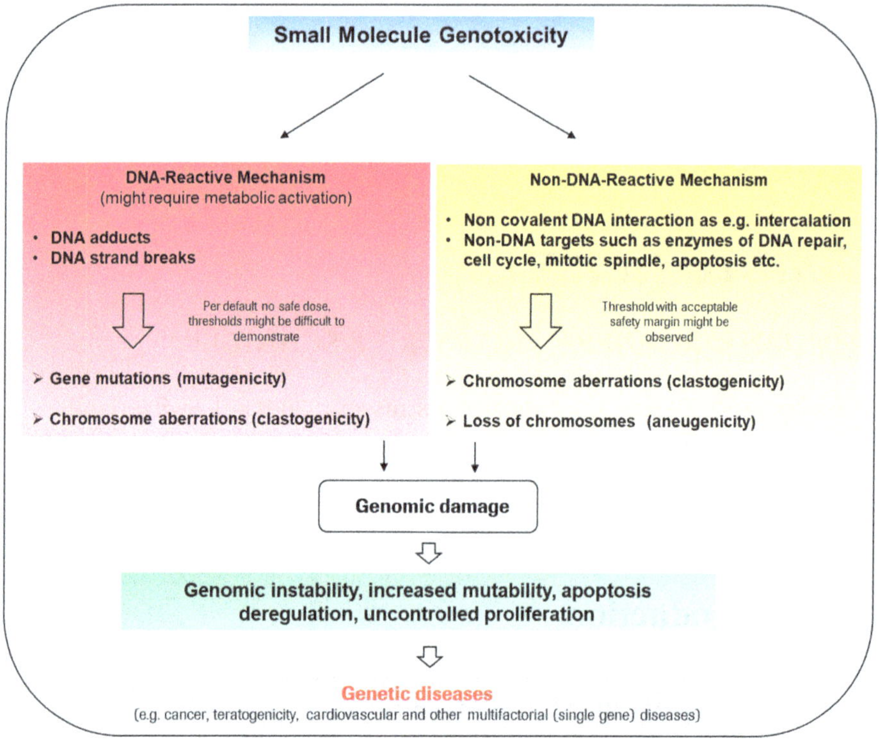

Figure 16.1 Small molecule genotoxic mechanisms and genetic risk.

Indirect, non-DNA-reactive mechanisms *e.g. via* interference with spindle fibers, inhibition of DNA repair enzymes, nucleotide pool imbalance or inhibition of topoisomerases may result in DNA strand breaks or impairment of the chromosome replication cycle. Any unrepaired DNA strand break or any impairment of the chromosome segregation can translate into chromosome breaks (clastogenicity[3]) or mis-segregation of whole chromosomes (aneugenicity[4,5]). For indirect non-DNA-reactive mechanisms thresholds with acceptable safety margins might be observed.

The accumulation of genomic damage could ultimately result in genetic diseases, such as cancer, but also teratogenicity (the capability of producing fetal malformations),[6] cardiovascular effects (by damage of mitochondrial DNA[7,8]) and other multifactorial diseases, such as diabetes.[9] Genomic instability phenotypes with increased mutability may accumulate initial genomic damage which might further lead to uncontrolled cell proliferation and cancer.[10,11]

16.1.2.1 DNA Reactive Mechanisms

Induction of DNA damage as a consequence of covalent modification of the purine and pyrimidine bases or the phosphodiester backbone by reaction with an electrophile is one of the common processes leading to positive

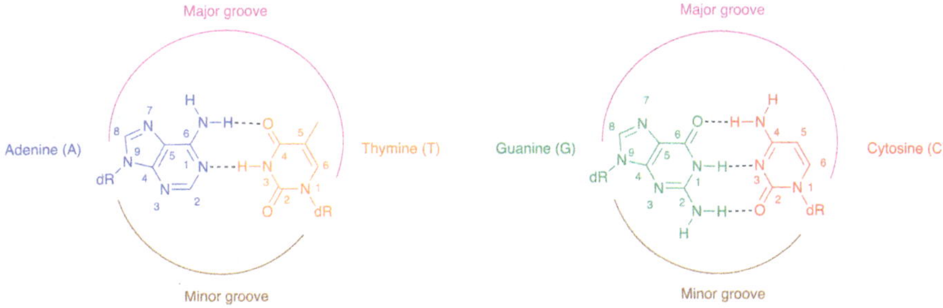

Figure 16.2 Numbering of the atoms and pairing of DNA nucleobases.

results in a genotoxicity assay in the course of drug discovery programs. The preferred sites of reaction in DNA have been explained by the nature of the electrophile and the nucleophilic nitrogen and oxygen atoms of the nucleobases according to the hard and soft acids and bases (HSAB) concept as well as by steric factors.[12,13]

Endocyclic nitrogens are more nucleophilic than exocyclic oxygens and nitrogens. In particular the N7 atom of guanine, which is also sterically exposed in the major groove, is often found to be the site which is modified most extensively (Figure 16.2). Alkylation of endocyclic nitrogens can lead to hydrolytic deglycosylation or ring-opening following nucleophilic attack of water at C8.[12,13] On the other hand, alkylation of the exocyclic nitrogen atoms N^2 of guanine, N^6 of adenine and N^4 of cytosine, the amidic endocyclic N^1 nitrogen atoms of guanine and thymine, and the oxygen atoms O^6 of guanine and O^4 of thymine gives rise to stable adducts.[12] The stability and position of alkylation have an influence on the likelihood of such a modification to evade the body's DNA repair mechanisms and result in a mutagenic outcome.[1,13]

Electrophilic DNA-reactive agents are commonly classified into direct-acting and indirect-acting agents.[13] The latter term does not refer to their mechanism of causing damage to DNA but to the requirement for metabolic activation to turn them into electrophiles. John Ashby and colleagues made a prominent contribution to popularizing alerts of structural moieties which can react as electrophiles with DNA directly or after metabolic activation by making up a hypothetical molecule featuring the structural alerts known at the time (Figure 16.3).[14,15]

16.1.2.1.1 Direct-acting Agents. Direct-acting electrophiles, which comprise mainly alkylating and acylating agents and which tend to be of concern mainly as impurities of reagents used during synthesis,[16] include

- small alkyl, benzyl and allyl esters of sulfuric, sulfonic, phosphoric and phosphonic acid,
- alkyl chlorides, bromides and iodides,
    - primary alkyl halides may also act as indirect-acting agents after undergoing metabolic oxidation to an aldehyde

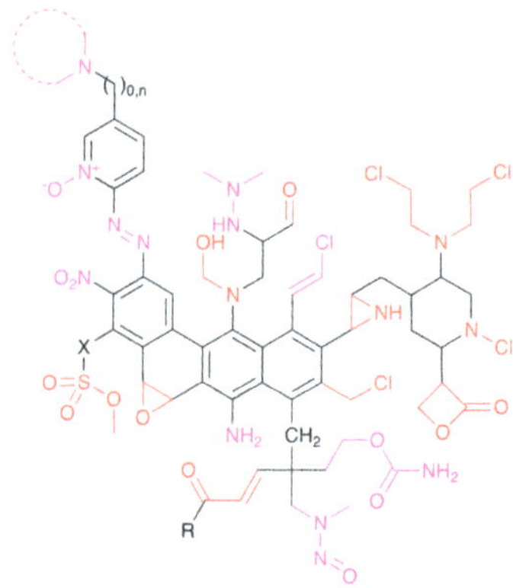

Figure 16.3 'Super mutagen', adapted from Ashby.[14,15] Direct-acting functional groups are colored in red, indirect-acting ones requiring metabolic activation in magenta.

- sulfur and nitrogen mustards [bis(2-haloethyl) sulfide and bis (2-haloethyl)amine derivatives], which can form highly reactive episulfonium and aziridinium ions, respectively,
- epoxides and aziridines,
- alkyl nitrites,
- *N*-hydroxymethyl derivatives, which can be formed metabolically by cytochrome P450 catalyzed oxidation of *N*-methyl amines,
- Michael acceptors, *i.e.* electron-deficient double or triple bonds which are substituted with one or more electron-withdrawing substituents such as carbonyl, cyano or nitro groups,
- acylating agents, including acyl halides, carbamoyl halides, haloformates, isocyanates, isothiocyanates, β-lactones and γ-sultones,
- aldehydes, which can react with primary amino groups of nucleobases and proteins to form Schiff bases, giving rise to DNA adducts and DNA interbase as well as DNA protein crosslinks.

16.1.2.1.2 Indirect-acting Agents. Electrophilic agents which are formed upon metabolic activation are termed indirect-acting agents.[13] The probably most prominent class of indirect-acting agents in medicinal chemistry comprises compounds which can be metabolized to form highly electrophilic *aryl nitrenium ions*. This includes aryl rings which are substituted with an amino, nitro, nitroso, hydroxylamino or diazo group (Figure 16.4). Arylamination occurs predominantly at the C8 position of guanine bases. Experimental evidence indicates that these thermodynamically stable adducts are formed after rearrangement of an initial N7 adduct (Figure 16.4).[17]

Metabolic hydroxylation at the α-position of a *tertiary aliphatic amine* followed by elimination of water can give rise to a reactive iminium ion. The formation of such stabilized iminium ions is often observed for cyclic amines.[19]

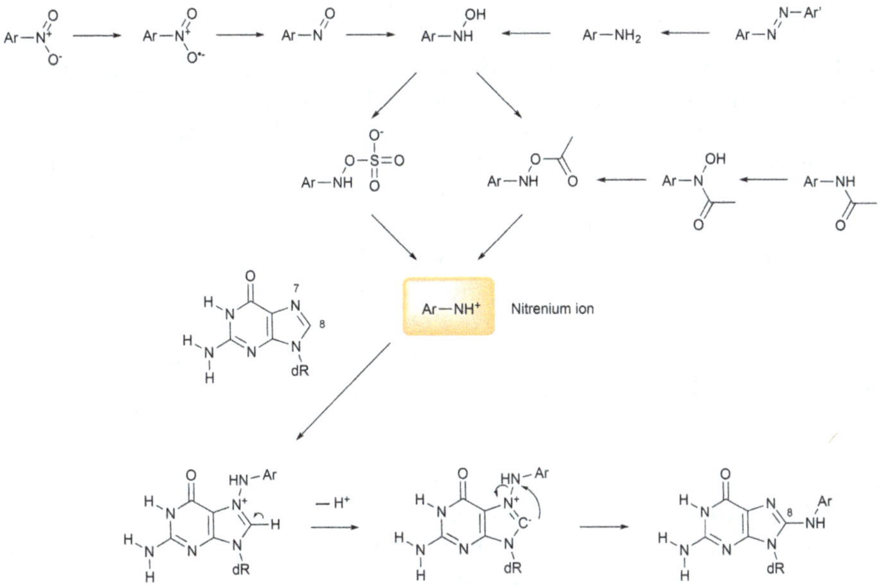

Figure 16.4 Mechanisms of formation of an aryl nitrenium ion and addition to a guanine base. Aryl nitrenium ions are formed from *N*-arylhydroxylamine derivatives following enzymatic conjugation and elimination. *N*-Arylhydroxylamines and their conjugates can be formed by enzymatic reduction of nitro and nitroso derivatives or cytochrome P450 mediated *N*-oxidation of aromatic amines, which can also be formed from azo compounds by an azoreductase.[13] Bioactivation of *N*-acetylated aromatic amines has also been described to lead to reactive aryl nitrenium ions.[18]

Hydrazines, *N-nitrosamines*, *N-nitrosamides* and *triazines* can undergo multi-step metabolic and chemical transformations to give highly reactive diazonium ions or, in the case of hydrazines, carbon radicals.[13]

Compounds with *electron-rich carbon–carbon double bonds* including halo- and alkoxy-substituted olefins and aromatic rings, such as furans and thiophenes or benzene rings substituted with electron-donating groups, can metabolically be activated to electrophilic epoxides.[13]

Certain *N-oxides of aromatic azaheterocycles* have been reported to be mutagenic. The mechanism by which *N*-oxides exert this effect will depend on the structure of the heterocycle as well as its substitution and may in some cases involve metabolic activation.[13,20]

Hydroxamic acids can undergo a Lossen rearrangement to form isocyanates. It has been proposed that the mutagenic effect observed for many hydroxamic acids is due to the formation of such highly reactive acylating agents.[21,22]

16.1.2.2 Non-DNA-reactive Mechanisms

16.1.2.2.1 Non-covalent Interactions with DNA: Intercalation and Groove Binding. The two major types of non-covalent and thus reversible interactions with double-stranded DNA, which can result in a disturbance of replication and transcription of DNA, are intercalation and groove-binding. Electrostatic interactions conferred by positive charges often account for a significant contribution to free binding energy of non-covalent binders to the polyanionic DNA. It is noteworthy that it has been concluded that the highest concentration of the negative potentials is located in the grooves despite the formal negative charge of the phosphate backbones.[23]

DNA intercalation is a non-covalent insertion into the hydrophobic space between two adjacent base pairs, which requires a transient widening of this space by a partial unwinding of the double helix.[23] Intercalation complexes are stabilized by interactions between the lowest unoccupied molecular orbital (LUMO) of the intercalator and the highest occupied molecular orbitals (HOMOs) of the adjacent purine bases. Enhancement of the HOMO–LUMO interaction by lowering of the LUMO can be effected by introduction of a positive charge to the intercalating moiety. This is exemplified by the classical intercalator ethidium, whereas its noncationic analog 3,8-diamino-6-phenylphenanthridine is not an intercalator (Figure 16.5).[24] While highly planar and electron-deficient structures such as aromatic polycycles are predisposed best to intercalation, the space between base-pairs can widen up to 7–8 Å to accommodate also less planar compounds.[25]

Other examples of intensely studied DNA intercalators include berberine, which used to be employed as a dye, the bacteriostatic disinfectant proflavine, 9-aminoacridine and its derivatives such as the antineoplastic amsacrine,

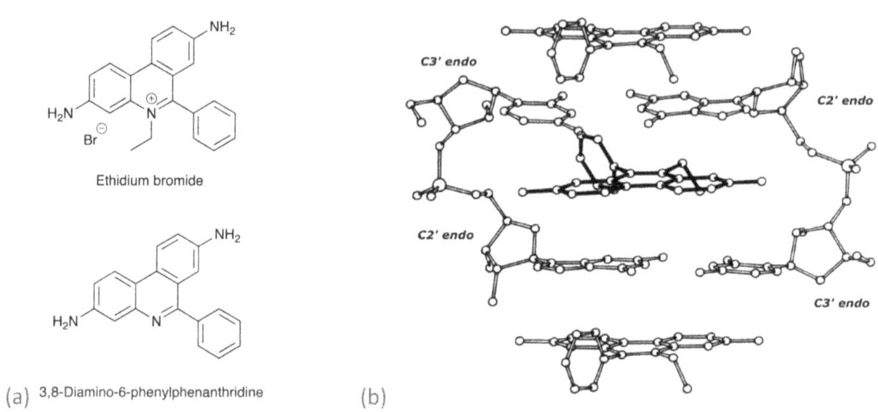

Figure 16.5 (a) Chemical structures of ethidium bromide and 3,8-diamino-6-phenanthridine. (b) Visualization of an X-ray crystal structure of a complex of ethidium with the dinucleoside monophosphate cytidylyl-(3′–5′)-guanosine (CpG) as a model of double-helical DNA and RNA: An intercalative ethidium molecule is shown with dark bonds, while stacked ethidium molecules are drawn with light open bonds.[26,27] Reproduced from ref. 26, https://doi.org/10.1007/s10969-016-9202-4, under the terms of the CC BY 4.0 license, http://creativecommons.org/licenses/by/4.0/.

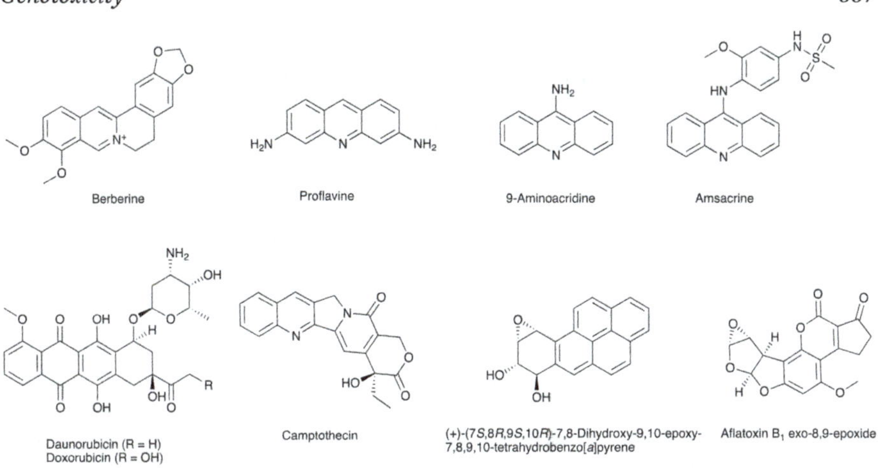

Figure 16.6 Chemical structures of DNA intercalators.

Figure 16.7 (a) Nuclear magnetic resonance (NMR) solution structure of an undecanucleotide duplex, d(CGGTC**A**CGAGG),d(CCTCGTGACCG), in which (+)-(7*S*,8*R*,9*S*,10*R*)-7,8-dihydroxy-9,10-epoxy-7,8,9,10-tetrahydrobenzo[*a*]pyrene is covalently bonded to the exocyclic N^6 amino group of the central deoxyadenosine, d**A**6, through *trans* addition at C10 of the epoxide and epoxide ring opening (PDB ID 1JDG; the image was created using Mol*).[30] (b) (+)-(7*S*,8*R*,9*S*,10*R*)-7,8-Dihydroxy-9,10-epoxy-7,8,9,10-tetrahydrobenzo[*a*]pyrene bonded to the central deoxyadenosine.[29]

anthracycline derivatives including doxo- and daunorubicin (also known as daunomycin), the anticancer agent camptothecin, (+)-(7*S*,8*R*,9*S*,10*R*)-7,8-dihydroxy-9,10-epoxy-7,8,9,10-tetrahydrobenzo[*a*]pyrene, the most carcinogenic metabolite of benzo[*a*]pyrene, and the 8,9-exo epoxide metabolite of aflatoxin B_1, a highly carcinogenic food contaminant produced by the molds *Aspergillus flavus* and *Aspergillus parasiticus* (Figure 16.6).[23,28] It should be pointed out that the epoxide metabolites of aflatoxin B_1 and benzo[*a*]pyrene are of course not pure intercalators but are also able to specifically alkylate DNA nucleobases as visualized for the latter in Figure 16.7.[29]

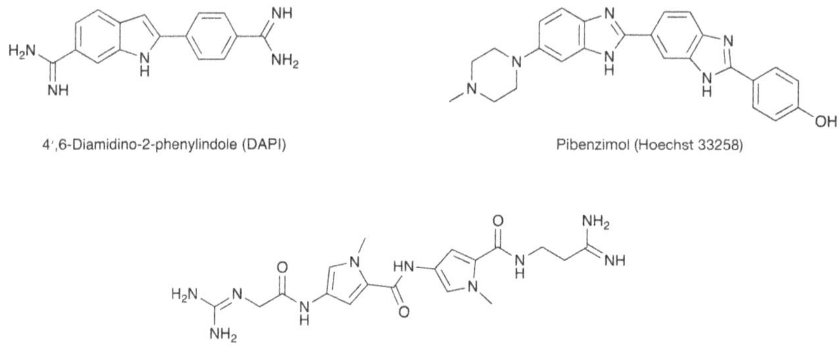

Figure 16.8 Chemical structures of three well-studied DNA minor groove binders.

Many, but not all, DNA intercalators are able to inhibit or "poison" topoisomerase I or II. Topoisomerases are enzymes that catalyze the transient breaking and rejoining of a single (topoisomerase I) or double strand (topoisomerase II) during the unwinding of DNA, thus facilitating replication and transcription. Poisoning of topoisomerase enzymes is a consequence of a stable intercalator–DNA–topoisomerase complex, which impedes religation of the DNA breaks. Examples of topoisomerase II inhibitors include anthracyclines and amsacrine, while camptothecin is a topoisomerase I inhibitor.[23,28]

The second major mode of non-covalent interaction of small molecules with a DNA double-helix is binding to the major or the minor groove. Structural features which enable binding to the latter, including planarity and positively charged groups, are, in part, similar to those of intercalators as exemplified by 4′,6-diamidino-2-phenylindole (DAPI), which can interact with DNA either as intercalator or as minor-groove binder (Figure 16.8).[23] Another prominent group of minor-groove binders are bisbenzimidazoles, which are useful as fluorescent dyes to stain DNA. Bisbenzimidazoles such as pibenzimol, which is also known as Hoechst 33258 and which had been evaluated in a phase 1 clinical trial as an anti-cancer treatment, as well as other minor groove binders have been shown to also have effects on topoisomerase enzymes (Figure 16.8).[31,32]

The natural product netropsin, which was first isolated from the actinobacterium *Streptomyces netropsis* and which has antibiotic as well as antiviral activity, is a representative of polyamide minor groove binders to DNA with sequence selectivity elicited by specific hydrogen bonding interactions (Figure 16.9).[31,32]

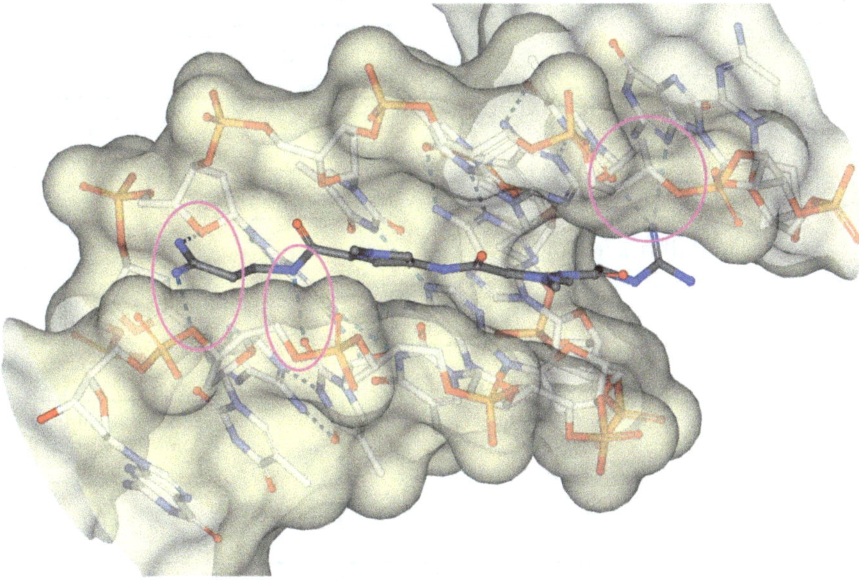

Figure 16.9 X-ray crystal structure of the DNA dodecamer d(CGCAAATTTGCG) complexed with the minor groove binding drug netropsin, which forms hydrogen bonds (blue dotted lines, highlighted with ovals in magenta) with O^2 of thymines and N^3 of adenine as well as backbone oxygens (PDB ID 121D; the image was created using NGL viewer and RCSB PDB).[33,34] Water molecules have been omitted.

16.1.2.2.2 Mechanisms Not Involving a Direct Interaction With DNA. Mechanisms which elicit genotoxicity without directly interacting with DNA are expected to show threshold dose–response curves.[35] Major mechanisms of non-direct DNA damage are:

- Inhibition of kinases regulating the cell cycle[36]
- Inhibition of topoisomerases[37]
- Inhibition of tubulin polymerization[38,39]
- Inhibition of DNA synthesis (nucleotide pool imbalance, inhibition of DNA polymerases)[40–43]
- Inhibition of histone deacetylases (HDACs), DNA methyl transferases[44,45]
- Overloading of oxidative defense mechanisms[46]
- Metal ion chelation; disturbance of metal homeostasis[47]
- Metabolic overload (phase II enzymes *etc.*)[48,49]

16.1.3 Screening Strategies and Regulatory Guidelines

Figure 16.10 depicts a schematic overview of a standard testing approach from early discovery stages to regulatory testing for assessment and optimization against genotoxic liabilities. The standard test battery consists of a bacterial

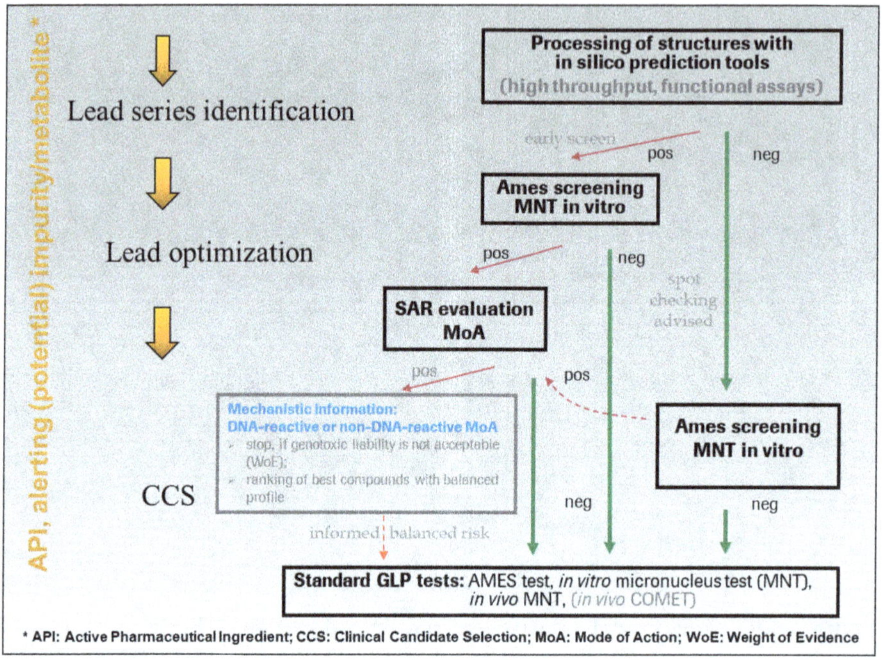

Figure 16.10 Strategy for early genotoxicity assessment and compound optimization.

mutation assay (the Ames test[50,51]), an *in vitro* mammalian cell assay measuring chromosomal damage (*e.g.* a chromosomal aberration test (CA),[52] a mouse lymphoma gene mutation assay (MLA)[53] or an *in vitro* micronucleus test (*in vitro* MNT)[54]) and an *in vivo* rodent assay, typically a micronucleus test using either bone marrow or peripheral blood cells. In case of a positive result in an *in vitro* mammalian cell assay with unknown biological relevance, *e.g.* if caused by *in vitro* cytotoxicity, a second *in vivo* test, such as a single cell gel electrophoresis (comet) assay[55] in a second tissue would be needed to assess and clarify the genotoxic potential of drug candidates. Alternatively two *in vivo* assays would be mandatory in case no *in vitro* mammalian cell assay is conducted.[56–58,104] These regulatory assays are expensive, time-consuming and require large quantities of test material. Therefore, early genotoxicity screening approaches have been introduced prior to the application of regulatory testing.

Considering the critical implications of genotoxicity for drug development and a general genotoxic liability rate of about 20% (based on Roche internal data), reliable screening methods predicting the good laboratory practice (GLP) regulatory assays are essential for efficient compound optimization. Genotoxicity screening should be both sensitive and specific. Throughput and predictability need to be balanced and suitable for the respective discovery stages, *i.e.* from lead series identification and compound optimization to clinical candidate selection. These typically include miniaturized versions of the regulatory Ames and *in vitro* MNT assays requiring much lower amounts

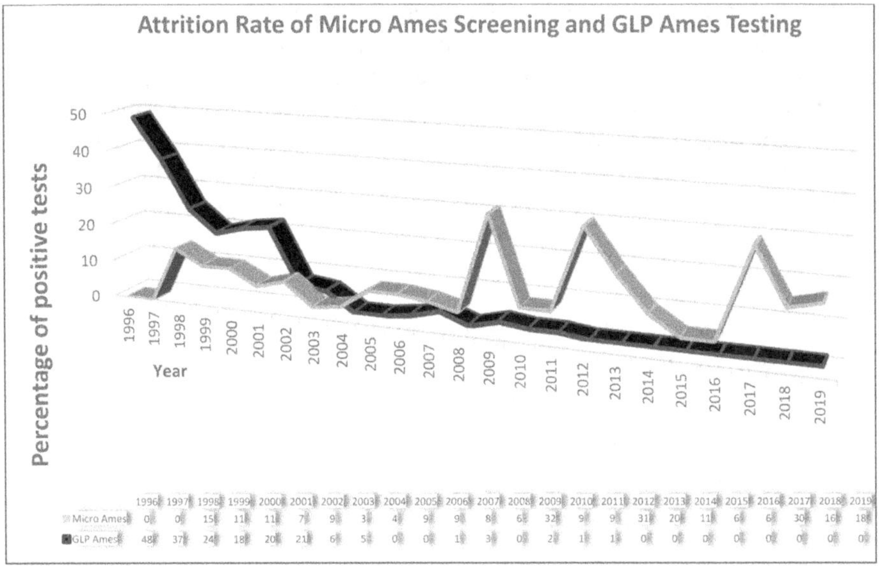

Figure 16.11 Highly effective prediction of regulatory GLP Ames test by a miniaturized Micro Ames screening version (Roche internal data). Positive GLP Ames assays after 2003 are due to a bacteria-specific mechanism and tested positive in the screening assay as well.

of test compound and time but predicting regulatory GLP assays very effectively (Figure 16.11). Higher throughput assays, such as bacterial reporter assays (*e.g.* VITOTOX[59]) and indicator assays measuring DNA damage [*e.g.* fluorometric analysis of DNA unwinding (FADU[60])] or DNA damage response (*e.g.* Greenscreen[61]) require even less compound and might be more appropriate in early phases than *in silico* approaches, which are successfully used in very early phases to highlight potential genotoxic liabilities and prioritization of testing.[62] Emerging flow cytometric and imaging analysis based methods for the micronucleus test (MNT) could increase throughput as well.[63,64]

16.1.4 Relevance

Genetic toxicology data are highly critical for further development of drug candidates, and late-stage genotoxicity failures are very costly. In general mutagenic activity by a positive Ames test might only be considered acceptable in case of a bacteria-specific mechanism (*e.g.* by azides[65]) or for life threatening indications where efficacy would relate to a genotoxic mechanism (mainly for late-stage cancer indications). There is no safe dose *per* default for such directly DNA-reactive compounds, and any potential identification of a possible threshold concentration would require very labor- and time-consuming additional (animal) testing with movement of the risk of failure to later development stages.[66,67]

Thresholds with acceptable safety margins might be observed for compounds with a positive *in vitro* mammalian cell test result in cases of chromosomal breakage (clastogenic activity) or chromosome loss (aneugenic activity) resulting from an indirect non-DNA-reactive mechanism interfering with the cell division or chromosome replication cycle. If such a compound is considered for progression, additional studies for mechanic proof of an indirect mode of action and demonstration of acceptable *in vivo* safety margins are warranted.[56]

Key References

- S. M. Galloway, *Environ. Mol. Mutagen.*, 2017, **58**, 296–324.
  - *Review of the international guidelines and regulatory requirements for genotoxicity testing of pharmaceuticals as well as their impurities and metabolites.*
- *ICH guideline S2 (R1) on genotoxicity testing and data interpretation for pharmaceuticals intended for human use*, EMA/CHMP/ICH/126642/2008, https://www.ema.europa.eu/en/news/european-medicines-agency-issues-revised-guidance-genotoxicity-testing-medicines.
- *ICH guideline M7(R1) on assessment and control of DNA reactive (mutagenic) impurities in pharmaceuticals to limit potential carcinogenic risk*, EMA/CHMP/ICH/83812/2013, https://www.ema.europa.eu/en/ich-m7-assessment-control-dna-reactive-mutagenic-impurities-pharmaceuticals-limit-potential.
- R. Benigni and C. Bossa, *Chem. Rev.*, 2011, **111**, 2507–2536.
  - *Comprehensive review of mechanisms of chemical carcinogenicity and mutagenicity and implications for predictive toxicology.*
- J. Bentzien, E. R. Hickey, R. A. Kemper, M. L. Brewer, J. D. Dyekjaer, S. P. East and M. Whittaker, *J. Chem. Inf. Model.*, 2010, **50**, 274–297.
  - *Quantum mechanical calculations of the stabilities of the nitrenium ions of 257 primary aromatic amines were used to predict Ames activities. The results are also compared to the predictions of commercial software packages DEREK, MULTICASE, and the MOE-Toxicophore descriptor.*
- C. Hasselgren *et al.*, *Regul. Toxicol. Pharmacol.*, 2019, **107**, 104403.
  - *Outlines key effects/mechanisms and their relationship to genetic toxicity and details a framework for assessment of genetic toxicity based on available data and in silico models.*
- L. Strekowski and B. Wilson, *Mutat. Res., Fundam. Mol. Mech. Mutagen.*, 2007, **623**, 3–13.
  - *A concise overview of non-covalent interactions of small organic molecules with DNA.*

16.2 Mitigation Strategies

Due to the numerous mechanisms which may disrupt the cell cycle and the fact that DNA damage is often caused by reactive metabolites, the formation of which may, in some cases, be hard to predict, there are no universal rules to mitigate genotoxicity. Avoidance of substructures which have previously been associated with genotoxic effects and which have been described in Section 16.1.2 is among the most successful strategies.[13] A number of *in silico* tools are available to assist a thorough assessment of a compound's potential to cause genetic toxicity.[62] However, avoidance or replacement of toxophores may not in all cases be possible without substantial loss of on-target activity. Therefore, strategies to mitigate some of the most common causes of genotoxicity including the formation of DNA-reactive nitrenium ions and alkylating agents as well as DNA intercalators and inhibitors of kinases regulating the cell cycle are discussed below.

A focus on fine-tuning physicochemical properties is often too simplistic to successfully mitigate a genotoxicity liability. However, reducing planarity, modification of lipophilicity and increasing molecular weight will be promising strategies in certain cases:

Planarity is possibly the most prominent risk factor of genotoxicity since it is a prerequisite of DNA intercalation and minor groove binding (see Section 16.2.3.1). Furthermore, an extended conjugated planar system may favor the formation of a reactive nitrenium ion from an aromatic amine (see 16.2.1.1). *In silico* descriptors of planarity can be used to guide optimization efforts. Analysis of more than 1500 *in vitro* micronucleation test results at Roche showed that plane-of-best-fit (PBF),[68] a measure of three-dimensionality of a molecule which is based on calculated conformations, or also simple proxy measures of planarity such as percent or fraction of heavy atoms in planar or aromatic rings or of aromatic bonds are good predictors of chromosome damage activity of molecules. *E.g.* data display a clear trend and greater risk of a positive or weakly positive MNT response for compounds with >60% aromatic bonds compared with those with <60% (Figure 16.12).

Lipophilicity is not generally correlated with a positive read-out in *in vitro* assays of genotoxicity. However, reducing lipophilicity may decrease the rate of metabolism (see Chapter 9) and thus attenuate the formation of a

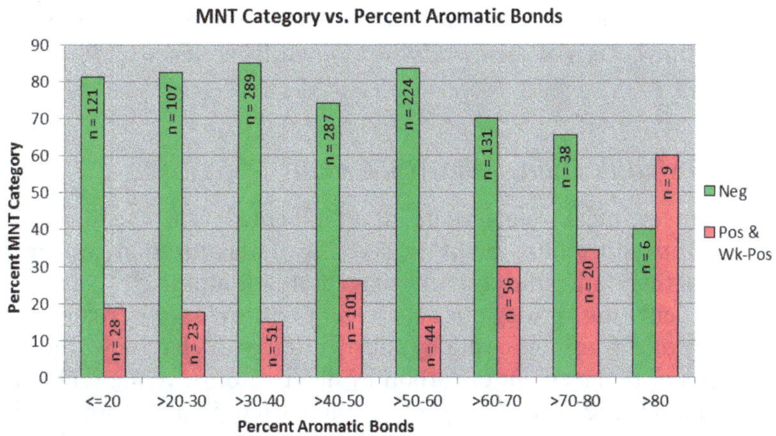

Figure 16.12 Percentage of aromatic bonds as a predictor of chromosome damage measured by *in vitro* micronucleation activity.

genotoxic reactive metabolite. On the other hand, modifications which disfavor the formation of a reactive metabolite electronically or sterically may lead to an increase in lipophilicity (this section and Section 16.3). Furthermore, increasing electron-density is a means of disrupting intercalation (see Section 16.2.3.2), which may entail an increase in lipophilicity. Last but not least, lipophilicity is a key determinant of cellular permeability (Chapter 3) and may thus affect the potential of a compound to exert a genotoxic effect within cells.

Molecular weight is another key property that determines permeability (Chapter 3). Good permeability due to low molecular weight may thus favor genotoxicity. Analysis of a Roche-internal data set indeed revealed an increasing likelihood of a positive readout with decreasing molecular weight in the micronucleus as well as the Ames test. The effect is particularly pronounced for aromatic amines below a molecular weight of 300 Daltons in the Ames test under conditions including metabolic activation. This may be due to efficient metabolism by certain cytochrome P450 enzymes, such as isoforms 1A1, 1A2, 1B1 and 2A6, which have a preference for smaller and aromatic compounds.[69–71]

16.2.1 Avoiding the Formation of Aryl Nitrenium Ions

Due to its propensity to form a highly reactive nitrenium metabolite the primary aromatic amine motif is often avoided in the design of active pharmaceutical ingredients (APIs). However, aromatic amines are commonly used as building blocks for the synthesis of heterocycles or amides and might still be released upon (metabolic) hydrolysis or be present as residual impurities. Mutagenic metabolites are generally not tolerated, and mutagenic impurities need to be limited according to the International Council for Harmonisation of Technical Requirements for Pharmaceuticals for Human Use (ICH) M7 guideline.[72] Therefore, compounds for which metabolic liberation of a mutagenic aromatic amine cannot be excluded are usually not considered for further development. The two main mitigation strategies are replacement of the mutagenic amine by a non-mutagenic isostere or prevention of the metabolic or chemical release of the aromatic amine.

16.2.1.1 Modification of the Aryl Group

Quantum mechanical calculations have indicated that the dissociation energy to form a nitrenium ion or the stability of the nitrenium ion are good predictors of the mutagenic potential of an aromatic amine.[73–76] Consistently with such calculations mutagenic activity may be decreased by reducing electron density of the aromatic ring by substitution with electron-withdrawing substituents or replacement of carbon by hetero atoms. While replacement of the aromatic ring by a non-aromatic bioisostere is often not feasible,[77] breaking conjugation can be a successful strategy, *e.g.* by removal of a ring in the case of polycyclic aromatic amines or increase of biaryl dihedral angles by introduction of *ortho* substituents.[76]

For nitroarenes modification of the one-electron reduction potential of the nitro group has been proposed as a promising approach to modulate mutagenicity.[78]

An alternative strategy to mitigate mutagenic activity of an aromatic amine or nitro compound is steric shielding of the amino or nitro group with one or, more effectively, two ortho substituents to prevent metabolic activation.[78,79] It has been shown for alkyl substituents that mutagenicity decreases with increasing steric demand of the alkyl group.[79] With the exception of bis-*ortho*-substitution, the effect of smaller substituents, such as methyl, ethyl or chlorine, tends not to be sufficient. Mutagenic activity may also be abolished by introduction of a remote sterically demanding substituent, in particular a tertiary butyl or adamantyl group.[80,81] The effectiveness of introducing bulky alkyl substituents to mitigate mutagenic activity might be due to a disruption of the interaction of a compound with metabolizing enzymes or redirection of metabolism to the alkyl group as a metabolic soft spot. In cases where an electrophilic quinone imine metabolite causes mutagenicity, the issue may be resolved by blocking the position *ortho* or *para* to the amino group. It should be kept in mind, though, that methyl or ethyl substituted aromatic amines may give rise to electrophilic imine methide metabolites.[19]

16.2.1.2 Preventing the Release of an Aromatic Amine

While hydrolysis of acyclic carboxylic amide bonds could also be influenced by modulating electronic and steric properties of the aromatic ring, it may be elusive to completely rule out in preclinical development that this process might happen in humans. Instead successful mitigation strategies of amide hydrolysis include stabilization by embedding the amide bond in a cycle, *N*-alkylation of secondary amides, amide replacement by a bioisostere[82] or inversion of the amide moiety. Furthermore, it should be pointed out that release of an aromatic amine due to hydrolytic instability is not restricted to amides but has also been observed for certain *N*-arylated heterocycles.[83] In such cases replacement or modification of the heterocycle may be a viable strategy to avoid genotoxicity.

16.2.2 Avoiding Alkylating Agents

With the exception of treatments for severe life-threatening diseases[84–87] highly reactive direct acting alkylating agents are usually avoided in candidate drugs. However, attention to thorough purification needs to be paid when such reagents are used in synthesis.[16] In drug discovery genotoxicity caused by DNA alkylation is therefore often due to reactive metabolites. Three main classes of metabolites stand out; their formation is discussed in detail in Chapter 15 – Reactive Metabolites:

- Epoxides of – in particular electron-rich – aromatic rings and olefins
- Michael acceptors, including quinones, quinone imines, diimines and quinone as well as quinone imine methides
- Iminium ions and aldehydes

16.2.3 Avoiding Intercalation and Minor Groove Binding

16.2.3.1 Reducing Planarity

One way of abolishing the capacity of a compound to insert between adjacent base pairs is introduction of appropriately placed substituents. While interaction with DNA may already be disturbed distinctly by even a small substituent, more sterically demanding groups tend to be more effective.

Another well-proven mitigation strategy is to break planarity by conformational changes, *e.g.* by increasing the torsion angle of a biaryl system by ortho substitution or (partial) saturation of planar rings.

16.2.3.2 Disfavoring Intercalation Electronically

Since intercalation complexes are stabilized by interactions between the LUMO of the intercalator and the HOMOs of the adjacent purine bases, these interactions are disfavored by increasing electron density of an aromatic system of an intercalator, thus raising the energy of the LUMO. This is accomplished by removing electron-withdrawing substituents or replacing them by electron-donating ones, reducing the number of fused aromatic rings, or replacing an electron-poor heteroaryl ring with a more electron-rich system.

16.2.3.3 Disrupting Binding to the Minor Groove

Reducing planarity and positive charge are not only effective strategies to disfavor intercalation but also binding to the minor groove. Furthermore, minor groove binding may be abrogated by disruption of hydrogen bonding.

16.2.4 Optimization Against Binding to the ATP Site of Kinases Regulating the Cell Cycle

Kinase inhibition often accounts for positive results in chromosome damage assays.[36] Most small molecule kinase inhibitors target the ATP binding site by mimicking adenine, which forms three hydrogen bonds with the amide groups of the backbone of the hinge (Figure 16.13). The vast majority of kinase inhibitors form at least one hydrogen bond with the hinge backbone. The most common feature is a hydrogen bond acceptor, which is often flanked by one or two hydrogen bond donors. Off-target kinase activity may thus be mitigated by modification of such a hydrogen bond acceptor(-donor) motif, which many planar – mostly aromatic – azaheterocycles feature, in particular when substituted with a primary or secondary amino group (Figure 16.13c).

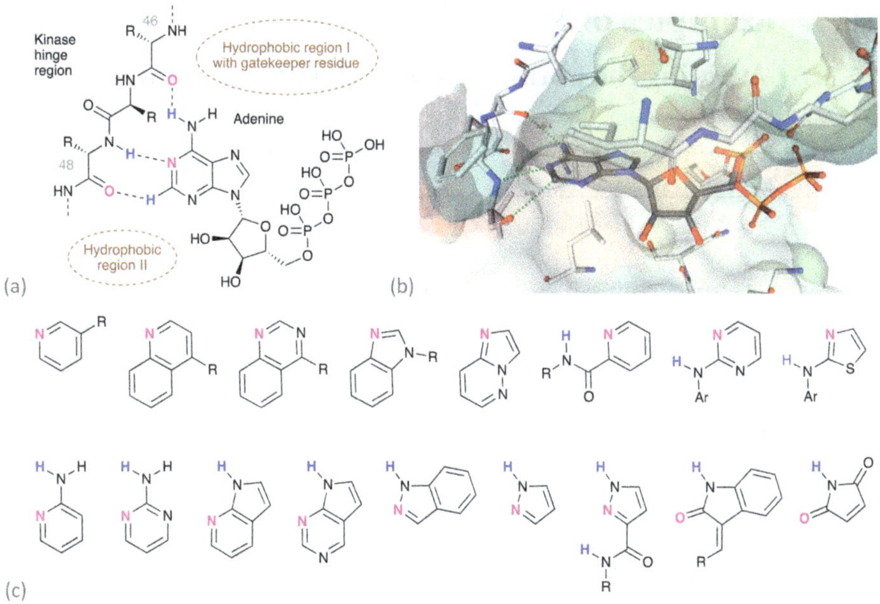

Figure 16.13 (a) Hydrogen bonding interactions of adenine with amino acids 46 and 48 of the hinge region of the ATP binding pocket of kinases.[88] (b) X-ray crystal structure of ATP binding to CDK2 (PDB ID 1FIN; the image was created using NGL viewer and RCSB PDB).[33,89] (c) Examples of hinge-binding motifs of kinase inhibitors confirmed by X-ray crystal structures. Top row: motifs interacting only with amino acid 48; bottom row: binders forming hydrogen bonds with amino acids 46 and 48.[88]

However, it should be emphasized that selectivity against unwanted kinases can also be achieved by disrupting interactions in other areas of the ATP binding site, *i.e.* the gate area, the back cleft or regions of the front cleft beside the hinge.[88] Decreasing interactions with the hydrophobic pockets of the hinge region can be applied to any other non-DNA reactive enzyme target containing a conserved ATP binding site and involved in the chromosome replication cycle, such as histone deacetylases (HDACs), topoisomerases, tubulin polymerases and mitotic kinesins, to eliminate any chromosome damaging activity.

Critical players of genetic stability are especially the CDK1, CDK2, CDK4, CDK6, AURKA, AURKB, PLK1, BUB1, and MPS1 kinases by regulating key steps of the chromosome replication cycle, such as mitotic signaling, centrosome condensation and segregation, spindle formation, and chromosome segregation and cytokinesis.[36]

16.3 Examples of Mitigation Strategies

16.3.1 Avoiding the Formation of Aryl Nitrenium Ions

Examples

1

	R	H	Me	Et	*i*-Pr	*n*-Bu	*t*-Bu	Ad
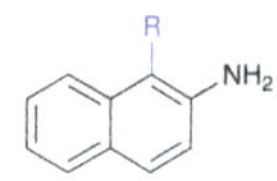	TA98 / TA100	− / +		+ / ++	(+) / +	(+) / ++	− / −	
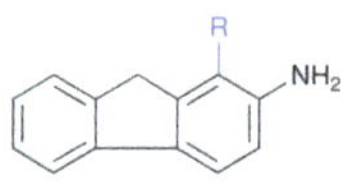	TA98 / TA100	++ / ++		+++ / ++	+++ / ++	+++ / ++	+ / (+)	
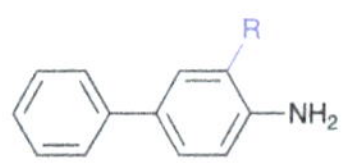	TA98 / TA100	+ / ++		+ / ++	+ / ++	(+) / +	(+) / +	
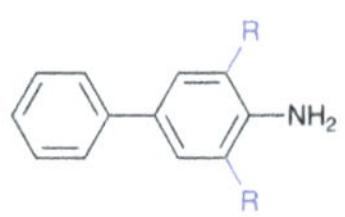	TA98 / TA100	+ / ++	− / (+)	− / (+)	− / −			
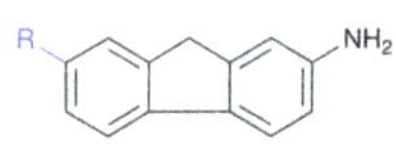	TA98 / TA100	+++ / ++	+++ / ++				+ / +	− / −
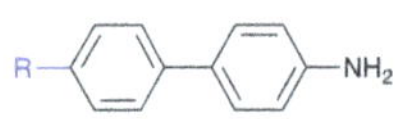	TA98 / TA100	++ / ++	++ / ++	+ / +	− / −	− / −	− / −	

In two independent studies Boche and colleagues explored the effect of alkyl substitution on the genotoxicity of 2-aminonaphthalene, 2-aminofluorene and 4-aminobiphenyl in the Ames test with *Salmonella typhimurium* strains TA98 and TA100.[79,80] Genotoxicity was mainly observed with metabolic activation in the presence of S9 (+S9), while most compounds did not show any genotoxicity without S9 (−S9). Steric shielding of the amino group of these systems by *ortho*-bis-substitution with a methyl, ethyl or isopropyl group effectively decreased mutagenicity. However, *ortho*-substitution with only one primary or secondary alkyl group increased mutagenicity. This may be rationalized by insufficient steric shielding, a tendency to increased metabolic turnover due to an increase in lipophilicity and/or the formation of imine methides. The latter cannot be formed from *ortho*-*t*-butyl substituted anilines, which showed no or strongly reduced mutagenicity. Remote *t*-butyl substitution in the 7-position of 2-aminofluorene and in the 4′-position of 4-aminobiphenyl effectively mitigated mutagenicity as well. Mutagenicity of 4-aminobiphenyl could also be abolished by 4′-substitution with *i*-propyl and *n*-butyl groups, which are potential metabolic soft spots.

Examples

2

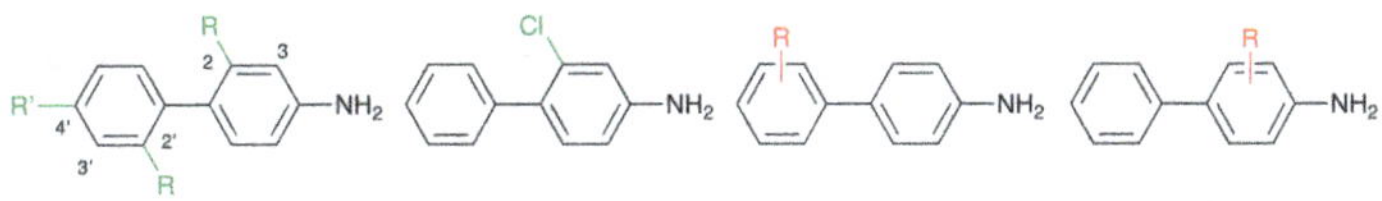

R = F, Cl, CN, R′ = H, F, Cl, CN — R = F, Cl — R = F, Cl

Ames TA98 / TA100	– / –	– / –	+ / +	+ / +

Using quantum mechanical calculations Leach and colleagues found that the dissociation of substituted *N*-acetoxy-4-aminobiphenyl derivatives to form the nitrenium ion correlates with a likelihood of the corresponding 4-aminobiphenyl being positive in the Ames test in *Salmonella typhimurium* strains TA98 and TA100.[76] Mutagenicity was only observed with metabolic activation. Substitution with electron-withdrawing substituents will destabilize the nitrenium ion and thus disfavor dissociation. It was found that single fluorine or chlorine substitution in any position does not abolish mutagenicity in the Ames test with the notable exception of the 2-chloro derivative, which effects a significant distortion of the biphenyl system in addition to its electron-withdrawing inductive effect. All derivatives with 2,2′-substitution with fluorine, chlorine or cyano were Ames negative.

3

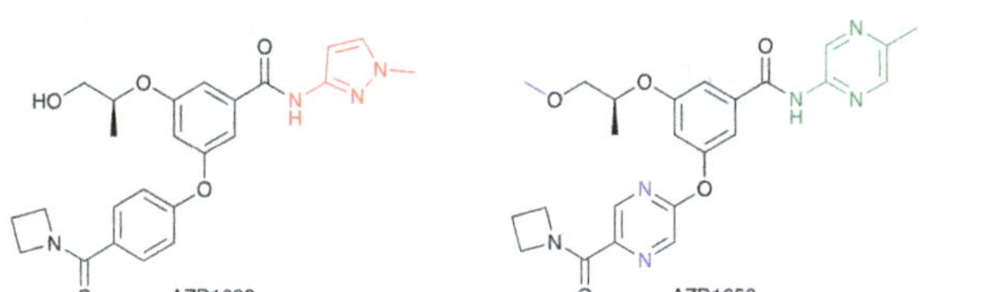

AZD1092 — AZD1656

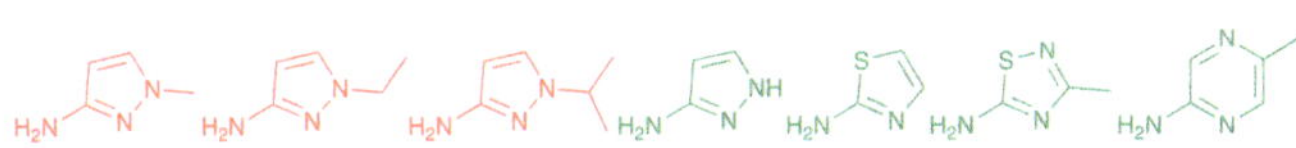

Ames test	positive	positive	positive	negative	negative	negative	negative
Calc. dE_{diss} (g) ArNHOAc → ArNH+ kcal/mol	149	147	144	157	180	170	157

In search of novel glucokinase activators a team at AstraZeneca had identified the secondary carboxylic acid amide AZD1092 as a development candidate. The potentially releasable 1-methyl-3-aminopyrazole fragment was found to be positive in the Ames assay. Although this aromatic amine was not observed in any preclinical pharmacokinetic studies and testing in other genotoxicity assays, including the rat liver unscheduled DNA synthesis and rat bone marrow micronucleus tests, did not reveal any concerns, the positive Ames result would have necessitated front-loading of more extensive genotoxicity characterization throughout development, which triggered the search for a new development candidate which is devoid of any genotoxicity liability.[90]

A number of aromatic amines were identified which tested negative in the Ames test. Using the same quantum mechanical method as in example 2 above to calculate the gas-phase dissociation energies of the corresponding *N*-acetoxy derivatives[73] revealed that the Ames-positive amines have a lower dissociation energy compared with those that were negative. Amides with the Ames-negative 2-amino-5-methylpyrazine showed a good potential to give rise to derivatives with a balanced pharmacodynamic and ADMET profile. Further optimization led to the selection of AZD1656 as a new development candidate.

(continued)

Examples

4 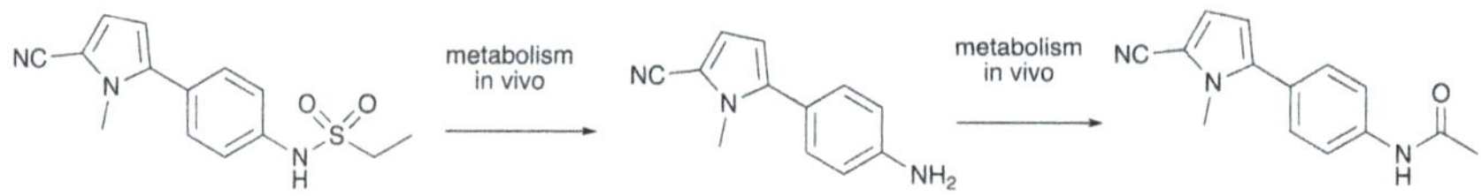

"cpd 19"

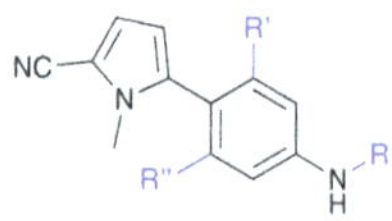

Ames (+S9)	R' = R" = H	R' = F, R" = H	R' = R" = F	R' = CF_3, R" = H
R = H	positive	negative	negative	negative
R = SO_2Et	positive	negative	**ND**	negative

As a representative example of a series of sulfonamides as novel progesterone receptor modulators explored at Pfizer "cpd 19" was selected to assess the risk of liberation of a potentially mutagenic aromatic amine.[91] After oral administration of a single dose to male rats, significant levels of the aniline and the secondary *N*-acetyl metabolite were found in plasma. The sulfonamide as well as the aniline tested positive in the Ames test with S9 activation. Mutagenic activity could be abolished by 3-trifluoromethyl, 3-fluoro or 3,5-difluoro substitution of the phenyl ring with retention of potency. Electronic as well as conformational effects may contribute to the mitigation of mutagenicity.

5 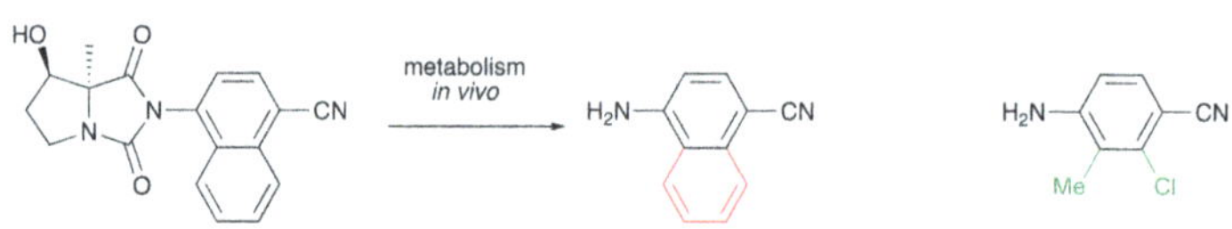

Ames (TA98 & TA100)	negative	positive	negative

As part of a program aimed at the optimization of selective androgen receptor modulators, researchers at Bristol–Myers Squibb found traces of the Ames-positive hydrolytic degradant 4-cyano-1-naphthylamine of a bicyclic *N*-aryl hydantoin lead compound in the urine of cynomolgus monkeys and dogs but not in plasma of any species.[83] As a replacement of the mutagenic 4-cyano-1-naphthylamine two dozen 2,3-substituted 4-aminobenzonitrile derivatives were explored, of which 14 were found to be less mutagenic than 4-cyano-1-naphthylamine in the SOS Chromotest assay.[92] The 2- chloro-3-methyl- and 2-bromo-3-methyl-4-aminobenzonitrile hydantoin derivatives were the most potent compounds in the series, and 2-chloro-3-methyl-4-aminobenzonitrile tested negative in the Ames test. The parent hydantoin subsequently advanced to human clinical trials.

Examples

6 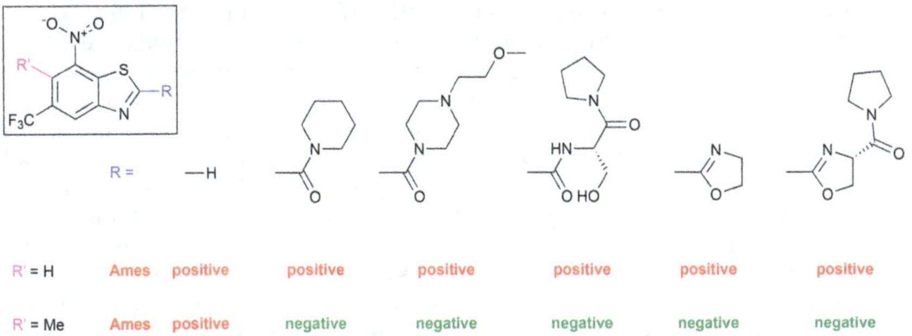

A series of antitubercular 7-nitrobenzothiazoles were found to be positive in the Ames test in *Salmonella typhimurium* TA98 and TA100 strains.[78] The nitro group was shown to be reduced by bacterial nitroreductase to its active nitroso form and thus critical for antitubercular activity, but its step-wise reduction, culminating in the formation of a highly reactive nitrenium ion, was also associated with the mutagenicity observed. It was found that the mutagenicity could be mitigated by introduction of a methyl substituent ortho to the nitro group, while the 2-unsubstituted derivative was still Ames positive, albeit to a lesser degree. The methyl group inductively decreases the single-electron affinity of the nitro group as shown computationally, thus disfavoring the first step of its metabolic reduction. In addition, steric hindrance may further decrease reactivity of the nitro group.

16.3.2 Avoiding Alkylating Agents

Examples

1 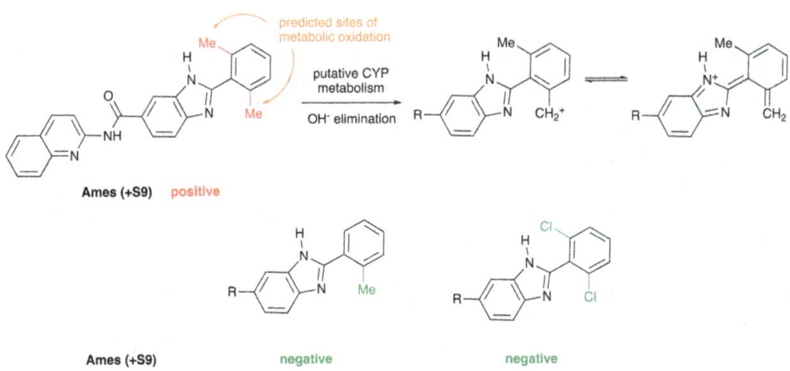

(continued)

Examples

Novartis researchers reported on an interesting observation on the effects of *ortho*-phenyl substituents on the mutagenicity of 2-phenylbenzimidazoles.[93] 2,6-Dimethyl substitution was found to consistently give rise to Ames-positive compounds in the presence of S9 metabolic activation. The methyl groups were predicted to be likely sites of CYP mediated metabolism by an *in silico* metabolism model. Methyl hydroxylation followed by hydroxide elimination would thus lead to a reactive methide. Consistent with this hypothesis the 2,6-dichloro derivative did not show any mutagenicity in the Ames test. More surprisingly it was found that the deletion of only one of the two *ortho*-methyl groups abolished mutagenicity as well. The 2,6-dimethyl derivative was also found to be a moderate time-dependent inhibitor of CYP3A4 [concentration giving 50% of maximum inhibition (IC_{50}) = 7.3 μM], whereas the mono-methyl analog was a much weaker inhibitor (IC_{50} > 50 μM). The authors proposed that this difference may be rationalized by the stronger twist of the biaryl system elicited by the *ortho*-bis-substitution, which renders the benzimidazole nitrogens more accessible to bind to the CYP3A4 heme iron, thus facilitating subsequent oxidation of one the methyl groups.

2

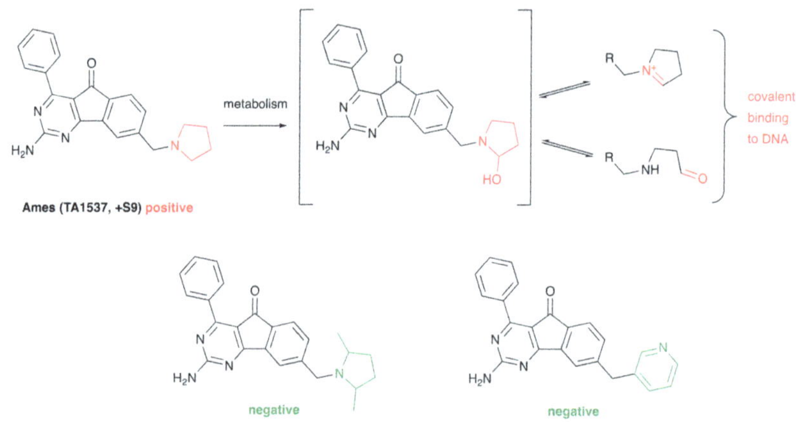

The A2A/A1 adenosine receptor antagonist 2-amino-4-phenyl-8-pyrrolidin-1-ylm ethyl-indeno[1,2-*d*]pyrimidin-5-one was determined to be genotoxic in the Ames as well as the mouse lymphoma L5178Y assays following metabolic activation.[94] The compound was positive only in the *Salmonella typhimurium* tester strain TA1537, indicating DNA intercalation. However, the fact that mutagenicity was dependent on metabolic activation indicated that covalent binding to DNA was the key driver of the effect. Results from detailed metabolism studies indicated extensive metabolism of the pyrrolidinylmethyl residue. It was concluded that the endocyclic iminium ion and the ring-opened aldehyde reactive metabolites are likely candidates for causing the genotoxicity, since covalent binding to DNA could be prevented in the presence of cyanide, which is the nucleophile of choice to trap iminium ions, and methoxylamine, which had been confirmed to form an oxime ether with the aldehyde. Genotoxicity could be abolished by replacing the unsubstituted pyrrolidinyl ring by 2,5-dimethyl-pyrrolidinyl or 3-pyridyl, which further corroborated the hypothesis that reactive intermediates resulted from metabolism of the pyrrolidinyl group.

Examples

3

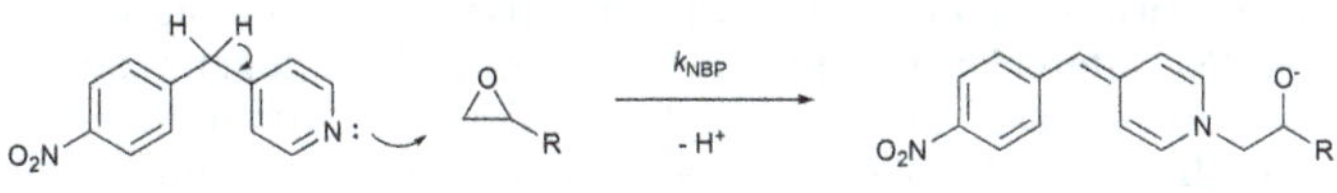

4-(4-Nitrobenzyl)pyridine (NBP)

NBP reactivity (log k_{NBP}) correlates well with experimental *in vitro* mutagenicity in Ames test / SOS Chromotest

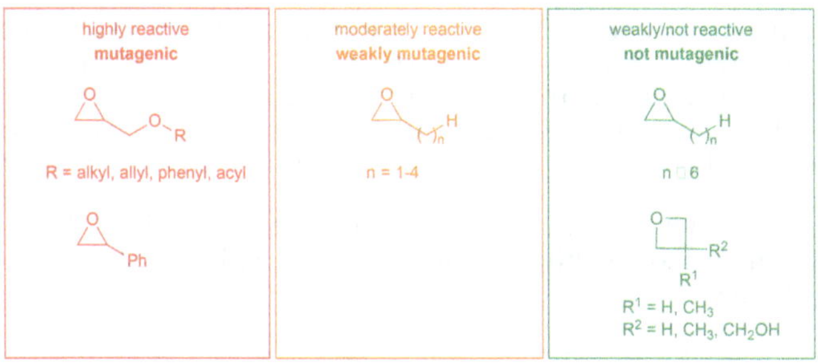

The N7 atom of guanine is a common site of DNA alkylation. As a model Thaens *et al.* used 4-(4-nitrobenzyl)pyridine (NBP) to develop a photometric assay to probe the reactivity of epoxides and oxetanes.[95] They found that the logarithm of the second-order rate constant of the NBP-electrophile reaction correlated well with the experimental mutagenicity determined in the Ames test or SOS Chromotest. They found that epoxides which are substituted with a moderately electron-withdrawing substituent were highly reactive and were reported to be mutagenic, while reactivity was markedly reduced with pure alkyl substituents. The three non-mutagenic oxetanes tested did not react with NBP.

16.3.3 Avoiding Intercalation

Examples

1

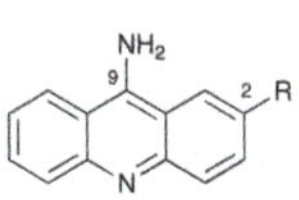

DNA binding constant: Me < H < NO_2 ≈ Cl < Br < I

Mutagenicity (Ames TA1537, -S9): Me ≈ H < Cl < NO_2

Toxicity against TA1537: H < Me < Cl < NO_2 << Br < I

(*continued*)

Examples

Acridine derivatives, including the topical antiseptic 9-aminoacridine, are known as DNA intercalators, which act as frameshift mutagens causing insertion or deletion of base pairs of DNA. Tomosaka *et al.* showed that binding affinity to calf thymus DNA was increased markedly by introduction of electron-withdrawing substituents.[96] Interestingly, intercalation gets stronger with increasing size of the halide substituents; the authors presume that the bulky bromide and iodide substituents are positioned outside the space of the base pairs in the intercalated state.

While mutagenicity of the 2-methyl substituted derivative, assessed in the Ames test in *Salmonella typhimurium* tester strain TA1537 in the absence of metabolic activation, was similar to that of the parent 9-aminoacridine, it was strongly increased for the chlorine and nitro analogs. Mutagenicity of the iodo and bromo derivatives could not be determined due to their strong toxicity.

9-Amino-2,7-dimethylacridine was found to have an even lower binding affinity than its monomethyl analog and, in line with this, showed very little mutagenicity.

2

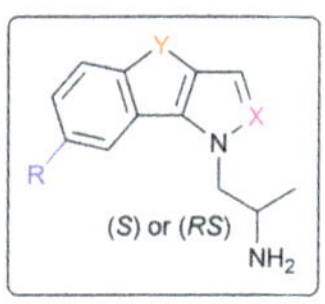

R	X	Y	Ames (TA1537, -S9)
OMe / Me	N	CH_2	+++ / +
OMe / Me	N	CMe_2	— / —
OMe / Me	CH	CH_2	+ / +
OMe / Me	CH	CMe_2	— / —
OMe	N	CH=CH	+++
OMe	CH	CH=CH	++
OMe	N	CH_2—CH_2	++
OMe	CH	CH_2—CH_2	—

A team at Roche had observed that a number of potent tricyclic 5-HT2C receptor agonists increased the number of revertant colonies in the Ames test in strain TA1537 without metabolic activation.[97] Using a DNA unwinding assay evidence was provided that the genotoxic effects were significantly attributable to the intercalating properties of the compounds.

The highest mutagenic responses were observed with two fully aromatic and consequently highly planar 1*H*-benz[*g*]indazole and -indole derivatives. Partial saturation of the central ring to reduce planarity did not abolish mutagenic activity but led to a clear reduction. It is worth noting that the indazole derivatives produced stronger mutagenic responses than their indole analogs. The same trend was observed for the 1,4-dihydro-indeno[1,2-*b*]pyrazole and -pyrrole scaffolds. However, introduction of three-dimensionality very effectively mitigated mutagenicity in this series: While all derivatives with an unsubstituted methylene bridge (Y = CH_2) showed mutagenicity, none of those with a geminal dimethyl substitution did.

Examples

3

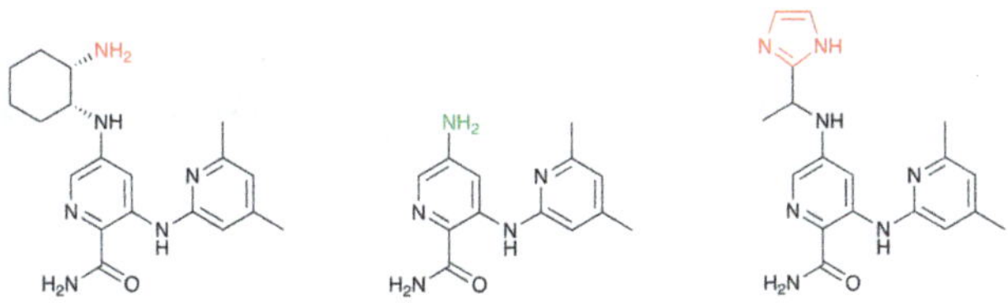

1 2 3

Ames (TA97a, +S9) positive negative positive

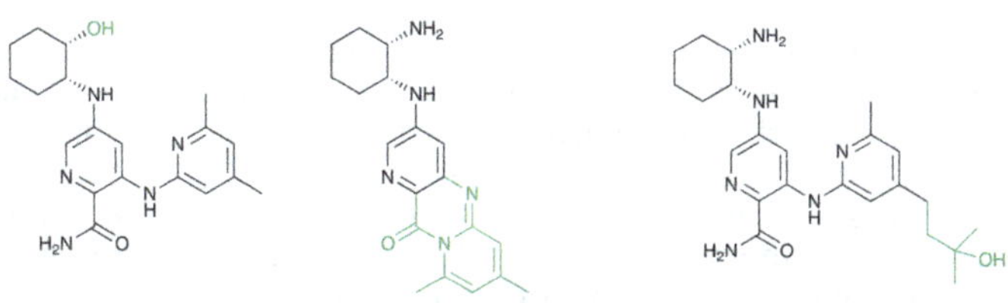

4 5 6

negative negative negative

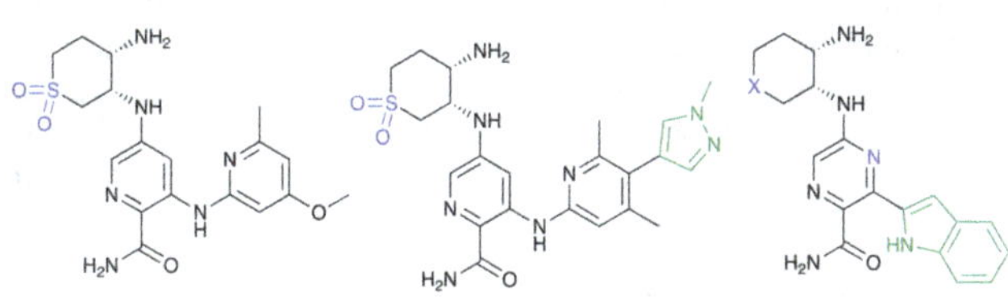

7 8 9 (X = CH_2) 10 (X = SO_2)

positive negative negative

In a program in search of selective spleen tyrosine kinase (Syk) inhibitors the highly potent lead compound **1** showed mutagenicity in the Ames test in the *Salmonella* TA97a strain.[98] Mutations in this strain are often indicative of DNA intercalation. Using a DNA unwinding assay, it was shown for several derivatives that compounds from this chemical class exert their mutagenic effect by means of intercalation. This was further corroborated by the observation that the *N*-dealkylation metabolite **2**, which had been identified in rat and human hepatocytes, tested negative in the Ames test, which was routinely run in the presence of metabolic activation with rat S9. A DNA binding model indicated that the protonated primary amino group of the 1,2-diaminocyclohexane motif forms a critical hydrogen bond with the tetrahydrofuran oxygen of the DNA backbone. While replacement of this strongly basic amine by the less basic imidazole did not abolish mutagenicity (**3**), this was achieved by substituting it with a hydroxy group (**4**). However, this modification led to a large drop in potency against Syk, necessitating the exploration of alternative tactics to mitigate mutagenicity. An interesting observation toward that end was that the tricyclic compound **5**, which had been observed as a degradation product upon standing with acid, tested Ames-negative, indicating that growing the intercalating moiety could be a successful mitigation strategy. It was shown that this could be achieved by substitution of the aminopyridine (compounds **6** and **8**) or by replacement of this group by a 2-indolyl residue, which fills a completely different space (compounds **9** and **10**). Replacement of the pyridine core by pyrazine was not critical to eliminate mutagenicity but contributed to an optimal overall profile.

(*continued*)

Examples

4

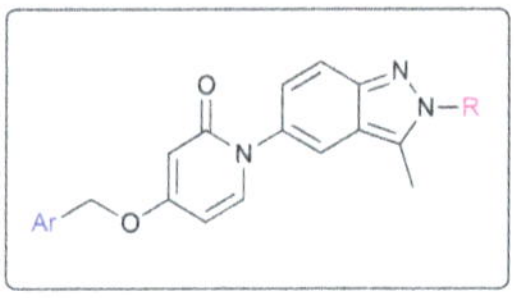

Ar	R = Methyl: hMCHR1 IC_{50} (nM)	R = Methyl: **Ames (TA1537, -S9)**	R = Cyclopropyl: hMCHR1 IC_{50} (nM)	R = Cyclopropyl: **Ames (TA1537, -S9)**
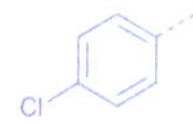	35	positive	31	negative
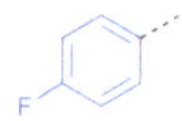	110	negative	43	negative
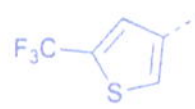	90	positive	38	negative

In a program in search of melanin-concentrating hormone receptor 1 antagonists researchers at Takeda found that lead compound 4-[(4-chlorobenzyl)oxy]-1-(2,3-dimethyl-2*H*-indazol-5-yl)pyridin-2(1*H*)-one exhibited mutagenicity in the Ames test in *Salmonella typhimurium* strain TA1537 without metabolic activation but not with metabolic activation, indicating that this compound acted as a DNA intercalator.[99] It was found that mutagenicity could be mitigated by replacement of 4-chlorophenyl by 4-fluorophenyl but not 5-(trifluoromethyl)thiophen-3-yl. Both replacements led to a marked drop in inhibitory activity against the human melanin-concentrating hormone receptor 1 (MCHR1). In contrast, potency was retained for derivatives with these three aryl groups in which the *N*-methyl substituent of the 1-(2*H*-indazole-5-yl)pyridin-2(1*H*)-one core, which was suggested to be the intercalating moiety, was substituted by cyclopropyl. All three analogues carrying this sterically more demanding group effectively breaking planarity were devoid of mutagenicity in the Ames test in TA1537 without S9.

5

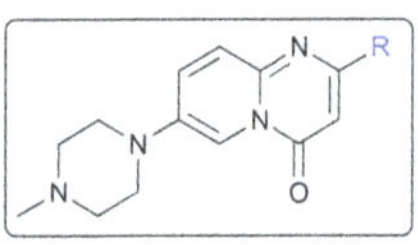

R =	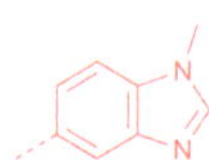	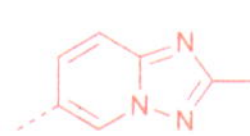	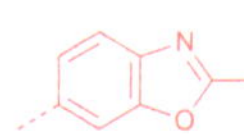	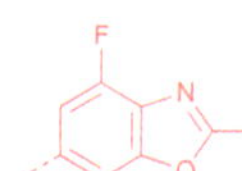
Ames (-S9)	positive	positive	positive	positive

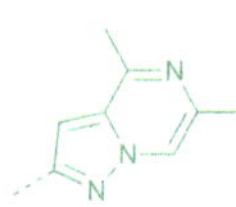	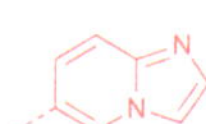	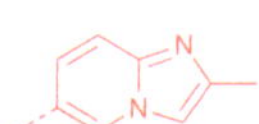	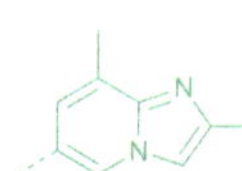	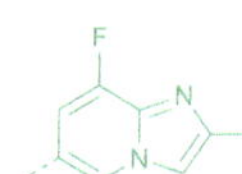
negative	positive	positive	negative	negative

In a program run jointly by F. Hoffmann-La Roche Ltd. and PTC Therapeutics directed towards the identification of small molecule survival motor neuron 2 (SMN2) splicing modifiers for the treatment of spinal muscular atrophy, genotoxicity was one of the toxicity liabilities the team had to overcome.[100] In an advanced pyridopyrimidinone lead series variation of the right-hand-side bicyclic heteroaryl substituent, which had been shown to be critically associated with the mutagenic effect observed in the Ames test, gave rise to substituted pyrazolopyrazine and imidazopyridine derivatives which were found to be Ames-negative as well as potent splicing modifiers. While there does not seem to be an evident structure–property relationship within this series of compounds, the results indicate that both electronic and steric effects may play a role.

16.3.4 Optimization Against Binding to the ATP Site of Kinases

Examples

1

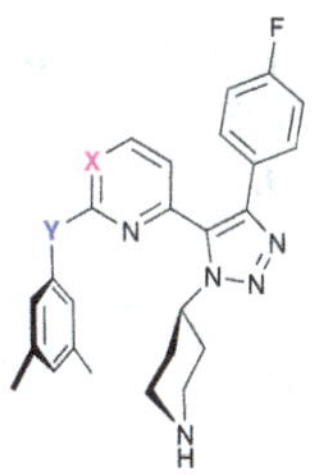

X	Y	p38α K_d (nM)	BRD4-D1 IC50 (nM)
N	NH	2.2	27'000
N	O	120	
CH	NH	6500	
CH	O	5000	

In an endeavor aimed at identifying selective inhibitors of bromodomain-containing protein 4 (BRD4) a double-digit micromolar *N*-aryl-2-aminopyrimidine derivative was used as a starting point.[101] The prototypical *N*-aryl-2-aminopyrimidine hinge binding motif[88] rendered this compound a potent kinase inhibitor, with >10 000-fold selectivity for p38α over BRD4-D1. While the p38α activity was not linked to a genotoxicity finding in this study, mitigating kinase activity was essential for the identification of selective BRD4 inhibitors to investigate their physiological and pharmacological effects.

It was found that kinase activity could be reduced substantially by disruption of the 2-aminopyrimidine donor–acceptor hinge binding motif: Replacing NH by O led to a 50-fold reduction in p38α inhibition. An even more dramatic drop of more than three orders of magnitude was achieved by replacing one of the pyrimidine N atoms by CH. The *syn* conformation of the resulting 2-substituted pyridine derivatives as drawn is the preferred conformation.[102] A further reduction of p38α inhibition was achieved by replacement of the 4-fluorophenyl group, which was suggested to form interactions with a hydrophobic pocket of the kinase, by a methyl group.

2

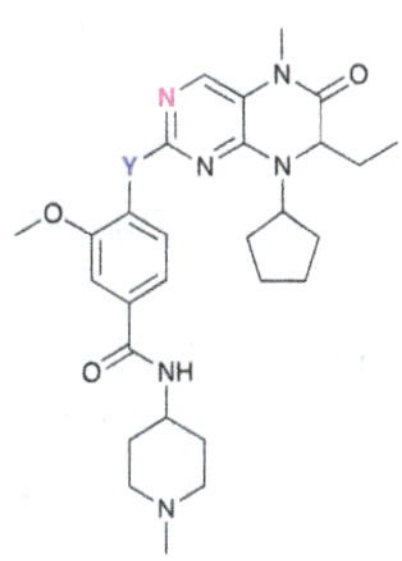

cpd	Y	PLK1 K_i (nM)
BI-2536	NH	0.22
39q	O	240

Polo-like kinase 1 (PLK1) plays an essential role in spindle formation and chromosome segregation during mitosis. Its inhibition can therefore induce genotoxicity (aneuploidy and polyploidy) by disrupting the cell cycle. Since it is often overexpressed in tumor cells, this effect may be of therapeutic use in the treatment of cancer. During SAR studies around BI-2536, a picomolar PLK1 inhibitor which also inhibits BRD4 with an inhibition constant (K_i) of 56 nM, it was found that replacement of the NH donor, which forms a hydrogen bond with the hinge region, by O led to a 1000-fold reduction of PLK1 inhibition, while BRD4 activity decreased less than twofold.[103]

References

1. O. D. Schärer, *Angew. Chem., Int. Ed.*, 2003, **42**, 2946–2974.
2. C. E. Weeden and M.-L. Asselin-Labat, *Biochim. Biophys. Acta, Mol. Basis Dis.*, 2018, **1864**, 89–101.
3. L. P. Bignold, *Mutat. Res., Rev. Mutat. Res.*, 2009, **681**, 271–298.
4. J. M. Parry, E. M. Parry, S. Ellard, T. Warr, J. O'Donovan and A. Lafi, in *Chromosome Segregation and Aneuploidy*, 1993.
5. M. Mishima, *Front. Biosci., Scholar Ed.*, 2017, **9**, 1–16.
6. E. Rencüzoğulları and M. Aydın, *Drug Chem. Toxicol.*, 2019, **42**, 409–429.
7. D. Lebrecht and U. A. Walker, *Cardiovasc. Toxicol.*, 2007, **7**, 108–113.
8. Z. V. Varga, P. Ferdinandy, L. Liaudet and P. Pacher, *Am. J. Physiol.: Heart Circ. Physiol.*, 2015, **309**, H1453–H1467.
9. I. J. Martins, in *Frontiers in Clinical Drug Research - CNS and Neurological Disorders*, ed. A. ur-Rahman, Bentham Science Publishers, 2015, pp. 158–192.
10. L. A. Loeb, *Cancer Res.*, 2001, **61**, 3230.
11. L. A. Loeb, *Nat. Rev. Cancer*, 2011, **11**, 450–457.
12. K. S. Gates, *Chem. Res. Toxicol.*, 2009, **22**, 1747–1760.
13. R. Benigni and C. Bossa, *Chem. Rev.*, 2011, **111**, 2507–2536.
14. J. Ashby, *Environ. Mutagen.*, 1985, **7**, 919–921.
15. J. Ashby and D. Paton, *Mutat. Res., Fundam. Mol. Mech. Mutagen.*, 1993, **286**, 3–74.
16. G. Szekely, M. C. Amores de Sousa, M. Gil, F. Castelo Ferreira and W. Heggie, *Chem. Rev.*, 2015, **115**, 8182–8229.
17. W. G. Humphreys, F. F. Kadlubar and F. P. Guengerich, *Proc. Natl. Acad. Sci. U. S. A.*, 1992, **89**, 8278–8282.
18. L. Verna, J. Whysner and G. M. Williams, *Pharmacol. Ther.*, 1996, **71**, 83–105.
19. A. S. Kalgutkar, I. Gardner, R. S. Obach, C. L. Shaffer, E. Callegari, K. R. Henne, A. E. Mutlib, D. K. Dalvie, J. S. Lee, Y. Nakai, J. P. O'Donnell, J. Boer and S. P. Harriman, *Curr. Drug Metab.*, 2005, **6**, 161–225.
20. M. Gabay, M. Cabrera, R. Di Maio, J. A. Paez, N. Campillo, M. L. Lavaggi, H. Cerecetto and M. Gonzalez, *Curr. Top. Med. Chem.*, 2014, **14**, 1374–1387.
21. M.-S. Lee and M. Isobe, *Cancer Res.*, 1990, **50**, 4300–4307.
22. S. Shen and A. P. Kozikowski, *ChemMedChem*, 2016, **11**, 15–21.
23. L. Strekowski and B. Wilson, *Mutat. Res., Fundam. Mol. Mech. Mutagen.*, 2007, **623**, 3–13.
24. S. E. Patterson, J. M. Coxon and L. Strekowski, *Bioorg. Med. Chem.*, 1997, **5**, 277–281.
25. L. B. Hendry, V. B. Mahesh, E. D. Bransome and D. E. Ewing, *Mutat. Res., Fundam. Mol. Mech. Mutagen.*, 2007, **623**, 53–71.
26. H. M. Sobell, *J. Struct. Funct. Genomics*, 2016, **17**, 17–31.
27. S. C. Jain and H. M. Sobell, *J. Biomol. Struct. Dyn.*, 1984, **1**, 1179–1194.
28. L. R. Ferguson and W. A. Denny, *Mutat. Res., Fundam. Mol. Mech. Mutagen.*, 2007, **623**, 14–23.

29. P. Pradhan, S. Tirumala, X. Liu, J. M. Sayer, D. M. Jerina and H. J. C. Yeh, *Biochemistry*, 2001, **40**, 5870–5881.
30. D. Sehnal, A. S. Rose, J. Koča, S. K. Burley and S. Velankar, in *Proceedings of the Workshop on Molecular Graphics and Visual Analysis of Molecular Data*, Eurographics Association, Goslar, DEU, 2018, pp. 29–33.
31. S. M. Nelson, L. R. Ferguson and W. A. Denny, *Mutat. Res., Fundam. Mol. Mech. Mutagen.*, 2007, **623**, 24–40.
32. P. R. Turner and W. A. Denny, *Mutat. Res., Fundam. Mol. Mech. Mutagen.*, 1996, **355**, 141–169.
33. A. S. Rose, A. R. Bradley, Y. Valasatava, J. M. Duarte, A. Prlić and P. W. Rose, *Bioinformatics*, 2018, **34**, 3755–3758.
34. L. Tabernero, N. Verdaguer, M. Coll, I. Fita, G. A. van der Marel, J. H. van Boom, A. Rich and J. Aymami, *Biochemistry*, 1993, **32**, 8403–8410.
35. M. Kirsch-Volders, A. Vanhauwaert, U. Eichenlaub-Ritter and I. Decordier, *Toxicol. Lett.*, 2003, **140–141**, 63–74.
36. S. Kirchner, in *Polypharmacology in Drug Discovery*, ed. J.-U. Peters, John Wiley & Sons, Inc., Hoboken, New Jersey, 2012, pp. 63–81.
37. L. C. Backer, J. W. Allen, K. Harrington-Brock, J. A. Campbell, D. M. Demarini, C. L. Doerr, D. R. Howard, A. D. Kligerman and M. M. Moore, *Mutagenesis*, 1990, **5**, 541–547.
38. M. Brunner, S. Albertini and F. E. Würgler, *Mutagenesis*, 1991, **6**, 1–1001.
39. Z. Cheng, X. Lu and B. Feng, *A Review of Research Progress of Antitumor Drugs Based on Tubulin Targets*, 2020.
40. D. E. Amacher and G. N. Turner, *Mutat. Res., Fundam. Mol. Mech. Mutagen.*, 1987, **176**, 123–131.
41. B. A. Kunz, *Mutat. Res., Fundam. Mol. Mech. Mutagen.*, 1988, **200**, 133–147.
42. S. M. Galloway, J. E. Miller, M. J. Armstrong, C. L. Bean, T. R. Skopek and W. W. Nichols, *Mutat. Res., Fundam. Mol. Mech. Mutagen.*, 1998, **400**, 169–186.
43. O. Rechkoblit, R. E. Johnson, A. Buku, L. Prakash, S. Prakash and A. K. Aggarwal, *Sci. Rep.*, 2019, **9**, 16400.
44. A. J. Olaharski, Z. Ji, J. Y. Woo, S. Lim, A. E. Hubbard, L. Zhang and M. T. Smith, *Toxicol. Sci.*, 2006, **93**, 341–347.
45. C. Stresemann, B. Brueckner, T. Musch, H. Stopper and F. Lyko, *Cancer Res.*, 2006, **66**, 2794–2800.
46. Z. Zhang, H. M. Shen, Q. F. Zhang and C. N. Ong, *Am. J. Physiol.: Lung Cell. Mol. Physiol.*, 1999, **277**, L743–L748.
47. W. Bal, A. Maria Protas and K. S. Kasprzak, in *Metal Ions in Toxicology: Effects, Interactions, Interdependencies*, The Royal Society of Chemistry, 2011, vol. 8, pp. 319–373.
48. W. W. Ku, A. Bigger, G. Brambilla, H. Glatt, E. Gocke, P. J. Guzzie, A. Hakura, M. Honma, H. J. Martus, R. S. Obach and S. Roberts, *Mutat. Res., Genet. Toxicol. Environ. Mutagen.*, 2007, **627**, 59–77.
49. D. J. Kirkland, M. L. Sheil, M. A. Streicker and G. E. Johnson, *Regul. Toxicol. Pharmacol.*, 2021, **119**, 104838.

50. D. M. Maron and B. N. Ames, *Mutat. Res., Environ. Mutagen. Relat. Subj.*, 1983, **113**, 173–215.
51. OECD, *Test No. 471: Bacterial Reverse Mutation Test*, 2020, https://www.oecd-ilibrary.org/content/publication/9789264071247-en.
52. OECD, *Test No. 473: In Vitro Mammalian Chromosomal Aberration Test*, 2016, https://www.oecd-ilibrary.org/content/publication/9789264264649-en.
53. OECD, *Test No. 490: In Vitro Mammalian Cell Gene Mutation Tests Using the Thymidine Kinase Gene*, 2016, https://www.oecd-ilibrary.org/content/publication/9789264264908-en.
54. OECD,*Test No. 487: In Vitro Mammalian Cell Micronucleus Test*, 2016, https://www.oecd-ilibrary.org/content/publication/9789264264861-en.
55. OECD, *Test No. 489: In Vivo Mammalian Alkaline Comet Assay*, 2016, https://www.oecd-ilibrary.org/content/publication/9789264264885-en.
56. *ICH Guideline S2 (R1) on Genotoxicity Testing and Data Interpretation for Pharmaceuticals Intended for Human Use*, EMA/CHMP/ICH/126642/2008, https://www.ema.europa.eu/en/news/european-medicines-agency-issues-revised-guidance-genotoxicity-testing-medicines, accessed January 2019.
57. OECD, *Test No. 474: Mammalian Erythrocyte Micronucleus Test*, 2016, https://www.oecd-ilibrary.org/content/publication/9789264264762-en.
58. D. Kirkland, Y. Uno, M. Luijten, C. Beevers, J. van Benthem, B. Burlinson, S. Dertinger, G. R. Douglas, S. Hamada, K. Horibata, D. P. Lovell, M. Manjanatha, H.-J. Martus, N. Mei, T. Morita, W. Ohyama and A. Williams, *Mutat. Res., Genet. Toxicol. Environ. Mutagen.*, 2019, **847**, 403035.
59. D. Van Der Lelie, L. Regniers, B. Borremans, A. Provoost and L. Verschaeve, *Mutat. Res., Genet. Toxicol. Environ. Mutagen.*, 1997, **389**, 279–290.
60. M. Moreno-Villanueva, T. Eltze, D. Dressler, J. Bernhardt, C. Hirsch, P. Wick, G. Von Scheven, K. Lex and A. Bürkle, *Altex*, 2011, **28**, 295–303.
61. P. W. Hastwell, T. W. Webster, M. Tate, N. Billinton, A. M. Lynch, J. S. Harvey, R. W. Rees and R. M. Walmsley, *Mutagenesis*, 2009, **24**, 455–463.
62. C. Hasselgren, E. Ahlberg, Y. Akahori, A. Amberg, L. T. Anger, F. Atienzar, S. Auerbach, L. Beilke, P. Bellion, R. Benigni, J. Bercu, E. D. Booth, D. Bower, A. Brigo, Z. Cammerer, M. T. D. Cronin, I. Crooks, K. P. Cross, L. Custer, K. Dobo, T. Doktorova, D. Faulkner, K. A. Ford, M. C. Fortin, M. Frericks, S. E. Gad-McDonald, N. Gellatly, H. Gerets, V. Gervais, S. Glowienke, J. Van Gompel, J. S. Harvey, J. Hillegass, M. Honma, J.-H. Hsieh, C.-W. Hsu, T. S. Barton-Maclaren, C. Johnson, R. Jolly, D. Jones, R. Kemper, M. O. Kenyon, N. L. Kruhlak, S. A. Kulkarni, K. Kümmerer, P. Leavitt, S. Masten, S. Miller, C. Moudgal, W. Muster, A. Paulino, E. Lo Piparo, M. Powley, D. P. Quigley, M. V. Reddy, A.-N. Richarz, B. Schilter, R. D. Snyder, L. Stavitskaya, R. Stidl, D. T. Szabo, A. Teasdale, R. R. Tice, A. Trejo-Martin, A. Vuorinen, B. A. Wall, P. Watts, A. T. White, J. Wichard, K. L. Witt, A. Woolley, D. Woolley, C. Zwickl and G. J. Myatt, *Regul. Toxicol. Pharmacol.*, 2019, **107**, 104403.

63. P. A. Escobar, R. A. Kemper, J. Tarca, J. Nicolette, M. Kenyon, S. Glowienke, S. G. Sawant, J. Christensen, T. E. Johnson, C. McKnight, G. Ward, S. M. Galloway, L. Custer, E. Gocke, M. R. O'Donovan, K. Braun, R. D. Snyder and B. Mahadevan, *Mutat. Res., Rev. Mutat. Res.*, 2013, **752**, 99–118.
64. A. V. Thougaard, J. Christiansen, T. Mow and J. J. Hornberg, *Environ. Mol. Mutagen.*, 2014, **55**, 704–718.
65. W. M. Owais, A. Kleinhofs and R. A. Nilan, *Mutat. Res., Genet. Toxicol.*, 1979, **68**, 15–22.
66. M. Guérard, M. Baum, A. Bitsch, G. Eisenbrand, A. Elhajouji, B. Epe, M. Habermeyer, B. Kaina, H. J. Martus, S. Pfuhler, C. Schmitz, A. Sutter, A. D. Thomas, C. Ziemann and R. Froetschl, *Mutat. Res., Rev. Mutat. Res.*, 2015, **763**, 181–201.
67. A. Hartwig, M. Arand, B. Epe, S. Guth, G. Jahnke, A. Lampen, H.-J. Martus, B. Monien, I. M. C. M. Rietjens, S. Schmitz-Spanke, G. Schriever-Schwemmer, P. Steinberg and G. Eisenbrand, *Arch. Toxicol.*, 2020, **94**, 1787–1877.
68. N. C. Firth, N. Brown and J. Blagg, *J. Chem. Inf. Model.*, 2012, **52**, 2516–2525.
69. R. J. Turesky and L. Le Marchand, *Chem. Res. Toxicol.*, 2011, **24**, 1169–1214.
70. D. F. V. Lewis and M. Dickins, *Drug Discovery Today*, 2002, **7**, 918–925.
71. D. F. Lewis and Y. Ito, *Expert Opin. Drug Metab. Toxicol.*, 2010, **6**, 661–674.
72. *ICH Guideline M7(R1) on Assessment and Control of DNA Reactive (Mutagenic) Impurities in Pharmaceuticals to Limit Potential Carcinogenic Risk*, EMA/CHMP/ICH/83812/2013, https://www.ema.europa.eu/en/ich-m7-assessment-control-dna-reactive-mutagenic-impurities-pharmaceuticals-limit-potential.
73. A. G. Leach, R. Cann and S. Tomasi, *Chem. Commun.*, 2009, 1094–1096.
74. J. Bentzien, E. R. Hickey, R. A. Kemper, M. L. Brewer, J. D. Dyekjaer, S. P. East and M. Whittaker, *J. Chem. Inf. Model.*, 2010, **50**, 274–297.
75. P. McCarren, G. R. Bebernitz, P. Gedeck, S. Glowienke, M. S. Grondine, L. C. Kirman, J. Klickstein, H. F. Schuster and L. Whitehead, *Bioorg. Med. Chem.*, 2011, **19**, 3173–3182.
76. A. M. Birch, S. Groombridge, R. Law, A. G. Leach, C. D. Mee and C. Schramm, *J. Med. Chem.*, 2012, **55**, 3923–3933.
77. T. M. Sodano, L. A. Combee and C. R. J. Stephenson, *ACS Med. Chem. Lett.*, 2020, **11**, 1785–1788.
78. S. Landge, V. Ramachandran, A. Kumar, J. Neres, K. Murugan, C. Sadler, M. D. Fellows, V. Humnabadkar, P. Vachaspati, A. Raichurkar, S. Sharma, S. Ravishankar, S. Guptha, V. K. Sambandamurthy, T. S. Balganesh, B. G. Ugarkar, V. Balasubramanian, B. S. Bandodkar and M. Panda, *ChemMedChem*, 2016, **11**, 331–339.
79. C. Glende, H. Schmitt, L. Erdinger, G. Engelhardt and G. Boche, *Mutat. Res., Genet. Toxicol. Environ. Mutagen.*, 2001, **498**, 19–37.

80. C. Glende, M. Klein, H. Schmitt, L. Erdinger and G. Boche, *Mutat. Res., Genet. Toxicol. Environ. Mutagen.*, 2002, **515**, 15–38.
81. M. Klein, L. Erdinger and G. Boche, *Mutat. Res.*, 2000, **467**, 69–82.
82. N. A. Meanwell, *J. Med. Chem.*, 2011, **54**, 2529–2591.
83. L. G. Hamann, M. C. Manfredi, C. Sun, S. R. Krystek, Y. Huang, Y. Bi, D. J. Augeri, T. Wang, Y. Zou, D. A. Betebenner, A. Fura, R. Seethala, R. Golla, J. E. Kuhns, J. A. Lupisella, C. J. Darienzo, L. L. Custer, J. L. Price, J. M. Johnson, S. A. Biller, R. Zahler and J. Ostrowski, *Bioorg. Med. Chem. Lett.*, 2007, **17**, 1860–1864.
84. M. H. Potashman and M. E. Duggan, *J. Med. Chem.*, 2009, **52**, 1231–1246.
85. J. Singh, R. C. Petter, T. A. Baillie and A. Whitty, *Nat. Rev. Drug Discovery*, 2011, **10**, 307–317.
86. R. A. Bauer, *Drug Discovery Today*, 2015, **20**, 1061–1073.
87. A. K. Ghosh, I. Samanta, A. Mondal and W. R. Liu, *ChemMedChem*, 2019, **14**, 889–906.
88. O. P. J. van Linden, A. J. Kooistra, R. Leurs, I. J. P. de Esch and C. de Graaf, *J. Med. Chem.*, 2014, **57**, 249–277.
89. P. D. Jeffrey, A. A. Russo, K. Polyak, E. Gibbs, J. Hurwitz, J. Massagué and N. P. Pavletich, *Nature*, 1995, **376**, 313–320.
90. M. J. Waring, D. S. Clarke, M. D. Fenwick, L. Godfrey, S. D. Groombridge, C. Johnstone, D. McKerrecher, K. G. Pike, J. W. Rayner, G. R. Robb and I. Wilson, *MedChemComm*, 2012, **3**, 1077–1081.
91. C. McComas, J. Cohen, C. Huselton, M. Marella, E. Melenski, C. Mugford, O. Slayden, R. Winneker, J. Wrobel, M. R. Yudt and A. Fensome, *Bioorg. Med. Chem. Lett.*, 2012, **22**, 7119–7122.
92. M. Hofnung and P. Quillardet, *Ann. N. Y. Acad. Sci.*, 1988, **534**, 817–825.
93. Y. Kwak, G. Coppola, C. J. Forster, T. A. Gilmore, Y. Gong, A. Kanter, A. Neubert, B. Stroup, P. Szklennik, S. Glowienke, P. Stadelmann, L. Bell, S. Bickford, E. Gangl, M. Gunduz, M. Jain, J. Zhan and M. H. Serrano-Wu, *Bioorg. Med. Chem. Lett.*, 2011, **21**, 1422–1424.
94. H. K. Lim, J. Chen, C. Sensenhauser, K. Cook, R. Preston, T. Thomas, B. Shook, P. F. Jackson, S. Rassnick, K. Rhodes, V. Gopaul, R. Salter, J. Silva and D. C. Evans, *Chem. Res. Toxicol.*, 2011, **24**, 1012–1030.
95. D. Thaens, D. Heinzelmann, A. Bohme, A. Paschke and G. Schuurmann, *Chem. Res. Toxicol.*, 2012, **25**, 2092–2102.
96. H. Tomosaka, S. Omata, E. Hasegawa and K. Anzai, *Biosci., Biotechnol., Biochem.*, 1997, **61**, 1121–1125.
97. S. Albertini, M. Bös, E. Gocke, S. Kirchner, W. Muster and J. Wichmann, *Mutagenesis*, 1998, **13**, 397–403.
98. J. M. Ellis, M. D. Altman, A. Bass, J. W. Butcher, A. J. Byford, A. Donofrio, S. Galloway, A. M. Haidle, J. Jewell, N. Kelly, E. K. Leccese, S. Lee, M. Maddess, J. R. Miller, L. Y. Moy, E. Osimboni, R. D. Otte, M. V. Reddy, K. Spencer, B. Sun, S. H. Vincent, G. J. Ward, G. H. Woo, C. Yang, H. Houshyar and A. B. Northrup, *J. Med. Chem.*, 2015, **58**, 1929–1939.

99. H. Igawa, M. Takahashi, M. Ikoma, H. Kaku, K. Kakegawa, A. Kina, J. Aida, S. Okuda, Y. Kawata, T. Noguchi, N. Hotta, S. Yamamoto, M. Nakayama, Y. Nagisa, S. Kasai and T. Maekawa, *Bioorg. Med. Chem.*, 2016, **24**, 2504–2518.
100. H. Ratni, G. M. Karp, M. Weetall, N. A. Naryshkin, S. V. Paushkin, K. S. Chen, K. D. McCarthy, H. Qi, A. Turpoff, M. G. Woll, X. Zhang, N. Zhang, T. Yang, A. Dakka, P. Vazirani, X. Zhao, E. Pinard, L. Green, P. David-Pierson, D. Tuerck, A. Poirier, W. Muster, S. Kirchner, L. Mueller, I. Gerlach and F. Metzger, *J. Med. Chem.*, 2016, **59**, 6086–6100.
101. A. S. Carlson, H. Cui, A. Divakaran, J. A. Johnson, R. M. Brunner, W. C. K. Pomerantz and J. J. Topczewski, *ACS Med. Chem. Lett.*, 2019, **10**, 1296–1301.
102. K. A. Brameld, B. Kuhn, D. C. Reuter and M. Stahl, *J. Chem. Inf. Model.*, 2008, **48**, 1–24.
103. L. Chen, J. L. Yap, M. Yoshioka, M. E. Lanning, R. N. Fountain, M. Raje, J. A. Scheenstra, J. W. Strovel and S. Fletcher, *ACS Med. Chem. Lett.*, 2015, **6**, 764–769.
104. S. M. Galloway, *Environ. Mol. Mutagen.*, 2017, **58**, 296–324.

CHAPTER 17

Drug-induced Photosensitivity

JEAN-FRANÇOIS FOURNIER*

Galapagos, 102 Avenue Gaston Roussel, 93230 Romainville, France
*E-mail: jean-francois.fournier@umontreal.ca

17.1 Introduction

17.1.1 Definition

Drug-induced photosensitivity (DIPS) refers to a toxicological event caused by the combination of a drug and UV or visible light irradiation of the affected organ, usually skin, but occasionally eyes.[1] DIPS is further categorized by mechanisms depending on the exact pathogenesis and clinical and histological features. This chapter will focus on the two main types: phototoxicity (common) and photoallergy (much less common). These can be clinically difficult to distinguish because: (1) the same drug can give rise to both mechanisms and (2) clinical and histological features can be similar.[2] The first corresponds to a dose-dependent undesired pharmacological effect and is similar to a severe sunburn.[1] The latter corresponds to a T-cell mediated immune reaction with an onset that can be delayed up to three days following exposure and is similar to eczema. Photoallergies are not dose-dependent and as such can be caused by minute amounts of drug. Other DIPS types, such as pellagra, pseudoporphyria or lichenoid, will not be discussed here.

Drug Discovery Series No. 79
The Medicinal Chemist's Guide to Solving ADMET Challenges
Edited by Patrick Schnider
© The Royal Society of Chemistry 2021
Published by the Royal Society of Chemistry, www.rsc.org

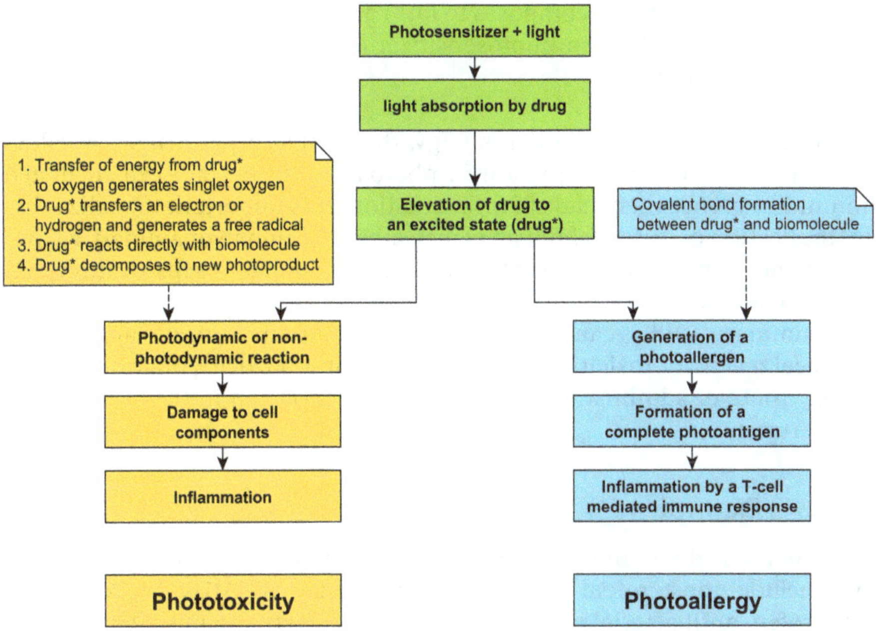

Figure 17.1 General mechanisms for phototoxicity and photoallergy.

17.1.2 Mechanisms

Both phototoxicity and photoallergy follow a similar generic path, albeit with a completely different inflammation mechanism (Figure 17.1). A photosensitizer drug, following irradiation by (1) UVA (most commonly since it penetrates skin deeper than UVB), (2) UVB (especially in the case of topical administration because this leads to high drug concentration in the epidermis) or (3) visible light, absorbs light and is elevated to an excited state. This unstable state induces a chemical reaction that ultimately leads to inflammation.

In the case of phototoxicity, four types of chemical reactions can occur.

1. The excited drug transfers energy to oxygen to produce singlet oxygen (photodynamic reaction), which is the ultimate perpetrator causing damage to cells (*e.g.* lomefloxacin and ciprofloxacin).[3]
2. The excited drug transfers an electron or hydrogen and generates a free radical, which directly (non-photodynamic reaction) or indirectly (formation of reactive oxygen species after reaction with oxygen; photodynamic reaction) reacts with biomolecules, thus causing damage to cells (*e.g.* naproxen, chlorpromazine and the fluoroquinolones).[3,4]
3. The excited drug directly reacts with biomolecules, which leads to damage to a cell component (*e.g.* fentichlor, psoralen, chlorpromazine and the fluoroquinolones).[3–6]

4. The excited drug decomposes to form at least one new molecule, which is either a toxin or a photosensitizer (*e.g.* naproxen, chlorpromazine and nifedipine).[3,5]

In contrast, in the case of photoallergy, the excited drug forms a covalent bond with a biomolecule to form a photoantigen, and a T-cell mediated immune response then leads to inflammation. It is important to realize that an implication of this mechanism is that subjects (human or animal) will not present any symptoms following their first exposure, as a sensitization step is a prerequisite.[2,7]

In summary, DIPS is caused by a mixture of light exposure and absorption, chemical reactivity, toxicity and, in the case of phototoxicity, pharmacokinetics (*i.e.* route of administration and drug concentration in the various skin compartments).[8]

17.1.3 Examples

Drugs in many different classes have been shown to cause DIPS. Some examples include nonsteroidal anti-inflammatory drugs (NSAIDs) naproxen and ketoprofen; antibacterial or antifungal fentichlor, tetracyclines (doxycycline, tetracycline) and fluoroquinolones (*e.g.* lomefloxacin, ciprofloxacin); and cholesterol-lowering drugs simvastatin and fenofibrate (Scheme 17.1).[1,3,4,9]

17.1.4 Screening Strategies

Compounds that do not absorb light between 290 and 700 nm with a molar extinction coefficient (MEC) of at least 1000 M^{-1} cm^{-1} are considered to be photoinactive and do not require any further testing.[10] Furthermore, both *in vitro* and *in vivo* tests are reputed to be extremely sensitive, therefore as soon as a negative result is reached in a screening cascade, compounds are considered photosafe and further evaluation is not warranted. This holds true even though *in vitro* evaluation and acute (single-dose) *in vivo* evaluation do not test for photoallergy. Indeed, photoallergy is rare and predictivity of current nonclinical photoallergy tests, whether it be following systemic or topical administration, is unknown and therefore not recommended[10] (Figure 17.2).

The most common and recognized *in vitro* test for phototoxicity is the neutral red uptake (NRU) assay in 3T3 cells recommended by the Organisation for Economic Co-operation and Development (OECD).[11] A variation of this assay conducted in normal human epidermal keratinocytes (NHEK) was recently published by Nestlé Skin Health.[12] Both assays measure cell viability in the presence and absence of daylight UV simulation, and a phototoxicity irritation factor (PIF) is calculated from the concentration giving 50% lethality (LC_{50}) ratio. The OECD guideline stipulates that compounds are considered phototoxic when the PIF is ≥ 5 whereas compounds with a PIF <2 are not phototoxic. However, a team from Novartis recently published a comparative study on 34 drug candidates evaluated both *in vitro* (3T3) and

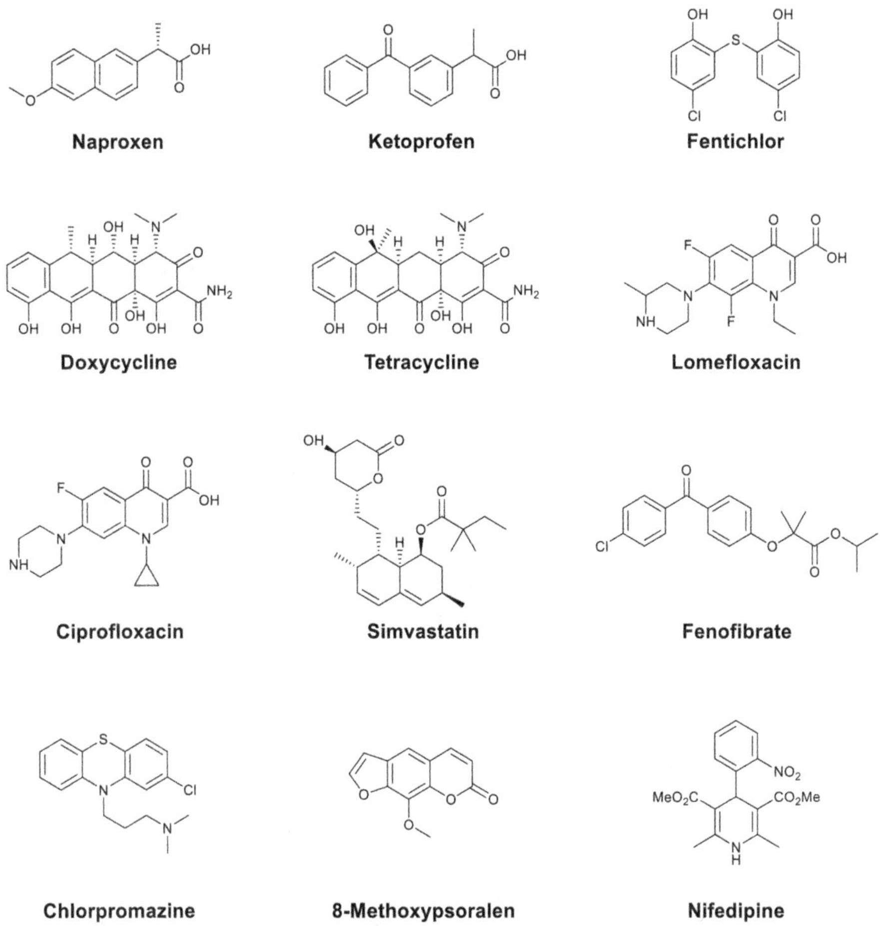

Scheme 17.1 Examples of known photosensitizing drugs.

in vivo [systemic mouse local lymph node assay (LLNA)] in which they showed that none of the compounds with an *in vitro* PIF <25 were phototoxic *in vivo*.[13]

In vivo tests can be done acutely (single-dose; only phototoxicity is assessed) or for three days (phototoxicity and photoallergy are assessed), on guinea pigs or mice, topically or systematically.[7,13–16] It is important to select a model with the correct route of administration as drug distribution in skin is fundamentally different in a systemic (high concentration in dermis, low in epidermis) *vs.* topical (high concentration in epidermis, low in dermis) approach.

17.1.5 Factors Promoting Photosensitivity

Two fundamental characteristics of photosensitizers are their ability to absorb light, especially in the UVA region, and cause toxicity. For example, MEC in excess of 20 000 M^{-1} cm^{-1}, high UV spectrum area under the curve (AUC) and small highest occupied molecular orbital–lowest unoccupied

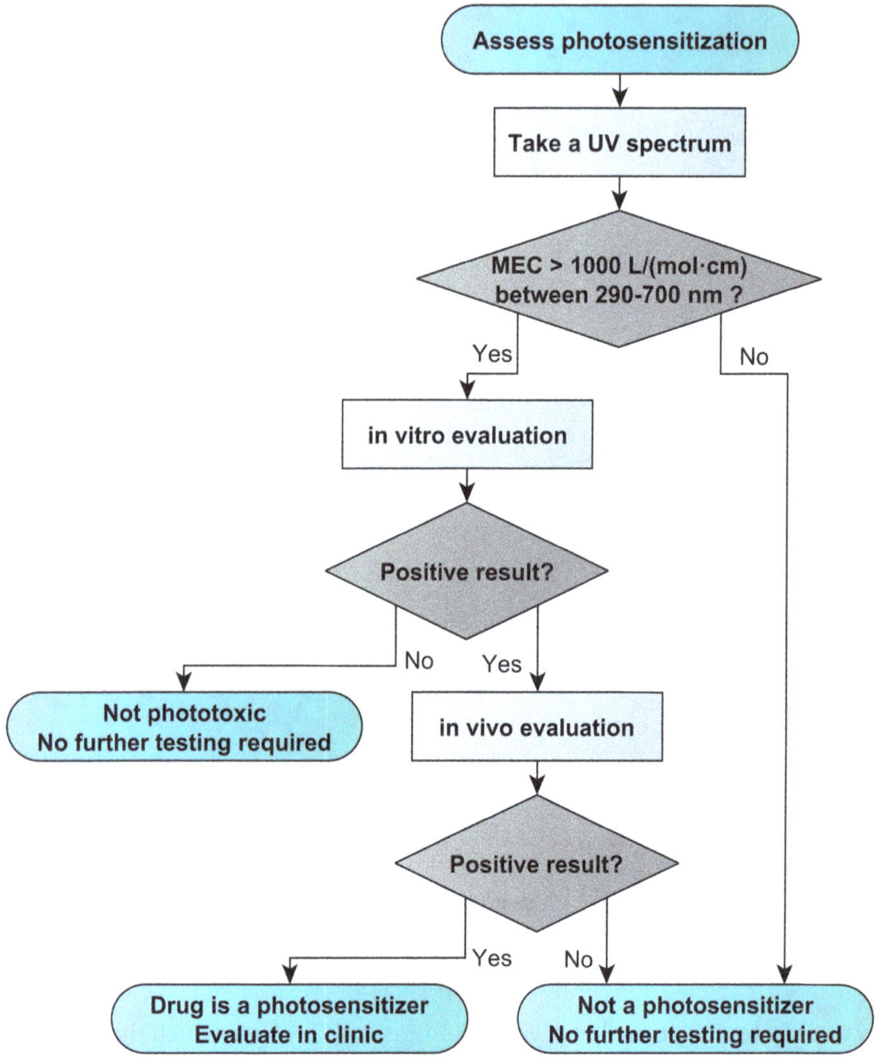

Figure 17.2 Workflow diagram of a photosensitization screening strategy.

molecular orbital (HOMO–LUMO) gap (6.7–7.7 eV) have all been linked with DIPS.[17,18] As such, factors affecting these properties, such as the number of conjugated π electrons, the number of aromatic rings and lipophilicity – a general marker of toxicity[19] – are known to favour DIPS.[12,20–22] Notably, the last two taken together add up to the intrinsic property forecast index (iPFI),[23] a general marker of compound quality in general and of phototoxicity in particular. Furthermore, as was found by Young and co-workers concerning human ether-à-go-go-related gene (hERG) and promiscuity, DIPS is better correlated with intrinsic [chromatographic log of the partition coefficient (ChromLogP)] than effective [chromatographic log of the distribution coefficient (ChromLogD)] lipophilicity (Figure 17.3).

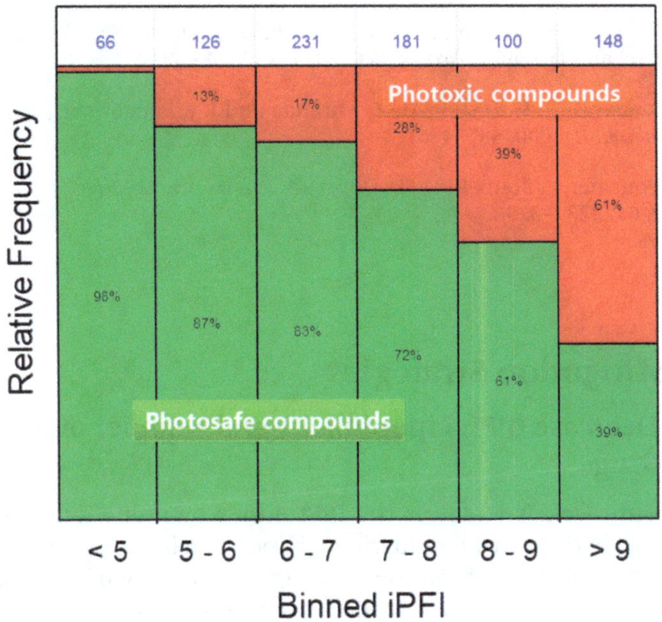

Figure 17.3 Proportion of phototoxic compounds as a function of the intrinsic property forecast indicates a good correlation. Reproduced from ref. 12 with permission from American Chemical Society, Copyright 2018.

In addition, some structural elements, such as diarylketone (fenofibrate, ketoprofen), halogenated arenes, such as thio- or amido- halogenated phenols (fentichlor), and fluoroquinolones (lomefloxacin or ciprofloxacin) are known to favour DIPS.[24] It is probably best to avoid such phototoxicophores, but if it cannot be helped, then one of the mitigation strategies presented below might prove useful in derisking DIPS.

Key References

- F. Monteiro, M. Rato and C. Martins, *Clin. Dermatol.*, 2016, **34**, 571–581.
- R. Mang, H. Stege and J. Krutmann, in *Contact Dermatitis*, ed. J. Johansen, P. Frosch and J. P. Lepoittevin, Springer, Berlin, Heidelberg, 2011, pp. 97–104.
- Department of Health and Human Services (US), Food and Drug Administration (US), Center for Biologics Evaluation and Research (US) and Center for Drug Evaluation and Research (US), in *Guidance for Industry*, Center for Drug Evaluation and Research, Silver Spring, MD, 2015, p. 1, online resource (1 PDF file (18 pages)), http://www.fda.gov/downloads/drugs/guidancecomplianceregulatoryinformation/guidances/ucm337572.pdf.
- OCDE, *Test No. 432: In Vitro 3T3 NRU Phototoxicity Test*, 2004.

(*continued*)

- J. Schümann, S. Boudon, P. Ulrich, N. Loll, D. Garcia, R. Schaffner, J. Streich, B. Kittel and D. Bauer, *Toxicol. Sci.*, 2014, **139**, 245–256.
- M. D. Barratt, J. V. Castell, M. A. Miranda and J. J. Langowski, *J. Photochem. Photobiol., B*, 2000, **58**, 54–61.
- J.-F. Fournier, C. Bouix-Peter, D. Duvert, A. P. Luzy and G. Ouvry, *J. Med. Chem.*, 2018, **61**, 3231–3236.

17.2 Mitigation Strategies

17.2.1 Decrease iPFI: Lipophilicity and Number of Aromatic Rings

A simple approach is to lower the iPFI either by reduction of the number of aromatic rings or by reduction of lipophilicity. This strategy probably works through a combination of decreased toxicity of photoproducts, decreased light absorption and decreased association of the photoactivated drug with biomolecules (for the same reasons that a drug is less bound to plasma proteins than a more lipophilic analog) which should decrease the probability of a photodynamic or non-photodynamic reaction from occurring.

Despite the utility of this mitigation strategy, it is not recommended to use a generic threshold value as other structural factors should affect where this value lies in any given chemical series.

17.2.2 Break Conjugation

Another approach is to break π-electrons conjugation between different groups. For example, between an aryl group and an acyl group, or simply between an aryl group and a heteroatom group. This can, for example, be achieved by the insertion of a methylene group between the two π systems or by the introduction of a group that sterically gears one of the π systems out of plane *vs.* the other. This strategy should lower the MEC, increase the HOMO–LUMO gap and therefore lower light absorption.

17.2.3 Remove an Aryl Halogen Atom

Several DIPS structural alerts incorporate aryl halides. This is especially true for the aryl chloride, bromide and iodide which can readily form radicals through C–X bond scission following UV irradiation, but can also be the case for aryl fluoride.[24,25] A third approach therefore simply consists of the removal of an aryl halogen atom.

17.2.4 Introduce Intramolecular Scavenger

A less conventional approach that requires a mechanistic understanding at the structural level of which photosensitization process depicted in Figure 17.1 is occurring is the introduction of an intramolecular scavenger that prevents the formation of the involved reactive radical species. In this approach, either the excited state of the molecule following UV excitation or the radical that ensues is trapped intramolecularly to yield a more stable product, thus blocking the photosensitization pathway. A typical example is preventing a photoexcited benzophenone species from abstracting hydrogen from cell membrane components by the introduction of an appropriate H-donor within 3.1 Å of the carbonyl.[26]

17.2.5 Change Positional Isomers

Because there are many different mechanisms leading to DIPS, each with their own factors, a variety of rational mitigation strategies exist. However, simply making a subtle change will sometimes suffice to thwart this threat if the change prevents one of the steps leading to DIPS from occurring. In this regard, if all rational approaches failed, one may attempt to try different positional isomers. Subtle changes in MEC, the HOMO–LUMO gap or in the fate of the photoexcited species, including potential photodegradation products, may be surprisingly efficacious in derisking a chemical series in spite of the absence of a clear mechanistic rationale.

17.3 Examples of Mitigation Strategies

In the following section, an attempt is made to classify under different strategies numerous previously published and unpublished examples of how a DIPS flag was removed through small structural modifications. The astute reader will recognize that, in some cases, a given structural change can have opposite effects on two DIPS promoting factors (*e.g.* inserting a lipophilic group to break conjugation increases iPFI), with one effect predominating over the other in terms of photosensitization outcome *for that particular example*. Conversely, other structural changes positively affect more than one promoting factor, and categorization as representing a single strategy in the absence of more data becomes somewhat arbitrary.

17.3.1 Decrease iPFI: Lipophilicity and Number of Aromatic Rings

Examples

1

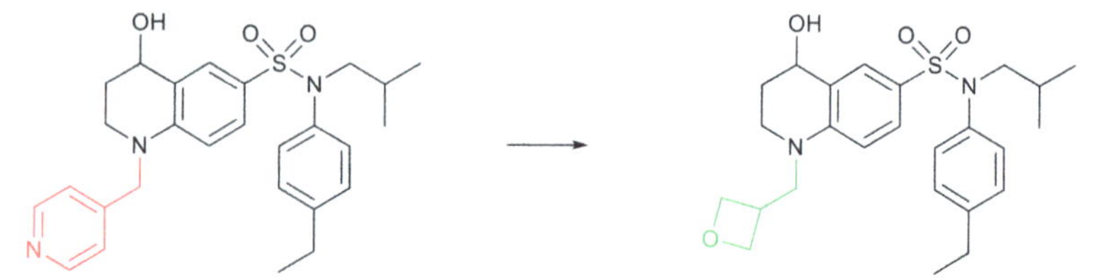

iPFI (ChromLogP + N_{Ar}) = 6.40 + 3 = 9.40
Phototoxic *in vitro*

iPFI (ChromLogP + N_{Ar}) = 6.40 + 2 = 8.40
Non-phototoxic *in vitro*

Replacing an aromatic heterocycle with an aliphatic heterocycle of similar lipophilicity reduced iPFI by 1 and removed the *in vitro* phototoxicity liability.[27]

2

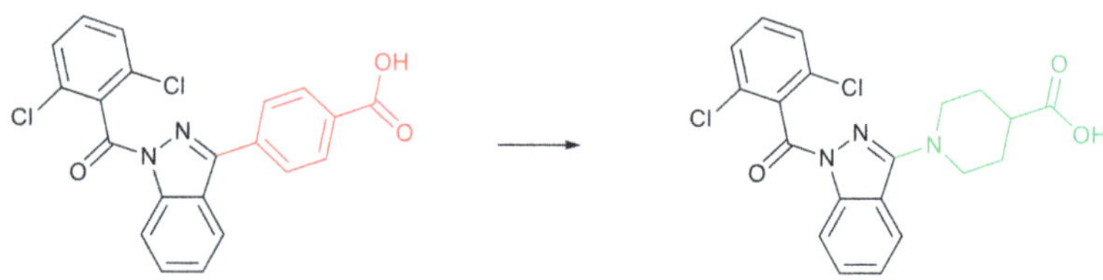

iPFI (ChromLogP + N_{Ar}) = 7.54 + 3 = 10.54
Phototoxic *in vitro* (PIF = 714)

iPFI (ChromLogP + N_{Ar}) = 7.18 + 2 = 9.18
Non-phototoxic *in vitro* (PIF = 1)

Replacing a benzoic acid aromatic group with an aliphatic heterocycle of lower lipophilicity reduced iPFI by 1.36 and removed the *in vitro* phototoxicity liability.[27]

3

cLogP + N_{Ar} = -0.03
Severe *in vivo* phototoxicity

cLogP + N_{Ar} = -0.65
Mild *in vivo* phototoxicity

Insertion of a single ethereal oxygen atom on a fluoroquinolone side-chain decreased cLogP by 0.7 and significantly reduced the severity of *in vivo* phototoxicity.[28]

Examples

4

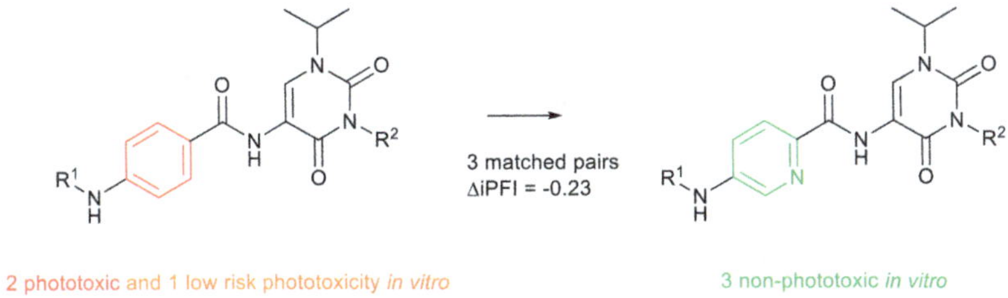

Replacing a benzene ring with a more hydrophilic pyridine ring in three matched pairs reduced iPFI by 0.23 on average and removed (in two cases) or reduced (in one case) the *in vitro* phototoxicity liability.[27]

5

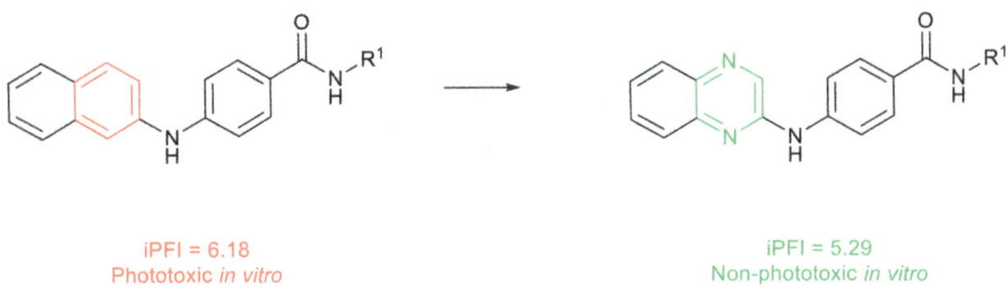

Replacing a naphthalene ring with a more hydrophilic quinoxaline ring reduced iPFI by 0.89 and removed the *in vitro* phototoxicity liability.[29]

6

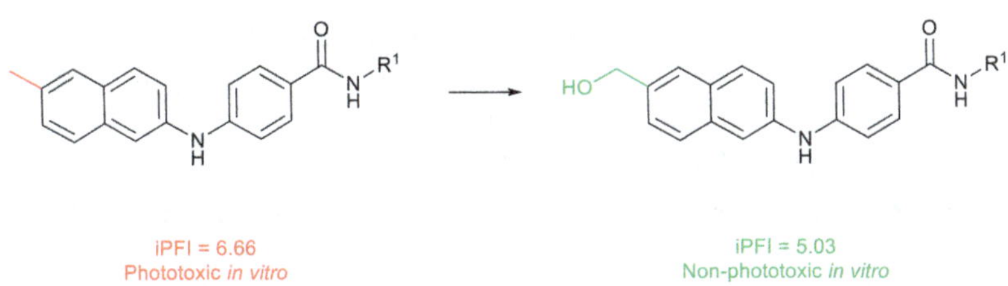

The addition of a hydrophilic hydroxy group reduced iPFI by 1.63 and removed the *in vitro* phototoxicity liability.[27]

(*continued*)

Examples	

7

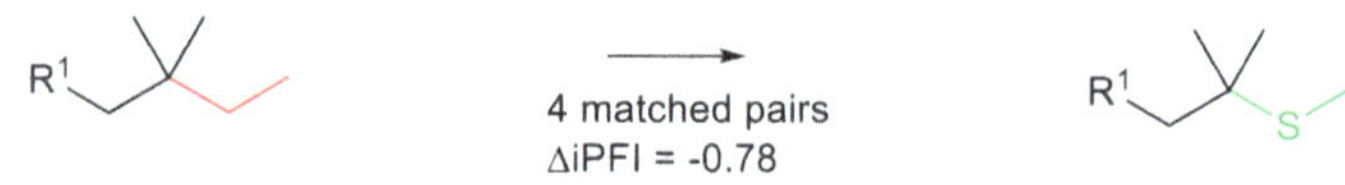

Replacement of a methylene group by sulfur in four matched pairs decreased iPFI by 0.78 on average and removed (in three cases) or reduced (in one case) the *in vitro* phototoxicity liability.[27]

8

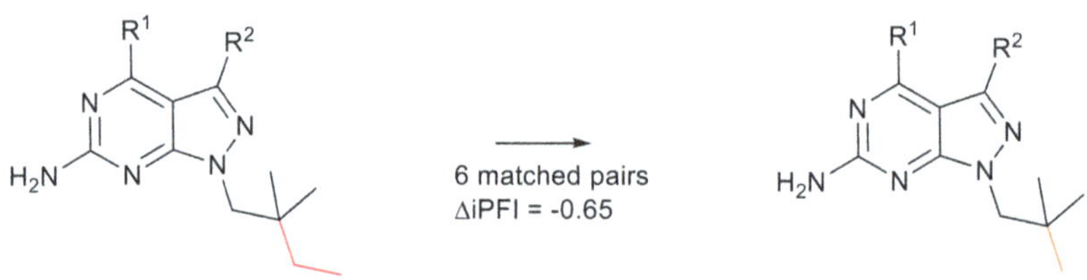

Deletion of a single lipophilic methylene group in six matched pairs decreased iPFI by 0.65 on average and removed (in two cases) or reduced (in three cases) the *in vitro* phototoxicity liability.[27]

9

	Phototoxicity potential			
	Increase	Stable	Reduction	ΔiPFI
4789 matched pairs	335	3143	1311	<0
1715 matched pairs	57	1005	653	<= −1

Reproduced from ref. 12 with permission from American Chemical Society, Copyright 2018.

A matched pair analysis on 926 compounds gave a total of 4789 matched pairs with negative ΔiPFI values, of which 1311 led to a reduction in phototoxicity risk, and 1715 matched pairs with ΔiPFI values below or equal to −1, of which 653 led to a reduction in phototoxicity risk. A larger iPFI reduction is thus accompanied by a stronger probability of phototoxicity potential reduction.

17.3.2 Break Conjugation

Examples

1

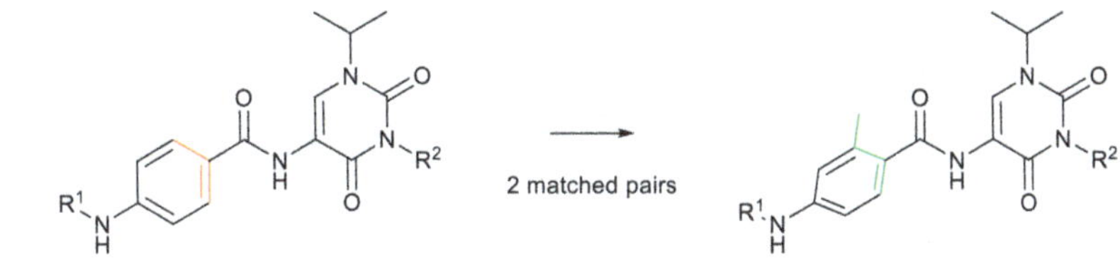

1 phototoxic and 1 low risk phototoxicity *in vitro* — 1 low risk phototoxicity and 1 non-phototoxic *in vitro*

Addition of an *ortho*-methyl group on a phenyl ring in two matched pairs forced the ring to twist out of plane *vs.* the acyl moiety, which broke conjugation and reduced the *in vitro* phototoxicity liability in both matched pairs.[29]

2

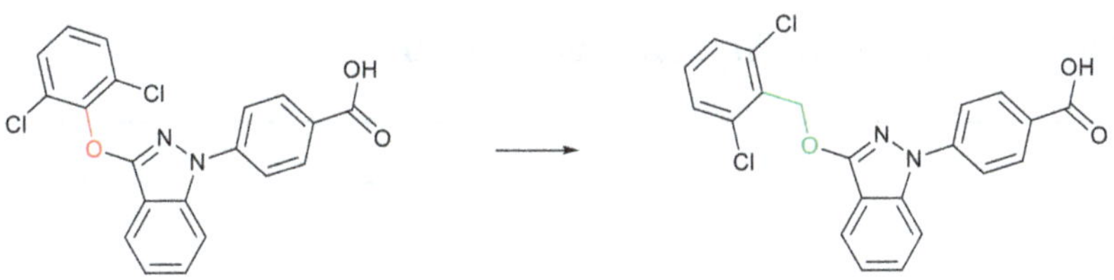

Phototoxic *in vitro* — Non-phototoxic *in vitro*

Insertion of a methylene group between an oxygen atom and an arene ring prevented delocalisation of the oxygen's doublet of electrons into the ring and removed the *in vitro* phototoxicity liability.[27]

3

Phototoxic *in vitro* — Non-phototoxic *in vitro*

Replacing the oxygen atom of an aryl carbamate by a methylene broke conjugation between the aryl and the acyl π electrons and removed the *in vitro* phototoxicity liability.[27]

(*continued*)

Examples

4

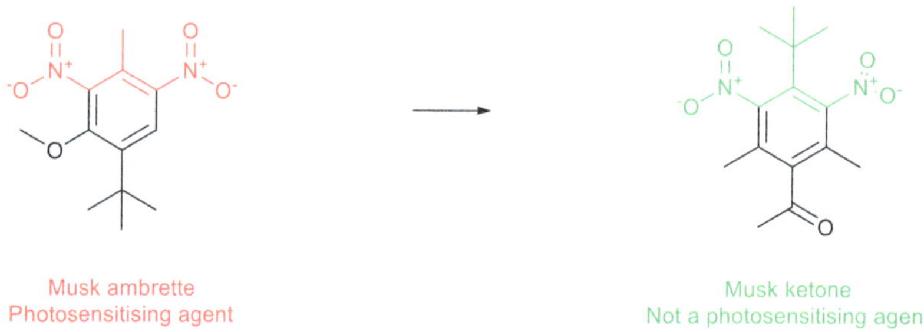

1,3-Dinitroaryl, such as in musk ambrette, is a structural alert for photosensitization unless a large 2-substituent is present, as in musk ketone: an in-plane nitro radical anion cannot form because of the steric hindrance.[30]

17.3.3 Remove an Aryl Halogen Atom

Examples

1

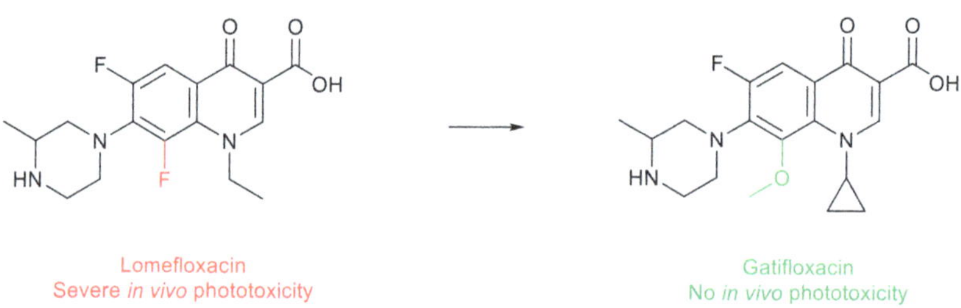

In the quasi-matched pair lomefloxacin and gatifloxacin, replacing the 8-fluoro group with a methoxy group removed the severe *in vivo* phototoxicity liability.[28]

2

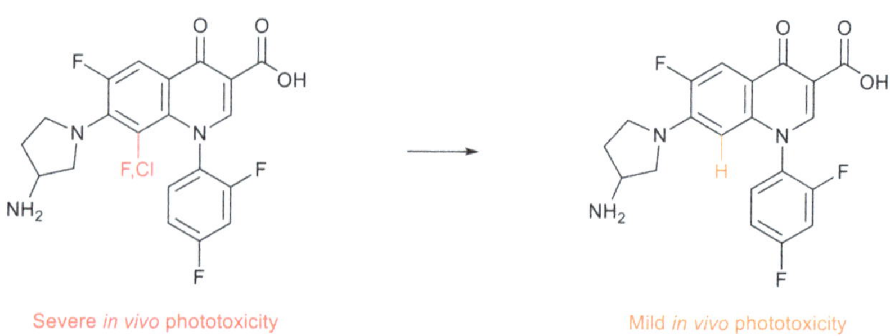

In a related fluoroquinolone example, removing the 8-halo group significantly reduced the *in vivo* phototoxicity liability from severe to mild.[28]

Examples

3

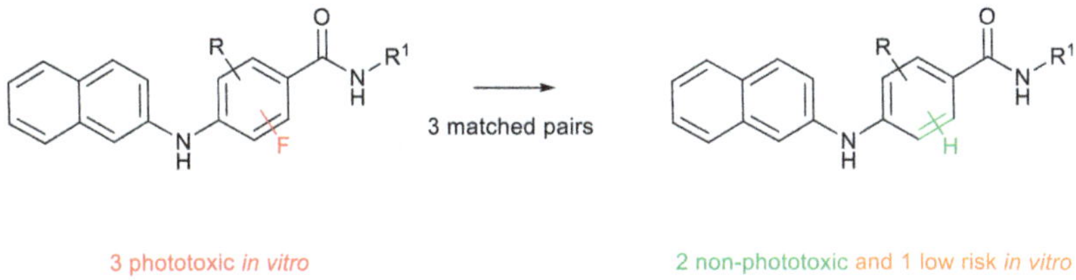

Removal of a fluorine atom in three matched pairs removed (in two cases) or reduced (in one case) the *in vitro* phototoxicity liability.[27]

4

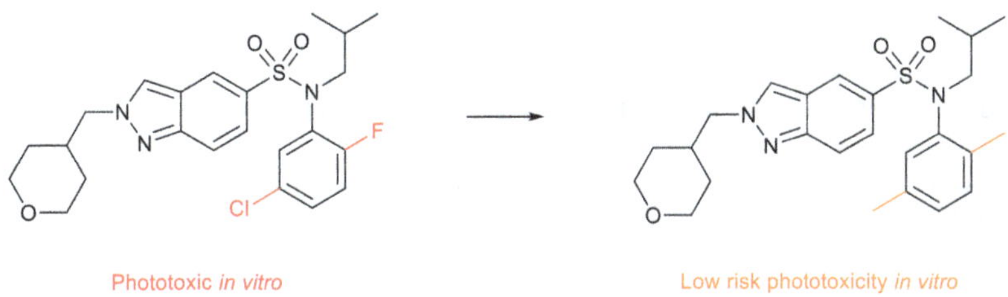

Replacing two halogens by methyls on a phenyl group reduced the *in vitro* phototoxicity liability.[27]

5

Replacing a bromine or fluorine atom by a methyl group on a benzoic acid ring removed the potential *in vitro* phototoxicity liability.[27]

6

Removal of a fluorine atom on a phenyl reduced the *in vitro* phototoxicity liability.[27]

(*continued*)

Examples

7

Replacing a chlorine atom by a methyl on an aminopyrimidine ring in two matched pairs removed the *in vitro* phototoxicity liability.[27]

17.3.4 Introduce an Intramolecular Scavenger

Examples

1

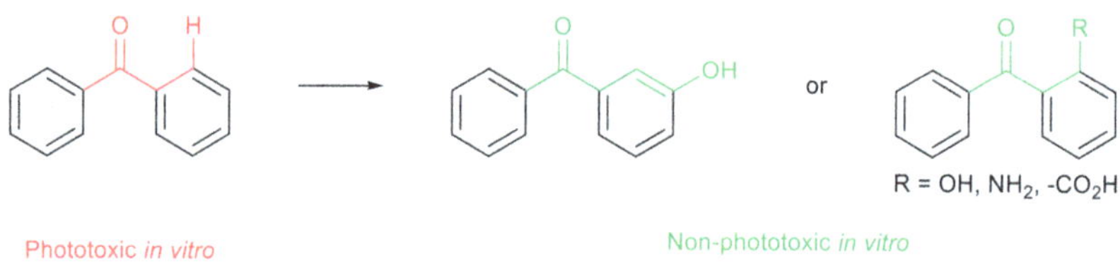

In the phototoxicophore benzophenone, the introduction of an *ortho* H-donor groups such as OH, NH_2 and CO_2H or even a *meta* H-donor group such as OH yielded photosafe products by permitting the intramolecular transfer of a hydrogen to the excited ketones. This, in turn, prevented them from abstracting a hydrogen from a cell component and forming singlet oxygen through energy transfer to ground state triplet oxygen. In support of the intramolecular quenching mechanism, the analogous 4-hydroxybenzophenone is phototoxic.[26]

2

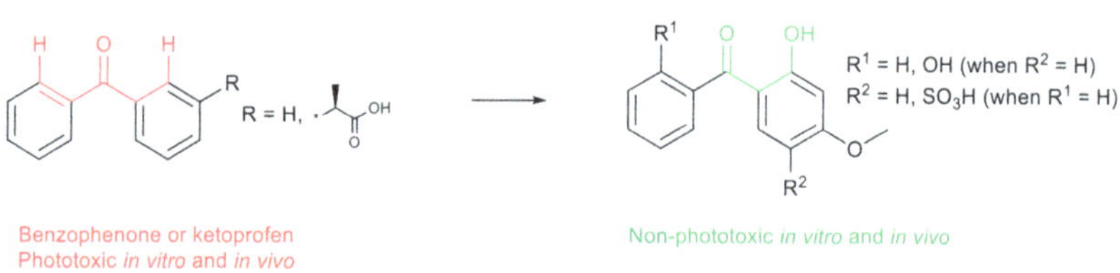

In a related example, benzophenone and ketoprofen were shown to be phototoxic *in vivo* and *in vitro* while the related ortho-hydroxy analogs were shown to be photosafe both *in vitro* and *in vivo*.[31]

17.3.5 Change Positional Isomers

Examples

1

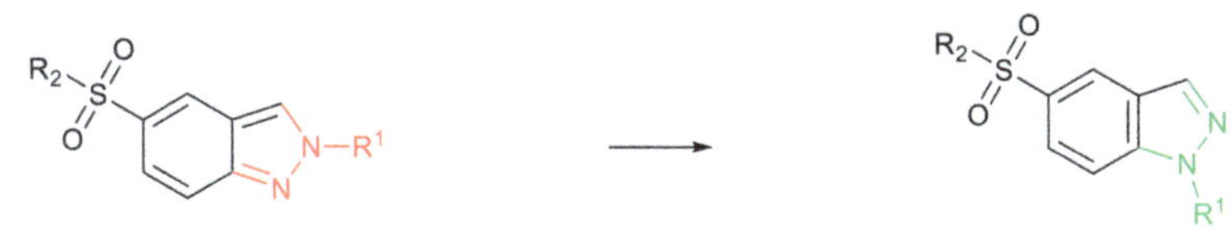

13 phototoxic *in vitro*
4 low risk phototoxicity *in vitro*
2 non-phototoxic *in vitro*

13 non-phototoxic *in vitro*
4 low risk phototoxicity *in vitro*

In a retinoic acid receptor-related orphan receptor gamma (RORγ) program, 2-substituted 2*H*-indazoles were found to be predominantly phototoxic whereas 1-substituted 1*H*-indazoles all had a PIF <2.[32]

2

Phototoxic *in vitro*

Non-phototoxic *in vitro*

The 5-aminoaryl-2-methyl phenol is found to be phototoxic whereas the matched pair 4-aminoaryl-2-methyl phenol where the two electron donating groups are *para* to each other was not.[27]

3

Low risk phototoxicity *in vitro*

Non-phototoxic *in vitro*

Switching from a 4- to a 2-methylpyridin-3-yl analog in two matched pairs removed the potential *in vitro* phototoxicity liability.[27]

(continued)

Examples

4

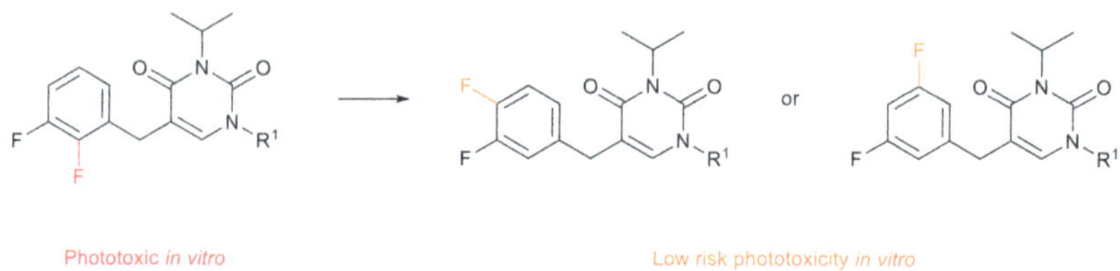

Switching a 2-fluoroaryl to a 4- or 5-fluoroaryl reduced the *in vitro* phototoxicity liability.[27]

References

1. E. B. Zuba, S. Koronowska, A. Osmola-Mańkowska and D. Jenerowicz, *Acta Dermatovenerol. Croat.*, 2016, **24**, 55–64.
2. R. Mang, H. Stege and J. Krutmann, in *Contact Dermatitis*, ed. J. Johansen, P. Frosch and J. P. Lepoittevin, Springer, Berlin, Heidelberg, 2011, pp. 97–104.
3. B. Quintero and M. A. Miranda, *Ars Pharm.*, 2000, **41**, 27–46.
4. K. E. Kochevar, *J. Invest. Dermatol.*, 1981, **77**, 59–64.
5. J. H. Epstein, *J. Am. Acad. Dermatol.*, 1983, **8**, 141–147.
6. R. U. Pendlington and M. D. Barratt, *Int. J. Cosmet. Sci.*, 1990, **12**, 91–103.
7. N. J. Neumann, A. Blotz, G. Wasinska-Kempka, M. Rosenbruch, P. Lehmann, H. J. Ahr and H. W. Vohr, *J. Photochem. Photobiol., B*, 2005, **79**, 25–34.
8. In our experience, drug distribution into skin following systemic administration is of the same order of magnitude as that in plasma in the majority of cases. For this reason, attempting to exclude the drug from skin might be a futile approach *vs* efficacy, Galderma R&D, unpublished results.
9. A. F. Monteiro, M. Rato and C. Martins, *Clin. Dermatol.*, 2016, **34**, 571–581.
10. Department of Health and Human Services (US), Food and Drug Administration (US), Center for Biologics Evaluation and Research (US) and Center for Drug Evaluation and Research (US), in *Guidance for Industry*, Center for Drug Evaluation and Research, Silver Spring, MD, 2015, p. 1 online resource (1 PDF file (18 pages)).
11. OCDE, *Test No. 432: In Vitro 3T3 NRU Phototoxicity Test*, 2004.
12. J.-F. Fournier, C. Bouix-Peter, D. Duvert, A. P. Luzy and G. Ouvry, *J. Med. Chem.*, 2018, **61**, 3231–3236.
13. J. Schümann, S. Boudon, P. Ulrich, N. Loll, D. Garcia, R. Schaffner, J. Streich, B. Kittel and D. Bauer, *Toxicol. Sci.*, 2014, **139**, 245–256.
14. E. V. Buehler, *Curr. Probl. Dermatol.*, 1985, **14**, 39–58.
15. I. Kimber and C. Weisenberger, *Arch. Toxicol.*, 1989, **63**, 274–282.
16. B. Homey, C. von Schilling, J. Blümel, H. C. Schuppe, T. Ruzicka, H. J. Ahr, P. Lehmann and H. W. Vohr, *Toxicol. Appl. Pharmacol.*, 1998, **153**, 83–94.

17. B. M. Verdel, P. C. Souverein, R. H. Meyboom, S. H. Kardaun, H. G. Leufkens and A. C. Egberts, *Pharmacoepidemiol. Drug Saf.*, 2009, **18**, 602–609.
18. S. Peukert, J. Nunez, F. He, M. Dai, N. Yusuff, A. DiPesa, K. Miller-Moslin, R. Karki, B. Lagu, C. Harwell, Y. Zhang, D. Bauer, J. F. Kelleher and W. Egan, *MedChemComm*, 2011, **2**, 973–976.
19. M. J. Waring, *Expert Opin. Drug Discovery*, 2010, **5**, 235–248.
20. O. G. Mekenyan and G. D. Veith, *SAR QSAR Environ. Res.*, 1994, **2**, 129–143.
21. S. Ringeissen, L. Marrot, R. Note, A. Labarussiat, S. Imbert, M. Todorov, O. Mekenyan and J. R. Meunier, *Toxicol. In Vitro*, 2011, **25**, 324–334.
22. Y. Haranosono, M. Kurata and H. Sakaki, *J. Toxicol. Sci.*, 2014, **39**, 655–664.
23. R. J. Young, D. V. Green, C. N. Luscombe and A. P. Hill, *Drug Discovery Today*, 2011, **16**, 822–830.
24. M. D. Barratt, J. V. Castell, M. A. Miranda and J. J. Langowski, *J. Photochem. Photobiol., B*, 2000, **58**, 54–61.
25. L. J. Martinez, G. Li and C. F. Chignell, *Photochem. Photobiol.*, 1997, **65**, 599–602.
26. M. Placzek, M. Dendorfer, B. Przybilla, K. P. Gilbertz and B. Eberlein, *Acta Derm.-Venereol.*, 2013, **93**, 30–32.
27. J.-F. Fournier, C. Bouix-Peter, D. Duvert, A. P. Luzy and G. Ouvry, unpublished work.
28. N. Hayashi, Y. Nakata and A. Yazaki, *Antimicrob. Agents Chemother.*, 2004, **48**, 799–803.
29. J.-F. Fournier, L. Clary, S. Chambon, L. Dumais, C. S. Harris, C. Millois, R. Pierre, S. Talano, E. Thoreau, J. Aubert, M. Aurelly, C. Bouix-Peter, A. Brethon, L. Chantalat, O. Christin, C. Comino, G. El-Bazbouz, A. L. Ghilini, T. Isabet, C. Lardy, A. P. Luzy, C. Mathieu, K. Mebrouk, D. Orfila, J. Pascau, K. Reverse, D. Roche, V. Rodeschini and L. F. Hennequin, *J. Med. Chem.*, 2018, **61**, 4030–4051.
30. A. G. Motten and C. F. Chignell, *Photochem. Photobiol.*, 1983, **37**, 17–26.
31. Y. Seto, H. Ohtake, M. Kato and S. Onoue, *J. Pharmacol. Exp. Ther.*, 2015, **354**, 195–202.
32. G. Ouvry, N. Atrux-Tallau, F. Bihl, A. Bondu, C. Bouix-Peter, I. Carlavan, O. Christin, M. J. Cuadrado, C. Defoin-Platel, S. Deret, D. Duvert, C. Feret, M. Forissier, J.-F. Fournier, D. Froude, F. Hacini-Rachinel, C. S. Harris, C. Hervouet, H. Huguet, G. Lafitte, A. P. Luzy, B. Musicki, D. Orfila, B. Ozello, C. Pascau, J. Pascau, V. Parnet, G. Peluchon, R. Pierre, D. Piwnica, C. Raffin, P. Rossio, D. Spiesse, N. Taquet, E. Thoreau, R. Vatinel, E. Vial and L. F. Hennequin, *ChemMedChem*, 2018, **13**, 321–337.

CHAPTER 18

Drug-induced Phospholipidosis

LAURA GORACCI*[a] AND GABRIELE CRUCIANI[a]

[a]University of Perugia, Department of Chemistry, Biology and Biotechnology, Via Elce di Sotto, 8, Perugia, 06123, Italy
*E-mail: laura.goracci@unipg.it

18.1 Introduction

18.1.1 Definition

Drug-induced phospholipidosis (DIPL) is a lipid storage disorder characterized by an accumulation of phospholipids in lysosomes induced by drug administration.[1–3] A number of drugs are known to induce phospholipidosis (PL), and a few examples are shown in Figure 18.1. From a morphological perspective, affected tissues show characteristic multilamellar inclusion bodies, usually detected by transmission electron microscopy.[2] Whether DIPL has to be considered a real toxic effect or just an adverse event is still under debate. On one hand, DIPL is generally a reversible process, with lipids returning to normal regulation upon drug therapy termination; on the other hand, there is evidence of toxicity in several organs connected to DIPL occurrence. In addition, DIPL lesions in tissues closely resemble those of patients affected by the rare genetic disorder Niemann–Pick type C disease, which is often fatal in early childhood.[1] Since the 1970s, when concerns about DIPL were first raised,[4,5] scientists have therefore put a lot of effort in studying the mechanism of PL induction and the relationship between the chemical structure of drugs and their PL-inducing potential. In 2004 the Food and Drug Administration (FDA) established

Drug Discovery Series No. 79
The Medicinal Chemist's Guide to Solving ADMET Challenges
Edited by Patrick Schnider
© The Royal Society of Chemistry 2021
Published by the Royal Society of Chemistry, www.rsc.org

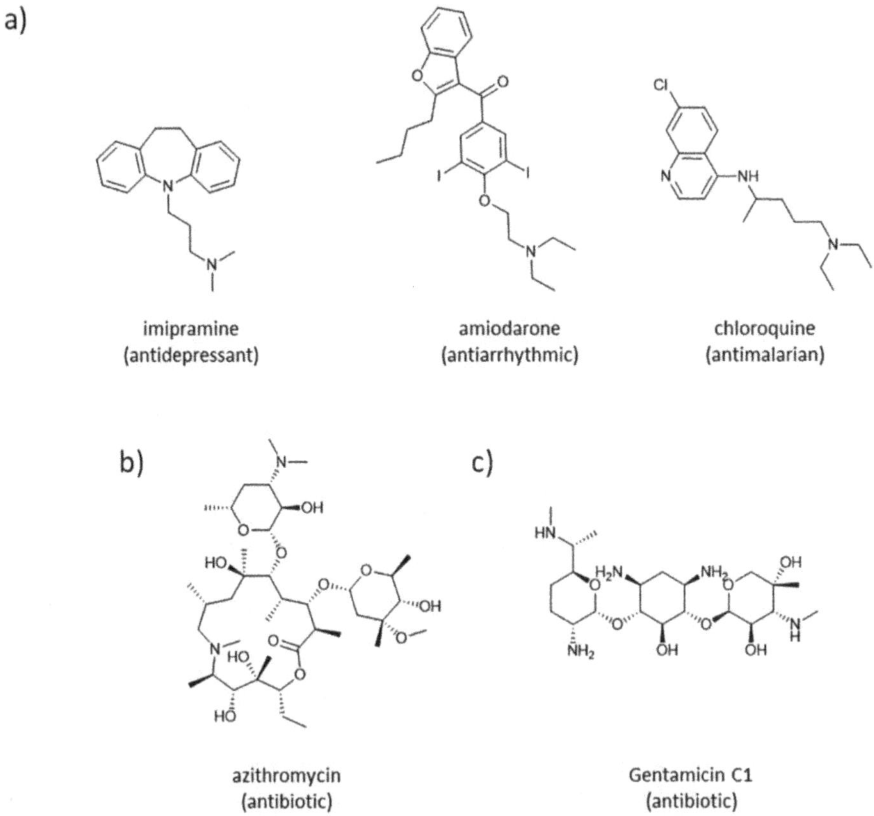

Figure 18.1 Examples of drugs inducing PL. (a) Cationic amphiphilic drugs (CADs); (b) macrolide antibiotic; (c) aminoglycoside antibiotic.

the Phospholipidosis Working Group (PLWG), aiming at clarifying the risk behind DIPL and related regulatory policies, which significantly boosted research in the field[6] and resulted in the recommendation to evaluate PL inducers for additional toxicity risks including QT prolongation, neurotoxicity, myopathy and hepatotoxicity.[7]

18.1.2 Proposed Mechanisms

Cationic amphiphilic drugs (CADs, Figure 18.1a) are referred to as the main class of PL inducers.[8] CADs cover many therapeutic areas, such as antidepressants, antiarrhythmics and antihistaminics, and PL induced by CADs can affect different tissues, rendering the detection of DIPL outcomes a challenging task. Their amphiphilic structure arises from the coexistence of a basic group (generally an amine group) and a relatively rigid, typically aromatic hydrophobic moiety. As a consequence, for many years it was generally assumed that two physico-chemical parameters, the negative log of the acid dissociation constant (pK_a) and log of the partition coefficient (LogP), could be sufficient to predict DIPL.[9,10] Over time, it emerged that additional parameters could

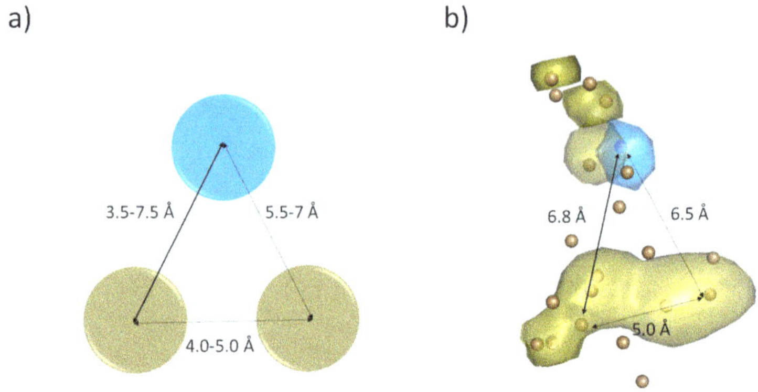

Figure 18.2 Proposed toxicophores for DIPL inducers. (a) Model by Slavov *et al.*[14] (b) model by Goracci *et al.*[15] The yellow regions represent hydrophobic moieties, and the blue region represents the protonated hydrogen-bond donor region corresponding to the amine group. Part (a) reproduced from ref. 14 with permission from Elsevier, Copyright 2014. Part (b) reproduced from ref. 15 with permission from Elsevier, Copyright 2015.

improve predictions.[11,12] The need for a more complex description arises from the observation that not all CADs are PL inducers.[13]

Slavov *et al.*[14] provided the first toxicophore model for DIPL from a three-dimensional quantitative spectral data–activity relationship (3D-QSDAR) study. The common structural pattern associated with the DIPL effect is shown in Figure 18.2a. According to this model, the distance between the amino-group and the centroid of an aromatic ring is between 3.5 Å and 7.5 Å in PL inducers (*two-center model*). A second aromatic ring can also be included in the toxicophore model, at a distance of 4–5 Å from the centroid of the first ring and at a distance of between 5.5 Å and 7 Å from the amino group (*three-center model*). A similar pattern was also confirmed a few years later by a different *in silico* approach, in which the toxicophore was generated upon alignment of a number of PL inducers using the FLAP software (Figure 18.2b).[15] It should be pointed out that the toxicophore models for DIPL are closely related to those for human ether-à-go-go-related gene (hERG) channel blockers; hence, most PL inducers are also hERG inhibitors, while the reverse is not always true.[16,17]

It is noteworthy that PL inducers can sometimes bear two amino groups, as the presence of two strong basic centers seems still a favorable feature for interaction with the lysosomal membrane.[18,19]

The generally accepted hypothesis on the origin of CAD-induced lipid accumulation is schematically depicted in Figure 18.3.

DIPL is a three-step process:

1. The drug enters the cells,
2. accumulates in lysosomes, and
3. causes phospholipid accumulation and formation of multilamellar inclusion bodies.[20]

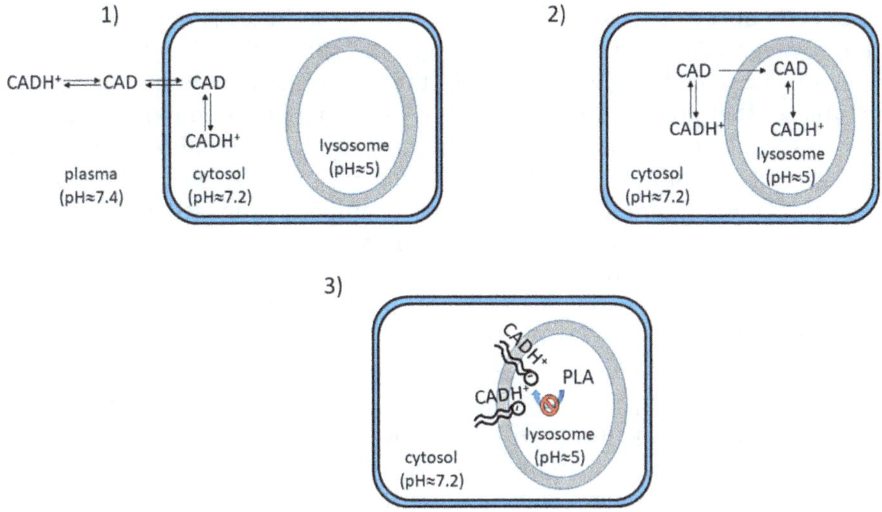

Figure 18.3 General scheme on the basis of DIPL mechanism for CADs. (1) The CAD enters the cell, maintaining an equilibrium between protonated and neutral form; (2) the CAD enters the lysosome becoming mostly protonated and inducing the '*lysosomal trapping*'; (3) phospholipase inhibition according to the '*negative charge hypothesis*'.

The mechanisms behind the first two steps are currently well-characterized for CADs, on the basis of biophysical and biological data. With a $pK_a > 8$, CADs still possess a neutral fraction at physiological pH (7.2–7.4), so that they can easily cross cell membranes by passive diffusion. However, when they enter lysosomes, they become fully protonated due to the acidic environment, and thus trapped inside these organelles. Due to their amphiphilic nature, CADs start to interact with the inner membrane of the lysosome. In particular, CADs seem to have a preference for interaction with acidic phospholipids (phosphatidylinositol, phosphatidylserine) through electrostatic interactions.[21,22]

For the third step, phospholipid accumulation and formation of multilamellar inclusion bodies, four mechanisms have been suggested on the basis of toxicogenomic data:[23]

1. inhibition of lysosomal phospholipase activity;
2. inhibition of lysosomal enzyme transport;
3. enhanced phospholipid biosynthesis;
4. enhanced cholesterol biosynthesis (a potential indirect trigger[24]).

A reversible inhibition of phospholipase activity is commonly considered to be the primary mechanism of DIPL. Depending on the CAD, dose-dependent inhibition of phospholipase A1 (PLA1), calcium-dependent or -independent phospholipase A2 (PLA2) and/or calcium-dependent phospholipase (PLC) have been observed.[25] One of the most cited hypotheses is that the interaction between CADs and negatively charged phospholipids, the abundance of which is estimated at about 9–25% in the lysosomal membrane, generates complexes that prevent the anchorage of lysosomal phospholipases to acidic lipids at the inner membrane (activation mechanism of phospholipases), causing activity inhibition.[20] This mechanism is usually referred to as the *negative charge hypothesis*.[26] This hypothesis, initially proposed for aminoglycosides inducing nephrotoxicity,[26,27] has also been proposed for CADs[1] and macrolides.[28] Other hypotheses are that DIPL-inducing drugs and phospholipids form complexes that are resistant to degradation, or that these complexes are enzyme inhibitors themselves.[29,30]

However, other studies raise doubts about the phospholipase inhibition theory. In particular, the lysosomal lipid bis(monoacylglycero)phosphate (BMP) is produced by phospholipase A2 action, and thus its concentration should decrease with the occurrence of DIPL; contrarily, an increased BMP concentration is nowadays considered a DIPL biomarker.[31] Therefore, further studies are needed for a full elucidation of DIPL mechanisms, so that DIPL risk prediction still remains a challenging task.

Beyond CADs several aminoglycoside and macrolide antibiotics are known to induce DIPL in a single tissue, mainly kidney or liver.[20,26,28,32,33] The tissue-specific effect in kidney seems to be related to accumulation following endocytosis by renal cells.[26] If this explains the drug accumulation into lysosomes, the mechanism of accumulation of phospholipids to give lamellar inclusion bodies, is still uncertain.[3,23]

18.1.3 Screening Strategies

Assessment of morphological changes in cells or tissue by transmission electron microscopy (TEM) still represents the gold standard method for DIPL detection. However, since this expensive and time-consuming method is used at a relatively late stage of drug discovery, it is unsuitable for screening during lead optimization. Nowadays, *in silico* methods represent the first tier for DIPL risk assessment. Such methods have been commonly generated according to structure–activity relationship (SAR) analyses, although a bioinformatics method based on reverse pharmacogenomics and toxicogenomics has also been proposed.[24,34]

In silico methods based on SAR studies can be classified as "equation-based methods" or "more complex methods". Table 18.1 briefly summarizes these approaches.

Concerning *equation-based methods*, the pioneering work by Ploemen *et al.*[9] in 2004 for the first time demonstrated that the DIPL potential could be predicted by means of pK_a and calculated LogP (CLogP) values, although the approach was tested in a limited dataset of 41 compounds. Since then, a number of equation-based methods have been developed, as modification of the model of Ploemen *et al.*[10,11,33] or as original new equations.[12]

Table 18.1 Overview of *in silico* methods for DIPL risk assessment.[a]

In silico		
Reference	Year	Description
Equation-based		
Ploemen *et al.*[9]	2004	DIPL+: pK_a ≥ 8, CLogP ≥ 1 and (pK_a-MB)^2 + (CLogP)^2 ≥ 90 DIPL−: pK_a < 8 or CLogP < 1 or (pK_a-MB)^2 + (CLogP)^2 < 90
Tomizawa *et al.*[10]	2006	DIPL+(low): NC ≈ 1 and CLogP < 1.61 DIPL+(medium): NC ≈ 1 and 1.61 ≤ CLogP < 2.75 DIPL+(high): (a) NC ≈ 1 and CLogP ≥ 2; (b) 1 < NC ≤ 2 DIPL−: NC < 1
Pelletier *et al.*[33]	2007	DIPL+: pK_a ≥ 6, CLogP ≥ 2 and (pK_a-MB)^2 + (CLogP)^2 ≥ 50 DIPL−: pK_a < 6 or CLogP < 2 or (pK_a-MB)^2 + (CLogP)^2 < 50
Hanumegowda *et al.*[11]	2010	DIPL+: CLogP ≥ 2 and (pK_a-MB × ClogP × V_d) ≥ 180 DIPL−: CLogP < 2 or (pK_a-MB × CLogP × V_d) < 180
Fischer et *al.*[12]	2012	DIPL+: pK_a ≥ 6.3 but ≤11 and $\Delta\Delta G_{AM}$ ≤ 6 kJ mol^{-1} DIPL−: pK_a < 6.3 or >11 and $\Delta\Delta G_{AM}$ > 6 kJ mol^{-1}
Models		
Pelletier *et al.*[33]	2007	Bayesian
Ivanciuc[35]	2008	Weka machine learning
Kruhlak *et al.*[36]	2008	QSAR approaches
Lowe *et al.*[38]	2010	Machine learning
Przybylak and Cronin[40]	2011	Molecular fragments (SMARTS strings)
Orogo *et al.*[37]	2012	QSAR approaches
Sun *et al.*[45]	2012	Support vector machine (SVM) models
Choi *et al.*[13]	2013	Neural network
Goracci *et al.*[41]	2013	PLS-DA approach
Slavov *et al.*[14]	2014	Toxicophore, 3D-QSDAR
Goracci *et al.*[15]	2015	Toxicophore, PLS-DA

[a]DIPL+: DIPL inducers; DIPL−: DIPL non-inducers; pK_a-MB: pK_a value of the most basic group; NC: sum of the charge of all dissociable functional groups; V_d: volume of distribution; $\Delta\Delta G_{AM}$: difference in free energy between the partitioning of a compound into the air–water interface, ΔG_{AW}; and free energy of the critical micellar concentration, ΔG_{MIC}.

The *more complex methods* are characterized by the use of large datasets and include Bayesian, neural networks and (quantitative) SAR [(Q)SAR] approaches.[13,15,33,35–41] A comparison of the performances of a large part of the methods in Table 18.1 has been provided.[7] It must be noticed that the statistical performances of the methods strongly depend on the dataset used. Available databases in most cases mix *in vivo* and *in vitro* data, often from different assays and tissues.[33] Moreover, drug exposures and treatment times differ strongly. An additional complication is that biotransformation can either increase or decrease the potential of a drug to induce PL. Since limited information on the DIPL–metabolism relationship is available,[42–44] the effect of drug metabolism is therefore still rarely considered for modelling purposes.[41,45–48] Therefore, slight statistical differences in prediction accuracy, sensitivity, specificity or concordance between the methods are probably not the best criteria for method selection. The use of multiple approaches may therefore be a cost-effective way to improve predictions.

In addition to *in silico* methods, a number of *in vitro* assays for DIPL potential evaluation have been developed, some of which can be applied in high-throughput workflows. Table 18.2 summarizes a number of published *in vitro* methods, which can be classified in two main categories, 'biophysical methods' and 'cell-based assays'.

The *biophysical methods* can usually identify cationic amphiphilic PL inducers. The method developed by Vitovic *et al.*[49] correlates the potential variation of the critical micellar concentration of a short-chain phosphatidylserine in the presence of the tested compound with its PL-inducing potential in a high-throughput assay. Ceccarelli *et al.*[50] demonstrated that the increase of the lateral pressure applied to a lipid monolayer containing zwitterionic and acidic phospholipids soon after the injection of a drug solution into the subphase correlates well with the drug's DIPL potential, also for non-CAD-like compounds. However, this method cannot be considered a high-throughput method until new instrumentation to automate the process is developed. Recently, the LogD in brain polar lipids ($LogD_{BPL}$) assay developed by Roche for the determination of drug distribution coefficients between an aqueous phase and porcine brain polar lipids extract was tested for DIPL risk evaluation, and the $LogD_{BPL}$ parameter was found to be an efficient descriptor to assess a PL-inducing potential of drugs, especially when corrected using the pK_a value.[51]

Table 18.2 Overview of *in vitro* methods for DIPL risk assessment.

		In vitro
Reference	Year	Description
Biophysical methods		
Vitovic *et al.*[49]	2008	Variation of the critical micelle concentration of a short chain acidic phospholipid
Ceccarelli *et al.*[50]	2015	Perturbation of a lipid monolayer
Ceccarelli *et al.*[51]	2017	Combination of $LogD_{BPL}$ and pK_a
Cell-based assays		
Casartelli *et al.*[52]	2003	Nile red staining
Sawada *et al.*[24,34]	2005, 2006	Toxicogenomic screening
Morelli *et al.*[53]	2006	NBD-PE accumulation
Kasahara *et al.*[54]	2006	NBD-PC accumulation
Nioi *et al.*[55,56]	2007, 2008	LipidTox accumulation
Mesens *et al.*[57,58]	2009, 2010	LipidTox accumulation (high throughput)
Shahane *et al.*[59]	2014	LipidTox accumulation (high-throughput)

Numerous *cell-based assays* have been developed to predict PL-inducing potential. Phospholipid accumulation within lysosomes is commonly monitored by using fluorescent dyes or probes, such us Nile red,[52] 7-nitrobenz -2-oxa-1,3-diazol-4-yl- phosphatidylethanolamine (NBD-PE),[60] 7-nitrobenz -2-oxa-1,3-diazol-4-yl-phosphatidylcholine (NBD-PC)[54] and, more recently, LipidTOX, which is probably the best option.[55,56,58,59] Several cell lines were tested and analyzed using various imaging platforms. Further improvements include the use of *in vitro* flow cytometry[57,58] and high-content screening.[60,61] Of course, the nature of the cell line also influences the potential effect of metabolism in the detected DIPL effect. In addition to traditional cell-based assays, a toxicogenomic approach was proposed by Sawada *et al.*[24]

Taking into consideration the uncertainties about the mechanism of PL induction on one hand and the large variety of *in silico* and *in vitro* methods available on the other hand, it appears reasonable to combine different approaches to reduce the risk of misprediction. In the last decade, a number of decision trees have been proposed,[3,6,11,20,62] variably including combinations of *in silico* and *in vitro* methods. Here, a decision tree to be applied in drug discovery and development is proposed, highlighting the importance of detecting the activating or deactivating effect of drug metabolism towards DIPL-inducing potential (Scheme 18.1).

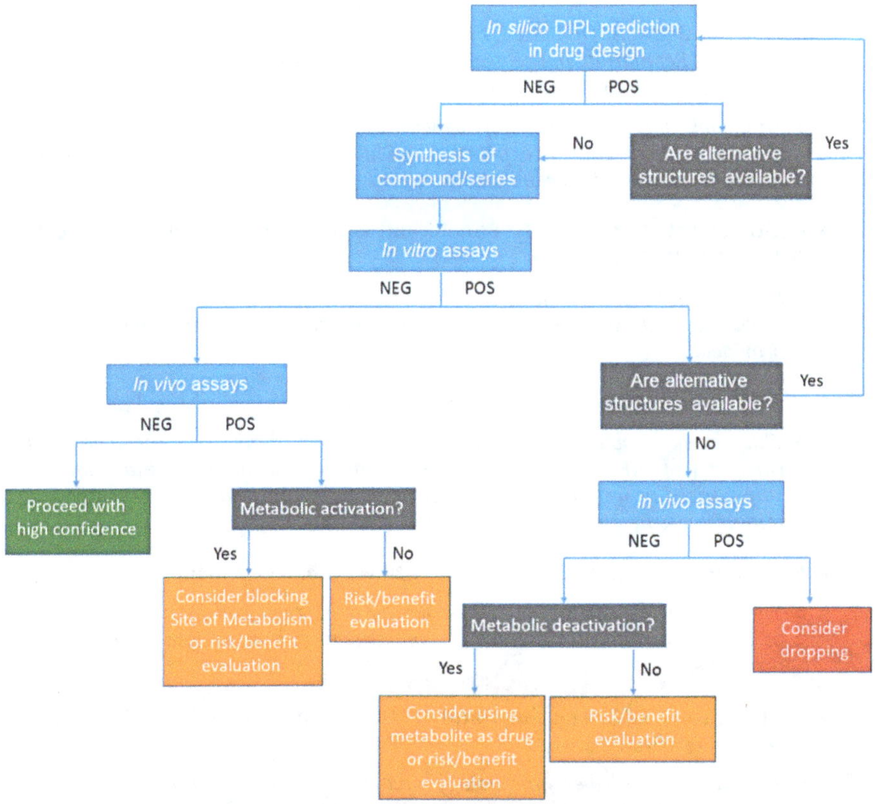

Scheme 18.1 Proposed decision tree for DIPL risk assessment.

The use of *in silico* DIPL prediction tools is recommended already in drug design to reduce costs and time for synthesis. When a designed compound is predicted to be a likely DIPL inducer, alternative structures should be considered. After synthesis, one or more *in vitro* assays should be performed to evaluate the DIPL-inducing potential. If results of *in vivo* studies are also negative, the compound is considered not to be a PL inducer with high confidence. In contrast, if results of *in vitro* studies are negative and those of *in vivo* studies are positive, a mechanism of metabolic activation should be considered, evaluating the metabolic profile of the compound to consider blocking the site of metabolism responsible for DIPL activation. In cases where a compound is found to be a PL-inducer in *in vitro* assays, progression of alternative structures, if available, should be considered. However, there is still the possibility that such a compound will turn out not to be a PL-inducer *in vivo*. If the negative outcome *in vivo* is related to a metabolic deactivation process, the use of the DIPL-inactive metabolite instead of the parent compound can also be evaluated. However, when DIPL is seen *in vivo* many additional parameters have to be taken into account for an assessment of the risk : benefit ratio, such as the therapeutic index, the duration of the therapy and the affected organ(s).

Key References

- J. M. Willard and A. De Felice, *Lysosomes and Phospholipidosis in Drug Development and Regulation, in Lysosomes: Biology, Diseases, and Therapeutics*, John Wiley & Sons, Inc., 1st edn, 2016, pp. 487–551.
    - *This book chapter covers many aspects of DIPL, from physical chemistry, to clinical and non-clinical expression, QSAR and the authors' perspective on future directions and recommendations.*
- L. Goracci, M. Ceccarelli, D. Bonelli and G. Cruciani, *J. Chem. Inf. Model.*, 2013, **53**, 1436–1446.
    - *In this paper the importance of using curated databases in modelling DIPL is discussed, and examples of the effects of metabolism in model generation are provided.*
- A. J. Ratcliffe, *Curr. Med. Chem.*, 2009, **16**, 2816–2823.
    - *This review probably represents the first paper focused on the medicinal chemistry of DIPL optimization. In silico methods available at that time are summarized as well as the role of metabolism in DIPL induction. Examples of mitigation strategies are also discussed.*
- N. Anderson and J. Borlak, *FEBS Lett.*, 2006, **580**, 5533–5540.
    - *This minireview mainly describes the morphological and biological features of DIPL, including an interesting comparison with other lysosomal storage disorders.*
- M. J. Reasor, K. L. Hastings and R. G. Ulrich, *Expert Opin. Drug Saf.*, 2006, **5**, 567–583.
    - *This review describes morphology and mechanisms of DIPL, toxicological implications, potential biomarkers and screening methods.*

18.2 Mitigation Strategies

18.2.1 Basicity Reduction

A basic pK_a greater than 8 is commonly considered a key parameter for PL-inducing potential,[9] although a pK_a > 6 has also been associated with lysosomal trapping.[12,63,64] Therefore, decreasing the basicity of a compound can reduce the risk of DIPL, especially for those compounds that fit the toxicophore models in Figure 18.2. Basicity reduction is probably the most used strategy for lowering the DIPL potential of a lead compound if applicable. Strategies include: (1) modification of the chemical structure in proximity to the basic center, to decrease its pK_a; (2) elimination of the basic center; (3) reduction of the number of basic centers (*e.g.*, for dibasic compounds).

18.2.2 Lipophilicity Reduction

Together with basicity reduction, the decrease of the overall lipophilicity of a lead compound is also widely used to reduce the DIPL risk. Introduction of polar substituents together with a reduction of the molecular weight proved to be useful to this aim. While polar substituents often do effect a drop in intrinsic lipophilicity (LogP), concomitant reduction of effective lipophilicity (LogD) and basicity can be a challenging task, though, since decreasing basicity and thus charge will increase logD. Yet there are examples in which the introduction of polar groups (*e.g.* hydroxyl or carbonyl moieties) in proximity to the basic center can decrease both the pK_a and logD.[65]

18.2.3 Amphiphilicity Reduction

The correlation between DIPL potential and amphiphilicity was first proposed by Fischer *et al.* at Roche, who developed the CAFCA program[66] able to quantify the amphiphilicity of a molecule by means of its amphiphilic moment ($\Delta\Delta G_{AM}$). Indeed, a $\Delta\Delta G_{AM} < 6$ has been associated with high risk of PL induction when pK_a is between 6.3 and 11.[12]

18.2.4 Modulation of Metabolism

Metabolism can decrease the potential of a drug to induce PL by reduction of overall lipophilicity and basicity. Several phase I metabolism reactions that might decrease the DIPL risk include aromatic and aliphatic hydroxylation by cytochromes P450 (CYP450), *N*-oxidation and *N*-hydroxylation by CYP450 or flavin containing monooxygenase (FMO3) or alpha-oxidation of a nitrogen-containing heteroaromatic ring by aldehyde oxidase (AOX).[46] Conversely, metabolic *N*-dealkylation of a tertiary amine to a more basic secondary amine[42,67] by CYP450 or *N*-deacylation to generate a basic amine metabolite by deacetylase, as observed for ketoconazole, can increase or even generate the potential of a drug to induce PL.[48,68]

18.3 Examples of Mitigation Strategies

18.3.1 Reduction of Basicity

Examples

1

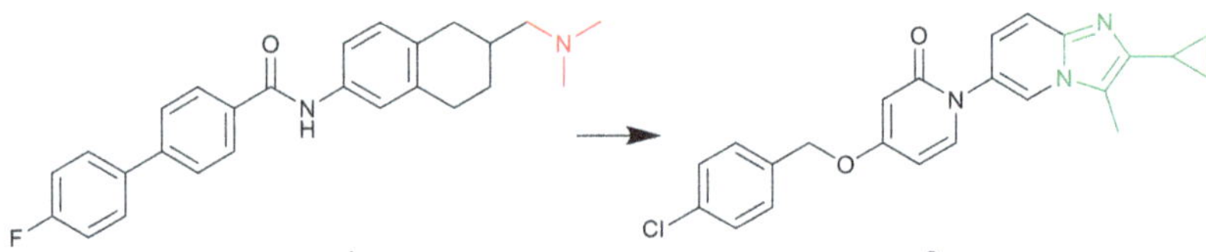

1 2

A number of small-molecule melanin concentrating hormone receptor 1 (MCHR1) antagonists exerted potent anorectic effects on obese animal models but raised toxicological concerns, including DIPL. In the optimization efforts starting from compound **1**, the authors set a cut-off value of basicity ($pK_a < 8$) in agreement with the model of Ploemen *et al.*[9] The optimization process led to the amine-free compound **2**, in which the weakly basic imidazo[1,2-*a*]pyridine ring (predicted $pK_a = 7.9$) gives a favorable interaction with MCHR1. The optimized compound proved to be negative towards DIPL *in vitro* on HepG2 cells.[69]

2

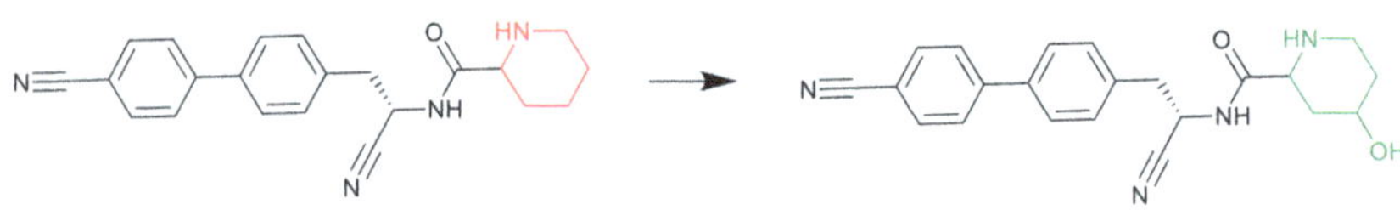

3 4

Compound **3**, a cathepsin C inhibitor with a pK_a of 8.4 and LogD of 2.7, was found to induce DIPL in rats *in vivo*, targeting mainly liver and lung. During optimization, compound **4** ($pK_a = 7.6$, LogD = 1.4) was predicted to have a lower propensity to induce PL, which was confirmed *in vitro*. Here, both basicity and lipophilicity reduction probably play a role, representing an example of a dual effect.[70]

3

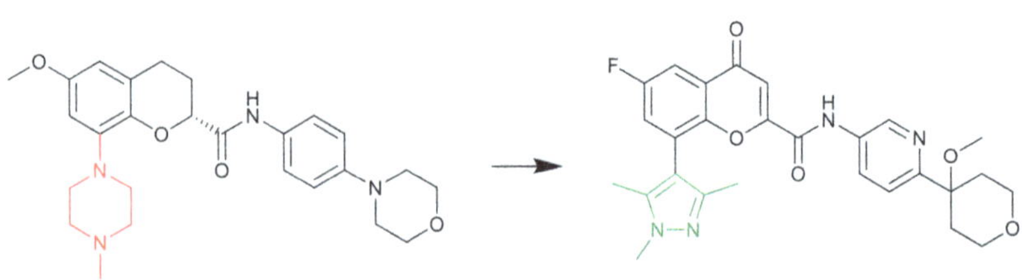

5 6

A chromane template (**5**) was used as the starting point for the design of novel 5-5-hydroxytryptamine receptor 1B (HT1B) receptor antagonists. To reduce adverse events, including DIPL potential, the basic piperazine ring was replaced with a pyrazole moiety, as in compound **6**. This strategy proved to be successful in getting rid of the DIPL-inducing effect on the basis of the results of *in vitro* assays. This optimization also demonstrated, for the first time to our knowledge, that a basic amine is not essential for *in vitro* and *in vivo* activity at the 5-HT1B receptor.[71]

Examples

4

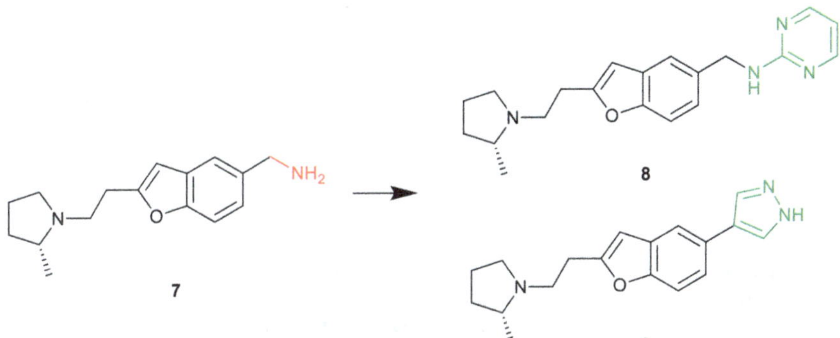

7 → 8, 9

Compound **7** is a histamine H_3 receptor antagonist with excellent binding potency, but it also represents an example of a dibasic compound inducing PL *in vitro*. The optimization strategy required keeping the pyrrolidine amine, which was found to be essential for H_3 binding, and modifying the primary amine. Substitution with a pyrimidine ring to give **8** or replacement of the amine group with a pyrazole ring to give **9** proved to be successful approaches in reducing the DIPL potential *in vitro*.[72]

5

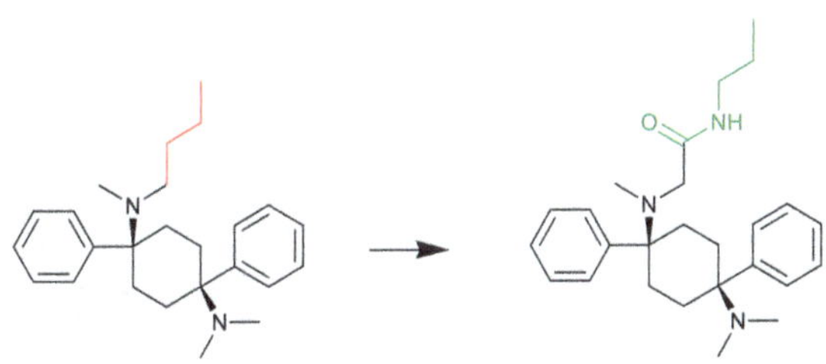

10	**11**
pK_a= 7.06/9.73	pK_a= 3.75/9.33
CLogP= 5.6	CLogP= 4.7
CLogD= 3.2	CLogD=2.5

In this study the authors compared the DIPL potential of two structurally related diprotic amines: GRT1 (**10**, measured pK_a = 7.06, 9.73) and GRT2 (**11**, measured pK_a = 3.75, 9.33). As with other dibasic rigid compounds, GRT1 induced PL both *in vitro* and *in vivo* (lung), while no signs of DIPL were associated with GRT2. Introduction of the amide moiety not only decreased basicity of one of the amino groups but also intrinsic as well as effective lipophilicity.[18]

(*continued*)

Examples

6

pK$_a$= 7.5/9.2
LogP= 3.7
LogD$_{7.4}$=1.3

12

pK$_a$= 4.9/8.5
LogP= 2.2
LogD$_{7.4}$=0.4

13

A series of chemical optimizations for the discovery of novel histamine H_3 receptor inverse agonists was recently published, taking into account also the DIPL effect of the tested compounds. In the present case, compounds **12** and **13** only differ by the nature of one basic center (piperidine and morpholine, respectively). This chemical modification not only reduces basicity of compound **13** (SUVN-G3031), but it also results in its lower lipophilicity. As a result, *in vitro* tests demonstrated the reduced DIPL potential of **13** with respect to **12**.[73]

7

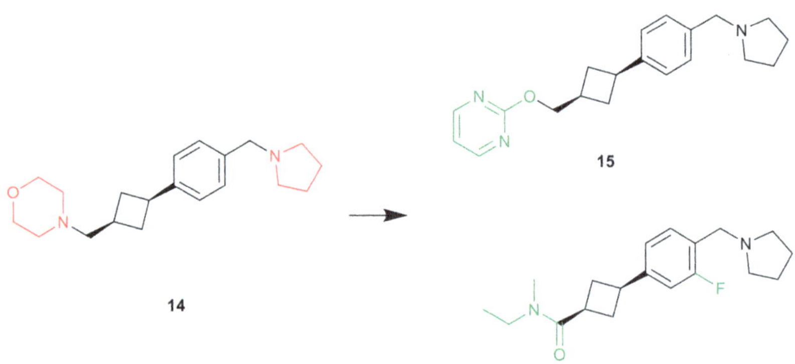

14 **15** **16**

In the optimization process to identify novel H_3 antagonists, the initial dibasic lead compound **14** turned out to induce PL *in vivo* (lung and kidney). Its calculated physicochemical properties are: pK_a = 9.8 and 7.7; CLogP = 3.6; CLogD = −0.2. Aiming at reducing basicity, compounds **15** (pK_a = 9.8; CLogP = 4.0; CLogD = 0.8) and **16** (pK_a = 9.3; CLogP = 3.3; CLogD = 0.8) were synthesized and tested. The DIPL potential was drastically reduced by the elimination of one basic center according to results from an *in vitro* assay.[74]

Examples

8

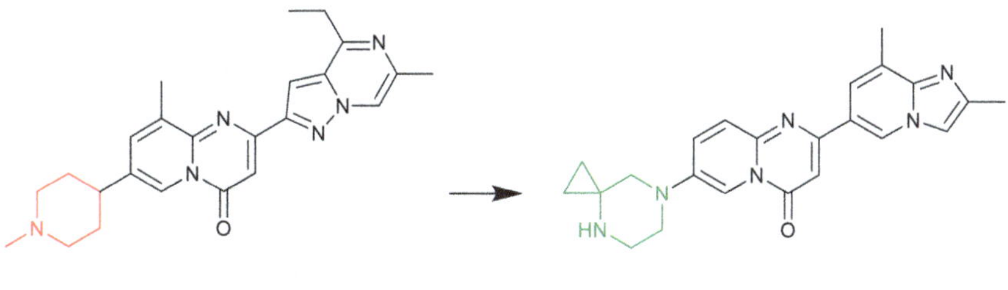

17 18

The survival of motor neuron-2 (SMN2) gene splicing modifier **17** (RG7800, RO6885247) was the first molecule from a joint Roche/PTC program which entered clinical trials for the treatment of spinal muscular atrophy. However, due to its large volume of distribution and its basic charge, it was found to induce PL in several tissues in rats. The optimization of compound **17** (measured pK_a = 10.9; LogD = 2.3), lead to the discovery of risdiplam (**18**), currently undergoing pivotal clinical studies. *In vivo* risdiplam (measured pK_a = 6.8; LogD = 2.5) did not show any evidence for PL in any tissues at the doses tested.[75]

18.3.2 Lipophilicity Reduction

Examples

1

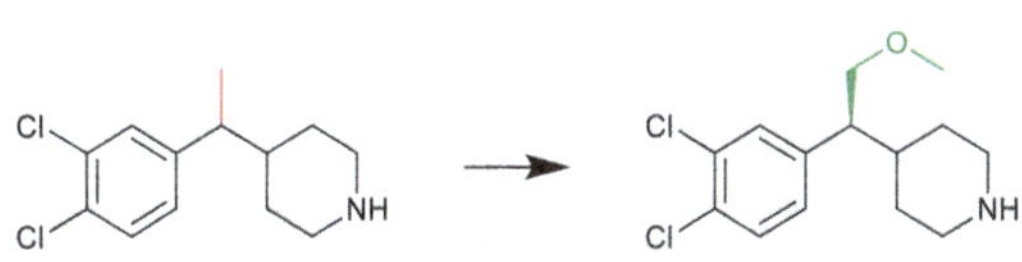

19 20

Monoamine reuptake inhibitors have been associated with a number of safety liabilities attributed to their cationic amphiphilic structure, including a potential to cause PL. Therefore, new compounds were designed setting constraints on MW (<300), CLogP (3.5) and number of aromatic rings (maximum 1). In addition, lipophilicity was modulated by modifying the substitution at the benzylic position. Compound **20** was discovered as a novel triple reuptake inhibitor with lower DIPL potential compared with compound **19**.[76]

(continued)

Examples

2

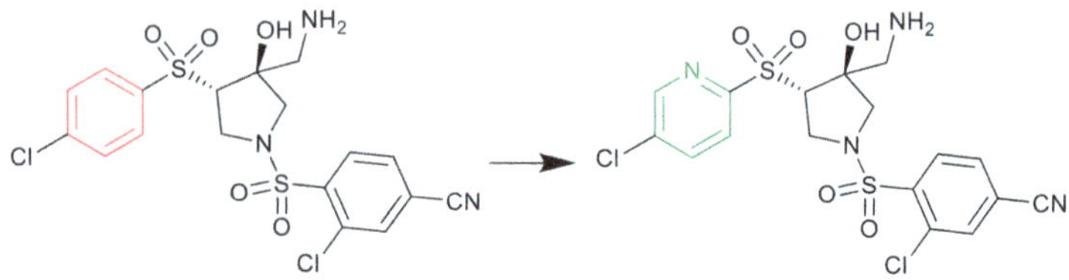

21 22

In a project aiming at identifying small molecule antagonists of the transient receptor potential vanilloid-4, concerns about a DIPL potential were raised for compound **21** on the basis of results from *in vitro* tests. The isosteric replacement made to introduce a pyridine ring (compound **22**) lowered the lipophilicity of the molecule, going from a chromatographic hydrophobicity index on immobilized artificial membrane (CHI IAM) of 49 to 42. *In vitro* assessment demonstrated that compound **22** displays a lower DIPL potential.[77]

3

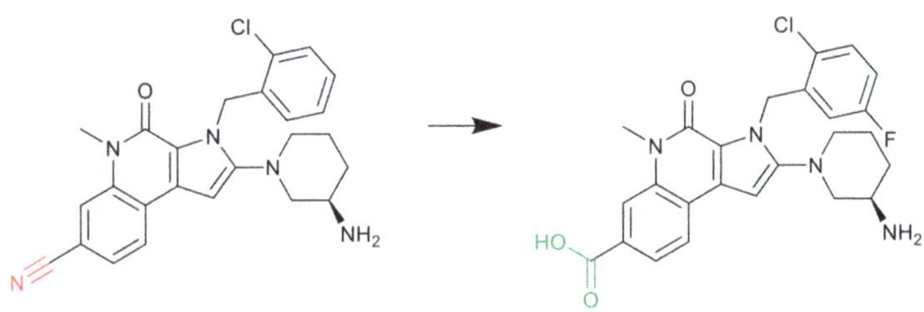

23 24

Compound **23** is a potent dipeptidyl peptidase-IV (DPP-IV) inhibitor but also a DIPL inducer in isolated human hepatocytes. The replacement of the –CN moiety with a carboxylic acid group (*i.e.* reducing the amphiphilicity and the lipophilicity) led to a more potent inhibitor with lower DIPL potential (**24**).[78]

18.3.3 Amphiphilicity Reduction

Examples

1

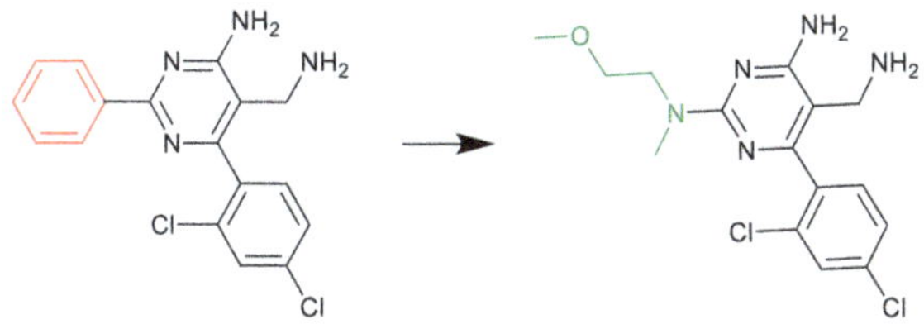

25 **26**

The aminomethylpyrimidine **25**, identified as a lead compound in the discovery of novel DPP-IV inhibitors, induced PL in cultured fibroblasts. The authors used the CAFCA program to design novel compounds devoid of a DIPL effect. Thus, starting from a $\Delta\Delta G_{AM}$ of −6.6 kJ mol^{-1} for compound **25**, the study led to compound **26** with a $\Delta\Delta G_{AM}$ of −5.6 kJ mol^{-1} which did not cause PL, in agreement with Fischer's method of DIPL prediction.[79]

2

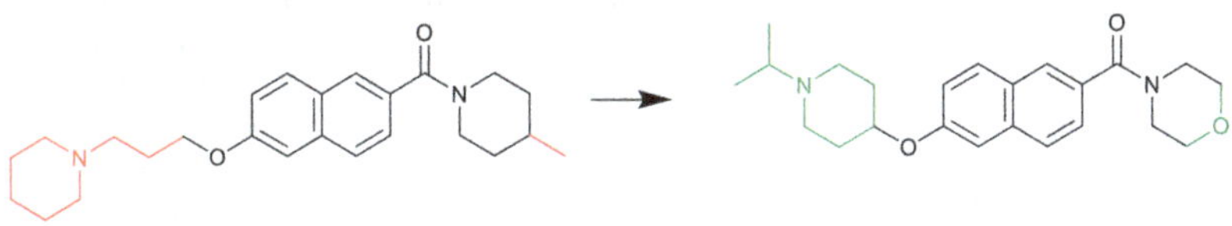

27 **28**

Compound **27** ($\Delta\Delta G_{AM} = -10.4$ kJ mol^{-1}) was identified as a lead compound for novel histamine 3 receptor (H_3R)-inverse agonists containing a naphthalene scaffold, but it was found to induce DIPL in cultured fibroblasts in a dose-dependent manner. The CAFCA program was used to support the design of the less amphiphilic compound **28** that, possessing a $\Delta\Delta G_{AM} = -4.27$ kJ mol^{-1}, was not flagged for DIPL.[80]

3

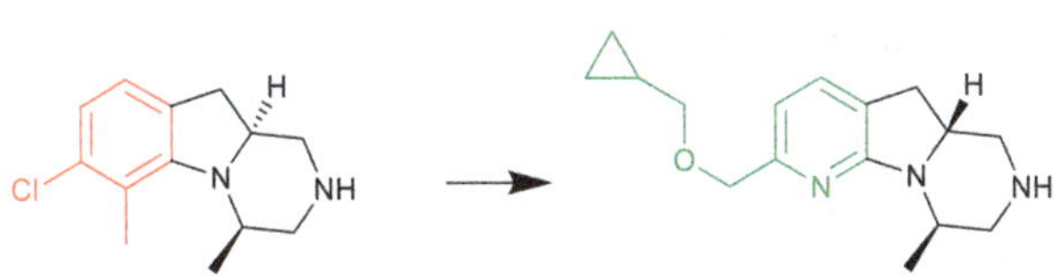

29 **30**

Compound **29** is a potent full agonist at the human 5-HT2C receptor, but also a strong PL inducer in cultured fibroblasts at a concentration of 7.5 μM. Therefore, less amphiphilic compounds were designed. Replacement of the indoline with an azoindole moiety and additional optimization led to compound **30**, which proved to be still a full agonist at the human 5-HT2C receptor and did not show PL at 20 μM.[81]

18.3.4 Modulation of Metabolism

Examples

1

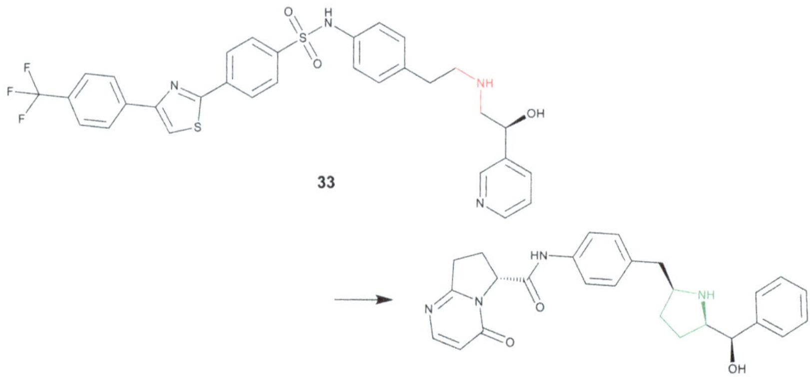

33

34

The development of the beta-3 adrenergic receptor (β3-AR) agonist MK-0634 (**33**) was discontinued due to structure-based toxicity in preclinical model species and PL induction. In addition, high levels of metabolites bearing a secondary amine moiety, generated by *N*-dealkylation at either side of the secondary amine, were observed in preclinical studies. Replacement of the amine group by a pyrrolidine scaffold to block the major site of metabolism and replacement of the lipophilic triaryl moiety by a smaller and less lipophilic group, resulting in reduced lipophilicity, finally gave rise to vibegron (**34**), a non-PL-inducing drug.[67]

References

1. N. Anderson and J. Borlak, *FEBS Lett.*, 2006, **580**, 5533–5540.
2. M. J. Reasor, K. L. Hastings and R. G. Ulrich, *Expert Opin. Drug Saf.*, 2006, **5**, 567–583.
3. J. M. Willard and A. De Felice, in *Lysosomes: Biology, Diseases, and Therapeutics*, ed. F. R. Maxfield, J. M. Willard and S. Lu, John Wiley & Sons, Inc., Weinheim, 2016, pp. 487–512.
4. H. Lullmann, R. Lullmann-Rauch and O. Wassermann, *Ger. Med.*, 1973, **3**, 128–135.
5. T. Shikata, T. Kanetaka, Y. Endo and K. Nagashima, *Acta Pathol. Jpn.*, 1972, **22**, 517–531.
6. L. A. Chatman, D. Morton, T. O. Johnson and S. D. Anway, *Toxicol. Pathol.*, 2009, **37**, 997–1005.
7. U. M. Hanumegowda and A. Regueiro-Ren, in *Tactics in Contemporary Drug Design*, ed. N. Meanwell, Springer, Berlin, Heidelberg, 2013, pp. 261–281.
8. W. H. Halliwell, *Toxicol. Pathol.*, 1997, **25**, 53–60.

9. J. P. Ploemen, J. Kelder, T. Hafmans, H. van de Sandt, J. A. van Burgsteden, P. J. Saleminki and E. van Esch, *Exp. Toxicol. Pathol.*, 2004, **55**, 347–355.
10. K. Tomizawa, K. Sugano, H. Yamada and I. Horii, *J. Toxicol. Sci.*, 2006, **31**, 315–324.
11. U. M. Hanumegowda, G. Wenke, A. Regueiro-Ren, R. Yordanova, J. P. Corradi and S. P. Adams, *Chem. Res. Toxicol.*, 2010, **23**, 749–755.
12. H. Fischer, E. A. Atzpodien, M. Csato, L. Doessegger, B. Lenz, G. Schmitt and T. Singer, *J. Med. Chem.*, 2012, **55**, 126–139.
13. S. S. Choi, J. S. Kim, L. G. Valerio Jr and N. Sadrieh, *Toxicol. Appl. Pharmacol.*, 2013, **269**, 195–204.
14. S. H. Slavov, J. G. Wilkes, D. A. Buzatu, N. L. Kruhlak, J. M. Willard, J. P. Hanig and R. D. Beger, *Bioorg. Med. Chem.*, 2014, **22**, 6706–6714.
15. L. Goracci, S. Buratta, L. Urbanelli, G. Ferrara, R. Di Guida, C. Emiliani and S. Cross, *Eur. J. Med. Chem.*, 2015, **92**, 49–63.
16. H. Sun, M. Xia, S. A. Shahane, A. Jadhav, C. P. Austin and R. Huang, *Bioorg. Med. Chem. Lett.*, 2013, **23**, 4587–4590.
17. S. Slavov, I. Stoyanova-Slavova, S. Li, J. Zhao, R. Huang, M. Xia and R. Beger, *Arch. Toxicol.*, 2017, **91**, 3885–3895.
18. I. Loryan, E. Hoppe, K. Hansen, F. Held, A. Kless, K. Linz, V. Marossek, B. Nolte, P. Ratcliffe, D. Saunders, R. Terlinden, A. Wegert, A. Welbers, O. Will and M. Hammarlund-Udenaes, *Mol. Pharm.*, 2017, **14**, 4362–4373.
19. P. Bonaventure, M. Letavic, C. Dugovic, S. Wilson, L. Aluisio, C. Pudiak, B. Lord, C. Mazur, F. Kamme, S. Nishino, N. Carruthers and T. Lovenberg, *Biochem. Pharmacol.*, 2007, **73**, 1084–1096.
20. J. M. Alakoskela, P. Vitovic and P. K. Kinnunen, *ChemMedChem*, 2009, **4**, 1224–1251.
21. A. Abe, M. Hiraoka and J. A. Shayman, *Drug Metab. Lett.*, 2007, **1**, 49–53.
22. M. Ceccarelli, B. Wagner, R. Alvarez-Sánchez, G. Cruciani and L. Goracci, *Chem. Res. Toxicol.*, 2017, **30**, 1145–1156.
23. M. D. Arbo, S. Melega, R. Stober, M. Schug, E. Rempel, J. Rahnenfuhrer, P. Godoy, R. Reif, C. Cadenas, M. de Lourdes Bastos, H. Carmo and J. G. Hengstler, *Arch. Toxicol.*, 2016, **90**, 3045–3060.
24. H. Sawada, K. Takami and S. Asahi, *Toxicol. Sci.*, 2005, **83**, 282–292.
25. T. Nonoyama and R. Fukuda, *J. Toxicol. Pathol.*, 2008, **21**, 9–24.
26. M. P. Mingeot-Leclercq, R. Brasseur and A. Schanck, *J. Toxicol. Environ. Health*, 1995, **44**, 263–300.
27. M. P. Mingeot-Leclercq, G. Laurent and P. M. Tulkens, *Biochem. Pharmacol.*, 1988, **37**, 591–599.
28. F. Van Bambeke, J. P. Montenez, J. Piret, P. M. Tulkens, P. J. Courtoy and M. P. Mingeot-Leclercq, *Eur. J. Pharmacol.*, 1996, **314**, 203–214.
29. R. Lullmann-Rauch, *Front. Biol.*, 1979, **48**, 49–130.
30. U. P. Kodavanti, V. G. Lockard and H. M. Mehendale, *J. Biochem. Toxicol.*, 1990, **5**, 245–251.
31. E. A. Tengstrand, G. T. Miwa and F. Y. Hsieh, *Expert Opin. Drug Metab. Toxicol.*, 2010, **6**, 555–570.

32. J. P. Montenez, F. Van Bambeke, J. Piret, R. Brasseur, P. M. Tulkens and M. P. Mingeot-Leclercq, *Toxicol. Appl. Pharmacol.*, 1999, **156**, 129–140.
33. D. J. Pelletier, D. Gehlhaar, A. Tilloy-Ellul, T. O. Johnson and N. Greene, *J. Chem. Inf. Model.*, 2007, **47**, 1196–1205.
34. H. Sawada, K. Taniguchi and K. Takami, *Toxicol. In Vitro*, 2006, **20**, 1506–1513.
35. O. Ivanciuc, *Curr. Top. Med. Chem.*, 2008, **8**, 1691–1709.
36. N. L. Kruhlak, S. S. Choi, J. F. Contrera, J. L. Weaver, J. M. Willard, K. L. Hastings and L. F. Sancilio, *Toxicol. Mech. Methods*, 2008, **18**, 217–227.
37. A. M. Orogo, S. S. Choi, B. L. Minnier and N. L. Kruhlak, *Mol. Inf.*, 2012, **31**, 725–739.
38. R. Lowe, R. C. Glen and J. B. Mitchell, *Mol. Pharm.*, 2010, **7**, 1708–1714.
39. R. Lowe, H. Y. Mussa, F. Nigsch, R. C. Glen and J. B. Mitchell, *J. Cheminf.*, 2012, **4**, 2.
40. K. R. Przybylak and M. T. Cronin, *Mol. Inf.*, 2011, **30**, 415–429.
41. L. Goracci, M. Ceccarelli, D. Bonelli and G. Cruciani, *J. Chem. Inf. Model.*, 2013, **53**, 1436–1446.
42. D. Quaglino, H. R. Ha, E. Duner, D. Bruttomesso, L. Bigler, F. Follath, G. Realdi, A. Pettenazzo and A. Baritussio, *Am. J. Physiol.*, 2004, **287**, L438–L447.
43. I. Smyej, S. De Jonghe, A. Looszova, G. Mannens, T. Verhaeghe, S. Thijssen, S. Starckx, A. Lampo and M. C. Rouan, *Toxicol. Pathol.*, 2017, **45**, 663–675.
44. G. Pomponio, C. C. Savary, C. Parmentier, F. Bois, A. Guillouzo, L. Romanelli, L. Richert, E. Di Consiglio and E. Testai, *Toxicol. In Vitro*, 2015, **30**, 36–51.
45. H. Sun, S. Shahane, M. Xia, C. P. Austin and R. Huang, *J. Chem. Inf. Model.*, 2012, **52**, 1798–1805.
46. L. Zhou, G. Geraci, S. Hess, L. Yang, J. Wang and U. Argikar, *Anal. Chem.*, 2011, **83**, 6980–6987.
47. K. R. Przybylak, A. R. Alzahrani and M. T. Cronin, *J. Chem. Inf. Model.*, 2014, **54**, 2224–2232.
48. A. J. Ratcliffe, *Curr. Med. Chem.*, 2009, **16**, 2816–2823.
49. P. Vitovic, J. M. Alakoskela and P. K. Kinnunen, *J. Med. Chem.*, 2008, **51**, 1842–1848.
50. M. Ceccarelli, R. Germani, S. Massari, C. Petit, A. Nurisso, J. L. Wolfender and L. Goracci, *Colloids Surf., B*, 2015, **136**, 175–184.
51. M. Ceccarelli, B. Wagner, R. Alvarez-Sanchez, G. Cruciani and L. Goracci, *Chem. Res. Toxicol.*, 2017, **30**, 1145–1156.
52. A. Casartelli, M. Bonato, P. Cristofori, F. Crivellente, G. Dal Negro, I. Masotto, C. Mutinelli, K. Valko and V. Bonfante, *Cell Biol. Toxicol.*, 2003, **19**, 161–176.
53. J. K. Morelli, M. Buehrle, F. Pognan, L. R. Barone, W. Fieles and P. J. Ciaccio, *Cell Biol. Toxicol.*, 2006, **22**, 15–27.
54. T. Kasahara, K. Tomita, H. Murano, T. Harada, K. Tsubakimoto, T. Ogihara, S. Ohnishi and C. Kakinuma, *Toxicol. Sci.*, 2006, **90**, 133–141.

55. P. Nioi, B. K. Perry, E. J. Wang, Y. Z. Gu and R. D. Snyder, *Toxicol. Sci.*, 2007, **99**, 162–173.
56. P. Nioi, I. D. Pardo and R. D. Snyder, *Drug Chem. Toxicol.*, 2008, **31**, 515–528.
57. N. Mesens, M. Steemans, E. Hansen, G. R. Verheyen, F. Van Goethem and J. Van Gompel, *Toxicol. In Vitro*, 2010, **24**, 1417–1425.
58. N. Mesens, M. Steemans, E. Hansen, P. Annelieke, G. Verheyen and P. Vanparys, *Toxicol. In Vitro*, 2009, **23**, 217–226.
59. S. A. Shahane, R. Huang, D. Gerhold, U. Baxa, C. P. Austin and M. Xia, *J. Biomol. Screening*, 2014, **19**, 66–76.
60. F. M. van de Water, J. Havinga, W. T. Ravesloot, G. J. Horbach and W. G. Schoonen, *Toxicol. In Vitro*, 2011, **25**, 1870–1882.
61. W. G. Schoonen, W. M. Westerink and G. J. Horbach, *EXS*, 2009, **99**, 401–452.
62. L. Fusani, M. Brown, H. Chen, E. Ahlberg and T. Noeske, *Mol. Pharm.*, 2017, **14**, 4346–4352.
63. F. Kazmi, T. Hensley, C. Pope, R. S. Funk, G. J. Loewen, D. B. Buckley and A. Parkinson, *Drug Metab. Dispos.*, 2013, **41**, 897–905.
64. B. L. Yano, D. M. Bond, M. N. Novilla, L. G. McFadden and M. J. Reasor, *Toxicol. Sci.*, 2002, **65**, 288–298.
65. P. Schnider, C. Dolente, H. Stalder, R. E. Martin, V. Reinmüller, R. Marty, C. Wyss Gramberg, B. Wagner, H. Fischer, A. M. Alker and K. Müller, *ChemBioChem*, 2020, **21**, 212–234.
66. H. K. Fischer, M. Kansy and D. Bur, *Chimia*, 2000, **54**, 640–645.
67. S. D. Edmondson, C. Zhu, N. F. Kar, J. Di Salvo, H. Nagabukuro, B. Sacre-Salem, K. Dingley, R. Berger, S. D. Goble, G. Morriello, B. Harper, C. R. Moyes, D. M. Shen, L. Wang, R. Ball, A. Fitzmaurice, T. Frenkl, L. N. Gichuru, S. Ha, A. L. Hurley, N. Jochnowitz, D. Levorse, S. Mistry, R. R. Miller, J. Ormes, G. M. Salituro, A. Sanfiz, A. S. Stevenson, K. Villa, B. Zamlynny, S. Green, M. Struthers and A. E. Weber, *J. Med. Chem.*, 2016, **59**, 609–623.
68. L. W. Whitehouse, A. Menzies, R. Mueller and R. Pontefract, *Toxicology*, 1994, **94**, 81–95.
69. H. Igawa, M. Takahashi, K. Kakegawa, A. Kina, M. Ikoma, J. Aida, T. Yasuma, Y. Kawata, S. Ashina, S. Yamamoto, M. Kundu, U. Khamrai, H. Hirabayashi, M. Nakayama, Y. Nagisa, S. Kasai and T. Maekawa, *J. Med. Chem.*, 2016, **59**, 1116–1139.
70. M. Furber, A. K. Tiden, P. Gardiner, A. Mete, R. Ford, I. Millichip, L. Stein, A. Mather, E. Kinchin, C. Luckhurst, S. Barber, P. Cage, H. Sanganee, R. Austin, K. Chohan, R. Beri, B. Thong, A. Wallace, V. Oreffo, R. Hutchinson, S. Harper, J. Debreczeni, J. Breed, L. Wissler and K. Edman, *J. Med. Chem.*, 2014, **57**, 2357–2367.
71. D. A. Nugiel, J. R. Krumrine, D. C. Hill, J. R. Damewood Jr, P. R. Bernstein, C. D. Sobotka-Briner, J. Liu, A. Zacco and M. E. Pierson, *J. Med. Chem.*, 2010, **53**, 1876–1880.

72. M. Sun, C. Zhao, G. A. Gfesser, C. Thiffault, T. R. Miller, K. Marsh, J. Wetter, M. Curtis, R. Faghih, T. A. Esbenshade, A. A. Hancock and M. Cowart, *J. Med. Chem.*, 2005, **48**, 6482–6490.
73. R. Nirogi, A. Shinde, A. R. Mohammed, R. K. Badange, V. Reballi, T. R. Bandyala, S. K. Saraf, K. Bojja, S. Manchineella, P. K. Achanta, K. K. Kandukuri, R. Subramanian, V. Benade, R. C. Palacharla, P. Jayarajan, S. Pandey and V. Jasti, *J. Med. Chem.*, 2019, **62**, 1203–1217.
74. T. T. Wager, B. A. Pettersen, A. W. Schmidt, D. K. Spracklin, S. Mente, T. W. Butler, H. Howard, D. J. Lettiere, D. M. Rubitski, D. F. Wong, F. M. Nedza, F. R. Nelson, H. Rollema, J. W. Raggon, J. Aubrecht, J. K. Freeman, J. M. Marcek, J. Cianfrogna, K. W. Cook, L. C. James, L. A. Chatman, P. A. Iredale, M. J. Banker, M. L. Homiski, J. B. Munzner and R. Y. Chandrasekaran, *J. Med. Chem.*, 2011, **54**, 7602–7620.
75. H. Ratni, M. Ebeling, J. Baird, S. Bendels, J. Bylund, K. S. Chen, N. Denk, Z. Feng, L. Green, M. Guerard, P. Jablonski, B. Jacobsen, O. Khwaja, H. Kletzl, C. P. Ko, S. Kustermann, A. Marquet, F. Metzger, B. Mueller, N. A. Naryshkin, S. V. Paushkin, E. Pinard, A. Poirier, M. Reutlinger, M. Weetall, A. Zeller, X. Zhao and L. Mueller, *J. Med. Chem.*, 2018, **61**, 6501–6517.
76. Y. Ishichi, E. Kimura, E. Honda, M. Yoshikawa, T. Nakahata, Y. Terao, A. Suzuki, T. Kawai, Y. Arakawa, H. Ohta, N. Kanzaki, H. Nakagawa and J. Terauchi, *Bioorg. Med. Chem.*, 2013, **21**, 4600–4613.
77. J. E. Pero, J. M. Matthews, D. J. Behm, E. J. Brnardic, C. Brooks, B. W. Budzik, M. H. Costell, C. A. Donatelli, S. H. Eisennagel, K. Erhard, M. C. Fischer, D. A. Holt, L. J. Jolivette, H. Li, P. Li, J. J. McAtee, B. W. McCleland, I. Pendrak, L. M. Posobiec, K. L. K. Rivera, R. A. Rivero, T. J. Roethke, M. R. Sender, A. Shu, L. R. Terrell, K. Vaidya, X. Xu and B. G. Lawhorn, *J. Med. Chem.*, 2018, **61**, 11209–11220.
78. Y. Ikuma, H. Hochigai, H. Kimura, N. Nunami, T. Kobayashi, K. Uchiyama, Y. Furuta, M. Sakai, M. Horiguchi, Y. Masui, K. Okazaki, Y. Sato and H. Nakahira, *Bioorg. Med. Chem.*, 2012, **20**, 5864–5883.
79. J. U. Peters, D. Hunziker, H. Fischer, M. Kansy, S. Weber, S. Kritter, A. Muller, A. Wallier, F. Ricklin, M. Boehringer, S. M. Poli, M. Csato and B. M. Loeffler, *Bioorg. Med. Chem. Lett.*, 2004, **14**, 3575–3578.
80. R. M. Rodriguez Sarmiento, M. H. Nettekoven, S. Taylor, J. M. Plancher, H. Richter and O. Roche, *Bioorg. Med. Chem. Lett.*, 2009, **19**, 4495–4500.
81. H. G. Richter, D. R. Adams, A. Benardeau, M. J. Bickerdike, J. M. Bentley, T. J. Blench, I. A. Cliffe, C. Dourish, P. Hebeisen, G. A. Kennett, A. R. Knight, C. S. Malcolm, P. Mattei, A. Misra, J. Mizrahi, N. J. Monck, J. M. Plancher, S. Roever, J. R. Roffey, S. Taylor and S. P. Vickers, *Bioorg. Med. Chem. Lett.*, 2006, **16**, 1207–1211.

CHAPTER 19

Cardiac Ion Channel Inhibition

CINZIA BORDONI[a], DANIEL J. BROUGH[a], GEMMA DAVISON[a], JAMES H. HUNTER[a], J. DANIEL LOPEZ-FERNANDEZ[a], KATE MCADAM[a], DUNCAN C. MILLER[a], PASQUALE A. MORESE[a], ALEXIA PAPAIOANNOU[a], MÉLANIE UGUEN[a], PAUL RATCLIFFE*[b], NIKOLAY SITNIKOV*[c] AND MICHAEL J. WARING*[a]

[a]Northern Institute for Cancer Research, School of Chemistry, School of Natural and Environmental Sciences, Newcastle University, Newcastle upon Tyne, NE1 7RU, UK; [b]Eurofins Discovery One™, Fyfield Business and Research Park, Fyfield Road, Ongar, Essex, CM5 0GS, UK; [c]Innovation Campus Berlin, Nuvisan ICB GmbH, Life Science Chemistry, Müllerstraße 178, 13353 Berlin, Germany
*E-mail: nikolay.sitnikov@nuvisan.com, Mike.Waring@newcastle.ac.uk, paulratcliffe@eurofins.com

19.1 Introduction

This chapter is focused on medicinal chemistry strategies to mitigate the risk of functional cardiotoxicity of a drug candidate resulting from direct interaction with cardiac ion channels, in particular human ether-à-go-go-related gene (hERG) potassium channels ($K_V11.1$), L-type voltage-gated calcium channels 1.2 ($Ca_V1.2$) and voltage-gated sodium channels 1.5 ($Na_V1.5$). Functional (*e.g.* contractile) or structural cardiotoxicity,[1] resulting from indirect modulation of channel activity or from interaction with other target classes (demonstrated *e.g.* for anticancer drugs[2–5] including kinase inhibitors[6,7]) is left outside of the

Drug Discovery Series No. 79
The Medicinal Chemist's Guide to Solving ADMET Challenges
Edited by Patrick Schnider
© The Royal Society of Chemistry 2021
Published by the Royal Society of Chemistry, www.rsc.org

scope of this chapter. For an overview of possible cardiac effects of drugs and preclinical cardiac safety assessment the reader is referred to a general review.[8]

19.1.1 Cardiac Ion Channels and Arrythmias – the Background of Counter-screen

Recent years have witnessed important advances in our understanding of the electrophysiological mechanisms underlying the development of cardiac arrhythmias. Mutations in genes encoding several cardiac ion channels were associated with inherited channelopathies: Brugada, long QT (time from the start of the Q wave to the end of the T wave) and short QT syndromes, all of which can lead to potentially fatal cardiac arrythmias in the absence of structural heart disease.[9] It was recognized that not only genetic mutations but also pharmacological inhibition of cardiac ion channel conductance can lead to life-threatening proarrhythmic effects.[10] For the hERG channel, this was first demonstrated in the late 1980s–1990s for the histamine H1 receptor antagonist terfenadine. In 1989, overdoses of the drug were reported to prolong the QT interval[11] and its use was linked to torsades de pointes.[12] Terfenadine was subsequently shown to inhibit the delayed rectifier potassium current in isolated myocytes[13] and hERG channels expressed in *Xenopus* oocytes.[14] During the following years, a significant number of drugs from various therapeutic classes (from the antihistamine astemizole and the class III antiarrhythmic dofetilide to the antipsychotic droperidol and antibiotic clarithromycin), were determined to prolong the QT interval and were either withdrawn from the market or received black box warning labels (based on increased proarrhythmic risk).[15] Although the correlation between blocking of a particular cardiac ion channel (or even multiple ones) and *in vivo* effects on cardiac function is not straightforward, the potential for putting patients at risk means that off-target activity on cardiac ion channels must be carefully evaluated. The most prominent of these channels, hERG, $Na_V1.5$ and $Ca_V1.2$, which traffic major ion currents shaping cardiac action potential, are recognized as primary counter-screen targets. These channels possess relatively large inner pores with multiple binding sites and can accommodate a variety of structurally diverse ligands. For example, a study that involved a high-throughput screening assay of 300 000 diverse compounds for hERG liability reported a hit rate of 27% at 10 μM.[16] Similarly, testing of a heterogenous set of >3000 compounds for inhibition of a customized $Na_V1.5$ channel revealed a hit rate (>50% inhibition) of around 13% at 10 μM.[17] Thus, historical experience and a strong desire to avoid late-stage compound failures motivate pharmaceutical companies to test compounds for cardiac ion channel liabilities early on in the lead optimization process.

19.1.2 Cardiac Action Potential

The systematic propagation of the cardiac action potential (AP) is essential for coordinated contraction of the heart. AP is a change of potential (depolarization followed by repolarization) across membranes of cardiac cells.

It is generated by pacemaker cells (found primarily in the sinoatrial node) and is transmitted throughout the sections of myocardium by cardiomyocytes due to the presence of intercellular gap junctions. Changes in polarization state of cardiomyocytes are mediated by movements of ions (Na^+, Ca^{2+}, K^+) across the cell membrane through a multitude of ion channels. Approximately 24 ion channel subtypes are known to contribute to around 13 different cardiac ion currents.[18] Major ion currents that shape the cardiac AP and correlation of AP phases with QT interval of surface electrocardiogram (ECG) are illustrated in Figure 19.1.[19] Generally, the instantaneous slope of the AP waveform is determined by the relative balance of total outward potassium and inward sodium and calcium currents.

In a simplified representation of AP, fast inward Na^+ current I_{NaF} (through $Na_V1.5$ channels) leads to initial membrane depolarization (upstroke of AP, phase 0). After rapid inactivation of $Na_V1.5$, transient outward K^+ current I_{to} (through $K_V1.4/1.7/3.4/4.2/4.3$) initiates rapid membrane repolarization (AP notch, phase 1). This event is followed by a slow repolarization phase (AP plateau, phase 2), resulting from superimposition of delayed rectifier outward K^+ current I_{Kr} (through hERG channels) and inward Ca^{2+} current I_{Ca} ($Ca_V1.2$ channels).

A short data overview on expression and cardiac function of $Ca_V1.2$, $Na_V1.5$ and hERG channels as well as associated channelopathies and ECG effects of channel blocking is given in Table 19.1.

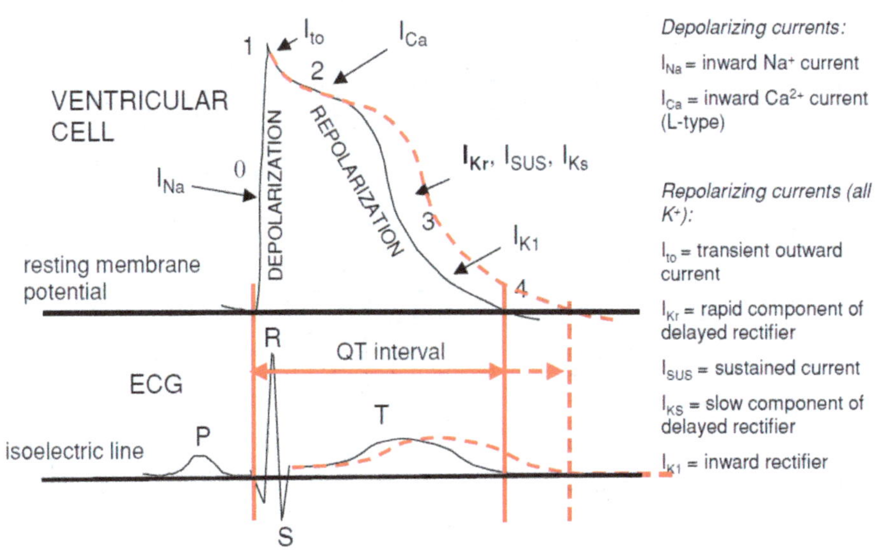

Figure 19.1 Ventricular cell action potential (AP) and surface ECG. I_{Na} current (AP phase 0) is mediated through $Na_V1.5$ channel, I_{to} (AP phase 1) – through $K_V1.4/1.7/3.4/4.2/4.3$, I_{Ca} (AP phase 2) – through $Ca_V1.2/3.1/3.2$, I_{Kr} and I_{Ks} (AP phase 3) – through hERG and K_VLQT1 respectively, I_{K1} (AP phase 4) – through Kir 2.1/2.2. Reproduced from ref. 19 with permission from Springer Nature, Copyright 2006.

Table 19.1 Major cardiac ion channels: general overview.

Channel	Isoform expression (human protein atlas)	Cardiac function	ECG effects of channel block	Channelopathies
$Ca_V1.2$ (CACNA1C gene)	Heart (major Ca^{2+}-channel isoform), smooth muscle, brain, endocrine tissue, male and female reproductive tissues, gastrointestinal tract, gallbladder, lung, kidney and bladder	(1) Cellular depolarization and pacemaker activity in the SA node. (2) Generation and conduction of influx from sinoatrial (SA) to atrioventricular (AV) node. (3) Initiation of contraction–relaxation cycle (triggering calcium-induced calcium release from sarcoplasmic reticulum). (4) Maintaining ventricular AP plateau phase.[20]	*RR prolongation* (bradycardia); *PR prolongation* (slowed conduction between atrial and ventricular myocardium); *Deep $Ca_V1.2$ block can result in: AV block; left bundle branch block.*	Arterial hypertension, Long QT syndrome type 8 (GoF[a]), Timothy syndrome (GoF), short QT syndrome (LoF[a]), Brugada syndrome type 3 (LoF) + mental disorders: *e.g.* bipolar disorder, schizophrenia.[21]
$Na_V1.5$ (SCN5A gene)	Predominantly heart (major Na^+-channel isoform, 77–88% of total sodium channel staining), minor: brain, gastrointestinal smooth muscle, male and female reproductive tissues[22]	(1) Upstroke of action potentials in atrial and ventricular myocardium (fast Na^+ current, $I_{Na\ fast}$). (2) Sustaining the action potential plateau, defining the action potential duration, modulation of intracellular sodium (hence contractility) (late sodium current, $I_{Na\ late}$).[23]	*QRS prolongation* (slowed intraventricular conduction); *PR prolongation.*	Long QT syndrome type 3, Brugada syndrome type 1, cardiac conduction disease, dilated cardiomyopathy, sick sinus node syndrome, atrial fibrillation.[24–26]
$K_V11.1$ (hERG, KCNH2 gene)	Heart and smooth muscle tissue, brain, bone marrow, endocrine tissue, gallbladder, male and female reproductive tissues, gastrointestinal tract	Conduction of the rapid component of the delayed rectifier potassium current, I_{Kr}, which is crucial for repolarization of cardiac action potentials.[27]	*QT prolongation.*	Long QT syndrome type 2 (LoF), short QT syndrome (GoF), both can lead to torsades de pointes (TdP).[28]

[a]GoF: gain-of-function mutation; LoF: loss-of-function mutation.

19.1.3 Ion Channel Function and Channel Blocking

Ion channels are membrane proteins that selectively regulate ion transport across lipophilic membranes of cells and organelles. The two fundamental properties of ion channels are *ion permeation* and *gating*. Ion permeation is the movement of specific ions (Na^+, K^+ or Ca^{2+}) through the channel pore. Ion selectivity forms the basis for channel classification. Ion current is generally non-linearly dependent on the electrochemical potential, and the magnitude of current at a given potential is also dependent on the direction (in or out of the cell). The latter property is called rectification. The second fundamental property, *gating*, is the mechanism of opening and closing of the ion channel. Generally, ion channels can interconvert between closed, open and inactivated states. This gating can be voltage- or ligand dependent or mechanosensitive. The majority of the cardiac channels, including $Na_V1.5$, $Ca_V1.2$ and hERG, are voltage-gated, meaning that conformational changes in these channels (state transitions) are triggered by changes in electrical fields.

Gating strongly effects the binding of ion channel blockers under physiological conditions. Tonic and phasic blockade refers to inhibition of static and gating channels, respectively. A *state-dependent* (*or voltage-dependent*) *block* indicates that the drug binds to open and/or inactivated channels with different (generally higher) affinity than to resting channels.[29,30] As voltage (or membrane polarization) is directly affecting channel conformation, voltage-dependence can be a result of altering accessibility of the binding site as well as changing affinity of the ligand. Drug dissociation from the ion channel complex can also be "state-dependent" (taking place only from the open state), as demonstrated *e.g.* for the hERG channel.[31] Ligand dissociation kinetics thus determines *use-dependence* (or *frequency-dependence*) of the block. For use-dependent blockers steady state levels of channel block increase with increasing pulse frequency. Slow drug dissociation kinetics can lead to *drug trapping and channel block accumulation*, a feature that significantly affects pharmacodynamic outcome. One should note that the effects of channel gating on drug binding diminish under *in vitro* conditions, as the gating frequency slows down below physiological levels and open channel times increase beyond physiological levels in electrophysiology assays and even to infinity in a binding assay (see Section 19.1.4.2 *In Vitro* Cardiac Safety Screening). In conclusion, *not only the affinity but also binding kinetics determine the propensity of the channel blocker to affect cardiovascular (CV) parameters in vivo*.

19.1.4 Preclinical Cardiotoxicity Screening

19.1.4.1 Regulatory Guidelines

Inhibition of I_{Kr} current is recognized as the predominant mechanism for drug-induced QT-prolongation that is linked to torsades de pointes (TdP). Thus, in line with current regulatory S7B International Committee on

Harmonization (ICH) guidelines,[32] *in vitro* screening for hERG antagonism is in focus during the early stages of drug discovery and development. The guidelines describe a core study battery (an *in vitro* test and *in vivo* experimental model) for preclinical assessment of QT interval prolongation liability. In particular, patch clamp experiments in cells (HEK293 or CHO) transfected with the hERG channel-encoding *KCNH2* gene or in native cardiac cells are recommended as an *in vitro* screening method, while a CV telemetric study in conscious, freely moving non-rodent species is specified for *in vivo* screening. Supplementary studies are also defined in S7B, which should be conducted in case core studies are inconclusive. Examples are AP duration in papillary muscle,[33] Purkinje fibres or Langendorff isolated heart.[34]

Figure 19.2 shows an example of a CV safety screening cascade aimed at identification and de-risking of candidate molecules advancing into GLP toxicity studies. The evolution of early *in vitro* screening paradigms and the matter of relevant safety margins is discussed below. *Ex vivo* and advanced *in vivo* models used for later stage CV risk assessment as well as their translatability into humans are left outside of the scope of this example.

19.1.4.2 In Vitro *Cardiac Safety Screening*

The overview of established *in vitro* techniques for cardiotoxicity evaluation is provided in Table 19.2. These fall into two major categories: single channel and phenotypic screening assays, both being briefly discussed below.

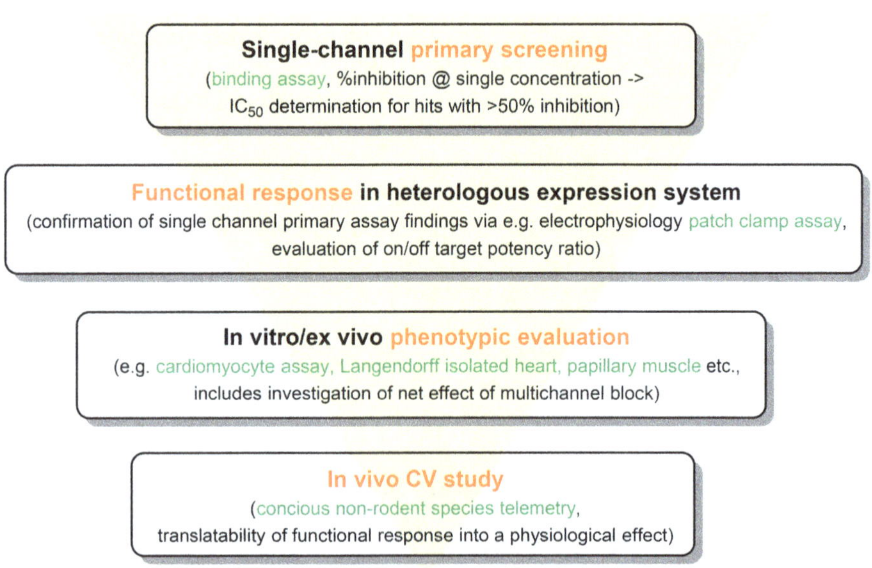

Figure 19.2 Possible preclinical non-good laboratory practice (non-GLP) CV safety evaluation cascade.

Detailed analysis of advantages and limitations of particular assays as well as comparison of their technical characteristics can be found in a dedicated review.[35]

19.1.4.2.1 Screening on Individual Cardiac Channels. The primary goal of an early *in vitro* screen on individual cardiac ion channels is to identify potential hazards and rank compound series and, importantly, to establish a structure–activity relationship (SAR) allowing compound optimization. The following information should be considered upon interpretation of *in vitro* data:

1. *Concentrations giving 50% of maximum inhibition (IC_{50}) values may vary significantly depending on the assay type.* In a study involving 49 drugs Huang *et al.* compared hERG potencies of 49 drugs across three different assays (radioligand binding, automated patch clamp (APC), Tl^+-flux).[36] In general, hERG inhibitors were most potent in APC assay, of intermediate potency in [^{3}H]dofetilide binding assays and least potent in Tl^+-flux assay [*e.g.* astemizole log of the inhibitory constant (pK_i) 8.50 ± 0.06 (binding), negative log of the half maximal inhibitory concentration (pEC_{50}) 7.67 ± 0.12 (APC) and 5.68 ± 0.01 (Tl^+-flux)]. Jenkinson and colleagues recently compared outcomes of various *in vitro* assays for three structurally distinct $Ca_V1.2$ inhibitors[37] as well as for 22 known $Na_V1.5$ blockers.[38] As discussed above for hERG, for multiple compounds more than one order of magnitude differences between IC_{50} values were observed. One should note that the difference in the outcome of a binding assay and an electrophysiology assay can be, at least in part, explained by existence of distinct binding sites (as well-evidenced for the $Ca_V1.2$ channel, see Section 19.1.5.2).
2. *IC_{50} values may vary significantly depending on the assay conditions.* The latter reflects not only the data fluctuation but also the complex nature of ligand–channel interaction, where ligand affinity and receptor accessibility are functions of several variables: pulse protocol (duration and amplitude),[39–41] temperature and frequency. For example, Stork *et al.* have demonstrated frequency-dependent hERG current inhibition by amiodarone [patch clamp IC_{50} (0.03 Hz) = 7.2 ± 2.05 μM, IC_{50} (0.3 Hz) = 1 ± 0.65 μM].[31] For $Na_V1.5$ a prominent example is lidocaine [IC_{50} (0.1 Hz) = 335.2 μM, IC_{50} (10 Hz) = 41.3 μM].[42]
3. *Kinetic aspects of drug–channel interaction should be taken into account when drawing conclusions on CV liability.* Using *in silico* modelling, Di Veroli *et al.*[43] concluded that slow-binding, non-trappable drugs induce less AP prolongation and minimal QT interval prolongation (approximately 5%) at a concentration equal to the electrophysiological hERG IC_{50}, whereas trappable drugs possess maximal pro-arrhythmic risk (>20% QT prolongation at IC_{50}) (for a detailed discussion see also Pearlstein *et al.*[44]). $Na_V1.5$ inhibitors with slow dissociation kinetics (k_{off}) have been shown to induce more pronounced QRS prolongation.

Table 19.2 Summary of *in vitro* cardiotoxicity screening assays.

Assay class	Single-channel primary	Single-channel functional			Phenotypic	
System	Isolated membrane	Heterologous expression system (HEK293 or CHO cells)			Primary cardiomyocytes[47]	Stem cell (SC)-derived cardiomyocytes (iPSC-CMs or ESC-CMs)[a].[48]
Technology	Binding assay (displacement of radiolabelled or photolabelled ligand)	Fluorescence-based	Manual patch clamp[49]	Automated patch clamp[50]	APC[a] (single-cell); MEA[a], RTCA[a] and Ca^{2+} FLIP[a] (cellular networks)[35]	
Advantages	High throughput, low costs	Low costs, rapid data generation (up to 50 000 data points per day)	High data accuracy; block potency, binding kinetics, voltage dependence can be studied	High throughput (up to 10 000 data points per day); blocking potency, binding kinetics,[51] voltage dependence can be studied	Recording of net endogenous currents under native cellular conditions→ significantly improved predictivity of *in vivo* outcome	Recording of net endogenous currents under conditions close to physiological (multiple channels and cardiac co-factors)

Table 19.2 Summary of *in vitro* cardiotoxicity screening assays. (*continued*)

Limitations	Measures affinity to specific binding site; no information on functional response	Less sensitive than patch clamp (up to two orders of magnitude IC_{50} shift), thus most useful for early identification of potent inhibitors	• Very limited throughput (1–3 days full time equivalent per IC_{50}); • Requires advanced technical skills	• Commonly high costs (approximately 400 US$ per plate) • Lower data accuracy than in manual patch clamp	Low proliferative capacity thus limited availability	• Correlation with adult cardiac physiology not well understood (immature phenotype); • No standardized protocol exists for evaluation of phenotype
	Biophysical channel properties non-identical to endogenous					
Comment	Radioligands: hERG: [^{3}H]-dofetilide; $Na_V1.5$: [^{3}H]-batrachotoxinin; $Ca_V1.2$: [^{3}H]nitrendipine (DHP site),[a] [^{3}H]verapamil (PAA site),[a] [^{3}H]diltiazem (BTZ site)[a]	Tl^+-flux FLIPR (hERG),[52] membrane potential dye FLIPR (hERG, $Na_V1.5$); Ca^{2+}-flux ($Ca_V1.2$)	CiPA[a] standardized protocol for hERG reported[53]	hERG,[54] $Ca_V1.2$,[55] $Na_V1.5$ channel[56]		hSC-CMs[a] preparations represent a mix of atrial, nodal and ventricular (predominant) cells with different distribution patterns depending on the vendor

[a]APC, automated patch clamp; BTZ, benzothiazepine; CiPA, comprehensive *in vitro* proarrhythmia assay; DHP, 1,4-dihydropyridine; ESC, embryonic stem cell; FLIPR, fluorometric imaging plate reader; hSC, human stem cells; iPSC, induced pluripotent stem cell; MEA, multi-electrode array; PAA, phenylalkylamine; RTCA, real-time cell analysis.

For example, lidocaine minimally affects QRS duration (due to rapid unblocking kinetics), while flecainide causes prominent QRS prolongation at physiological heart rates due to greater accumulation of blocked channels.[45,46]

19.1.4.2.2 Evaluation of Safety Window on the Basis of Single Channel Data

hERG. Redfern *et al.* correlated *in vivo* QTc prolongation and occurrence of clinical TdP reports with data on free plasma exposures and hERG IC_{50} values across a set of 52 drugs.[57] Basing on this analysis a provisional safety margin of 30 (defined as IC_{50}/C_{max}, where C_{max} is the maximum free plasma concentration a drug achieves clinically) was generally suggested as predictive of cardiac safety (although a higher margin of 100 was advised for indications with low disease burden). The general applicability of these safety margins was critically addressed, as they, for example, do not differentiate between trappable (higher risk) and non-trappable (lower risk) blockers (for details see 14.1.6 of Pearlstein *et al.*[44]). There is evidence that almost any appreciable block (>3%, approximately 1/30 of IC_{50}) of hERG current can be associated with corrected QT interval (QTc) prolongation and arrhythmia (see Section 2.3 of Leishman and Rankovic[27]). A pharmacokinetic–pharmacodynamic (PK–PD) study quantifying the relationship between *in vitro* hERG blockade and the magnitude of QT prolongation in humans for dofetilide revealed that 10% hERG block corresponds to a QT prolongation of 20 ms[†] (95% confidence interval, 12–32 ms) and that 50% of the maximal effect occurs at concentrations which only block 20% of hERG current *in vitro*.[58]

Na_V1.5. A thorough study of Na_V1.5 *in vitro* data predictivity for clinical QRS prolongation[‡] using a set of 22 clinical compounds was reported by Jenkinson *et al.*[38] The authors correlated *in vitro* concentration–response curves with published clinical data on human unbound plasma exposures and QRS prolongation. On the basis of the results from that analysis, safety thresholds were determined to be at IC_{20} for binding assay, IC_{10} for membrane potential assay and IC_{20} for patch clamp assay (as exposures leading to 10% QRS prolongation in humans). Predictivity of all three *in vitro* assays was rendered similar, while no safety margins were suggested. In line with the conclusions of Jenkinson *et al.* is an earlier study by Harmer *et al.*,[59] in which patch clamp Na_V1.5 data were correlated with clinical QRS prolongation across a heterogenous drug panel. The authors found that for 42 class I anti-arrhythmics and other QRS prolonging drugs, 67% had $IC_{50}:C_{max}$ free ratios <30. For 55 non-QRS prolonging drugs tested, 72% had

[†]The current FDA draft guidance suggests that an increased risk for TdP exists at 20 ms of QT prolongation or more [ICH E14 guidance, https://www.fda.gov/media/71372/download].

[‡]QRS duration can be considered as a rough surrogate marker for conduction-based proarrhythmia (analogous to QT prolongation being a surrogate marker for TdP proarrhythmia). Arguably, there is however no consensus on which extent of QRS prolongation can be reliably associated with occurrence of arrhythmia in humans.

ratios >100. Finally, for 37 drugs, QRS complex prolongation was observed at free plasma concentrations that were about 15-fold lower than the corresponding IC_{50} at $hNa_V1.5$ channels. On the basis of that data, a margin of 30- to 100-fold between $hNa_V1.5$ IC_{50} and $C_{max\ free}$ was suggested to confer an acceptable degree of safety from QRS prolongation (this study was critically reviewed by Gintant *et al.*[39]). Other studies also support the finding that exposures significantly below IC_{50} are sufficient to cause 10–20% QRS prolongation in humans[60] as well as in preclinical model species [dogs and non-human primates (NHP)].[61]

$Ca_V1.2$. In contrast to blocking of hERG and $Na_V1.5$ current, no systematic study of the translatability of *in vitro* $Ca_V1.2$ inhibition to CV effects in higher preclinical model species has so far been reported. Jenkinson *et al.* have correlated L-type Ca_V activity from *in vitro* assays (radioligand binding, patch clamp, $Ca2^+$-flux) and *ex vivo* tissue preparations (rat aorta and guinea pig Langendorff isolated heart) with CV observations in conscious telemetered rats[37] for three marketed antagonists from different structural classes: nifedipine (DHP site), verapamil (PAA site) and diltiazem (BTZ site) (see Figure 19.7). It was observed that nifedipine requires *in vivo* exposures corresponding to approximately maximal effects *in vitro* or *ex vivo* in order to result in an observable change in either blood pressure (BP) or heart rate (HR). In contrast, a significant change in BP was seen with verapamil at, or below, the pK_i/pIC_{50} values for patch clamp, FLIPR, aortic ring and Langendorff contractility.

19.1.4.2.3 Phenotypic Screening and the CiPA Initiative. The net outcome of a drug's integrated effect on multiple ion channels rather than on a single channel should be considered for more reliable CV safety assessment.[62] The most prominent example in this context is verapamil, which, despite being a potent blocker of repolarizing potassium current I_{Kr} (through hERG), does not induce significant QT prolongation due to concomitant blockade of the depolarizing calcium current *I*CaL (through $Ca_V1.2$).[63] *In vitro* phenotypic screening using primary cardiomyocytes,[47] human embryonic stem cell derived cardiomyocytes (hESC-CMs) or induced pluripotent stem cell derived cardiomyocytes (hiPSC-CMs)[48] represent possible approaches to get a more comprehensive view of drug effects on cardiac currents (Table 19.2). Electrophysiological approaches for evaluating drug effects with these cellular networks (microelectrode array, real-time cell analysis, Ca^{2+} dynamics) have been recently reviewed.[35] Arguably, data produced with SC-CMs still require validation with large sets of positive and negative reference compounds, because the correlation with adult cardiac physiology is not well understood. The electrophysiological phenotype of hSC-CMs is generally characterized as being immature (see Jiang *et al.* for a review,[64] Goodrow *et al.* for an example Na^+ current[65]). No standardized guidelines currently exist for the evaluation of hSC-CM phenotypes and functionality that may affect drug responses.

In 2013 the *Comprehensive in vitro Proarrhythmia Assay* (CiPA) initiative[66] was launched to elaborate an improved paradigm for preclinical evaluation of

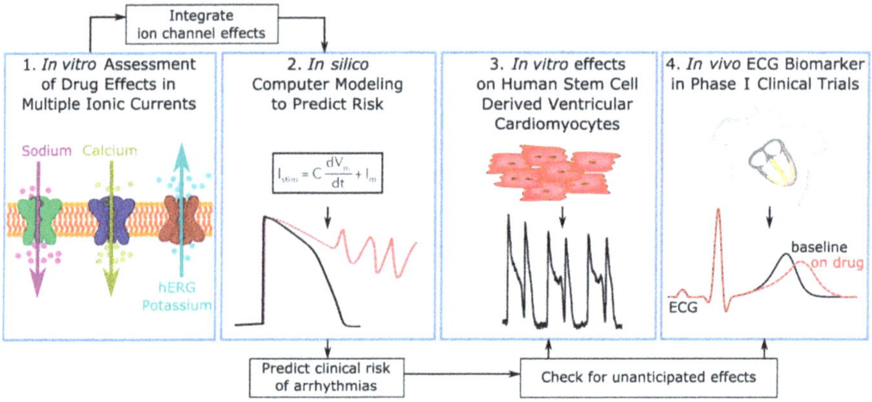

Figure 19.3 Stages of cardiac safety profiling according to CiPA. Adapted from ref. 69 with permission from John Wiley and Sons, Copyright © 2017 The Authors. Clinical Pharmacology & Therapeutics published by Wiley Periodicals, Inc. on behalf of American Society for Clinical Pharmacology and Therapeutics.

CV liabilities. The three main elements of the proposed schema are depicted in Figure 19.3.

1. *In vitro* evaluation of drug effects (concentration response curves and binding kinetics) on three ion currents ($Ca_V1.2$, $Na_V1.5$ and hERG) using an APC assay under standardized conditions.[53,67] These standard protocols open up an opportunity for the pharmaceutical industry to generate electrophysiology data that can be compared across laboratories, thus simplifying potential risk assessment.
2. Integration of *in vitro* information into an *in silico* computational model (modified O'Hara–Rudy myocyte model[68]) of a human ventricular myocyte to assess proarrhythmic liability.
3. Confirmation of drug effects on ionic currents using an hSC-CM-based electrophysiological assay.

A clinical stage that includes assessment of ECGs from Phase I studies represents the final component of the CiPA paradigm.[70] For further information on goals of this initiative as well as advances and challenges associated with it the reader is referred to the most recent report[71] and reviews.[69,72]

19.1.5 Structural Data and Models to Support MedChem Programs

19.1.5.1 Na_V1.5 Channel

Na_V channels are protein constructs built of a transmembrane α-subunit (nine alfa-subunit isoforms in humans are known: $Na_V1.1$–$Na_V1.9$), associated β-subunits and partner proteins. The large transmembrane α-subunit consists of about 2000 amino acid residues (approximately 260 kDa) and forms the channel pore. It is capable of building a functional channel which generates electrical pulses, independent of the β-subunits. The latter modulate the channel's gating kinetics and voltage dependence, as well as controlling cellular localization and interaction with other biomolecules.[73] While the α-subunit of many voltage-gated ion channels is a homotetramer, the polypeptide chain of eukaryotic sodium channels folds into four non-identical repeats I–IV (Figure 19.4).[74] Each repeat can be subdivided into six transmembrane segments S1–S6. Among those, S1–S4 form the voltage-sensing domain (VSD) (S4 carries several sensing positively charged residues), while S5 and S6 enclose the pore domain (PD). Conformational changes in the VSD in response to membrane potential changes lead to conformational transitions of the PD (channel gating). The sequences between the S5 and S6 segments form the ion selectivity filter (SF) (located about 40% of the way through the pore from the extracellular side). Four SF residues of each repeat [Asp/Glu/Lys/Ala (DEKA)] confer Na^+ selectivity. The inner pore is composed of neutral, mostly hydrophobic amino acid residues.

A complete experimental structure of the $Na_V1.5$ channel is not available to date. The first experimental structure of a voltage gated sodium channel – Na_VAb (from *Arcobacter butzleri*) – was disclosed in 2011,[75] and several additional structures have been determined since then (Table 19.3). Very recently,

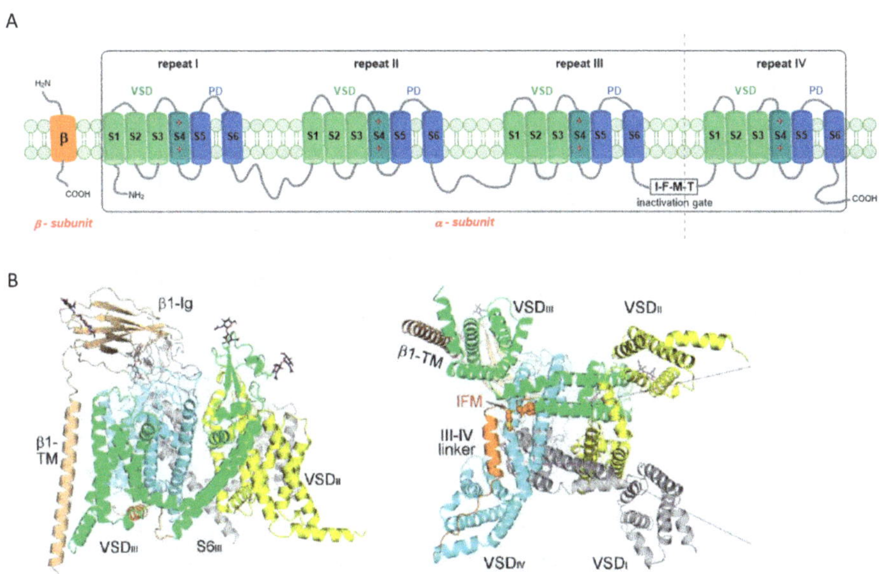

Figure 19.4 (A): Topology of eukaryotic voltage-gated sodium channel showing four homologous repeats. (B): Structure of the human $Na_V1.4$-β1 complex: side view (left) and bottom (intracellular) view (right). The structure is domain coloured. The IFM (fast inactivation motif) is shown as spheres and the III–IV linker is coloured orange. The glycosyl moieties are shown as sticks. (B) Reproduced from ref. 74 with permission from the AAAS, Copyright © 2018 The Authors.

Table 19.3 Published structures of voltage-gated sodium channels and $Na_V1.5$ homology models[a] on their basis.

	Channel	State	Reference	Method	Resolution [Å]	PDB codes	Respective $Na_V1.5$ homology models
Prokaryotic	Mutant **Na_VAb**	Closed	Payandeh *et al.* 2011[75]	X-Ray	2.7–2.95	3RVY, 3RVZ, 3RW0	Ahmed *et al.* 2017[79] (closed), Zablocki *et al.* 2016[80] (open, closed and inactivated), Wang *et al.* 2015,[81] Poulin *et al.* 2014[82] (closed)
	Na_VAb	Inactivated	Payandeh *et al.* 2012[83]		3.2	4EKW	—
	Na_VAb	Open/closed	Lenaeus *et al.* 2017[84]		2.85/3.2	5VB8, 5VB2	—
	WT/mutant **Na_VAb**		Irie *et al.* 2017[85]		2.8–3.4	5YUC, 5YUA, 5YUB	—
	Na_VMs	Open	McCusker *et al.* 2012[86]		3.49	4F4L	—
	Na_VMs	Open	Bagneris *et al.* 2013[87]		2.92	3ZJZ	Poulin *et al.* 2014[82] (open)
	Na_VMs	Open/with blockers	Bagneris *et al.* 2014[78]		2.66–3.43	4PA9, 4PA7, 4PA6, 4PA4, 4PA3, 4P9P, 4P9O, 4P30, 4OXS, 4CBC	Ji *et al.* 2018[88] (open)
	Na_VRh	Inactivated	Zhang *et al.* 2012[89]		3.05	4DXW	—
Eukaryotic	**Na_VPaS**	Undefined	Shen *et al.* 2017[90]	Cryo-EM	3.8	5X0M	—
	Na_VPaS	Toxin-bound	Shen *et al.* 2018[91]		2.6–3.2	6A90, 6A91, 6A95	—
	Ee**$Na_V1.4$**-β1	Undefined	Yan *et al.* 2017[92]		4.0	5XSY	—
	h**$Na_V1.4$**-β1	Undefined	Pan *et al.* 2018[74]		3.2	6AGF	—

[a]For studies on binding interactions of sodium channel blockers using $Na_V1.4$ homology models see *e.g.* Lipkind and Fozzard and Tikhonov and Zhorov (2017 and 2012).[93–95]

the first experimental structure of a human isoform Na$_V$1.4 was obtained *via* cryo-electron microscopy[74] (Figure 19.4). These data stimulated studies on molecular dynamics of state transitions (*e.g.* Bagneris *et al.*[76]), ion transport (*e.g.* Ulmschneider *et al.*[77]) as well as structural aspects of channel-blocker interactions.[78] The latter aspect is discussed in more detail below.

Toxins are known to bind to at least seven different sites on Na$_V$ channels, most of which are located or accessed from the extracellular side.[96,97] In contrast, structurally diverse *small molecules*, such as basic antiarrhythmic drugs and local anaesthetics (LAs) as well as non-basic anticonvulsants (Figure 19.5), are assumed to *interact with a common region*[23,93,98–102] formed by S6 segments of repeats III and IV in the central pore[93,103] (one should note that for non-cardiac isoform Na$_V$1.7 the existence of an antagonist binding site outside of the channel pore – in voltage sensing domain IV – was recently demonstrated,[104] however antagonists binding to this site are isoform-selective). The structural basis for voltage- and use-dependent channel block by LA drugs has been studied in most detail.[103] Single-point mutagenesis studies revealed that two residues on S6 of repeat IV have a major effect on high-affinity (open-state) blocking by LAs, for the Na$_V$1.5 isoform these are F1760 and Y1767.[105–107] LAs can access their binding site either through the inner pore (hydrophilic intracellular pathway, open state) or through membrane fenestrations (hydrophobic extracellular pathway, both open and closed state).[108]

In the absence of experimentally derived structures of ligand-bound eukaryotic Na$_V$ channels, homology modelling is a valuable tool to study ligand–channel interactions. Lipkind and Fozzard created an *open state* Na$_V$1.4 inner pore model to study binding interactions of several lidocaine-class LAs.[93] A common structural feature of LA drugs is a basic alkylamine,

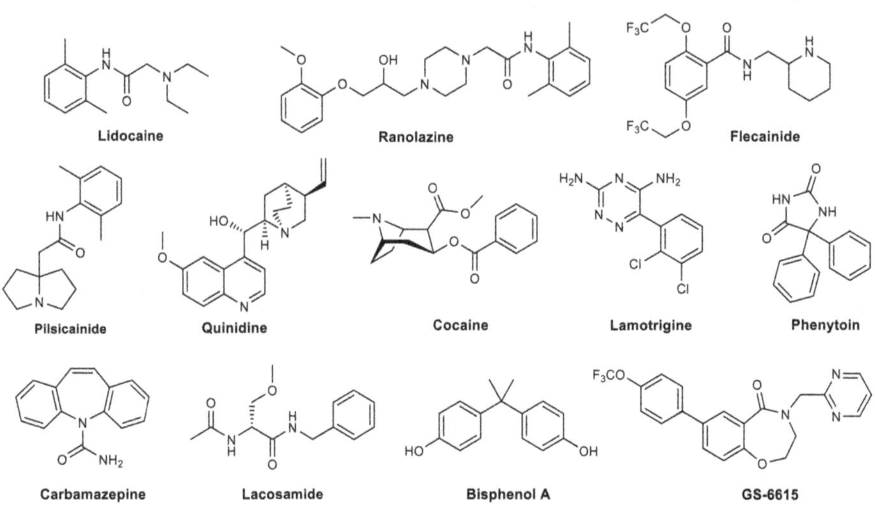

Figure 19.5 Structures of selected small molecules known to bind to the inner pore of Na$_V$1.5.

linked to a hydrophobic aromatic ring *via* a short amide or ether linker. The authors suggested the presence of a cation–Pi interaction between this alkylamine group and F1760 as well as a hydrophobic interaction between the aromatic ring and Y1767. Very recently Ji *et al.*[88] also proposed an *open-state* homology model of the $Na_V1.5$ pore domain (Na_VMs-based) and addressed the characterization of the pilsicainide, flecaidine, cocaine and ranolazine binding sites. The key role of F1760 and Y1767 in binding of the above mentioned LAs was also observed. However, in contrast to the previously suggested cation–Pi interaction, the model indicated strong Pi–Pi and Pi–sigma interactions of LAs with these amino acid residues (Figure 19.6, F526 and Y533 correspond to F1760 and Y1767).

Ahmed *et al.* published a Na_VAb-based homology model of a complete *closed-state* $Na_V1.5$.[79] The model was validated using a set of 35 known ligands with different affinities, and a certain correlation between binding energies and measured IC_{50} values was observed ($r_{pearson} = 0.7$). Binding poses for marketed antiarrhythmic $Na_V1.5$ blockers lidocaine, ranolazine and flecainide

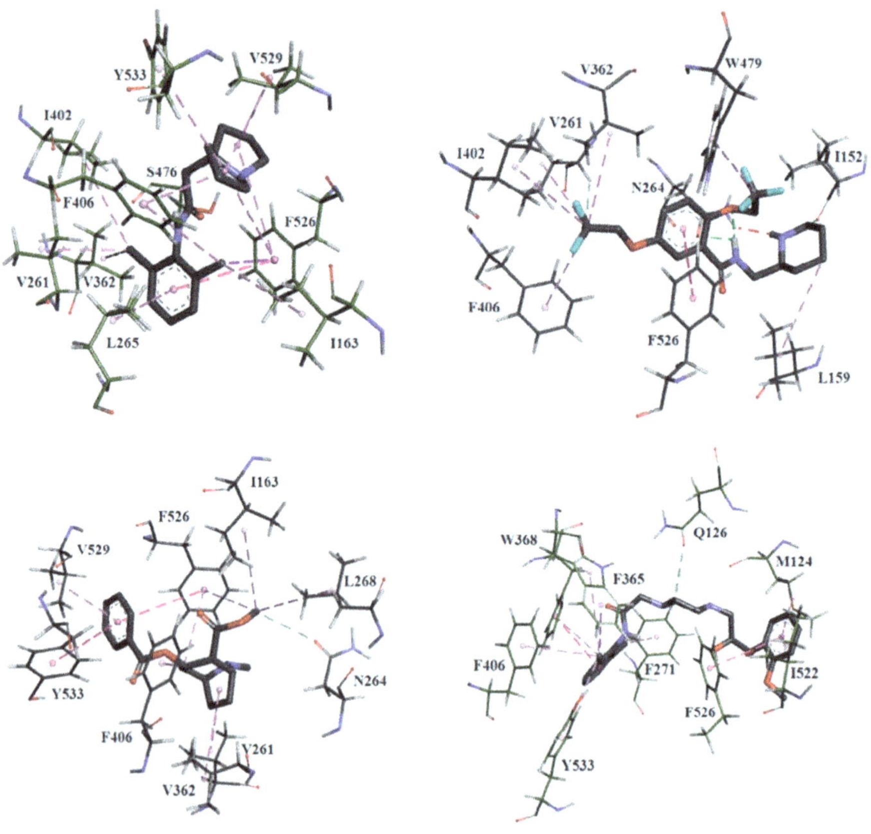

Figure 19.6 Interaction modes of pilsicainide, flecainide, cocaine and ranolazine (left to right) with the $hNa_V1.5$ pore domain according to Ji *et al.* Adapted from ref. 88 with permission from Taylor & Francis Ltd, http://www.tandfonline.com.

were analysed in detail. All three ligands were suggested to bind to the same region in proximity to F1760, however calculated binding poses were distinct with no obvious conserved interactions.

On the basis of the Na_VAb structure, scientists from Gilead Sciences also built a $Na_V1.5$ model which they manipulated into open, closed and inactivated states.[80] They demonstrated that the antiarrhythmic compound GS-6615 (Figure 19.5), although non-basic, also binds to the LA site, with its biphenyl motif forming stacking interactions with F1760 and Y1767 and the polar part pointing in the direction of the DEKA locus of the selectivity filter.

Several other molecular models of $hNa_V1.5$ have been reported (see Table 19.3). The recently disclosed $hNa_V1.4$ cryo-electron microscopy structure will facilitate further work on accurate $Na_V1.5$ modelling.

19.1.5.2 *$Ca_V1.2$ Channel*

Ca_V1 channels (L-type calcium channels or high-voltage activated calcium channels) are composed of an α-subunit[109] associated with several auxiliary subunits: β,[110] γ[111] and α2δ.[112] β-Subunits are key regulators of channel expression, voltage dependence, gating kinetics and modulation by other proteins. The α-subunit forms the ion-conducting pore and possesses all the key characteristics that define channel properties, including gating, ion selectivity and permeation. Within the L-type family four α-subunit isoforms are known: α1S, α1C, α1D and α1F, corresponding to $Ca_V1.1$, $Ca_V1.2$, $Ca_V1.3$ and $Ca_V1.4$, respectively.[113] These channel isoforms differ in their expression pattern, with $Ca_V1.2$ being the major cardiac isoform.[114,115] In contrast to $Na_V1.5$, which is a heterotetramer, the $Ca_V1.2$ α1C subunit consists of four homologous repeats I–IV with six transmembrane segments S1–S6 in each repeat. S1–S4 form the voltage sensing gating machinery, while S5, S6 and intervening P-loops form the pore. Four conserved Glu residues of each repeat constitute the EEEE locus, acting as selectivity filter for Ca^{2+} ions.

The experimental structure of the $hCa_V1.2$ channel has not yet been disclosed. However, the architecture of the mammalian skeletal muscle $Ca_V1.1$ channel complex has been determined by cryo-electron microscopy at 4.2 Å[116] and later at 3.6 Å resolution.[117] Additionally, a 2.7 Å-resolution crystal structure of artificial bacterial Ca_VAb channel has been reported in an unbound form and in complex with several antagonists (*e.g.* pdb id: 5KLG).[118]

Three known chemotypes of L-type Ca_V blockers (Figure 19.7) *interact with distinct, but allosterically connected, binding sites*[119] *– phenylalkylamine (PAA),*[120,121] *benzothiazepine (BTZ)*[122] *and 1,4-dihydropyridine (DHP) sites*[123–125] (Table 19.4). DHPs are thought to be allosteric modulators of Ca_V channel activation (both agonists and antagonists have been reported[126]), while PAAs and BTZs are assumed to physically block the pore. Several L-type calcium channel agonists which are structurally unrelated to DHPs, PAAs and BTZs have also been reported, for example. Compound **8** from Kang *et al.* and Ortner *et al.*[127,128] and FPL 64176[129–131] (Figure 19.5). Experimental knowledge on the location and structure of drug binding sites is primarily based on photoaffinity labelling, site-directed mutagenesis and studies with chimeric proteins.[119]

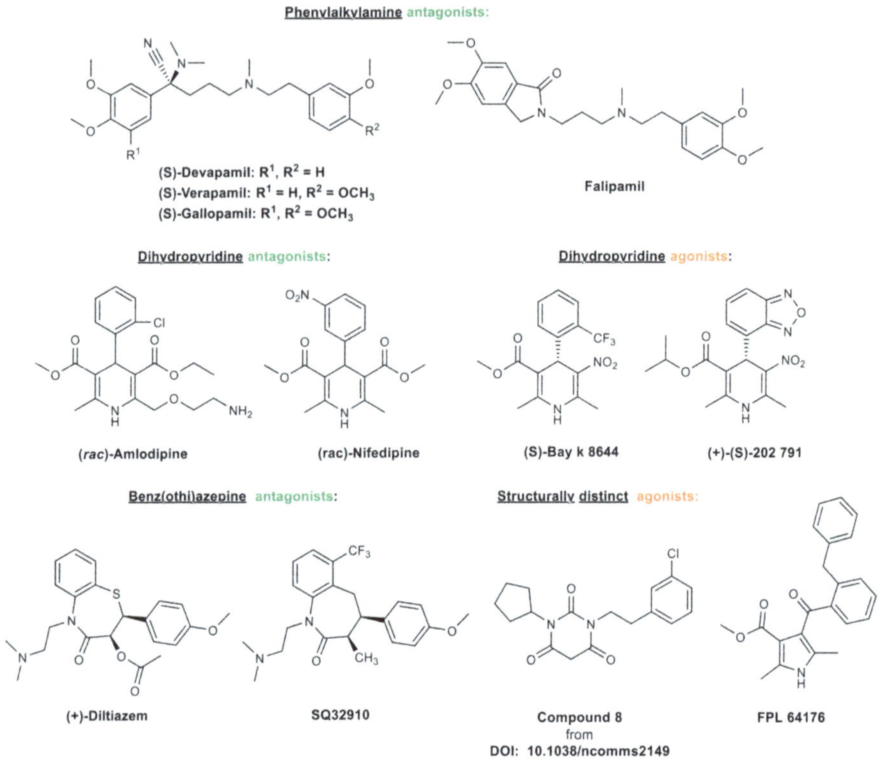

Figure 19.7 Structures of prominent Ca$_V$1.2 ligands.

In addition, on the basis of recent progress in structure determination of related voltage-gated ion channels, several Ca$_V$1.2 homology models have been developed. The experimental and computational insights into ligand–Ca$_V$1.2 channel interactions for three classes of blockers are briefly summarized below.

19.1.5.2.1 Phenylalkylamine Site. The characteristic structural elements of PAAs (*e.g.* verapamil) are two polymethoxylated aromatic rings, linked to the tertiary amine *via* two flexible aliphatic chains, one of which has a Lewis-basic substituent (*e.g.* nitrile or carbonyl) next to the aromatic fragment. PAAs induce voltage- and frequency-dependent blocking of L-type Ca^{2+} current upon entering the channel pore from the intracellular side (after penetration through the cell membrane in an unprotonated form followed by re-protonation in the cytosol).[119]

Lipkind and Fozzard developed a model of Ca$_V$1.2.[132] Docking of (*S*)-devapamil [(*S*)-D888] in a half-folded conformation and suggested that its two aromatic rings interact with the side chains of Y1508 and I1515 [here and subsequently the hCa$_V$1.2 (α1c) numbering is used (Q13936 UniProt sequence), which might differ from the numbering in the original publication],

Table 19.4 Characterization data for $Ca_V1.2$ binding sites.

Binding site	Location	Access from	Ca^{2+} effect on blocking	Residues contributing to binding (experimental data)[a]	Homology modelling studies of binding interactions
PAA	In the pore, at the interface between DIII and DIV	Intracellular side	Potentiation[133]	IIIS5: T1056; IIIS6: Y1169, I1170, F1181, V1182; IVS6: Y1508, A1512, I1515; P-loops: T704, T1133, F1134, E1135, E1464[119,120,133,134]	• Lipkind and Fozzard 2003 (KcsA-based, $Ca_V1.2$ inner pore);[132] • Cheng *et al.* 2009 (4 models, MthK, K_VAP, $K_V1.2$ and KcsA-based, $Ca_V1.2$ open/closed)[120]
BTZ	In the pore, at the interface between DIII and DIV	Extracellular side	Biphasic: inhibition at high concentrations, potentiation at low concentrations.[135,136]	IIIS6: I1167, I1170, I1173, M1177, F1181, V1182; IVS6: I1505, **Y1508**, M1509, **A1512**, **I1515**, P-loops: F1134, E1135, E1464[119,122]	Tikhonov and Zhorov. 2008 (K_VAP & KcsA-based, $Ca_V1.2$ pore, open/closed)[122]
DHP	Within the lipid bilayer at the interface between DIII and DIV	Extracellular side	Biphasic[137]	IIIS5: T1056, Q1060; IIIS6: **Y1169**, I1170, I1173, M1177, M1178; V1182 IVS6: I1505, **Y1508**, **M1509**, I1515, **I1516**, N1517; P-loops: F1129 & S1132 (agonists), F1134, E1135, E1464[119,123,126] and references therein	• Zhorov *et al.* 2001 (KcsA-based; rabbit $Ca_V1.2$ pore);[138] • Tikhonov and Zhorov. 2009 (K_VAP and KcsA-based, $Ca_V1.2$ pore, open/closed)[123] • Lipkind and Fozzard 2003 (KcsA-based, rabbit $Ca_V1.2$ pore);[132] • Cosconati *et al.* 2007 (KcsA-based, $hCa_V1.2$ central pore);[139] • Xu *et al.* 2016 (Ca_VAb-based, $hCa_V1.2$);[140] • Schaller *et al.* 2018 ($Ca_V1.1$-based, rat $Ca_V1.2$ pore)[141]

[a] $hCa_V1.2$ (α1c) numbering used; residues, mutations of which are causing the most pronounced affinity reduction, are highlighted in bold type.

while the central amine interacts directly with the carboxylate of E1135 of the SF, thereby blocking Ca^{2+} permeation. Zhorov and colleagues later built four K^+-channel-based molecular models of the $Ca_V1.2$ T1056Y mutant and proposed another (*S*)-devapamil binding mode (Figure 19.8).[120] The authors suggested the nitrile group to be coordinating to the Ca^{2+}-ion (bound to glutamate residues of the SF) and the protonated amino group to be located in the nucleophilic region at the focus of the P-helices. The first assumption is supported by experimental data on Ca^{2+} potentiation of PAA action.[133]

Recently, a co-crystal structure of Br-verapamil bound to the bacterial channel Ca_VAb was published.[118] A single molecule of the ligand was shown to be bound on the intracellular side of the ion selectivity filter, in agreement with experimental evidence for the intracellular access of PAAs to the binding site.

19.1.5.2.2 Benzothiazepine Site. The BTZ-binding site on L-type $Ca2^+$ channels is located in close proximity to the PAA site and is allosterically linked to both the PAA (negative modulation) and DHP (positive modulation) sites.[119] The channel blocking induced by the prototypic BTZ ligand diltiazem is voltage- and frequency-dependent due to preferential binding to the inactivated state and rapid access to the binding site in the open state, respectively. In contrast to PAAs, BTZs have been shown to access their binding site from the extracellular side. Raising concentrations of Ca^{2+} and Ba^{2+} are known to antagonize diltiazem-induced channel blocking,[135] but at the same time the presence of Ca^{2+} in the SF is required for maximum blocking efficacy.[136] Using mutagenesis studies BTZ-sensing residues were localized (Table 19.4). However, application of experimental data for mapping of the BTZ binding site is complicated due to unequal effects of mutations on different blockers and different characteristics of the block,

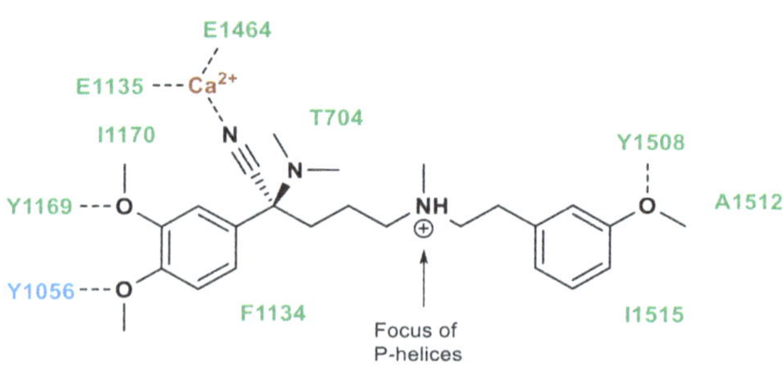

Figure 19.8 $Ca_V1.2$ binding interactions of (*S*)-devapamil according to molecular modelling studies by Zhorov and colleagues.[120] Residues highlighted in green are PAA-sensing according to results of mutational experiments. The T1056Y mutation highlighted in blue was introduced in order to increase the number of anchoring interactions and was experimentally shown to increase the potency of devapamil.

altered channel properties and possible allosteric effects of mutations on drug binding.[142]

A molecular modelling study of BTZ ligand–channel interactions was done by Tikhonov and Zhorov.[122] Figure 19.9 shows the most conserved binding interactions amongst 42 identified low-energy complexes of the highly potent benzazepine SQ32910 with the open-state channel. The key binding features reported by the authors were: (1) the interaction of the carbonyl group with Ca^{2+} coordinated to E1135 and E1464 of the SF (similar to PAAs); (2) location of the ammonium group in the inner pore at the focus of the P-helices; (3) interaction of the aromatic groups with Y1508, F1181 and F1134 (known to affect the potency of BTZs). In the absence of a Ca^{2+} ion the ammonium group is assumed to interact directly with the SF glutamates of DIII and DIV.

19.1.5.2.3 Dihydropyridine Site. DHPs are allosteric channel modulators and can act either as agonists [*e.g.* (*S*)-Bay k 8644,[143] (*S*)-202 791[144]] or antagonists (the majority, *e.g.* amlodipine, nifedipine, (*R*)-Bay k 8644, (*R*)-202 791). The type of action depends primarily on the structure but also on the membrane potential and stimulation frequency. Thus, (*S*)-Bay k 8644[143] and (*S*)-202-791[145] have been shown to act as both agonists and antagonists depending on experimental conditions. The affinity for DHP antagonists measured in binding assays with isolated muscle membranes and the activity measured electrophysiologically in intact muscle tissue can differ by 100- to 1000-fold.[119] This difference results from the divergent ligand affinities for the resting and open or inactivated states. DHP antagonists bind to the inactivated state of the calcium channel with the highest affinity, resulting in blocking that is steeply voltage dependent. Agonists favour open state binding and stabilize the open state. The DHP binding site is located within the

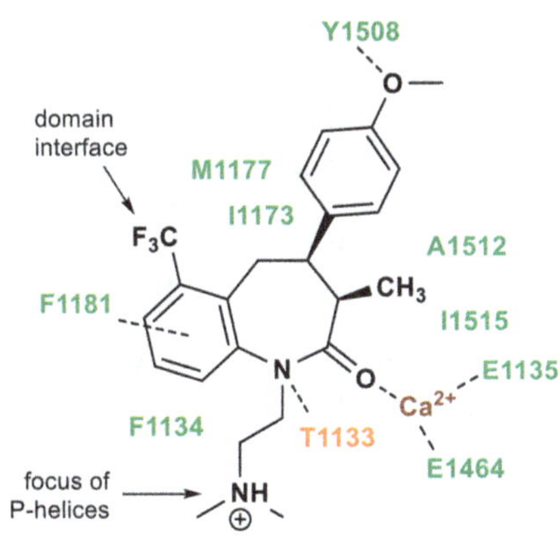

Figure 19.9 $Ca_V1.2$ binding interactions of SQ32910 according to molecular modelling studies by Tikhonov and Zhorov.[122] Residues highlighted in green are PAA-sensing according to the results from mutational experiments.

lipid bilayer approximately 11–14 Å away from the extracellular surface. The results of site-directed mutagenesis studies revealed residues essential for DHP binding (Table 19.4). Coordination of Ca^{2+} to the Glu residues in the SF stabilizes the DHP site in its high-affinity state.[137] Generally, the effects of divalent cations on DHP binding are biphasic, with enhanced binding at intermediate concentrations followed by reduced binding as the concentration is increased.

In order to study $Ca_V1.2$–DHP interactions, several homology models based on X-ray structures of bacterial potassium channels KscA and KvAP as well as the more closely related bacterial Ca_VAb channel have been published (Table 19.4). Most recently, a homology model of the rat $Ca_V1.2$ pore domain based on a crystal structure of the rabbit $Ca_V1.1$ channel has been reported. Docking poses of isradipine were used to build a pharmacophore model (Figure 19.10), characterized by H-bond interactions with S1132 and Y1508. The model allowed rationalization of the SAR of a series of novel DHPs possessing $Ca_V1.2$ or $Ca_V3.2$ blocking activity.

19.1.5.3 hERG Channel

The hERG channel is a homotetrameric protein consisting of four identical α subunits of 1159 amino acid residues.[146] Each subunit contains two membrane-spanning domains: the pore domain (PD) and the voltage sensing domain (VSD), with two cytoplasmic terminal domains, the C-terminal cyclic nucleotide-binding domain and the N-terminal period circadian protein–aryl hydrocarbon receptor nuclear translocator protein–single-minded protein [Per–Arnt–Sim (PAS)] domain (Figure 19.11). The VSD of each subunit contains four membrane-spanning α-helices (S1–S4), while the PD contains a further two transmembrane helices (S5–S6), along with a shorter pore

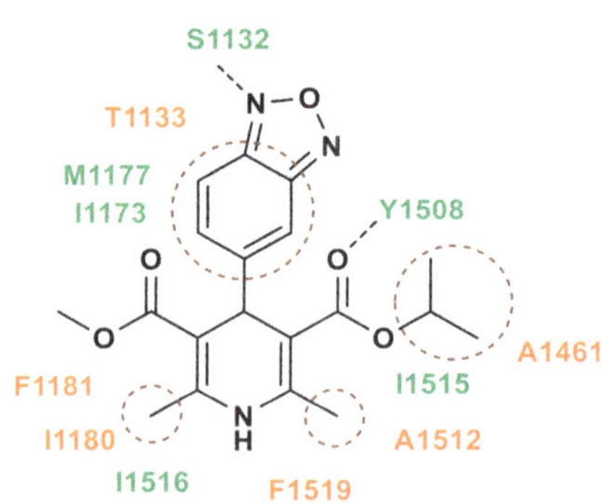

Figure 19.10 $Ca_V1.2$ binding interactions of Isradipine according to results from molecular modelling studies by Schaller *et al.*[141] Residues highlighted in green are PAA-sensing according to the results of mutational experiments. Brown circles indicate hydrophobic interactions.

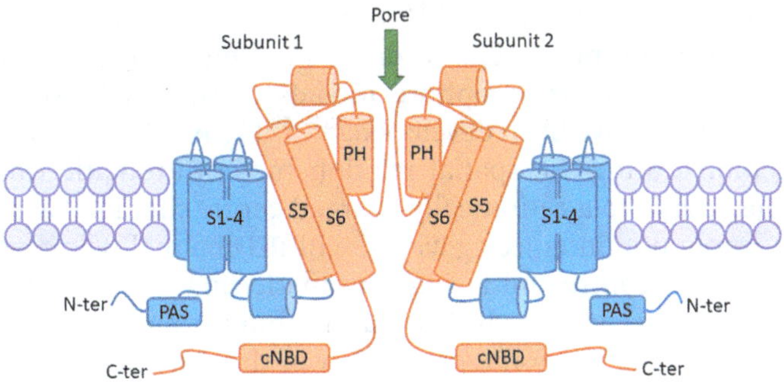

Figure 19.11 Topology diagram of the hERG channel in a lipid bilayer (only two subunits are shown for clarity).

helix (PH) which partially spans the membrane. A cryo-electron microscopy (Cryo-EM) structure of a truncated hERG channel construct, with residues 141–350 and 871–1005 deleted, but with similar functional activity to the wild-type channel was published in 2017 at a resolution of 3.8 Å (PDB accession code 5va1).[147]

The complete homotetrameric channel has fourfold symmetry around a conical pore to allow the passage of potassium ions. The S5 helices of the pore domain associate with the VSD, positioning the pore helices. Ion selectivity is associated with the carbonyl groups of a region of five amino acid residues (Ser–Val–Gly–Phe–Gly) on the SF-region toward the extracellular end of each pore domain that provide an electrostatic environment conducive to the passage of K^+ ions. Ion passage back out of the cell is thought to be prevented by the crossing over of the S6 helices at the cytoplasmic end of the pore. The VSD of the hERG channel senses changes in membrane potential. The exact mechanism of voltage gating continues to be investigated.[148] However, it is known that several positively charged (Arg and Lys) residues in the S4 helix are involved in the voltage-sensing activity.[149,150] Helices S1–S3 possess a number of negatively charged residues which help to stabilize the S4 helix in the membrane. The S4 helix moves outward in response to changes in the membrane potential, taking several gating charged residues out of the membrane bilayer. This movement is coupled to opening of the channel *via* the S4–S5 linker to open the gate located at the bottom of the S6 transmembrane helix.[148,151]

The gating kinetics of the hERG channel are significantly slower than those of the other members of the Kv channel family.[152] Inactivation onset and subsequent recovery are much faster than the kinetics for activation and deactivation.[153] This is important for the physiological role of the hERG channel conducting repolarizing current during the late phase of the cardiac action potential.

Several allosteric binding sites for drugs and toxins have been identified[42] and drug interference with trafficking of the hERG channels to the cell membrane has also been reported,[154] but the effects of drugs on the hERG channel are most commonly due to direct binding of drugs in the pore of the channel. Several pharmacophore models to predict hERG activity of drug-like molecules have been developed.[155–159] Figure 19.12 shows two such models, demonstrating the spatial relationships of aromatic rings and/or hydrophobic groups to a cationically-charged centre that increase the probability of a compound interacting with the hERG channel. The presence of a protonatable basic centre and a single aromatic ring can be sufficient for a compound to interact with the hERG channel, but the more pharmacophoric features present, the greater the probability of the drug interacting with the hERG channel.

Representative examples of drugs discontinued due to prolongation of QT interval can be seen in Figure 19.13, with their pharmacophoric features highlighted. Type III antiarrhythmic drugs such as dofetilide[160] have also been developed that deliberately target the hERG ion channel and also conform to the hERG pharmacophore models in Figure 19.12.

Weakly basic, neutral, acidic and zwitterionic compounds may also cause hERG channel blocking[161–163] but generally with much lower frequency than with basic compounds. An analysis of a hERG compound dataset by ionisation class and lipophilicity (Figure 19.14)[164] demonstrates the probabilites of a compound exhibiting hERG blockade in each ionisation class.

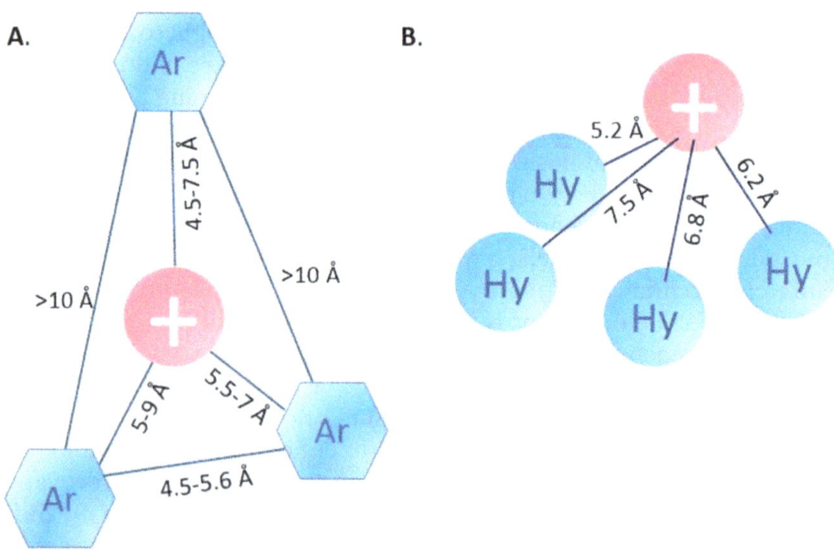

Figure 19.12 Representative pharmacophore models for hERG inhibition.[155] (A): Four-point pharmacophore showing positioning of aromatic groups and basic centre; (B): Five-Point pyramidal pharmacophore showing positioning of hydrophobic group and protonated centre.[156]

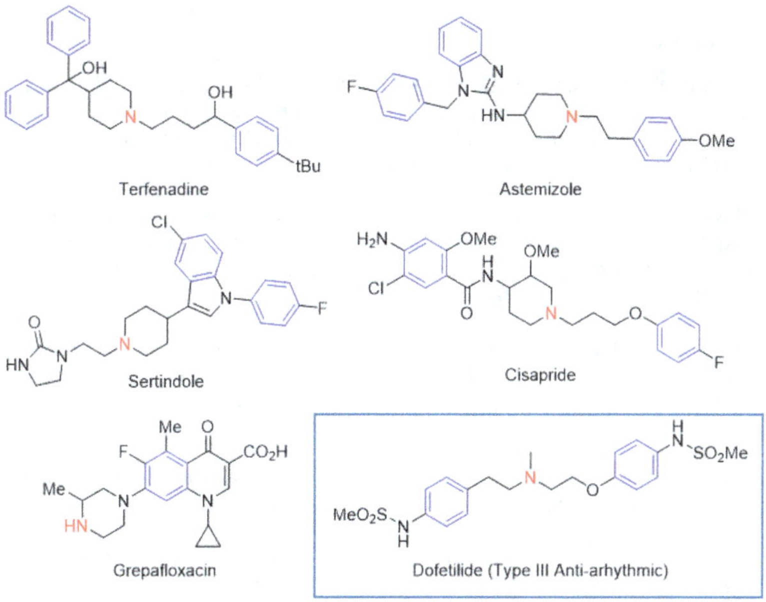

Figure 19.13 Examples of compounds that bind to the hERG channel, with pharmacophoric features coloured: Red, ionisable centre; Blue, hydrophobic and/or aromatic group.

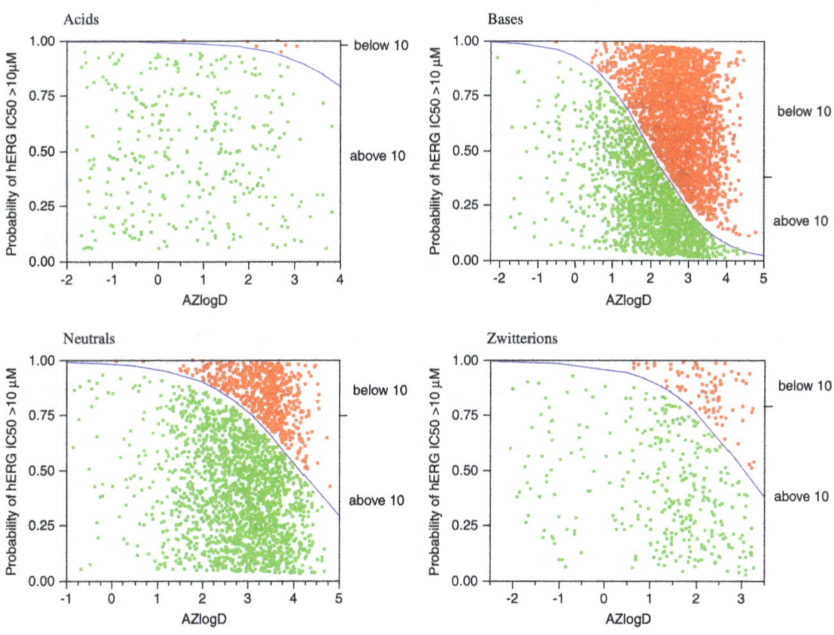

Figure 19.14 Logistic regressions showing how the probability of a compound achieving a hERG IC_{50} of >10 μM changes with logD for each ionisation class. Those compounds with IC_{50} values above 10 μM are shown in green; those below 10 μM are in red. Reproduced from ref. 164 with permission from Elsevier, Copyright 2007.

Key References

- C. Antzelevitch and A. Burashnikov, *Card. Electrophysiol. Clin.*, 2011, **3**, 23–45. *The review summarizes mechanisms underlying different types of cardiac arrythmias and provides a tabular overview of known ion channel gene mutation-associated cardiac diseases (channelopathies).*

- O. Grant, *Circ.: Arrhythmia Electrophysiol.*, 2009, **2**, 185–195. *The publication provides an overview of known cardiac ion channels including their general properties, structural information and role in the formation of ion currents shaping the cardiac action potential.*

- https://www.ich.org/fileadmin/Public_Web_Site/ICH_Products/Guidelines/Safety/S7B/Step4/S7B_Guideline.pdf. *The link directs to the current regulatory guidelines for the preclinical cardiac safety risk evaluation.*

- G. Gintant, *et al.*, *Nat. Rev. Drug Discovery*, 2016, **15**, 457–471. *The review contains analysis of the evolution of the preclinical safety assessment paradigm and describes the concepts and goals of the CiPA initiative.*

- X. Li, *et al.*, *Arch. Toxicol.*, 2016, 1803–1816. *The paper provides a concise overview of rapid-throughput in vitro approaches to cardiotoxicity screening.*

- G. A. Gintant, in *Antitargets and Drug Safety*, ed. L. Urbán, V. F. Patel and R. J. Vaz, Wiley-VCH Verlag GmbH & Co. KGaA, 2015, ch. 12, pp. 253–277. *The book chapter highlights basic aspects of $Na_V1.5$ structure, function and electrophysiological effects in relation to the evaluation of potential off-target actions of novel pharmacological agents.*

- S. Jenkinson, *et al.*, *J. Pharmacol. Toxicol. Methods*, 2018, **89**, 9–18. *The study analyses predictivity of several $Na_V1.5$ in vitro assays for clinical QRS prolongation using a set of 22 clinical compounds.*

- B. Dumotier and M. Traebert, in *Antitargets and Drug Safety*, ed. L. Urbán, V. F. Patel and R. J. Vaz, Wiley-VCH Verlag GmbH & Co. KGaA, 2015, ch. 11, pp. 235–251. *The book chapter highlights basic aspects of $Ca_V1.2$ structure, function and electrophysiological effects in relation to the evaluation of potential off-target actions of novel pharmacological agents.*

- B. Fermini *et al.*, *J. Pharmacol. Toxicol. Methods*, 2017, **84**, 86–92. *The study correlates the results from several L-type Ca_V in vitro assays and ex vivo tissue preparations with CV observations in conscious telemetered rats for three marketed L-type Ca_V antagonists.*

- J. Hancox, et al., *Pharm. Ther.*, 2008, **119**, 118–132. *A general review discussing the physiological role of hERG, mechanistic aspects of channel blockade and its association with TdP, hERG screening possibilities and matter of relevant safety margins.*

- R. A. Pearlstein, *et al.*, in *Antitargets and Drug Safety*, ed. L. Urbán, V. F. Patel and R. J. Vaz, Wiley-VCH Verlag GmbH & Co. KGaA, 2015, ch. 14, pp. 295–328. *The book chapter provides detailed analysis of the mechanistic and kinetic aspects of pharmacological hERG blockade in relation to the potential risk of proarrhythmic effects and relevant safety margins.*
- X. P. Huang, *et al.*, *Assay Drug Dev. Technol.*, 2010, **8**, 727–742. *The publication provides a comparative analysis of the outcome of several in vitro hERG screening assays using a set of 49 drugs.*
- M. Perry, *et al.*, *J. Physiol.*, 2010, **588**, 3157–3167.*Structural determinants for the action of hERG blockers and activators are discussed.*
- *The open-state hERG channel structure determined via cryo-EM at 3.8 Å resolution is disclosed.*
- S. Kalyaanamoorthy and K. H. Barakat, *Med. Res. Rev.*, 2018, **38**, 525–555. *The review summarizes the spectrum of reported binding modes for hERG blockers, the various in silico models developed for predicting hERG affinity and the known optimization strategies to avoid hERG off-target interactions.*
- M. Waring and C. Johnstone, *Bioorg. Med. Chem. Lett.*, 2007, **17**, 1759–1764. *The article provides quantitative assessment of hERG liability as a function of lipophilicity.*

19.2 Mitigation Strategies

19.2.1 $Na_V1.5$ Channel

Modification of heterocycles (*e.g.* repositioning of heteroatoms) or aromatic substitution pattern (removal, substitution or repositioning of fluoroalkyl or fluoroalkoxy groups and halogen atoms – often leading to decreased lipophilicity) are the most powerful strategies to avoid $Na_V1.5$ off-target potency. As exemplified below, even subtle modifications can lead to significant decrease or loss of channel blocking. Reduction of substrate basicity (pK_a) as well as introduction of structural elements leading to steric disfavouring of binding represent alternative approaches to diminishing $Na_V1.5$ affinity.

19.2.1.1 Modification of (Hetero)Aromatic Rings and/or(Hetero) Aromatic Substitution Pattern

In many of the examples shown this structural modification is accompanied by a significant drop in lipophilicity.

	Examples	Reference
1	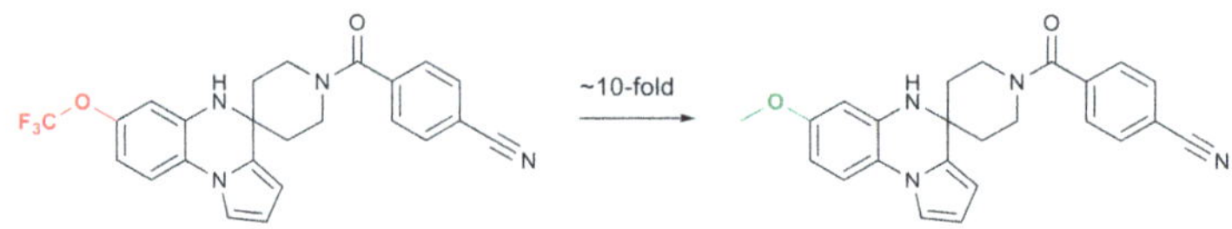 clogP = 4.2; $Na_V1.5$ pIC_{50} (FLIPR) = 5.8 → clogP = 2.6; $Na_V1.5$ pIC_{50} (FLIPR) = 4.9	165
2	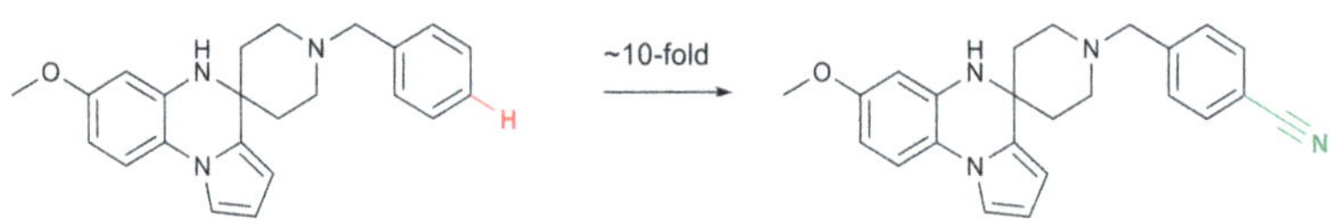 clogP = 3.1; $Na_V1.5$ pIC_{50} (FLIPR) = 6.3 → clogP = 3.1; $Na_V1.5$ pIC_{50} (FLIPR) = 5.3	165
3	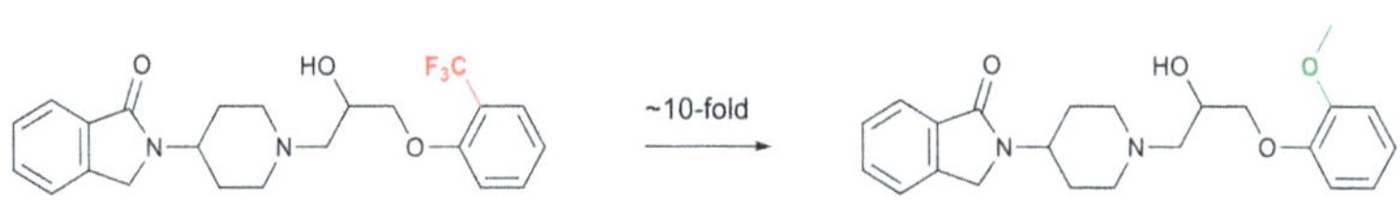 clogP = 3.1; $Na_V1.5$ pIC_{50} (PC) = 5.1 → clogP = 2.1; $Na_V1.5$ pIC_{50} (PC) = 4.1	166
4	 clogP = 3.1; $Na_V1.5$ pIC_{50} (PC) = 5.6 → clogP = 2.2; $Na_V1.5$ pIC_{50} (PC) = 4.4	166
5	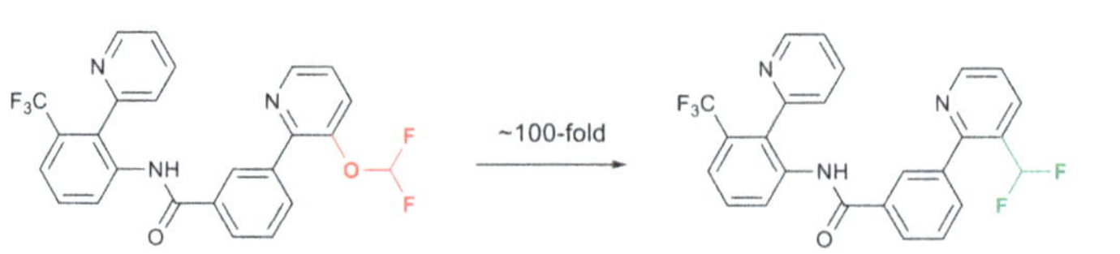 clogP = 5.9; $Na_V1.5$ pIC_{50} (APC) = 7.7 → clogP = 5.6; $Na_V1.5$ pIC_{50} (APC) = 5.7	167

	Examples	Reference
6	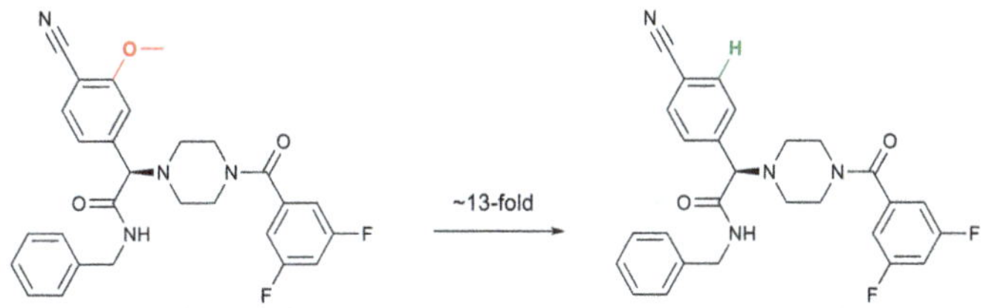 clogP = 3.9 → clogP = 4.0 $Na_V1.5$ pIC_{50} (APC) = 6.3 → $Na_V1.5$ pIC_{50} (APC) = 5.2	168
7	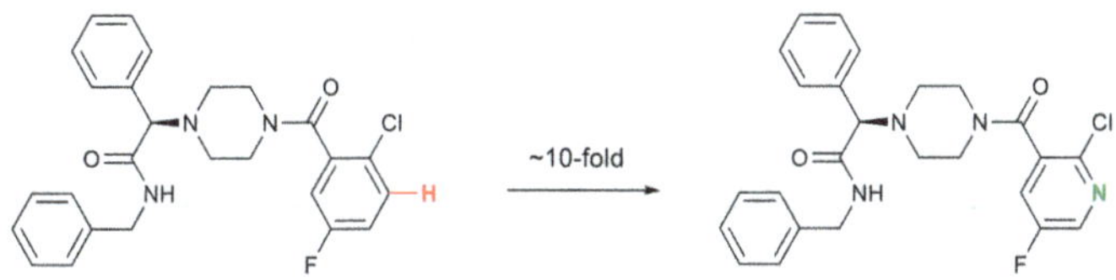 clogP = 4.3 → clogP = 3.4 $Na_V1.5$ pIC_{50} (APC) = 6.0 → $Na_V1.5$ pIC_{50} (APC) = 5.0	168
8	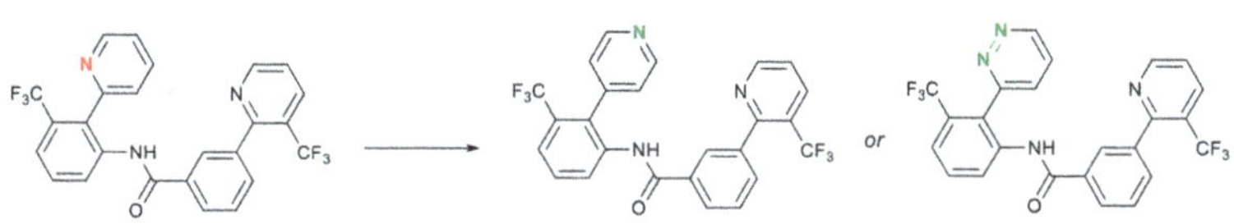 **A** → **B** or **C** A: clogP = 6.2; $Na_V1.5$ pIC_{50} (APC) = 6.3 B: clogP = 5.8; $Na_V1.5$ pIC_{50} (APC) = 4.9 C: clogP = 5.8; $Na_V1.5$ pIC_{50} (APC) = 5.2 Rabbit Langendorff isolated heart study with compound **A** showed no effect at 0.3 μM but dose-dependent QRS prolongation at 1–10 μM.	167
9	 clogP = 3.2 → clogP = 2.8 $Na_V1.5$ pIC_{50} (APC) = 5.1 → $Na_V1.5$ pIC_{50} (APC) < 4.5	169

(*continued*)

Examples		Reference
10	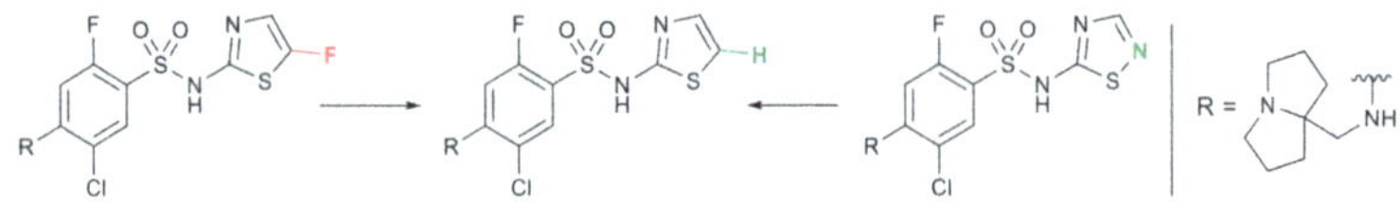 clogP = 3.0; Na$_V$1.5 pIC$_{50}$ (APC) = 5.7 → clogP = 2.9; Na$_V$1.5 pIC$_{50}$ (APC) < 4.5 ← clogP = 3.2; Na$_V$1.5 pIC$_{50}$ (APC) = 5.1	169
11	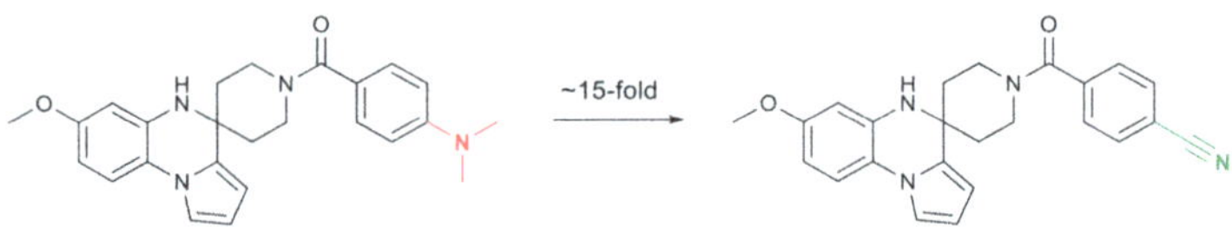 clogP = 2.8; NaV1.5 pIC$_{50}$ (FLIPR) = 6.0 → (~15-fold) clogP = 2.6; NaV1.5 pIC$_{50}$ (FLIPR) = 4.9	165

19.2.1.2 Reduction or Elimination of Basic pK_a

Examples		Reference
1	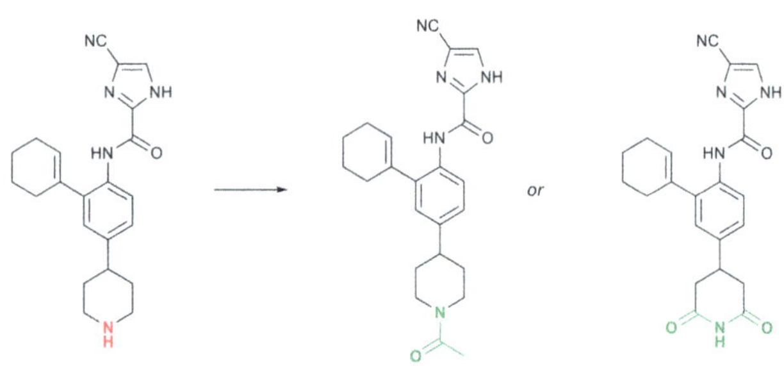 clogP = 2.9 → clogP = 2.5 *or* clogP = 1.6 Na$_V$ inhibition @10μM (binding assay): 96% → 27% / 11%	170
2	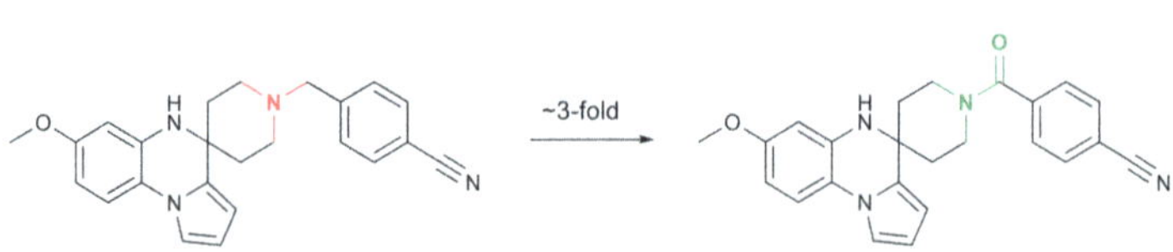 clogP = 3.1; Na$_V$1.5 pIC$_{50}$ (FLIPR) = 5.3 → (~3-fold) clogP = 2.6; Na$_V$1.5 pIC$_{50}$ (FLIPR) = 4.9	165

19.2.1.3 Introduction of Steric Clashes

Diverse types of structural modifications which are assumed to lead to steric disfavouring of binding are exemplified in this section.

Examples		Reference
1	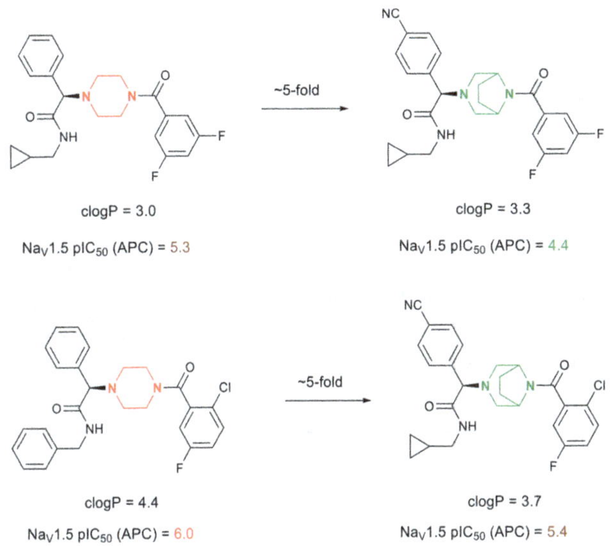	168
2	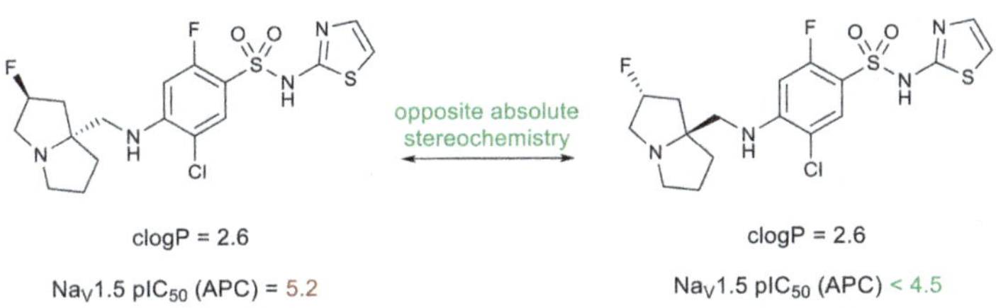	169
3	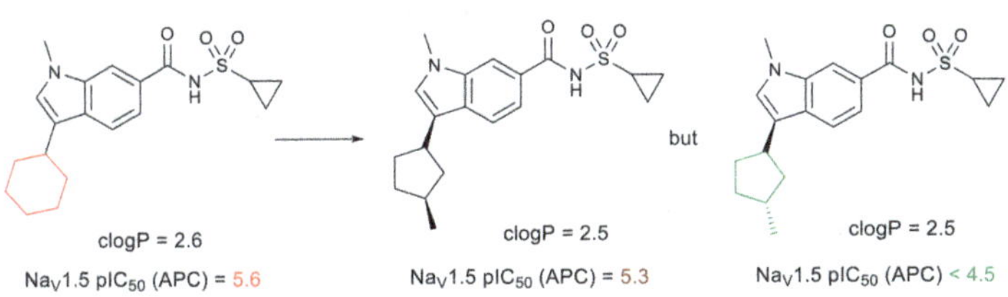	171

(*continued*)

Examples	Reference
4	168

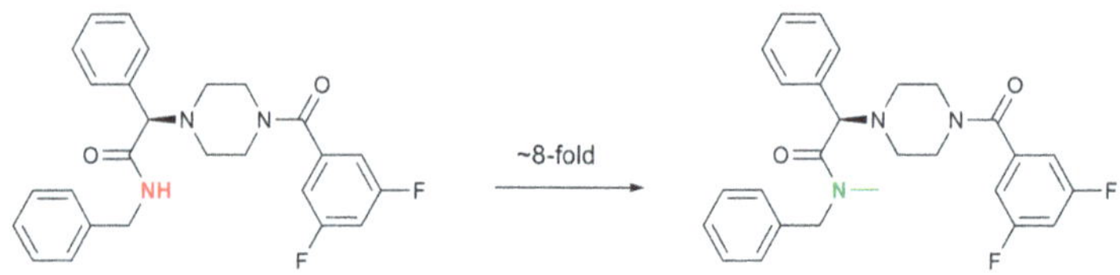

clogP = 4.0 → clogP = 4.2

$Na_V1.5$ pIC_{50} (APC) = 5.9 → $Na_V1.5$ pIC_{50} (APC) = 5.0

19.2.1.4 *Lipophilicity Reduction*

Diverse types of structural modifications leading to reduced lipophilicity are exemplified in this section. Another efficient strategy for the reduction of lipophilicity is modification of aromatic substituents – for respective examples see Section 19.2.1.1.

Examples	Reference
1	166

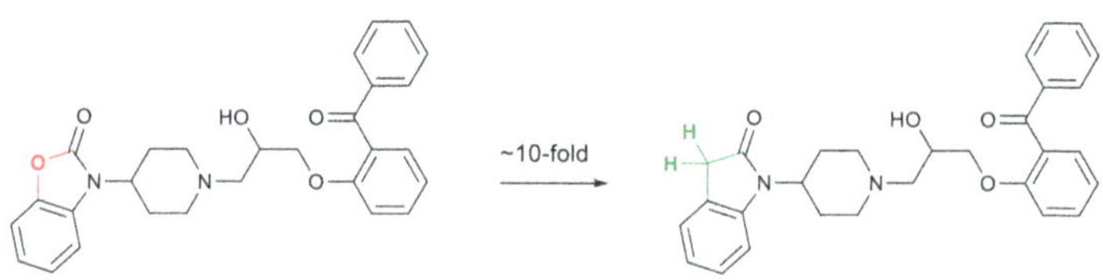

clogP = 3.7 → clogP = 3.3

clogD (pH 7) = 2.9 → clogD (pH 7) = 2.3

$Na_V1.5$ pIC_{50} (PC) = 5.5 → $Na_V1.5$ pIC_{50} (PC) = 4.5

The lactam is about 0.4 units less lipophilic than the carbamate. In addition, it is less electron-withdrawing,
rendering the piperidine nitrogen more basic, which further decreases logD.

2	172

clogP = 3.5 → clogP = 1.7

$Na_V1.5$ pIC_{50} (APC) = 5.3 → $Na_V1.5$ pIC_{50} (APC) < 4.5

Examples		Reference
3	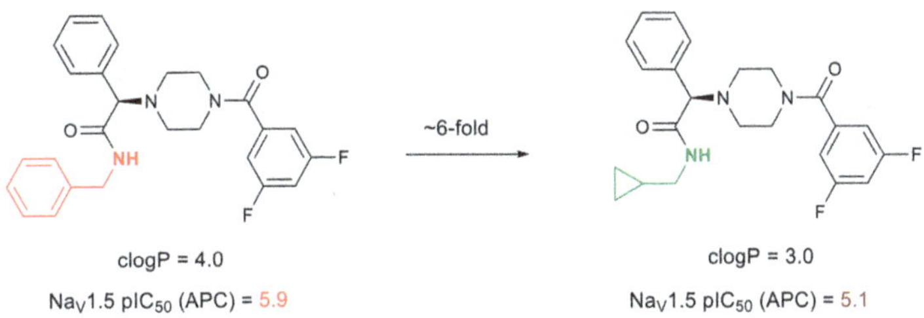	168
4	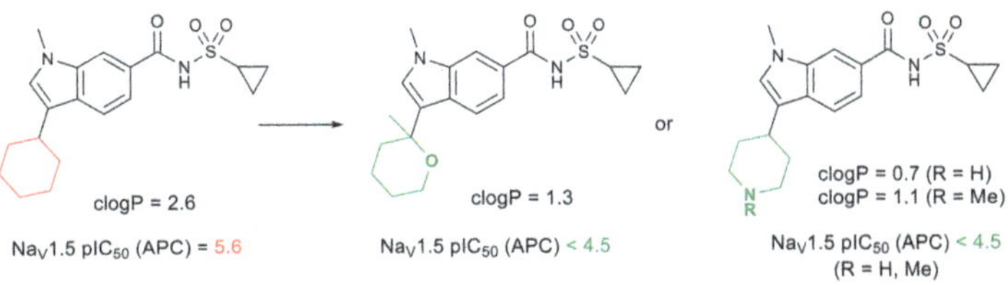	171

19.2.2 $Ca_V1.2$ Channel

Several research groups have demonstrated that reduction of compound basicity or number of basic centres, when tolerated by on-target potency, can be a successful strategy to reduce L-type calcium channel off-target activity. Lipophilicity reduction can be considered as another general approach. Similar to $Na_V1.5$ channels, very steep SAR is often observed in Ca_V channel inhibitors with regards to aromatic substitution pattern.

19.2.2.1 Modification of (Hetero)Aromatic Rings and/or (Hetero) Aromatic Substitution Pattern

In many of the examples shown this structural modification is accompanied by a significant drop in lipophilicity.

Examples		Reference
1	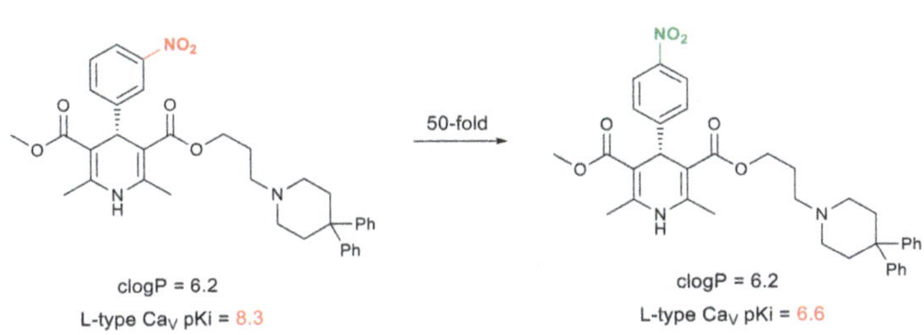	173

(*continued*)

Examples		Reference
2	 clogP = 5.2 → clogP = 2.5; $Ca_V1.2$ inhibition @1μM (FLIPR assay): 73% → 0%	174
3	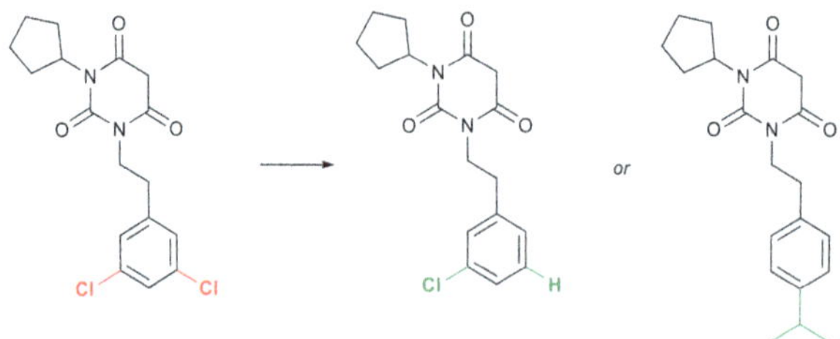 clogP = 3.4, $Ca_V1.2$ pIC$_{50}$ (FLIPR) = 5.5; clogP = 2.8, $Ca_V1.2$ pIC$_{50}$ (FLIPR) < 4; clogP = 3.5, $Ca_V1.2$ pIC$_{50}$ (FLIPR) = 4.2	175
4	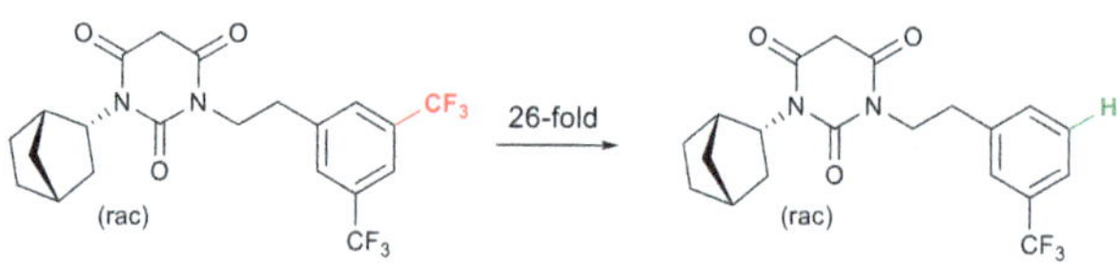 clogP = 4.3, $Ca_V1.2$ pIC$_{50}$ (FLIPR) = 5.4; clogP = 3.4, $Ca_V1.2$ pIC$_{50}$ (FLIPR) = 4.0	175
5	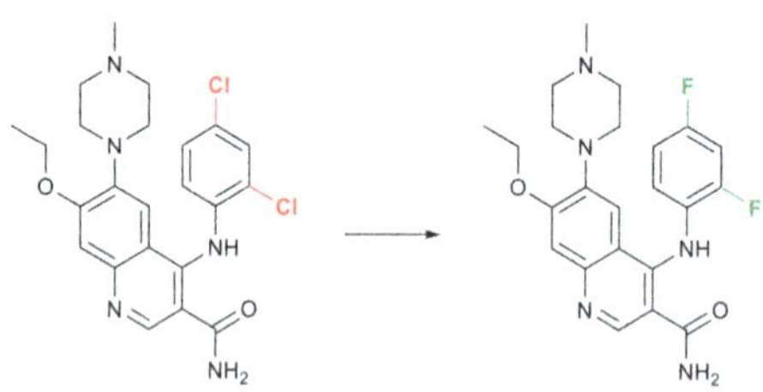 clogP = 3.9 → clogP = 3.1; L-type Ca_V %inhibition @10μM (binding, PAA site): 80% → 41%	176

Examples	Reference

6 — Reference 177

R = piperidin-1-yl

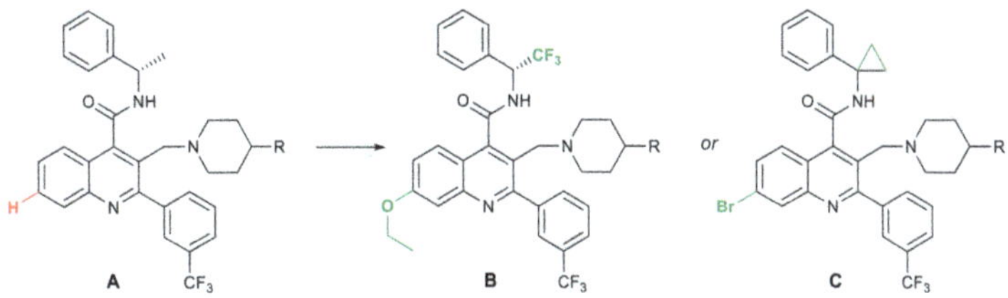

clogP = 7.0 clogP = 7.8 clogP = 7.6

L-type Ca_V pIC_{30} (Patch Clamp):
2 Hz (120 beats/min) frequency is accepted as being more physiologically relevant

5.8 (0.1 Hz), 6.9 (2 Hz) 5.0 (0.1 Hz), 4.9 (2 Hz) 5.0 (0.1 Hz), 5.0 (2 Hz)

Data from anesthetized dog CV study with compound **B** are reported in the original publication.

19.2.2.2 Elimination or Reduction of Basic pK$_a$

Examples	Reference

1 — Reference 178

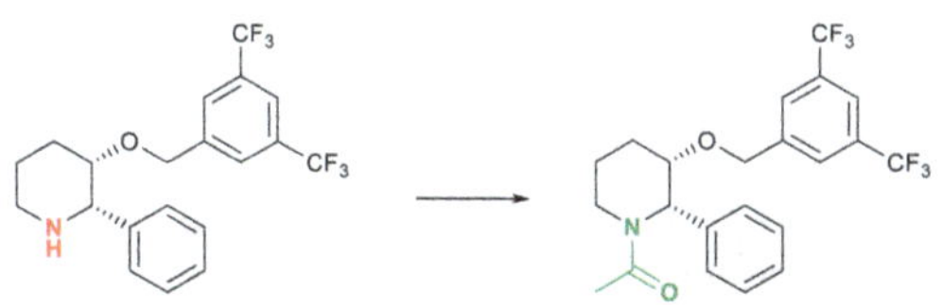

clogP = 5.3 clogP = 4.9

L-type Ca_V inhibition (binding assay):

pIC_{50} = 6.1 0% Inh @ 5μM

2 — Reference 178

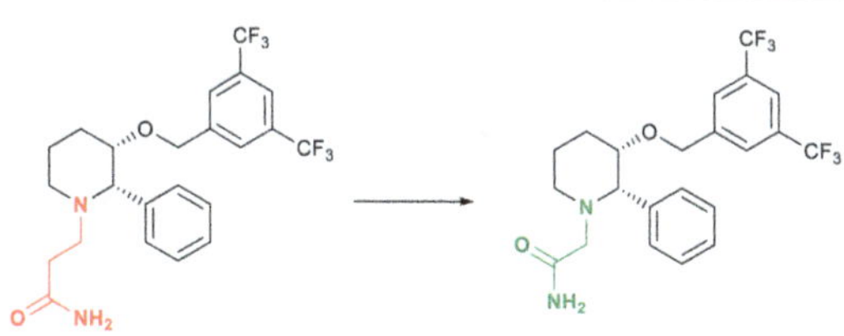

clogP = 4.6 clogP = 4.3

L-type Ca_V inhibition (binding assay):

98% @ 5μM 16% @ 30μM

(continued)

Examples	Reference

3

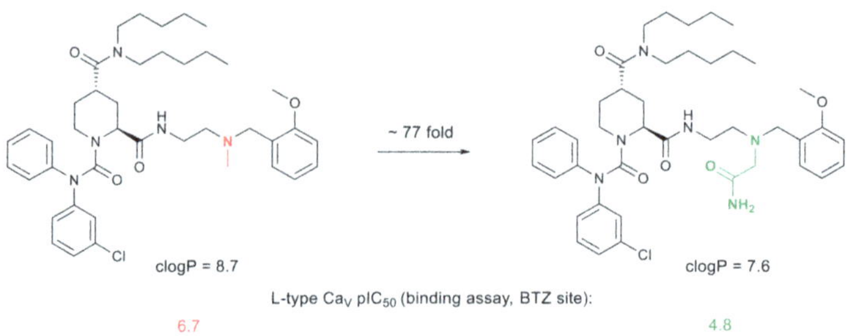

179

4

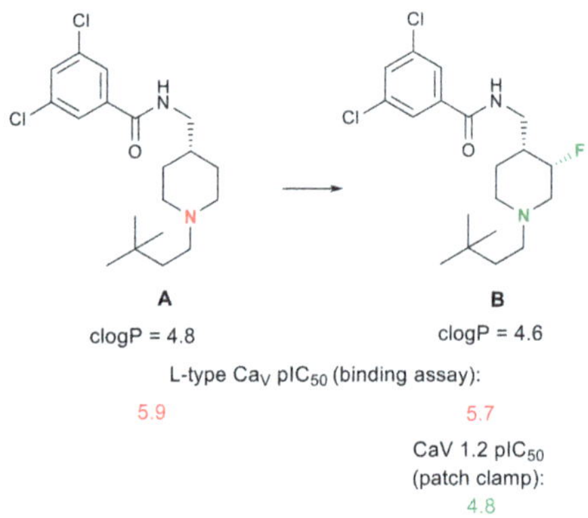

180

Data from anesthetized dog CV study with compound B are reported in the original publication.

5

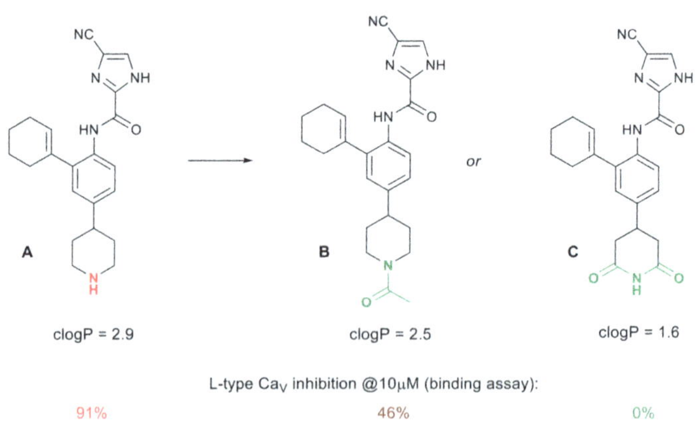

170

Examples	Reference
Exact assay type and channel isoform are not specified in the original publication. The CV liability potential of compounds **A** and **B** was investigated in the guinea pig right atrium assay and is reported in the original publication.	

6 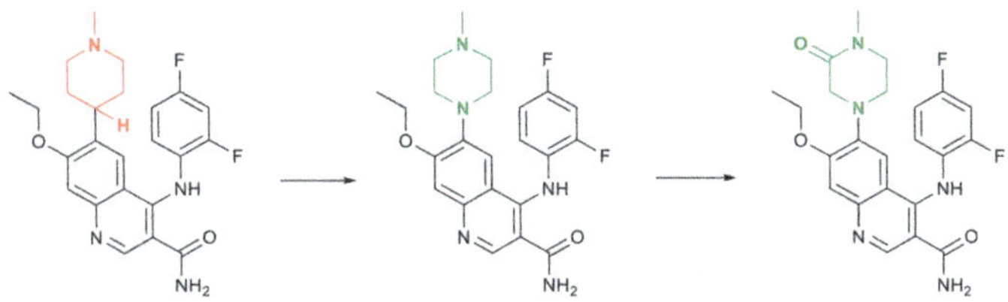 176

clogP = 3.5 clogP = 3.1 clogP = 2.1

L-type Ca_V %inhibition @10μM (binding, PAA site):

75% 41% 14%

7 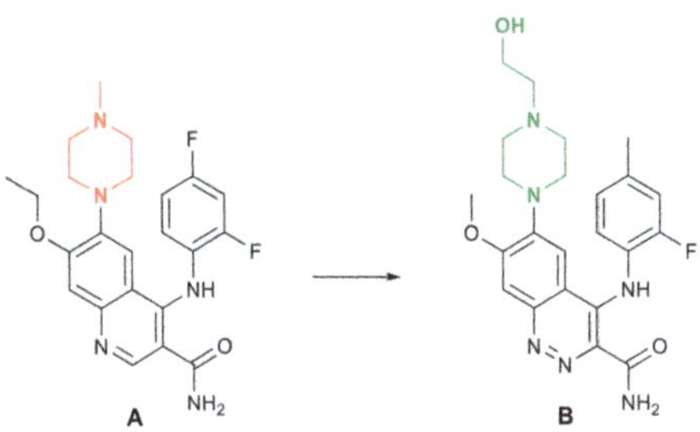 176

A **B**

clogP = 3.1 clogP = 2.6

L-type Ca_V pIC_{50} (canine cardiomyocyte L-type CaV current assay):

5.5 < 4.7

Data from dog cardiomyocyte assay, guinea pig monophasic action potential duration assay and dog telemetry study with compounds **A** and **B** are reported in the original publication.

8 **twisted** **more planar** 181

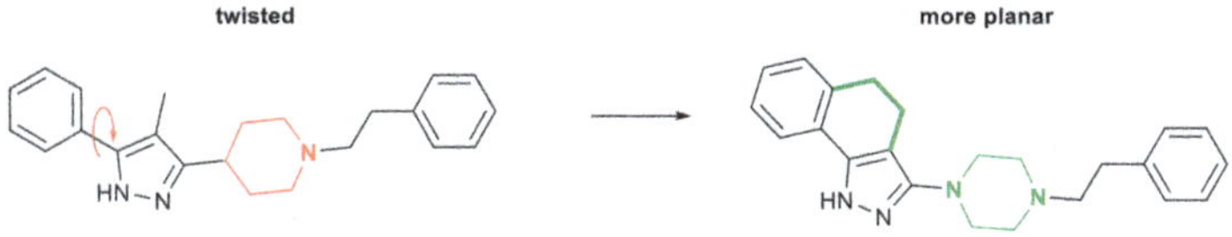

A **B**

clogP = 5.2 clogP = 5.1

L-type Ca_V %inhibition @10μM (binding assay):

77% 17%

In addition to reduced pK_a compound **B** has increased structural rigidity, which may lead to further disruption of binding

19.2.2.3 Lipophilicity Reduction

Examples		Reference
1	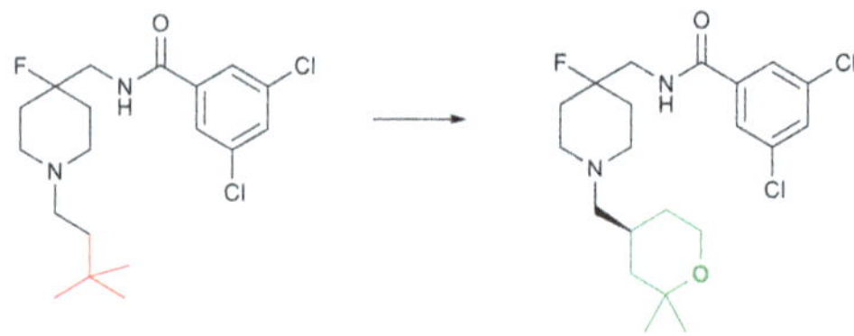 clogP = 4.4 → clogP = 3.1 L-type Ca_V pIC_{50} (binding assay): 6.0 → 5.2	182
2	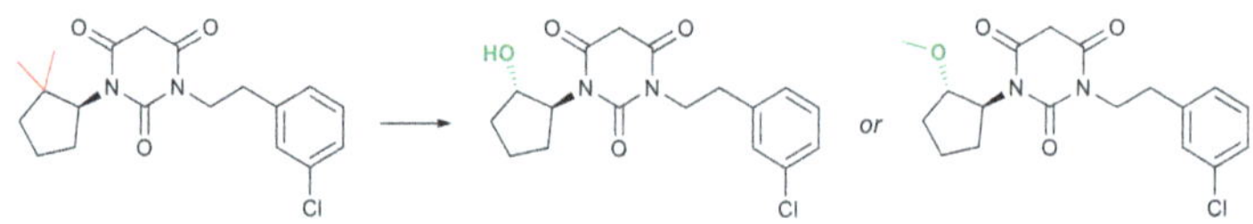 clogP = 3.8 → clogP = 1.9 or clogP = 2.2 $Ca_V1.2$ pIC_{50} (FLIPR) = 5.7 → $Ca_V1.2$ pIC_{50} (FLIPR) < 4.0 or $Ca_V1.2$ pIC_{50} (FLIPR) = 4.5	175
3	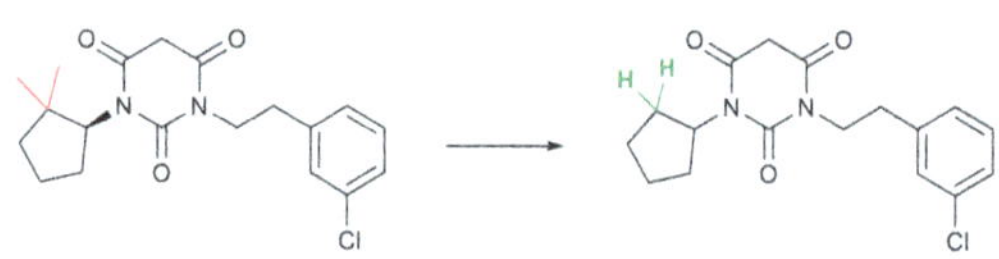 clogP = 3.8 → clogP = 2.8 $Ca_V1.2$ pIC_{50} (FLIPR) = 5.7 → $Ca_V1.2$ pIC_{50} (FLIPR) < 4.0	175
4	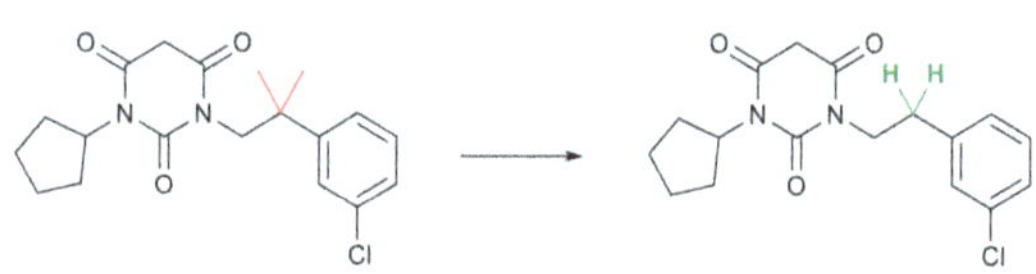 clogP = 3.8 → clogP = 2.8 $Ca_V1.2$ pIC_{50} (FLIPR) = 4.9 → $Ca_V1.2$ pIC_{50} (FLIPR) < 4.0	175
5	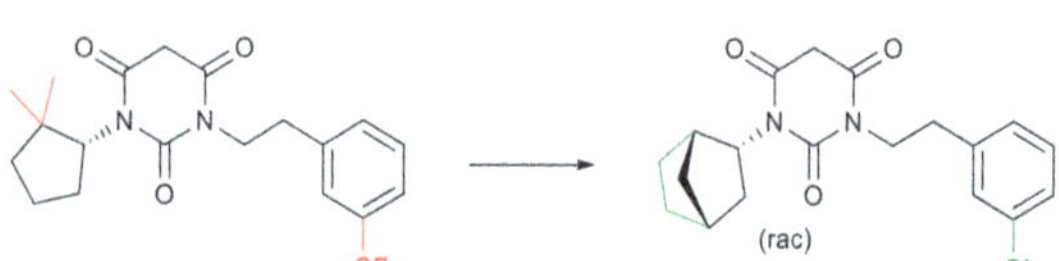 clogP = 4.1 → clogP = 3.0 $Ca_V1.2$ pIC_{50} (FLIPR) = 5.5 → $Ca_V1.2$ pIC_{50} (FLIPR) = 4.0	175

19.2.3 hERG Channel

Binding to the hERG channel is generally driven by charge, lipophilic and pi-stacking interactions.[183] Potent hERG blockers are generally found to be lipophilic bases, often containing multiple aromatic systems, while less lipophilic and/or non-basic compounds are less likely to carry hERG activity.[164] Accordingly, the majority of strategies to reduce hERG binding can be broadly classified into four approaches, which are often not mutually exclusive:

1. Removal or reduction of basicity and lipophilicity.
2. Introduction of acidic centres.
3. Reduction of lipophilicity.
4. Disruption of binding by introduction of steric clashes and changes in conformation.

Due to the promiscuity and prominence of the hERG channel as a safety liability, there is a vast array of studies on reducing activity.[184] The following examples are selected from a systematic search of prominent primary medicinal chemistry journals from the period 2012 to 2018. To provide greater structural granularity within the categories of approach outlined above and to give prospective applicability, the examples are classified into removal or manipulation of basicity, introduction of acidity, manipulation of aromatic substitution, replacement of aryl rings, modifications of heteroaryl systems, aliphatic ring or chain modifications and larger changes to core scaffolds. Many of these serve to modify basicity and lipophilicity, in some cases in an interdependent manner. The examples are generally selected from examples of overall optimisations with a positive trajectory and so highlight examples in which hERG liabilities were reduced in a manner that was generally compatible with maintaining primary activity and overall desirable properties.

To allow assessment of the context of the change in activity relative to concomitant changes to lipophilicity and basicity, logD and pK_a values have been provided where they have been reported, calculated logP values are provided in other cases. Whilst calculated logP values are not generally accurate enough to assess SAR,[185] relative changes within the reported compound pairs can illustrate whether a change in potency is outwith the change in lipophilicity for each pair with a reasonable level of confidence.

19.2.3.1 Replacement of Basic Centres

Examples		Reference
1	 pIC$_{50}$ = 6.7, clogP = 4.7 → pIC$_{50}$ = 5.2, clogP = 4.1	186
2	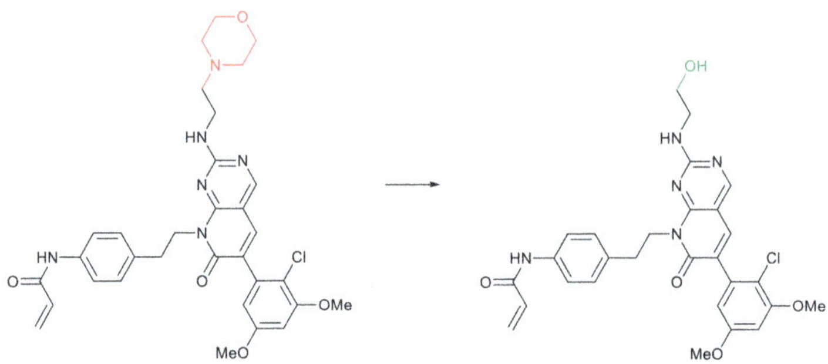 81% inhibition at 1 μM, clogP = 4.1 → 5% inhibition at 1 μM, clogP = 3.1	187
3	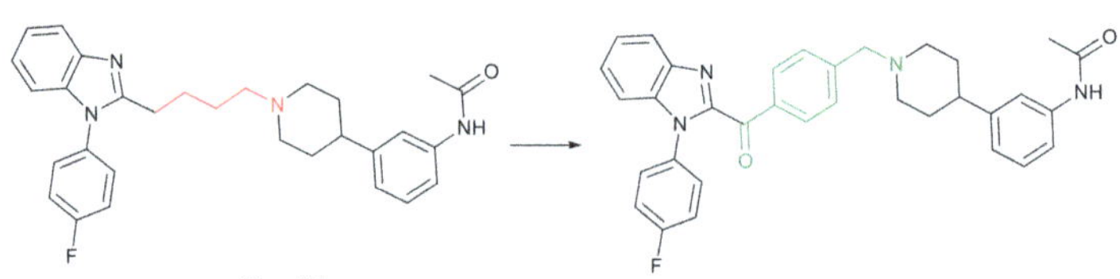 pIC$_{50}$ = 7.5, clogP = 6.1 → pIC$_{50}$ = 5.7, clogP = 7.1	188
4	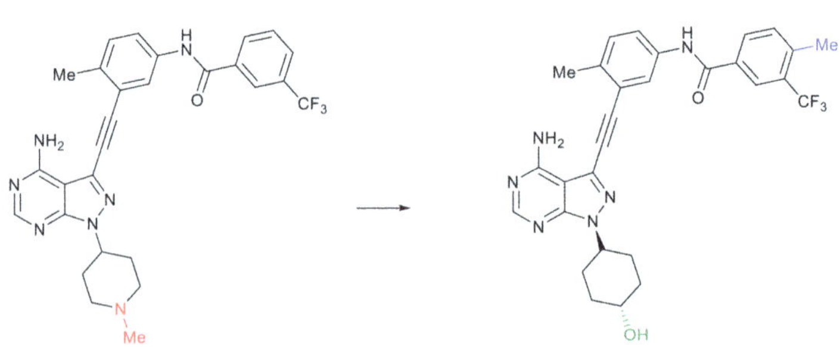 pIC$_{50}$ = 6.2, cLogP = 4.5 → pIC$_{50}$ < 4.5, cLogP = 4.9	189

Examples				Reference
5	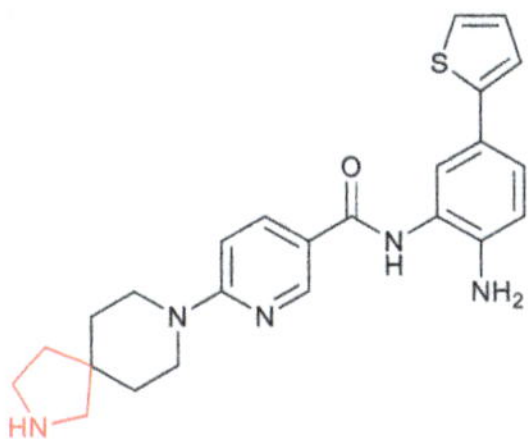 pIC_{50} = 5.0 clogP = 2.9	→	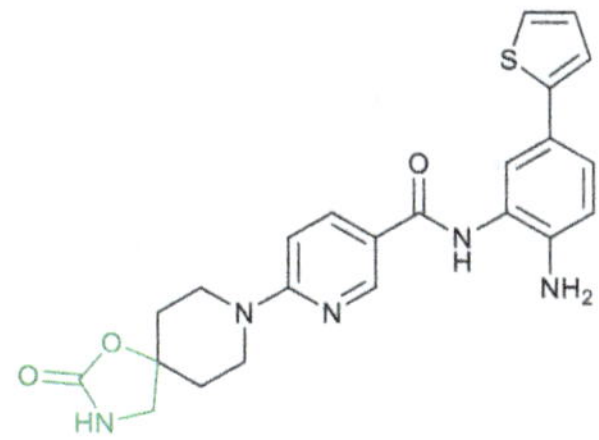 pIC_{50} < 4.5 clogP = 1.7	190
6	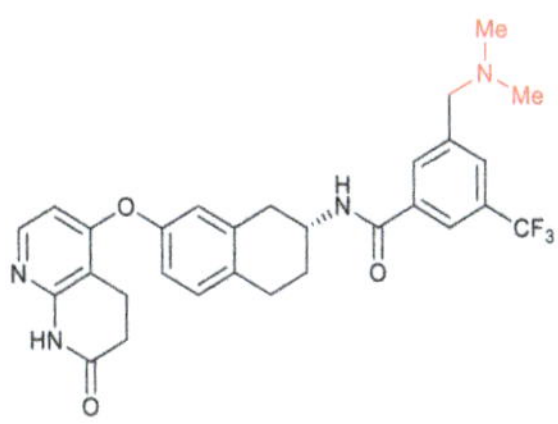 pIC_{50} = 5.2 clogP = 5.1	→	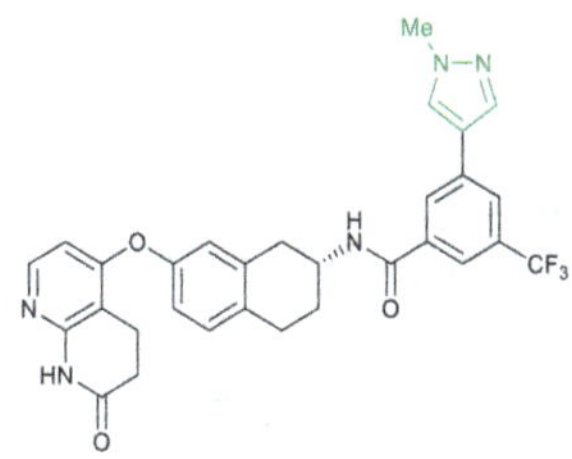 pIC_{50} < 4.0 clogP = 5.3	191
7	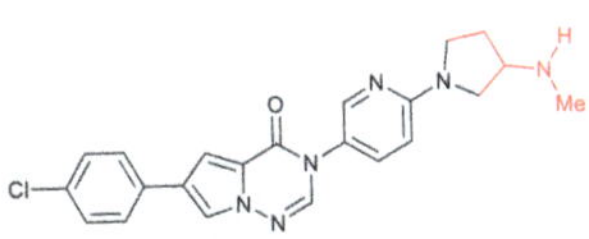 pIC_{50} = 5.5 clogP = 2.8	→	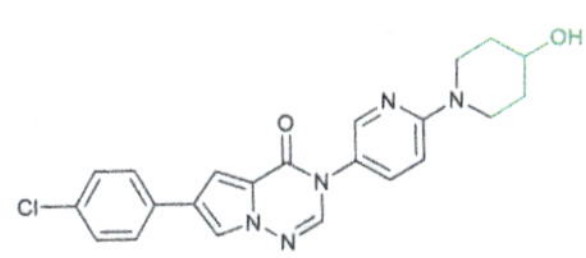 pIC_{50} < 4.1 clogP = 2.2	192
8	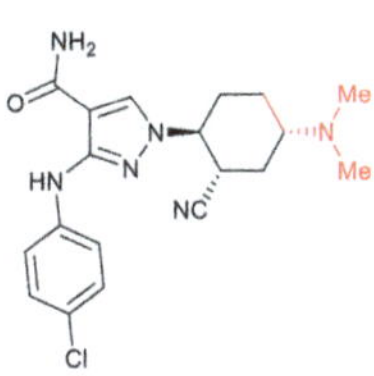 pIC_{50} = 4.9 clogP = 2.2	→	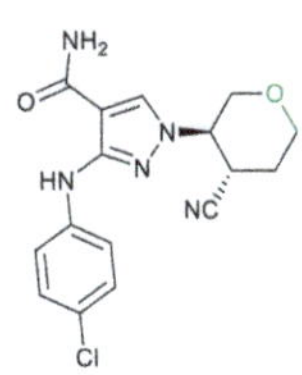 pIC_{50} = 4.4 clogP = 0.9	193

(*continued*)

Examples			Reference
9			194
	pIC_{50} = 5.7 clogP = 4.6	pIC_{50} < 4.5 clogP = 5.0	
10	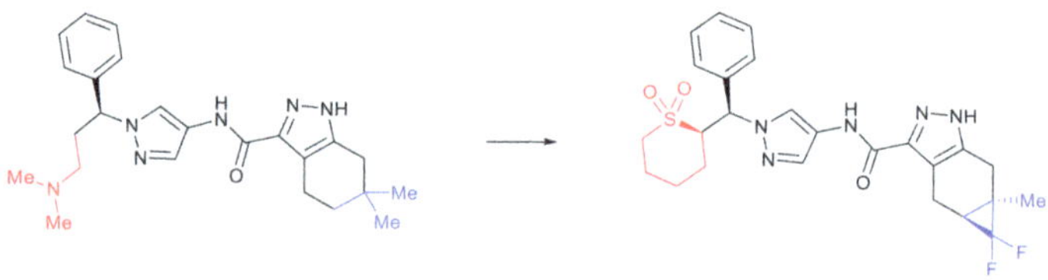		195
	88% inhibition at 10 μM clogP = 4.2	6.8% inhibition at 10 μM clogP = 1.9	

19.2.3.2 Modification of Basic Centres

Examples			Reference
1	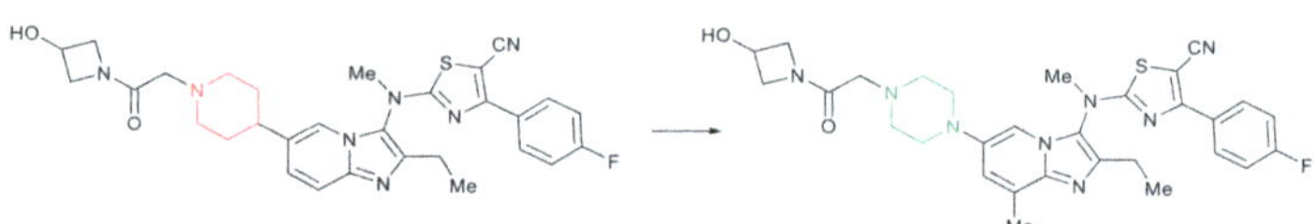		196
	pIC_{50} = 5.5 clogP = 6.5 calcd pK_a = 6.6	pIC_{50} < 5.0 clogP = 5.8 calcd pK_a = 5.5	
2	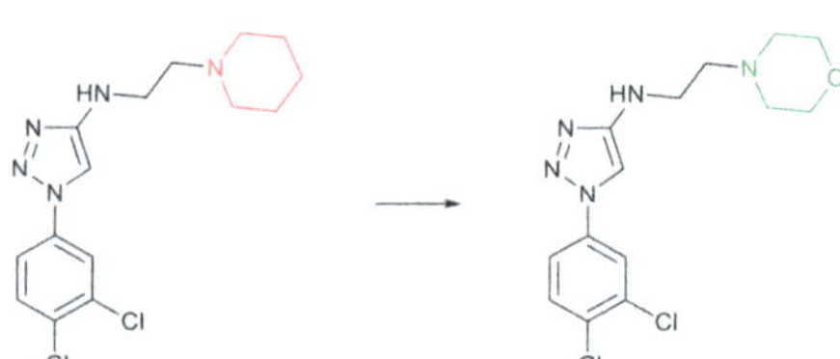		197
	pIC_{50} = 5.8 clogP = 4.6	pIC_{50} < 5 clogP = 3.4	

Examples				Reference
3	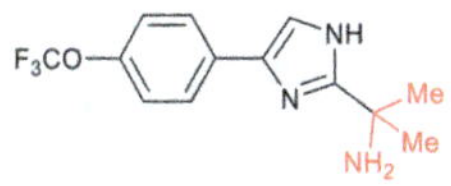 pIC$_{50}$ = 5.6 clogP = 2.9 pK$_a$ = 7.9	→	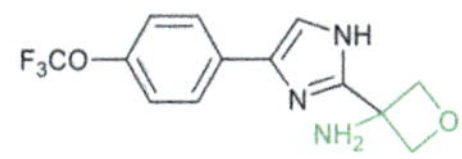 pIC$_{50}$ <5.0 clogP = 2.3 pK$_a$ = 6.4	198
4	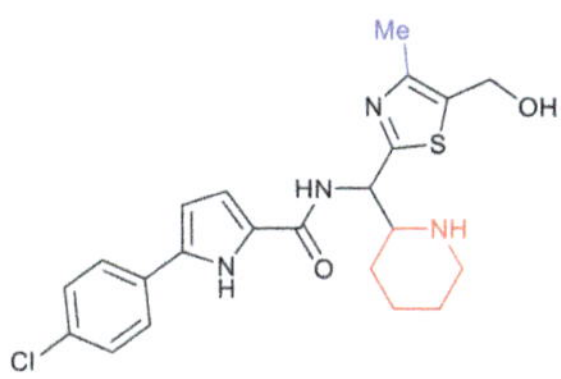 pIC$_{50}$ = 5.4 clogP = 3.4	→	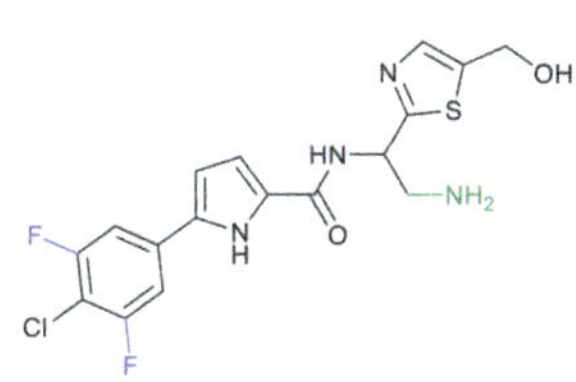 pIC$_{50}$ = 4.9 clogP = 1.7	199
5	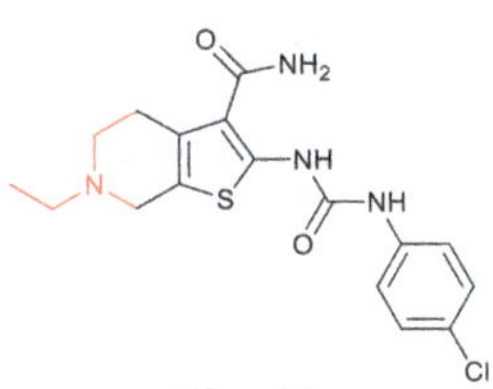 pIC$_{50}$ = 5.5 clogP = 3.6	→	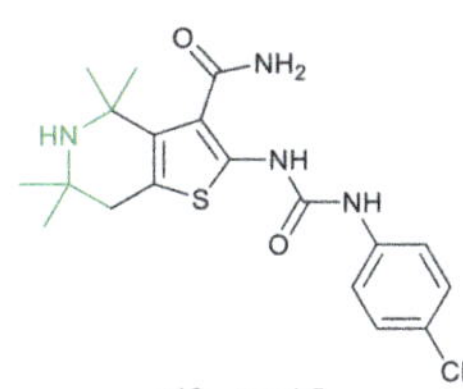 pIC$_{50}$ = < 4.5 clogP = 4.7	200
6	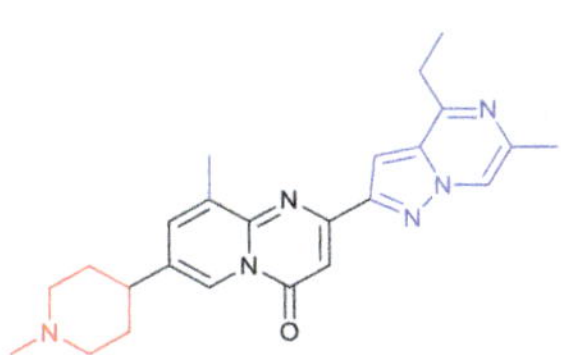 pIC$_{50}$ = 5.8 logD = 2.3 pK$_a$ = 10.9	→	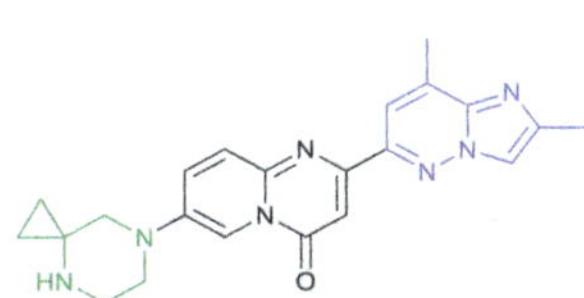 pIC$_{20}$ < 5.3 logD = 2.5 pK$_a$ = 6.8	201
7	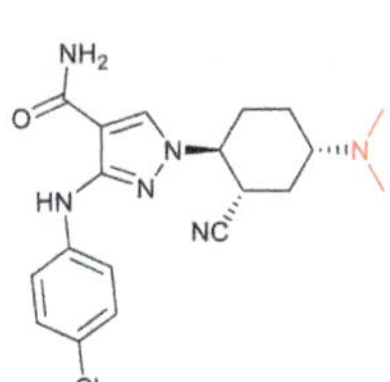 pIC$_{50}$ = 4.9 clogP = 2.2	→	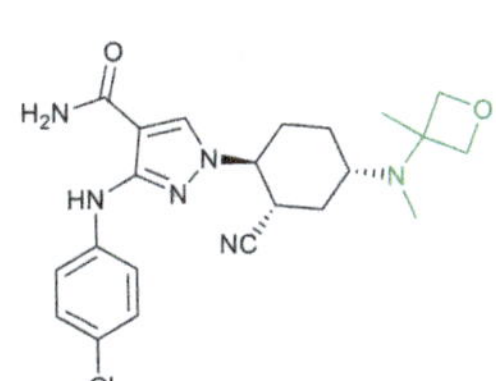 pIC$_{50}$ = 4.3 clogP = 3.1	193

(*continued*)

Examples		Reference
8	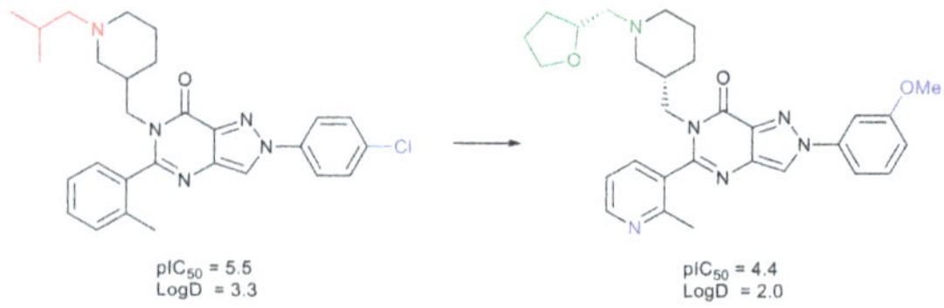	202
9	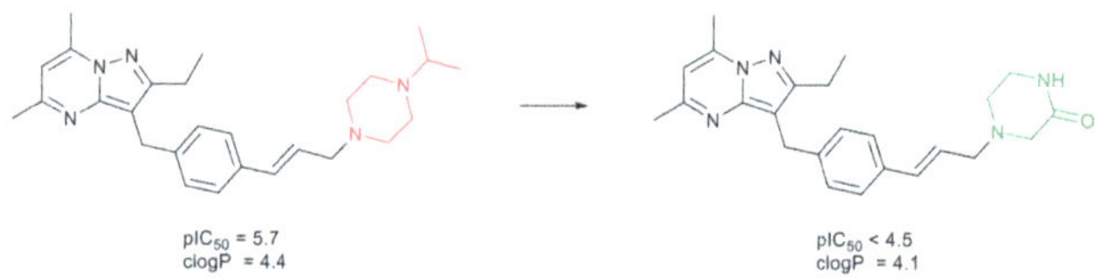	203
10	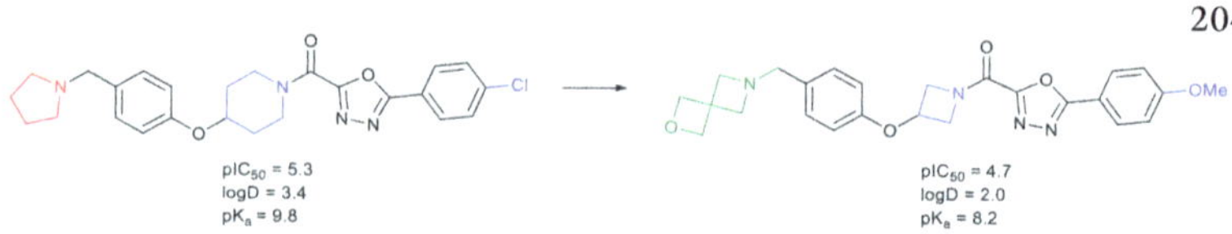	204
11	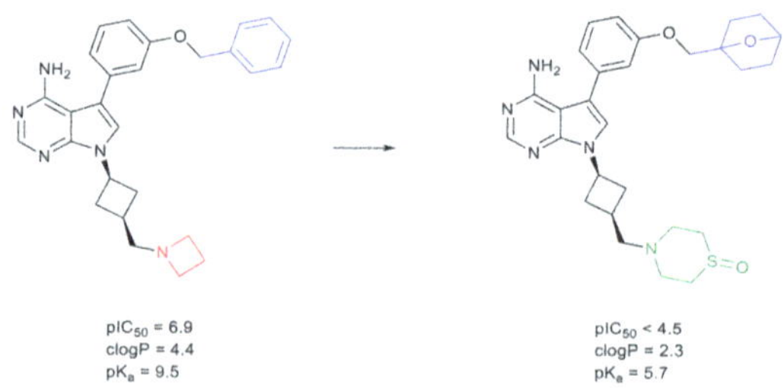	205
12	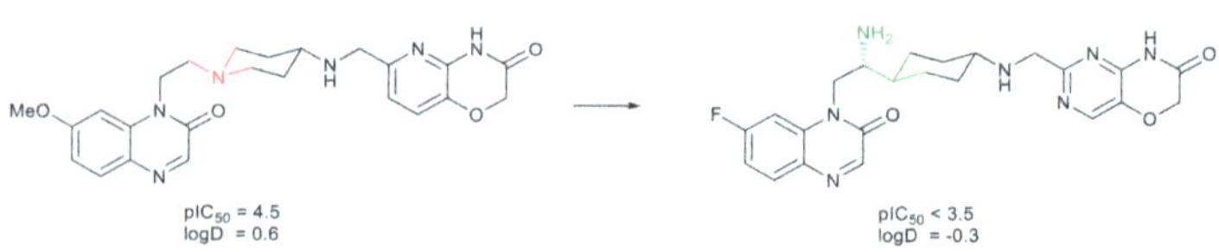	206

19.2.3.3 Addition of Acids

	Examples	Reference
1	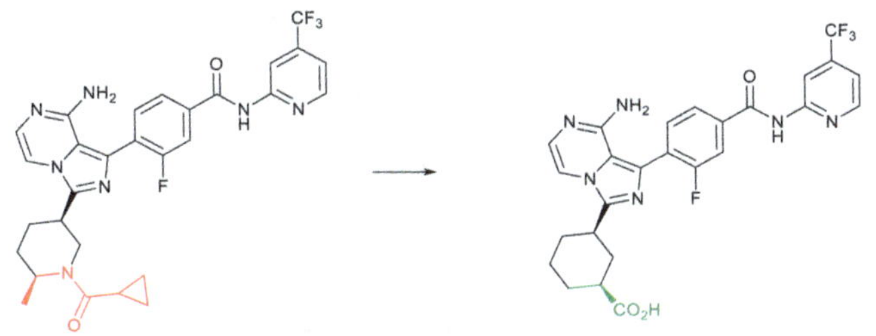 pIC_{50} = 5.4, clogP = 3.7 → pIC_{50} < 4.2, clogP = 4.3	208
2	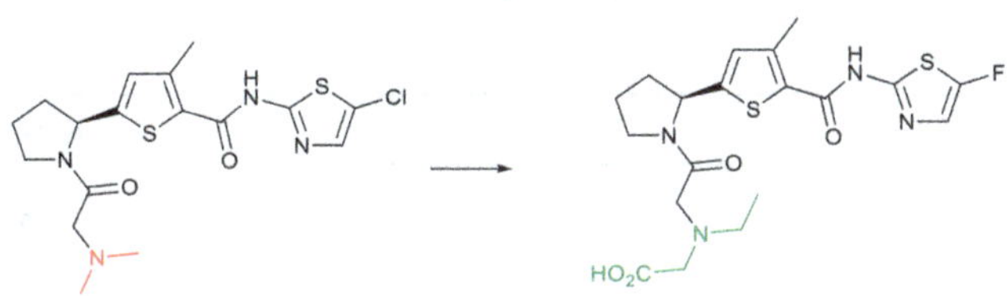 pIC_{50} > 6.5, clogP = 2.0 → pIC_{50} < 3.5, clogP = 0.0	207
3	 pIC_{50} = 5.7, logD = 3.3 → pIC_{50} > 4.0, logD = 1.2	208
4	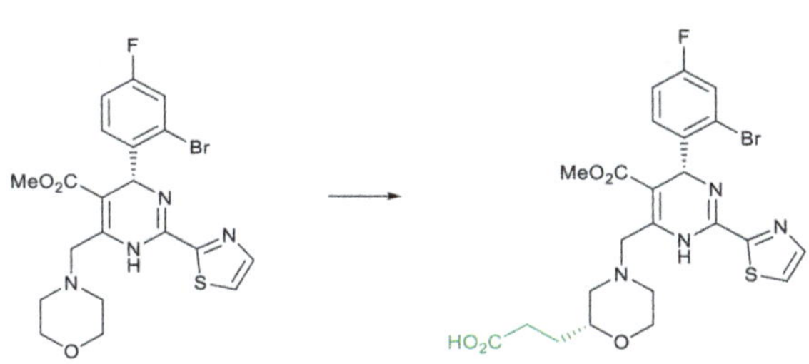 pIC_{50} = 5.9, clogP = 4.7 → pIC_{50} < 4.5, clogP = 2.5	209

(*continued*)

Examples		Reference
5	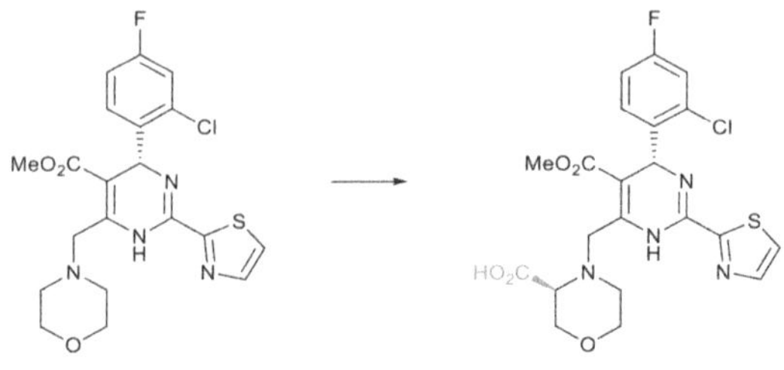 pIC_{20} = 6.3, clogP = 3.2 → pIC_{20} < 5.0, logD = 0.1	210

19.2.3.4 Addition of (Hetero)Aromatic Substituents

Examples		Reference
1	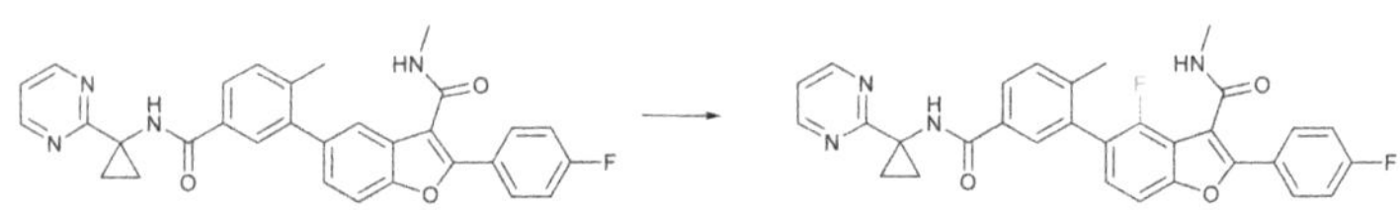 pIC_{50} = 5.3, clogP = 4.8 → pIC_{50} = 4.6, LogD = 5.1	211
2	 pIC_{50} = 5.2, clogP = 2.9 → pIC_{50} < 5, clogP = 3.0	212
3	 pIC_{50} = 4.9, clogP = 2.3 → pIC_{50} < 4.5, clogP = 2.5	213

Examples			Reference
4	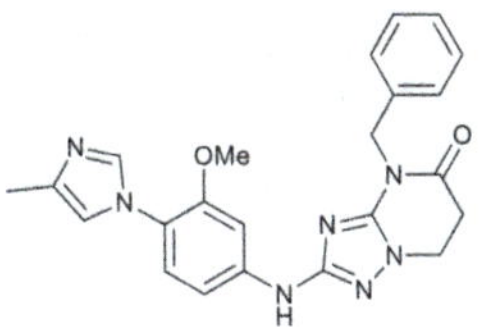 pIC_{50} = 5.2 logD = 2.7	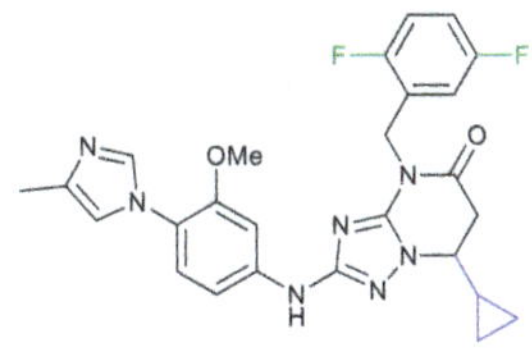 pIC_{50} < 4.5 logD = 4.0	214
5	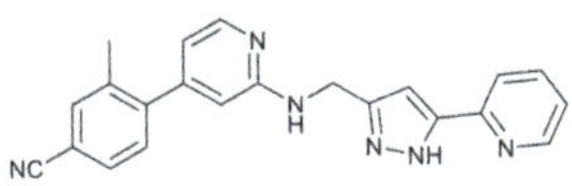 pIC_{50} = 6.0 clogP = 3.1	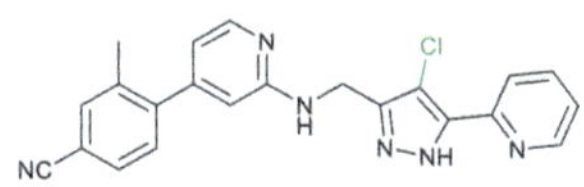 pIC_{50} < 4.5 clogP = 3.7	215
6	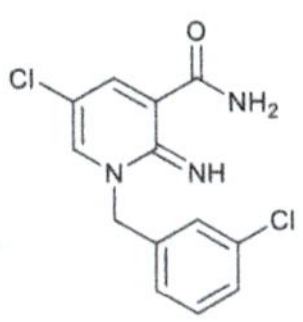 47% inhibition at 10 μM clogP: 1.8	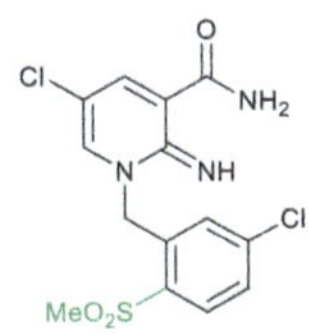 16% inhibition at 10 μM clogP: 0.3	216
7	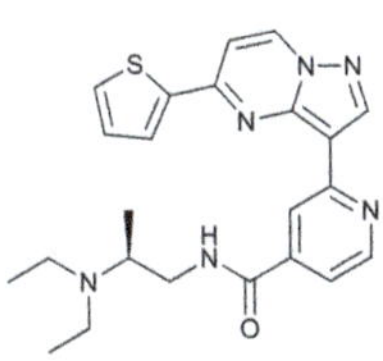 pIC_{50} = 6.1 clogP = 4.3	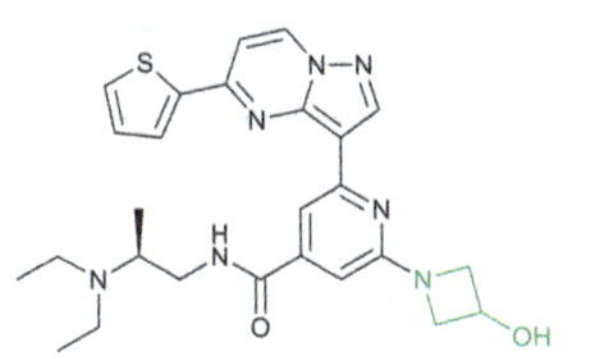 pIC_{50} < 4.5 clogP = 4.0	217
8	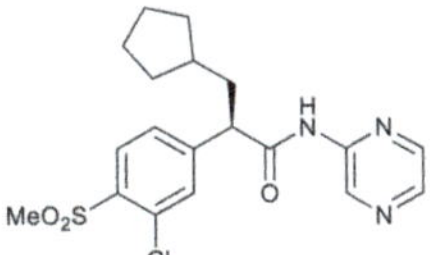 59% inhibiton at 10 μM clogP = 3.0	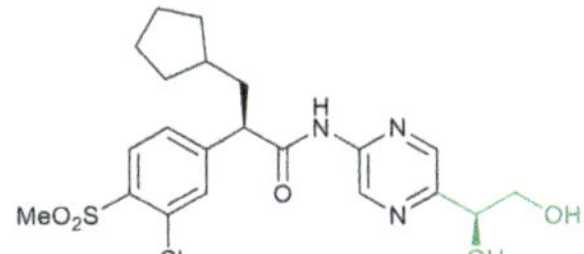 22% inhibiton at 10 μM clogP = 1.0	218

(continued)

Examples	Reference
9  70% inhibition at 30 μM, clogP = 3.5 → 30% inhibition at 30 μM, clogP = 2.9	219
10  pIC_{50} = 5.8, clogP = 5.4 → pIC_{50} < 4.5, clogP = 7.4	220

19.2.3.5 *Modification of (Hetero)Aromatic Substituents*

Examples	Reference
1  pIC_{50} = 4.9, LogD = 3.1 → pIC_{50} = 4.2, LogD = 2.6	221
2  75% inhibition at 30 μM, clogP = 3.0 → 20% inhibition at 30 μM, clogP = 2.4	222

Examples		Reference
3	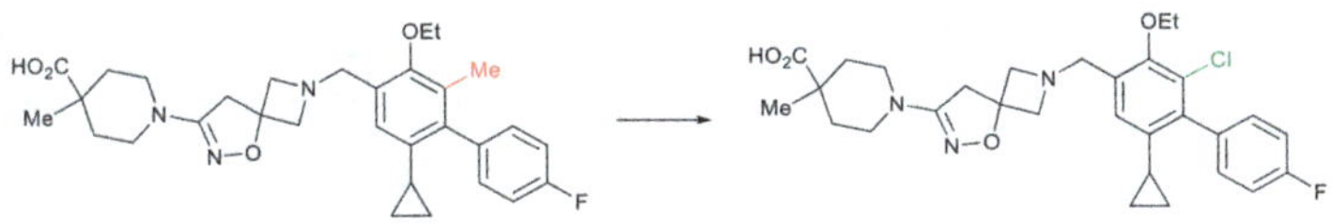 76% inhibition at 30 μM, logD = 4.1 → 4% inhibition at 30 μM, logD = 4.2	223
4	 $pIC_{50} = 5.9$, logD = 2.7 → $pIC_{50} < 4.5$, logD = 1.0	224
5	 $piC_{50} = 5.4$, clogP = 3.0 → 45% inhibition at 30 μM, clogP = 2.5	225
6	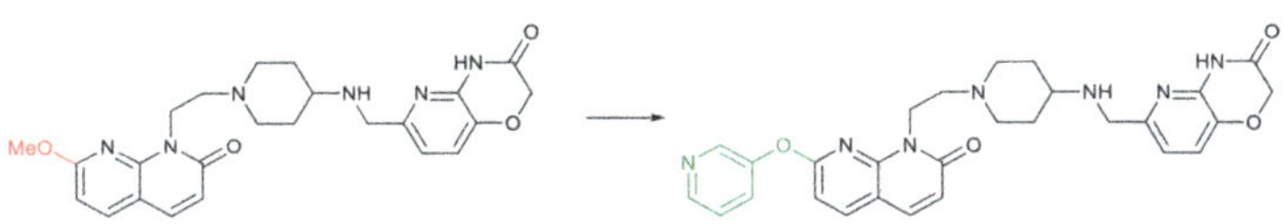 $pIC_{50} = 4.6$, clogP = 1.0 → $pIC_{50} = 3.8$, clogP = 0.7	226
7	 $pIC_{50} = 4.7$, logD = 2.1 → $pIC_{50} < 4.5$, logD = 2.0	227
8	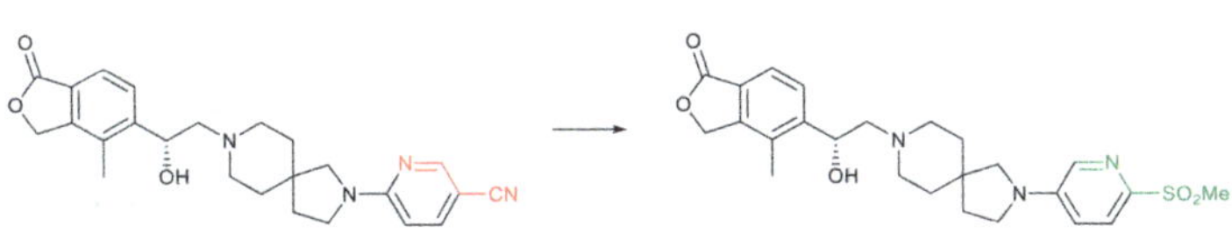 $pIC_{50} = 5.6$, clogP = 1.2 → $pIC_{50} < 4.2$, clogD = 1.4	228

(*continued*)

Examples		Reference
9	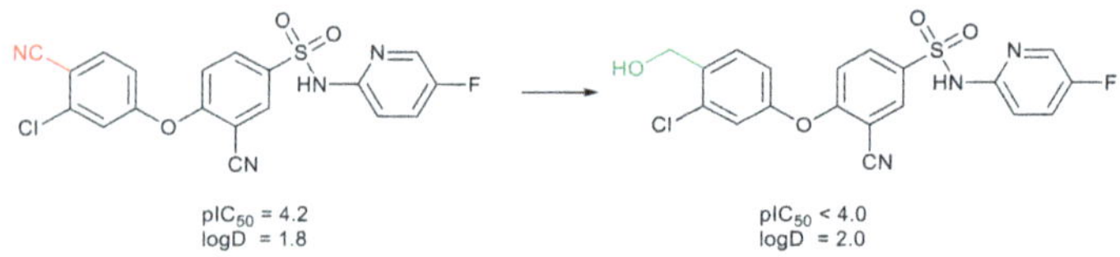	229
10	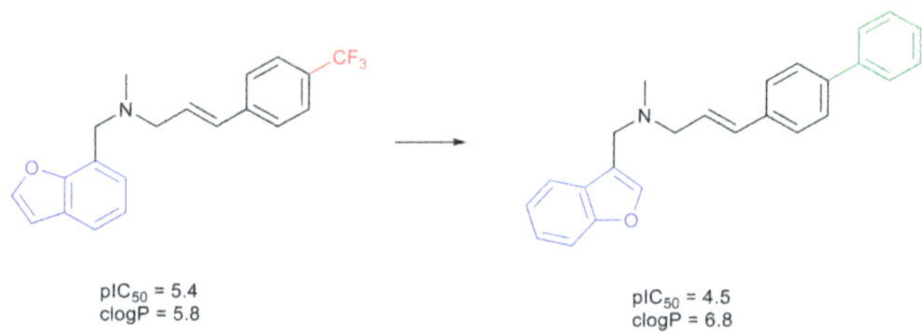	230
11	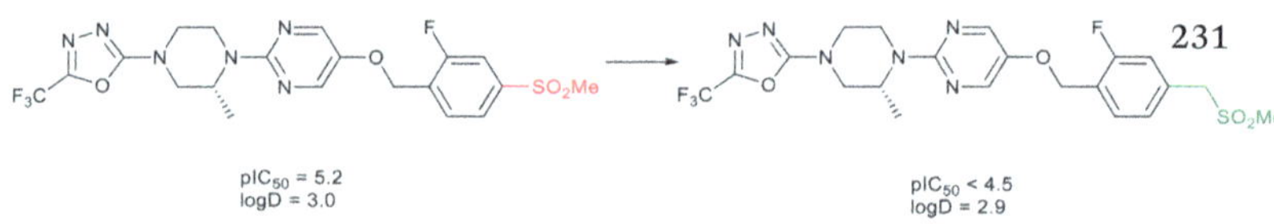	231
12	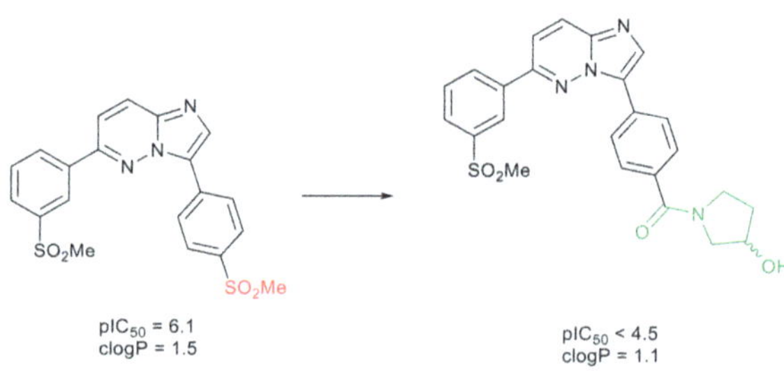	232
13	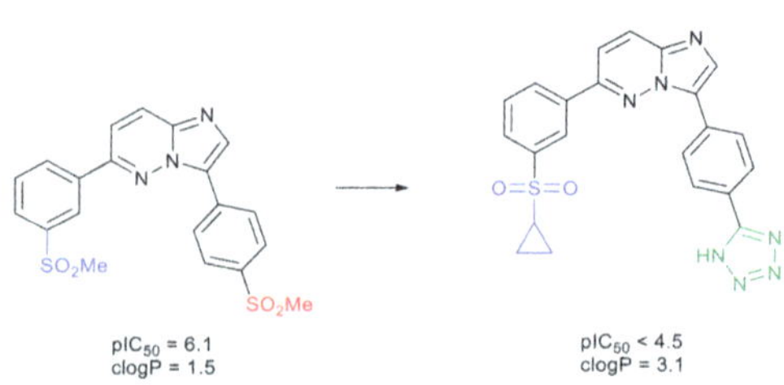	232

Examples		Reference
14	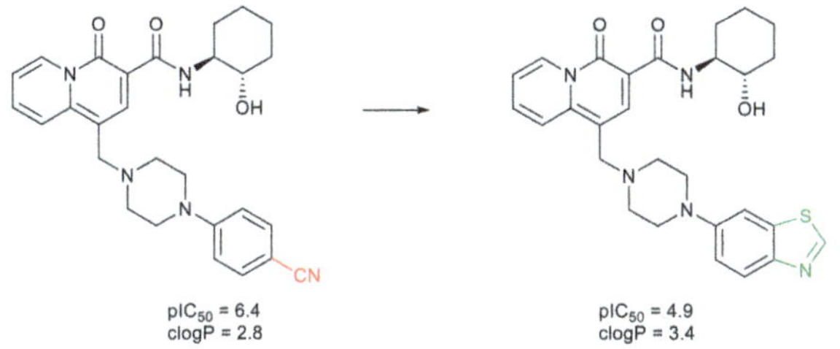	233
15	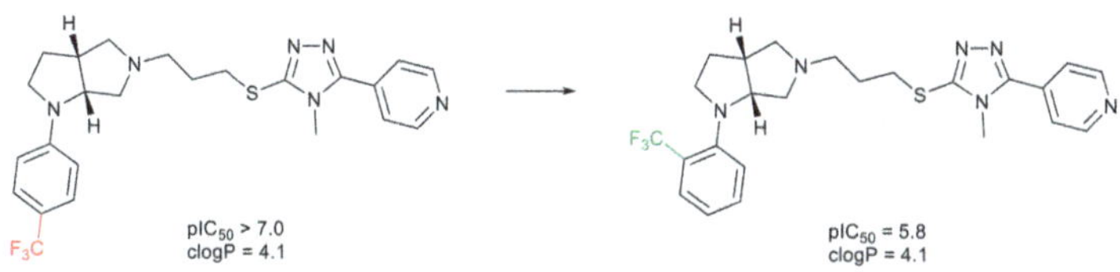	234
16	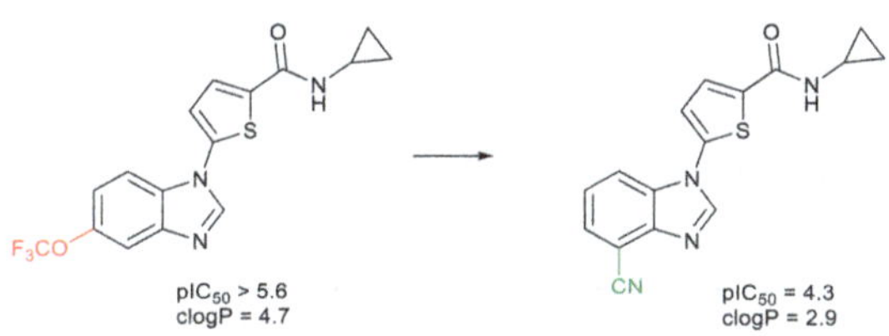	235

19.2.3.6 *Removal of Aryl Rings*

Examples		Reference
1	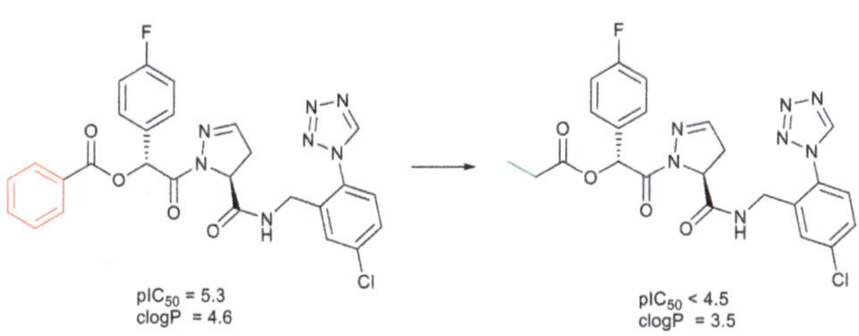	236

(continued)

Examples		Reference
2	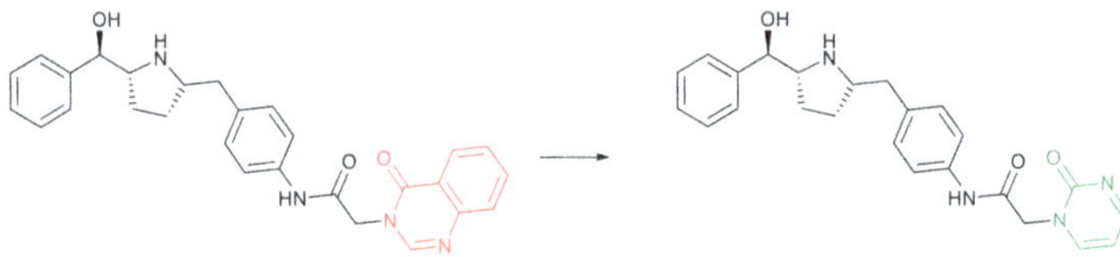 pIC_{50} = 5.7, logD = 2.0 → pIC_{50} < 4.5, logD = 0.6	237
3	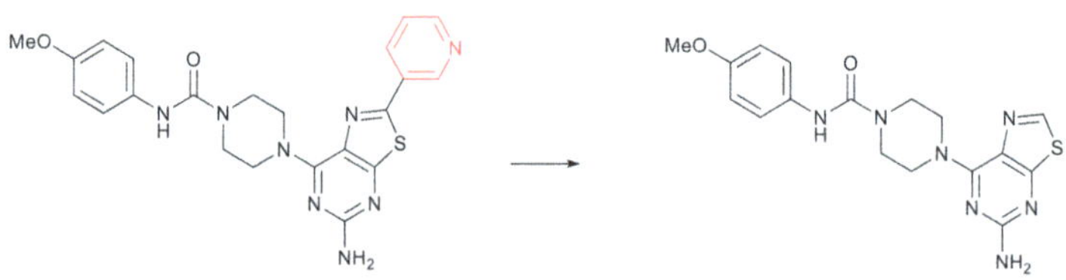 pIC_{50} = 5.7, logD = 3.1 → pIC_{50} = 4.6, logD = 1.9	238

19.2.3.7 Replacement of Phenyl Rings with Heteroaryl Systems

Examples		Reference
1	 25% inhibition at 10 μM, clogP = 1.8 → 0% inhibition at 10 μM, clogP = 0.3	239
2	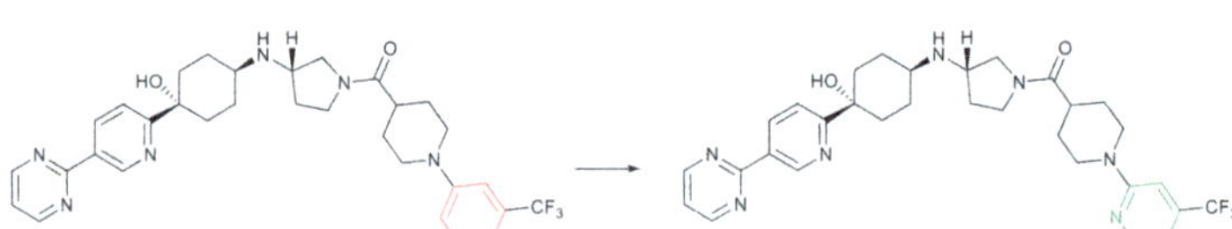 pIC_{50} = 5.0, clogP = 4.1 → pIC_{50} = 4.6, clogP = 3.5	240
3	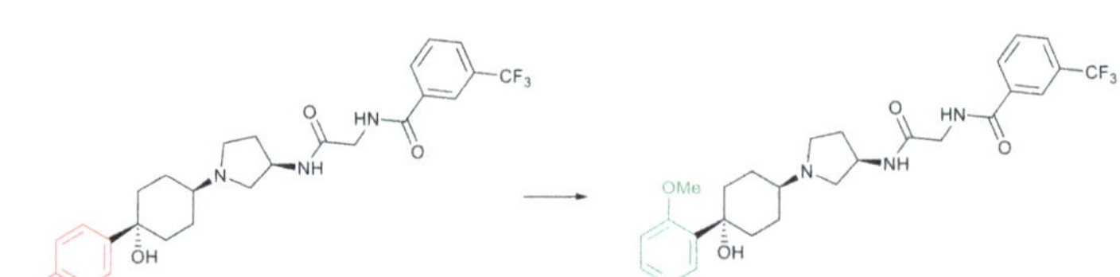 pIC_{50} = 4.9, clogP = 3.2 → pIC_{50} < 4.5, clogP = 1.3	241

Examples		Reference
4	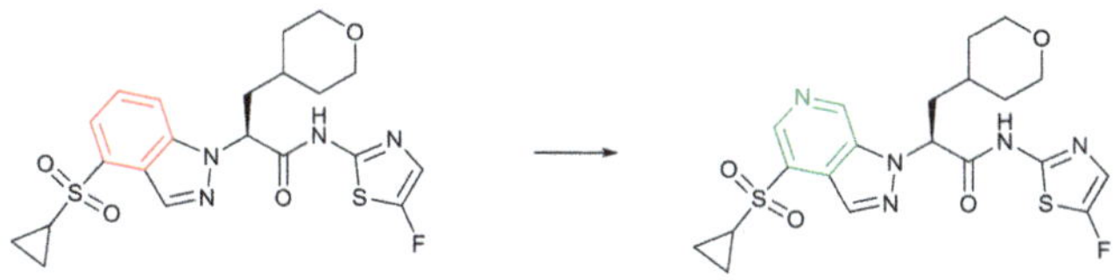 84% inhibition at 10 μM, clogP = 1.8 → 46% inhibition at 10 μM, clogP = 1.0	242
5	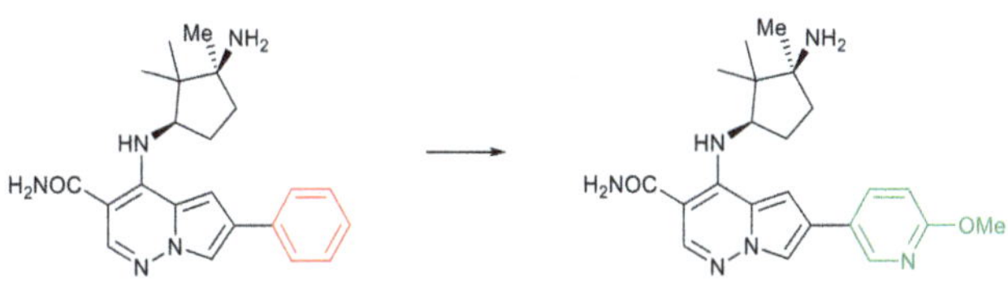 pIC_{50} = 5.4, clogP = 3.9 → pIC_{50} < 4.1, clogP = 3.2	243
6	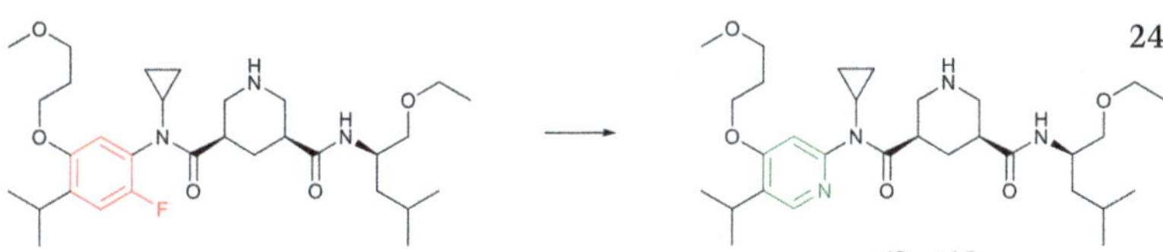 pIC_{50} = 5.3, clogP = 5.8 → pIC_{50} < 4.5, clogP = 4.7	244
7	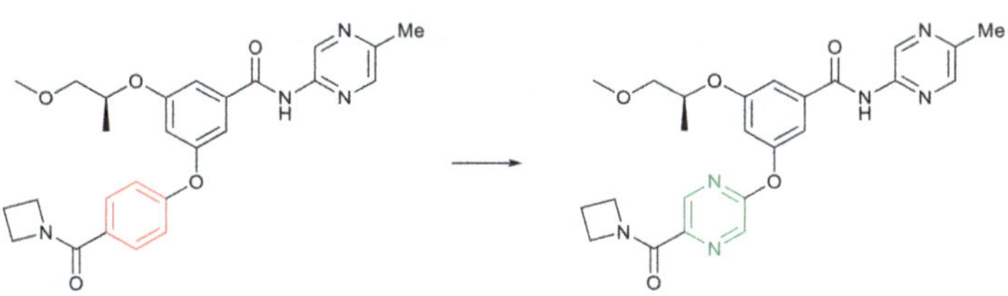 pIC_{50} = 4.7, logD = 2.9 → pIC_{50} < 4.0, logD = 2.4	245
8	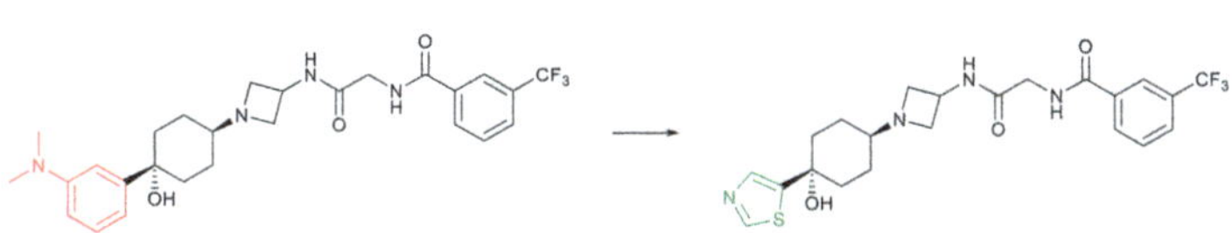 pIC_{50} = 4.8, clogP = 2.9 → pIC_{50} < 4.3, clogP = 1.4	246

(*continued*)

Examples			Reference
9	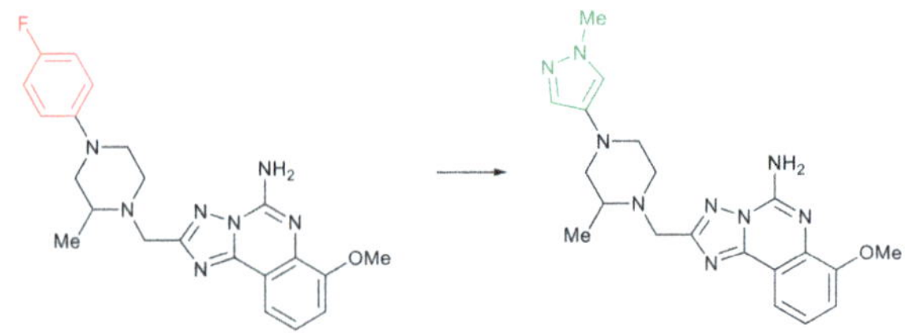		247
	pIC_{50} = 5.8 clogP = 4.0	pIC_{50} < 4.2 clogP = 2.1	
10	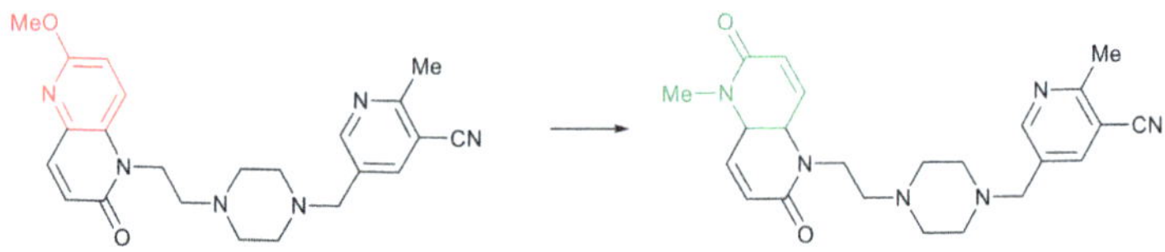		248
	pIC_{50} = 5.0 logD = 1.1	pIC_{50} < 4.5 logD = -0.6	
11	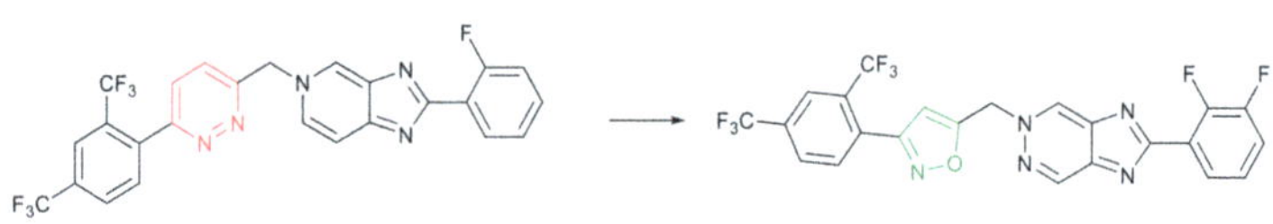		249
	pIC_{50} = 6.1 clogP = 5.5	pIC_{50} < 4.3 clogP = 6.3	
12	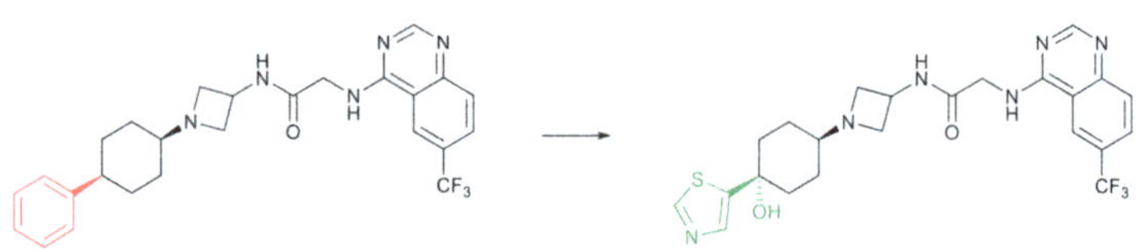		250
	pIC_{50} = 5.3 clogP = 6.2	pIC_{50} = 4.3 clogP = 3.0	
13			251
	pIC_{50} = 4.6 logD = 2.2	pIC_{50} = 3.9 logD = 2.1	

Examples		Reference
14	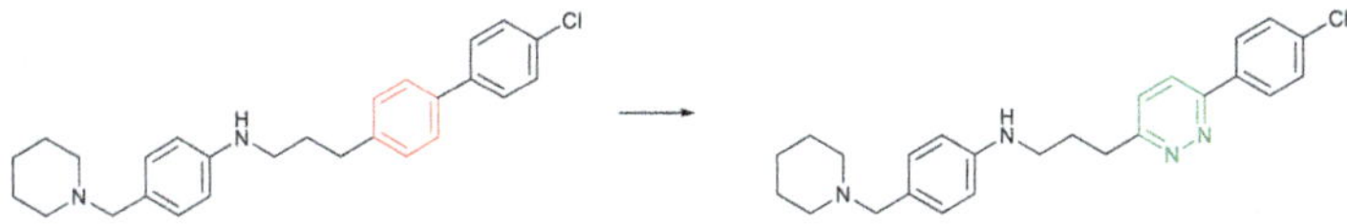 76% inhibition at 1 μM clogP = 6.9 → 8% inhibition at 1 μM clogP = 5.8	252
15	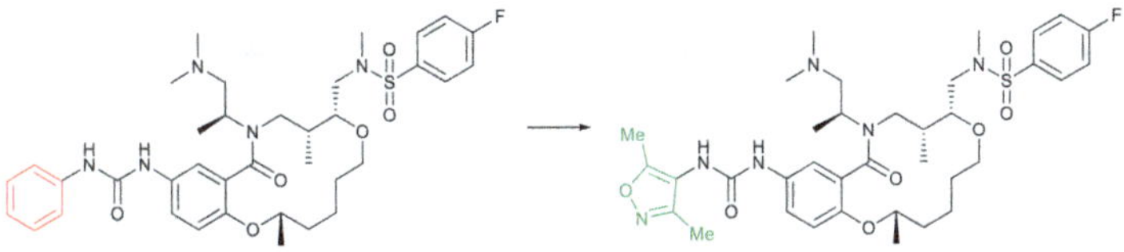 95% inhibition at 10 μM logD = 4.1 → 13% inhibition at 10 μM logD = 2.9	253

19.2.3.8 Modification of Heteroaryl Ring Systems

Examples		Reference
1	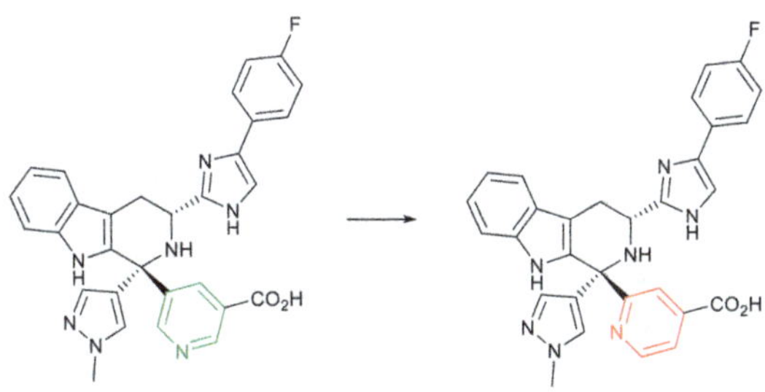 70% inhibition at 30 μM clogP = 0.8 → 26% inhibition at 30 μM clogP = 0.9	254
2	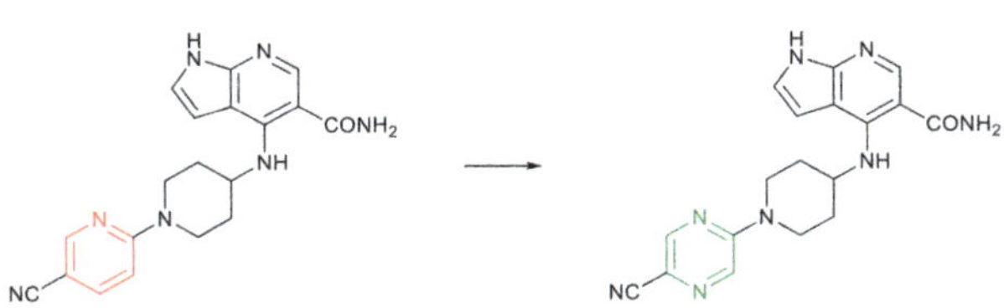 pIC_{50} = 4.9 clogP = 2.8 → pIC_{50} < 4.0 clogP = 2.4	255

(continued)

Examples				Reference
3	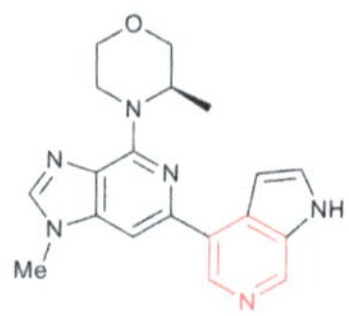 pIC_{50} = 4.9 clogP = 0.9	→	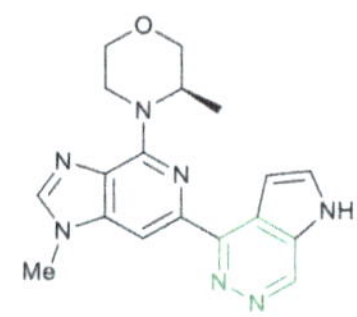 pIC_{50} < 4.5 clogP = 1.0	213
4	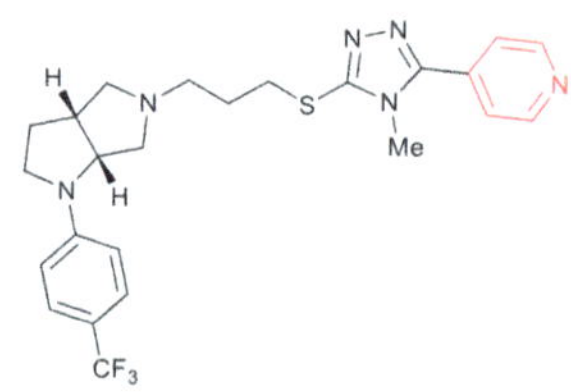 pIC_{50} > 7.0 clogP = 4.1	→	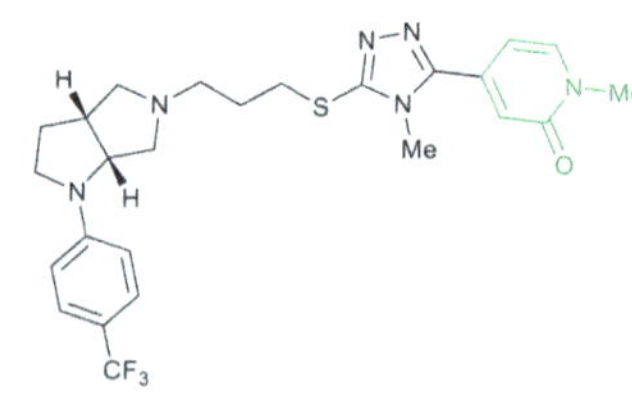 pIC_{50} = 5.9 clogP = 3.4	234
5	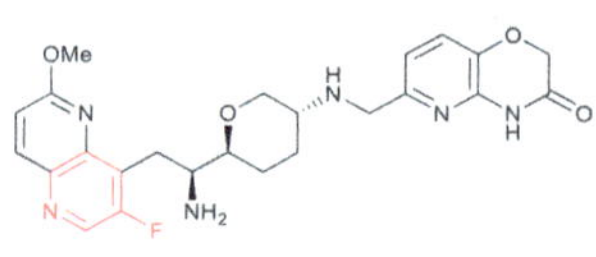 pIC_{50} = 5.3 logD = 0.9	→	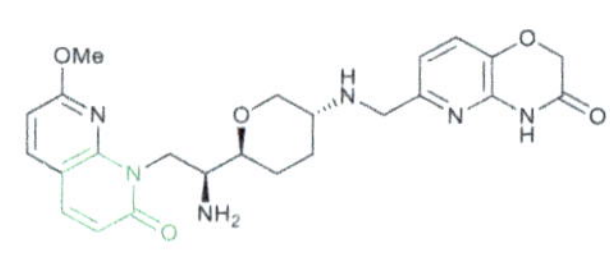 pIC_{50} = 4.3 logD = 0.3	256
6	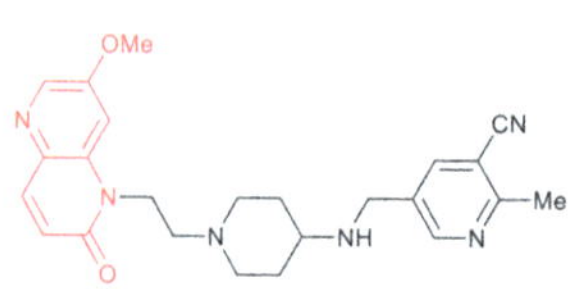 pIC_{50} = 4.3 logD = 1.1	→	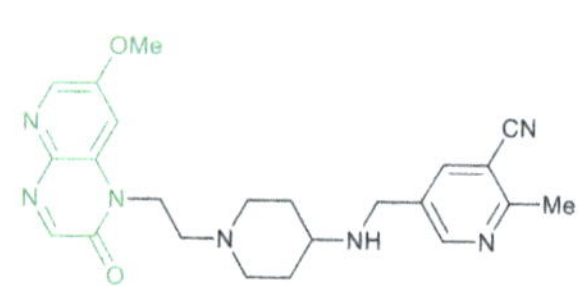 pIC_{50} < 4.0 logD = 0.4	248
7	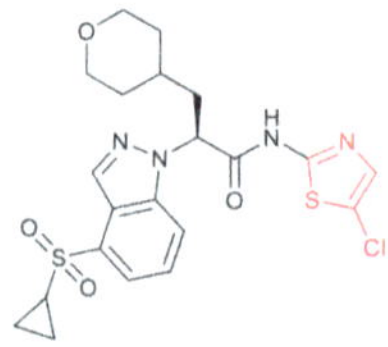 93% inhibition at 10 μM clogP = 2.4	→	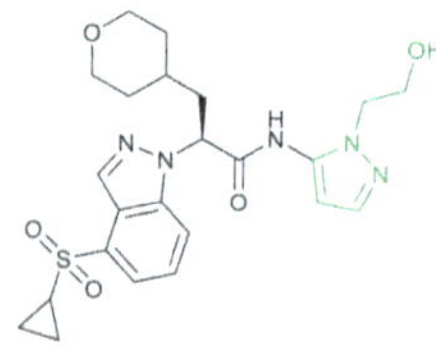 13% inhibition at 10 μM clogP = 0.8	242

Examples		Reference
8	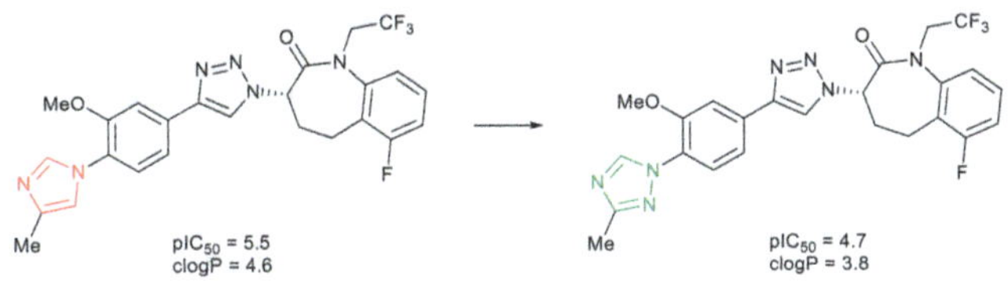	257
9	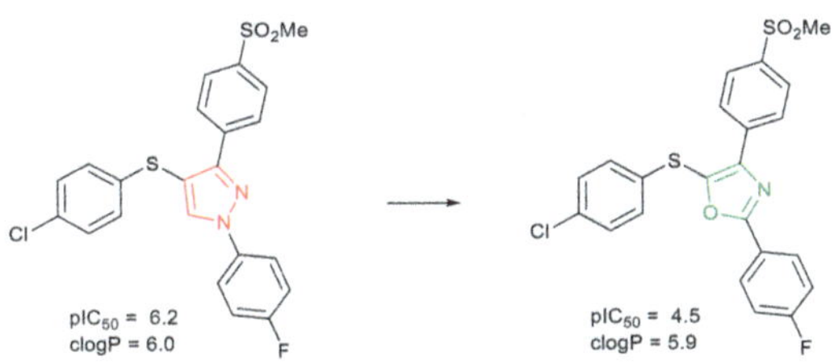	220
10	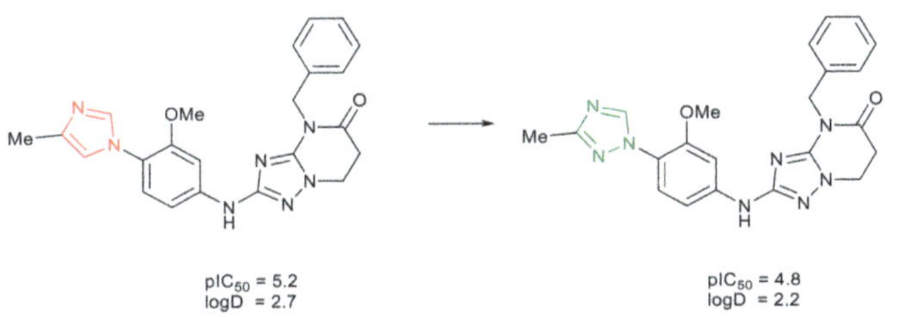	214
11	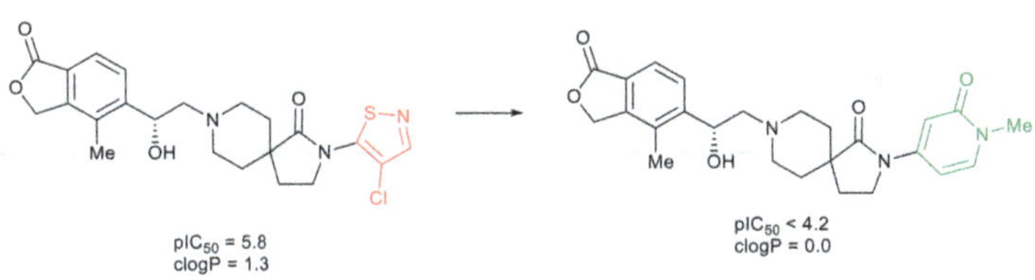	258
12	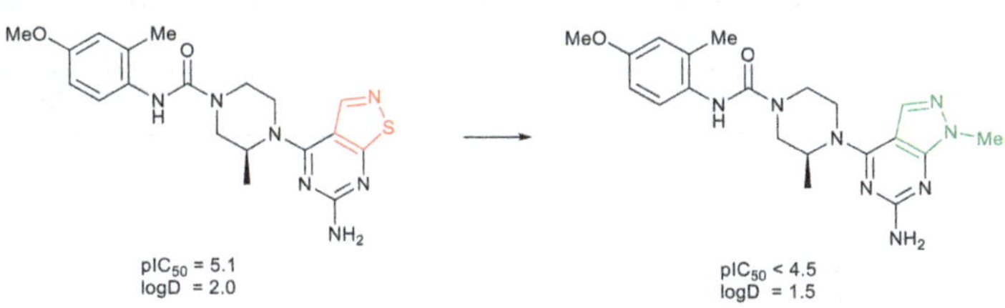	238

(*continued*)

Examples		Reference
13	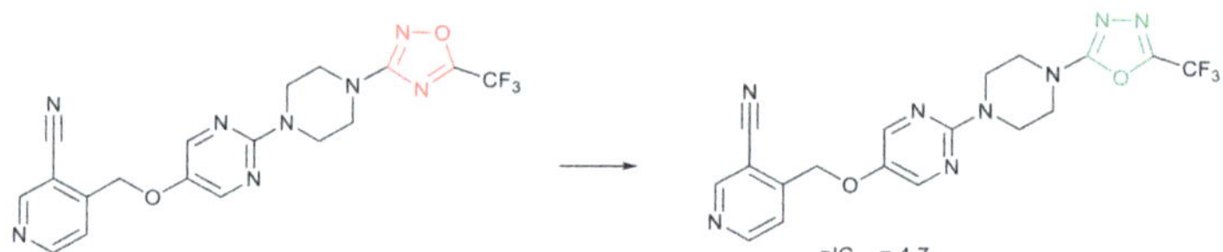 Left: pIC_{50} = 5.1, logD = 4.0 Right: pIC_{50} = 4.7, logD = 2.7	259
14	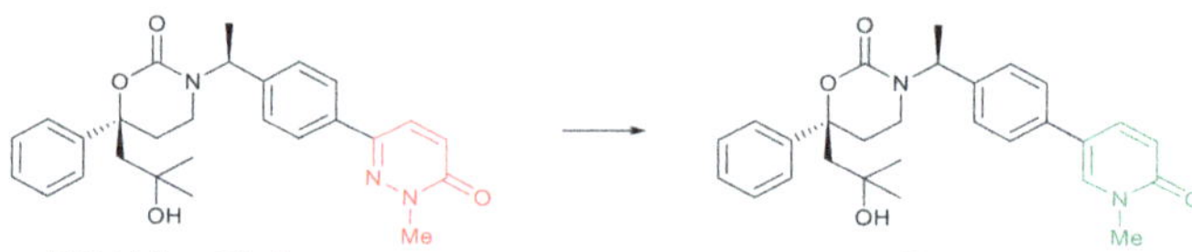 Left: 50% inhibition at 10 μM, clogP = 3.7 Right: pIC_{50} < 4.0, clogP = 4.2	260
15	 Left: pIC_{50} = 5.5, clogP = 1.4 Right: pIC_{50} < 4.5, clogP = 0.3	261
16	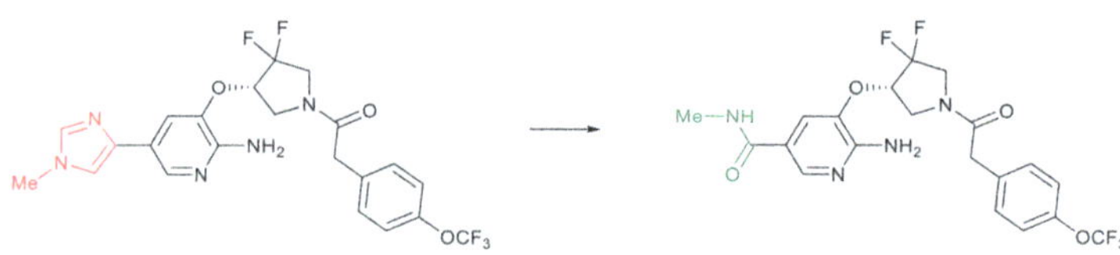 Left: pIC_{50} = 5.0, logD = 2.8 Right: pIC_{50} = 4.1, logD = 2.5	262

19.2.3.9 Replacement of Phenyl Rings with Non-aromatic Systems

Examples		Reference
1	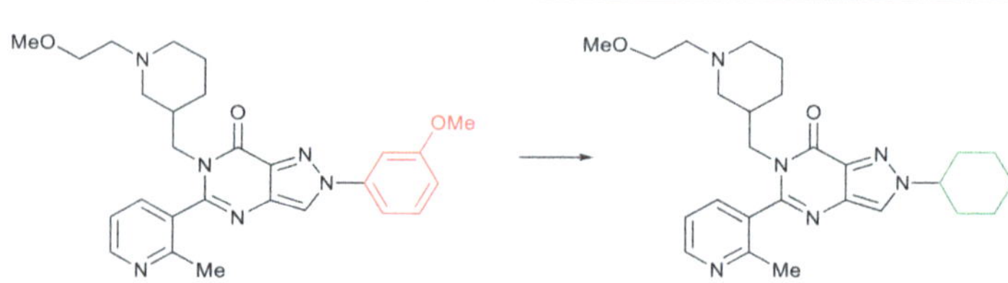 Left: pIC_{50} = 4.4, logD = 1.3 Right: pIC_{50} < 4.0, logD = 1.4	202

Examples		Reference
2	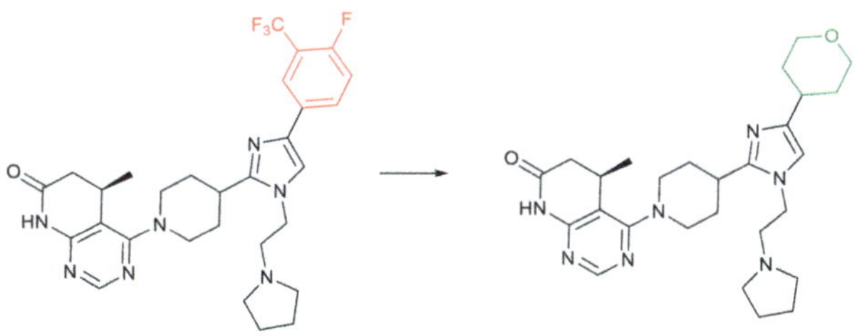 pIC_{50} = 5.8, clogP = 4.8 → pIC_{50} < 4.0, clogP = 1.0	263
3	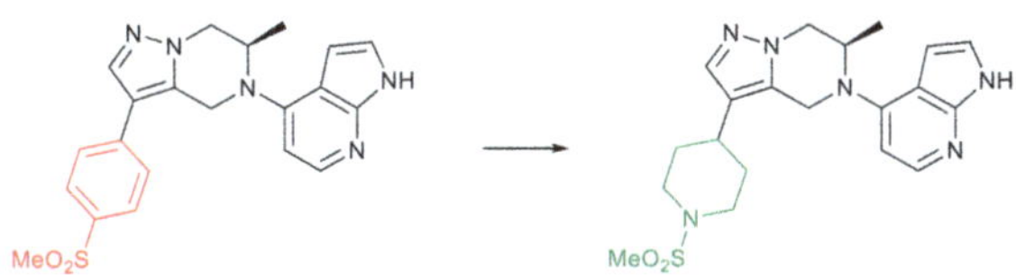 pIC_{50} = 5.8, logD = 2.3 → pIC_{50} < 4.5, logD = 2.4	264
4	 pIC_{50} = 5.9, LogD = 5.0 → pIC_{50} = 4.6, LogD = 3.1	265
5	 pIC_{50} = 5.1, clogP = 5.3 → pIC_{50} = 4.1, clogP = 4.3	266
6	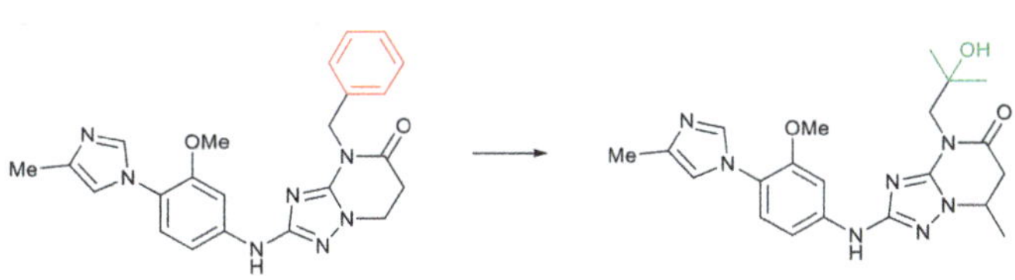 pIC_{50} = 5.2, logD = 2.7 → pIC_{50} = 4.5, logD = 1.8	214

(continued)

Examples		Reference
7	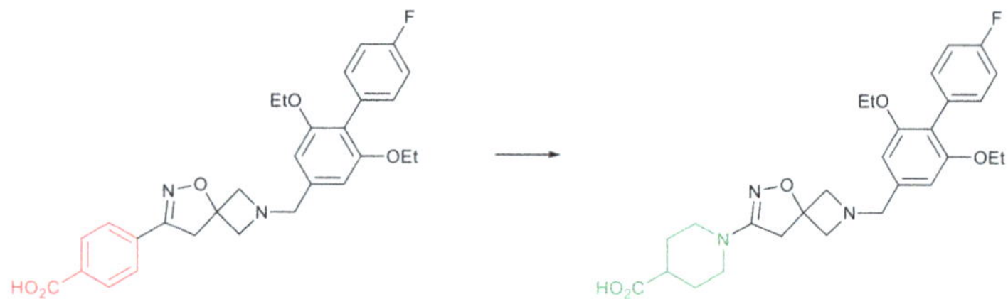 65% inhibition at 30 μM logD = 3.3 → 5.6% inhibition at 30 μM logD = 3.0	267
8	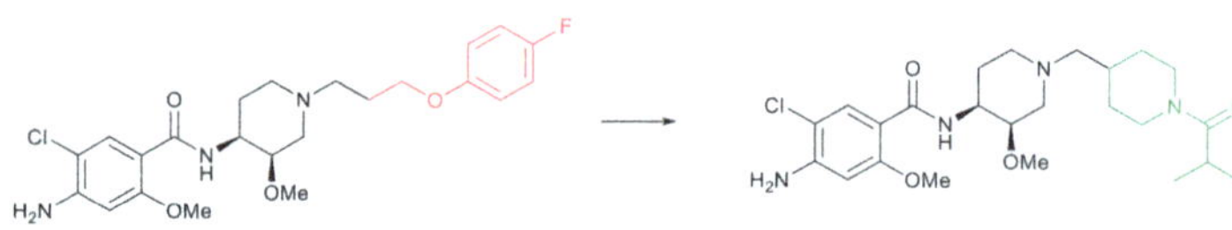 $pIC_{50} > 6.0$ clogP = 3.8 → $pIC_{50} < 5.0$ clogP = 1.3	268

19.2.3.10 *Aliphatic Carbon Substitution*

Examples		Reference
1	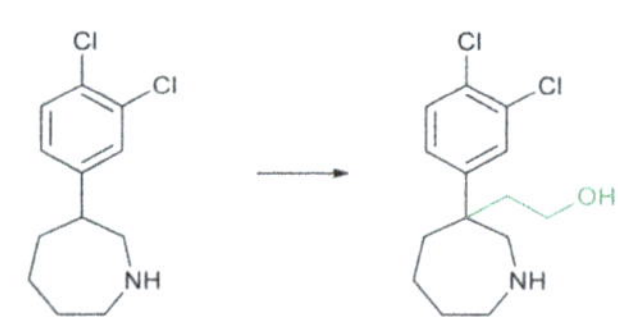 $pIC_{50} = 5.2$ clogP = 4.4 → $pIC_{50} < 4.5$ clogP = 3.4	269
2	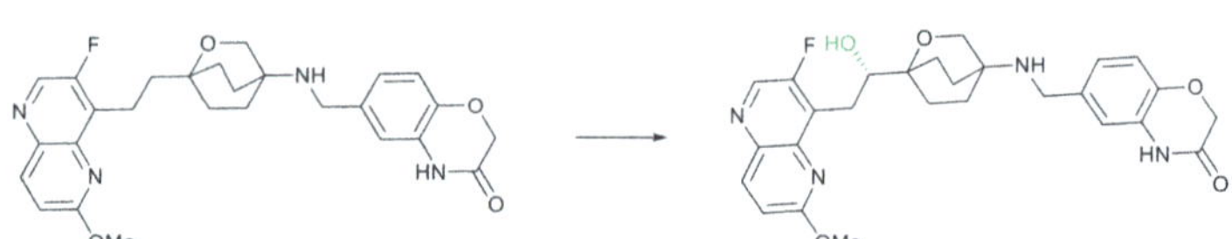 $pIC_{50} = 6.2$ clogP = 3.5 → $pIC_{50} = 4.7$ clogP = 2.2	270

Examples		Reference
3	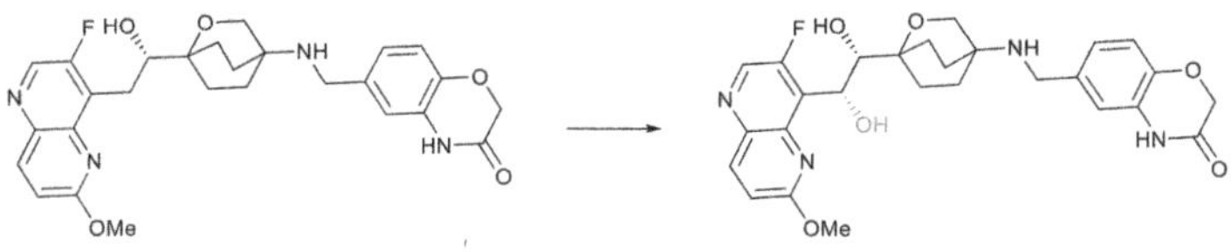 pIC_{50} = 4.7; clogP = 1.9 → pIC_{50} = 4.2; clogP = 0.7	271
4	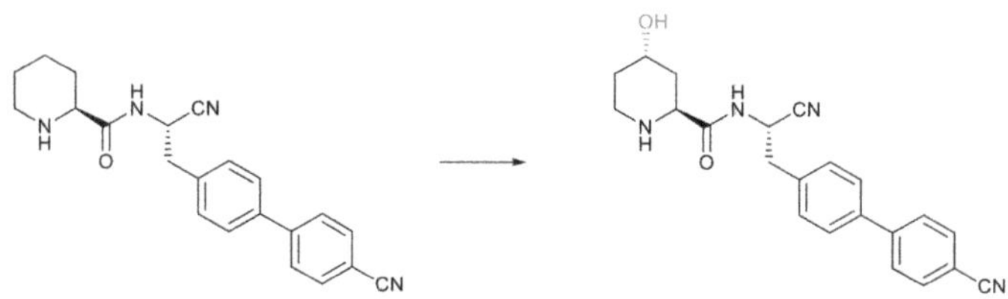 pIC_{50} = 5.9; logD = 2.7; pK_a = 8.4 → pIC_{50} < 4.5; logD = 1.4; pK_a = 7.6	224
5	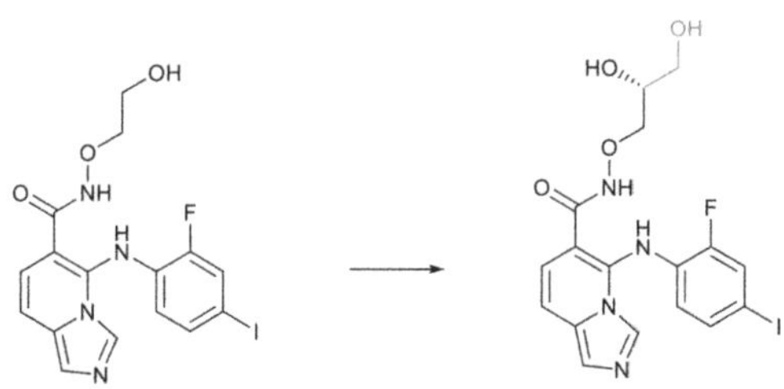 pIC_{50} = 5.4; logD = 2.1 → pIC_{50} = 4.9; logD = 1.1	272
6	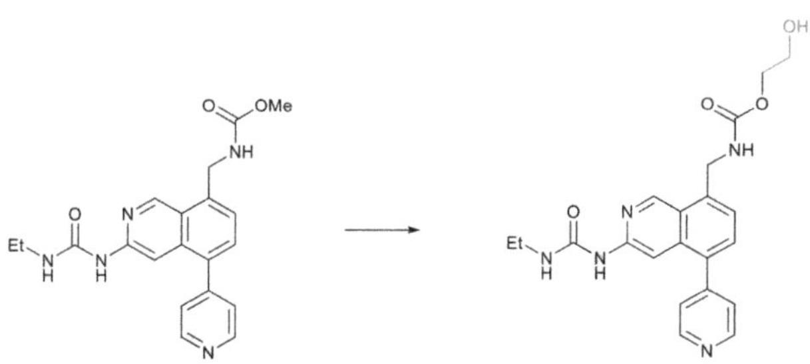 pIC_{50} = 41% inhibition at 10 μM; logD = 2.7 → pIC_{50} = 0% inhibition at 10 μM; logD = 2.0	273

(*continued*)

Examples				Reference
7	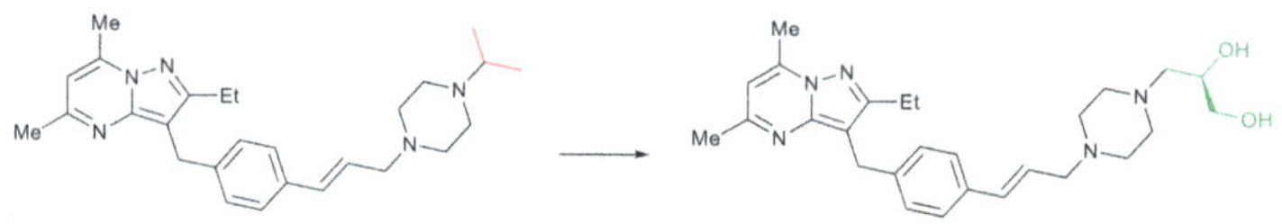	pIC_{50} = 5.7 clogP = 4.4	pIC_{50} < 4.5 clogP = 2.8	203
8	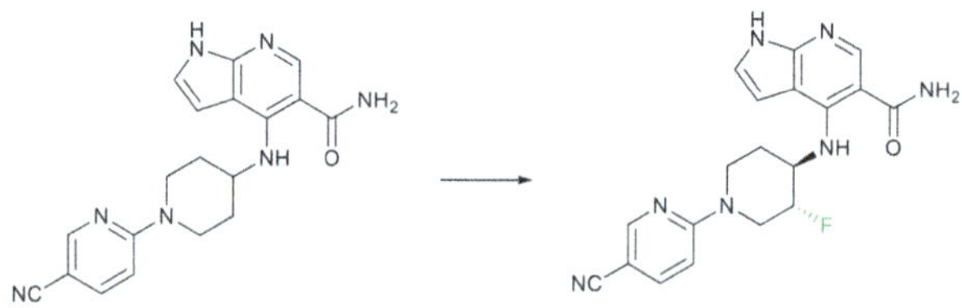	pIC_{50} = 4.9 clogP = 2.8	pIC_{50} < 4.0 clogP = 2.7	255
9	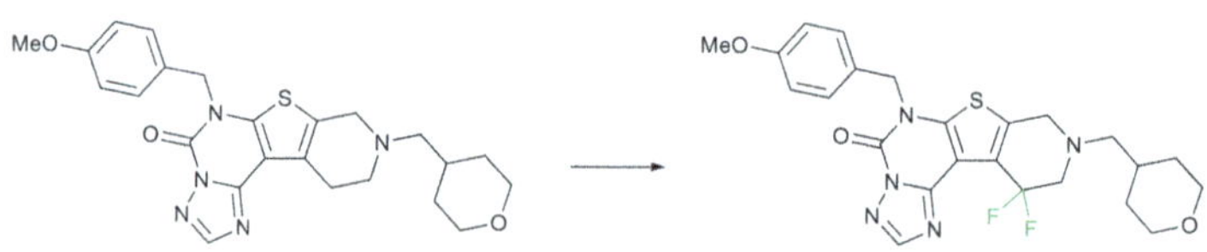	pIC_{50} = 5.7 clogP = 1.3	pIC_{50} = 5.0 clogP = 2.0	274
10		pIC_{50} = 6.4 clogP = 3.9 pK_a = 6.8	pIC_{50} = 4.6 clogP = 3.8 pK_a = 4.4	275
11	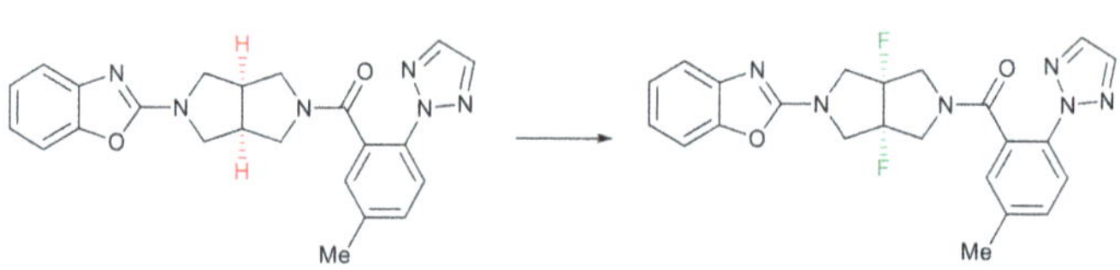	pIC_{50} = 5.3 clogP = 3.4	pIC_{50} < 4.5 clogP = 4.2	276

Examples — **Reference**

12 — 248

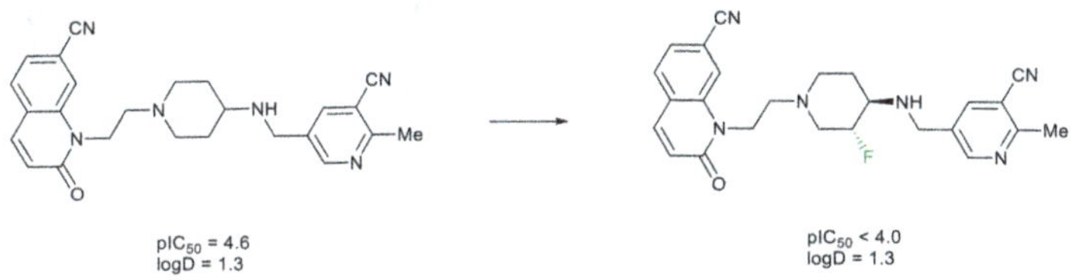

13 — 277

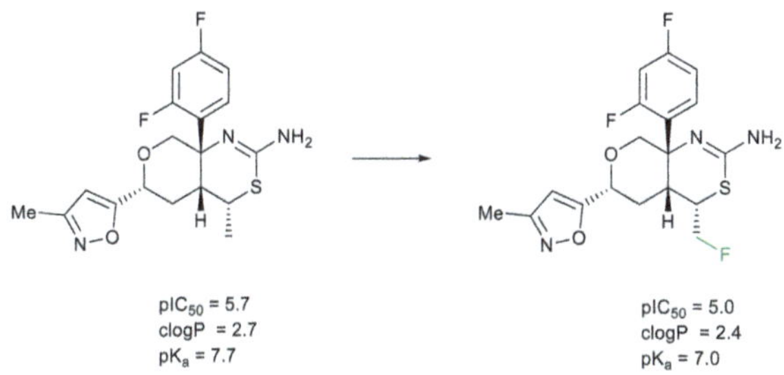

14 — 278

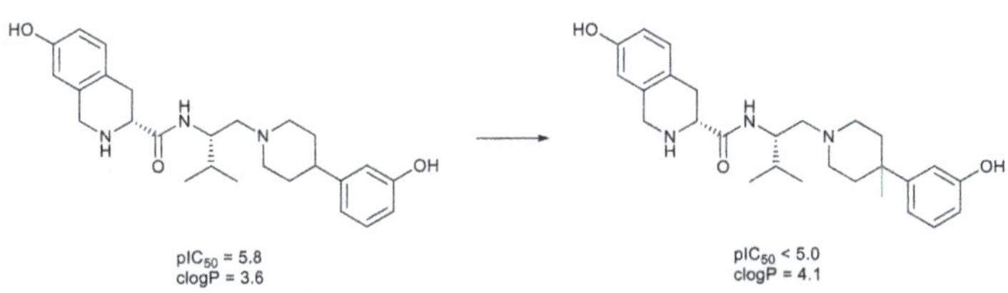

15 — 279

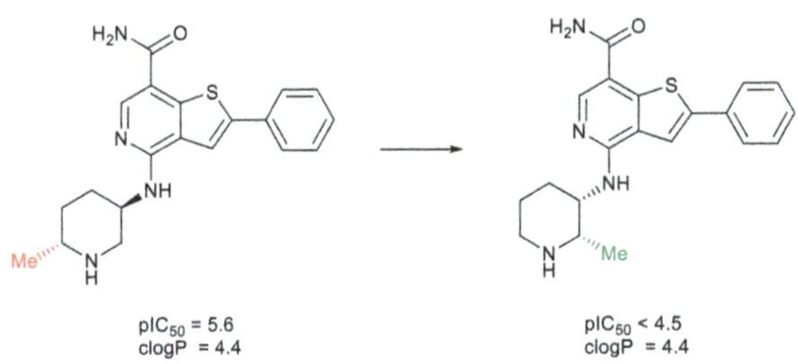

(continued)

Examples		Reference
16	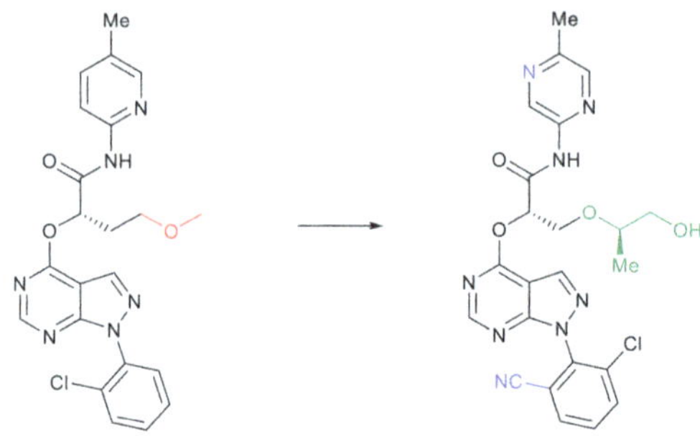 pIC_{50} = 4.7 / logD = 3.6 → pIC_{50} < 4.0 / logD = 1.9	280 and 281
17	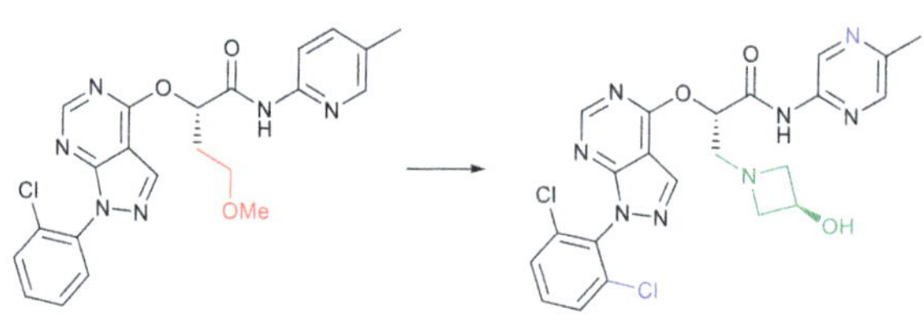 pIC_{50} = 4.7 / logD = 3.6 → pIC_{50} = 4.2 / logD = 2.3	280 and 281
18	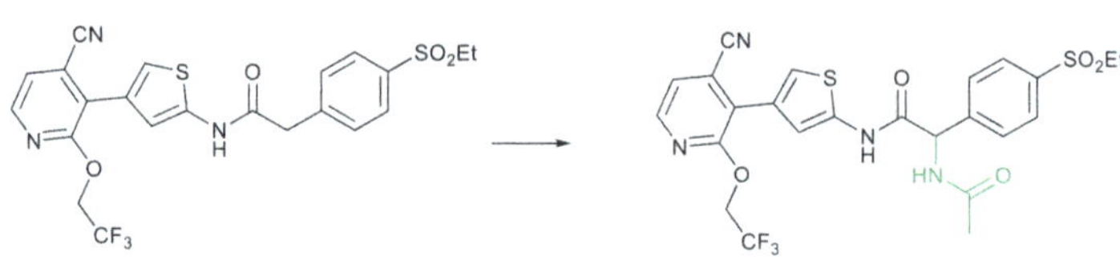 47% inhibition at 11 μM / clogP = 2.5 → 3% inhibition at 11 μM / clogP = 1.7	282
19	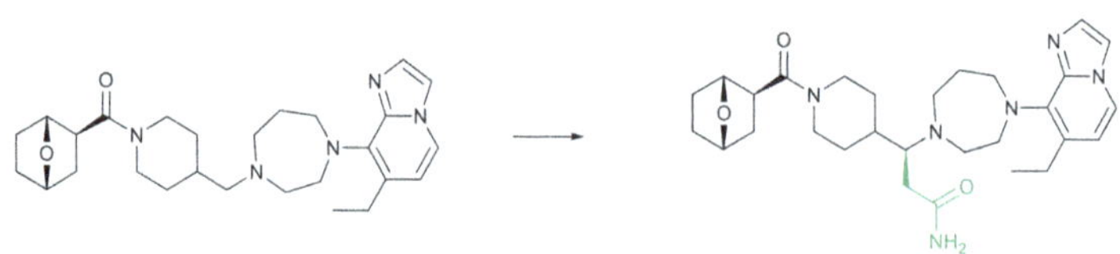 pIC_{50} = 5.9 / LogD = 0.8 → pIC_{50} = 5.1 / LogD = 1.1	283

Examples		Reference
20	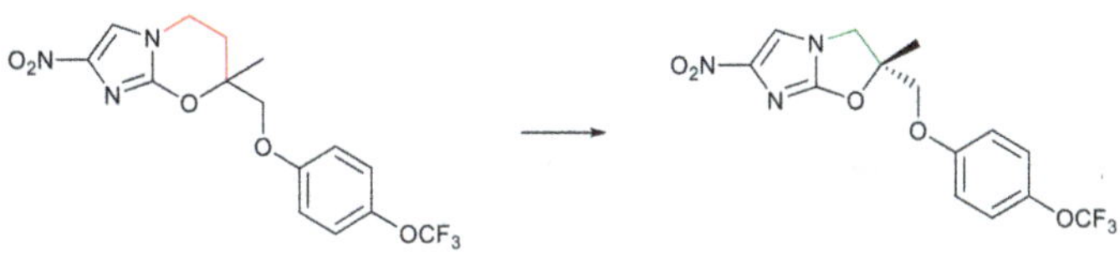 pIC_{50} = 5.4, clogP = 3.8 → pIC_{50} = 4.9, clogP = 3.5	284
21	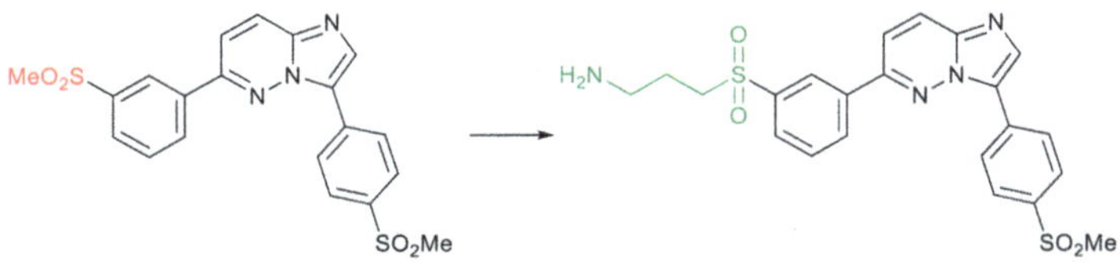 pIC_{50} = 6.1, clogP = 1.5 → pIC_{50} < 4.5, clogP = 1.4	232
22	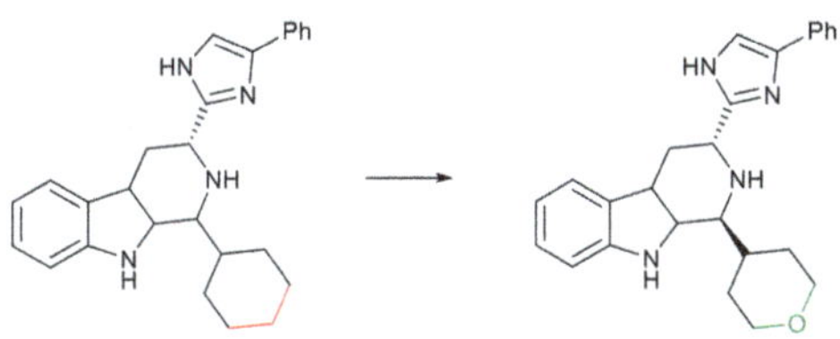 pIC_{50} = 7.1, logD = 4.7 → pIC_{50} = 6.1, logD = 3.0	285
23	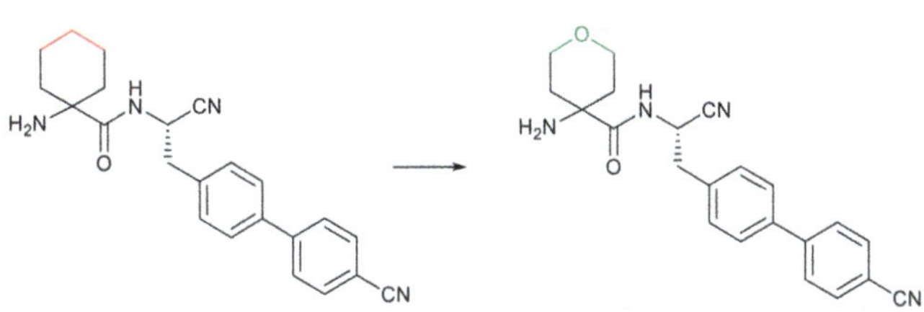 pIC_{50} = 5.0, logD = 2.9, pK_a = 7.1 → pIC_{50} = 4.6, logD = 1.9, pK_a = 5.8	224

(*continued*)

Examples		Reference
24	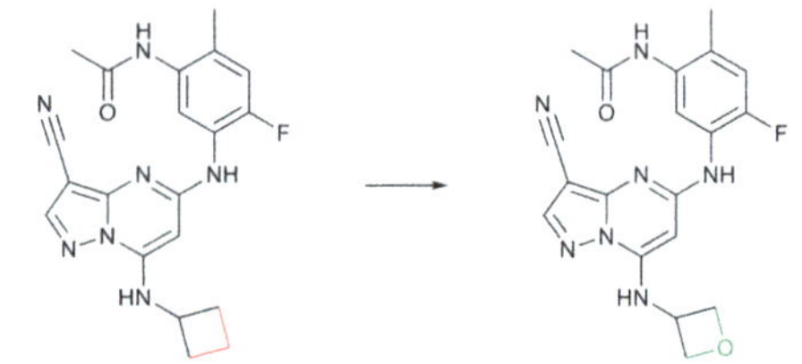 pIC_{50} = 4.8, logD = 3.0 → pIC_{50} < 4.5, logD = 1.5	286
25	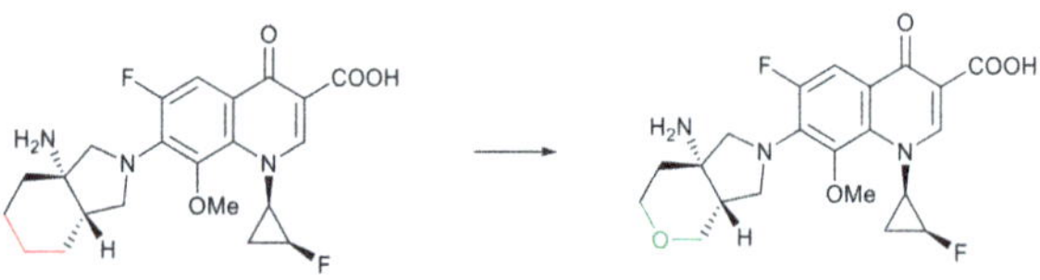 36% inhibition @ 100 μM, clogP = 0.9 → 0% inhibition @ 100 μM, clogP = -0.6	287
26	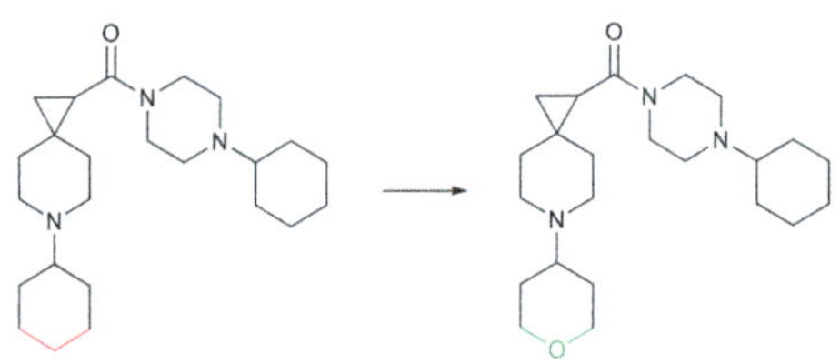 pIC_{50} = 4.7, clogP = 3.9 → pIC_{50} < 4.5, clogP = 1.5	288
27	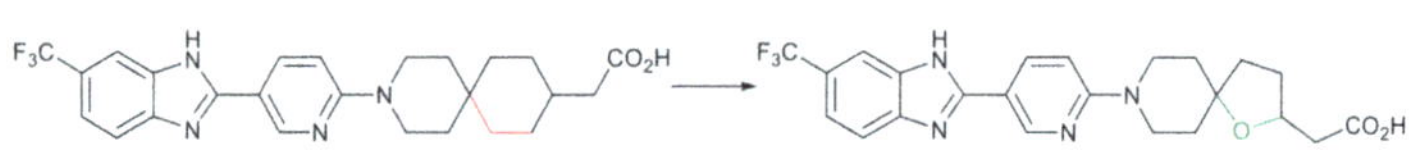 pIC_{50} = 5.3, clogP = 5.8 → pIC_{50} < 4.2, clogP = 3.6	289
28	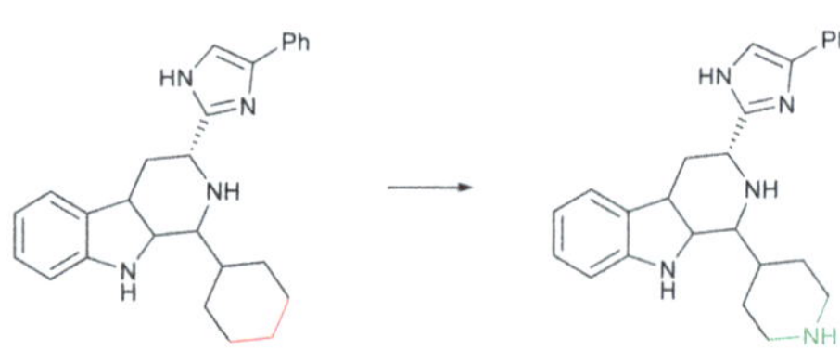 pIC_{50} = 7.1, logD = 4.7 → pIC_{50} = 5.3, logD = 2.8	285

Examples		Reference
29	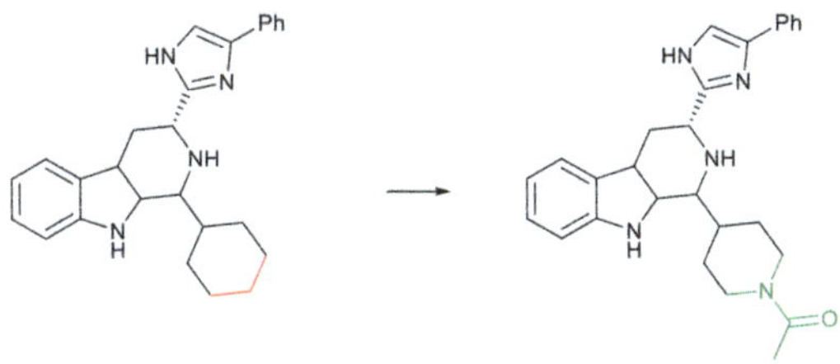 pIC_{50} = 7.1, logD = 4.7 → pIC_{50} = 5.0, logD = 2.5	285
30	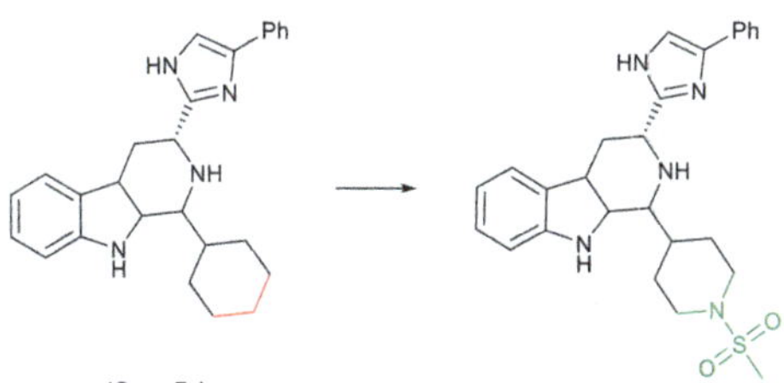 pIC_{50} = 7.1, logD = 4.7 → pIC_{50} = 5.5, logD = 2.1	285
31	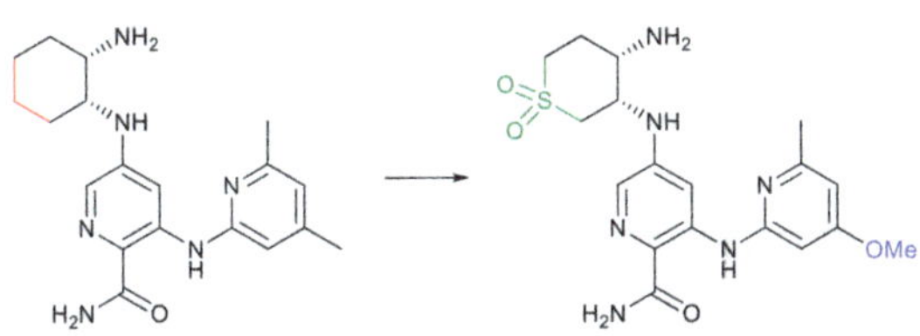 pIC_{50} = 5.8, log D = 2.1 → pIC_{50} < 4.5, log D = 1.5	290
32	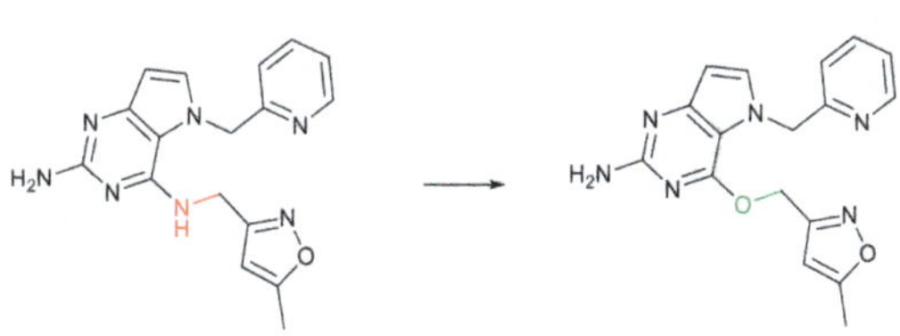 pIC_{50} = 5.9, logD = 1.1 → pIC_{50} < 4.3, logD = 1.1	291

(*continued*)

Examples		Reference
33	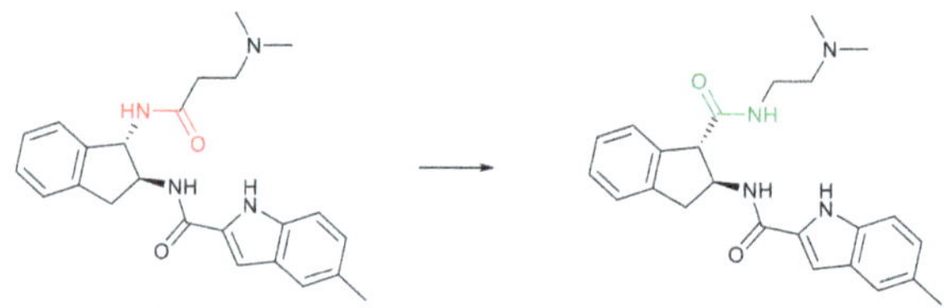 pIC_{50} = 5.1 clogP = 3.7 → pIC_{50} = 4.5 clogP = 3.6	292
34	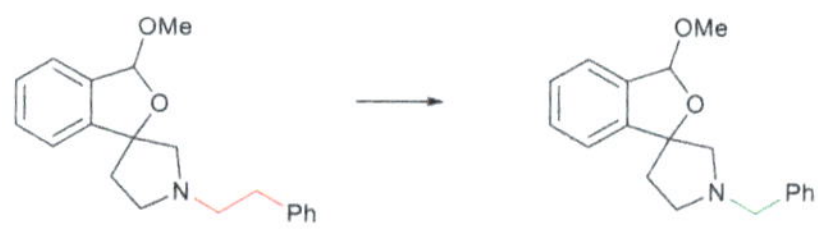 77% inhibition at 10 μM clogP = 3.2 → 7% inhibition at 10 μM clogP = 2.9	293
35	 pIC_{50} = 5.0 logD = 3.1 → pIC_{50} < 4.5 logD = 2.6	294
36	 pIC_{50} = 6.7 clogP = 3.0 → pIC_{50} = 4.9 clogP = 3.0	295
37	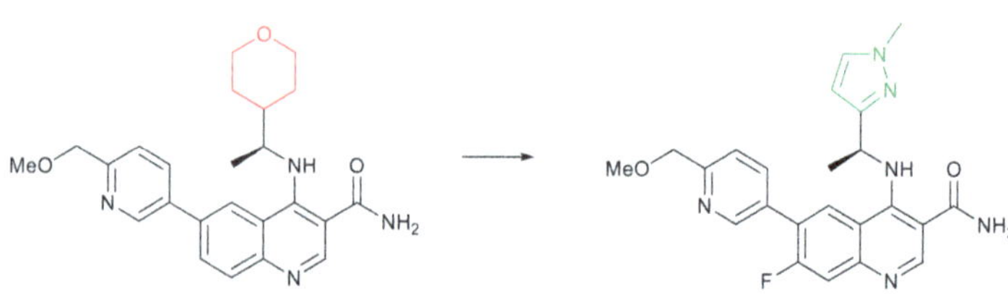 pIC_{50} = 5.3 logD = 2.5 → pIC_{50} = 4.7 logD = 2.7	227

19.2.3.11 Scaffold Changes

Examples		Reference
1	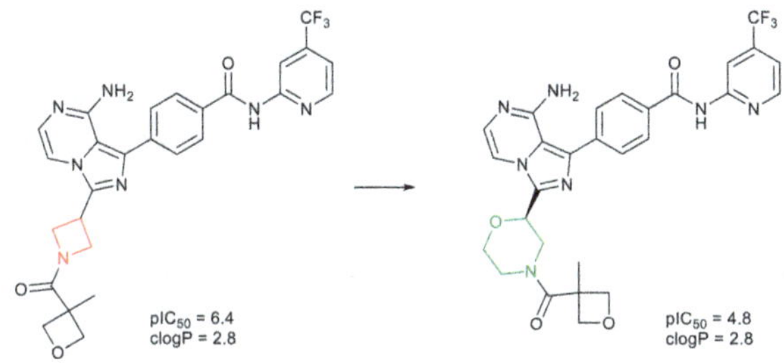	296
2	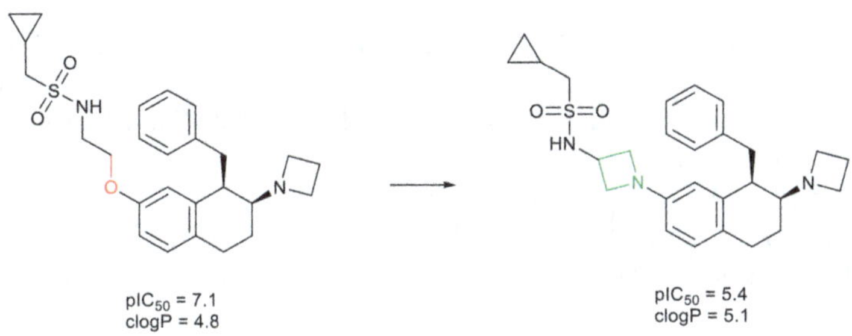	297
3	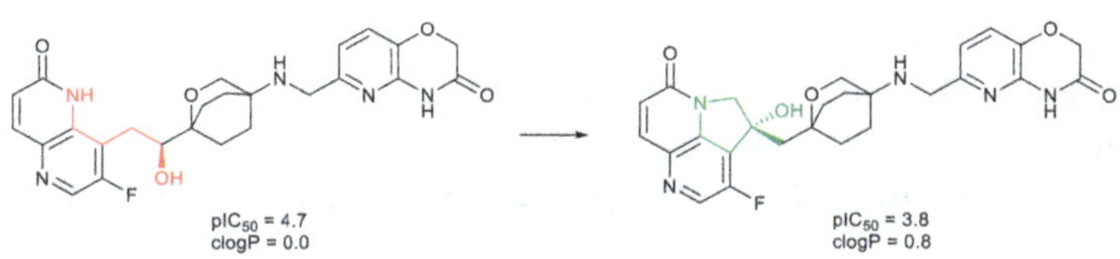	271
4	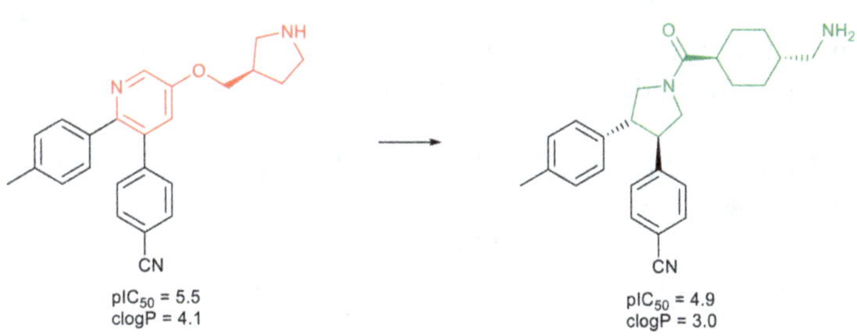	298

(*continued*)

Examples		Reference
5	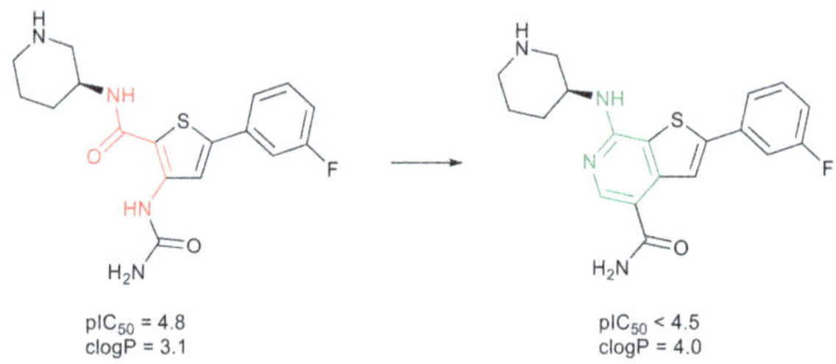	279

References

1. M. J. Cross, B. R. Berridge, P. J. M. Clements, L. Cove-Smith, T. L. Force, P. Hoffmann, M. Holbrook, A. R. Lyon, H. R. Mellor, A. A. Norris, M. Pirmohamed, J. D. Tugwood, J. E. Sidaway and B. K. Park, *Br. J. Pharmacol.*, 2015, **172**, 957–974.
2. C. Jesse and N. Anju, *Curr. Oncol. Rep.*, 2018, **20**, 61.
3. J. Tamargo, R. Caballero and E. Delpon, *Drug Saf.*, 2015, **38**, 129–152.
4. C. E. Hamo and M. W. Bloom, *Clin. Med. Insights: Cardiol.*, 2015, **9**, 47–51.
5. J. Alexandre, J. J. Molsehi, K. R. Bersell, C. Funck-Brentano, D. M. Roden and J. E. Salem, *Pharmacol. Ther.*, 2018, **189**, 89–103.
6. B. Yang and T. Papoian, *J. Appl. Toxicol.*, 2018, **38**, 790–800.
7. T. Papoian and B. Yang, *J. Appl. Toxicol.*, 2012, **32**, 945–951.
8. G. Hanton, *Drugs R&D*, 2007, **8**, 213–228.
9. C. Antzelevitch and A. Burashnikov, *Card. Electrophysiol. Clin.*, 2011, **3**, 23–45.
10. R. R. Fenichel, M. Malik, C. Antzelevitch, M. Sanguinetti, D. M. Roden, S. G. Priori, J. N. Ruskin, R. J. Lipicky, L. R. Cantilena and Independent Academic Task Force, *J. Cardiovasc. Electrophysiol.*, 2004, **15**, 475–495.
11. A. J. Davies, V. Harindra, A. McEwan and R. R. Ghose, *BMJ*, 1989, **298**, 325.
12. D. R. Mathews, B. McNutt, R. Okerholm, M. Flicker and G. McBride, *JAMA*, 1991, **266**, 2375–2376.
13. R. L. Woosley, *JAMA*, 1993, **269**, 1532.
14. M.-L. Roy, R. Dumaine and A. M. Brown, *Circulation*, 1996, **94**, 817–823.
15. Y. G. Yap and A. J. Camm, *Heart*, 2003, **89**, 1363–1372.
16. H. B. Yu, B. Y. Zou, X. L. Wang and M. Li, *Acta Pharmacol. Sin.*, 2016, **37**, 111–123.
17. H. Zhang, B. Zou, F. Du, K. Xu and M. Li, *Mol. Pharmacol.*, 2014, **87**, 207–217.
18. A. O. Grant, *Circ.: Arrhythmia Electrophysiol.*, 2009, **2**, 185–195.
19. C. Arrigoni and P. Crivori, *Cell Biol. Toxicol.*, 2007, **23**, 1–13.

20. B. Dumotier and M. Traebert, in *Antitargets and Drug Safety*, ed. L. Urbán, V. F. Patel and R. J. Vaz, Wiley-VCH Verlag GmbH & Co. KGaA, 2015.
21. P. Liao and T. W. Soong, *Pflügers Arch.*, 2010, **460**, 353–359.
22. M. B. Rook, M. M. Evers, M. A. Vos and M. F. Bierhuizen, *Cardiovasc. Res.*, 2012, **93**, 12–23.
23. G. A. Gintant, in *Antitargets and Drug Safety*, ed. L. P. Urbán, V. F. Patel and R. J. Vaz, Wiley-VCH Verlag GmbH & Co. KGaA, 2015, pp. 253–277.
24. E. Savio-Galimberti, M. Argenziano and C. Antzelevitch, *Handb. Exp. Pharmacol.*, 2018, **246**, 331–354.
25. M. Liu, K. C. Yang and S. C. Dudley Jr, *Curr. Top. Membr.*, 2016, **78**, 513–559.
26. Y. Ruan, N. Liu and S. G. Priori, *Nat. Rev. Cardiol.*, 2009, **6**, 337–348.
27. D. J. Leishman and Z. Rankovic, in *Tactics in Contemporary Drug Design*, ed. N. A. Meanwell, Springer Berlin Heidelberg, Berlin, Heidelberg, 2015, pp. 225–259.
28. P. L. Hedley, P. Jorgensen, S. Schlamowitz, R. Wangari, J. Moolman-Smook, P. A. Brink, J. K. Kanters, V. A. Corfield and M. Christiansen, *Hum. Mutat.*, 2009, **30**, 1486–1511.
29. C. Starmer and A. O. Grant, *Mol. Pharmacol.*, 1985, **28**, 348–356.
30. C. F. Starmer, *Biophys. J.*, 1987, **52**, 405–412.
31. D. Stork, E. N. Timin, S. Berjukow, C. Huber, A. Hohaus, M. Auer and S. Hering, *Br. J. Pharmacol.*, 2007, **151**, 1368–1376.
32. Anonymous, 2005, https://www.ich.org/fileadmin/Public_Web_Site/ICH_Products/Guidelines/Safety/S7B/Step4/S7B_Guideline.pdf.
33. J. Ducroq, in *Principles of Safety Pharmacology*, ed. M. K. Pugsley and M. J. Curtis, Springer Berlin Heidelberg, Berlin, Heidelberg, 2015, pp. 205–219.
34. J. P. Valentin, P. Hoffmann, F. De Clerck, T. G. Hammond and L. Hondeghem, *J. Pharmacol. Toxicol. Methods*, 2004, **49**, 171–181.
35. X. Li, R. Zhang, B. Zhao, C. Lossin and Z. Cao, *Arch. Toxicol.*, 2016, **90**, 1803–1816.
36. X. P. Huang, T. Mangano, S. Hufeisen, V. Setola and B. L. Roth, *Assay Drug Dev. Technol.*, 2010, **8**, 727–742.
37. B. Fermini, D. S. Ramirez, S. Sun, A. Bassyouni, M. Hemkens, T. Wisialowski and S. Jenkinson, *J. Pharmacol. Toxicol. Methods*, 2017, **84**, 86–92.
38. S. Jenkinson, A. Bassyouni, J. Cordes, B. Fermini, D. Guo, D. M. Potter, D. S. Ramirez, J. Steidl-Nichols, S. Sun and T. Wisialowski, *J. Pharmacol. Toxicol. Methods*, 2018, **89**, 9–18.
39. G. A. Gintant, D. J. Gallacher and M. K. Pugsley, *Br. J. Pharmacol.*, 2011, **164**, 254–259.
40. J. T. Milnes, H. J. Witchel, J. L. Leaney, D. J. Leishman and J. C. Hancox, *J. Pharmacol. Toxicol. Methods*, 2010, **61**, 178–191.
41. G. E. Kirsch, E. S. Trepakova, J. C. Brimecombe, S. S. Sidach, H. D. Erickson, M. C. Kochan, L. M. Shyjka, A. E. Lacerda and A. M. Brown, *J. Pharmacol. Toxicol. Methods*, 2004, **50**, 93–101.

42. Z. Yu, E. Klaasse, L. H. Heitman and A. P. Ijzerman, *Toxicol. Appl. Pharmacol.*, 2014, **274**, 78–86.
43. G. Y. Di Veroli, M. R. Davies, H. Zhang, N. Abi-Gerges and M. R. Boyett, *J. Cardiovasc. Electrophysiol.*, 2014, **25**, 197–207.
44. R. A. Pearlstein, K. Andrew MacCannell, Q. Y. Hu, R. Farid and J. S. Duca, in *Antitargets and Drug Safety*, ed. L. P. Urbán, V. Patel and R. J. Vaz, Wiley-VCH Verlag GmbH & Co. KGaA, 2015, pp. 295–328.
45. A. W. Nathan, K. J. Hellestrand, R. S. Bexton, R. A. Spurrell and A. J. Camm, *Drugs*, 1985, **29**(Suppl 4), 45–53.
46. J. Sallstrom, A. Al-Saffar and R. Pehrson, *J. Pharmacol. Toxicol. Methods*, 2014, **69**, 24–29.
47. C. Davie, J. Pierre-Valentin, C. Pollard, N. Standen, J. Mitcheson, P. Alexander and B. Thong, *J. Cardiovasc. Electrophysiol.*, 2004, **15**, 1302–1309.
48. G. Gintant, B. Fermini, N. Stockbridge and D. Strauss, *Cell Stem Cell*, 2017, **21**, 14–17.
49. O. P. Hamill, A. Marty, E. Neher, B. Sakmann and F. J. Sigworth, *Pflügers Arch.*, 1981, **391**, 85–100.
50. A. Obergrussberger, T. A. Goetze, N. Brinkwirth, N. Becker, S. Friis, M. Rapedius, C. Haarmann, I. Rinke-Weiss, S. Stolzle-Feix, A. Bruggemann, M. George and N. Fertig, *Expert Opin. Drug Discovery*, 2018, **13**, 269–277.
51. G. Y. Di Veroli, M. R. Davies, H. Zhang, N. Abi-Gerges and M. R. Boyett, *Am. J. Physiol.*, 2013, **304**, H104–H117.
52. T. Haisa, K. Matsumiya, N. Yoshimasu and N. Kuribayashi, *J. Neurosurg.*, 1990, **72**, 292–294.
53. M. J. Windley, N. Abi-Gerges, B. Fermini, J. C. Hancox, J. I. Vandenberg and A. P. Hill, *J. Pharmacol. Toxicol. Methods*, 2017, **87**, 99–107.
54. S. Houtmann, B. Schombert, C. Sanson, M. Partiseti and G. A. Bohme, *Methods Mol. Biol.*, 2017, **1641**, 187–199.
55. B. Balasubramanian, J. P. Imredy, D. Kim, J. Penniman, A. Lagrutta and J. J. Salata, *J. Pharmacol. Toxicol. Methods*, 2009, **59**, 62–72.
56. T. Li, G. Lu, E. Y. Chiang, T. Chernov-Rogan, J. L. Grogan and J. Chen, *PLoS One*, 2017, **12**, e0180154.
57. W. S. Redfern, L. Carlsson, A. S. Davis, W. G. Lynch, I. MacKenzie, S. Palethorpe, P. K. Siegl, I. Strang, A. T. Sullivan, R. Wallis, A. J. Camm and T. G. Hammond, *Cardiovasc. Res.*, 2003, **58**, 32–45.
58. D. M. Jonker, L. A. Kenna, D. Leishman, R. Wallis, P. A. Milligan and E. N. Jonsson, *Clin. Pharmacol. Ther.*, 2005, **77**, 572–582.
59. A. R. Harmer, J. P. Valentin and C. E. Pollard, *Br. J. Pharmacol.*, 2011, **164**, 260–273.
60. B. M. Heath, Y. Cui, S. Worton, B. Lawton, G. Ward, E. Ballini, C. P. Doe, C. Ellis, B. A. Patel and N. C. McMahon, *J. Pharmacol. Toxicol. Methods*, 2011, **63**, 258–268.
61. J. Cordes, C. Li, J. Dugas, R. Austin-LaFrance, I. Lightbown, M. Engwall, M. Sutton and J. Steidl-Nichols, *J. Pharmacol. Toxicol. Methods*, 2009, **60**, 221.

62. R. L. Martin, J. S. McDermott, H. J. Salmen, J. Palmatier, B. F. Cox and G. A. Gintant, *J. Cardiovasc. Pharmacol.*, 2004, **43**, 369–379.
63. S. Zhang, Z. Zhou, Q. Gong, J. C. Makielski and C. T. January, *Circ. Res.*, 1999, **84**, 989–998.
64. Y. Jiang, P. Park, S. M. Hong and K. Ban, *Mol. Cells*, 2018, **41**, 613–621.
65. R. J. Goodrow Jr, S. Desai, J. A. Treat, B. K. Panama, M. Desai, V. V. Nesterenko and J. M. Cordeiro, *J. Pharmacol. Toxicol. Methods*, 2018, **90**, 19–30.
66. P. T. Sager, G. Gintant and N. Stockbridge, *Nat. Rev. Drug Discovery*, 2016, **15**, 457–471.
67. B. Fermini, J. C. Hancox, N. Abi-Gerges, M. Bridgland-Taylor, K. W. Chaudhary, T. Colatsky, K. Correll, W. Crumb, B. Damiano, G. Erdemli, G. Gintant, J. Imredy, J. Koerner, J. Kramer, P. Levesque, Z. Li, A. Lindqvist, C. A. Obejero-Paz, D. Rampe, K. Sawada, D. G. Strauss and J. I. Vandenberg, *J. Biomol. Screening*, 2016, **21**, 1–11.
68. Z. Li, S. Dutta, J. Sheng, P. N. Tran, W. Wu, K. Chang, T. Mdluli, D. G. Strauss and T. Colatsky, *Circ.: Arrhythmia Electrophysiol.*, 2017, **10**, e004628.
69. J. Vicente, R. Zusterzeel, L. Johannesen, J. Mason, P. Sager, V. Patel, M. K. Matta, Z. Li, J. Liu, C. Garnett, N. Stockbridge, I. Zineh and D. G. Strauss, *Clin. Pharmacol. Ther.*, 2018, **103**, 54–66.
70. J. Vicente, R. Zusterzeel, L. Johannesen, R. Ochoa-Jimenez, J. W. Mason, C. Sanabria, S. Kemp, P. T. Sager, V. Patel, M. K. Matta, J. Liu, J. Florian, C. Garnett, N. Stockbridge and D. G. Strauss, *Clin. Pharmacol. Ther.*, 2019, **105**, 943–953.
71. D. G. Strauss, G. Gintant, Z. Li, W. Wu, K. Blinova, J. Vicente, J. R. Turner and P. T. Sager, *Ther. Innov. Regul. Sci.*, 2018, **53**, 519–525.
72. R. Wallis, C. Benson, B. Darpo, G. Gintant, Y. Kanda, K. Prasad, D. G. Strauss and J. P. Valentin, *J. Pharmacol. Toxicol. Methods*, 2018, **93**, 15–25.
73. W. A. Catterall, A. L. Goldin and S. G. Waxman, *Pharmacol. Rev.*, 2005, **57**, 397–409.
74. X. Pan, Z. Li, Q. Zhou, H. Shen, K. Wu, X. Huang, J. Chen, J. Zhang, X. Zhu, J. Lei, W. Xiong, H. Gong, B. Xiao and N. Yan, *Science*, 2018, **362**, eaau2486.
75. J. Payandeh, T. Scheuer, N. Zheng and W. A. Catterall, *Nature*, 2011, **475**, 353–358.
76. C. Bagneris, C. E. Naylor, E. C. McCusker and B. A. Wallace, *J. Gen. Physiol.*, 2015, **145**, 5–16.
77. M. B. Ulmschneider, C. Bagneris, E. C. McCusker, P. G. Decaen, M. Delling, D. E. Clapham, J. P. Ulmschneider and B. A. Wallace, *Proc. Natl. Acad. Sci. U. S. A.*, 2013, **110**, 6364–6369.
78. C. Bagneris, P. G. DeCaen, C. E. Naylor, D. C. Pryde, I. Nobeli, D. E. Clapham and B. A. Wallace, *Proc. Natl. Acad. Sci. U. S. A.*, 2014, **111**, 8428–8433.

79. M. Ahmed, H. Jalily Hasani, A. Ganesan, M. Houghton and K. Barakat, *Drug Des., Dev. Ther.*, 2017, **11**, 2301–2324.
80. J. A. Zablocki, E. Elzein, X. Li, D. O. Koltun, E. Q. Parkhill, T. Kobayashi, R. Martinez, B. Corkey, H. Jiang, T. Perry, R. Kalla, G. T. Notte, O. Saunders, M. Graupe, Y. Lu, C. Venkataramani, J. Guerrero, J. Perry, M. Osier, R. Strickley, G. Liu, W. Q. Wang, L. Hu, X. J. Li, N. El-Bizri, R. Hirakawa, K. Kahlig, C. Xie, C. H. Li, A. K. Dhalla, S. Rajamani, N. Mollova, D. Soohoo, E. I. Lepist, B. Murray, G. Rhodes, L. Belardinelli and M. C. Desai, *J. Med. Chem.*, 2016, **59**, 9005–9017.
81. L. Wang, X. Meng, Z. Yuchi, Z. Zhao, D. Xu, D. Fedida, Z. Wang and C. Huang, *Cell. Physiol. Biochem.*, 2015, **36**, 2250–2262.
82. H. Poulin, I. Bruhova, Q. Timour, O. Theriault, J. M. Beaulieu, D. Frassati and M. Chahine, *Mol. Pharmacol.*, 2014, **86**, 378–389.
83. J. Payandeh, T. M. Gamal El-Din, T. Scheuer, N. Zheng and W. A. Catterall, *Nature*, 2012, **486**, 135–139.
84. M. J. Lenaeus, T. M. Gamal El-Din, C. Ing, K. Ramanadane, R. Pomes, N. Zheng and W. A. Catterall, *Proc. Natl. Acad. Sci. U. S. A.*, 2017, **114**, E3051–E3060.
85. K. Irie, Y. Haga, T. Shimomura and Y. Fujiyoshi, *FEBS Lett.*, 2018, **592**, 274–283.
86. E. C. McCusker, C. Bagneris, C. E. Naylor, A. R. Cole, N. D'Avanzo, C. G. Nichols and B. A. Wallace, *Nat. Commun.*, 2012, **3**, 1102.
87. C. Bagneris, P. G. Decaen, B. A. Hall, C. E. Naylor, D. E. Clapham, C. W. Kay and B. A. Wallace, *Nat. Commun.*, 2013, **4**, 2465.
88. X. Ji, Y. Xiao and S. Liu, *J. Biomol. Struct. Dyn.*, 2018, **36**, 2268–2278.
89. X. Zhang, W. Ren, P. DeCaen, C. Yan, X. Tao, L. Tang, J. Wang, K. Hasegawa, T. Kumasaka, J. He, J. Wang, D. E. Clapham and N. Yan, *Nature*, 2012, **486**, 130–134.
90. H. Shen, Q. Zhou, X. Pan, Z. Li, J. Wu and N. Yan, *Science*, 2017, **355**, eaal4326.
91. H. Shen, Z. Li, Y. Jiang, X. Pan, J. Wu, B. Cristofori-Armstrong, J. J. Smith, Y. K. Y. Chin, J. Lei, Q. Zhou, G. F. King and N. Yan, *Science*, 2018, **362**, eaau2596.
92. Z. Yan, Q. Zhou, L. Wang, J. Wu, Y. Zhao, G. Huang, W. Peng, H. Shen, J. Lei and N. Yan, *Cell*, 2017, **170**, 470–482 e411.
93. G. M. Lipkind and H. A. Fozzard, *Mol. Pharmacol.*, 2005, **68**, 1611–1622.
94. D. B. Tikhonov and B. S. Zhorov, *J. Gen. Physiol.*, 2017, **149**, 465–481.
95. D. B. Tikhonov and B. S. Zhorov, *Mol. Pharmacol.*, 2012, **82**, 97–104.
96. M. Stevens, S. Peigneur and J. Tytgat, *Front. Pharmacol.*, 2011, **2**, 71.
97. R. Chen and S. H. Chung, *Biochem. Biophys. Res. Commun.*, 2014, **446**, 370–374.
98. D. S. Ragsdale, J. C. McPhee, T. Scheuer and W. A. Catterall, *Proc. Natl. Acad. Sci. U. S. A.*, 1996, **93**, 9270–9275.
99. C. C. Kuo, *Mol. Pharmacol.*, 1998, **54**, 712–721.
100. G. Liu, V. Yarov-Yarovoy, M. Nobbs, J. J. Clare, T. Scheuer and W. A. Catterall, *Neuropharmacology*, 2003, **44**, 413–422.

101. M. E. O'Leary and M. Chahine, *J. Physiol.*, 2002, **541**, 701–716.
102. A. O. O'Reilly, E. Eberhardt, C. Weidner, C. Alzheimer, B. A. Wallace and A. Lampert, *PLoS One*, 2012, **7**, e41667.
103. H. A. Fozzard, M. F. Sheets and D. A. Hanck, *Front. Pharmacol.*, 2011, **2**, 68.
104. S. Ahuja, S. Mukund, L. Deng, K. Khakh, E. Chang, H. Ho, S. Shriver, C. Young, S. Lin, J. P. Johnson Jr, P. Wu, J. Li, M. Coons, C. Tam, B. Brillantes, H. Sampang, K. Mortara, K. K. Bowman, K. R. Clark, A. Estevez, Z. Xie, H. Verschoof, M. Grimwood, C. Dehnhardt, J. C. Andrez, T. Focken, D. P. Sutherlin, B. S. Safina, M. A. Starovasnik, D. F. Ortwine, Y. Franke, C. J. Cohen, D. H. Hackos, C. M. Koth and J. Payandeh, *Science*, 2015, **350**, aac5464.
105. K. W. Chan, D. Tabuena, C. Xie, J. Shryock, L. Belardinelli and C. Smith-Maxwell, *Biophys. J.*, 2012, **102**, 324a.
106. D. S. Ragsdale, J. C. McPhee, T. Scheuer and W. A. Catterall, *Science*, 1994, **265**, 1724–1728.
107. V. Yarov-Yarovoy, J. C. McPhee, D. Idsvoog, C. Pate, T. Scheuer and W. A. Catterall, *J. Biol. Chem.*, 2002, **277**, 35393–35401.
108. I. Bruhova, D. B. Tikhonov and B. S. Zhorov, *Mol. Pharmacol.*, 2008, **74**, 1033–1045.
109. X. G. Zhen, C. Xie, A. Fitzmaurice, C. E. Schoonover, E. T. Orenstein and J. Yang, *J. Gen. Physiol.*, 2005, **126**, 193–204.
110. Z. Buraei and J. Yang, *Physiol. Rev.*, 2010, **90**, 1461–1506.
111. M. G. Kang and K. P. Campbell, *J. Biol. Chem.*, 2003, **278**, 21315–21318.
112. A. Davies, J. Hendrich, A. T. Van Minh, J. Wratten, L. Douglas and A. C. Dolphin, *Trends Pharmacol. Sci.*, 2007, **28**, 220–228.
113. E. A. Ertel, K. P. Campbell, M. M. Harpold, F. Hofmann, Y. Mori, E. Perez-Reyes, A. Schwartz, T. P. Snutch, T. Tanabe, L. Birnbaumer, R. W. Tsien and W. A. Catterall, *Neuron*, 2000, **25**, 533–535.
114. G. W. Zamponi, J. Striessnig, A. Koschak and A. C. Dolphin, *Pharmacol. Rev.*, 2015, **67**, 821–870.
115. A. Mikami, K. Imoto, T. Tanabe, T. Niidome, Y. Mori, H. Takeshima, S. Narumiya and S. Numa, *Nature*, 1989, **340**, 230–233.
116. J. Wu, Z. Yan, Z. Li, C. Yan, S. Lu, M. Dong and N. Yan, *Science*, 2015, **350**, aad2395.
117. J. Wu, Z. Yan, Z. Li, X. Qian, S. Lu, M. Dong, Q. Zhou and N. Yan, *Nature*, 2016, **537**, 191–196.
118. L. Tang, T. M. Gamal El-Din, T. M. Swanson, D. C. Pryde, T. Scheuer, N. Zheng and W. A. Catterall, *Nature*, 2016, **537**, 117–121.
119. G. H. Hockerman, B. Z. Peterson, B. D. Johnson and W. A. Catterall, *Annu. Rev. Pharmacol. Toxicol.*, 1997, **37**, 361–396.
120. R. C. Cheng, D. B. Tikhonov and B. S. Zhorov, *J. Biol. Chem.*, 2009, **284**, 28332–28342.
121. G. H. Hockerman, B. D. Johnson, M. R. Abbott, T. Scheuer and W. A. Catterall, *J. Biol. Chem.*, 1997, **272**, 18759–18765.
122. D. B. Tikhonov and B. S. Zhorov, *J. Biol. Chem.*, 2008, **283**, 17594–17604.
123. D. B. Tikhonov and B. S. Zhorov, *J. Biol. Chem.*, 2009, **284**, 19006–19017.

124. B. Z. Peterson, T. N. Tanada and W. A. Catterall, *J. Biol. Chem.*, 1996, **271**, 5293–5296.
125. K. J. Schleifer, *J. Med. Chem.*, 1999, **42**, 2204–2211.
126. S. Yamaguchi, B. S. Zhorov, K. Yoshioka, T. Nagao, H. Ichijo and S. Adachi-Akahane, *Mol. Pharmacol.*, 2003, **64**, 235–248.
127. S. Kang, G. Cooper, S. F. Dunne, B. Dusel, C.-H. Luan, D. J. Surmeier and R. B. Silverman, *Nat. Commun.*, 2012, **3**, 1–7.
128. N. J. Ortner, G. Bock, D. H. Vandael, R. Mauersberger, H. J. Draheim, R. Gust, E. Carbone, P. Tuluc and J. Striessnig, *Nat. Commun.*, 2014, **5**, 3897.
129. A. J. Baxter, J. Dixon, F. Ince, C. N. Manners and S. J. Teague, *J. Med. Chem.*, 1993, **36**, 2739–2744.
130. D. Rampe and A. E. Lacerda, *J. Pharmacol. Exp. Ther.*, 1991, **259**, 982–987.
131. S. I. McDonough, Y. Mori and B. P. Bean, *Biophys. J.*, 2005, **88**, 211–223.
132. G. M. Lipkind and H. A. Fozzard, *Mol. Pharmacol.*, 2003, **63**, 499–511.
133. N. Dilmac, N. Hilliard and G. H. Hockerman, *Mol. Pharmacol.*, 2004, **66**, 1236–1247.
134. I. G. Huber, E. Wappl-Kornherr, M. J. Sinnegger-Brauns, J. C. Hoda, D. Walter-Bastl and J. Striessnig, *J. Biol. Chem.*, 2004, **279**, 55211–55217.
135. K. S. Lee and R. W. Tsien, *Nature*, 1983, **302**, 790–794.
136. N. Dilmac, N. Hilliard and G. H. Hockerman, *Mol. Pharmacol.*, 2003, **64**, 491–501.
137. B. Z. Peterson and W. A. Catterall, *Mol. Pharmacol.*, 2006, **70**, 667–675.
138. B. S. Zhorov, E. V. Folkman and V. S. Ananthanarayanan, *Arch. Biochem. Biophys.*, 2001, **393**, 22–41.
139. S. Cosconati, L. Marinelli, A. Lavecchia and E. Novellino, *J. Med. Chem.*, 2007, **50**, 1504–1513.
140. L. Xu, D. Li, L. Tao, Y. Yang, Y. Li and T. Hou, *Mol. BioSyst.*, 2016, **12**, 379–390.
141. D. Schaller, M. G. Gunduz, F. X. Zhang, G. W. Zamponi and G. Wolber, *Eur. J. Med. Chem.*, 2018, **155**, 1–12.
142. I. Bodi, S. E. Koch, H. Yamaguchi, G. P. Szigeti, A. Schwartz and G. Varadi, *J. Biol. Chem.*, 2002, **277**, 20651–20659.
143. R. S. Kass, *Circ. Res.*, 1987, **61**, I1–I5.
144. J. S. Williams, I. L. Grupp, G. Grupp, P. L. Vaghy, L. Dumont, A. Schwartz, A. Yatani, S. Hamilton and A. M. Brown, *Biochem. Biophys. Res. Commun.*, 1985, **131**, 13–21.
145. T. J. Kamp, M. C. Sanguinetti and R. J. Miller, *Circ. Res.*, 1989, **64**, 338–351.
146. S. Kalyaanamoorthy and K. H. Barakat, *Med. Res. Rev.*, 2018, **38**, 525–555.
147. W. Wang and R. MacKinnon, *Cell*, 2017, **169**, 422–430 e410.
148. P. de la Pena, P. Dominguez and F. Barros, *Pflügers Arch.*, 2018, **470**, 517–536.

149. D. J. Elliott, N. Y. Dondas, T. S. Munsey and A. Sivaprasadarao, *Mol. Membr. Biol.*, 2009, **26**, 435–447.
150. M. Perry, M. Sanguinetti and J. Mitcheson, *J. Physiol.*, 2010, **588**, 3157–3167.
151. V. Vardanyan and O. Pongs, *Front. Pharmacol.*, 2012, **3**, 145.
152. Y. M. Cheng and T. W. Claydon, *Front. Pharmacol.*, 2012, **3**, 83.
153. S. Wang, S. Liu, M. J. Morales, H. C. Strauss and R. L. Rasmusson, *J. Physiol.*, 1997, **502**(Pt 1), 45–60.
154. S. Rajamani, L. L. Eckhardt, C. R. Valdivia, C. A. Klemens, B. M. Gillman, C. L. Anderson, K. M. Holzem, B. P. Delisle, B. D. Anson, J. C. Makielski and C. T. January, *Br. J. Pharmacol.*, 2006, **149**, 481–489.
155. A. Cavalli, E. Poluzzi, F. De Ponti and M. Recanatini, *J. Med. Chem.*, 2002, **45**, 3844–3853.
156. S. Ekins, *J. Pharmacol. Exp. Ther.*, 2002, **301**, 427–434.
157. J. M. Kratz, D. Schuster, M. Edtbauer, P. Saxena, C. E. Mair, J. Kirchebner, B. Matuszczak, I. Baburin, S. Hering and J. M. Rollinger, *J. Chem. Inf. Model.*, 2014, **54**, 2887–2901.
158. S. Durdagi, H. J. Duff and S. Y. Noskov, *J. Chem. Inf. Model.*, 2011, **51**, 463–474.
159. L. Du-Cuny, L. Chen and S. Zhang, *J. Chem. Inf. Model.*, 2011, **51**, 2948–2960.
160. T. L. Lenz and D. E. Hilleman, *Pharmacotherapy*, 2000, **20**, 776–786.
161. A. M. Aronov, *J. Med. Chem.*, 2006, **49**, 6917–6921.
162. P. K. Agarwal, F. Du, J. J. Babcock, H. Yu, B. Zou and M. Li, *PLoS One*, 2015, **10**, e0118324.
163. N. G. Nikolov, M. Dybdahl, S. Ó. Jónsdóttir and E. B. Wedebye, *Bioorg. Med. Chem.*, 2014, **22**, 6004–6013.
164. M. J. Waring and C. Johnstone, *Bioorg. Med. Chem. Lett.*, 2007, **17**, 1759–1764.
165. G. D. Ho, D. Tulshian, A. Bercovici, Z. Tan, J. Hanisak, S. Brumfield, J. Matasi, C. R. Heap, W. G. Earley, B. Courneya, R. Jason Herr, X. Zhou, T. Bridal, D. Rindgen, S. Sorota and S.-W. Yang, *Bioorg. Med. Chem. Lett.*, 2014, **24**, 4110–4113.
166. S. Suzuki, T. Kuroda, H. Kimoto, Y. Domon, K. Kubota, Y. Kitano, T. Yokoyama, A. Shimizugawa, R. Sugita, R. Koishi, D. Asano, K. Tamaki, T. Shinozuka and H. Kobayashi, *Bioorg. Med. Chem. Lett.*, 2015, **25**, 5419–5423.
167. L. B. Schenkel, E. F. DiMauro, H. N. Nguyen, N. Chakka, B. Du, R. S. Foti, A. Guzman-Perez, M. Jarosh, D. S. La, J. Ligutti, B. C. Milgram, B. D. Moyer, E. A. Peterson, J. Roberts, V. L. Yu and M. M. Weiss, *Bioorg. Med. Chem. Lett.*, 2017, **27**, 3817–3824.
168. B. A. Sparling, S. Yi, J. Able, H. Bregman, E. F. DiMauro, R. S. Foti, H. Gao, A. Guzman-Perez, H. Huang, M. Jarosh, T. Kornecook, J. Ligutti, B. C. Milgram, B. D. Moyer, B. Youngblood, V. L. Yu and M. M. Weiss, *MedChemComm*, 2017, **8**, 744–754.

169. A. J. Roecker, M. Egbertson, K. L. G. Jones, R. Gomez, R. L. Kraus, Y. Li, A. J. Koser, M. O. Urban, R. Klein, M. Clements, J. Panigel, C. Daley, J. Wang, E. N. Finger, J. Majercak, V. Santarelli, I. Gregan, M. Cato, T. Filzen, A. Jovanovska, Y.-H. Wang, D. Wang, L. A. Joyce, E. C. Sherer, X. Peng, X. Wang, H. Sun, P. J. Coleman, A. K. Houghton and M. E. Layton, *Bioorg. Med. Chem. Lett.*, 2017, **27**, 2087–2093.
170. K. J. Wilson, C. R. Illig, J. Chen, M. J. Wall, S. K. Ballentine, R. L. DesJarlais, Y. Chen, C. Schubert, R. Donatelli, I. Petrounia, C. S. Crysler, C. J. Molloy, M. A. Chaikin, C. L. Manthey, M. R. Player, B. E. Tomczuk and S. K. Meegalla, *Bioorg. Med. Chem. Lett.*, 2010, **20**, 3925–3929.
171. Y.-J. Wu, B. Venables, J. Guernon, J. Chen, S.-Y. Sit, R. Rajamani, R. J. Knox, M. Matchett, R. L. Pieschl, J. Herrington, L. J. Bristow, N. A. Meanwell, L. A. Thompson and C. Dzierba, *Bioorg. Med. Chem. Lett.*, 2019, **29**, 659–663.
172. J. M. Frost, D. A. DeGoey, L. Shi, R. J. Gum, M. M. Fricano, G. L. Lundgaard, O. F. El-Kouhen, G. C. Hsieh, T. Neelands, M. A. Matulenko, J. F. Daanen, M. Pai, N. Ghoreishi-Haack, C. Zhan, X.-F. Zhang and M. E. Kort, *J. Med. Chem.*, 2016, **59**, 3373–3391.
173. J. M. Wetzel, S. W. Miao, C. Forray, L. A. Borden, T. A. Branchek and C. Gluchowski, *J. Med. Chem.*, 1995, **38**, 1579–1581.
174. P. P. Shao, F. Ye, P. K. Chakravarty, D. J. Varughese, J. B. Herrington, G. Dai, R. M. Bugianesi, R. J. Haedo, A. M. Swensen, V. A. Warren, M. M. Smith, M. L. Garcia, O. B. McManus, K. A. Lyons, X. Li, M. Green, N. Jochnowitz, E. McGowan, S. Mistry, S.-Y. Sun, C. Abbadie, G. J. Kaczorowski and J. L. Duffy, *J. Med. Chem.*, 2012, **55**, 9847–9855.
175. S. Kang, G. Cooper, S. F. Dunne, C.-H. Luan, D. J. Surmeier and R. B. Silverman, *J. Med. Chem.*, 2013, **56**, 4786–4797.
176. D. A. Scott, L. A. Dakin, K. Daly, D. J. Del Valle, R. B. Diebold, L. Drew, J. Ezhuthachan, T. W. Gero, C. A. Ogoe, C. A. Omer, S. P. Redmond, G. Repik, K. Thakur, Q. Ye and X. Zheng, *Bioorg. Med. Chem. Lett.*, 2013, **23**, 4591–4596.
177. M. Cheung, W. Bao, D. J. Behm, C. A. Brooks, M. J. Bury, S. E. Dowdell, H. S. Eidam, R. M. Fox, K. B. Goodman, D. A. Holt, D. Lee, T. J. Roethke, R. N. Willette, X. Xu, G. Ye and K. S. Thorneloe, *ACS Med. Chem. Lett.*, 2017, **8**, 549–554.
178. T. Harrison, A. P. Owens, B. J. Williams, C. J. Swain, R. Baker, P. H. Hutson, S. Sadowski and M. A. Cascieri, *Bioorg. Med. Chem. Lett.*, 1995, **5**, 209–212.
179. S. G. Mills, M. MacCoss, M. A. Cascieri, S. Sadowski, S. Patel, K. L. Chapman and P. H. Hutson, *Bioorg. Med. Chem. Lett.*, 1995, **5**, 599–604.
180. Z.-Q. Yang, J. C. Barrow, W. D. Shipe, K.-A. S. Schlegel, Y. Shu, F. V. Yang, C. W. Lindsley, K. E. Rittle, M. G. Bock, G. D. Hartman, V. N. Uebele, C. E. Nuss, S. V. Fox, R. L. Kraus, S. M. Doran, T. M. Connolly, C. Tang, J. E. Ballard, Y. Kuo, E. D. Adarayan, T. Prueksaritanont, M. M. Zrada, M. J. Marino, V. K. Graufelds, A. G. DiLella, I. J. Reynolds, H. M. Vargas, P. B. Bunting, R. F. Woltmann, M. M. Magee, K. S. Koblan and J. J. Renger, *J. Med. Chem.*, 2008, **51**, 6471–6477.

181. I. Collins, M. Rowley, W. B. Davey, F. Emms, R. Marwood, S. Patel, S. Patel, A. Fletcher, I. C. Ragan, P. D. Leeson, A. L. Scott and T. Broten, *Bioorg. Med. Chem.*, 1998, **6**, 743–753.
182. W. D. Shipe, J. C. Barrow, Z.-Q. Yang, C. W. Lindsley, F. V. Yang, K.-A. S. Schlegel, Y. Shu, K. E. Rittle, M. G. Bock, G. D. Hartman, C. Tang, J. E. Ballard, Y. Kuo, E. D. Adarayan, T. Prueksaritanont, M. M. Zrada, V. N. Uebele, C. E. Nuss, T. M. Connolly, S. M. Doran, S. V. Fox, R. L. Kraus, M. J. Marino, V. K. Graufelds, H. M. Vargas, P. B. Bunting, M. Hasbun-Manning, R. M. Evans, K. S. Koblan and J. J. Renger, *J. Med. Chem.*, 2008, **51**, 3692–3695.
183. A. Aronov, *Curr. Top. Med. Chem.*, 2008, **8**, 1113–1127.
184. C. Jamieson, E. M. Moir, Z. Rankovic and G. Wishart, *J. Med. Chem.*, 2006, **49**, 5029–5046.
185. I. V. Tetko, G. I. Poda, C. Ostermann and R. Mannhold, *QSAR Comb. Sci.*, 2009, **28**, 845–849.
186. S. Hayashi, K. Ohashi, S. Mihara, E. Nakata, C. Emoto and A. Ohta, *Eur. J. Med. Chem.*, 2016, **114**, 345–364.
187. K. A. Brameld, T. D. Owens, E. Verner, E. Venetsanakos, J. M. Bradshaw, V. T. Phan, D. Tam, K. Leung, J. Shu, J. LaStant, D. G. Loughhead, T. Ton, D. E. Karr, M. E. Gerritsen, D. M. Goldstein and J. O. Funk, *J. Med. Chem.*, 2017, **60**, 6516–6527.
188. P. K. Sasmal, S. Sasmal, C. Abbineni, B. Venkatesham, P. T. Rao, M. Roshaiah, I. Khanna, V. Sebastian, J. Suresh and M. P. Singh, *MedChemComm*, 2011, **2**, 385–389.
189. C. H. Zhang, K. Chen, Y. Jiao, L. L. Li, Y. P. Li, R. J. Zhang, M. W. Zheng, L. Zhong, S. Z. Huang, C. L. Song, W. T. Lin, J. Yang, R. Xiang, B. Peng, J. H. Han, G. W. Lu, Y. Q. Wei and S. Y. Yang, *J. Med. Chem.*, 2016, **59**, 9788–9805.
190. J. L. Methot, D. M. Hoffman, D. J. Witter, M. G. Stanton, P. Harrington, C. Hamblett, P. Siliphaivanh, K. Wilson, J. Hubbs, R. Heidebrecht, A. M. Kral, N. Ozerova, J. C. Fleming, H. Wang, A. A. Szewczak, R. E. Middleton, B. Hughes, J. C. Cruz, B. B. Haines, M. Chenard, C. M. Kenific, A. Harsch, J. P. Secrist and T. A. Miller, *ACS Med. Chem. Lett.*, 2014, **5**, 340–345.
191. S. C. Huang, S. Adhikari, R. Afroze, K. Brewer, E. F. Calderwood, J. Chouitar, D. B. England, C. Fisher, K. M. Galvin, J. Gaulin, P. D. Greenspan, S. J. Harrison, M. S. Kim, S. P. Langston, L. T. Ma, S. Menon, H. Mizutani, M. Rezaei, M. D. Smith, D. M. Zhang and A. E. Gould, *Bioorg. Med. Chem. Lett.*, 2016, **26**, 1156–1160.
192. P. Devasthale, W. Wang, J. Mignone, K. Renduchintala, S. Radhakrishnan, J. Dhanapal, J. Selvaraj, R. Kuppusamy, M. A. Pelleymounter, D. Longhi, N. Huang, N. Flynn, A. V. Azzara, K. Rohrbach, J. Devenny, S. Rooney, M. Thomas, S. Glick, H. Godonis, S. Harvey, M. J. Cullen, H. Zhang, C. Caporuscio, P. Stetsko, M. Grubb, C. Huang, L. Zhang, C. Freeden, B. J. Murphy, J. A. Robl and W. N. Washburn, *Bioorg. Med. Chem. Lett.*, 2015, **25**, 4412–4418.

193. T. Siu, J. Brubaker, P. Fuller, L. Torres, H. Zeng, J. Close, D. M. Mampreian, F. Shi, D. Liu, X. Fradera, K. Johnson, N. Bays, E. Kadic, F. He, P. Goldenblatt, L. Shaffer, S. B. Patel, C. A. Lesburg, C. Alpert, L. Dorosh, S. V. Deshmukh, H. Yu, J. Klappenbach, F. Elwood, C. J. Dinsmore, R. Fernandez, L. Moy, J. R. Young, P. Devasthale, W. Wang, J. Mignone, K. Renduchintala, S. Radhakrishnan, J. Dhanapal, J. Selvaraj, R. Kuppusamy, M. A. Pelleymounter, D. Longhi, N. Huang, N. Flynn, A. V. Azzara, K. Rohrbach, J. Devenny, S. Rooney, M. Thomas, S. Glick, H. Godonis, S. Harvey, M. J. Cullen, H. Zhang, C. Caporuscio, P. Stetsko, M. Grubb, C. Huang, L. Zhang, C. Freeden, B. J. Murphy, J. A. Robl and W. N. Washburn, *J. Med. Chem.*, 2017, **60**, 9676–9690.
194. Y. Xu, B. G. Brenning, S. G. Kultgen, J. M. Foulks, A. Clifford, S. Lai, A. Chan, S. Merx, M. V. McCullar, S. B. Kanner and K. K. Ho, *ACS Med. Chem. Lett.*, 2015, **6**, 63–67.
195. J. D. Burch, K. Barrett, Y. Chen, J. DeVoss, C. Eigenbrot, R. Goldsmith, M. H. Ismaili, K. Lau, Z. Lin, D. F. Ortwine, A. A. Zarrin, P. A. McEwan, J. J. Barker, C. Ellebrandt, D. Kordt, D. B. Stein, X. Wang, Y. Chen, B. Hu, X. Xu, P. W. Yuen, Y. Zhang and Z. Pei, *J. Med. Chem.*, 2015, **58**, 3806–3816.
196. N. Desroy, C. Housseman, X. Bock, A. Joncour, N. Bienvenu, L. Cherel, V. Labeguere, E. Rondet, C. Peixoto, J. M. Grassot, O. Picolet, D. Annoot, N. Triballeau, A. Monjardet, E. Wakselman, V. Roncoroni, S. Le Tallec, R. Blanque, C. Cottereaux, N. Vandervoort, T. Christophe, P. Mollat, M. Lamers, M. Auberval, B. Hrvacic, J. Ralic, L. Oste, E. van der Aar, R. Brys and B. Heckmann, *J. Med. Chem.*, 2017, **60**, 3580–3590.
197. J. L. Diaz, U. Christmann, A. Fernandez, A. Torrens, A. Port, R. Pascual, I. Alvarez, J. Burgueno, X. Monroy, A. Montero, A. Balada, J. M. Vela and C. Almansa, *J. Med. Chem.*, 2015, **58**, 2441–2451.
198. S. K. Bagal, M. I. Kemp, P. J. Bungay, T. L. Hay, Y. Murata, C. E. Payne, E. B. Stevens, A. Brown, D. C. Blakemore and M. S. Corbett, *MedChemComm*, 2016, **7**, 1925–1931.
199. F. Curreli, Y. D. Kwon, D. S. Belov, R. R. Ramesh, A. V. Kurkin, A. Altieri, P. D. Kwong and A. K. Debnath, *J. Med. Chem.*, 2017, **60**, 3124–3153.
200. S. Chowdhury, K. N. Owens, R. J. Herr, Q. Jiang, X. Chen, G. Johnson, V. E. Groppi, D. W. Raible, E. W. Rubel and J. A. Simon, *J. Med. Chem.*, 2018, **61**, 84–97.
201. H. Ratni, M. Ebeling, J. Baird, S. Bendels, J. Bylund, K. S. Chen, N. Denk, Z. Feng, L. Green, M. Guerard, P. Jablonski, B. Jacobsen, O. Khwaja, H. Kletzl, C. P. Ko, S. Kustermann, A. Marquet, F. Metzger, B. Mueller, N. A. Naryshkin, S. V. Paushkin, E. Pinard, A. Poirier, M. Reutlinger, M. Weetall, A. Zeller, X. Zhao and L. Mueller, *J. Med. Chem.*, 2018, **61**, 6501–6517.
202. W. McCoull, P. Barton, A. Broo, A. J. Brown, D. S. Clarke, G. Coope, R. D. Davies, A. G. Dossetter, E. E. Kelly and L. Knerr, *MedChemComm*, 2013, **4**, 456–462.
203. J. Velcicky, W. Miltz, B. Oberhauser, D. Orain, A. Vaupel, K. Weigand, J. Dawson King, A. Littlewood-Evans, M. Nash, R. Feifel and P. Loetscher, *J. Med. Chem.*, 2017, **60**, 3672–3683.

204. A. Johansson, C. Lofberg, M. Antonsson, S. von Unge, M. A. Hayes, R. Judkins, K. Ploj, L. Benthem, D. Linden, P. Brodin, M. Wennerberg, M. Fredenwall, L. Li, J. Persson, R. Bergman, A. Pettersen, P. Gennemark and A. Hogner, *J. Med. Chem.*, 2016, **59**, 2497–2511.
205. R. A. Fairhurst, T. H. Marsilje, S. Stutz, A. Boos, M. Niklaus, B. Chen, S. Jiang, W. Lu, P. Furet, C. McCarthy, F. Stauffer, V. Guagnano, A. Vaupel, P. Y. Michellys, C. Schnell and S. Jeay, *Bioorg. Med. Chem. Lett.*, 2016, **26**, 2057–2064.
206. F. Reck, D. E. Ehmann, T. J. Dougherty, J. V. Newman, S. Hopkins, G. Stone, N. Agrawal, P. Ciaccio, J. McNulty, H. Barthlow, J. O'Donnell, K. Goteti, J. Breen, J. Comita-Prevoir, M. Cornebise, M. Cronin, C. J. Eyermann, B. Geng, G. R. Carr, L. Pandarinathan, X. Tang, A. Cottone, L. Zhao and N. Bezdenejnih-Snyder, *Bioorg. Med. Chem.*, 2014, **22**, 5392–5409.
207. H. Fujieda, M. Kogami, M. Sakairi, N. Kato, M. Makino, N. Takahashi, T. Miyazawa, S. Harada and T. Yamashita, *Eur. J. Med. Chem.*, 2018, **156**, 269–294.
208. R. L. Dow, M. Ammirati, S. W. Bagley, S. K. Bhattacharya, L. Buckbinder, C. Cortes, A. F. El-Kattan, K. Ford, G. B. Freeman, C. R. W. Guimaraes, S. Liu, M. Niosi, A. Skoura and D. Tess, *J. Med. Chem.*, 2018, **61**, 3114–3125.
209. Q. Ren, X. Liu, G. Yan, B. Nie, Z. Zou, J. Li, Y. Chen, Y. Wei, J. Huang, Z. Luo, B. Gu, S. Goldmann, J. Zhang and Y. Zhang, *J. Med. Chem.*, 2018, **61**, 1355–1374.
210. Z. Qiu, X. Lin, W. Zhang, M. Zhou, L. Guo, B. Kocer, G. Wu, Z. Zhang, H. Liu, H. Shi, B. Kou, T. Hu, Y. Hu, M. Huang, S. F. Yan, Z. Xu, Z. Zhou, N. Qin, Y. F. Wang, S. Ren, H. Qiu, Y. Zhang, Y. Zhang, X. Wu, K. Sun, S. Zhong, J. Xie, G. Ottaviani, Y. Zhou, L. Zhu, X. Tian, L. Shi, F. Shen, Y. Mao, X. Zhou, L. Gao, J. A. T. Young, J. Z. Wu, G. Yang, A. V. Mayweg, H. C. Shen, G. Tang and W. Zhu, *J. Med. Chem.*, 2017, **60**, 3352–3371.
211. K. S. Yeung, B. R. Beno, K. Parcella, J. A. Bender, K. A. Grant-Young, A. Nickel, P. Gunaga, P. Anjanappa, R. O. Bora, K. Selvakumar, K. Rigat, Y. K. Wang, M. Liu, J. Lemm, K. Mosure, S. Sheriff, C. Wan, M. Witmer, K. Kish, U. Hanumegowda, X. Zhuo, Y. Z. Shu, D. Parker, R. Haskell, A. Ng, Q. Gao, E. Colston, J. Raybon, D. M. Grasela, K. Santone, M. Gao, N. A. Meanwell, M. Sinz, M. G. Soars, J. O. Knipe, S. B. Roberts and J. F. Kadow, *J. Med. Chem.*, 2017, **60**, 4369–4385.
212. W. Embrechts, F. Herschke, F. Pauwels, B. Stoops, S. Last, S. Pieters, V. Pande, G. Pille, K. Amssoms, I. Smyej, D. Dhuyvetter, A. Scholliers, W. Mostmans, K. Van Dijck, B. Van Schoubroeck, T. Thone, D. De Pooter, G. Fanning, T. H. M. Jonckers, H. Horton, P. Raboisson and D. McGowan, *J. Med. Chem.*, 2018, **61**, 6236–6246.
213. P. A. Barsanti, Y. Pan, Y. Lu, R. Jain, M. Cox, R. J. Aversa, M. P. Dillon, R. Elling, C. Hu, X. Jin, M. Knapp, J. Lan, S. Ramurthy, P. Rudewicz, L. Setti, S. Subramanian, M. Mathur, L. Taricani, G. Thomas, L. Xiao and Q. Yue, *ACS Med. Chem. Lett.*, 2015, **6**, 42–46.

214. U. Yngve, K. Paulsen, I. Macsari, M. Sundström, E. Santangelo, C. Linde, K. Bogar, F. Lake, Y. Besidski and J. Malmborg, *MedChemComm*, 2013, **4**, 422–431.
215. J. Kim, D. Lee, C. Park, W. So, M. Jo, T. Ok, J. Kwon, S. Kong, S. Jo, Y. Kim, J. Choi, H. C. Kim, Y. Ko, I. Choi, Y. Park, J. Yoon, M. K. Ju, J. Kim, S. J. Han, T. H. Kim, J. Cechetto, J. Nam, P. Sommer, M. Liuzzi, J. Lee and Z. No, *ACS Med. Chem. Lett.*, 2012, **3**, 678–682.
216. N. Sakauchi, Y. Kohara, A. Sato, T. Suzaki, Y. Imai, Y. Okabe, S. Imai, R. Saikawa, H. Nagabukuro, H. Kuno, H. Fujita, I. Kamo and M. Yoshida, *J. Med. Chem.*, 2016, **59**, 2989–3002.
217. D. O. Koltun, E. Q. Parkhill, R. Kalla, T. D. Perry, E. Elzein, X. Li, S. P. Simonovich, C. Ziebenhaus, T. R. Hansen, B. Marchand, W. K. Hung, L. Lagpacan, M. Hung, R. G. Aoyama, B. P. Murray, J. K. Perry, J. R. Somoza, A. G. Villasenor, N. Pagratis and J. A. Zablocki, *Bioorg. Med. Chem. Lett.*, 2018, **28**, 541–546.
218. Y. Qian, W. L. Corbett, S. J. Berthel, D. S. Choi, M. T. Dvorozniak, W. Geng, P. Gillespie, K. R. Guertin, N. E. Haynes, R. F. Kester, F. A. Mennona, D. Moore, J. Racha, R. Radinov, R. Sarabu, N. R. Scott, J. Grimsby and N. L. Mallalieu, *ACS Med. Chem. Lett.*, 2013, **4**, 414–418.
219. C. N. Johnson, J. S. Ahn, I. M. Buck, E. Chiarparin, J. E. H. Day, A. Hopkins, S. Howard, E. J. Lewis, V. Martins, A. Millemaggi, J. M. Munck, L. W. Page, T. Peakman, M. Reader, S. J. Rich, G. Saxty, T. Smyth, N. T. Thompson, G. A. Ward, P. A. Williams, N. E. Wilsher and G. Chessari, *J. Med. Chem.*, 2018, **61**, 7314–7329.
220. H. R. Chobanian, Y. Guo, P. Liu, M. D. Chioda, S. Fung, T. J. Lanza, L. Chang, R. K. Bakshi, J. P. Dellureficio, Q. Hong, M. McLaughlin, K. M. Belyk, S. W. Krska, A. K. Makarewicz, E. J. Martel, J. F. Leone, L. Frey, B. Karanam, M. Madeira, R. Alvaro, J. Shuman, G. Salituro, J. L. Terebetski, N. Jochnowitz, S. Mistry, E. McGowan, R. Hajdu, M. Rosenbach, C. Abbadie, J. P. Alexander, L. L. Shiao, K. M. Sullivan, R. P. Nargund, M. J. Wyvratt, L. S. Lin and R. J. DeVita, *ACS Med. Chem. Lett.*, 2014, **5**, 717–721.
221. D. C. Beshore, C. N. Di Marco, R. K. Chang, T. J. Greshock, L. Ma, M. Wittmann, M. A. Seager, K. A. Koeplinger, C. D. Thompson, J. Fuerst, G. D. Hartman, M. T. Bilodeau, W. J. Ray and S. D. Kuduk, *ACS Med. Chem. Lett.*, 2018, **9**, 652–656.
222. H. Wan, G. M. Schroeder, A. C. Hart, J. Inghrim, J. Grebinski, J. S. Tokarski, M. V. Lorenzi, D. You, T. McDevitt, B. Penhallow, R. Vuppugalla, Y. Zhang, X. Gu, R. Iyer, L. J. Lombardo, G. L. Trainor, S. Ruepp, J. Lippy, Y. Blat, J. S. Sack, J. A. Khan, K. Stefanski, B. Sleczka, A. Mathur, J. H. Sun, M. K. Wong, D. R. Wu, P. Li, A. Gupta, P. N. Arunachalam, B. Pragalathan, S. Narayanan, C. K. Nanjundaswamy, P. Kuppusamy and A. V. Purandare, *ACS Med. Chem. Lett.*, 2015, **6**, 850–855.
223. T. Yamasaki, H. Hirose, T. Yamashita, N. Takakura, S. Morimoto, T. Nakahata, A. Kina, Y. Nakano, Y. Okano Tamura, J. Sugama, T. Odani, Y. Shimizu, S. Iwasaki, M. Watanabe, T. Maekawa and S. Kasai, *Bioorg. Med. Chem.*, 2017, **25**, 4153–4162.

224. M. Furber, A. K. Tiden, P. Gardiner, A. Mete, R. Ford, I. Millichip, L. Stein, A. Mather, E. Kinchin, C. Luckhurst, S. Barber, P. Cage, H. Sanganee, R. Austin, K. Chohan, R. Beri, B. Thong, A. Wallace, V. Oreffo, R. Hutchinson, S. Harper, J. Debreczeni, J. Breed, L. Wissler and K. Edman, *J. Med. Chem.*, 2014, **57**, 2357–2367.
225. J. Cheng, P. M. Giguere, O. K. Onajole, W. Lv, A. Gaisin, H. Gunosewoyo, C. M. Schmerberg, V. M. Pogorelov, R. M. Rodriguiz, G. Vistoli, W. C. Wetsel, B. L. Roth and A. P. Kozikowski, *J. Med. Chem.*, 2015, **58**, 1992–2002.
226. P. S. Hameed, P. Manjrekar, A. Raichurkar, V. Shinde, J. Puttur, G. Shanbhag, M. Chinnapattu, V. Patil, S. Rudrapatana, S. Sharma, C. N. Kumar, R. Nandishaiah, P. Madhavapeddi, D. Sriram, S. Solapure and V. K. Sambandamurthy, *ACS Med. Chem. Lett.*, 2015, **6**, 741–746.
227. K. G. Pike, B. Barlaam, E. Cadogan, A. Campbell, Y. Chen, N. Colclough, N. L. Davies, C. de-Almeida, S. L. Degorce, M. Didelot, A. Dishington, R. Ducray, S. T. Durant, L. A. Hassall, J. Holmes, G. D. Hughes, P. A. MacFaul, K. R. Mulholland, T. M. McGuire, G. Ouvry, M. Pass, G. Robb, N. Stratton, Z. Wang, J. Wilson, B. Zhai, K. Zhao and N. Al-Huniti, *J. Med. Chem.*, 2018, **61**, 3823–3841.
228. H. R. Chobanian, Y. Guo, B. Pio, H. Tang, N. Teumelsan, M. Clements, J. Frie, R. Ferguson, Z. Guo, B. S. Thomas-Fowlkes, J. P. Felix, J. Liu, M. Kohler, B. Priest, C. Hampton, L. Y. Pai, A. Corona, J. Metzger, V. Tong, E. M. Joshi, L. Xu, K. Owens, K. Maloney, K. Sullivan and A. Pasternak, *Bioorg. Med. Chem. Lett.*, 2017, **27**, 1109–1114.
229. R. I. Storer, R. M. Owen, A. Pike, C. L. Benn, E. Armstrong, D. C. Blakemore, M. Bictash, K. Costelloe, E. Impey and P. H. Milliken, *MedChemComm*, 2016, **7**, 1587–1595.
230. B. Li, S. Ni, F. Mao, F. Chen, Y. Liu, H. Wei, W. Chen, J. Zhu, L. Lan and J. Li, *J. Med. Chem.*, 2018, **61**, 224–250.
231. J. S. Scott, S. S. Bowker, K. J. Brocklehurst, H. S. Brown, D. S. Clarke, A. Easter, A. Ertan, K. Goldberg, J. A. Hudson, S. Kavanagh, D. Laber, A. G. Leach, P. A. MacFaul, E. A. Martin, D. McKerrecher, P. Schofield, P. H. Svensson and J. Teague, *J. Med. Chem.*, 2014, **57**, 8984–8998.
232. C. Le Manach, T. Paquet, D. Gonzalez Cabrera, Y. Younis, D. Taylor, L. Wiesner, N. Lawrence, S. Schwager, D. Waterson, M. J. Witty, S. Wittlin, L. J. Street and K. Chibale, *J. Med. Chem.*, 2014, **57**, 8839–8848.
233. S. D. Kuduk, R. K. Chang, T. J. Greshock, W. J. Ray, L. Ma, M. Wittmann, M. A. Seager, K. A. Koeplinger, C. D. Thompson, G. D. Hartman and M. T. Bilodeau, *ACS Med. Chem. Lett.*, 2012, **3**, 1070–1074.
234. F. Micheli, A. Bernardelli, F. Bianchi, S. Braggio, L. Castelletti, P. Cavallini, P. Cavanni, S. Cremonesi, M. Dal Cin, A. Feriani, B. Oliosi, T. Semeraro, L. Tarsi, S. Tomelleri, A. Wong, F. Visentini, L. Zonzini and C. Heidbreder, *Bioorg. Med. Chem.*, 2016, **24**, 1619–1636.
235. R. T. Skerlj, C. M. Bastos, M. L. Booker, M. L. Kramer, R. H. Barker Jr, C. A. Celatka, T. J. O'Shea, B. Munoz, A. B. Sidhu, J. F. Cortese, S. Wittlin, P. Papastogiannidis, I. Angulo-Barturen, M. B. Jimenez-Diaz and E. Sybertz, *ACS Med. Chem. Lett.*, 2011, **2**, 708–713.

236. K. Abrahamsson, P. Andersson, J. Bergman, U. Bredberg, J. Brånalt, A.-C. Egnell, U. Eriksson, D. Gustafsson, K.-J. Hoffman and S. Nielsen, *MedChemComm*, 2016, **7**, 272–281.
237. S. D. Edmondson, C. Zhu, N. F. Kar, J. Di Salvo, H. Nagabukuro, B. Sacre-Salem, K. Dingley, R. Berger, S. D. Goble, G. Morriello, B. Harper, C. R. Moyes, D. M. Shen, L. Wang, R. Ball, A. Fitzmaurice, T. Frenkl, L. N. Gichuru, S. Ha, A. L. Hurley, N. Jochnowitz, D. Levorse, S. Mistry, R. R. Miller, J. Ormes, G. M. Salituro, A. Sanfiz, A. S. Stevenson, K. Villa, B. Zamlynny, S. Green, M. Struthers and A. E. Weber, *J. Med. Chem.*, 2016, **59**, 609–623.
238. J. Reuberson, H. Horsley, R. J. Franklin, D. Ford, J. Neuss, D. Brookings, Q. Huang, B. Vanderhoydonck, L. J. Gao, M. Y. Jang, P. Herdewijn, A. Ghawalkar, F. Fallah-Arani, A. R. Khan, J. Henshall, M. Jairaj, S. Malcolm, E. Ward, L. Shuttleworth, Y. Lin, S. Li, T. Louat, M. Waer, J. Herman, A. Payne, T. Ceska, C. Doyle, W. Pitt, M. Calmiano, M. Augustin, S. Steinbacher, A. Lammens and R. Allen, *J. Med. Chem.*, 2018, **61**, 6705–6723.
239. T. Aoki, I. Hyohdoh, N. Furuichi, S. Ozawa, F. Watanabe, M. Matsushita, M. Sakaitani, K. Morikami, K. Takanashi, N. Harada, Y. Tomii, K. Shiraki, K. Furumoto, M. Tabo, K. Yoshinari, K. Ori, Y. Aoki, N. Shimma and H. Iikura, *ACS Med. Chem. Lett.*, 2014, **5**, 309–314.
240. C. B. Xue, A. Wang, Q. Han, Y. Zhang, G. Cao, H. Feng, T. Huang, C. Zheng, M. Xia, K. Zhang, L. Kong, J. Glenn, R. Anand, D. Meloni, D. J. Robinson, L. Shao, L. Storace, M. Li, R. O. Hughes, R. Devraj, P. A. Morton, D. J. Rogier, M. Covington, P. Scherle, S. Diamond, T. Emm, S. Yeleswaram, N. Contel, K. Vaddi, R. Newton, G. Hollis and B. Metcalf, *ACS Med. Chem. Lett.*, 2011, **2**, 913–918.
241. C. B. Xue, H. Feng, G. Cao, T. Huang, J. Glenn, R. Anand, D. Meloni, K. Zhang, L. Kong, A. Wang, Y. Zhang, C. Zheng, M. Xia, L. Chen, H. Tanaka, Q. Han, D. J. Robinson, D. Modi, L. Storace, L. Shao, V. Sharief, M. Li, L. G. Galya, M. Covington, P. Scherle, S. Diamond, T. Emm, S. Yeleswaram, N. Contel, K. Vaddi, R. Newton, G. Hollis, S. Friedman and B. Metcalf, *ACS Med. Chem. Lett.*, 2011, **2**, 450–454.
242. Z. S. Cheruvallath, S. L. Gwaltney 2nd, M. Sabat, M. Tang, H. Wang, A. Jennings, D. Hosfield, B. Lee, Y. Wu, P. Halkowycz and C. E. Grimshaw, *Bioorg. Med. Chem. Lett.*, 2017, **27**, 2678–2682.
243. J. Hynes Jr, H. Wu, J. Kempson, J. J. Duan, Z. Lu, B. Jiang, S. Stachura, J. S. Tokarski, J. S. Sack, J. A. Khan, J. S. Lippy, R. F. Zhang, S. Pitt, G. Shen, K. Gillooly, K. McIntyre, P. H. Carter, J. C. Barrish, S. G. Nadler, L. M. Salter-Cid, A. Fura, G. L. Schieven, W. J. Pitts and S. T. Wrobleski, *Bioorg. Med. Chem. Lett.*, 2017, **27**, 3101–3106.
244. T. Ehara, O. Irie, T. Kosaka, T. Kanazawa, W. Breitenstein, P. Grosche, N. Ostermann, M. Suzuki, S. Kawakami, K. Konishi, Y. Hitomi, A. Toyao, H. Gunji, F. Cumin, N. Schiering, T. Wagner, D. F. Rigel, R. L. Webb, J. Maibaum and F. Yokokawa, *ACS Med. Chem. Lett.*, 2014, **5**, 787–792.

245. M. J. Waring, D. S. Clarke, M. D. Fenwick, L. Godfrey, S. D. Groombridge, C. Johnstone, D. McKerrecher, K. G. Pike, J. W. Rayner and G. R. Robb, *MedChemComm*, 2012, **3**, 1077–1081.
246. X. Zhang, C. Hou, H. Hufnagel, M. Singer, E. Opas, S. McKenney, D. Johnson and Z. Sui, *ACS Med. Chem. Lett.*, 2012, **3**, 1039–1044.
247. Q. Deng, Y. H. Lim, R. Anand, Y. Yu, J. H. Kim, W. Zhou, J. Zheng, P. Tempest, D. Levorse, X. Zhang, S. Greene, D. Mullins, C. Culberson, B. Sherborne, E. M. Parker, A. Stamford and A. Ali, *Bioorg. Med. Chem. Lett.*, 2015, **25**, 2958–2962.
248. P. S. Hameed, V. Patil, S. Solapure, U. Sharma, P. Madhavapeddi, A. Raichurkar, M. Chinnapattu, P. Manjrekar, G. Shanbhag, J. Puttur, V. Shinde, S. Menasinakai, S. Rudrapatana, V. Achar, D. Awasthy, R. Nandishaiah, V. Humnabadkar, A. Ghosh, C. Narayan, V. K. Ramya, P. Kaur, S. Sharma, J. Werngren, S. Hoffner, V. Panduga, C. N. Kumar, J. Reddy, K. N. M. Kumar, S. Ganguly, S. Bharath, U. Bheemarao, K. Mukherjee, U. Arora, S. Gaonkar, M. Coulson, D. Waterson, V. K. Sambandamurthy and S. M. de Sousa, *J. Med. Chem.*, 2014, **57**, 4889–4905.
249. M. Leivers, J. F. Miller, S. A. Chan, R. Lauchli, S. Liehr, W. Mo, T. Ton, E. M. Turner, M. Youngman, J. G. Falls, S. Long, A. Mathis and J. Walker, *J. Med. Chem.*, 2014, **57**, 1964–1975.
250. N. L. Subasinghe, J. Lanter, T. Markotan, E. Opas, S. McKenney, C. Crysler, C. Hou, J. O'Neill, D. Johnson and Z. Sui, *Bioorg. Med. Chem. Lett.*, 2013, **23**, 1063–1069.
251. S. Brand, E. J. Ko, E. Viayna, S. Thompson, D. Spinks, M. Thomas, L. Sandberg, A. F. Francisco, S. Jayawardhana, V. C. Smith, C. Jansen, M. De Rycker, J. Thomas, L. MacLean, M. Osuna-Cabello, J. Riley, P. Scullion, L. Stojanovski, F. R. C. Simeons, O. Epemolu, Y. Shishikura, S. D. Crouch, T. S. Bakshi, C. J. Nixon, I. H. Reid, A. P. Hill, T. Z. Underwood, S. J. Hindley, S. A. Robinson, J. M. Kelly, J. M. Fiandor, P. G. Wyatt, M. Marco, T. J. Miles, K. D. Read and I. H. Gilbert, *J. Med. Chem.*, 2017, **60**, 7284–7299.
252. G. J. Roth, A. Heckel, J. T. Kley, T. Lehmann, S. G. Muller, T. Oost, K. Rudolf, K. Arndt, R. Budzinski, M. Lenter, R. R. Lotz, M. Schindler, L. Thomas and D. Stenkamp, *Bioorg. Med. Chem. Lett.*, 2015, **25**, 3270–3274.
253. E. Comer, J. A. Beaudoin, N. Kato, M. E. Fitzgerald, R. W. Heidebrecht, t. Lee Md, D. Masi, M. Mercier, C. Mulrooney, G. Muncipinto, A. Rowley, K. Crespo-Llado, A. E. Serrano, A. K. Lukens, R. C. Wiegand, D. F. Wirth, M. A. Palmer, M. A. Foley, B. Munoz, C. A. Scherer, J. R. Duvall and S. L. Schreiber, *J. Med. Chem.*, 2014, **57**, 8496–8502.
254. S. He, Z. Lai, Z. Ye, P. H. Dobbelaar, S. K. Shah, Q. Truong, W. Du, L. Guo, J. Liu, T. Jian, H. Qi, R. K. Bakshi, Q. Hong, J. Dellureficio, M. Reibarkh, K. Samuel, V. B. Reddy, S. Mitelman, S. X. Tong, G. G. Chicchi, K. L. Tsao, D. Trusca, M. Wu, Q. Shao, M. E. Trujillo, G. Fernandez, D. Nelson, P. Bunting, J. Kerr, P. Fitzgerald, P. Morissette, S. Volksdorf, G. J. Eiermann, C. Li, B. Zhang, A. D. Howard, Y. P. Zhou, R. P. Nargund and W. K. Hagmann, *ACS Med. Chem. Lett.*, 2014, **5**, 748–753.

255. Y. Nakajima, T. Inoue, K. Nakai, K. Mukoyoshi, H. Hamaguchi, K. Hatanaka, H. Sasaki, A. Tanaka, F. Takahashi, S. Kunikawa, H. Usuda, A. Moritomo, Y. Higashi, M. Inami and S. Shirakami, *Bioorg. Med. Chem.*, 2015, **23**, 4871–4883.
256. J. P. Surivet, C. Zumbrunn, T. Bruyere, D. Bur, C. Kohl, H. H. Locher, P. Seiler, E. A. Ertel, P. Hess, M. Enderlin-Paput, S. Enderlin-Paput, J. C. Gauvin, A. Mirre, C. Hubschwerlen, D. Ritz and G. Rueedi, *J. Med. Chem.*, 2017, **60**, 3776–3794.
257. J. L. Methot, C. Fischer, C. Li, A. Rivkin, S. P. Ahearn, W. C. Brown, S. Kattar, E. Kelley, D. M. Mampreian, A. Schell, A. Rosenau, H. Zhou, R. Ball, S. V. Deshmukh, V. V. Jeliazkova-Mecheva, D. Diaz, L. Y. Moy, C. M. Kenific, C. Moxham, S. Shah, H. Nuthall, A. A. Szewczak, A. Hill, B. Hughes, N. Smotrov, B. Munoz, T. A. Miller and M. S. Shearman, *Bioorg. Med. Chem. Lett.*, 2015, **25**, 3495–3500.
258. S. Dong, K. VanGelder, Z. C. Shi, Y. Yu, Z. Wu, R. Ferguson, Z. Z. Guo, H. Tang, J. Frie, Q. Fu, X. Gu, B. T. Priest, B. Thomas-Fowlkes, A. Weinglass, M. Margulis, J. Liu, L. Y. Pai, C. Hampton, R. E. Haimbach, K. Owens, V. Tong, S. Xu, M. Hu, G. J. Zingaro, P. Morissette, J. Ehrhart, S. Roy, K. Sullivan and A. Pasternak, *Bioorg. Med. Chem. Lett.*, 2017, **27**, 2559–2566.
259. K. Goldberg, S. Groombridge, J. Hudson, A. G. Leach, P. A. MacFaul, A. Pickup, R. Poultney, J. S. Scott, P. H. Svensson and J. Sweeney, *MedChemComm*, 2012, **3**, 600–604.
260. L. Zhuang, C. M. Tice, Z. Xu, W. Zhao, S. Cacatian, Y. J. Ye, S. B. Singh, P. Lindblom, B. M. McKeever, P. M. Krosky, Y. Zhao, D. Lala, B. A. Kruk, S. Meng, L. Howard, J. A. Johnson, Y. Bukhtiyarov, R. Panemangalore, J. Guo, R. Guo, F. Himmelsbach, B. Hamilton, A. Schuler-Metz, H. Schauerte, R. Gregg, G. M. McGeehan, K. Leftheris and D. A. Claremon, *Bioorg. Med. Chem.*, 2017, **25**, 3649–3657.
261. D. King, C. Iwuagwu, J. Cook, I. M. McDonald, R. Mate, F. C. Zusi, M. D. Hill, H. Fang, R. Zhao, B. Wang, A. E. Easton, R. Miller, D. Post-Munson, R. J. Knox, L. Gallagher, R. Westphal, T. Molski, J. Fan, W. Clarke, Y. Benitex, K. A. Lentz, R. Denton, D. Morgan, R. Zaczek, N. J. Lodge, L. J. Bristow, J. E. Macor and R. E. Olson, *ACS Med. Chem. Lett.*, 2017, **8**, 366–371.
262. S. K. Bagal, M. Andrews, B. M. Bechle, J. Bian, J. Bilsland, D. C. Blakemore, J. F. Braganza, P. J. Bungay, M. S. Corbett, C. N. Cronin, J. J. Cui, R. Dias, N. J. Flanagan, S. E. Greasley, R. Grimley, K. James, E. Johnson, L. Kitching, M. L. Kraus, I. McAlpine, A. Nagata, S. Ninkovic, K. Omoto, S. Scales, S. E. Skerratt, J. Sun, M. Tran-Dube, G. J. Waldron, F. Wang and J. S. Warmus, *J. Med. Chem.*, 2018, **61**, 6779–6800.
263. S. Parthasarathy, K. Henry, H. Pei, J. Clayton, M. Rempala, D. Johns, O. De Frutos, P. Garcia, C. Mateos, S. Pleite, Y. Wang, S. Stout, B. Condon, S. Ashok, Z. Lu, W. Ehlhardt, T. Raub, M. Lai, S. Geeganage and T. P. Burkholder, *Bioorg. Med. Chem. Lett.*, 2018, **28**, 1887–1891.
264. P. A. Barsanti, R. J. Aversa, X. Jin, Y. Pan, Y. Lu, R. Elling, R. Jain, M. Knapp, J. Lan, X. Lin, P. Rudewicz, J. Sim, L. Taricani, G. Thomas, L. Xiao and Q. Yue, *ACS Med. Chem. Lett.*, 2015, **6**, 37–41.

265. H. R. Hoveyda, G. L. Fraser, M. O. Roy, G. Dutheuil, F. Batt, M. El Bousmaqui, J. Korac, F. Lenoir, A. Lapin, S. Noel and S. Blanc, *J. Med. Chem.*, 2015, **58**, 3060–3082.
266. T. Ogiyama, M. Inoue, S. Honda, H. Yamada, T. Watanabe, T. Gotoh, T. Kiso, A. Koakutsu, S. Kakimoto and J. Shishikura, *Bioorg. Med. Chem.*, 2014, **22**, 6899–6907.
267. H. Hirose, T. Yamasaki, M. Ogino, R. Mizojiri, Y. Tamura-Okano, H. Yashiro, Y. Muraki, Y. Nakano, J. Sugama, A. Hata, S. Iwasaki, M. Watanabe, T. Maekawa and S. Kasai, *Bioorg. Med. Chem.*, 2017, **25**, 4175–4193.
268. J. S. Park, W. Im, S. Choi, S. J. Park, J. M. Jung, K. S. Baek, H. P. Son, S. Sharma, I. S. Kim and Y. H. Jung, *Eur. J. Med. Chem.*, 2016, **109**, 75–88.
269. D. G. Brown, P. R. Bernstein, Y. Wu, R. A. Urbanek, C. W. Becker, S. R. Throner, B. T. Dembofsky, G. B. Steelman, L. A. Lazor, C. W. Scott, M. W. Wood, S. S. Wesolowski, D. A. Nugiel, S. Koch, J. Yu, D. E. Pivonka, S. Li, C. Thompson, A. Zacco, C. S. Elmore, P. Schroeder, J. Liu, C. A. Hurley, S. Ward, H. J. Hunt, K. Williams, J. McLaughlin, V. Hoesch, S. Sydserff, D. Maier and D. Aharony, *ACS Med. Chem. Lett.*, 2013, **4**, 46–51.
270. S. B. Singh, D. E. Kaelin, J. Wu, L. Miesel, C. M. Tan, P. T. Meinke, D. Olsen, A. Lagrutta, P. Bradley, J. Lu, S. Patel, K. W. Rickert, R. F. Smith, S. Soisson, C. Wei, H. Fukuda, R. Kishii, M. Takei and Y. Fukuda, *ACS Med. Chem. Lett.*, 2014, **5**, 609–614.
271. S. B. Singh, D. E. Kaelin, J. Wu, L. Miesel, C. M. Tan, P. T. Meinke, D. B. Olsen, A. Lagrutta, C. Wei and Y. Liao, *MedChemComm*, 2015, **6**, 1773–1780.
272. K. D. Robarge, W. Lee, C. Eigenbrot, M. Ultsch, C. Wiesmann, R. Heald, S. Price, J. Hewitt, P. Jackson, P. Savy, B. Burton, E. F. Choo, J. Pang, J. Boggs, A. Yang, X. Yang and M. Baumgardner, *Bioorg. Med. Chem. Lett.*, 2014, **24**, 4714–4723.
273. P. Panchaud, T. Bruyere, A. C. Blumstein, D. Bur, A. Chambovey, E. A. Ertel, M. Gude, C. Hubschwerlen, L. Jacob, T. Kimmerlin, T. Pfeifer, L. Prade, P. Seiler, D. Ritz and G. Rueedi, *J. Med. Chem.*, 2017, **60**, 3755–3775.
274. B. Dyck, B. Branstetter, T. Gharbaoui, A. R. Hudson, J. G. Breitenbucher, L. Gomez, I. Botrous, T. Marrone, R. Barido, C. K. Allerston, E. P. Cedervall, R. Xu, V. Sridhar, R. Barker, K. Aertgeerts, K. Schmelzer, D. Neul, D. Lee, M. E. Massari, C. B. Andersen, K. Sebring, X. Zhou, R. Petroski, J. Limberis, M. Augustin, L. E. Chun, T. E. Edwards, M. Peters and A. Tabatabaei, *J. Med. Chem.*, 2017, **60**, 3472–3483.
275. L. R. Marcin, J. Warrier, S. Thangathirupathy, J. Shi, G. N. Karageorge, B. C. Pearce, A. Ng, H. Park, J. Kempson, J. Li, H. Zhang, A. Mathur, A. B. Reddy, G. Nagaraju, G. Tonukunuru, G. Gupta, M. Kamble, R. Mannoori, S. Cheruku, S. Jogi, J. Gulia, T. Bastia, C. Sanmathi, J. Aher, R. Kallem, B. N. Srikumar, K. K. Vijaya, P. S. Naidu, M. Paschapur, N. Kalidindi, R. Vikramadithyan, M. Ramarao, R. Denton, T. Molski, E. Shields, M. Subramanian, X. Zhuo, M. Nophsker, J. Simmermacher, M. Sinz, C. Albright, L. J. Bristow, I. Islam, J. J. Bronson, R. E. Olson, D. King, L. A. Thompson and J. E. Macor, *ACS Med. Chem. Lett.*, 2018, **9**, 472–477.

276. S. Wu, Y. Sun, Y. Hu, H. Zhang, L. Hou, X. Liu, Y. Li, H. He, Z. Luo, Y. Chen, Y. Wang, W. Shi, L. Shen, C. Cao, W. Liang, Q. Xu, Q. Lv, J. Lan, J. Li and S. Chen, *Bioorg. Med. Chem. Lett.*, 2017, **27**, 1458–1462.
277. B. T. O'Neill, E. M. Beck, C. R. Butler, C. E. Nolan, C. Gonzales, L. Zhang, S. D. Doran, K. Lapham, L. M. Buzon, J. K. Dutra, G. Barreiro, X. Hou, L. A. Martinez-Alsina, B. N. Rogers, A. Villalobos, J. C. Murray, K. Ogilvie, E. A. LaChapelle, C. Chang, L. F. Lanyon, C. M. Steppan, A. Robshaw, K. Hales, G. G. Boucher, K. Pandher, C. Houle, C. W. Ambroise, D. Karanian, D. Riddell, K. R. Bales and M. A. Brodney, *J. Med. Chem.*, 2018, **61**, 4476–4504.
278. F. I. Carroll, M. G. Gichinga, C. M. Kormos, R. Maitra, S. P. Runyon, J. B. Thomas, S. W. Mascarella, A. M. Decker and H. A. Navarro, *Bioorg. Med. Chem.*, 2015, **23**, 6379–6388.
279. B. Yang, M. M. Vasbinder, A. W. Hird, Q. Su, H. Wang, Y. Yu, D. Toader, P. D. Lyne, J. A. Read, J. Breed, S. Ioannidis, C. Deng, M. Grondine, N. DeGrace, D. Whitston, P. Brassil and J. W. Janetka, *J. Med. Chem.*, 2018, **61**, 1061–1073.
280. M. J. Waring, S. N. Bennett, S. Boyd, L. Campbell, R. D. Davies, S. Gerhardt, D. Hargreaves, N. G. Martin, G. R. Robb and G. Wilkinson, *MedChemComm*, 2013, **4**, 657–662.
281. M. J. Waring, S. N. Bennett, S. Boyd, L. Campbell, R. D. Davies, D. Hargreaves, P. MacFaul, N. G. Martin, D. J. Ogg and G. R. Robb, *MedChemComm*, 2013, **4**, 663–668.
282. F. Narjes, Y. Xue, S. von Berg, J. Malmberg, A. Llinas, R. I. Olsson, J. Jirholt, H. Grindebacke, A. Leffler, N. Hossain, M. Lepisto, L. Thunberg, H. Leek, A. Aagaard, J. McPheat, E. L. Hansson, E. Back, S. Tangefjord, R. Chen, Y. Xiong, G. Hongbin and T. G. Hansson, *J. Med. Chem.*, 2018, **61**, 7796–7813.
283. E. Menhaji-Klotz, K. D. Hesp, A. T. Londregan, A. S. Kalgutkar, D. W. Piotrowski, M. Boehm, K. Song, T. Ryder, K. Beaumont, R. M. Jones, K. Atkinson, J. A. Brown, J. Litchfield, J. Xiao, D. P. Canterbury, K. Burford, B. A. Thuma, C. Limberakis, W. Jiao, S. W. Bagley, S. Agarwal, D. Crowell, S. Pazdziorko, J. Ward, D. A. Price and V. Clerin, *J. Med. Chem.*, 2018, **61**, 3685–3696.
284. A. M. Thompson, P. D. O'Connor, A. J. Marshall, V. Yardley, L. Maes, S. Gupta, D. Launay, S. Braillard, E. Chatelain, S. G. Franzblau, B. Wan, Y. Wang, Z. Ma, C. B. Cooper and W. A. Denny, *J. Med. Chem.*, 2017, **60**, 4212–4233.
285. A. Pasternak, Z. Feng, R. de Jesus, Z. Ye, S. He, P. Dobbelaar, S. A. Bradley, G. G. Chicchi, K. L. Tsao, D. Trusca, G. J. Eiermann, C. Li, Y. Feng, M. Wu, Q. Shao, B. B. Zhang, R. Nargund, S. G. Mills, A. D. Howard, L. Yang and Y. P. Zhou, *ACS Med. Chem. Lett.*, 2012, **3**, 289–293.
286. J. E. Dowling, M. Alimzhanov, L. Bao, M. H. Block, C. Chuaqui, E. L. Cooke, C. R. Denz, A. Hird, S. Huang, N. A. Larsen, B. Peng, T. W. Pontz, C. Rivard-Costa, J. C. Saeh, K. Thakur, Q. Ye, T. Zhang and P. D. Lyne, *ACS Med. Chem. Lett.*, 2013, **4**, 800–805.

287. T. Odagiri, H. Inagaki, M. Nagamochi, T. Kitamura, S. Komoriya and H. Takahashi, *J. Med. Chem.*, 2018, **61**, 7234–7244.
288. D. G. Brown, P. R. Bernstein, A. Griffin, S. Wesolowski, D. Labrecque, M. C. Tremblay, M. Sylvester, R. Mauger, P. D. Edwards, S. R. Throner, J. J. Folmer, J. Cacciola, C. Scott, L. A. Lazor, M. Pourashraf, V. Santhakumar, W. M. Potts, S. Sydserff, P. Giguere, C. Levesque, M. Dasser and T. Groblewski, *J. Med. Chem.*, 2014, **57**, 733–758.
289. T. Cernak, N. J. Gesmundo, K. Dykstra, Y. Yu, Z. Wu, Z. C. Shi, P. Vachal, D. Sperbeck, S. He, B. A. Murphy, L. Sonatore, S. Williams, M. Madeira, A. Verras, M. Reiter, C. H. Lee, J. Cuff, E. C. Sherer, J. Kuethe, S. Goble, N. Perrotto, S. Pinto, D. M. Shen, R. Nargund, J. Balkovec, R. J. DeVita and S. D. Dreher, *J. Med. Chem.*, 2017, **60**, 3594–3605.
290. J. M. Ellis, M. D. Altman, A. Bass, J. W. Butcher, A. J. Byford, A. Donofrio, S. Galloway, A. M. Haidle, J. Jewell, N. Kelly, E. K. Leccese, S. Lee, M. Maddess, J. R. Miller, L. Y. Moy, E. Osimboni, R. D. Otte, M. V. Reddy, K. Spencer, B. Sun, S. H. Vincent, G. J. Ward, G. H. Woo, C. Yang, H. Houshyar and A. B. Northrup, *J. Med. Chem.*, 2015, **58**, 1929–1939.
291. D. C. McGowan, F. Herschke, F. Pauwels, B. Stoops, I. Smyej, S. Last, S. Pieters, W. Embrechts, M. D. Khamlichi, T. Thone, B. Van Schoubroeck, W. Mostmans, D. Wuyts, D. Verstappen, A. Scholliers, D. De Pooter, D. Dhuyvetter, H. Borghys, M. Tuefferd, E. Arnoult, J. Hong, G. Fanning, J. Bollekens, V. Urmaliya, A. Teisman, H. Horton, T. H. M. Jonckers and P. Raboisson, *J. Med. Chem.*, 2017, **60**, 6137–6151.
292. C. Gardelli, H. Wada, A. Ray, M. Caffrey, A. Llinas, I. Shamovsky, J. Tholander, J. Larsson, U. Sivars, L. Hultin, U. Andersson, H. J. Sanganee, K. Stenvall, B. Leidvik, K. Gedda, L. Jinton, M. Ryden Landergren and K. Karabelas, *J. Med. Chem.*, 2018, **61**, 5974–5987.
293. A. Jasper, D. Schepmann, K. Lehmkuhl, J. M. Vela, H. Buschmann, J. Holenz and B. Wunsch, *Eur. J. Med. Chem.*, 2012, **53**, 327–336.
294. A. M. Thompson, P. D. O'Connor, A. J. Marshall, A. Blaser, V. Yardley, L. Maes, S. Gupta, D. Launay, S. Braillard, E. Chatelain, B. Wan, S. G. Franzblau, Z. Ma, C. B. Cooper and W. A. Denny, *J. Med. Chem.*, 2018, **61**, 2329–2352.
295. A. T. Nchinda, C. Le Manach, T. Paquet, D. Gonzalez Cabrera, K. J. Wicht, C. Brunschwig, M. Njoroge, E. Abay, D. Taylor, N. Lawrence, S. Wittlin, M. B. Jimenez-Diaz, M. Santos Martinez, S. Ferrer, I. Angulo-Barturen, M. J. Lafuente-Monasterio, J. Duffy, J. Burrows, L. J. Street and K. Chibale, *J. Med. Chem.*, 2018, **61**, 4213–4227.
296. S. B. Boga, A. B. Alhassan, J. Liu, D. Guiadeen, A. Krikorian, X. Gao, J. Wang, Y. Yu, R. Anand, S. Liu, C. Yang, H. Wu, J. Cai, H. Zhu, J. Desai, K. Maloney, Y. D. Gao, T. O. Fischmann, J. Presland, M. Mansueto, Z. Xu, E. Leccese, I. Knemeyer, C. G. Garlisi, N. Bays, P. Stivers, P. E. Brandish, A. Hicks, A. Cooper, R. M. Kim and J. A. Kozlowski, *Bioorg. Med. Chem. Lett.*, 2017, **27**, 3939–3943.

297. W. Amberg, U. E. W. Lange, M. Ochse, F. Pohlki, B. Behl, A. L. Relo, W. Hornberger, C. Hoft, M. Mezler, J. Sydor, Y. Wang, H. Zhao, J. T. Brewer, J. Dietrich, H. Li, I. Akritopoulou-Zanze, Y. Lao, S. M. Hannick, Y. Y. Ku and A. Vasudevan, *J. Med. Chem.*, 2018, **61**, 7503–7524.
298. D. P. Mould, U. Bremberg, A. M. Jordan, M. Geitmann, A. E. McGonagle, T. C. P. Somervaille, G. J. Spencer and D. J. Ogilvie, *Bioorg. Med. Chem. Lett.*, 2017, **27**, 4755–4759.

Subject Index

References to figures are given in *italic* type. References to tables are given in **bold** type.

www.ingramcontent.com/pod-product-compliance
Lightning Source LLC
Jackson TN
JSHW070523180826
106902JS00003B/55